COLUMBIA UNIVERSITY
LIBRARIES

The Utility of Regional Gravity and Magnetic Anomaly Maps

William J. Hinze
Editor

Martin F. Kane
Norbert W. O'Hara
Mike S. Reford
James Tanner
Christian Weber
Associate Editors

SOCIETY OF EXPLORATION GEOPHYSICISTS

ISBN 0-931830-34-6

Library of Congress Catalog Card Number: 84-052447

Society of Exploration Geophysicists
P.O. Box 702740
Tulsa, Oklahoma 74170-2740

© 1985 Society of Exploration Geophysicists. All rights reserved.

Published 1985

Printed in the United States of America

CONTENTS

Preface   vii

Maps

Composite magnetic anomaly map of the conterminous United States   ix

Composite magnetic anomaly map of Alaska and Hawaii   x

Gravity anomaly map of the conterminous United States   xi

Residual Bouguer gravity anomaly map of the conterminous United States based on wavelengths less than 250 km   xii

Magnetic anomaly map of Canada   xiii

Gravity anomaly map of Canada   xiv

The composite magnetic-anomaly map of the conterminous United States   1
William J. Hinze and Isidore Zietz

Preparation of magnetic-anomaly maps of Alaska and Hawaii   25
Richard H. Godson

Preparation and overview of the gravity-anomaly map of the United States   33
Norbert W. O'Hara and Paul L. Lyons

Preparation of a digital grid of gravity anomaly values of the conterminous United States   38
Richard H. Godson

Features of a pair of long-wavelength (>250 km) and short-wavelength (< 250 km) Bouguer gravity maps of the United States   46
M. F. Kane and R. H. Godson

Evolution of Geological Survey of Canada magnetic-anomaly maps: A Canadian perspective   62
P. J. Hood, P. H. McGrath, and D. J. Teskey

The new series of 1:1 000 000-scale magnetic-anomaly maps of the Geological Survey of Canada: Compilation techniques and interpretation   69
S. D. Dods, D. J. Teskey, and P. J. Hood

Global gravity maps and the structure of the Earth   88
Carl Bowin

Block structure of continental crust derived from gravity and magnetic maps, with Australian examples   102
Peter Wellman

...dies of the Grenville province: Significance for Precambrian plate collision ...e origin of anorthosite 109
M. D. Thomas

Interpretation of gravity and magnetic data, central and eastern Brazil 124
Nicolau L. E. Haralyi and Yociteru Hasui

Applications of aeromagnetic data to mineral-resources exploration — Baluchistan, Pakistan 132
Allan Spector, Peter J. Hood, Abul Farah, and Waheeduddin Ahmed

The Kalahari Desert, central southern Africa: A case history of regional gravity and magnetic exploration 144
Colin V. Reeves

Geologic mapping of the basement of the Paris basin (France) by gravity- and magnetic- data interpretation 154
N. Debeglia and C. Weber

Crustal geology of Arizona as interpreted from magnetic, gravity, and geologic data 164
John S. Sumner

Mapping basement magnetization zones from aeromagnetic data in the San Juan basin, New Mexico 181
Lindrith E. Cordell and V. J. S. Grauch

Regional gravity and magnetic study of west Texas 198
G. R. Keller, R. A. Smith, W. J. Hinze, and C. L. V. Aiken

Kansas basement study using spectrally filtered aeromagnetic data 213
Harold L. Yarger

Model of the geothermal system in southwestern South Dakota from gravity and aeromagnetic studies 233
T. G. Hildenbrand and R. P. Kucks

Magnetic terranes in the central United States determined from the interpretation of digital data 248
Thomas G. Hildenbrand

Geologic interpretation of gravity and magnetic data for northern Michigan and Wisconsin 267
J. S. Klasner, E. R. King, and W. J. Jones

Geologic significance of regional gravity and magnetic anomalies in the east-central Midcontinent 287
E. G. Lidiak, W. J. Hinze, G. R. Keller, J. E. Reed, L. W. Braile, and R. W. Johnson

Studies of gravity anomalies in Georgia and adjacent areas of the southeastern
  United States   308
> Leland Timothy Long and Anton M. Dainty

Some effects of regional metamorphism and geologic structure on magnetic anomalies
  over the Carolina Slate belt near Roxboro, North Carolina   320
> E. S. Robinson, P. V. Poland, L. Glover, III, and J. A. Speer

Structure of the U.S. Atlantic continental margin from derivative and filtered maps
  of the magnetic field   325
> John C. Behrendt and Muriel S. Grim

The change in the magnetic anomaly pattern at the ocean – continent boundary   339
> J. R. Heirtzler

An isostatic residual gravity map of California — A residual map for interpretation
  of anomalies from intracrustal sources   347
> Robert C. Jachens and Andrew Griscom

Analysis of gravity data in volcanic terrain and gravity anomalies and subvolcanic
  intrusions in the Cascade Range, U.S.A., and at other selected volcanoes   361
> David L. Williams and Carol Finn

Interpretation of Precambrian geology in Minnesota using low-altitude, high-
  resolution aeromagnetic data   375
> Val W. Chandler

Mineral-exploration aspects of gravity and aeromagnetic surveys in the Sudbury –
  Cobalt area, Ontario   392
> V. K. Gupta and F. S. Grant

Geologic interpretation of a high-resolution aeromagnetic survey in the Amos –
  Barraute area of Quebec   413
> Sun Yunsheng, D. W. Strangway, and W. E. S. Urquhart

Interpretation of part of an aeromagnetic survey in the Matagami area of Quebec   426
> W. E. S. Urquhart and D. W. Strangway

Interpretation of an aeromagnetic survey of the Qian'an Archean metamorphic-rock
  series in China   439
> Sun Yunsheng, D. W. Strangway, W. E. S. Urquhart, and Sun Fengxing

Index   451

PREFACE

Mapping of the Earth's gravity and magnetic fields has a long and distinguished history as part of the investigations of the structure and petrologic variation within the Earth's lithosphere. Prior to 1950, observations of anomalies in these planetary fields had generally been focused on limited areas for specific mineral-resource and geological objectives. The increasing availability of portable gravimeters and the development of aeromagnetic-survey technology shortly after World War II, however, led to efficient and precise gravity and magnetic mapping of extensive regions. These "regional" surveys were conducted for the most part by governmental organizations most often as a means of evaluating or stimulating the exploitation of earth resources. The data from these federal surveys are generally available in the public domain and have been augmented by large amounts of data collected by academic institutions for basic-research investigations. Vast amounts of gravity- and magnetic-anomaly data also have been acquired by industrial firms in mineral-resource exploration, but with few exceptions these data are generally not available to the public. In the United States, the publicly available data have recently been composited into new or improved country-wide anomaly maps.

In 1975, the Society of Exploration Geophysicists (SEG) and the U.S. Geological Survey, recognizing the value of regional gravity- and magnetic-anomaly maps, jointly organized gravity- and magnetic-anomaly map committees to prepare anomaly maps of the United States. The immediate objective of the committees was to compile and publish a revised gravity-anomaly map of the conterminous United States and the first magnetic-anomaly maps of the conterminous United States and Alaska. These objectives were met in late 1982 with the publication of the Gravity Anomaly Map of the United States by the SEG and the release of the Composite Magnetic Anomaly Map of the Conterminous United States and the Magnetic Anomaly Map of Alaska by the U.S. Geological Survey. Small-scale versions of these maps are reproduced in color in this volume together with the most recently published gravity- and magnetic-anomaly maps of Canada. In recognition of the publication of these national maps, and to illustrate the many uses of gravity- and magnetic-anomaly maps, a series of special technical sessions was held at the 52nd Annual International Meeting of the SEG in the fall of 1982 in Dallas, Texas. A total of 33 of the 53 papers presented at the special sessions were accepted for publication in this volume, *The Utility of Regional Gravity and Magnetic Anomaly Maps.*

The subjects of the papers that make up the volume vary from the preparation of national maps to examples of the many uses of regional maps. The anomalies that are discussed range in areal dimension from hundreds of kilometers to tens of meters. The majority of the papers illustrate the utility of the maps in mapping structures and lithologic variations within the continental crust, the configuration of the crystalline basement rocks, zones of crustal weakness, distribution of extrusive and intrusive igneous rocks, and the geometry of sedimentary basins. Most cases are drawn from the United States and Canada, but examples from Europe, Africa, South America, and Asia are included. The uses of regional gravity and magnetic maps are illustrated by actual examples from petroleum, mineral, and geothermal exploration; earthquake-hazard evaluation; and general geologic mapping.

A secondary objective of the anomaly-map committees was to provide appropriate regional sets of digital data to the geophysical community. The availability of the data sets permits the rapid mechanical production of contour maps and other plots at a variety of scales, map projections, and contour intervals. Their availability also makes possible the employment of digital two-dimensional filtering to produce maps of selected attributes of the anomaly fields such as gradients, wavelength ranges, and strike directions as an aid in interpretation. The importance of digital filtering in the analysis of gravity and magnetic data is exemplified by the contents of the papers of this volume, more than one-half of which use some form of digital processing. As an example of these filtering procedures, the 250-km high-pass Gravity Anomaly Map of the Conterminous United States, which is discussed in this volume by M. F. Kane and R. H. Godson, is also reproduced in color in this Preface.

The increasing availability of regional- and continental-scale gravity- and magnetic-anomaly maps and their corresponding data sets, which makes possible the performance of digital filtering of these maps, is well documented in this volume. The majority of the papers included herein testify to the wide uses of these data, which range from investigations of broad crustal structure to exploration of mineral resources, to say nothing of the geodetic applications of the gravity data, which have not been considered in this volume. We envision the papers of this volume as constituting a benchmark from which we can expand the understanding of our Earth by improved regional anomaly maps and data sets, additional processing of regional digital-data sets by filtering and inversion, and enhanced integration of the anomaly data with available geological and collateral geophysical data.

Finally, we note that efforts are now under way to compile and publish magnetic and gravity maps for North America as a contribution to the Decade of North American Geology, a project sponsored by the Geological Society of America. In addition to contributions from the United States and Canada, we look forward to those of colleagues from Mexico and Central America, Greenland and Denmark, Iceland, and other countries covered by the maps. At this point, we would like to challenge our colleagues from other continents to embark on similar projects. For many regions there already exist sufficient data to undertake a project of continental dimension. For others, adequate data do not exist, and the challenge will be to mount the operations for the collection of the data. For still others, data exist but never have been publicly released, presumably for reasons of military security, at least in the case of gravity data. The loss to science and to mankind in general, owing to such restrictive measures, is enormous.

We, the editors of this volume, are indebted to the many individuals and groups who made this volume possible. Our appreciation goes to the authors of the papers for their efforts to present readable, authoritative, and well-illustrated manuscripts; Dr. David W. Strangway, who suggested this volume while serving as Vice-President of the SEG; the Gravity and Magnetic Anomaly Map Committees; Jerry Henry, Belynda Bland Morton, and their colleagues in the Publications Office of the SEG; William D. Rose, project editor; and the individuals who spent many hours reviewing and suggesting revisions to submitted manuscripts. The reviewers included Carlos L. V. Aiken, John C. Behrendt, Carl O. Bowin, William C. Brisbin, Val W. Chandler, Richard Coles, Lindrith E. Cordell, Duncan R. Cowan, David L. Daniels, N. Debeglia, Edward R. Decker, William H. Diment, T. Feininger, R. A. Gibb, Richard H. Godson, A. Goodacre, Frazer S. Grant, V.J.S. Grauch, Andrew Griscom, A. Hahn, Richard T. Haworth, Peter J. Hood, Thomas G. Hildenbrand, Robert C. Jachens, B. David Johnson, G. Randy Keller, Douglas P. Klein, Stephen Kumarapeli, Lawrence K. Law, Edward G. Lidiak, P. R. Louis, R. Niblett, Don R. Mabey, Kenneth McConnell, Lyle D. McGinnis, P. H. McGrath, Paul Morgan, Chris Nind, Howard W. Oliver, Don Plouff, R. Riddihough, Robert W. Simpson, Jr., William D. Stanley, Patrick Taylor, Dennis Teskey, Michael D. Thomas, and Isidore Zietz.

WILLIAM J. HINZE, Chairman
MARTIN F. KANE
NORBERT W. O'HARA
MIKE S. REFORD
JAMES TANNER
CHRISTIAN WEBER

Composite Magnetic Anomaly Map of the Conterminous United States. Color-contour interval is 200 nT. (Published by the U.S. Geological Survey as Map GP-954A, 1982.)

Composite Magnetic Anomaly Map of Alaska and Hawaii. Color-contour interval is 200 nT. (Published by the U.S. Geological Survey as Map GP-954B, 1984.)

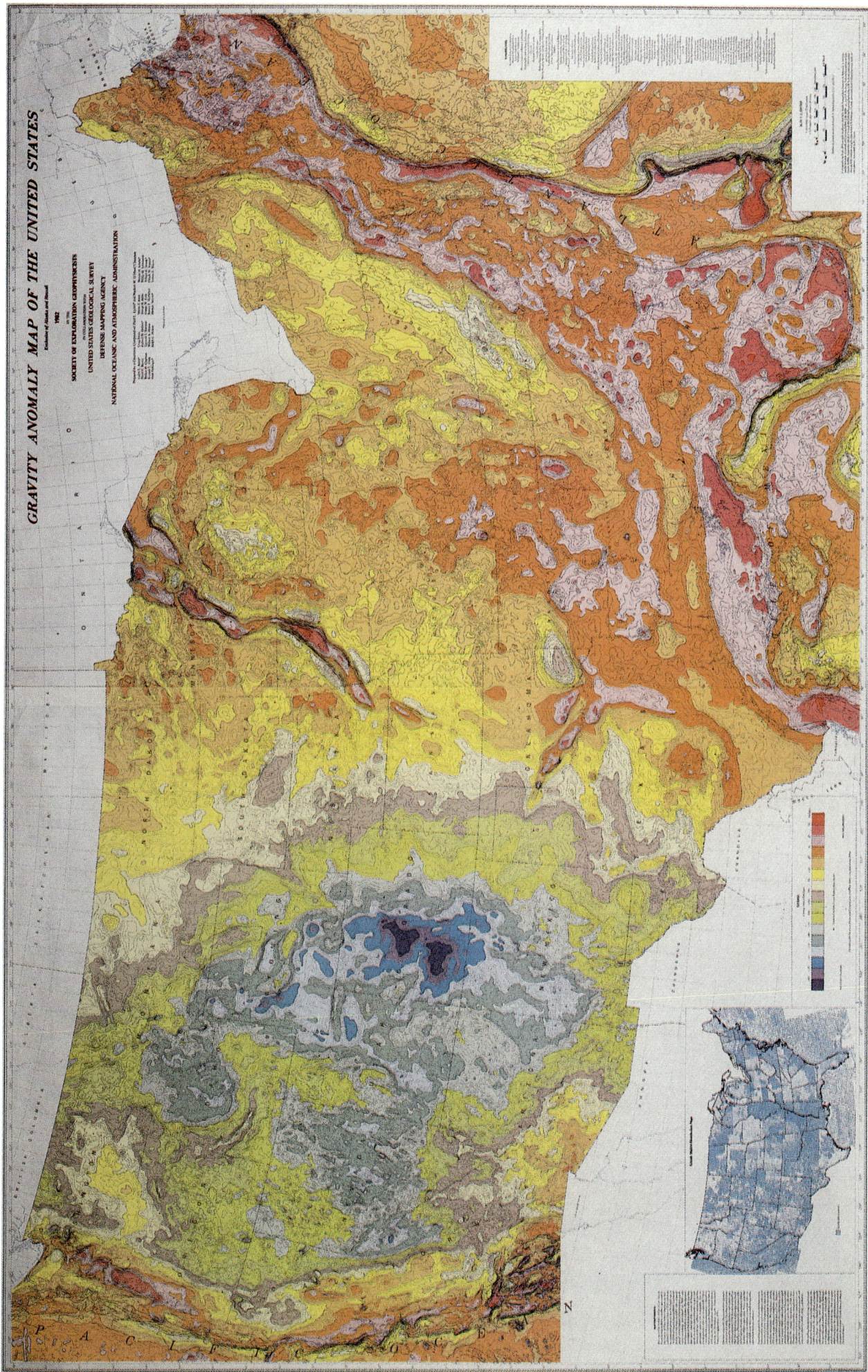

Gravity Anomaly Map of the Conterminous United States. Contour interval is 5 mGal; color-contour interval is 25 mGal. (Published by the Society of Exploration Geophysicists, 1982.)

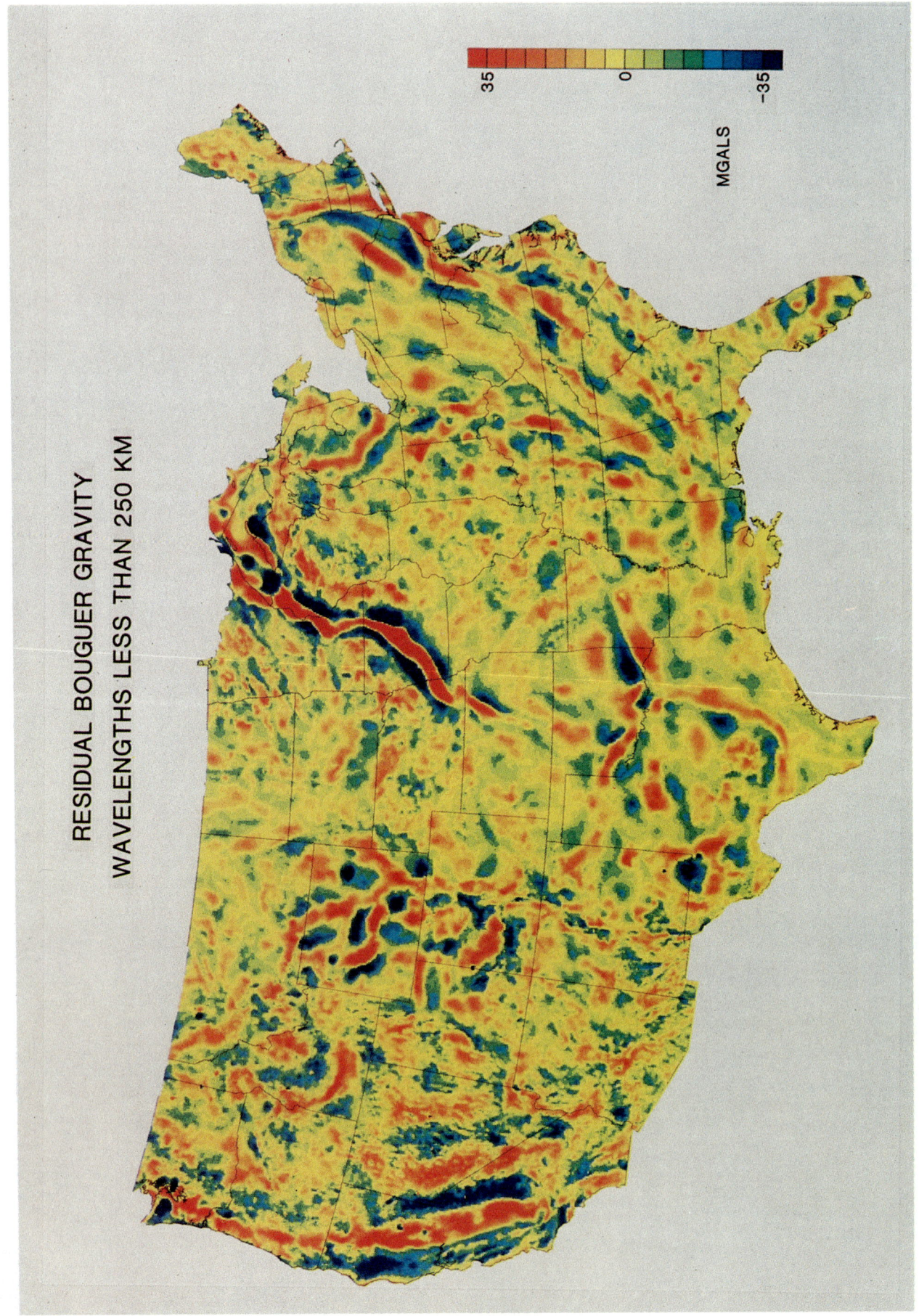

Residual Bouguer Gravity Anomaly Map of the Conterminous United States Based on Wavelengths Less than 250 Km. Color-contour interval is 5 mGal. (Published by the U.S. Geological Survey as Map GP-953A, 1982.)

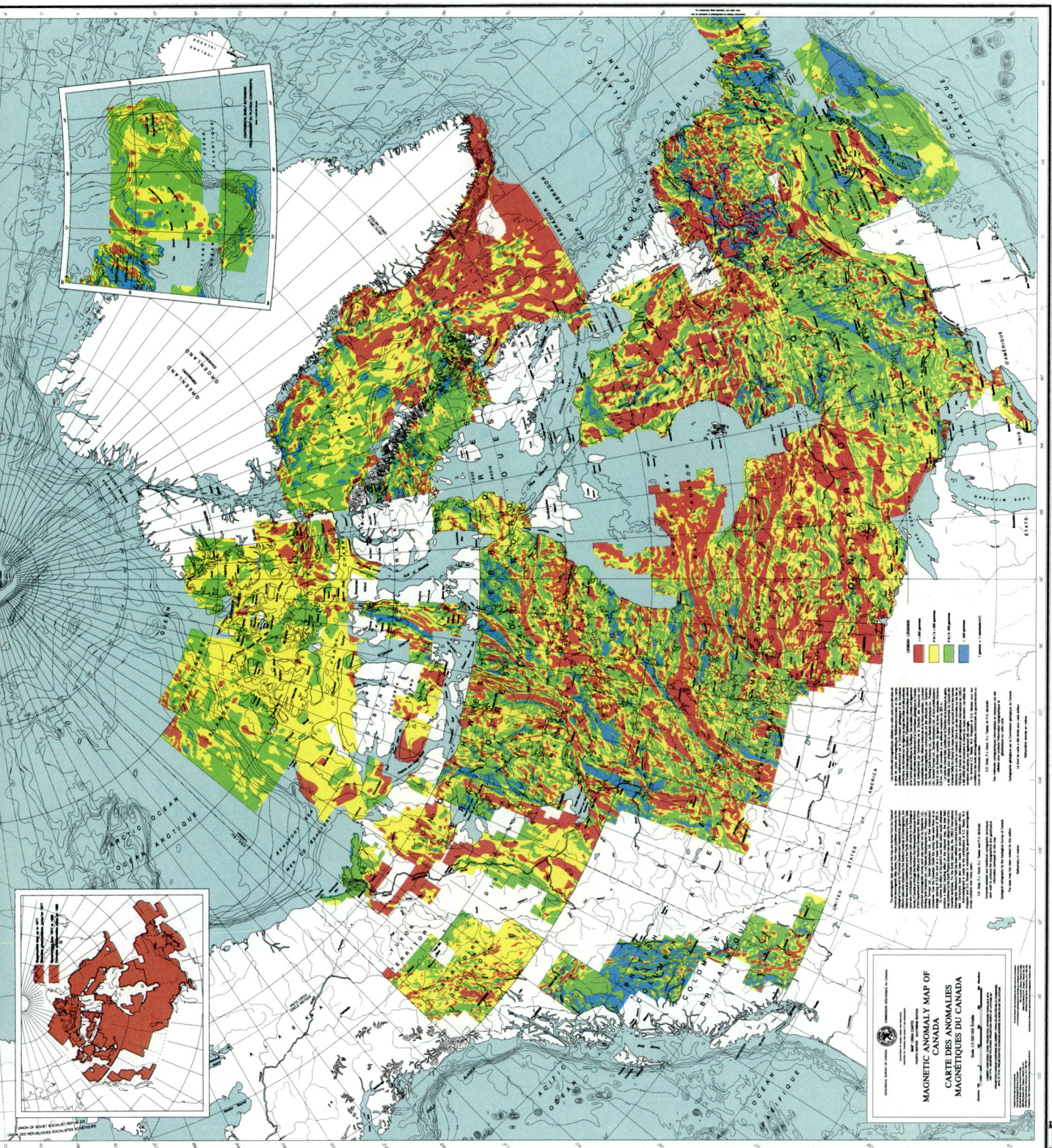

Magnetic Anomaly Map of Canada. Color-contour interval is 200 nT. (Published by the Geological Survey of Canada, Department of Energy, Mines and Resources, Canada, as Map 1255A, fourth edition, 1984.)

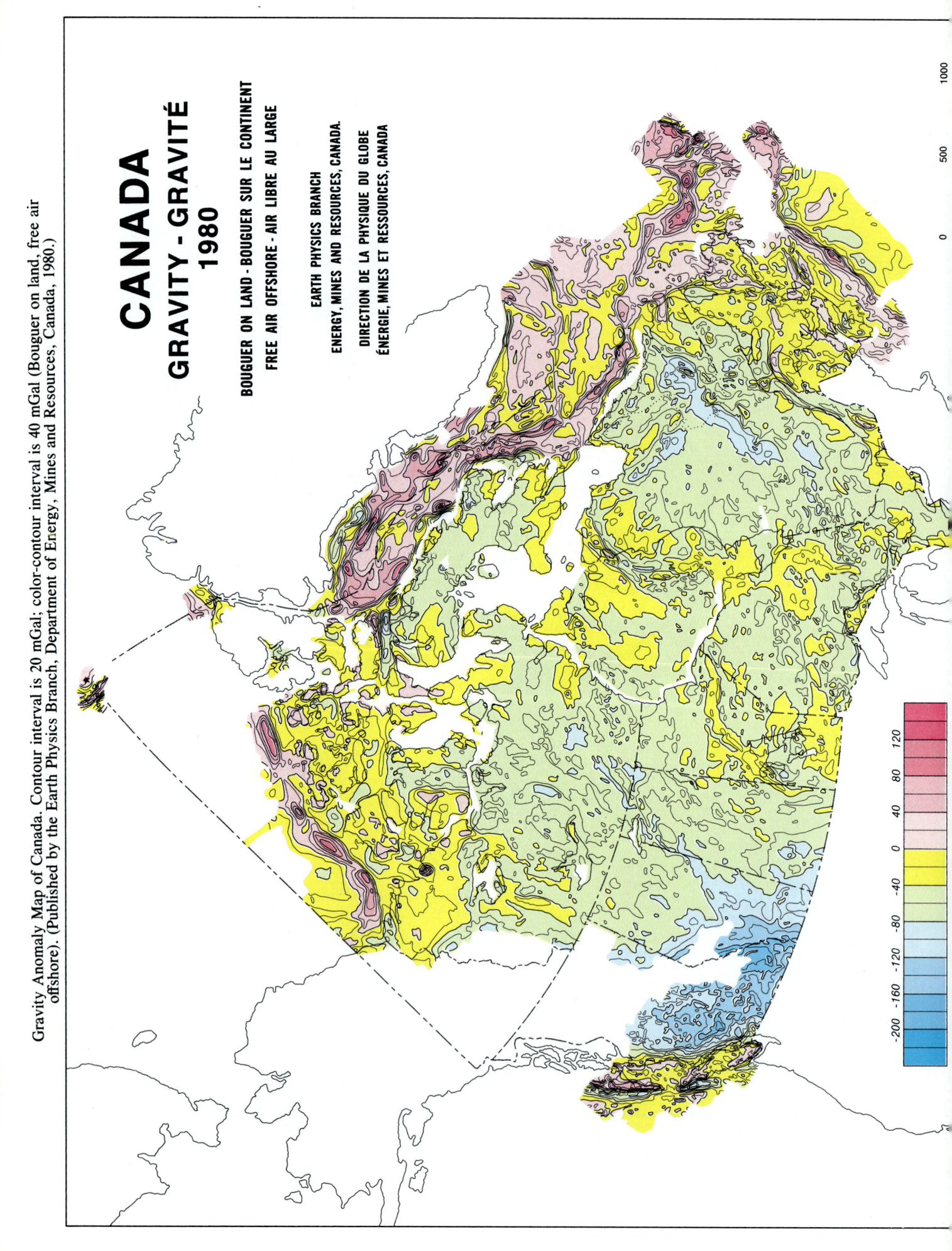

The composite magnetic-anomaly map of the conterminous United States

William J. Hinze* and Isidore Zietz‡

ABSTRACT

The first composite magnetic-anomaly map of the conterminous United States and adjacent offshore areas has been published at a color-contour interval of 200 gammas and at the scale and projection of other national geologic and geophysical maps for easy comparison. This map, despite the inconsistent characteristics of the surveys from which it was compiled, is useful in providing a regional framework for the interpretation of magnetic studies of limited areas, in selecting areas for more detailed magnetic investigations, and in studying the distribution and character of regional geologic features.

The map has a wide variation of magnetic-anomaly patterns, trends, and types, thus reflecting the diversity of the geologic terranes of the United States. In general, the anomaly pattern east of the Cordillera in the craton and in the Appalachian Mountains consists of more and greater intensity anomalies. The muted nature of the anomalies of much of the Cordillera is a result of several factors but appears to be primarily related to a decreased crustal magnetization caused by an abnormally shallow Curie isotherm. The anomalies of the Appalachian Mountains and the Cordilleran system primarily reflect the major structural patterns of the orogens, but important exceptions occur, such as those associated with rocks underlying thrust sheets in the Appalachian Mountains and westerly-striking anomaly trends in the Cordillera, which are correlated with igneous intrusives, faults, and mineral deposits.

The buried southern and eastern edges of the Precambrian craton are indicated by changes in the magnetic anomalies and their dominant trends. Within the central United States, numerous regional magnetic-anomaly provinces are observed that reflect the long, complex history of the Precambrian basement rocks of the craton. These provinces are transected by conspicuous, intense, long, generally linear anomalies that originate from mafic extrusive or shallow intrusive igneous bodies within failed rifts, such as the Midcontinent rift system, the Southern Oklahoma aulacogen, and the Reelfoot rift buried beneath the Mississippi embayment. These are only a few of the many interesting regional geologic features that are observed on the composite magnetic-anomaly map of the United States.

INTRODUCTION

The development of aeromagnetic-surveying technology in the United States in the mid-1940's paved the way for efficient, precise magnetic mapping of extensive regions (Balsley, 1952). Improvements in this technology soon permitted surveying to move from "anomaly hunting" to the preparation of comprehensive anomaly maps. The magnetic-anomaly map is an efficient tool in mapping the character and configuration of igneous and other crystalline rocks of the continental crust. As a result, aeromagnetic surveying has been used to map igneous rocks that occur within the sedimentary-rock column and, more importantly, to investigate the crystalline rocks that make up the Precambrian shields and the "basement" of continental regimes. The basement is largely hidden from direct observation in the United States by younger sediments and sedimentary rocks; it is poorly sampled by drill holes and, therefore, remains largely unknown to the geological community. Yet, the majority of the Earth's history is tied up in basement rocks, many significant mineral deposits occur within them, and there is increasing evidence (Flawn, 1965) that the basement—its structure, physical properties, and configuration—has a profound effect on lithologic variations, structural development, and fluid migration in overlying sedimentary rocks. Thus, magnetic-anomaly maps, in concert with geo-

*Department of Geosciences, Purdue University, West Lafayette, IN 47907.
‡Phoenix Corporation, 7643 Leesburg Pike, Falls Church, VA 22043.

logic information and other geophysical data, play an important role in deciphering Earth history and in characterizing crustal rocks.

In the United States, aeromagnetic surveying has been conducted largely on an ad-hoc basis aimed at the solution of particular geologic problems by either private industry or governmental organizations. The data acquired by private industry remain largely confidential, but aeromagnetic maps prepared by university, state, and federal organizations have been widely distributed in journal articles and various map series. However, the lack of a national aeromagnetic mapping program in the United States left many areas unsurveyed or data unavailable to the public domain. In addition, data of surveyed areas were not tied together, and maps were prepared on the basis of inconsistent survey and map specifications. As a result, the preparation of a comprehensive magnetic-anomaly map of the United States was thwarted.

In 1975, the U.S. Geological Survey and the Society of Exploration Geophysicists organized a joint effort to prepare a magnetic-anomaly map of North America (Hinze, 1976). The first step toward this goal was the preparation of magnetic-anomaly maps of the conterminous United States and Alaska. An inventory of the available magnetic maps showed that somewhat less than 50 percent of the United States was covered by published maps and that most of these were in U.S. Geological Survey publications. At approximately that time, an aeroradiometric survey of the United States was initiated by the National Uranium Resource Evaluation (NURE) program of the U.S. Department of Energy. Aeromagnetic observations were made concurrently with the radiometric-data acquisition, and the data were made available to the public. These data, together with maps obtained from ongoing federal and state programs, plus a minor quantity of data made available from private industry, were used to supplement the existing coverage to prepare the composite magnetic-anomaly map of the conterminous United States. The map has been published by the U.S. Geological Survey as Geophysical Map 954A. An estimated 2.5 to 3.0 million line kilometers of aeromagnetic data have been used to prepare this map. The complementary composite magnetic-anomaly maps of Alaska and Hawaii are published as Geophysical Map 954B. (These maps are reproduced in color in the Preface of this volume. Readers are encouraged to refer to the colored map of the conterminous United States.)

The composite magnetic-anomaly map of the conterminous United States and adjacent offshore areas is published in color on two sheets, showing magnetic-anomaly contours at an interval of 200 gammas (nT) with supplemental contours at an interval of 100 gammas in low-gradient areas, on an Albers equal-area projection at a scale of 1:2 500 000. The map can be compared directly with the tectonic (U.S. Geological Survey and American Association of Petroleum Geologists, 1961), basement-rock (Bayley and Muehlberger, 1968), geologic (King and Beikman, 1974), and gravity-anomaly (Society of Exploration Geophysicists, 1982) maps of the conterminous United States. A gray-shade version of the colored magnetic-anomaly map prepared by the Phoenix Corporation from a 2-minute digital-data set is shown in Figure 1.

The composite magnetic-anomaly map of the United States is useful in providing a regional framework for the interpretation of magnetic studies of limited areas, in selecting areas for more detailed investigations, and in studying the distribution and character of regional geologic features measured in tens or hundreds of kilometers. The map and its accompanying text also provide an index to publicly available magnetic-map coverage. Thus, the map is a valuable asset to both geologists and geophysicists, but caution needs to be used in the geologic interpretation of the map. We strongly recommend that the map be used only at the 1:2 500 000 publication scale or smaller scales because of the diversity of data types, data-acquisition specifications, and compilation techniques. The scale and contour interval of the map limit its use to investigation of anomalies that are broader than several kilometers and higher in amplitude than about 200 gammas. We recommend that original data sources be used for detailed studies at scales larger than the 1:2 500 000 publication scale. Care must also be exercised in interpreting anomalies at the other end of the spectrum. Long-wavelength anomalies—that is, anomalies that exceed roughly 1 000 km in horizontal dimension—may be erroneous, owing to limitations in the process of removing the core-derived, main magnetic field from the observed data and tying together maps prepared from surveys that have different base levels.

DATA SOURCES

The magnetic-anomaly map of the United States was compiled from roughly 660 magnetic maps. Most of these maps are based on total-intensity aeromagnetic-anomaly data; others were constructed from total-intensity ground and shipborne magnetic-anomaly data and vertical-intensity ground observations. The vertical-intensity anomaly data cover major parts of South Dakota and Missouri and a small area of New Mexico. The maps date back to 1943, but almost one-half are based on observations acquired since 1975. The aeromagnetic-observation tracks vary widely in direction, altitude, and spacing. However, most observations were made within 1 000 ft (305 m) of the ground surface, and 50 percent of the maps were constructed from surveys with spacings of 1 mile (1.6 km) or less while 20 percent are a result of 3-mile or greater track spacings. The critical specifications of the surveys are detailed and referenced in the text that accompanies the map (U.S. Geological Survey and Society of Exploration Geophysicists, 1982).

No attempt was made to continue analytically the anomaly data to a common altitude. However, the data were adjusted to a common magnetic-field datum, because data of the anomaly maps are based on numerous, often arbitrarily selected, magnetic-field levels. Comparison of the compiled magnetic-anomaly map with recent nationwide magnetic observations by the U.S. Naval Oceanographic Office and the NURE program suggests that the actual zero level of the compiled map is approximately 1 000 gammas higher than the zero level, based on the 1975 International Geomagnetic Reference Field (IGRF) (Barraclough and Fabiano, 1978).

In many areas, more than one map was available for compilation. Generally, the most detailed map that tied into adjacent maps with the least difference was selected for

FIG. 1. Composite magnetic-anomaly map of the conterminous United States. Gray-shade levels indicate anomaly intensity from white (lowest) to black (highest) at 200-gamma intervals. Solid-black blocked-out areas are no-data areas. (Courtesy of Phoenix Corporation.)

compilation. Major segments of the conterminous United States, particularly in the Gulf Coast and the Great Plains, were compiled from data and maps made available through the NURE program. The flight-line spacing of these surveys was usually 3 or 6 miles (4.8 or 9.6 km).

Magnetic maps covering several areas were unavailable for compilation at the time the draft copy of the magnetic-anomaly map was sent to the publisher in late 1981. These areas are left blank on the map. Subsequently, these areas in Florida, Ohio, Illinois, Texas, New Mexico, California, Washington, and Oregon and Lakes Erie and Ontario have been magnetically mapped and thus will be included in revised versions of the map.

The borders of the maps used in preparing the composite anomaly map are indicated by light lines on the original color map. They are shown to assist in identifying magnetic-anomaly patterns that are based on varying data types, data-acquisition specifications, and compilation techniques rather than on geologic-anomaly sources. An excellent example is shown in the State of Washington. A block of aeromagnetic data in central Washington, which extends to the southwest into eastern Oregon (Figure 2), is based on a survey flown at 1 000 ft (305 m) above mean terrain with a flight-line spacing of 0.5 mile (0.8 km). This portion of the map shows complex, short-wavelength magnetic anomalies derived from the surface Columbia Plateau basaltic rocks. In contrast, the areas immediately adjacent to this block in the southern third of the state do not have this complicated anomaly pattern, even though the surface rocks are similar in both areas. The rather simple, long-wavelength anomaly pattern of the area adjacent to the block reflects the high level (15 000 ft or 4 575 m above sea level) and broad flight-line spacing (5 miles or 8 km) of the survey conducted in this area. This is a dramatic example of the need for using the borders and specifications of the individual maps in interpreting the composite map.

MAP COMPILATION

Compilation of the magnetic-anomaly map of the conterminous United States involved the following steps: (1) the magnetic-anomaly map of a given survey was inspected and, as necessary, was referenced to the appropriate IGRF adjusted for the date of the survey, and the datum of the data; (2) contour lines at an interval of 100 or 200 gammas were selected; (3) the map of the selected contour lines was photographically adjusted to the 1:1 000 000 compilation scale and positioned on an Albers equal-area projection master base map of the conterminous United States and offshore areas; (4) near the boundaries of adjacent surveys, contour lines were visually joined as smoothly as possible; (5) where major discontinuities of anomaly values existed, contoured NURE data were used to guide the connecting of contour lines; and (6) the map at the 1:1 000 000 compilation scale was photographically reduced to the 1:2 500 000 publication scale.

The NURE data, acquired during a 7-year period over the conterminous United States and referenced primarily to the updated IGRF–1975 (Tinnel and Hinze, 1981), provided a base net for controlling the compilation of individual magnetic-anomaly maps. As an independent check on the compilation, profiles from the map were compared with a series of north–south aeromagnetic traverses of the U.S. Naval Oceanographic Office (NOO). These traverses were flown in 1976 and 1977 at approximately 1° longitude intervals across the conterminous United States. This comparison shows that the compiled data agree with the NOO data, after adjustment to the IGRF–1975, to within 100 gammas throughout the country. However, visual inspection of the composite map indicates a broad scale increase in average magnetic values from the eastern to the northwestern United States. A similar trend is observed in the magnetic-anomaly map (Sexton et al., 1982) prepared from the NOO data by removal of the IGS–75 geomagnetic-field model from the observed data. This long-wavelength anomaly may result from a continent-wide geologic variation or an error in the 1975 geomagnetic reference field. At the time the composite map was published, this problem was unresolved. Subsequently, errors have been identified in the IGRF–1975, and a revised provisional international geomagnetic reference field for the 1975–80 epoch has been defined (Peddie, 1982). The difference between this revised reference field and the IGRF–1975, used to reduce the majority of the NURE data as well as the NOO data and thus which played a critical role in tying together the compilation and checking its validity, has been found to vary from the order of −650 gammas along the eastern seaboard to −150 gammas in southern California and +150 gammas in northwestern Washington (Peddie, 1983). The effect of this error in the IGRF–1975 is to impose a long-wavelength trend upon the composite magnetic-anomaly map, decreasing from the western part of the country, particularly the northwestern corner, to the east coast. This is exactly the nature of the observed broad-scale variation in the composite map. A more local potential problem in leveling of the composite data has been identified for the State of Minnesota magnetic-anomaly map (Schnetzler et al., 1984). The data over Minnesota may be as much as +100 gammas in error in comparison with the surrounding magnetic-anomaly maps.

ANOMALY SOURCES

The diversity of magnetic-anomaly patterns and types observed over the United States is anticipated because of the broad spectrum of geologic terranes that occur within the mapped area. Magnetic-polarization (intensity) variations within the crust, which primarily reflect the nature and quantity of magnetite present, are the primary source of these anomalies. In general, the magnetic anomalies of maps of the scale and contour interval of the United States map originate within the igneous- and metamorphic-rock basement on which largely nonmagnetic sediments and sedimentary rocks have been deposited. Additional anomalies are derived from igneous rocks that have been intruded into or erupted onto the sedimentary-rock section. Measurable magnetic-polarization variations are anticipated to occur throughout the crust as a result of structure, igneous-rock differentiates, metamorphism, and alteration. The base of the magnetic layer may be the Moho (Wasilewski et al., 1979), if the upper mantle is non-ferrimagnetic as inferred to be, or the base may be within the crust, where the level of the Curie isotherm ($\cong 580°C$) is affected by increased heat transfer from the mantle into the crust. Magnetic anomalies

FIG. 2. Composite magnetic-anomaly map of the State of Washington and adjoining areas.

with dominant wavelengths measured in hundreds of kilometers are anticipated from the lower crust, because the magnetization of this layer is an order of magnitude greater than the upper crust (Shuey et al., 1973; Hall, 1974). In marked contrast, magnetic anomalies commonly mapped on larger scale, more detailed maps than the United States map, which are derived from sedimentary rocks or are due to relief on the basement surface, cannot be recognized on the United States map.

The magnetic polarization of crustal rocks is a result of induction in the Earth's current magnetic field or remanent magnetization that reflects ancient or modern geomagnetic fields. We anticipate that many of the magnetic anomalies on the composite magnetic-anomaly map of the United States are caused primarily by induction of magnetization in magnetite because, particularly in the central and eastern parts of the country, the anomalies are derived from ancient rocks, older than several hundred million years, that may have lost much of their remanent magnetization by long-term decay or alteration by subsequently superimposed processes. This conclusion is supported by the general nature of the magnetic anomalies, which are characteristic of anomalies caused by induction in the direction of the geomagnetic field over the United States. In general, the intense anomalies are positive; the magnetic minima associated with positive anomalies are weaker in amplitude than the maxima, and the minima are located on the north side of the anomalies. A line connecting the positive peak of an isolated anomaly with its complementary minimum to the north is commonly directed along the local magnetic meridian. Furthermore, as pointed out by Wasilewski et al. (1979), the ancient remanent magnetization in lower crustal rocks should be diminished by their great age, high ambient temperatures, and structural deformation. However, it is possible that these rocks carry a viscous remanent magnetization directed along the current geomagnetic field.

The magnetic anomaly resulting from the induced magnetic polarization of a specific geologic feature will differ over the United States because of the varying direction and intensity of the inducing geomagnetic field. However, these variations are limited. The azimuth of the geomagnetic field over the United States ranges from −20 to +20 degrees and the inclination from 60 to 75 degrees, and the intensity of the total-field component increases by roughly 20 percent from the southern to the northern limits of the country. It is noteworthy that an anomaly derived from a specific positive magnetic contact within the Earth will increase in amplitude and symmetry from southerly to northerly latitudes. In addition, the maximum of the anomaly will shift to the north, and the polarization minimum to the north of the maximum will decrease with increasing magnetic latitude. Furthermore, the direction of the line connecting the positive and negative extremes of anomalies will shift from roughly 20 degrees west in the northeast to 20 degrees east in the northwest.

Model magnetic-anomaly maps that illustrate the effect of varying inducing geomagnetic fields are given by Vacquier et al. (1951) and Andreasen and Zietz (1969); the latter also shows the effect of remanent magnetization. These and similar suites of theoretical anomalies also show the important effect of the geometric properties of the source upon magnetic anomalies. One of the more significant of these properties is the depth to the source. The amplitude of the anomaly will decrease with increasing depth to the source—approximately in a linear manner for long-wide sources, as the inverse square of the distance for long-narrow horizontal bodies, and as the inverse cube of distance for equidimensional, concentrated sources. As a result of this inverse distance function, shallow sources such as encountered in the Llano uplift in central Texas have much greater amplitudes and sharper gradients than the anomalies of central Louisiana, where the magnetic basement rocks are buried at a depth of roughly 10 km.

The amplitude of anomalies derived from induced magnetization sources is also a function of the quantity of magnetite and related magnetic minerals and in a complex manner to the nature of the magnetic minerals. Magnetite is the end member of a solid-solution series that decreases in magnetic characteristics and Curie temperature with increasing percentage of titanium. Magnetite is a minor component of the vast majority of rocks and thus is not directly related to lithology and rock-classification schemes. The relationship between magnetic-mineral content (and thus magnetic-anomaly intensity) and lithologies is tenuous at best, but certain broad generalities can be made that may be useful in identifying the source of anomalies. These generalities are apparent in compendia of magnetic-rock properties (e.g., Lindsley et al., 1966; Strangway, 1981; Carmichael, 1982). Both clastic and chemical sediments and sedimentary rocks carry only minor magnetite and thus are transparent in regional magnetic-anomaly mapping. However, there are exceptions, and some are striking. For example, the magnetite-bearing sedimentary banded iron formations are among the most magnetic rocks in the United States. Several of the intense, long, linear, positive magnetic anomalies in the Lake Superior region mark the outcrop or subcrop pattern of these Precambrian sedimentary units. In general, the magnetite content increases in crystalline rocks from felsic to mafic rocks. Thus, granitic rocks are usually related to magnetic minima and gabbros to magnetic maxima. Exceptions to these generalities do occur, however. For example, Proterozoic granite gneisses are commonly more magnetic than Archean metamorphosed mafic volcanic rocks and granites in Nebraska (Lidiak, 1972) and eastern Kansas (Yarger, 1981), which contain 2 percent or more magnetite and are associated with positive magnetic anomalies.

There are notable exceptions to the generalization that most magnetic anomalies on the United States map are due primarily to induction. For example, the linear, alternating positive and negative magnetic anomalies off the west coast in the Pacific and Juan de Fuca plates are caused by alternating normal and reversed magnetizations of the oceanic crust. In contrast, the featureless magnetic-anomaly pattern off the eastern continental shelf reflects either a consistent remanent magnetization resulting from a constant magnetic-field polarity during Jurassic time (Larson and Hilde, 1975), when this crust originated, or subsequent modification of preexisting remanent magnetization (Bleil and Petersen, 1983).

Perhaps the most dramatic examples of the effects of remanent magnetization on magnetic anomalies are observed in northwest Texas (Figure 3). The Crosbytown

Fig. 3. Composite magnetic-anomaly map of Texas and Oklahoma and adjoining states.

anomaly, located at 33.5° N, 101° W, has been shown to be related to an early Paleozoic mafic intrusive that has a strong reversed remanent polarization (Shurbet et al., 1976). Another example of an anomaly probably related to reversed remanent magnetization is observed northwest of the Crosbytown anomaly at 34.5° N, 102.5° W. This anomaly is strongly negative, suggesting an intense remanent magnetization. Other examples of anomalies derived largely from remanent magnetic polarization are observed over volcanic rocks of Tertiary and Quaternary age in the western United States, but even ancient volcanic rocks may carry a strong remanent-polarization component. For example, the anomalies associated with the 1 100-m.y.-old Keweenawan volcanic rocks of the Lake Superior region and the Midcontinent rift, which transects the general anomaly pattern and extends southward from western Lake Superior to Kansas, are highly distorted by a remanent component that is at least three times as great as the induced magnetic polarization and directed obliquely to it (King and Zietz, 1971). The long geologic history of these volcanic rocks and their low-grade, greenschist metamorphism have been ineffective in destroying the remanent magnetic component. We may assume from these examples that many other anomalies characteristic of sources having induced magnetization may be to some degree caused by remanent magnetizations having the same direction and polarity as the present geomagnetic field.

Thus, it is apparent that the vast majority of the anomalies on the magnetic-anomaly map of the United States are derived from intracrystalline rock variations within the upper crust. Localized high-intensity anomalies are anticipated from igneous rocks within the sedimentary-rock section, and long-wavelength anomalies measured in hundreds of kilometers originate within the lower crust. The amplitudes and gradients of these anomalies decrease with increasing depth. Although, in general, the intensity of anomalies is directly related to the relative mafic nature of their sources, sufficient exceptions require care in interpreting lithologies directly from the mapped anomalies. The highest confidence is placed on interpretations based on extrapolation from known geology. Auxiliary information, particularly gravity data, is also useful in the interpretation process. Despite problems in identifying lithology from the magnetic-anomaly map, the character and anomaly pattern can be useful in

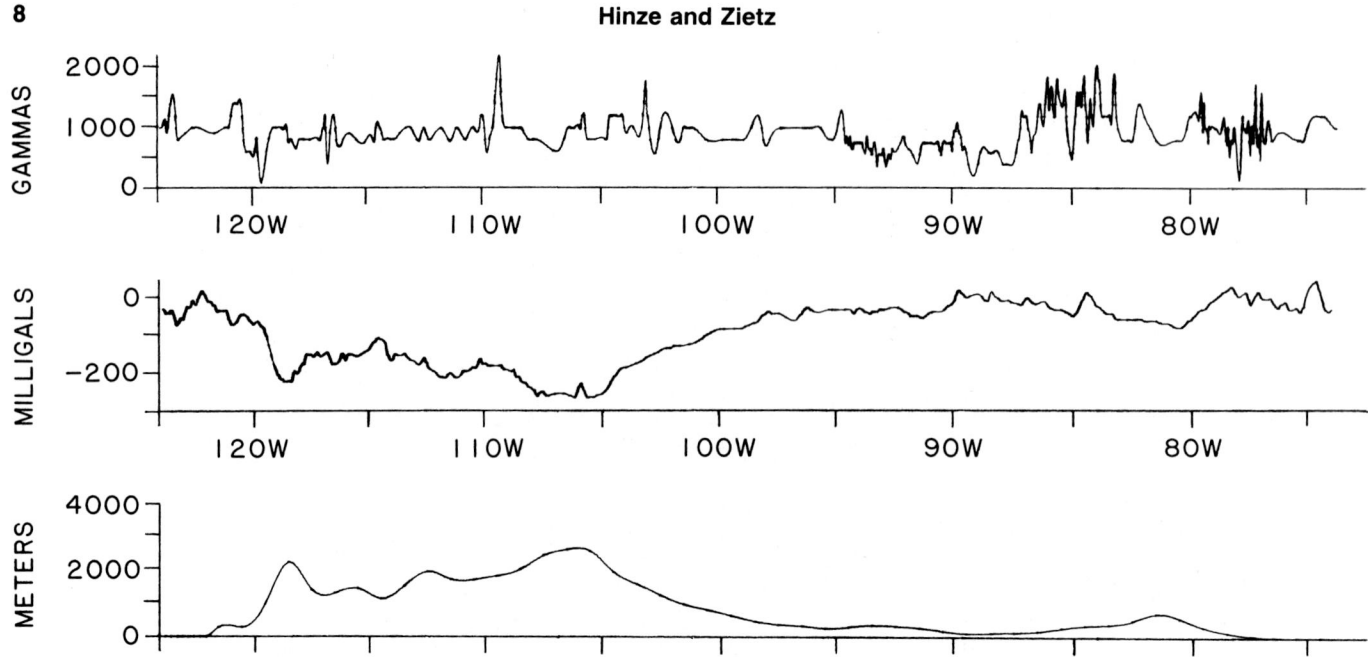

FIG. 4. Composite magnetic-anomaly, Bouguer gravity-anomaly (free-air anomaly in oceans), and elevation profiles along 37° N latitude.

identifying structural fabric and geometric properties of the source rocks.

GEOLOGIC SIGNIFICANCE OF SELECTED ANOMALIES

Utilizing the basic concepts regarding magnetic-anomaly sources discussed above, and expanding on them with more specific geology/magnetization information from particular regions of the country, it will be possible to interpret geologically the magnetic-anomaly pattern of the United States map. Of course, magnetic-anomaly maps and profiles of many regions of the country previously have been discussed and interpreted, but for the first time it is possible to view the anomalies for many parts of the country on a broad-scale, multi-state basis. This view allows extrapolation from and interpolation between observed anomaly sources, based on the magnetic anomalies and the study of long-wavelength anomalies characteristic of geologic provinces.

The importance of the United States magnetic-anomaly map is particularly evident upon comparison with the geologic map of the conterminous United States (King and Beikman, 1974). Direct correlation of anomalies and anomaly patterns with geology is only possible in areas of outcropping basement rocks, as in the Appalachian Mountains, and surface Tertiary and Quaternary volcanic rocks, as in the Snake River Plain of Idaho. In the vast majority of the United States, there is no indication of crystalline basement–anomaly sources in the surface geology. Thus, the magnetic-anomaly map provides information that is only available through geophysical studies and the occasional deep drill hole. This is evident on the magnetic profile across the United States along the 37° N parallel of latitude, shown in Figure 4 together with the Bouguer gravity-anomaly profile derived from the 1982 map (Society of Exploration Geophysicists, 1982) and a smoothed elevation profile. Geologic provinces based on surface geology show little correlation with the magnetic-anomaly profile, except that in deep basement provinces (e.g., the Great Plains at roughly 100° W) the magnetic anomalies have low gradients.

To illustrate the potential utility of the United States magnetic-anomaly map, the following overview is presented of the geologic significance of selected anomalies. Comprehensive analysis of the map awaits integration of available geological and geophysical data in selected regions and quantitative modeling based on anomaly information derived from larger scale maps.

Atlantic and Gulf Coasts

The magnetic-anomaly pattern along the Atlantic Coast of the United States (Figure 5) is dominated by the East Coast magnetic anomaly (ECMA), a linear positive magnetic anomaly that roughly coincides with the continental slope over much of its 3 000-km length from Newfoundland to east of Georgia (Figure 6(1)). The anomaly splits into two segments off the northern North Carolina coast and off southern Georgia, at roughly 31° N latitude; the oceanward segment turns into the continent and continues as an irregular series of positive magnetic anomalies (the Brunswick anomaly) in an arcuate fashion across southern Georgia into Alabama (Figure 6(2)). The ECMA has been known for a considerable

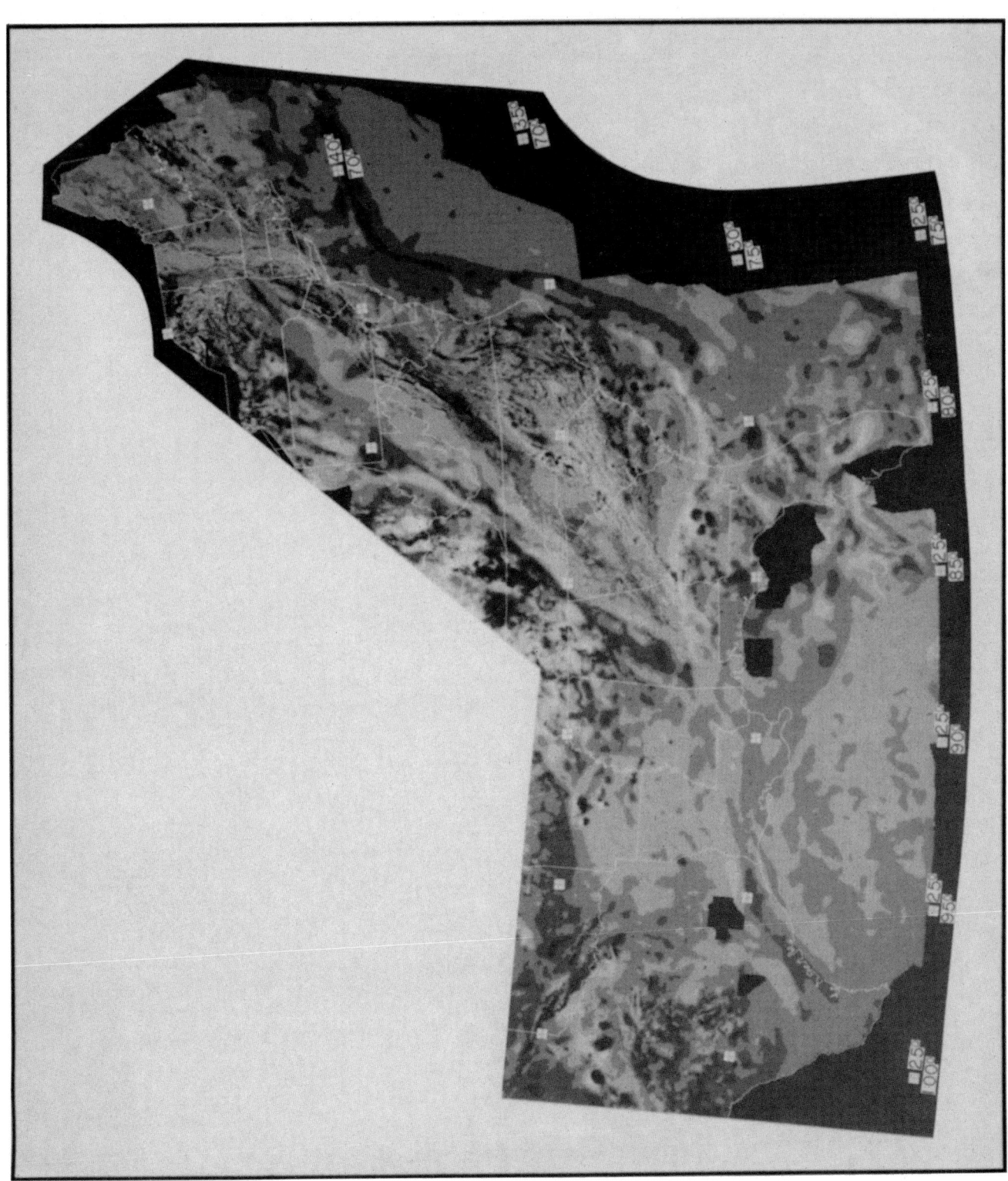

FIG. 5. Composite magnetic-anomaly map of the eastern United States.

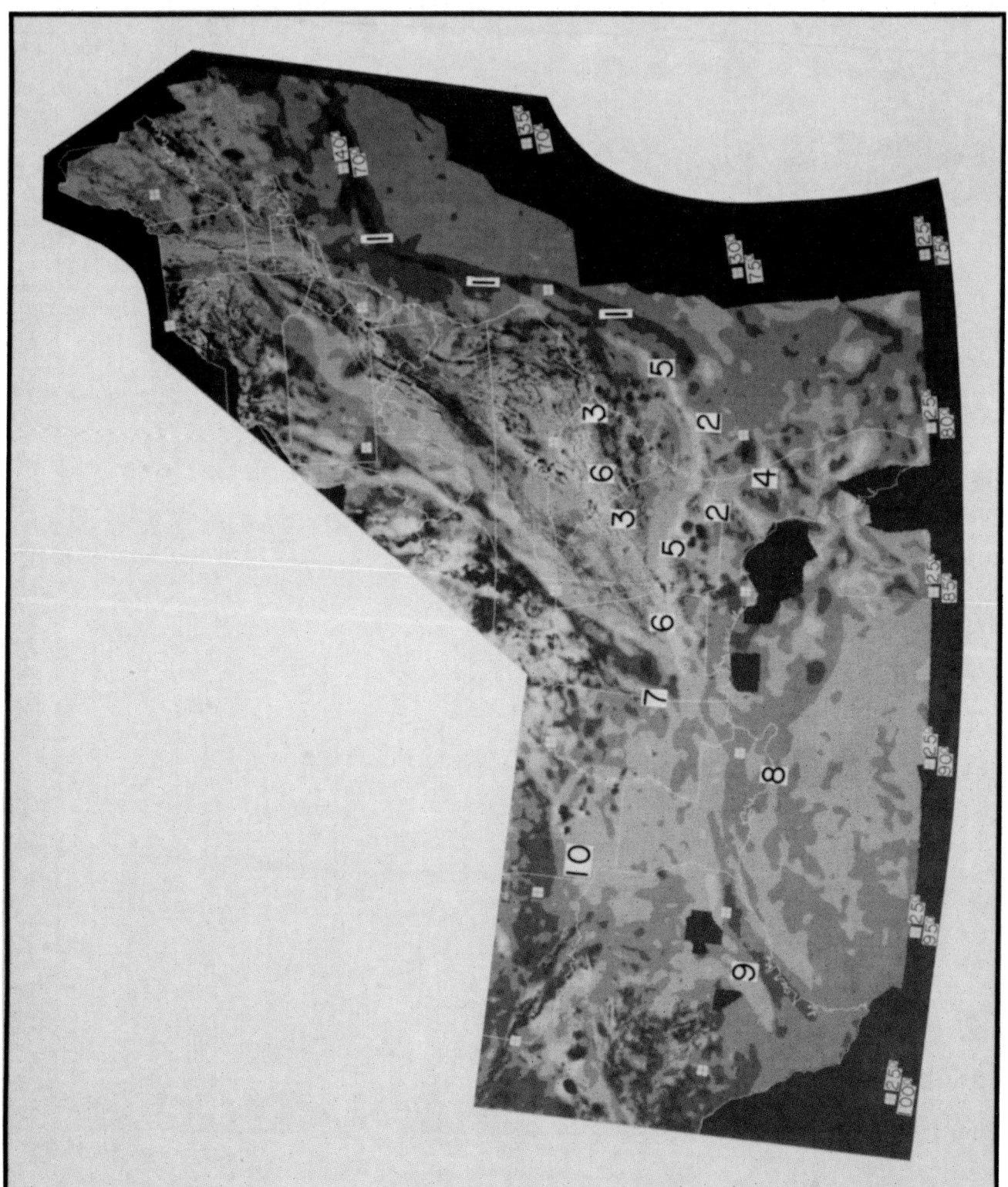

FIG. 6. Composite magnetic-anomaly map of the eastern United States with anomalies, anomaly patterns, and geologic features identified. 1 = East Coast magnetic anomaly; 2 = Brunswick anomaly; 3 = northern margin of Charleston magnetic terrane ("suspect" Brunswick terrane); 4 = Tallahassee/Suwannee terrane; 5 = Altamaha anomaly; 6 = Carolina–Mississippi fault; 7 = termination of northeast magnetic trends; 8 = Mississippi River delta area; 9 = Gulf Coast magnetic anomaly; 10 = Magnet Cove anomaly.

period of time (Keller et al., 1954) and has been the subject of extensive study and controversy. Taylor et al. (1968) reviewed the previous explanations for the anomaly and conducted modeling studies on it. Their favored origin for the anomaly is a felsic basement-intrusive body along the pre-Paleozoic continental–oceanic boundary. However, Folger et al. (1979), Behrendt and Klitgord (1980), and Schlee et al. (1979) support an origin related to the edge effect associated with oceanic–continental crustal transitions. More recently, Behrendt and Grim (this volume) suggest that the edge effect may account for most of the ECMA but that the shorter wavelength components of it are caused by more complex structures, perhaps igneous intrusions. Oceanward of the ECMA the magnetic-anomaly field is featureless. In contrast, shoreward and south of the ECMA the anomaly pattern is complex, reflecting the magnetic heterogeneity of the continental crust, but the anomalies are broadened and subdued owing to the oceanward thickening of the shelf sedimentary rocks, particularly in the basins, as discussed by Klitgord and Behrendt (1979) and Behrendt and Grim (this volume).

Taylor et al. (1968) interpreted the ECMA and its landward extension, the Brunswick anomaly, as the pre-Paleozoic continental–oceanic boundary. Using new geologic data and magnetic- and gravity-anomaly compilations, Higgins and Zietz (1983) place the boundary between the North American continent and what was part of the African or African/South American plate until late Paleozoic time along the northwestern margin of the Charleston magnetic terrane (Figure 6(3)). This terrane is a broad positive magnetic anomaly, with local magnetic maxima, which extends across southern South Carolina into Georgia and is equivalent to the "suspect" Brunswick terrane of Williams and Hatcher (1983). They differentiate the Brunswick terrane from the Tallahassee/Suwannee terrane (Figure 6(4)) of northern Florida and southern Georgia on the basis of magnetic and gravity patterns. The boundary between these terranes and the Charleston and the northern Florida terranes of Higgins and Zietz (1983) lies along the Brunswick anomaly and a major linear magnetic minimum that extends for more than 1 000 km in an arcuate manner from the continental shelf to southern Alabama. This minimum, the Altamaha anomaly (Figure 6(5)), is interpreted by Higgins and Zietz as a deep fault-bounded trough filled with nonmagnetic sedimentary rocks. It appears to be truncated in Alabama by the Carolina–Mississippi fault (Figure 6(6)), an extension of the northwestern boundary of the Charleston (Brunswick) terrane with the Avalon terrane to the north.

The magnetic trends of the Appalachian orogen are truncated in Alabama by the Altamaha anomaly and the Carolina–Mississippi fault of Higgins and Zietz (1983). Farther west in Mississippi, and extending to the northwest into Arkansas, a postulated transform fault (Cebull et al., 1976) connecting the Appalachian and Ouachita orogens terminates the northeast magnetic trends of the east-central United States (Figure 6(7)). The Mississippi River delta (Figure 6(8)) and embayment are characterized by long-wavelength anomalies that probably reflect both the thick section of Phanerozoic sedimentary rocks and a largely magnetically homogeneous crust. The western margin of this zone extends southwestward from central Arkansas to southeast Oklahoma and then south-southeastward to south Texas. This break in the magnetic-anomaly pattern marks the limit of the North American craton prior to subduction associated with the Appalachian and Ouachita orogenies at the end of the Paleozoic Period (Thomas, 1977). A positive magnetic anomaly extends along the shoreward side, parallel to the coastline in Texas and discontinuously into Louisiana (Figure 6(9)). A few local, intense, positive magnetic anomalies within the featureless magnetic terrane of the Gulf Coastal Plain mark the site of late Paleozoic and Mesozoic intrusives of alkalic and related affinities (Zartman, 1977). The relatively intense magnetic minima associated with the positive anomalies (e.g., the Magnet Cove complex; see Figure 6(10) at 34°30′ N latitude and 92°50′ W longitude) suggest strong contributions of remanent magnetizations to the total-rock magnetization.

Appalachian orogen

Aeromagnetic data have had a significant role in mapping the metamorphic and igneous rocks of the Appalachian orogen (Figure 7) and the associated basement rocks, especially in the southern Appalachian Mountains, because of sparse outcrops and a thick covering of saprolite. Many of the anomalies are linear in form and parallel the general northeasterly strike of the Appalachian orogenic structural trends. Linear anomalies originate from magnetic-polarization variations within bedded rocks and compositional layering in plutonic rocks. Cataclastic zones as well as dikes also cause linear anomalies. Roughly equidimensional magnetic maxima may originate from either mafic or felsic plutons, but magnetic minima may also be associated with felsic plutons. Northeast-striking linear magnetic minima whose widths are generally measured in a few tens of kilometers are related to Triassic sedimentary troughs. The correlation of magnetic signature with lithology is complicated by the wide range of metamorphic grade of the belts of rocks that make up the Appalachian orogen. Reed et al. (1968) and Popenoe and Zietz (1977) reported that the magnetite content is increased in medium-grade metamorphic rocks and thus these rocks have an enhanced magnetic susceptibility. In addition, Hatcher and Zietz (1980) pointed out that both regional aeromagnetic and gravity anomalies of the southern Appalachian Blue Ridge and Inner Piedmont are caused by deep crustal rocks rather than by near-surface rocks. The broad anomalies cannot be explained by the surficial rocks.

The most striking magnetic anomaly of the Appalachian region and one of the longest magnetic lineaments on the North American continent is the New York–Alabama lineament (Figure 8(1)) (King and Zietz, 1978), which extends for more than 1 600 km from the Mississippi embayment to the Green Mountains of Vermont. The lineament roughly coincides with the western limit of the regional Appalachian gravity minimum and the western edge of the Appalachian fold belt. The lineament is tangential to the arcuate fold-belt salients in Pennsylvania and Tennessee, which led King and Zietz to suggest that the lineament marks the southeast edge of a stable crustal block that acted as a buttress for the strong deformation of the Appalachian fold belt. The lineament is

FIG. 7. Composite magnetic-anomaly map of the Appalachian orogen, U.S.A.

also a boundary between the strike of prevailing regional magnetic and gravity anomalies. To the southeast, the anomalies tend to parallel the northeast Appalachian trend, while to the north they have the more northerly trend of the buried Grenville province.

East of the New York–Alabama lineament, the magnetic-anomaly pattern is generally negative. Anomalies that occur within the minimum are broadened and subdued, because the basement crystalline rocks are overlain by thick sedimentary rocks in the Appalachian fold belt (Figure 8(2)). East of the fold belt, short-wavelength anomalies are observed that are derived from the surface magnetite-bearing rocks of the Blue Ridge and Piedmont. The long-wavelength magnetic anomalies that occur in generally northeast-trending belts are unrelated to the surface rocks, suggesting that the Blue Ridge and Inner Piedmont are thin, crystalline thrust sheets overlying nonmagnetic sedimentary rocks (Hatcher and Zietz, 1980). Hatcher and Zietz interpret the variations in the long-wavelength pattern of the Inner Piedmont as inverse relationships to the thickness of sedimentary rocks beneath the thrust sheets. Thus, magnetic minima correspond to thickened sedimentary sections.

The buried edge of the Precambrian craton in the Appalachian orogen is generally placed along the prominent regional gravity gradient that separates the regional Appalachian gravity minimum from the eastern gravity positive anomaly (Rankin, 1975; Hatcher and Zietz, 1980; Cook and Oliver, 1981; Thomas, 1983). In the southern Appalachians, this buried Taconic suture lies along the Inner Piedmont–Charlotte slate belt transition. It is observed in the magnetic-anomaly pattern in Georgia, South and North Carolina, and southern Virginia as the western limit of the high-intensity, short-wavelength anomalies (Figure 8(4)) of the Charlotte–Carolina slate belts. From Virginia to New England, the suture lies along the eastern margin of a curvilinear zone of local anomalies derived from near-surface Piedmont rocks. Zen (1983) also showed the correlation of this Taconic suture with the gravity gradient and magnetic anomalies in northwestern Maine. His interpretation places the suture along the northwestern margin of a diffuse band of positive magnetic anomalies that strike northeast from immediately south of the common boundary point between Canada, Maine, and New Hampshire.

Craton

Geophysical surveying, particularly aeromagnetic studies, have had a profound impact on our knowledge of the Precambrian rocks of the craton of the United States. In the craton—the central part of the country from the Appalachian fold belt to the eastern margin of the Rocky Mountains and from the United States–Canadian border to the Gulf Coastal Plain—the dominant sources of magnetic anomalies are the Precambrian basement rocks that underlie the mildly deformed Phanerozoic strata. These basement rocks are difficult to study by direct methods because of the paucity of deep drilling through the sedimentary-rock cover; and even where exposed, such as in the Adirondack Mountains, the Lake Superior region, the St. Francois Mountains, the Llano uplift, and the Black Hills, outcrops are sparse. However, these regions of Precambrian crystalline-rock outcrop, plus scattered basement drill holes, are critical because they provide "benchmarks" from which geological information can be extrapolated using regional magnetic-anomaly maps.

Geologic studies and isotopic age determinations (Van Schmus and Bickford, 1981; Denison et al., 1984) testify to the long and complex Precambrian geologic history of the craton, which in turn is manifested in the wide variety of magnetic anomalies (Figure 9) observed over the region. Generally, these anomalies can be grouped into either units or terranes based on characteristic patterns of anomalies with similar dimensions, amplitudes, and strike direction or relatively linear anomaly patterns that transect prevailing widespread patterns.

The most prominent magnetic anomaly of the craton extends north-northeastward from eastern Kansas to the west end of Lake Superior. It corresponds to the Midcontinent gravity high, an outstanding feature of the United States gravity-anomaly map. The anomaly (Figure 10), which can be traced directly to outcropping Keweenawan basalts in the Lake Superior region, is interpreted as the geophysical expression of a 1 100-m.y.-old paleorift zone (Halls, 1978) that extends in an arcuate pattern from Kansas to Lake Superior and then to southeast Michigan. Interpretation of magnetic- and gravity-anomaly data along the Kansas to Lake Superior portion of the anomaly by King and Zietz (1971), in Lake Superior by Hinze et al. (1982), and in Michigan by Oray et al. (1973) and Hinze et al. (1975), supplemented by information obtained from outcrops and basement drill holes, suggests that the anomaly is related to a graben (basin) filled with mafic volcanic rocks and clastic sedimentary rocks. Clastic sedimentary basins occur locally over this graben. The magnetic anomaly on the eastern limb—i.e., beneath the Michigan basin—is neither as continuous nor as intense as the anomalies of the western limb, despite the continuity of the gravity anomaly. Hinze et al. (1975) suggest that the different nature of the magnetic anomaly in Michigan is related to the limited volume and extent of the mafic volcanism, even though deep-crustal intrusives associated with the rift retain the continuity of the gravity anomaly. The southern limit of the eastern arm of the rift has been continued south from southern Michigan on the basis of a discontinuous, north–south–trending belt of positive gravity and magnetic anomalies and related mafic basement rocks (Halls, 1978; Keller et al., 1982) across western Ohio, western Kentucky, and into central Tennessee (Figure 9). Also, Lidiak and Zietz (1976) recognized rifts in western Kentucky on the basis of magnetic anomalies.

Another prominent rift-related magnetic anomaly is associated with the Southern Oklahoma aulacogen (Figure 3), a linear trough that extended west-northwestward from the Ouachita fold belt in southern Oklahoma into Colorado. The magnetic anomaly correlates with the Wichita system (King, 1977) and its extension to the west and is presumably derived from Lower Cambrian basalts and Lower to Middle Cambrian gabbro, anorthosite, and diorite intrusives (Keller et al., 1983). However, not all rift zones are as obvious on the magnetic map as the Midcontinent rift system and the Southern Oklahoma aulacogen. Keller et al. (1983) show that ancient rift zones are widespread in the Midcontinent region. One of these rifts, which is expressed in only a subtle

Fig. 8. Composite magnetic-anomaly map of the Appalachian orogen, U.S.A., with anomalies, anomaly patterns, and geologic features identified. 1 = New York–Alabama lineament; 2 = magnetic-anomaly minimum over the Appalachian fold belt; 3 = magnetic anomalies associated with the Blue Ridge and Inner Piedmont; 4 = edge of the Precambrian craton.

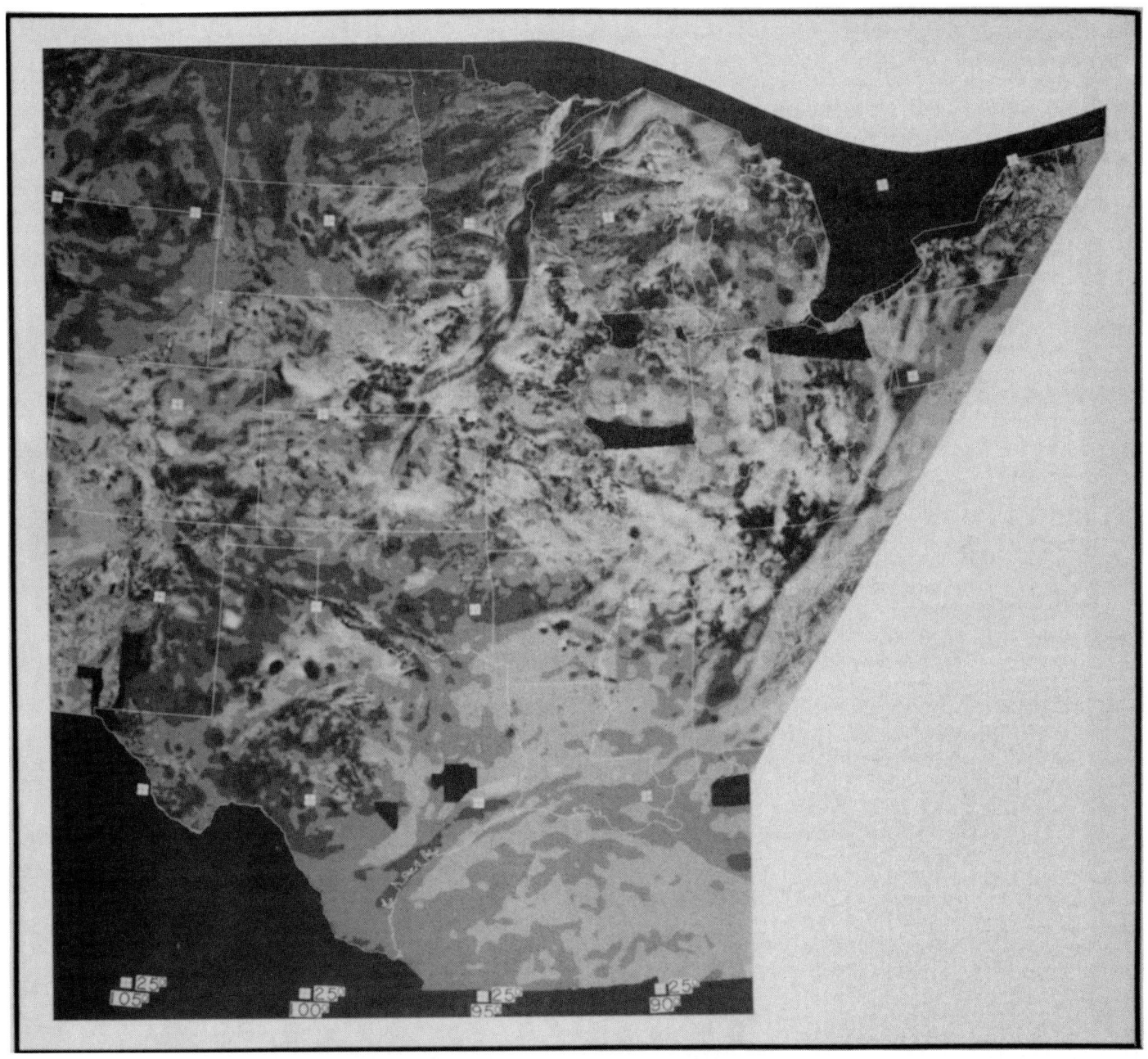

Fig. 9. Composite magnetic-anomaly map of the craton, U.S.A.

manner on the United States magnetic-anomaly map, is the Reelfoot rift (Ervin and McGinnis, 1975). It extends northeast form east-central Arkansas into western Kentucky and southeastern Missouri along the axis of the Mississippi embayment. The sedimentary-rock-filled graben associated with the rift was mapped by Hildenbrand et al. (1977) on the basis of two parallel, linear, northeast trends of magnetic anomalies that are interpreted as having originated from mafic intrusives along the graben-margin faults; long-wavelength anomalies occur over the graben itself. More recent studies utilizing detailed geophysical maps indicate that this feature continues to the northeast into Indiana and breaks up into a complex of similar features in southern Illinois (Braile et al., 1982). Braile and his co-workers suggest that geophysical signatures of selected arms of the New Madrid rift complex, plus the Rome trough in eastern Kentucky and West Virginia, make up the 38th-parallel lineament (Heyl, 1972; Lidiak and Zietz, 1976). This lineament is not a conspicuous feature of the United States magnetic-anomaly map because of the subtle nature of the related anomalies.

In the eastern craton, roughly from western Ohio eastward, the magnetic-anomaly belts strike from north to northnortheast. Over the shallower basement areas along the Cincinnati arch and its extensions and in the Adirondack Mountains region, the anomalies have an intricate "birdseye" pattern. The trend of the magnetic belts and the texture

FIG. 10. Composite magnetic-anomaly map of the Midcontinent gravity high.

of the anomalies are characteristic of the Grenville basement province, which crops out immediately to the north in Canada. The eastern limit of this anomaly pattern is the New York–Alabama lineament, although similar subdued anomaly trends are observed over the deeply buried basement rocks of the Appalachian fold belt. In New England, the Grenville anomaly pattern is terminated at the eastern margin of the Adirondack Mountains and the western limit of the Appalachian orogen. The western margin of this terrane coincides with the location of the Grenville front, which separates the roughly 1 100-m.y.-old rocks of the Grenville basement province from the older basement to the west. The Grenville terrane has been mapped on the basis of the lithology and age of basement rocks obtained from deep drill holes and the nature of magnetic and gravity anomalies from its outcrop in Canada across Lake Huron and southeastern Michigan, southward across western Ohio and south-southwestward across Kentucky to central Tennessee (Bass, 1960; Rudman et al., 1965; Lidiak et al., 1966; Lidiak and Zietz, 1976; O'Hara and Hinze, 1980).

In the Lake Superior region, the regional magnetic anomalies occur in east- to northeast-striking, alternating bands of maxima and minima. In northeast Minnesota, where the basement rocks crop out, the positive anomalies generally correlate with granite gneiss, and the negative anomalies are identified with greenstone/sedimentary-rock terrane (Zietz, 1980). Narrow, high-intensity anomalies associated with steeply dipping magnetic strata—e.g., iron formations, slates, and tuffs—are useful in mapping the complex structure of the Archean and Proterozoic basins. The regional bands of alternating-sign anomaly trends, which are discernible from the Canadian border south to roughly the 43° N parallel, are disrupted by the anomalies of the Midcontinent rift system. They are broadened and attenuated to the east as their source continues under cover of the Phanerozoic sedimentary rocks of the Michigan basin. The strike of the anomalies turns from the south-southeast to the southeast in the vicinity of Lake Michigan, and the anomalies are terminated at the Grenville front. The regional anomalies continue to the west into the Dakotas, where they are truncated at the boundary with the Churchill basement province, which extends from Canada into North Dakota (Green et al., 1979; Lidiak, 1982).

The south-southeastward-trending magnetic anomalies of the buried Churchill province in eastern Montana and western North Dakota arc to the southeast in South Dakota (Figure 9). These trends are on strike with narrower and generally less intense anomalies in a region of lower magnetic-level background in Missouri and southern Illinois that cut across Kentucky and Tennessee to the vicinity of the New York–Alabama lineament. North of these trends in Illinois and Indiana and adjacent states, the trends are not clear-cut, but the regional anomalies tend to strike northwest. The Churchill province trends of the Dakotas are truncated, or disrupted, if they do indeed continue to the New York–Alabama lineament, in Nebraska, Iowa, Kansas, and Missouri by the generally north-northeastward-striking central positive anomaly and marginal negative anomalies associated with the Midcontinent rift system. The northern margin of these trends is delimited by a dipolar lineament anomaly with the positive side to the northeast, which strikes northwest from the New York–Alabama lineament to eastern Missouri. The Midcontinent rift magnetic anomalies in turn are bordered on both sides by northeast-striking bands of local, roughly equidimensional, intense positive anomalies. The anomalies are at least in part derived from 1 350-m.y.-old magnetite-rich granitic plutons intruded into the older basement rocks in Kansas (Yarger, 1981) and somewhat older granitic plutons farther northeast along the same belt (Coates et al., 1983).

An arcuate zone of broad positive magnetic anomalies lies south of the northwestward-striking anomalies in Missouri and Tennessee. The zone continues to the west into Colorado and curves south through Oklahoma to the Mexican border. Its northern margin is marked by a discontinuous bordering negative anomaly, which is most intense in south-central Kansas. Yarger (1981) suggests that this band of negative anomalies in Kansas may correlate with a boundary between an older, mesozonal granitic terrane to the north and a younger, epizonal granitic terrane to the south. He postulates that the boundary may be a suture associated with convergent-plate processes in mid-Proterozoic time. The magnetic-anomaly pattern of south-central Texas (Figure 11) is dominated by positive magnetic anomalies related to the Proterozoic crystalline rocks of the Llano uplift and their extensions into the subsurface to the north and southwest. It is speculated that the southwestern edge of this block in the vicinity of 30° N, 103° W, may be related to a discontinuity in the Paleozoic orogenic pattern in south Texas that King (1975) associated with a northwest-trending pre-Paleozoic transform fault. It is interesting to note that the "birds-eye" pattern of magnetic anomalies in south-central Texas associated with the Grenville-age rocks of the Llano uplift is similar to parts of the magnetic terrane over the Grenville basement province in the northeastern United States and eastern Canada. These anomalies are truncated, and thus presumably the Grenville basement province is terminated, in west Texas along a northeast-striking magnetic minimum that extends from southwestern Oklahoma (the Southern Oklahoma aulacogen) into southwest Texas, where anomalies from surface Tertiary volcanic rocks complicate the magnetic pattern. The northeast-trending magnetic minimum coincides with the position of the Grenville front as suggested by King (1975).

Cordilleran system

The term Cordilleran system (King, 1977) refers to the mountains of the western United States that were largely developed during the Mesozoic Era and were subsequently modified by various thermal and tectonic events in Tertiary and Quaternary time. On the basis of the magnetic anomalies observed on transcontinental magnetic traverses and scattered magnetic-anomaly maps, Pakiser and Zietz (1965) noted a change in character of magnetic anomalies along the eastern edge of the Cordilleran system. Within the Cordillera (Figure 12), they observed that the magnetic anomalies are muted relative to the pervasive and intense anomalies of the craton and the Appalachian orogen. They showed that this change in magnetic character coincides with variations in attributes of the lithosphere as indicated by other geophysi-

FIG. 11. Composite magnetic-anomaly map of south-central United States.

FIG. 12. Composite magnetic-anomaly map of western United States.

cal measurements. This change in magnetic character is confirmed by the patterns observed in the United States magnetic-anomaly map. Mabey et al. (1978) define this zone of attenuated anomalies where basement magnetic anomalies are largely absent, which they name the "quiet basement zone," as an extensive region of the Cordillera from southern Nevada to northern Idaho. Mabey and his co-workers point out that although the quiet zone could in part be explained by the deep burial of the basement crystalline rocks by the thrust-thickened sedimentary rocks of the Cordilleran system, the change in anomaly pattern is probably in large measure due to a variation in the character of the crust. The character change may be related to magnetic-susceptibility variations in the crust, a thinned crust, or a thinned magnetic crust owing to an elevated Curie isotherm (Zietz et al., 1970; Shuey et al., 1973). Mayhew finds an inverse correlation between muted Magsat long-wavelength magnetic anomalies in the western United States (1982a) and crustal magnetization (1982b) with heat flow, suggesting that the regional quiet zone is related to a shallow Curie isotherm. The extent of this quiet zone closely corresponds to the extent of the high-heat-flow regions of the Cordillera (Blackwell, 1978). Low regional magnetic-anomaly values extend westward from the Rio Grande rift in central New Mexico and Colorado to the Sierra Nevada in eastern California. The minimum is mottled by numerous high-gradient positive anomalies arising from Cenozoic surface volcanic and near-surface intrusive igneous rocks. The Colorado Plateau, characterized by relatively low heat-flow values, is associated with a regional magnetic high (Figure 13(1)). The magnetic trends of the Plateau are predominantly northeast, reflecting intra-basement lithologic variations and Tertiary to Holocene volcanic rocks. The intense magnetic anomalies of central and eastern Montana, which are largely derived from Precambrian basement crystalline rocks (Mabey et al., 1978), strike primarily northeast. However, a strong cross-pattern of basement-derived anomalies trending northwest in eastern Montana grade into the anomalies of the western Dakotas associated with the Churchill basement province boundary (Figure 13(2)).

A series of west-trending magnetic-anomaly zones transects the Cordillera from eastern Nevada into Colorado and possibly across the Rocky Mountain front into Kansas. Within the Cordillera, they are recognized by interruption of north-trending magnetic anomalies associated with Basin and Range province structures (Fuller, 1964) and west-trending belts of igneous rocks (Mabey and Morris, 1967). Several significant mineral districts occur on these west-trending belts (Zietz et al., 1969; Stewart et al., 1977). These belts have been interpreted as the site of deeply extending faults that have exerted a profound effect on the structural development of the area and the concentration of magma injection into the upper crust (Mabey et al., 1978). A change in the character of magnetic anomalies along a southwestward-striking zone from southeastern Wyoming across Colorado and Utah coincides with the Cheyenne belt in Wyoming (Figure 13(3)). This belt was recognized by Houston (1971) and Hills and Houston (1979) as a shear zone that separates the Archean rocks to the north and the Proterozoic rocks to the south. Hills and Houston suggest that the boundary represents the site of a Proterozoic subduction zone which, they hypothesize, dipped toward the southeast. The change in magnetic pattern also coincides with the northern edge of the Colorado lineament (Warner, 1978).

The complex array of magnetic anomalies of Washington, Oregon, and adjacent states is derived from the Miocene basalts of the Columbia River Group. Similar magnetic anomalies extend into Idaho and over the Snake River Plain (Figure 13(4)) (Mabey et al., 1978) northeastward across southern Idaho to the magnetic minimum over the Yellowstone caldera in northwest Wyoming (Figure 13(5)). The minimum reflects the low magnetic polarization of the altered volcanic rocks and the shallow Curie isotherm (Bhattacharyya and Leu, 1975). The intense positive anomalies, immediately northeast of the caldera, are produced by Tertiary andesitic volcanic rocks and Precambrian crystalline basement rocks (Eaton et al., 1975). The northeast trend of anomalies of the eastern Snake River Plain, the Yellowstone Plateau, and eastern Montana has a continuation to the southwest across Nevada as a subtle magnetic expression that Mabey et al. (1978) termed the Humboldt zone (Figure 13(6)). A northeast alignment of geologic features north of the Snake River Plain in Idaho and continuing on into Montana is an extensive Cretaceous batholith, the Idaho batholith (Figure 13(7)), composed primarily of granite and granodiorite. The principal components of this magnetically poorly defined batholith (Zietz et al., 1971), which intrude Proterozoic metasedimentary and gneissic rocks, are "S-type" (Chappell and White, 1974), according to Hyndman (1983). In contrast, the Boulder batholith to the northeast (Figure 13(8)) (46° N, 112.5° W), which is well defined by a northeast-striking anomaly block, is an "I-type" granite (Hyndman, 1983), which contains more magnetite than the S-type units. According to Chappell and White, the S-type granitic rocks are derived from sedimentary source rocks, while the I-type are derived from igneous rocks.

In California, the extensive Sierra Nevada batholith (Figure 13(9)) is observed to be associated with a magnetic maximum extending southward and roughly parallel to the Nevada border from approximately 41° N to the Mexican border. The batholith is composed of numerous granitoids of the I-type (Bateman, 1983) and thus is anticipated to contain magnetite. In fact, Oliver (1977) explains that the major changes in the intensity of the magnetic anomalies over the batholith primarily reflect the abundance of magnetite in the intrusives. An intense, low-gradient, curvilinear magnetic-anomaly maximum (Figure 13(10)), which lies west of the Sierra Nevada anomaly, overlies the sedimentary rocks of the Great Valley. This anomaly correlates with an intense gravity maximum and originates from mafic rocks, perhaps as a remnant of oceanic crust involved in subduction along the west coast. The magnetic-anomaly pattern north of the Transverse Ranges ($\cong$35° N) between the California coast and the San Andreas fault is almost featureless. This zone coincides with the Salinian crustal block (Figure 13(11)), which has abnormally high heat flow (Blackwell, 1978). It is interesting to speculate that this terrane, which is made up of granitic and metamorphosed basement rock immersed in the Franciscan subduction assemblage, is magnetically featureless because of the shallow Curie point isotherm suggested by the high heat flow.

FIG. 13. Composite magnetic-anomaly map of western United States with anomalies, anomaly patterns, and geologic features identified. 1 = Colorado Plateau; 2 = Churchill province boundary; 3 = Cheyenne belt; 4 = Snake River Plain; 5 = Yellowstone caldera; 6 = Humboldt zone; 7 = Idaho batholith; 8 = Boulder batholith; 9 = Sierra Nevada batholith; 10 = Great Valley anomaly; 11 = Salinian crustal block.

SUMMARY

The first magnetic-anomaly map of the conterminous United States and adjacent offshore areas has been composited from several hundred magnetic maps. Even though the individual maps were prepared from largely aeromagnetic surveys with inconsistent characteristics and reduction parameters, the composite map is useful in providing a regional framework for the interpretation of magnetic surveys of limited areas, in selecting areas for more detailed magnetic investigations, and in studying the distribution and character of regional geologic features.

The wide variation of magnetic-anomaly patterns, trends, and types is evidence for the diversity of the geologic terranes of the United States. In general, there are fewer and less intense anomalies in the Cordillera, probably as a result of several factors but primarily related to a decreased crustal magnetization associated with an abnormally shallow Curie isotherm. The anomalies of the Appalachian Mountains and the Cordilleran system strongly reflect the principal structural patterns of the orogens, but important exceptions occur such as those related to rocks underlying thrust sheets in the Appalachian Mountains and westward-striking anomaly trends in the Cordillera correlated with faults, igneous intrusives, and mineral deposits. Numerous regional magnetic-anomaly provinces are observed over the sedimentary-rock cover of the central United States. They provide critical evidence of the poorly known but long, complex history of the Precambrian basement rocks of the craton. These provinces are transected by conspicuous, intense, generally linear anomalies that originate from mafic extrusive and shallow intrusive igneous bodies within failed rifts, such as the Midcontinent rift system, the Southern Oklahoma aulacogen, and the Reelfoot rift buried beneath the Mississippi embayment. Many additional anomalies are observed that have great significance in characterizing the continental crust and interpreting its geologic history.

ACKNOWLEDGMENTS

We express our deep appreciation to the many individuals of the Society of Exploration Geophysicists and the U.S. Geological Survey whose continuing efforts made possible the preparation of the first magnetic-anomaly map of the conterminous United States. We also acknowledge the Phoenix Corporation for making the gray-shade version of this map available to us for use in this paper.

REFERENCES

Andreasen, G. E., and Zietz, I., 1969, Magnetic fields for a 4×6 prismatic model: Prof. Pap. 666, U.S. Geol. Surv.

Balsley, J. R., 1952, Aeromagnetic surveying, in Advances in geophysics: Academic Press, 313–349.

Barraclough, D. R., and Fabiano, E. B., 1978, Grid values and charts for the International Geomagnetic Reference Field, 1975: Int. Assoc. Geomagn. and Aeronomy Bull., **38**.

Bass, M. N., 1960, Grenville boundary: J. Geol., **68**, 673–677.

Bateman, P. C., 1983, A summary of critical relations in the central part of the Sierra Nevada batholith, California, U.S.A., in Roddick, J. A., Circum-Pacific plutonic terranes: Mem. 159, Geol. Soc. Am., 213–240.

Bayley, R. W., and Muehlberger, W. R. (compilers), 1968, Basement rock map of the United States (exclusive of Alaska and Hawaii): U.S. Geol. Surv., scale 1:2 500 000, 2 sheets.

Behrendt, J. C., and Klitgord, K. D., 1980, High resolution aeromagnetic survey of the U.S. Atlantic continental margin: Geophysics, **45**, 1813–1846.

Bhattacharyya, B. K., and Leu, Lei-kuang, 1975, Analysis of magnetic anomalies over Yellowstone National Park: Mapping of Curie point isothermal surface for geothermal reconnaissance: J. Geophys. Res., **80**, 4461–4465.

Blackwell, D. D., 1978, Heat flow and energy loss in the western United States, in Smith, R. B., and Eaton, G. P., Eds., Cenozoic tectonics and regional geophysics of the western Cordillera: Mem. 152, Geol. Soc. Am., 175–208.

Bleil, V., and Petersen, N., 1983, Variations in magnetization intensity and low-temperature titanomagnetite oxidation of ocean floor basalts: Nature, **301**, 384–388.

Braile, L. W., Keller, G. R., Hinze, W. J., and Lidiak, E. G., 1982, An ancient rift complex and its relation to contemporary seismicity in the New Madrid seismic zone: Tectonics, **1**, 225–237.

Carmichael, R. S., 1982, Magnetic properties of minerals and rocks, in Carmichael, R. S., Ed., Handbook of physical properties of rocks: CRC Press, **1**.

Cebull, S. E., Shurbet, D. H., Keller, G. R., and Russell, L. R., 1976, Possible role of transform faults in the development of apparent offsets in the Ouachita–southern Appalachian tectonic belt: J. Geol., **84**, 107–114.

Chappell, B. W., and White, A.J.R., 1974, Two contrasting granite types: Pac. Geol., **8**, 173–174.

Coates, M. S., Haimson, B. C., Hinze, W. J., and Van Schmus, W. R., 1983, Introduction to the Illinois Deep Hole Project: J. Geophys. Res., **88**, 7267–7275.

Cook, F. A., and Oliver, J. E., 1981, The early Paleozoic continental edge in the Appalachian orogen: Am. J. Sci., **281**, 993–1008.

Denison, R. E., Lidiak, E. G., Bickford, M. E., and Kisvarsanyi, E. B., 1984, Geology and geochronology of Precambrian rocks in the Central Interior region of the United States: Econ. Geol.

Eaton, G. P., Christiansen, R. L., Pitt, A. M., Mabey, D. R., Blank, H. R., Zietz, I., and Gettings, M. E., 1975, Magma beneath Yellowstone National Park: Science, **188**, 787–796.

Ervin, C. P., and McGinnis, L. D., 1975, Reelfoot rift: reactivated precursor to the Mississippi embayment: Geol. Soc. Am. Bull., **86**, 1287–1295.

Flawn, P. T., 1965, Basement—not the bottom but the beginning: AAPG Bull., **49**, 883–886.

Folger, D. W., Dillon, W. P., Grow, J. A., Klitgord, K. D., and Schlee, J. S., 1979, Evolution of the Atlantic continental margin of the United States, in Talwani, M., Hay, W., and Ryan, W.B.F., Eds., Deep-drilling results in the Atlantic Ocean: Continent. Margins and Paleoenviron., Am. Geophys. Union, Maurice Ewing ser., **3**, 87–108.

Fuller, M. D., 1964, Expressions of E–W fractures in magnetic surveys in parts of the U.S.A.: Geophysics, **29**, 602–622.

Green, A. G., Cumming, G. L., and Cedarwell, D., 1979, Extension of the Superior–Churchill boundary zone into southern Canada: Can. J. Earth Sci., **16**, 1691–1701.

Hall, D. H., 1974, Long-wavelength aeromagnetic anomalies and deep crustal magnetization in Manitoba and northwestern Ontario, Canada: Pure and Appl. Geophysics, **40**, 403–430.

Halls, H. C., 1978, The late Precambrian central North American rift system—a survey of recent geological and geophysical investigations, in Neumann, E. R., and Ramberg, I., Eds., Tectonics and geophysics of continental rifts: Reidel, NATO Adv. Stud. Inst., ser. C, **37D**, 111–123.

Hatcher, R. D., Jr., and Zietz, I., 1980, Tectonic implications of regional aeromagnetic and gravity data from the southern Appalachians, in Wones, D. R., Ed., The Caledonides in the U.S.A.: Va. Polytech. Inst. and State Univ. Mem. 2, 235–244.

Heyl, A. V., 1972, The 38th Parallel lineament and its relationship to ore deposits: Econ. Geol., **67**, 879–894.

Higgins, M. W., and Zietz, I., 1983, Geologic interpretation of geophysical maps of the pre-Cretaceous "basement" beneath the Coastal Plain of southeastern United States, in Hatcher, R. D., Jr., et al., Eds., Contributions to the tectonics and geophysics of mountain chains: Mem. 158, Geol. Soc. Am., 125–130.

Hildenbrand, T. G., Kane, M. F., and Stauder, W., 1977, Magnetic and gravity anomalies in the northern Mississippi embayment and their spatial relation to seismicity: Misc. Field Stud. Map MF-914, U.S. Geol. Surv.

Hills, F. A., and Houston, R. S., 1979, Early Proterozoic tectonics of the central Rocky Mountains, North America: Contrib. Geol., Univ. Wyo., **17**, 89–109.

Hinze, W. J., 1976, Report of the SEG Committee for a National Magnetic Anomaly Map: Geophysics, **41**, 1055.

Hinze, W. J., Kellogg, R. L., and O'Hara, N. W., 1975, Geophysical studies of basement geology of Southern Peninsula of Michigan: AAPG Bull., **59**, 1562–1584.

Hinze, W. J., Wold, R. J., and O'Hara, N. W., 1982, Gravity and magnetic anomaly studies of Lake Superior, *in* Wold, R. J., and Hinze, W. J., Eds., Geology and tectonics of the Lake Superior basin: Mem. 156, Geol. Soc. Am., 203–221.

Houston, R. S., 1971, Regional tectonics of the Precambrian rocks of the Wyoming province and its relationship to Laramide structure: 23rd Annu. Field Conf. Guideb., Wyo. Geol. Assoc., 19–27.

Hyndman, D. W., 1983, The Idaho batholith and associated plutons, Idaho and western Montana, *in* Roddick, J. A., Circum-Pacific plutonic terranes: Mem. 159, Geol. Soc. Am., 213–240.

Keller, F., Jr., Meuschke, J. L., and Alldredge, L. R., 1954, Aeromagnetic surveys of the Aleutian, Marshall, and Bermuda Islands: Eos, **35**, 558–572.

Keller, G. R., Bland, A. E., and Greenberg, J. K., 1982, Evidence for a major late Precambrian tectonic event (rifting?) in the eastern midcontinent region, U.S.A.: Tectonics, **1**, 213–223.

Keller, G. R., Lidiak, E. G., Hinze, W. J., and Braile, L. W., 1983, The role of rifting in the tectonic development of the midcontinent, U.S.A.: Tectonophysics, **94**, 391–412.

King, E. R., and Zietz, I., 1971, Aeromagnetic study of the midcontinent gravity high of central United States: Geol. Soc. Am. Bull., **82**, 2187–2208.

——— 1978, The New York–Alabama lineament: Geophysical evidence for a major crustal break in the basement beneath the Appalachian basin: Geology, **6**, 312–318.

King, P. B., 1975, Ancient southern margin of North America: Geology, **3**, 732–734.

——— 1977, The evolution of North America: Princeton Univ. Press.

King, P. B., and Beikman, H. M. (compilers), 1974, Geologic map of the United States (exclusive of Alaska and Hawaii): U.S. Geol. Surv., scale 1:2 500 000, 3 sheets.

Klitgord, K. D., and Behrendt, J. C., 1979, Basin structure of the U.S. Atlantic margin, *in* Watkins, J. S., Montadert, L., and Dickerson, P. W., Eds., Geological and geophysical investigations of continental margins: Mem. 29, AAPG, 85–112.

Larson, R. L., and Hilde, T.W.C., 1975, A revised time scale of magnetic reversals for the Early Cretaceous and Late Jurassic: J. Geophys. Res., **80**, 2586–2594.

Lidiak, E. G., 1972, Precambrian rocks in the subsurface of Nebraska: Bull. 26, Nebr. Geol. Surv.

——— 1982, Basement rocks of the main interior basins of the midcontinent: J., Univ. Mo.–Rolla, **3**, 5–24.

Lidiak, E. G., Marvin, R. F., Thomas, H. H., and Bass, M. N., 1966, Geochronology of the midcontinent region, United States, Part 4, Eastern area: J. Geophys. Res., **71**, 5427–5438.

Lidiak, E. G., and Zietz, I., 1976, Interpretation of aeromagnetic anomalies between latitudes 37° N and 38° N in the eastern and central United States: Spec. Pap. 167, Geol. Soc. Am.

Lindsley, D. H., Andreasen, G. E., and Balsley, J. R., 1966, Magnetic properties of rocks and minerals, *in* Clark, S. P., Ed., Handbook of physical constants: Mem. 97, Geol. Soc. Am., 543–552.

Mabey, D. R., and Morris, H. T., 1967, Geologic interpretation of gravity and aeromagnetic maps of Tintic Valley and adjacent areas, Tooele and Juab Counties, Utah: Prof. Pap. 516-D, U.S. Geol. Surv.

Mabey, D. R., Zietz, I., Eaton, G. P., and Kleinkopf, M. D., 1978, Regional magnetic patterns in part of the Cordillera in the western United States, *in* Smith, R. B., and Eaton, G. P., Eds., Cenozoic tectonics and regional geophysics of the western Cordillera: Mem. 152, Geol. Soc. Am., 93–106.

Mayhew, M. A., 1982a, An equivalent layer magnetization model for the United States derived from satellite altitude magnetic anomalies: J. Geophys. Res., **87**, 4837–4845.

——— 1982b, Application of satellite magnetic anomaly data to Curie isotherm mapping: J. Geophys. Res., **87**, 4846–4854.

O'Hara, N. W., and Hinze, W. J., 1980, Regional basement geology of Lake Huron: Geol. Soc. Am. Bull., **91**, 348–358.

Oliver, H. W., 1977, Gravity and magnetic investigations of the Sierra Nevada batholith, California: Geol. Am. Bull., **88**, 445–461.

Oray, E., Hinze, W. J., and O'Hara, N. W., 1973, Gravity and magnetic evidence for the eastern termination of the Lake Superior syncline: Geol. Soc. Am. Bull., **84**, 2763–2780.

Pakiser, L. C., and Zietz, I., 1965, Transcontinental crustal and upper-mantle structure, Rev. Geophys., **3**, 505–520.

Peddie, N. W., 1982, International geomagnetic reference field: The third generation: J. Geomagn. Geoelect., **34**, 309–326.

——— 1983, International geomagnetic reference field—its evolution and the difference in total field intensity between new and old models for 1965–1980: Geophysics, **48**, 1961–1966.

Popenoe, P., and Zietz, I., 1977, The nature of the geophysical basement beneath the Coastal Plain of South Carolina and northeastern Georgia, *in* Rankin, D. W., Ed., Studies related to the Charleston, South Carolina earthquake of 1886—A preliminary report: Prof. Pap. 1028, U.S. Geol. Surv., 119–138.

Rankin, D. W., 1975, The continental margin of eastern North America in the southern Appalachians: The opening and closing of the proto-Atlantic Ocean: Am. J. Sci., **275–A**, 298–336.

Reed, J. C., Jr., Owens, J. P., and Stockard, H. P., 1968, Interpretation of basement rocks beneath the Atlantic Coastal Plain from reconnaissance aeromagnetic data [abstr.]: Spec. Pap. 115, Geol. Soc. Am., 182–183.

Rudman, A. J., Summerson, C. E., and Hinze, W. J., 1965, Geology of basement in midwestern United States: AAPG Bull., **84**, 2763–2780.

Schlee, J. S., Dillon, W. P., and Grow, J. A., 1979, Structure of the continental slope off the eastern United States: Spec. Publ. 27, SEPM, 95–117.

Schnetzler, C. C., Taylor, P. T., Langel, R. A., Hinze, W. J., and Phillips, J. D., in press, Comparison between the recent United States composite magnetic anomaly map and Magsat anomaly data: J. Geophys. Res., **89**.

Sexton, J. L., Hinze, W. J., von Frese, R.R.B., and Braile, L. W., 1982, Long-wavelength aeromagnetic anomaly map of the conterminous United States: Geology, **10**, 364–369.

Shuey, R. T., Schellinger, D. K., Johnson, E. H., and Alley, L. B., 1973, Aeromagnetics and the transition between the Colorado Plateau and the Basin and Range Provinces: Geology, **1**, 107–110.

Shurbet, D. H., Keller, G. R., and Friess, J. P., 1976, Remanent magnetization from comparison of gravity and magnetic anomalies: Geophysics, **41**, 56–61.

Society of Exploration Geophysicists, 1982, Gravity anomaly map of the United States (exclusive of Alaska and Hawaii): scale 1:2 500 000, 2 sheets.

Stewart, J. H., Moore, W. J., and Zietz, I., 1977, East–west patterns of Cenozoic igneous rocks, aeromagnetic anomalies, and mineral deposits, Nevada and Utah: Geol. Soc. Am. Bull., **81**, 67–77.

Strangway, D. W., 1981, Magnetic properties of rocks and minerals, *in* Touloukian, R. S., Judd, W. R., and Roy, R. F., Eds., Physical properties of rocks and minerals: McGraw-Hill, **II-2**.

Taylor, P. T., Zietz, I., and Dennis, L. S., 1968, Geologic implications of aeromagnetic data for the eastern continental margin of the United States: Geophysics, **33**, 755–780.

Thomas, M. D., 1983, Tectonic significance of paired gravity anomalies in the southern and central Appalachians, *in* Hatcher, R. D., Jr., et al., Eds., Contributions to the tectonics and geophysics of mountain chains: Mem. 158, Geol. Soc. Am., 113–124.

Thomas, W. A., 1977, Evolution of Appalachian–Ouachita salients and recesses from reentrants and promontories in the continental margin: Am. J. Sci., **277**, 1233–1278.

Tinnel, E. P., and Hinze, W. J., 1981, Preparation of magnetic anomaly profile and contour maps from DOE-NURE aerial survey data, Volume 1: Processing Procedures TM-155, Oak Ridge Natl. Lab.

U.S. Geological Survey and American Association of Petroleum Geologists, 1961, Tectonic map of the United States (exclusive of Alaska and Hawaii): U.S. Geol. Surv., scale 1:2 500 000, 2 sheets.

U.S. Geological Survey and Society of Exploration Geophysicists, 1982, Composite magnetic anomaly map of the United States, Part A: Conterminous United States: Map GP 954-A, U.S. Geol. Surv., scale 1:1 250 000, 2 sheets.

Vacquier, V., Steenland, N. C., Henderson, R. G., and Zietz, I., 1951, Interpretation of aeromagnetic maps: Mem. 47, Geol. Soc. Am.

Van Schmus, W. R., and Bickford, M. E., 1981, Proterozoic chronology and evolution of midcontinent region, North America, *in* Kroner, A., Ed., Precambrian plate tectonics: Elsevier, 261–296.

Warner, L. A., 1978, The Colorado lineament: A middle Precambrian wrench fault system: Geol. Soc. Am. Bull., **89**, 161–171.

Wasilewski, P. J., Thomas, H. H., and Mayhew, M. A., 1979, The Moho as a magnetic boundary: Geophys. Res. Lett., **6**, 541–544.

Williams, H., and Hatcher, R. D., Jr., 1983, Appalachian suspect terranes, *in* Hatcher, R. D., Jr., et al., Eds., Contributions to the tectonics and geophysics of mountain chains: Mem. 158, Geol. Soc. Am., 33–53.

Yarger, H. L., 1981, Aeromagnetic survey of Kansas: Eos, **62**, 173–178.

Zartman, R. E., 1977, Geochronology of some alkalic rock provinces in eastern and central United States: Earth Planet. Sci. Lett. Annu. Rev., **5**, 257–286.

Zen, E-an, 1983, Exotic terranes in the New England Appalachians—limits, candidates, and ages: A speculative essay, *in* Hatcher, R. D., Jr., et al., Eds., Contributions to the tectonics and geophysics of mountain chains: Mem. 158, Geol. Soc. Am., 55–82.

Zietz, I., 1980, Exploration of the continental crust using aeromagnetic data, *in* Continental tectonics: Stud. Geophys., Natl. Acad. Sci., 127–138.

Zietz, I., Andreasen, G. E., and Cain, J. C., 1970, Magnetic anomalies from satellite magnetometer: J. Geophys. Res., **75**, 4007–4015.

Zietz, I., Bateman, P. C., Case, J. E., Crittenden, M. D., Griscom, A., King, E. R., Roberts, R. J., and Lorentzen, G. R., 1969, Aeromagnetic investigation of crustal structure for a strip across the western United States: Geol. Soc. Am. Bull., **80**, 1703–1714.

Zietz, I., Hearn, B. C., Higgins, M. W., Robinson, G. D., and Swanson, D. H., 1971, Interpretation of an aeromagnetic strip across the northwestern United States: Geol. Soc. Am. Bull., **82**, 3347–3372.

Preparation of magnetic anomaly maps of Alaska and Hawaii

Richard H. Godson*

ABSTRACT

Regional magnetic-anomaly maps of the States of Alaska and Hawaii and adjacent offshore areas have been published at scales of 1:2 500 000. These maps were manually compiled from previously published maps of various scales and contour intervals by merging 200-gamma (200-nT) contours. A study of the International Geomagnetic Reference Field (IGRF) in Alaska revealed that this regional field would not cause a problem in merging data from surveys flown several years apart. Comparisons of data flown with different specifications over the same areas in Alaska show variations in anomaly trends and wavelengths; therefore, individual survey specifications should be considered when interpreting this map.

INTRODUCTION

As part of the cooperative effort between the Society of Exploration Geophysicists and the U.S. Geological Survey (Hinze, 1976), magnetic-anomaly maps of the States of Alaska and Hawaii and adjacent offshore areas were prepared at a scale of 1:2 500 000 with contour intervals of 200 gammas. These maps have been published in black and white (Godson, 1982) and also in color (Godson, 1983) in two sheets on Albers equal-area projections.

The compilation procedures used to produce the Hawaii map were different from the methods used to prepare the Alaska map, and therefore each state compilation is discussed separately.

HAWAII COMPILATION

The Hawaiian Islands and adjacent islets are scattered over a considerable area of nearly 3 000 km; only data on the eight principal islands and their adjacent offshore areas were used for this project. The map was compiled by assimilating maps originally published by Malahoff and Woollard (1968). The 200-gamma contours from these maps were traced, reduced to the publication scale, and then manually adjusted in sections to an Albers equal-area-projection base map. The following information about the field surveys and data-reduction procedures used for the original maps is derived from Malahoff and Woollard (1966, 1968, 1970).

Field surveys

The data used to produce the map were obtained from two surveys. The first survey in 1963–64 over the islands was flown at barometric elevations between 2 440 and 3 050 m, except over the peaks of Mauna Loa and Mauna Kea on the island of Hawaii, where the barometric flight elevation was 4 575 m. This survey was flown at a 1.6-km flight-line spacing at a sampling interval of 250 m, using a proton-precession magnetometer. Positioning was accomplished by plotting locations from film strips on 1:62 500-scale topographic maps on which were annotated cultural and topographic features observed during the flights. Horizontal accuracy is regarded to be better than 150 m.

The second survey in 1965–66 consisted of offshore flight lines that radiated out from the principal Hawaiian islands with spacings that varied from 5 to 15 km (Figure 1). The flight elevation was about 3 100 m above sea level, with observations recorded every 7 s with a proton-precession magnetometer. Navigation was based on ground control from a Loran A/Loran C recording system. Profile crossing errors indicated that navigation was good to only ±6.4 km.

Regional field

As these surveys were performed before the International Geomagnetic Reference Field (IGRF) was established in 1968, a regional field was constructed from recorded data by using the smooth field in areas of little magnetic disturbance and interpolating across areas of more intense magnetic anomalies. A planar field with a gradient of 3.8 gammas/km to the north resulted from this exercise. This regional field was then subtracted from the observed data to derive anomaly values used to produce contour maps.

ALASKA COMPILATION

The magnetic-anomaly map of Alaska was compiled from a variety of sources; data from about 90 surveys were used.

---

*U.S. Geological Survey, Box 25046, MS 964, Denver Federal Center, Denver, CO 80225.

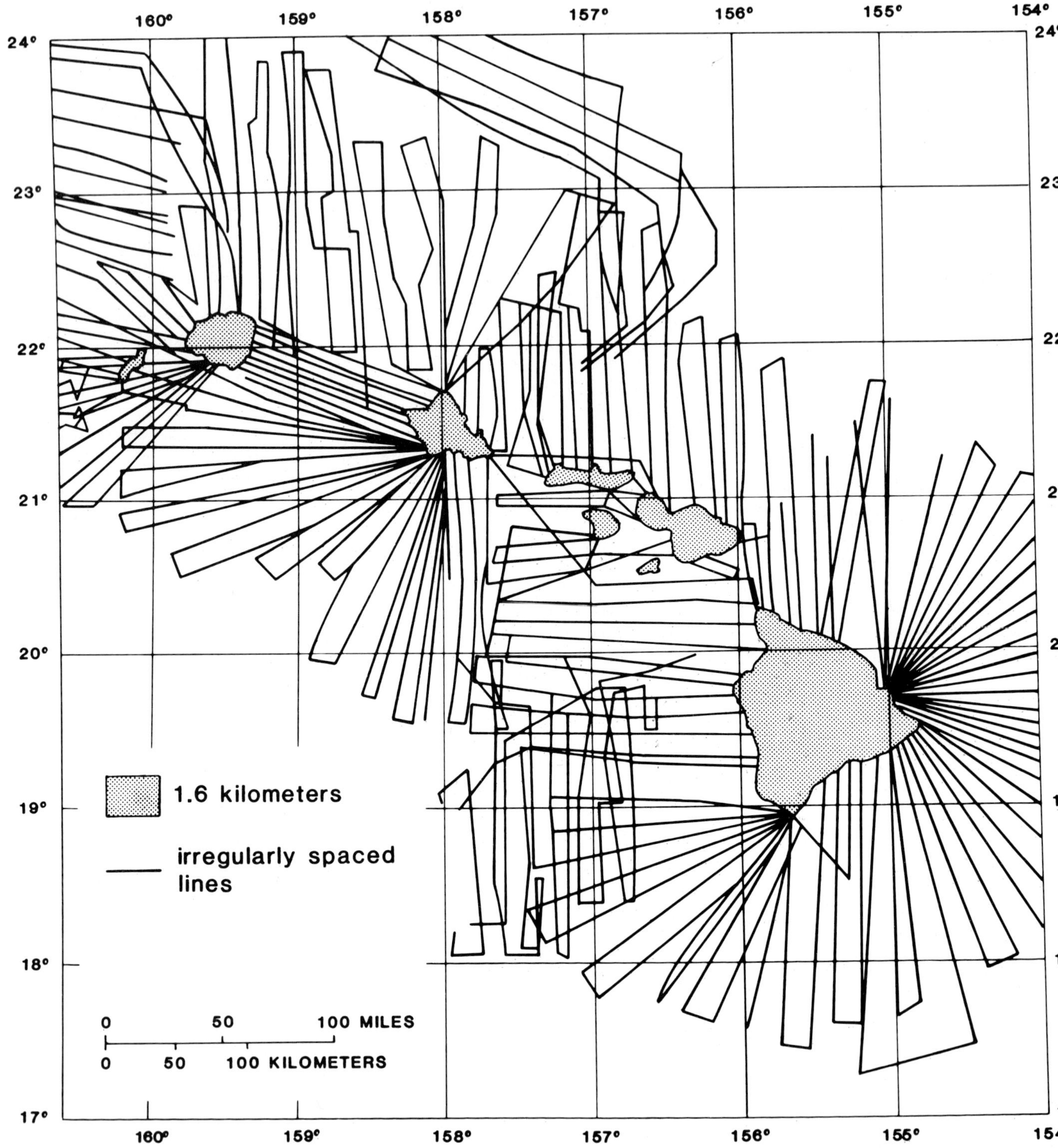

FIG. 1. Flight-line spacings of surveys of the Hawaiian Islands.

Aeromagnetic maps of large areas of the State have been released as U.S. Geological Survey Open-File Reports (Decker and Karl, 1977a–e; Decker, 1979) at a scale of 1:1 000 000, and athough the individual surveys were not merged on these maps, the contours from them were used wherever possible. Larger scale maps, usually at 1:250 000, were the most commonly used maps in the compilation.

The first step in the compilation was to join 200-gamma contours between adjacent maps and, where necessary, to reduce the maps to the compilation scale of 1:1 000 000. The contours were then placed on a 1:1 000 000-scale master base map prepared using an Albers equal-area projection. The map was then photographically reduced to the 1:2 500 000 publication scale and subsequently edited to eliminate closely spaced contours.

Most of these maps were previously reduced to anomaly form by removal of either the 1965 IGRF (Fabiano and Peddie, 1969) or the 1975 IGRF (Barraclough and Fabiano, 1978). Magnetic maps not previously reduced to anomaly form were digitized and the appropriate IGRF removed; machine-contoured anomaly maps were then prepared and compiled into the Alaskan map.

One of the reasons it is sometimes difficult to match contours on maps from surveys that were flown several years apart is that the continual change of the IGRF cannot be predicted accurately. The predicted change with time of both the 1965 and the 1975 IGRF is small in Alaska. In order to verify this, a study was made of readings taken at repeat stations, which are locations where magnetic measurements are taken every 5 years. By calculating the actual changes of the magnetic field over a period of several years at several stations, and comparing these changes with the secular variations using the IGRF models, the errors involved in using the IGRF can be determined. Figure 2 shows a contour map of these errors calculated from readings taken at 12 stations in 1965 and again in 1975. The errors are small when one considers making maps with 200-gamma intervals, and therefore no special compilation techniques were used to correct for the use of the IGRF.

Type of surveys

Figure 3 shows the different types of line spacings of the surveys used in the compilation. Most surveys with line spacings of 3 km or less were flown at 300 m above ground. Almost all of the surveys with 10-km spacings were conducted as part of the National Uranium Resources Evaluation (NURE) program and were flown at 120 m above ground.

The bold lines offshore in Figure 3 represent ship-track locations where the spacings between lines are irregular. Other marine areas are shaded, and even though the individual ship tracks are irregular, their spacing is adequate to cover most of the area. All of the data in the marine areas were collected by ships except in the eastern and southern parts of the Gulf of Alaska, where the data used were from an aeromagnetic survey conducted by the U.S. Naval Oceanographic Office.

NURE data

Processing.—As previously stated, published maps were used for the compilation, and generally they were prepared by standard methods with no special processing. This was not true for the NURE data, which make up about 40 percent of the onshore part of the map. Determining how to handle the NURE data was a special case because of the wide flight-line spacing of 10 km and the low altitude of 120 m above ground. These specifications were used because the surveys were of a reconnaissance nature concerned mainly with airborne gamma-ray measurements.

These surveys produced high-frequency/high-amplitude anomalies on many flight lines and made the data difficult to contour in some areas. To attempt to alleviate this problem, a 4-km low-pass filter was applied to data from each individual flight line before machine contouring (Tinnel and Hinze, 1981). The selection of a 4-km filter was empirical and was deemed sufficient to preserve anomalies that could be seen at the publication scale. This processing of NURE data was performed by Union Carbide Corporation in Oak Ridge, Tennessee, under a contract from the U.S. Department of Energy. A few contour maps of NURE data were provided by Exxon Company U.S.A. with no filtering, and some of these were used in this compilation.

Comparison with other data.—The filtering of the NURE data suppressed the high-frequency anomalies prevalent on individual flight lines and generally decreased the amplitude of all anomalies. Many 1° × 3° quadrangles were flown according to NURE specifications and also by surveys having closer line spacings and higher altitudes. In order to determine if the NURE data produced contour maps of sufficient quality for this project, comparisons at the compilation scale were made with maps from more detailed surveys. Figures 4 and 5 show representative examples of these comparisons. It is obvious from these figures that significant differences exist in the shapes and trends of many anomalies. A major discrepancy can be seen in Figure 5 in which a narrow anomaly of about 80 km in length is only partially shown on the NURE map. This lack of detection results from an anomaly trend that is nearly the same as the east–west flight-line direction used for both surveys.

In spite of these deficiencies, the NURE data were considered good enough for a regional map at the publication scale. In general, these data exhibit longer wavelengths than the closer spaced data. This is a result of the survey specifications and the data-processing procedures, and should not necessarily be interpreted as having geologic implications. Similar differences in wavelengths are observed between other types of data in other areas where the survey specifications, such as spacing, height, or direction, vary. Therefore, caution should be exercised when interpreting the Alaskan magnetic-anomaly map that was compiled from a wide variety of data sources.

ACKNOWLEDGMENTS

Helpful suggestions were received from members of the National Magnetic Anomaly Map (NMAM) editorial committee: Alan Cooper, John Corbett, Michael Fuller, William Hanna, James Heirtzler, William Hinze, James Schwartz,

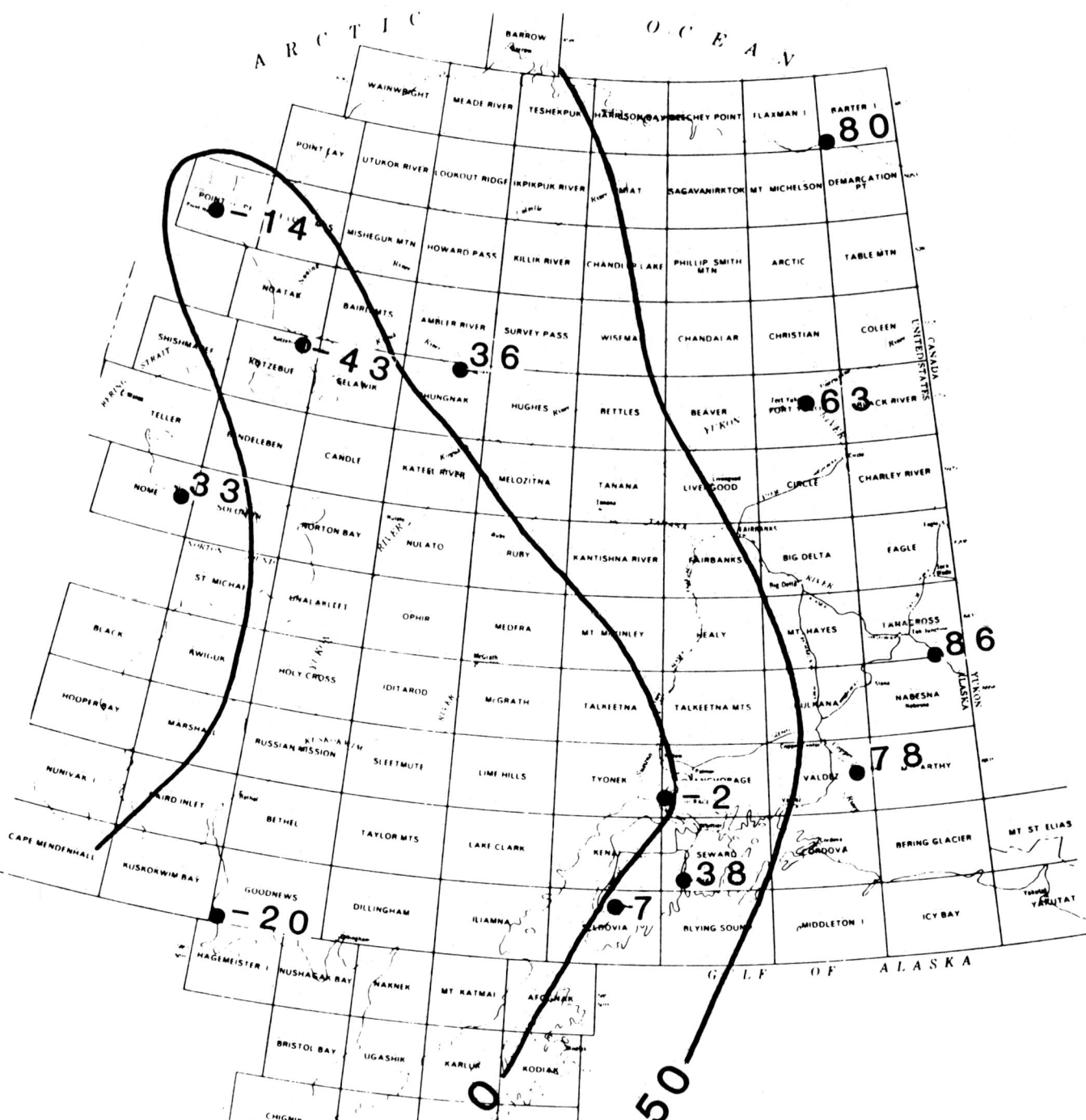

FIG. 2. Contours and values of differences in gammas between observed magnetic values and calculated values, using the IGRF models at 12 stations in Alaska between 1965 and 1975.

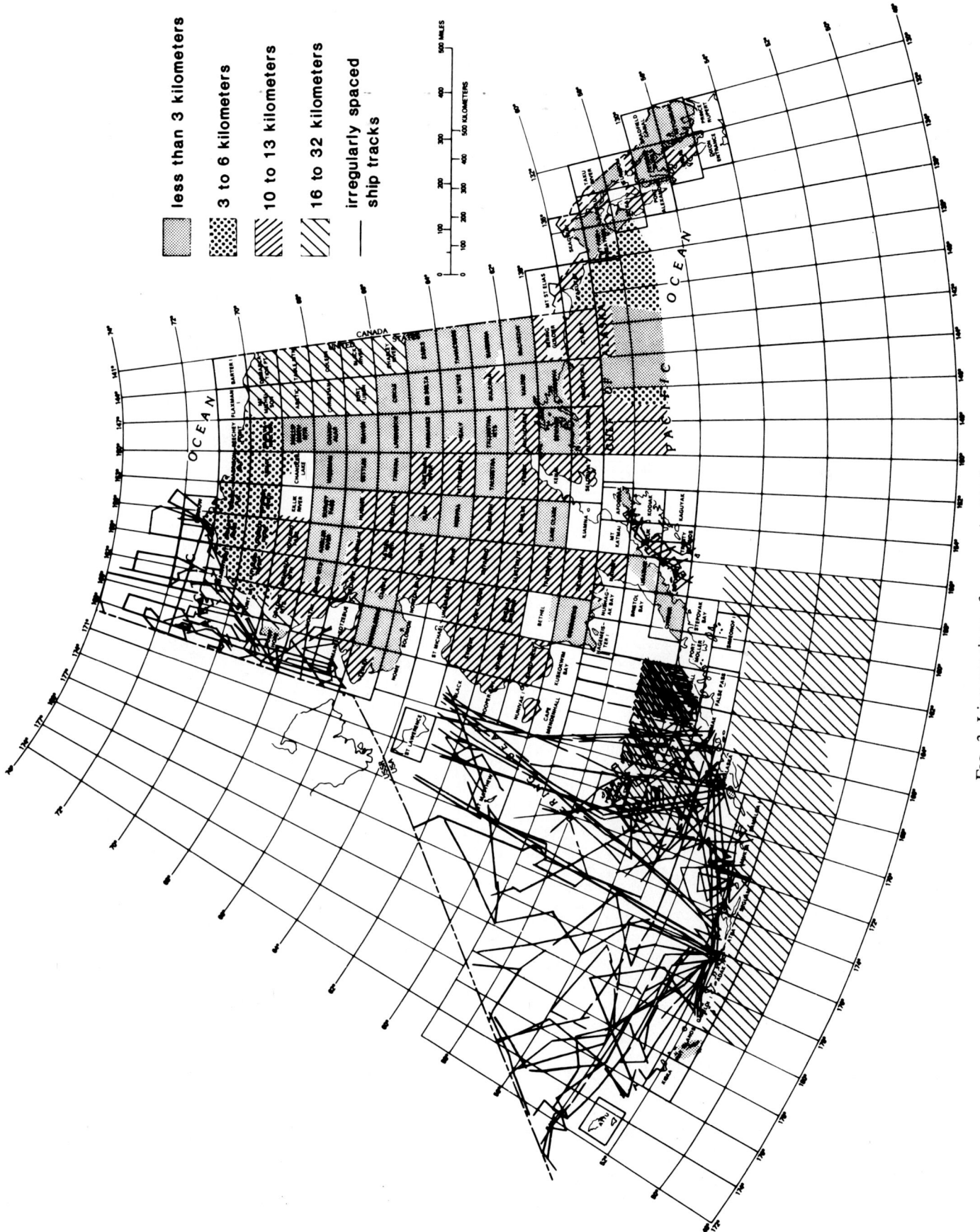

FIG. 3. Line spacings of surveys in Alaska.

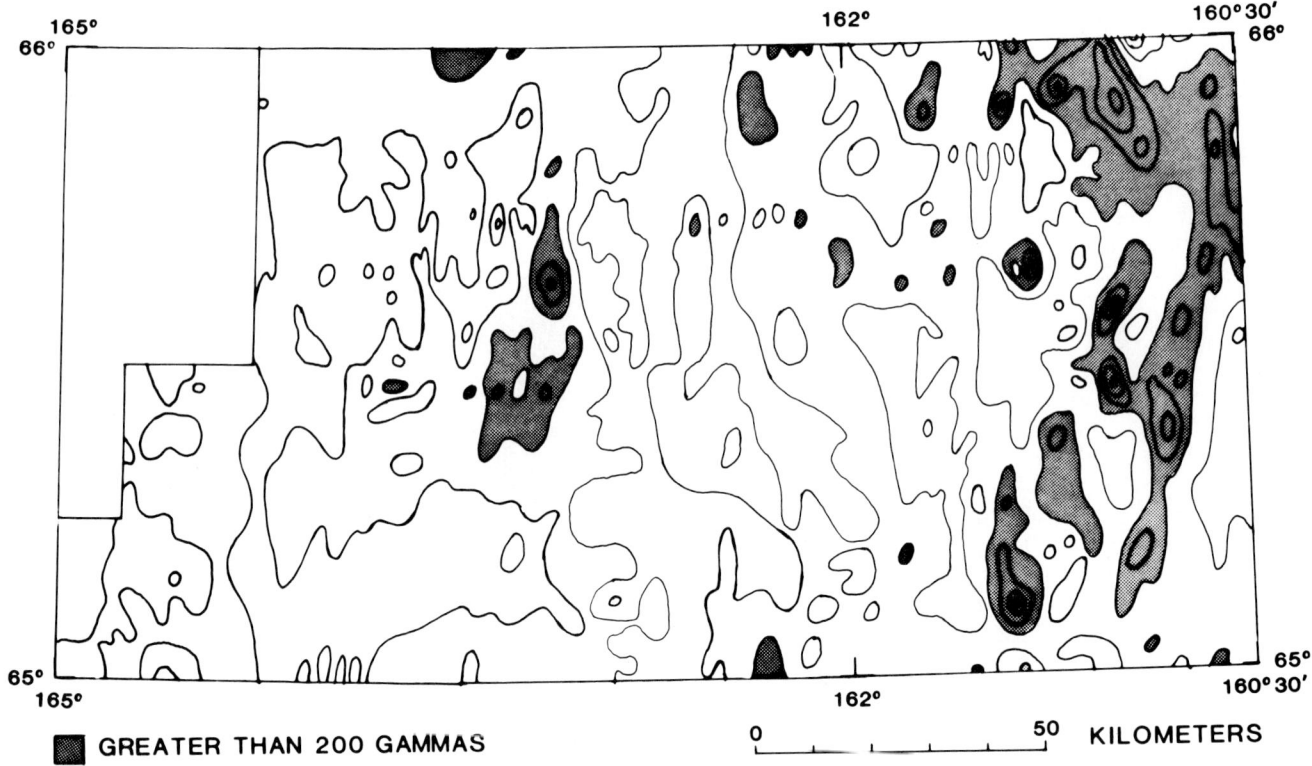

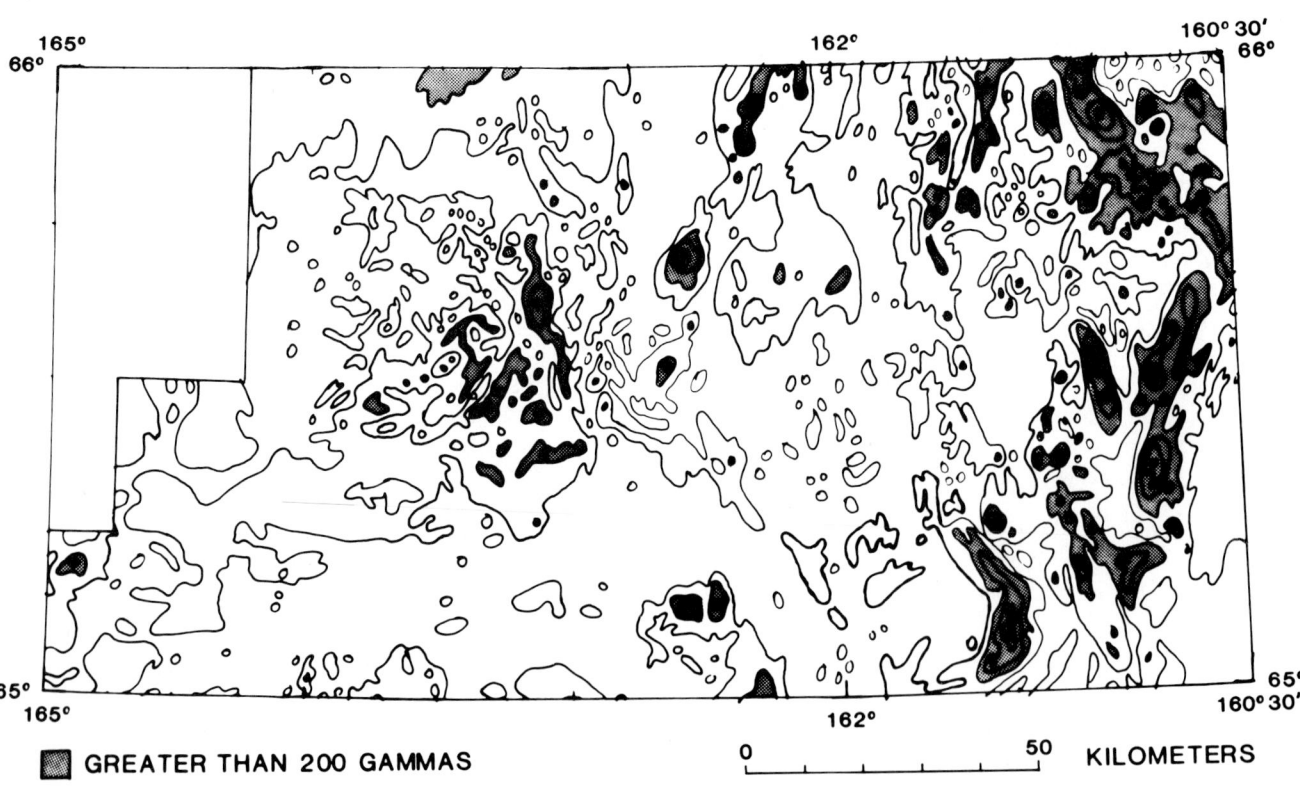

FIG. 4. Comparison of contours of data obtained from surveys with different line spacings and altitudes in the Bendeleben and Candle quadrangles, Alaska. Contour intervals are 200 gammas, and the flight direction for both surveys is east–west.

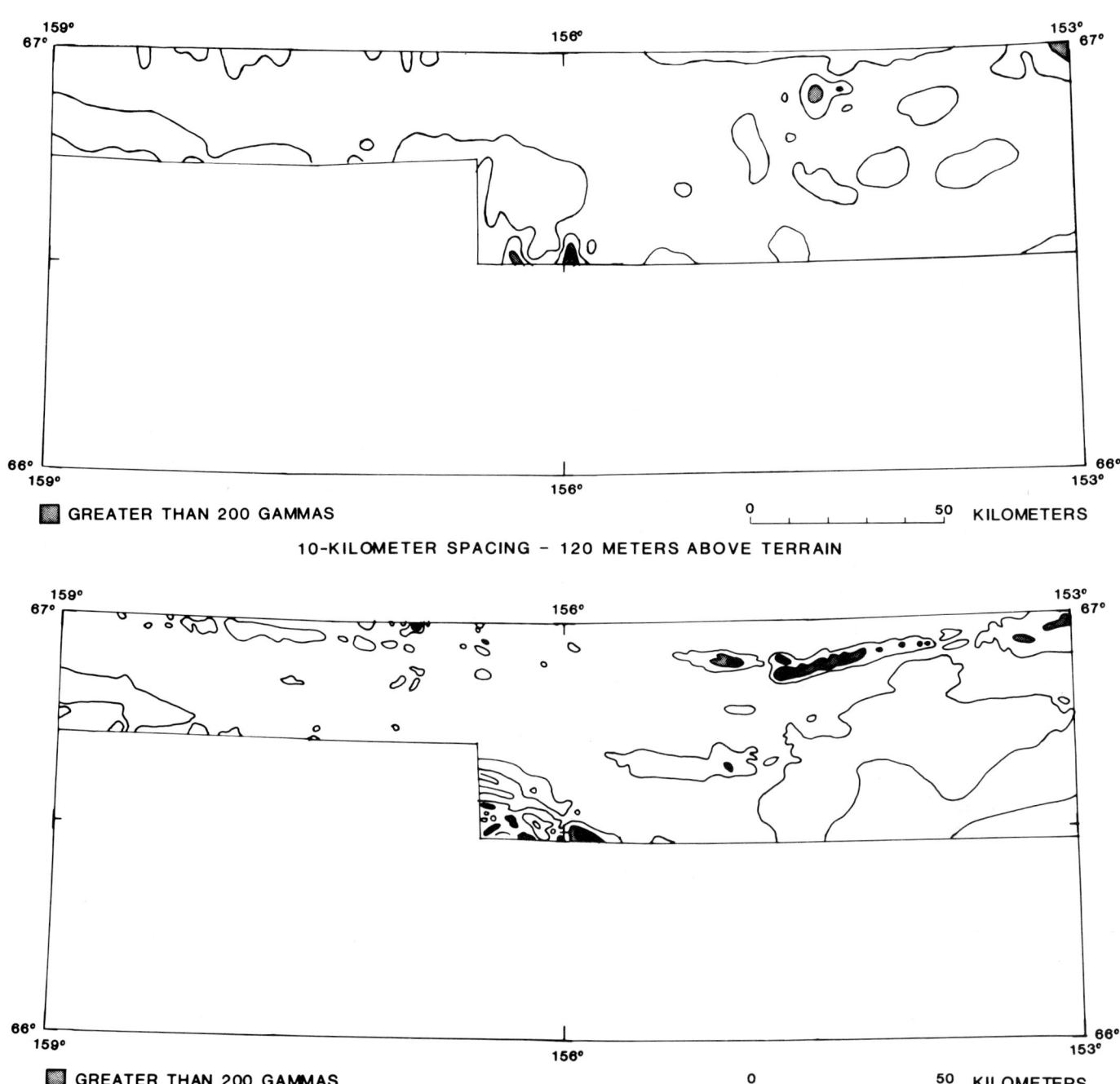

FIG. 5. Comparison of contours of data obtained from surveys with different line spacings and altitudes in the Shungnak and Hughes quadrangles, Alaska. Contour intervals are 200 gammas, and the flight direction for both surveys is east–west.

and Isidore Zietz. Special thanks are given to W. J. Hinze in obtaining the NURE maps from Union Carbide Corporation, to J. Heirtzler for assimilating the marine data, and to J. Schwartz for review of map contours.

Compilation of the maps was accomplished by U.S. Geological Survey personnel under the supervision of Frank Petrafeso. James Linton, Herbert Pierce, and Joseph Plesha assisted in the compilation of the Alaska map, and Patricia Hill and Jackie Williams assisted in the compilation of the Hawaii map.

REFERENCES

Barraclough, D. R., and Fabiano, E. B., 1978, Grid values and charts for the International Geomagnetic Reference Field: Nat. Tech. Inf. Serv. Rep. PB276 630, U.S. Dep. Commerce.

Decker, J., 1979, Preliminary aeromagnetic map of southeastern Alaska: Open-File Rep. 79-1694, U.S. Geol. Surv., scale 1:1 000 000.

Decker, J., and Karl, S., 1977a, Preliminary aeromagnetic map of central Alaska: Open-File Rep. 77-168E, U.S. Geol. Surv., scale 1:1 000 000.

———1977b, Preliminary aeromagnetic map of Seward Peninsula, Alaska: Open-File Rep. 77-796E, U.S. Geol. Surv., scale 1:1 000 000.

———1977c, Preliminary aeromagnetic map of the Brooks Range and Arctic Slope, Alaska: Open-File Rep. 77-166E, U.S. Geol. Surv., scale 1:1 000 000.

———1977d, Preliminary aeromagnetic map of the eastern part of southern Alaska: Open-File Rep. 77-169E, U.S. Geol. Surv., scale 1:1 000 000.

———1977e, Preliminary aeromagnetic map of the western part of southern Alaska: Open-File Rep. 77-169J, U.S. Geol. Surv., scale 1:1 000 000.

Fabiano, E. B., and Peddie, N. W., 1969, Grid values of total magnetic intensity, International Geomagnetic Reference Field—1965: Tech. Rep. C and GS 38, U.S. Environ. Sci. Serv. Adm.

Godson, R. H., 1982, compiler, Composite magnetic anomaly map of the United States, Part B—Alaska and Hawaii: Open-File Rep. 82-970, U.S. Geol. Surv., scale 1:2 500 000, 2 sheets.

———1984, compiler, Composite magnetic anomaly map of the United States, Part B—Alaska and Hawaii: Geophys. Invest. Map GP-954-B, U.S. Geol. Surv., scale 1:2 500 000, 2 sheets.

Hinze, W. J., 1976, Report of the SEG committee for a National Magnetic Anomaly Map: Geophysics, **41**, 1055.

Malahoff, A., and Woollard, G. P., 1966, Magnetic surveys over the Hawaiian Islands and their geologic implications: Pac. Sci. **20**, 285-311.

———1968, Magnetic and tectonic trends over the Hawaiian Ridge, in Knopoff, C., Drake, C. L., and Hart, P. J., Eds., The crust and upper mantle of the Pacific area: Monogr. 12, Am. Geophys. Union, 501-516.

———1970, Geophysical studies of the Hawaiian Ridge and Murray Fracture Zone, in Maxwell, W. E., Ed., The sea: Wiley Intersci., **4**, pt. 2, 73-131.

Tinnel, E. P., and Hinze, W. J., 1981, Preparation of magnetic anomaly profile and contour maps from DOE-NURE aerial survey data: Open-File Rep. GJBX-177(81), U.S. Dep. Energy.

Preparation and overview of the gravity anomaly map of the United States

Norbert W. O'Hara* and Paul L. Lyons‡

ABSTRACT

The Gravity Anomaly Map of the United States was published in October 1982 by the Society of Exploration Geophysicists. The regional map was compiled and edited in collaboration with the U.S. Geological Survey, the U.S. Defense Mapping Agency, and the National Oceanic and Atmospheric Administration.

Publication of the conterminous United States Bouguer and offshore free-air gravity data is at a scale of 1:2 500 000. Map scale, base, and projection are identical with the existing geologic, tectonic, basement-rock, and magnetic maps of the United States. The contour interval is 5 mGal, with gravity amplitudes depicted in color intervals of 25 mGal. Anomalies were calculated using the IGSN71 standard and the GRS67 ellipsoid with Bouguer values based on a rock density of 2.67 g/cm^3. Nearly 2 million digital gravity-data points were initially examined, then sorted to produce an equally gridded station spacing of 4 km. The screened data were next terrain-corrected, where appropriate, and machine-contoured. Detailed editing, assimilation of nondigital data where necessary, hand contouring, and final cartographic work completed the process.

The most obvious characteristic of the new national gravity map is the spectacular contrast between the generally high anomalous amplitudes observed on the eastern part of the map when compared with the predominance of low anomalous amplitudes illustrated on the western part. Typical regional geologic structural features recognizable from the gravity anomalies in the eastern half of the United States include the Midcontinent rift system, present and possibly past continental margins, Precambrian-basement trends, orogenic belts, buried basins, and more obvious mountainous terrain. Major structural features reflected by the anomalies in the western United States include the Southern Rockies, Colorado Plateau, Idaho batholith, Basin and Range pattern, Columbia Plateau volcanic-rock region, and indications of ancient plate collisions, subduction, and crustal uplift associated with the Pacific Cordillera.

Comparisons of specific geologic structures, represented on the geologic and tectonic maps, with corresponding gravity and magnetic anomalies provide essential information regarding the distribution and configuration of basement crystalline rocks, structural and lithologic provinces, zones of crustal weakness, and the distribution of mafic rocks and sedimentary basins.

INTRODUCTION

Gravity methods, used in the analysis of geologic structure, have had a long and successful history in helping to study the Earth's crust for scientific and applied objectives. Although many require detailed surveys with restricted areal coverage, most objectives first require a broad, regional approach to establish the basic scientific framework. Regional gravity-anomaly maps are particularly useful in mapping (1) geographic distribution and configuration of the basement crystalline rocks, (2) structural and lithologic provinces, (3) zones of crustal weakness, (4) mass imbalances within the lithosphere, (5) geometry of sedimentary basins, and (6) the distribution of extrusive and intrusive igneous rocks.

Realizing the need for and potential importance of regional gravity data, the Society of Exploration Geophysicists (SEG) and the U.S. Geological Survey (USGS) agreed in 1975 to co-sponsor the compilation and publication of the first gravity map of North America. However, the Gravity Anomaly Map (GAM) Committee, appointed to carry out this assignment, soon realized that it was first necessary to

---

*Scientific Evaluations & Applications, 1300 South Ramona Avenue, Indialantic, FL 32903.
‡Lyons & Lyons, Inc., 1519 South Baltimore Street, Tulsa, OK 74101.
Much of the material presented here was published previously in the December 1983 issue of Geotimes (American Geological Institute, v. 28, no. 12, p. 22–27).

prepare an updated gravity-anomaly map of the United States. Although the maps prepared by Lyons (1951) and Woollard and Joesting (1964) had served their purpose well, the increasing use of gravity data, in both theoretical and applied studies, dictated the need for a map capable of providing an order of magnitude more for areal distribution and data saturation. Even more evident and important to the GAM Committee was the need for a digital database accessible to the entire geophysical community.

Focusing on the above objectives, the GAM Committee directed its efforts toward (1) acquiring all non-proprietary United States and offshore gravity data available from industry, government, and academia through the U.S. Defense Mapping Agency/Aerospace Center (DMA/AC) in St. Louis; (2) providing a digital gravity database accessible to the public through the National Oceanic and Atmospheric Administration/National Geophysical Data Center (NOAA/NGDC) in Boulder, Colorado; (3) compiling a gravity-anomaly map of the United States with a station distribution of at least 5 miles × 5 miles (8 km or closer station spacing); and (4) promoting the acquisition and compilation of gravity data within all the countries being included in the North American (and Central American) gravity-anomaly map. With the publication and distribution of the new Gravity Anomaly Map of the United States by the SEG in October 1982, the first three goals of the GAM Committee were realized. This map is reproduced in color in the Preface to this volume. The GAM Committee's final goal to produce the Gravity Anomaly Map of North America should be completed by early 1987. The North American map will be a contribution to the series of maps being prepared for the Decade of North American Geology (DNAG) project sponsored by the Geological Society of America.

MAP SPECIFICATIONS AND DATA PROCESSING

Publication of the conterminous United States and offshore gravity data was at a scale of 1:2 500 000 on an Albers map projection. This scale and projection permit direct comparison and overlay with previously published geologic, tectonic, and basement-rock maps of the United States and with the new national magnetic map published in October 1982 by the USGS. The contour interval is 5 mGal, with larger intervals of 25 mGal in 16 colors, each representing an interval of 25 mGal. The International Gravity Standardization Net 1971 (IGSN71) and the Geodetic Reference System 1967 (GRS67) were employed in calculating the anomalies. Bouguer gravity anomalies based on a rock density of 2.67 g/cm^3 were calculated for onshore areas, while free-air gravity anomalies were computed for the offshore regions.

The gravity-anomaly map has been compiled through a hand-contoured integration of a machine-contoured map derived from the DMA/AC digital database and from a mosaic of published gravity maps. The machine-contoured map, produced through the joint efforts of the DMA and the USGS, was also reproduced as a machine-colored map and used to edit the digital database and the final anomaly map.

Initially, more than 1 000 000 onshore and 800 000 marine stations from the U.S. Department of Defense Gravity Library were examined by the USGS and the DMA. The onshore Bouguer data were processed first by sorting them into 4-km × 4-km cells. The values in each cell were qualified by comparing the first-encountered station with a second station. If the difference was less than 15 mGal, the first station was retained and the rest of the stations within the cell were eliminated. If the 15-mGal tolerance was exceeded, all stations within the cell were rejected. This screening procedure, conducted by the USGS, provided both first-order editing and a data reduction of about 70 percent.

The second step in the onshore data processing involved the application of terrain corrections to all stations west of the Kansas–Colorado border, in the Ouachita–Ozark Mountains, and in the Appalachian Mountains. Employing a modified computer program developed by Plouff (1977), each station was terrain-corrected for radial distances from 0.895 to 166.7 km, using a rock density of 2.67 g/cm^3. More details regarding data-quality control, the machine-contouring of the DMA digital database, and the integration of terrain corrections are discussed by Godson (this volume).

The offshore gravity data were processed in the same manner, except free-air rather than Bouguer anomaly values were used, primarily because acceptable bathymetric control was lacking. The reduced number of offshore stations (about 88 000) were then combined with the onshore stations. This is possible, because at the shoreline (zero elevation) both the onshore Bouguer calculation and the offshore free-air value are identical. The random data were next transferred to an equally spaced grid (4 km × 4 km), using a program based on a minimum-curvature method developed by Briggs (1974). The resulting data grid was then contoured with a computer program that used no smoothing techniques. It should be noted that many of the data used to compile the digital database came from the proprietary files of the DMA.

Final editing and drafting of the United States gravity map were focused on a close comparison of three maps: (1) the machine-contoured digital-data map, (2) an annotated map showing the location of each grid station, and (3) a mosaic map of published data. All drafting was done at the 1:1 750 000 scale and then photographically reduced to the publication scale of 1:2 500 000.

The final map represents about 95-percent digital data in which the machine contours have been hand-smoothed. In those areas deficient in digital data (no data points within the prescribed 5′ × 5′ grid) and where the map borders Canadian and Mexican contours, data from the mosaic map were given priority. Included on the United States gravity map are the locations of the official base stations provided by the National Ocean Survey group of NOAA. Also included is a separate map showing all data points initially analyzed (before screening), a list of contributing organizations, and the names of individuals involved in the preparation of the map.

DISCUSSION

Eastern half of map

This discussion will attempt to correlate briefly only the most obvious anomalies, shown on the Gravity Anomaly Map of the United States, with regional geologic structural-

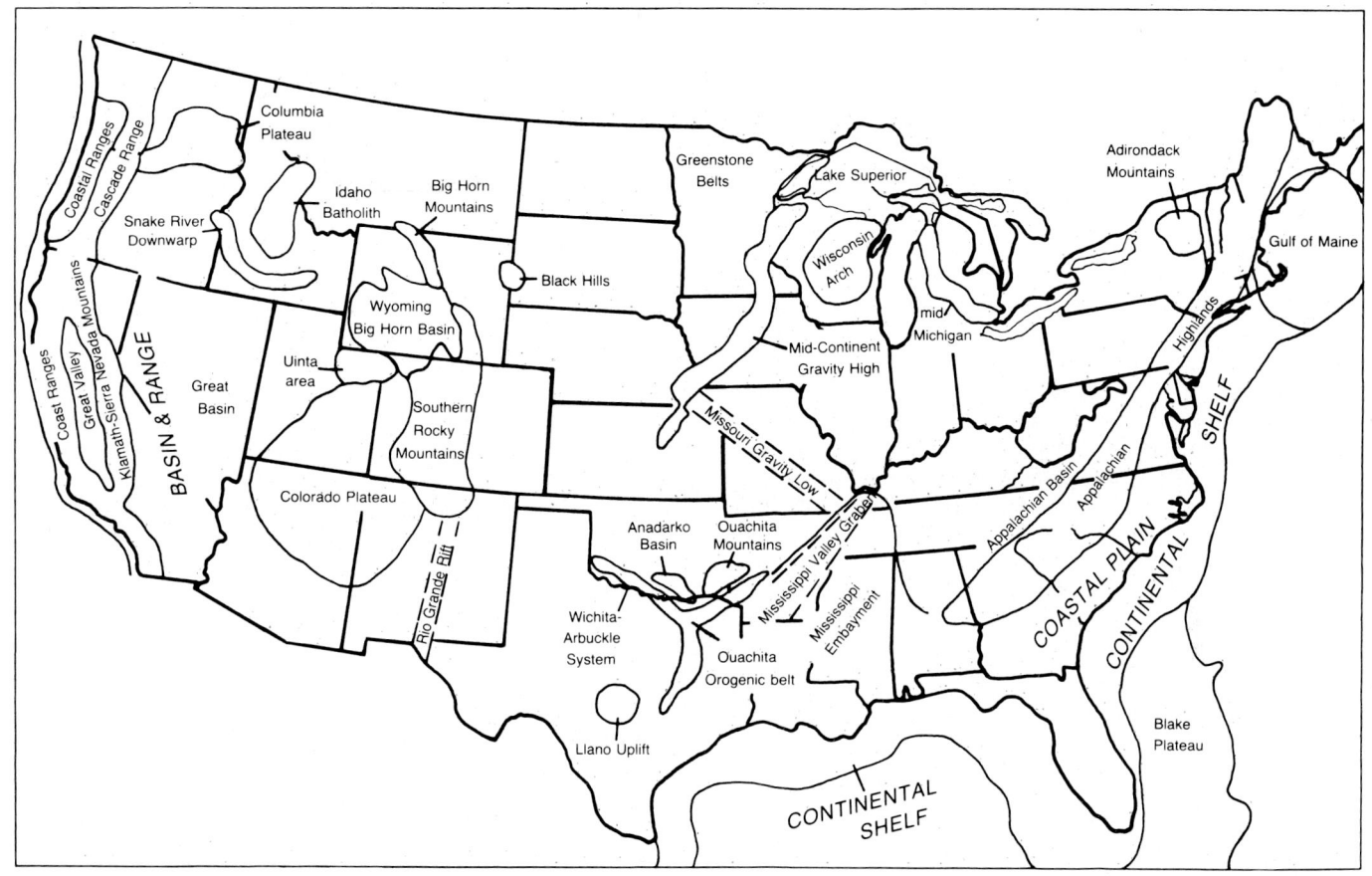

FIG. 1. Sketch map showing major geologic features observable on the gravity-anomaly map.

source features (Figure 1). State names referred to in the following discussion are indicated only on the Gravity Anomaly Map, which is given in the Preface of this volume. The reader should refer to that map in the following discussion.

In the eastern part of the map, the most spectacular gravity anomaly is the elongate Midcontinent gravity high extending from the western tip of Lake Superior to at least central Kansas (King and Zietz, 1971). These anomalous gravity maxima represent the western limb of an 1 100-m.y.-old rift system into which dense basaltic rock was accumulated to produce some of the highest Bouguer gravity anomalies observed in the United States. Flanking both sides of the elongate gravity maxima are a series of anomaly minima, produced by structural faults and thick deposits of downfaulted, sedimentary rock of low density.

Striking at a right angle to the Midcontinent gravity high lies a strip of relatively low gravity anomalies, known as the Missouri gravity low, extending for nearly 1 000 km from Nebraska southeastward through Missouri (Stauder et al., 1977). The geologic structure producing this faint band of anomalously low gravity could have originated in one of several ways: (1) as a failed arm of the Midcontinent rift, (2) as a suture between two pieces of collided crust, or (3) as the trace of a basement slip fault (Kane et al., 1981). Other anomalous gravity lineaments showing parallel orientation can be found throughout the Midcontinent area. This northwest–southeast pattern suggests an initial relationship with Precambrian fracturing of the basement.

Extending from the eastern margin of Lake Superior through the Southern Peninsula of Michigan is located a gravity maximum similar to the Midcontinent gravity anomaly. Although less pronounced, this gravity maximum, known as the Mid-Michigan geophysical anomaly, reflects another, more deeply buried arm of the rift system that produced the Midcontinent gravity high (Hinze et al., 1975). Through the center of Lake Superior, where the structurally related features merge, low gravity anomalies occur, possibly because of the deep burial and thick accumulation of basement complex and crustal rock (Hinze et al., 1982). Collectively, the more than 3 000-km-long geologic feature, stretching from southern Kansas to southeastern Michigan (and probably southward to northeastern Mississippi; Lyons, 1970), is known as the Midcontinent rift system. Centered within this rift system's horseshoe-shaped mass of extrusive basalt and intrusive mafic rock lies a region of anomalously low gravity, probably resulting from the less dense igneous rocks of the Wisconsin arch.

South of the Great Lakes region lies an extensive gravity maximum, reflecting what is generally accepted as a late Precambrian aulacogen that was reactivated during Mesozoic and Cenozoic time (Ervin and McGinnis, 1975). Located along the axis of the Mississippi River and at the eastern end of the Missouri gravity low previously mentioned is a rift known as the Mississippi Valley graben (Hildenbrand et al., 1977). The Mississippi embayment is associated with the broad, positive anomalous area and the rift. The embayment developed as a structural trough into which thick accumulations of Mesozoic and Cenozoic sediments were deposited to form a projection into the North American craton from the Gulf Coast area. The most recent major structural activity within the Mississippi embayment was during the Late Cretaceous and was probably related to the formation of the present Atlantic Ocean basin.

Also related to the formation of the Atlantic Ocean are the crustal displacements reflected by a series of free-air gravity anomalies extending along the outer margin of the continental shelf and the Blake Plateau. Similarly, the parallel series of positive Bouguer gravity anomalies striking northeastward from southern Alabama to Connecticut, along the crest of the Appalachian Mountains, may be related to crustal displacements initiated during the formation of a Proto-Atlantic Ocean. This series of positive Bouguer anomalies may, in addition, reflect a major rift system hidden beneath the Appalachian Mountains. Through the Blake Plateau, southern Florida, and the Gulf Coast region, an area of relatively positive gravity anomalies can be observed outlining the continental margin.

From northern Georgia to eastern New York, along the western flank of the Appalachian Mountains, lies an elongated, relatively low anomaly. The basement rocks underlying this area consist of sedimentary rock of Precambrian and early Paleozoic age, possibly indicating the position of a continental-shelf area related to the formation of a Proto-Atlantic Ocean. This northeastward-trending anomalous gravity feature appears to terminate, possibly by deep burial, the existing southeastward-trending pattern of anomalies to the west. The gravity-anomaly pattern is generally believed to be associated with the structural features responsible for the previously discussed Midcontinent rift system and the Precambrian basement complex.

The eastern New England area and the offshore Gulf of Maine show a rather mottled array of positive gravity anomalies, suggesting the presence of deformed crustal material accreted to the North American craton. This crustal remnant is an exotic terrane and may represent the remains of a former island arc that collided with the mainland.

Western half of map

In the western half of the United States, the most dominant structural features generate negative anomalies. In fact, the most obvious general characteristic of the Bouguer gravity map of the conterminous United States is the contrast between the relatively high anomalous gravity values found in the eastern half as compared to the low anomalous values observed in the western half. This gravity contrast reflects conditions of general isostatic equilibrium across the nation. The negative anomalies of the western United States result from the deficiency of mass at depth, which compensates for the higher elevations.

Located along the western margin of the Mississippi embayment, extending through eastern Texas with a branch protruding into southern Oklahoma, is a "Y"-shaped gravity maximum. The eastern arms of the "Y"-shaped anomaly reflect the Ouachita orogenic belt. The northeastern arm of the belt borders the southern flank of the Ouachita Mountains, which correlate with a low gravity anomaly in southern Oklahoma and west-central Arkansas. The western arm of the "Y"-shaped positive anomaly is associated with the Wichita–Arbuckle system. North of this system, another series of small positive anomalies pinpoints the location of the Anadarko basin. To the south, in south-central Texas, a circular, positive gravity anomaly represents the Llano uplift, which contains relatively dense rock of Precambrian age.

Extending from central Arizona–New Mexico northward through western Wyoming is an extensive negative anomalous gravity region with the largest amplitudes located in western Colorado, coincident with the Southern Rocky Mountains. Several geologic features occupy the eastern margin of the large negative anomalous region. Starting in south-central New Mexico, the Rio Grande rift system produces an observable gravity trend that extends northward through the state until it merges with the lower gravity anomalies of the Front Range through the central portion of Colorado. Continuing northward, the eastern margin of the negative anomalous region strikes northwestward through central Wyoming, where the low gravity values are related to thick sequences of granitic and sedimentary crustal rock composing the Laramie Range (and basin) and the Big Horn Mountains. Throughout the southern part of Wyoming, in the Big Horn basin area, and farther south in the Uinta area of Utah, extensive negative anomalies are associated with numerous uplifted blocks and downwarped basins.

To the northwest, in the Idaho region, there is a rather detached, circular gravity minimum. The western part of this negative anomaly is correlative with the Idaho Batholith.

West of and adjacent to the Colorado Plateau area in Utah and Nevada, an elongate series of north–south–oriented negative anomalies forms the gravity pattern over a large part of the Great Basin. The relatively thin crust of the Great Basin, together with negative gravity anomalies, high elevations, and tectonic activity, are believed to be the consequence primarily of lithospheric heating and thermal expansion (Eaton et al., 1978). South of the Great Basin, through the southern parts of California and Arizona, the gravity anomalies become much higher. Transition zones between the regional high and low anomalous Bouguer gravity expressions may help to delineate and explain the existence of such speculative linear features as the Walker lane in southwestern Nevada and the Texas lineament, which crosses Arizona and New Mexico.

The relatively high gravity anomaly observed in southern Idaho can be attributed to the dense, more recent volcanism of the Snake River downwarp. Similarly, the much larger, more circular positive anomalous gravity region to the northwest, which covers the eastern Washington area, is associated with the flood basalts of the Columbia Plateau.

The western margin of the map exhibits north–south–oriented belts of positive Bouguer gravity anomalies onshore, with parallel belts of positive free-air gravity anomalies offshore. The entire region is referred to as the Pacific Cordillera. Located along most of the western margin of the United States are the Coast Ranges, which exhibit relatively high Bouguer gravity anomalies. Paralleling the Coast Ranges offshore, north–south–oriented belts of negative and positive free-air gravity anomalies coincide with the structural trend of the continental margin. Through the central part of California a series of high Bouguer gravity anomalies coincides with a trough of sedimentary rock in the Great Valley of California. These anomalies originate from mafic rocks in the crust, perhaps a remnant of oceanic crust involved in subduction along the west coast. Extending along the eastern flank of the Great Valley, a minimum gravity anomaly with steep contour gradients marks the location of the Sierra Nevada batholith. To the north, separating the Sierra Nevada from the Cascade Mountains of Oregon and Washington (also indicated by steep gravity gradients), lie the Klamath Mountains. The relatively low gravity anomaly of this area, reflecting crystalline rocks of low density, extends over northern California and southern Oregon. Many of the geologic structural features of the Pacific Cordillera are believed to be the result of ancient subduction, plate collisions, and crustal uplift. More recent deformation and a mottled pattern of offshore gravity anomalies along the southern California region are related to further collisions that have produced the active dextral motion generally observed along the San Andreas fault system.

SUMMARY

The gravity-anomaly map of the United States recently published by the Society of Exploration Geophysicists was prepared from nearly 2 million gravity observations. The map, published at a scale of 1:2 500 000, consists of terrain-corrected Bouguer gravity anomalies onshore and free-air anomalies in adjacent marine areas. The most obvious characteristic of the map is the marked negative anomalies over the topographically high regions of the western United States. However, most anomalies can be directly related to geologic features of the Earth's crust. These anomalies provide useful information regarding the distribution and configuration of crystalline basement rocks, structural and lithologic provinces, zones of crustal weakness, and the distribution of intrusive igneous rocks and sedimentary basins.

REFERENCES

Briggs, I. C., 1974, Machine contouring using minimum curvature: Geophysics, **39**, 39–48.

Eaton, G. P., Wahl, R. R., Prostka, H. J., Mabey, D. R., and Kleinkopf, M. D., 1978, Regional gravity and tectonic patterns: Their relation to late Cenozoic epeirogeny and lateral spreading in western Cordillera, in Smith, R. B., and Eaton, G. P., Eds., Cenozoic tectonics and regional geophysics of the western Cordillera: Mem. 152, Geol. Soc. Am., 51–92.

Ervin, C. P., and McGinnis, L. D., 1975, Reelfoot rift: Reactivated precursor to the Mississippi embayment: Geol. Soc. Am. Bull., **86**, 1287–1295.

Godson, R. H., Preparation of a digital grid of gravity-anomaly values of the conterminous United States, this volume.

Hildenbrand, T. G., Kane, M. F., and Stauder, W., 1977, Magnetic and gravity anomalies in northern Mississippi embayment and their spatial relation to seismicity: Misc. Field Stud. Map MF-914, U.S. Geol. Surv.

Hinze, W. J., Kellogg, R. L., and O'Hara, N. W., 1975, Geophysical studies of basement geology of Southern Peninsula of Michigan: AAPG Bull., **59**, 1562–1584.

Hinze, W. J., Wold, R. J., and O'Hara, N. W., 1982, Gravity and magnetic studies of Lake Superior, in Wold, R. J., and Hinze, W. J., Eds., Geology and tectonics of the Lake Superior basin: Mem. 156, Geol. Soc. Am., 203–221.

Kane, M. F., Hildenbrand, T. G., and Hendricks, J. D., 1981, Model for the tectonic evolution of the Mississippi Embayment and its contemporary seismicity: Geology, **9**, 563–568.

King, E. R., and Zietz, I., 1971, Aeromagnetic study of the midcontinent gravity high of central United States: Geol. Soc. Am. Bull., **82**, 2187–2208.

Lyons, P. L., 1951, Bouguer gravity map of the United States: Tulsa Geol. Soc.

——— 1970, The megatectonics of continents and oceans, in Johnson, H., and Smith, B. L., Eds., Continental and oceanic geophysics: Rutgers Univ. Press, 147–166.

Plouff, D., 1977, Preliminary documentation for a fortran program to compute gravity-terrane corrections based on topography digitized on a geographic grid: Open-File Rep. 77–535, U.S. Geol. Surv.

Stauder, W., Kramer, M., Fischer, G., Schaeffer, S., and Morrissey, S. T., 1977, Seismic characteristics of southeast Missouri as indicated by a regional telemeter microearthquake array: Seismol. Soc. Am. Bull., **66**, 1953–1964.

Woollard, G. P., and Joesting, H. R., 1964, Bouguer gravity anomaly map of the United States: U.S. Geol. Surv.

Preparation of a digital grid of gravity-anomaly values of the conterminous United States

Richard H. Godson*

ABSTRACT

Irregularly spaced gravity data of the conterminous United States have been converted to a digital data set in gridded format suitable for contouring and analytical transformations. The data consist of Bouguer gravity-anomaly values on land and free-air gravity anomaly values offshore. The data have been terrain corrected in areas of substantial relief and extensively edited to remove erroneous values. Contours generated from this grid were used in the preparation of regional gravity-anomaly maps of the United States.

INTRODUCTION

As part of a U.S. Geological Survey (USGS)–Society of Exploration Geophysicists (SEG) cooperative effort to publish a regional gravity-anomaly map of the conterminous United States at a scale of 1:2 500 000, a grid of digital data was produced. The data set used in the reduction process was provided in digital form by the U.S. Defense Mapping Agency (DMA). These data consisted of approximately 1 million land stations and about 0.8 million marine stations. The onshore part of the grid consists of Bouguer gravity-anomaly values computed using a density of 2.67 g/cm^3, and the offshore data consist of free-air gravity-anomaly values. It was not possible to consider the use of Bouguer anomaly values offshore, because most marine stations did not have water-depth information. All gravity data were referenced to the International Gravity Standardization Net, 1971 (Morelli, 1974), and theoretical gravity was computed using the 1967 Geodetic Reference System formula (International Association of Geodesy, 1971).

SCREENING

The onshore and offshore data were processed separately, with the initial step being a screening of the onshore data,

*U.S. Geological Survey, Box 25046, MS 964, Denver Federal Center, Denver, CO 80225.

which involved sorting the data by latitude and examining data values in each 4-km by 4-km area. This examination consisted of comparing the Bouguer anomaly value of the first station encountered in a compartment with the Bouguer anomaly value of the second station found in the same compartment. If the difference between the two Bouguer anomalies was less than 15 mGal, the first value encountered was retained, and all other stations in that compartment were removed. When the tolerance of 15 mGal was exceeded, all stations in the compartment were eliminated. If only one value was found in a compartment, it was kept. The marine data were subsequently screened in the same manner as the land data, except that free-air anomaly values were used instead of Bouguer anomaly values.

The purpose of this screening effort was twofold: (1) it was an initial attempt at removing highly erroneous gravity values, and (2) it produced a substantial reduction of about 70 percent in the number of onshore stations and about 85 percent reduction in the number of marine stations. This reduction was important because of the limited disk space available on the computer and because of the central processing unit (CPU) time involved in the following data-processing steps.

TERRAIN CORRECTIONS

Those stations in land areas of substantial topographic relief were terrain-corrected. This included all stations west of longitude 102° W, the area of the Ouachita–Ozark Mountains (latitude 34° N–38° N, longitude 90° W–95° W), the southern Appalachian Mountains (latitude 34° N–39° N, longitude 78° W–86.5° W) and the area north of latitude 39° N and east of longitude 81° W. As the number of stations was too large to be corrected all at once, 25 subsets of about 2 750 stations each were created. Each subset was then terrain-corrected using a modified version of a program developed by Plouff (1977). This program allows the user to access the terrain data automatically from a disk pack on the computer for the area of interest. Terrain elevations averaged in cells of 30 seconds of latitude and longitude were used up to distances of 5 km from the stations. For distances from 5 to 21 km, elevations averaged in cells of 1 minute were used, and from 21 to 166.7 km, 3-minute averages were used.

Corrections were determined from 0.9 km (the outer edge of Hammer zone F) to 166.7 km (the end of Hayford zone O) for each station, using a density of 2.67 g/cm^3. The inner-zone corrections (<0.9 km) were not made. It is conceivable that more error would be introduced by computing the inner-zone correction using 30-second average terrain than by leaving this correction out. In general, errors of only 2 or 3 mGal might be introduced by not computing the inner-zone correction. Certain stations on sharp peaks or in deep canyons can have much larger corrections; however, the number of these stations is thought to be small.

GRIDDING

In preparation for gridding, the onshore and offshore data sets were combined and their geodetic coordinates converted to x–y coordinates, using an Albers equal-area projection with a central meridian of longitude 96° W and standard parallels of latitude 29.5° N and latitude 45.5° N. This projection is the same as used for other national maps published at a scale of 1:2 500 000 (U.S. Geological Survey and American Association of Petroleum Geologists, 1961; American Geophysical Union, 1964; Bayley and Muehlberger, 1968; King and Beikman, 1974; Zietz, 1982).

The data were then transformed to an equally spaced grid using a program (Webring, 1981) based on a minimum-curvature method (Briggs, 1974). This procedure honors the individual random values and generates a smooth surface, especially in areas of sparse data. The grid interval selected was 4 km, an empirical value based on previous experience in generating regional magnetic and gravity maps at scales of 1:1 000 000 and smaller. A search radius of 40 km, which is the maximum distance for interpolation from a known value, was used. This distance was sufficient to calculate values at all grid locations inside the boundaries of the data except for one area off the southeast coast of Florida. The grid size created with this interval was about 1 300 columns by 770 rows or approximately 1 million grid positions.

EDITING

A contour map was produced from this initial grid using a computer program (Godson and Webring, 1982) that uses a spline under tension technique (Cline, 1974) for a slight smoothing of the contours. Although some erroneous values were expected on this preliminary map, the number and extent of these were higher than expected.

Blocks of erroneous values

Figure 1 shows an example of clusters of poor data that are associated with particular sources—i.e., several erroneous values concentrated in one area or along ship tracks that can be traced to a single source. These types of errors are due to such things as wrong datum values, incorrect units of elevations, or location errors. Usually these sources could be identified by examining computer printouts for land data and by overlaying color-coded track plots on contour maps for marine data.

After identification of an erroneous source, it was often possible to determine the exact error and to correct it; at other times it was necessary to eliminate all values associated with a particular source.

Figure 2 shows as an example an area in the vicinity of New York City that did not contain many obviously incorrect gravity values. However, the contours appeared suspicious and did not correspond to published maps for the same area. The same type of situation existed in other areas off the East Coast, and so a detailed study of the entire eastern offshore area was made. This examination revealed six more poor data sources that were not evident in this first editing procedure. Figure 3 shows the same area as Figure 2 after removal of these sources.

Isolated erroneous values

In addition to blocks of incorrect values, a number of isolated erroneous values were present on the initial contour map (Figure 4). These errors appeared as short-wavelength anomalies that are attributed to a single station with no source relationship from one station to another. This type of error was difficult to remove because it requires time and expense to pinpoint on computer printouts or CRT's. This is because locations cannot be determined accurately enough, nor can one detect the erroneous Bouguer values from maps at a scale of 1:2 500 000. Larger scale maps could have revealed these incorrect values, but it still is necessary to determine the locations and then search through the data set to find and eliminate the incorrect values with an editing program. This editing approach was not used because of the large number of erroneous values present in this data set of several hundred thousand stations.

Polynomial surface fit.—To solve this editing problem, a computer program (Sweeney, 1981) was written that fits a polynomial surface to a set of data. The program is interactive and requests responses to several questions.

In one mode of operation, the program requests the order of the desired polynomial and the number of standard deviations to use for data rejection. The program then fits the specified polynomial surface to the data and eliminates those stations that are farther removed than the specified number of standard deviations from the calculated root-mean-square (RMS) value.

Alternatively, the program asks the user to specify a certain RMS value in milligals, and the program then iterates to calculate the order of polynomial necessary to fit the data to an RMS value less than that specified. The program then removes all values above the specified number of standard deviations from the RMS value. After some experience with the program, an RMS value of 8 mGal and 3 standard deviations was chosen for most areas.

A square or rectangular area encompassing isolated incorrect values was extracted from the screened data set, and these random values were used as input to the program. The advantage of this procedure is that the erroneous values are automatically located and removed from the data set.

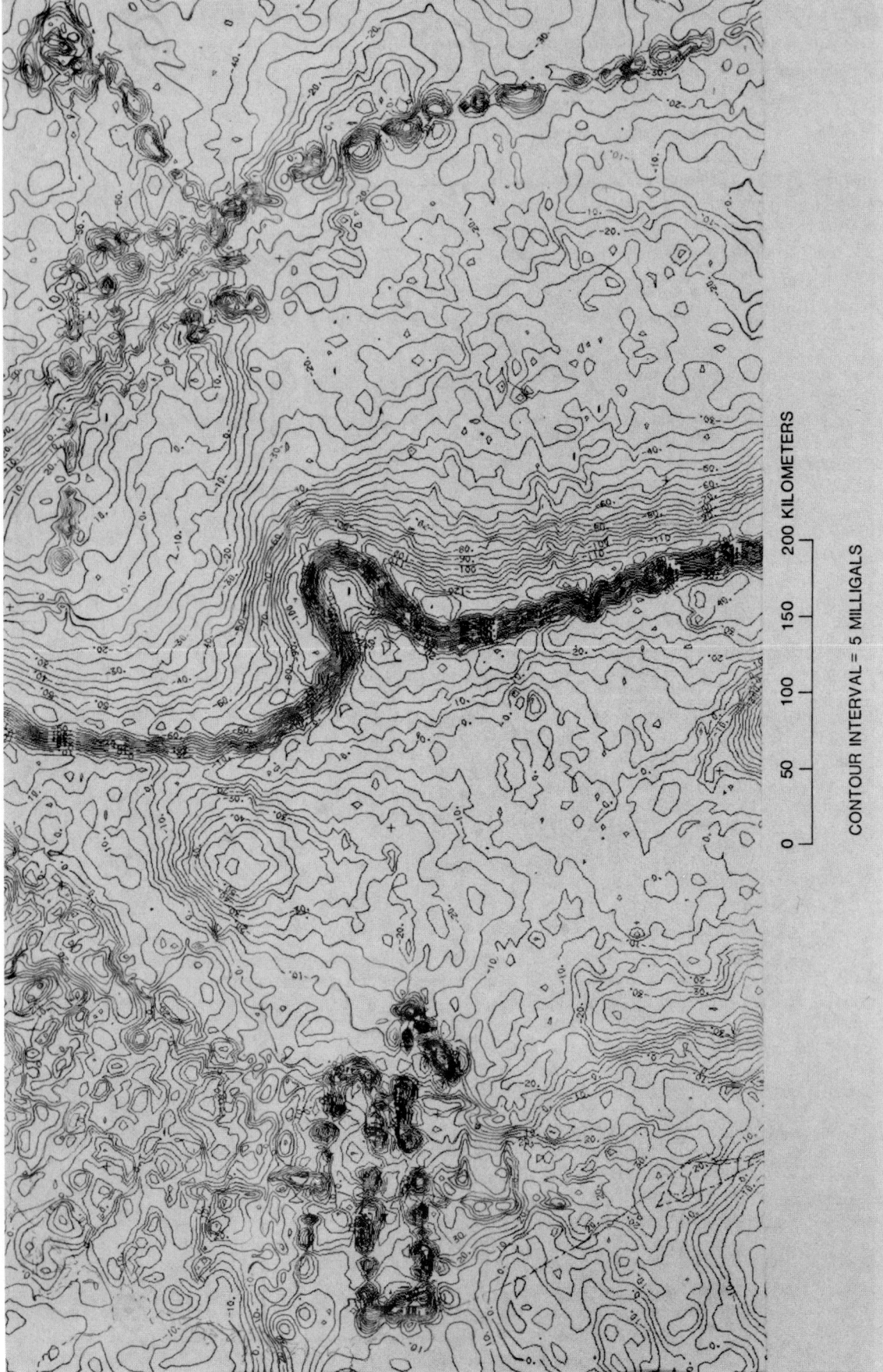

FIG. 1. Contours of free-air-gravity data off the east coast of Georgia and Florida, illustrating examples of blocks of obvious erroneous values.

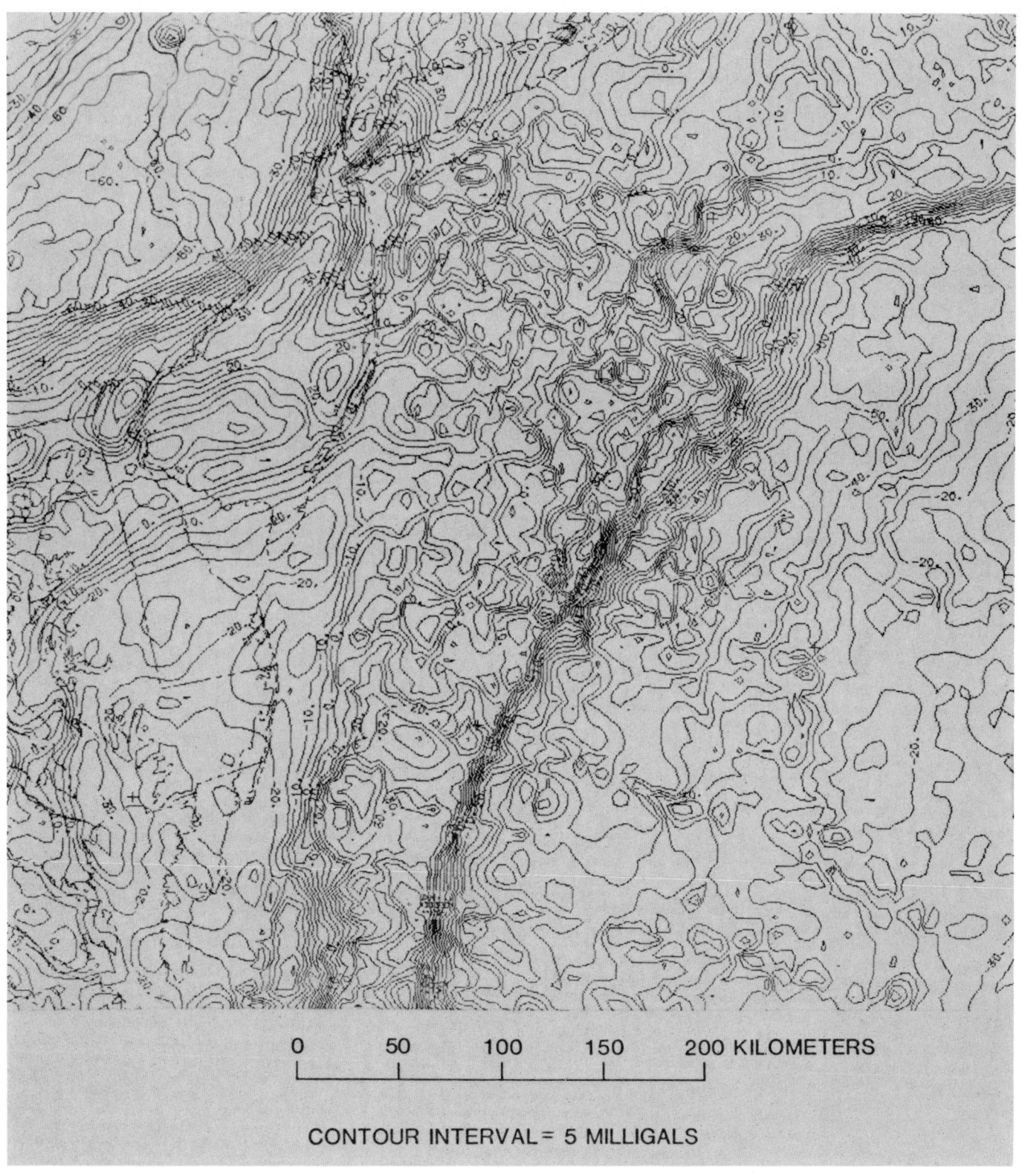

FIG. 2. Contours of gravity data centered off the coast of Delaware before editing.

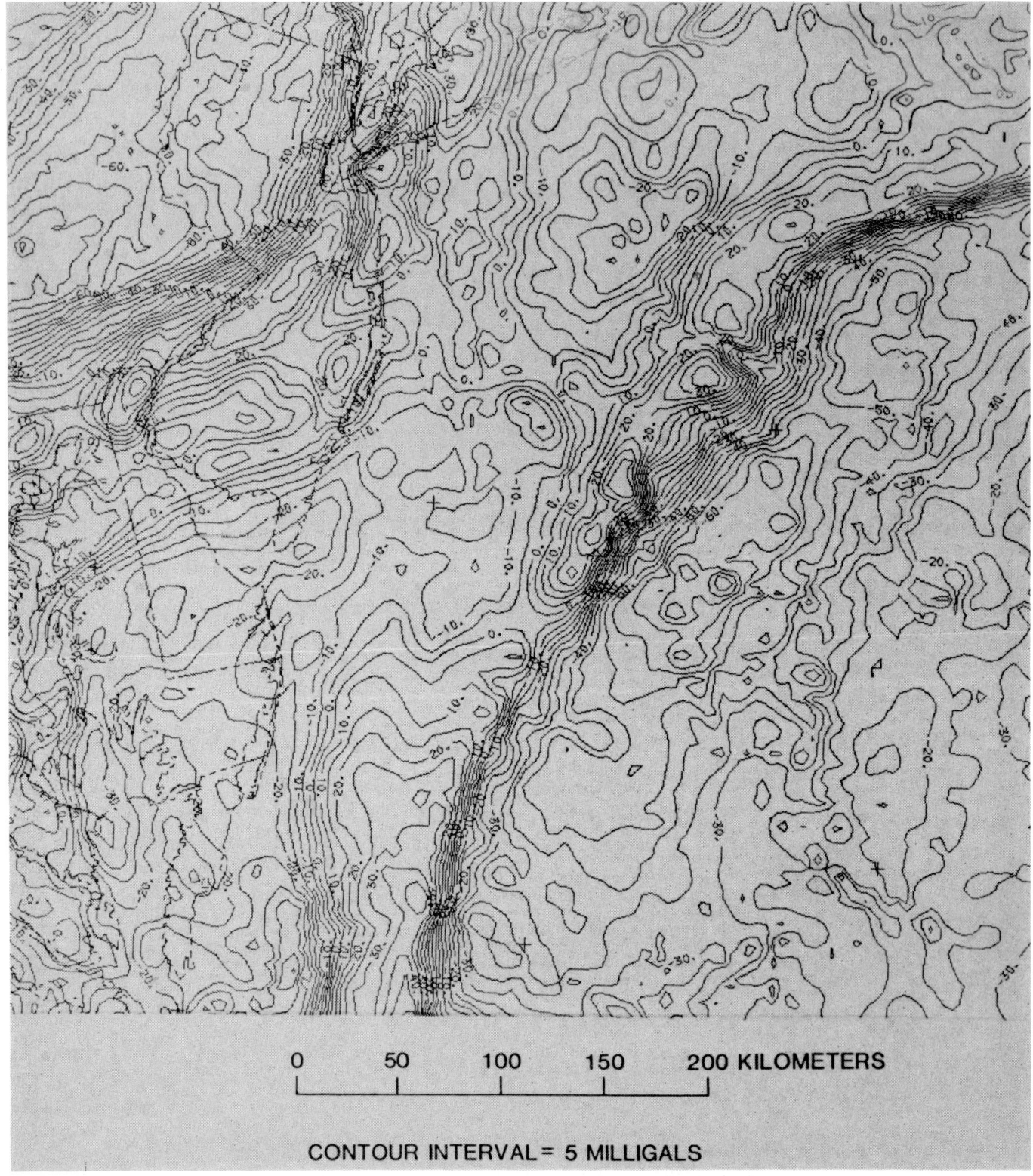

FIG. 3. Contours of gravity data centered off the coast of Delaware after editing.

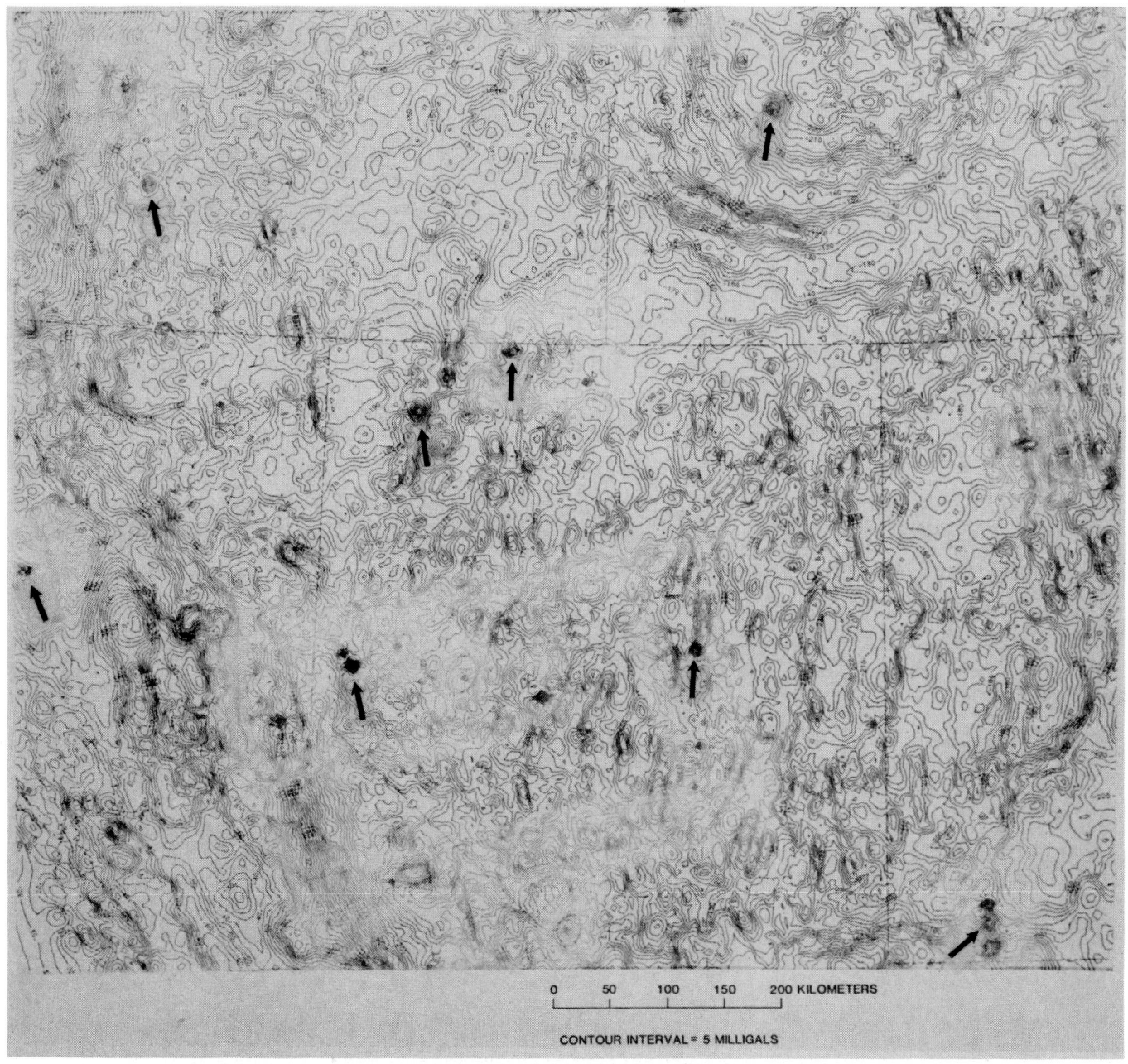

FIG. 4. Contours of gravity data in Nevada and adjacent states exhibiting numerous isolated erroneous values. Arrows show examples of small-wavelength anomalies that are probably incorrect.

FIG. 5. Distribution of gravity stations in the conterminous United States and offshore areas after final editing.

Many of these stations occurred within the data set and considerable time was spent in this editing process.

DMA editing

In addition to the type of editing explained above, the DMA used its standard procedures to provide a list of about 1 300 questionable stations that could be present in the final screened data set. These DMA methods consisted mainly of comparing station values in an area for consistency and checking station elevations against elevations shown on topographic maps. The final step in the editing process was to check for and eliminate these stations that remained in the data set. Figure 5 shows the distribution of stations after final editing.

SUMMARY

A grid of the data set was submitted to the SEG Gravity Anomaly Map Committee for use in producing a colored gravity-anomaly map of the United States (Society of Exploration Geophysicists, 1982). The grid has been used to produce contours for publication by the USGS at a later date of a transparent black and white map at a scale of 1:2 500 000. This map can be used as an overlay on previously published national maps that have been produced at the same scale and projection. Magnetic tapes and short descriptions of processing procedures used to produce the grid were sent to the National Technical Information Service (NTIS) in Springfield, Virginia (Godson and Scheibe, 1982), and to the Geophysical Data Services Center (GDSC) in Boulder, Colorado, for distribution on request to the scientific community.

ACKNOWLEDGMENTS

The author thanks the National Geodetic Survey for submitting a list of about 20 questionable stations it found in an independent editing process. Don Scheibe of the DMA supplied the initial data set and assisted greatly in the editing process.

REFERENCES

American Geophysical Union, Special Committee for the Geophysical and Geological Study of the Continents, and U.S. Geological Survey, 1964, Bouguer gravity anomaly map of the United States (exclusive of Alaska and Hawaii): U.S. Geol. Surv., scale 1:2 500 000, 2 sheets.

Bayley, R. W., and Muehlberger, W. R., compilers, 1968, Basement rock map of the United States (exclusive of Alaska and Hawaii): U.S. Geol. Surv., scale 1:2 500 000, 2 sheets.

Briggs, I. C., 1974, Machine contouring using minimum curvature: Geophysics, **39**, 39–48.

Cline, A. K., 1974, Scalar and planar-valued curve fitting using splines under tension: Commun. Assoc. for Comput. Mach., **17**, 218–224.

Godson, R. H., and Scheibe, D. M., 1982, Magnetic tape containing conterminous U.S. gravity anomaly data in a gridded format: Nat. Tech. Inf. Serv. Rep. PB82-254798, U.S. Dep. Commerce.

Godson, R. H., and Webring, M. W., 1982, CONTOUR: a modification of G. I. Evenden's general purpose contouring program: Open-File Rep. 82-797, U.S. Geol. Surv.

International Association of Geodesy, 1971, Geodetic Reference System, 1967: Spec. Publ. 3, Int. Assoc. Geod.

King, P. B., and Beikman, H. M., compilers, 1974, Geologic map of the United States (exclusive of Alaska and Hawaii): U.S. Geol. Surv., scale 1:2 500 000, 3 sheets.

Morelli, C., Ed., 1974, The International Gravity Standardization Net, 1971: Spec. Publ. 4, Int. Assoc. Geod.

Plouff, D., 1977, Preliminary documentation for a Fortran program to compute gravity terrain corrections based on topography digitized on a geographic grid: Open-File Rep. 77-535, U.S. Geol. Surv.

Society of Exploration Geophysicists, 1982, Gravity anomaly map of the United States (exclusive of Alaska and Hawaii): SEG, scale 1:2 500 000, 2 sheets.

Sweeney, R. E., 1981, FIT POST: a polynomial surface fitting program: U.S. Geol. Surv. unpubl. program.

U.S. Geological Survey and American Association of Petroleum Geologists, 1961, Tectonic map of the United States (exclusive of Alaska and Hawaii): scale 1:2 500 000, 2 sheets.

Webring, Michael, 1981, MINC: a gridding program based on minimum curvature: Open-File Rep. 81-1230, U.S. Geol. Surv.

Zietz, I., compiler, 1982, Composite magnetic anomaly map of the United States, Part A: Conterminous United States: Geophys. Invest. Map GP 954-A, U.S. Geol. Surv., scale 1:2 500 000, 2 sheets.

Features of a pair of long-wavelength (>250 km) and short-wavelength (<250 km) Bouguer gravity maps of the United States

M. F. Kane* and R. H. Godson‡

ABSTRACT

A complementary pair of wavelength-filtered regional and residual gravity maps of the United States has been compiled from a digital gravity base. Processing of simple model anomalies with the same filters as those used on the maps shows that the short-wavelength (SWL) anomalies, although exaggerated in shape and distorted in amplitude, give a clearer indication of source geometry than the total-field Bouguer anomalies. Side lobes created by the filtering process appear to be the most serious deficiency of the SWL maps, but they can often be distinguished by using simple criteria. The model studies also show that the horizontal-gradient amplitude map is a useful adjunct for interpreting the SWL anomalies. Maximum horizontal-gradient/amplitude relations indicate that most of the anomalies of the SWL map, particularly the prominent ones, have their source in the crust. There is considerable leakage of long-wavelength (LWL) energy from relatively narrow but intense anomalies into the LWL field so that moderate-wavelength anomalies in the LWL map must be interpreted with care. The SWL map defines a series of zones that are based on anomaly characteristics, primarily anomaly trend. These zones correspond in many places to previously identified geologic terranes. It is concluded that rift systems can account for many of the SWL gravity features of the eastern United States; a model of a rift deformed by compression is proposed to explain many of these features. Other gravity–geologic correlations, notably over igneous features, are discernible.

INTRODUCTION

The availability of high-quality digital gravity data for the entire continental United States (Godson and Scheibe, 1982) offers opportunities to develop many types of derived maps

*U.S. Geological Survey, USGS Mission, APO New York, NY 09697.
‡U.S. Geological Survey, Box 25046, MS 964, Denver Federal Center, Denver, CO 80225.

for study of continental structure (Arvidson et al., 1982; Hildenbrand et al., 1982; Kane et al., 1982; Simpson et al., 1982). The data set used for this study was extracted from the nonproprietary gravity holdings of the U.S. Defense Mapping Agency (DMA), St. Louis, Missouri, and provides coverage at about 5-km spacing for most of the country. An exception to this spacing is in North Dakota, South Dakota, and parts of Wyoming and Montana, where it may be as much as 15 km. The data were collected by DMA from public agencies, including universities, and from some private concerns. Terrain corrections were made, and accuracies are generally better than 1 mGal except where terrain effects caused by local topography are large.

We experimented with several kinds of derived maps and concluded that one composed of wavelengths less than 250 km is especially effective in emphasizing anomalies and anomaly patterns at a continental scale. The choice of the 250-km-wavelength cutoff is based on a comparison of maps using different wavelength cutoffs. Map pairs (SWL/LWL) using either a 200- or 300-km cutoff do not show great differences from ones using a 250-km cutoff, whereas those using either a 100-km or 400-km cutoff do. Our choice is supported by the observation that many SWL anomalies using the 250-km cutoff correspond closely with the outline of geologic features for which the geologic–gravity correlation is already established—for example, the large gravity low over the Sierra Nevada batholith.

Our analyses indicate that the sources of the anomalies of the 250-km SWL map reside principally in the crust and that the broad, lateral patterns formed by them provide a first-order depiction of the structural framework of the continental crust underlying the United States. We also discuss briefly the origins of anomalies of a more enigmatic nature that are shown on a map of wavelengths longer than 250 km—that is, the map complementary to the one described above. In the following paragraphs we examine some effects of the wavelength filtering and some characteristics of the resulting anomalies, describe the general distribution of the anomalies, and speculate on their sources.

DEPTH OF SOURCE AND EFFECTS OF FILTERING ON THE ANOMALIES OF THE SWL/LWL MAPS

The SWL residual Bouguer anomaly map was prepared by transforming the Bouguer gravity database (Godson and

Scheibe, 1982) to the frequency domain, filtering the transformed data with a LWL cutoff at 250 km, and subtracting the LWL data grid (the regional) from that of the original data set. The filter was sloped linearly from 0-percent gain at 300-km wavelength to 100-percent gain at 200-km wavelength. The regional LWL field is shown as Figure 1; the complementary residual SWL field is shown as Figure 2.

An insight into probable limits on source-depths of anomalies of the SWL map may be had by considering together the effects of the filtering and the effects of a change in source-depth. To evaluate the latter, we use the wavelength/amplitude-attenuation relation from upward continuation:

$$D_\lambda = A_\lambda e^{-2\pi \Delta Z/\lambda}$$

where: λ = a discrete wavelength;
A_λ = amplitude of λ at original source-depth;
D_λ = attenuated amplitude of λ with increased source-depth;
ΔZ = increase in source-depth.

Adopting 5 km as a shallow crustal-source depth and 40 km as a top-of-the-mantle depth, we calculate how an anomaly decreases in amplitude as its source is increased in depth from 5 to 40 km. As shown in Figure 3 (solid line from 50- to 200-km wavelength and dashed line from 200- to 300-km wavelength), the effect of the 35-km increase in depth is to attenuate wavelength components of an anomaly inversely with respect to the size of wavelength. The attenuation ranges from 99 percent for wavelengths shorter than 50 km to 52 percent for a wavelength of 300 km. The combined effect of the increase in source-depth and the filter with its sloped cutoff (shown by the solid line from 200- to 300-km wavelength) is to limit the maximum contribution of any anomaly component to 33 percent at 200-km wavelength. Since the total anomaly amplitude is the sum of components that have all been attenuated by 33 percent or more, then it follows that the maximum amplitude from a source at 40 km (top of the mantle) is less than 33 percent of an equivalent source at 5 km. Our initial conclusion that the sources of anomalies shown in Figure 2 are primarily crustal in origin was based on this large attenuation in anomaly amplitude for deeper sources.

A more precise estimate of source depths for specific anomalies of the SWL map (Figure 2) is given by the amplitude/maximum-gradient relation (Bott and Smith, 1958; Bancroft, 1960; Kane and Bromery, 1968) for point, line, and plate sources:

$$d_{max} = k\, A/G_{max}$$

where: G_{max} = maximum horizontal gradient;
A = anomaly amplitude;
d_{max} = maximum depth to source;
k = constant (point source, .32; line source, .65; edge source, .86).

Locations of gravity highs selected for depth calculations are shown on the index map (Figure 4); anomaly characteristics are shown in Table 1. Lengths and widths of anomalies listed in Table 1 were estimated from the locations of the maximum gradients. The amplitudes were estimated from the Bouguer Gravity Anomaly Map of the United States (Society of Exploration Geophysicists and U.S. Geological Survey, 1982). Maximum gradients for the California and Midcontinent gravity anomalies were also estimated from the map cited above, but the remainder were taken from the horizontal-gradient map (Figure 5). As will be discussed below, the position of the maximum gradient gives a reliable measure of the lengths of sources that are much longer than deep and, under some conditions, a measure of the widths of sources much wider than deep. The constant for a line source, 0.65, was used in the calculations because it gives a conservative estimate for linear anomalies where source width is greater than depth or thickness. Averages of maximum depths for three anomalies were used for the Appalachian and Rocky Mountain regions and represent depth ranges of 4 and 6 km, respectively. The depths given in Table 1 show that the deepest sources of this group are in the middle part of the crust. The horizontal-gradient map shows a preponderance of gradients of about 1 mGal per km associated with anomalies that are generally less than 30 mGal in amplitude, indicating that the sources of these smaller anomalies are also in the crust. It is known, of course, that many anomaly patterns like those of the Rocky Mountains can be directly associated with surface geologic features (bedrock ranges and intervening intermontane basins). The main purpose of the exercise, however, is to establish that the anomalies of the SWL map arise primarily from sources in the crust.

To further our analysis of the general nature of the sources of anomalies shown in Figure 2, we examined a model based on the anomaly characteristics given in Table 1. The steepest gradients and half-amplitude points of the anomalies listed in the table lie close together. Nettleton (1940, p. 109, 114) showed that the steepest gradients for line sources occur about midway between the half- and full-amplitude points, whereas the steepest gradients for a step-like edge coincide with the half-amplitude points. Thus, the lateral anomaly dimensions of Table 1 are also reasonable approximations of the lateral dimensions of the sources. For our model, we used a source 300 km long, 60 km wide, and 1 km thick at a depth of 20 km. The following considerations were made in choosing this model: source lengths more than 10 times the depth can be considered infinite in length; most of the anomaly widths are within 10 km of 60 km; thickness is not a critical dimension if it is much less than width or depth.

Figure 6 shows the total, SWL, and LWL anomalies of the horizontal rectangular plate, using the same filter that produced the maps of Figures 1 and 2. The steep gradient of the unfiltered total anomaly conforms closely, except for rounded corners, to the outline of the plate source. The position of the half amplitude conforms closely to the length of the source, but it lies slightly outside the width of the source. The maximum gradient in the contour plot of Figure 6b remains in the same location as in 6a, even though the contours are bowed slightly inward. As before, the steepest gradient location outlines the horizontal shape of the source. No shape information appears to be retained in the LWL plot, except that the source is elongate. The amplitude of the LWL plot is about 30 percent higher than that of the SWL plot. The amplitudes of the side lobes of Figure 6b are about 60 percent of that of the main peak.

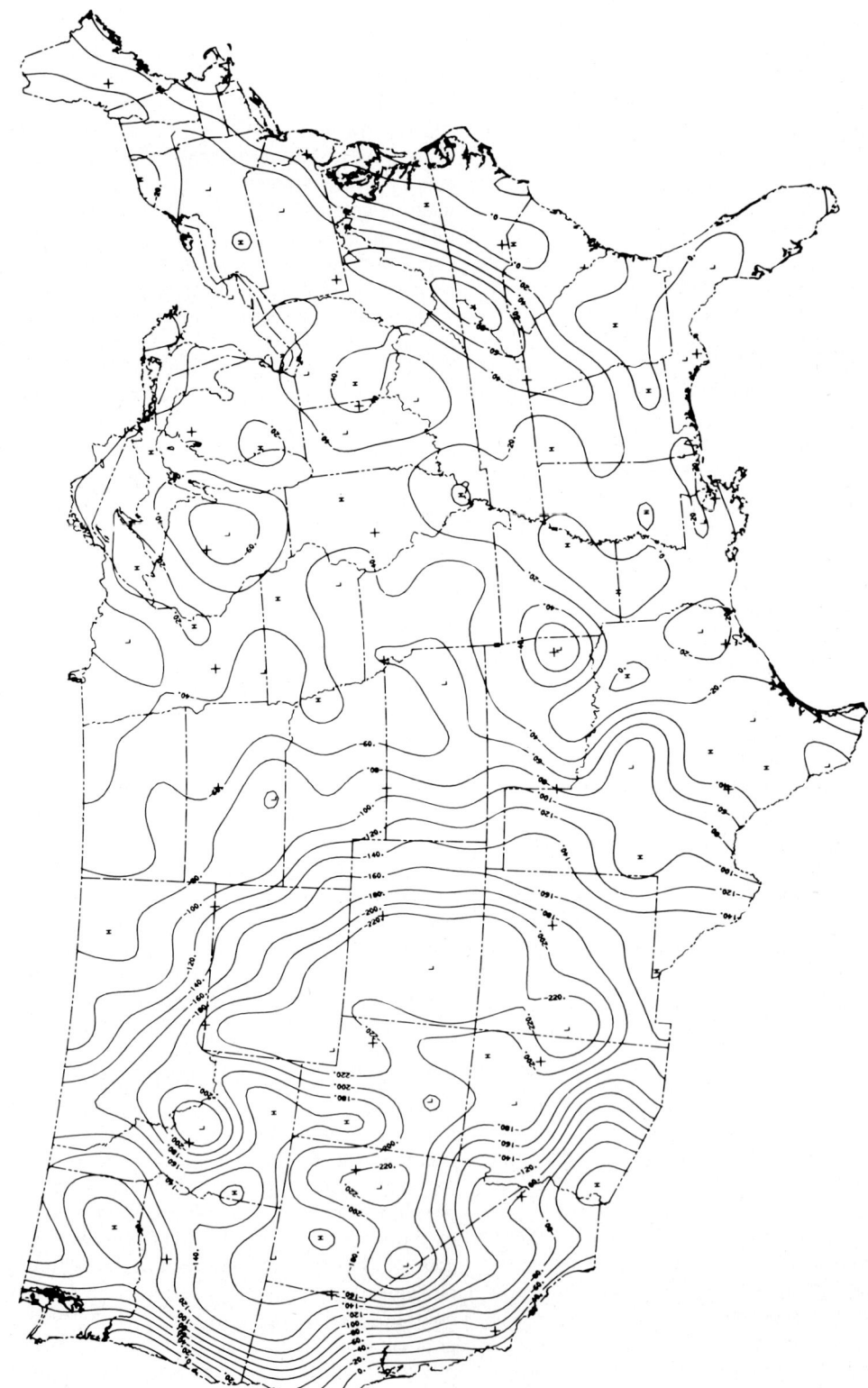

Fig. 1. Regional Bouguer gravity map of the United States composed of wavelengths longer than 250 km. Contour interval, 20 mGal.

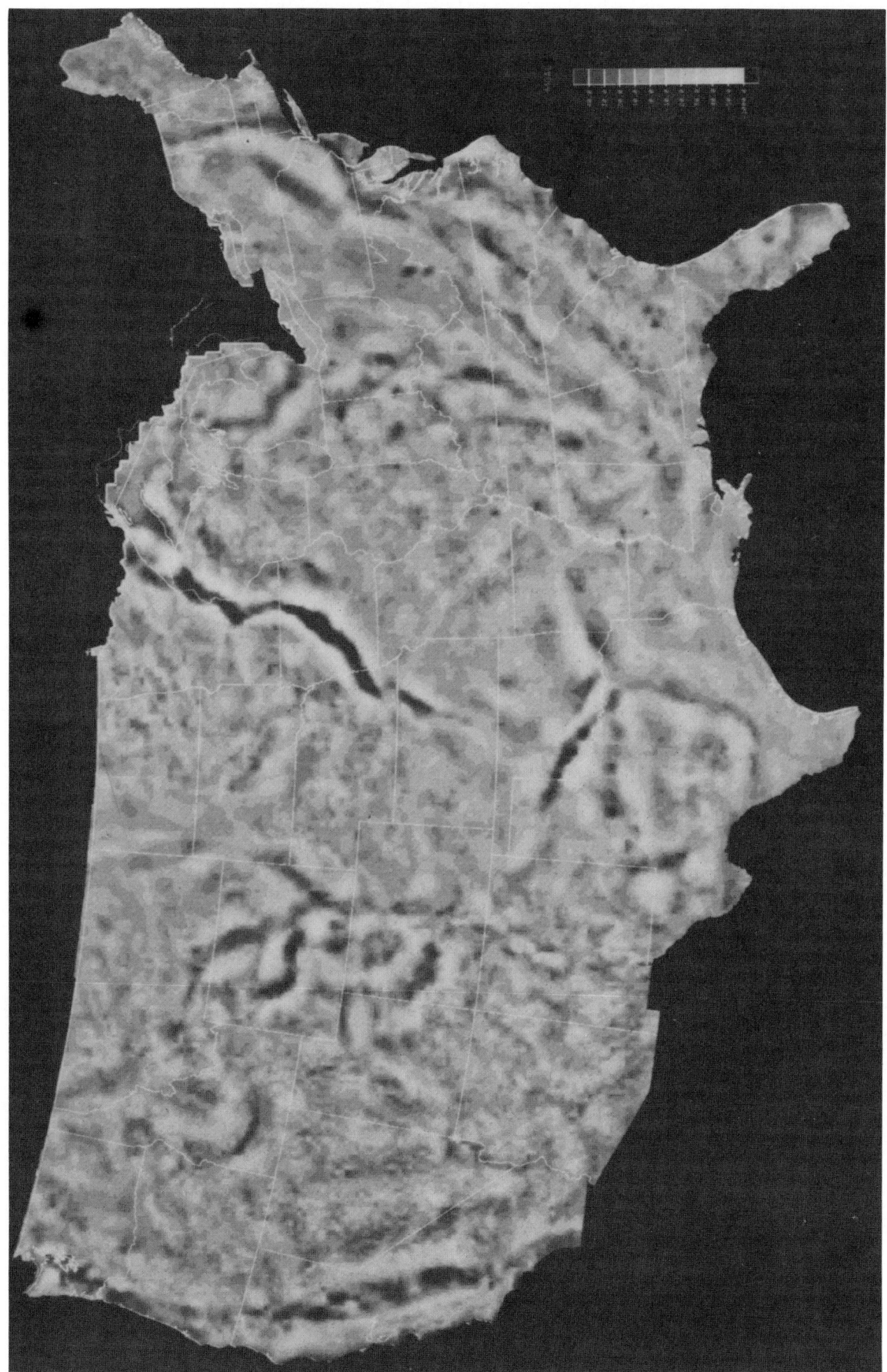

FIG. 2. Residual Bouguer gravity map of the United States composed of wavelengths shorter than 250 km. Shade interval, 10 mGal.

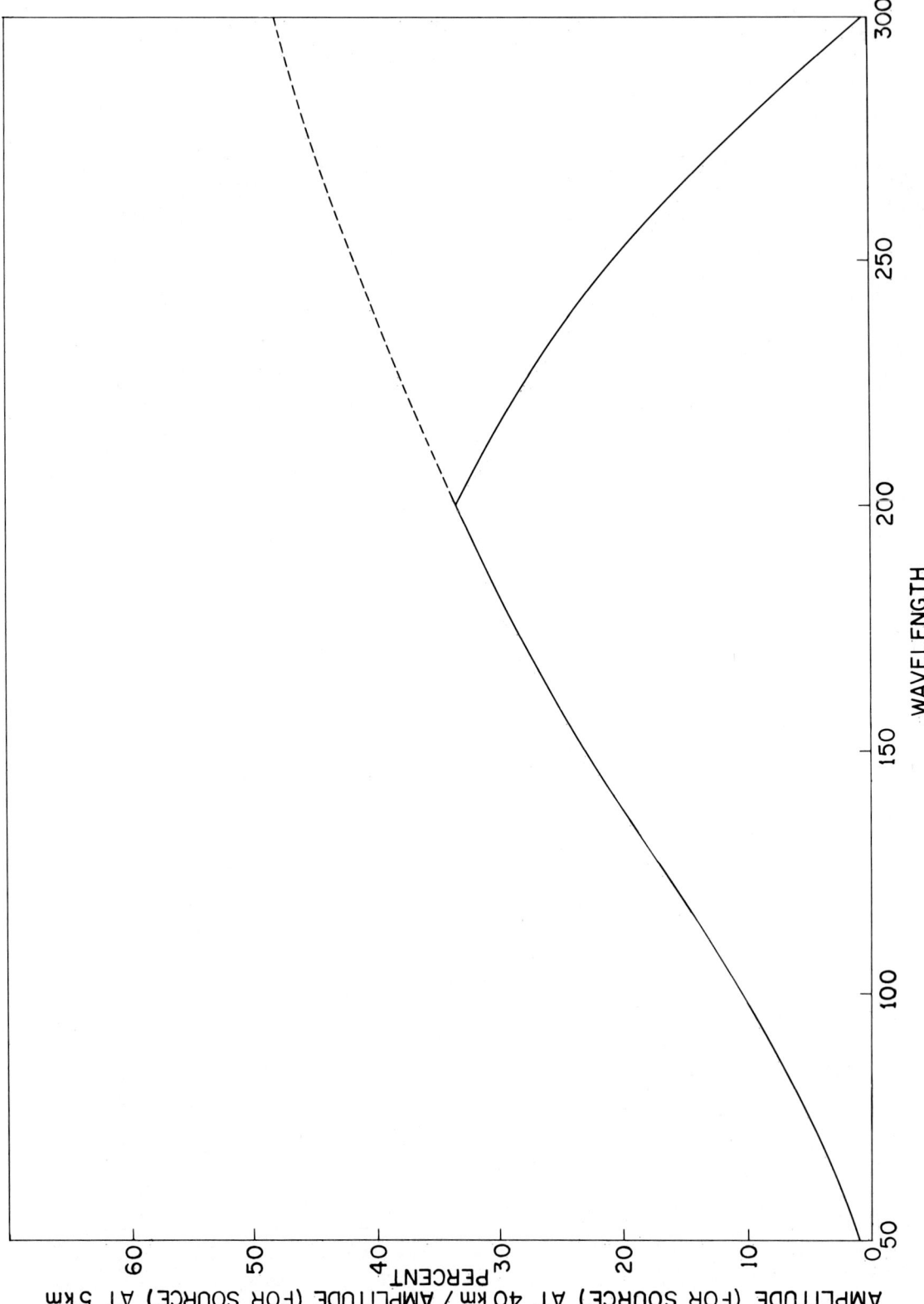

FIG. 3. Attenuation of component wavelengths of anomalies caused by combined effects of deepening of source (5 to 40 km) and filter cutoff; dashed line shows source-deepening effect only.

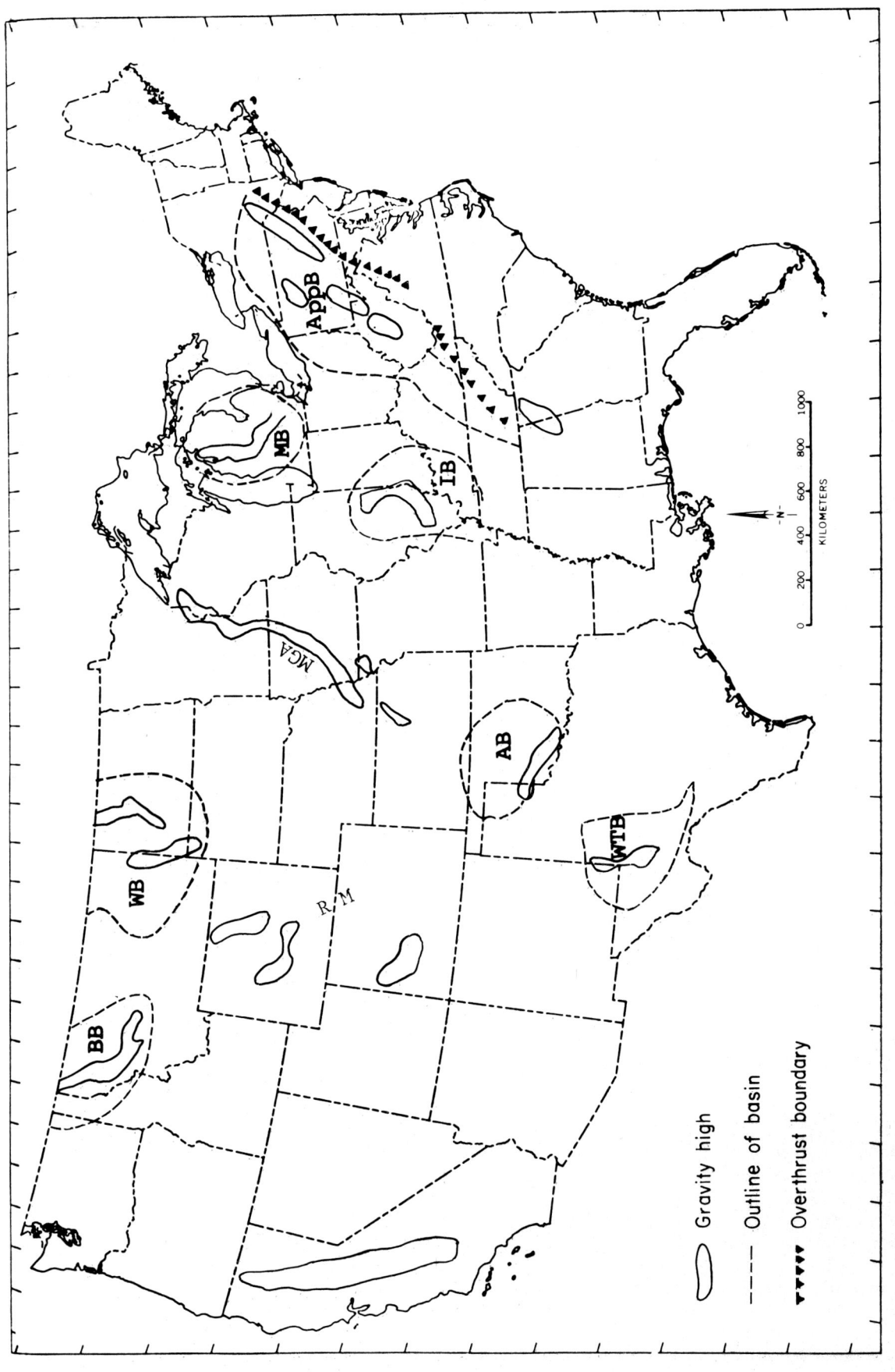

FIG. 4. Location map of the United States showing outlines of the states, gravity features, and sedimentary basins. Abbreviations are as follows: MGA = Midcontinent geophysical anomaly; AB = Anadarko basin; App. B = Appalachian basin; BB = Belt basin; IB = Illinois basin; MB = Michigan basin; WB = Williston basin; WTB = West Texas basin; RM = Rocky Mountains.

Table 1. Prominent anomalies of the SWL map and their characteristics.

Location	Width (km)	Length (km)	Amplitude (mGal)	Maximum gradient (mGal/km)	Maximum depths[1] (km)
California	110	800	50	1.4	23
Midcontinent geophysical anomaly	65	800	125	5.0	16
Pennsylvania–New York	60	350	50	1.6	20
Michigan basin	60	400	35	1.5	23
Belt basin	70	400	35	1.5	23
Williston basin	70	300	35	1.8	13
West Texas basin	60	250	40	1.0	26
Appalachian basin	80	150	20, 20, 35	0.8, 0.6, 1.1	19 (avg.)
Rocky Mountains	65	200	50, 60, 75	2.5, 2.5, 2.5	16 (avg.)

[1] Assumes line source.

We made a series of SWL and LWL plots for the 300- × 60- × 1-km plate source at various depths and found that the characteristics of the unfiltered anomalies described above are typical. Both the side lobes of the SWL plots and the high amplitudes of the LWL plots present serious drawbacks in using filtering as a complete residual/regional separation process. As shown in Figure 6, however, the side lobes have low gradients along their outer margins and horizontal shapes that form symmetrically about the main anomaly. These characteristics may be useful in distinguishing side lobes from true anomalies, which most often will be outlined by relatively steep gradients and which will have shapes that are obviously not related to those of nearby anomalies of opposite sign. The LWL anomalies are co-located on the plots with the SWL anomalies so that highs (or lows) on LWL maps that have SWL counterparts should not be regarded as valid features.

The SWL anomalies emphasize source shape. Because structure is generally associated with shape, it seems reasonable to conclude that the map is useful in discriminating different structural terranes. In this regard it should be noted that although anomaly shape is somewhat distorted, the location of the steepest gradient is relatively stable and is closely related to source shape. We conclude that the horizontal-gradient map (Figure 5) gives a truer indication of source boundaries. The SWL-anomaly map, however, provides additional structural information from anomaly elongation and width, distinguishes highs from lows (with due regard given to side lobes), and adds a relative measure of vertical dimension of the source from anomaly amplitude. The SWL-anomaly and horizontal-gradient maps thus appear to be complementary data plots for analysis of structure.

Although the preceding sections have established that the sources of the anomalies on the SWL map (Figure 2) are primarily in the Earth's crust, the map does not contain all crustal anomalies. Specifically, crustal-anomaly wavelengths greater than 300 km are completely removed, and wavelengths from 200 to 300 km are reduced in amplitude 0 to 100 percent, respectively. As examples, gravity lows caused by broad distributions of relatively low-density sedimentary strata are removed from the SWL map as well as those caused by broad regions of anomalous density in the crystalline part of the crust. The net effect is to constrain anomaly sources to features of the crystalline basement and to narrow sedimentary basins like the intermontane basins of the western United States. The removal of broad sources tends to emphasize the linear features that are usually indicative of structure.

ANOMALY PATTERNS ON THE SWL MAP

General

The dominant characteristic of the gravity map of Figure 2 is the elongate or linear aspect of the anomalies. In many regions anomalies of similar trend form patterns that characterize the area. In the easternmost part of the United States over terranes collectively identified as the Appalachians, the trends are pervasively northeast like those of the geologic structures (King, 1969). This direction is repeated by the Midcontinent gravity anomalies (MGA). In much of the central region between the Appalachians and the Rocky Mountains the trends are variable, but they tend to be similar in zones that lie on opposite sides of MGA. A most notable feature in the southernmost central region is a continuous narrow high that nearly encircles central and western Texas. Strongly expressed highs and lows clearly distinguish the northern and central Rocky Mountain region (western Montana, Wyoming, western Colorado, and eastern Utah). The anomalies display a variety of trends, including a strong western one. The anomaly pattern of southeastern Utah, western New Mexico, Arizona, and southernmost California is variable, but the pattern as a whole contrasts sharply with that of the regions to the north and east.

Trends over the westernmost United States are generally more variable than those to the east, but they form consistent patterns over regions many hundreds of kilometers in average diameter. The most prominent of these patterns are the linear lows and highs that parallel the west coast of the United States. In the north the pattern consists of a moderately well-defined high on the west and a less distinct low on the east. In the south the pattern persists much farther eastward and consists of seven parallel anomalies arranged in a pattern of alternating lows and highs. In the northwest, east of the coast-region anomalies, the anomaly trends and shapes are highly variable.

It is clear that the juxtaposition of highs and lows, which is a prominent characteristic of the map, is in part a result of the filtering process. But it is also clear that many of the example, the Wabigoon ("WB" in Figure 2) volcanic belt is associated in general with a magnetic low. Thus, it is within the broad, magnetic-low areas in which most of the economic ore deposits occur. Locally, magnetic highs in greenstone belts are associated with iron formation, iron-rich tholeiitic basalts, and iron-rich, mafic intrusive rocks. Volcanic rocks of calc-alkaline and komatiitic affinities are generally nonmagnetic. Other pattern changes also occur within the Superior province itself. The Kapuskasing lineament ("KL" in Figure 2), extending from James Bay to the eastern end of Lake Superior, is a structural zone within a geologic province in which both the magnetic base level and the magnetic-anomaly patterns change. On the other hand, the Quetico ("QS" in Figure 2) structural zone is associated with a linear magnetic low that transgresses through the magnetic-anomaly pattern associated with the western Superior province.

It should be realized that a 200-gamma magnetic-anomaly map is a form of filtered map in which the majority of anomalies less than 200 gammas in amplitude are eliminated. Because of the small scale of the 1:5 000 000-scale map and color-contour interval, the longer wavelength anomalies that have an amplitude greater than 200 gammas will tend to be emphasized. Features that produce anomalies less than about 0.5 mm in width at the map scale cannot be physically drawn on the map; 0.5 mm at the 1:5 000 000 scale represents 2 500 m, but only 500 m at the 1:1 000 000 scale. This means that many important narrow geologic features, such as dikes or dike swarms, will not be portrayed on the resulting end product. This is illustrated in Figure 3 for the Great Slave Lake area of the Northwest Territories, which shows a part of the 1:5 000 000-scale magnetic-anomaly map for the area and also the concomitant geologic map on which can be seen a number of diabase dikes. The magnetic expression of the dikes is readily apparent on the 1:1 000 000-scale magnetic-anomaly map of the area (see Figure 7 in Dods et al., this volume). Thus, the short-wavelength cutoff is directly proportional to the scale of the map. Hence the scale and color-contour interval represent the graphic-filtering parameters that may be varied to produce various filtering effects on the magnetic-anomaly map. Thus, inevitably, there is a considerable loss of detail in the smaller scale maps.

Consequently, although the 1:5 000 000-scale Magnetic Anomaly Map of Canada series serves a useful purpose, it is felt that the scale is not the optimum one for regional studies. The 1:1 000 000-scale Magnetic Anomaly Map series, which is the subject of the paper by Dods et al. in this volume, is much better suited for that purpose. However, there is no doubt that the Geological Survey of Canada will continue to produce a national magnetic-anomaly map at the 1:5 000 000 scale to accompany the concomitant geologic, tectonic, and gravity maps at the same scale.

Aeromagnetic coverage of the sedimentary areas of Saskatchewan and Alberta has been obtained by the petroleum industry. Negotiations to obtain these data are being undertaken at the present time both to complete the Magnetic Anomaly Map of Canada and also the Magnetic Anomaly Map of North America, which is being produced as part of the Geological Society of America centenary. The Geological Survey of Canada wishes to thank the Gulf and Mobil oil companies for their recent contributions of data from southern Saskatchewan, which will be a useful addition to contributions made in former years by other oil and mining companies in various parts of Canada. We hope that their example will be followed by the other oil companies that hold similar 30-year-old data in the Western Plains area of Canada.

The production of digitally compiled color magnetic-anomaly maps became feasible with the advent of the Applicon color plotter in the late 70's and has revolutionized the techniques of producing such maps. One of the first end products produced using these newly developed techniques has been a Magnetic Anomaly Map of Arctic Canada, compiled by P. H. McGrath of the Geological Survey of Canada and I. Fraser of the Department of Indian and Northern Affairs, at a scale of 1:3 500 000. Much of the data was supplied by private companies to the Department of Indian and Northern Affairs. The digitally compiled map, published in February 1982, utilized 33 shades extending over the spectrum of white light, each one representing 25 gammas except at the extremities. The map was printed using three color separations produced by the Applicon color plotter. The data from the Magnetic Anomaly Map of the Canadian Arctic have been incorporated into the fourth edition of the Magnetic Anomaly Map of Canada (shown at the front of this volume), although with the reduction of scale there is inevitably some loss of detail.

There are many features of regional interest on the map. For instance, the extension of the Slave province under the Paleozoic rocks of Victoria Island can be recognized. Distinctive patterns of elongated highs and lows occur adjacent to the Boothia Peninsula and in the Devon Island area. The latter area, which has an oval shape, may represent the pivotal area around which Greenland rotated to form Baffin Bay and the Labrador Sea. A number of grabens can also be recognized, such as the sediment-filled Melville Bay graben and one that strikes south-southwestward from Prince of Wales Island.

To summarize, Figure 4 shows the various scales of the airborne-magnetic maps published by the Geological Survey of Canada mostly in cooperation with Provincial government geological surveys. Each area outlined in Figure 4 indicates, in turn, the area covered by the adjacent larger scale map. The four scales are 1:50 000, 1:250 000, 1:1 000 000, and 1:5 000 000, which are roughly equally spaced steps of 5:1, 4:1, and 5:1. Thus, a given area on the 1:50 000-scale aeromagnetic map would be reduced by a factor of 10 000 on the 1:5 000 000-scale magnetic-anomaly map. It is common usage to call the maps at the smaller scales Magnetic Anomaly Maps because the Earth's core field has been removed in compiling them to leave only the magnetic anomalies from the crustal rocks; hence the geophysical parameter portrayed is the residual total field. Thus the colored Magnetic Anomaly Map of Canada demonstrates that features of large areal extent, such as geologic provinces, are readily delineated and proves the usefulness of producing such regional compilations.

Consequently, in carrying out aeromagnetic surveys overseas as contributions to Canadian aid programs of the Canadian International Development Agency (CIDA), the preparation of regional magnetic-anomaly maps has usually

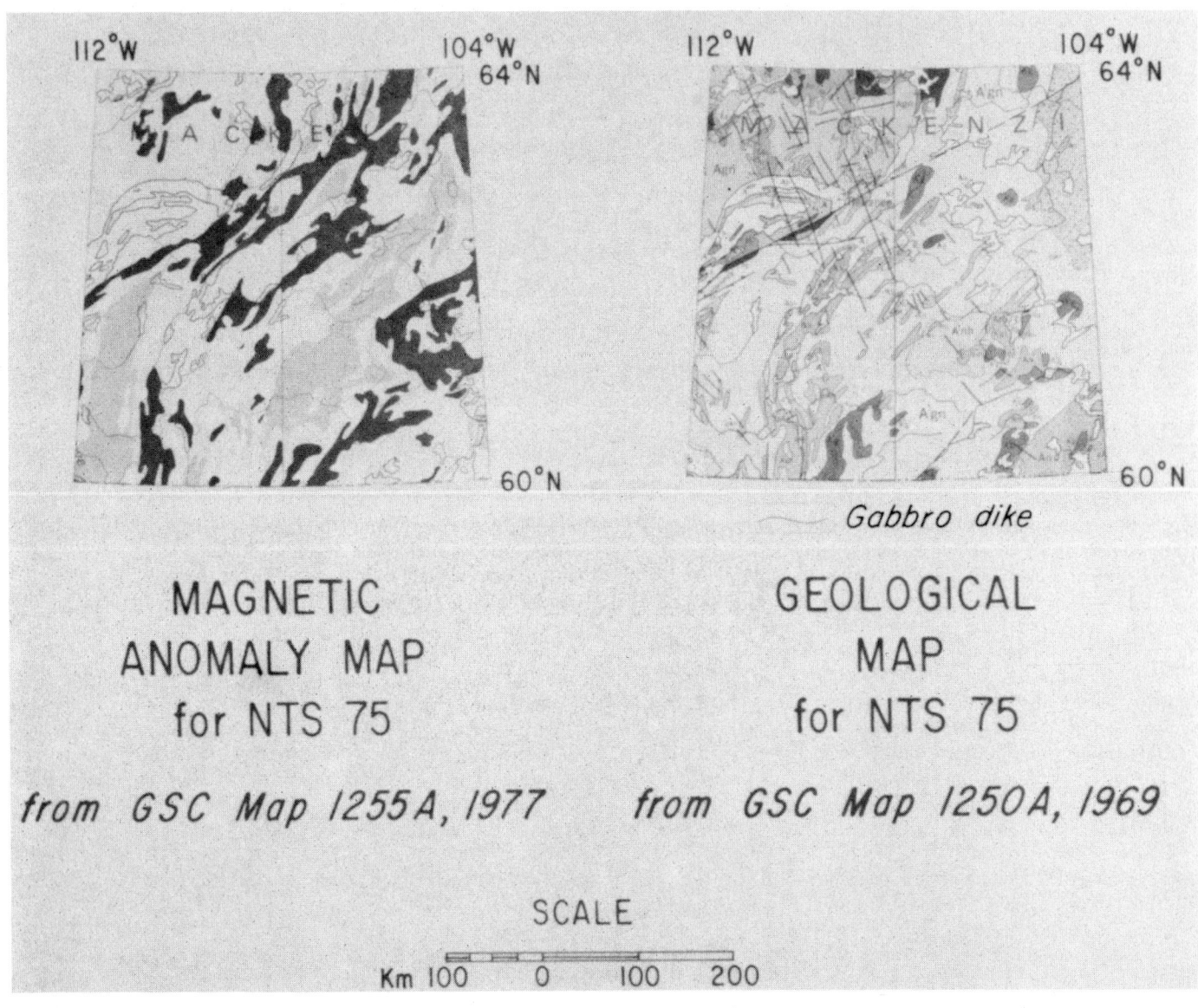

FIG. 3. Comparison of geologic and magnetic-anomaly maps for the Great Slave Lake area, Northwest Territories (NTS 75).

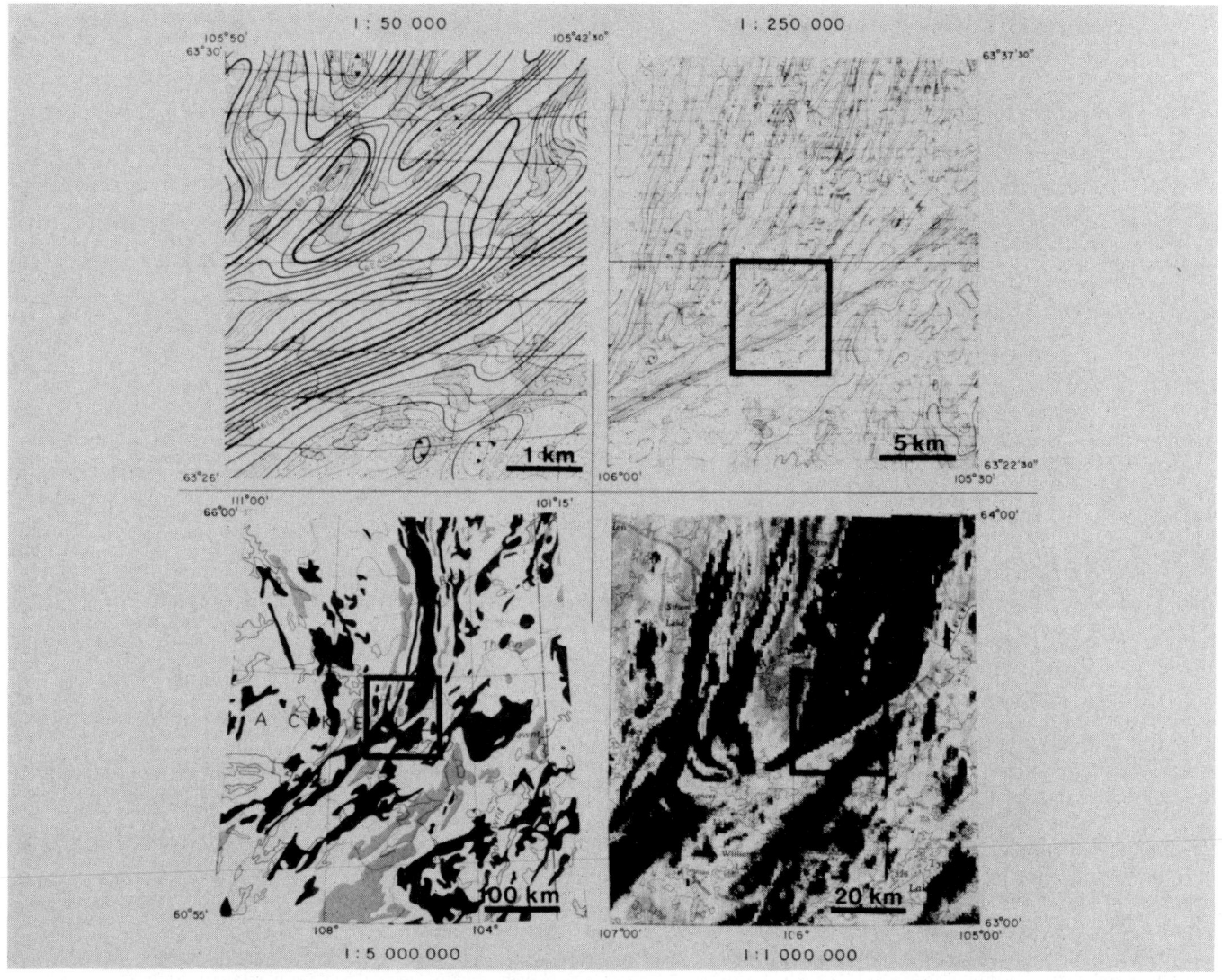

FIG. 4. Various scales used in Canadian government aeromagnetic maps. Each area outlined indicates, in turn, the area covered by the adjacent larger scale map.

been made part of the contract at the interpretation stage. The first such MAM published under contract in 1973 was a colored 1:1 000 000-scale MAM of Guyana, which was the first of its kind in South America (Hood and Tyl, 1977). It was compiled from two aeromagnetic surveys that covered about two-thirds of the country—one by the United Nations and the second by Canada. The IGRF 1965.0 extrapolated to the year of the aeromagnetic surveys was subtracted from the total-field data, using a graphic-subtraction technique. The resulting 1:1 000 000-scale published map was produced by Terra Surveys, Ltd., of Ottawa, using six colors, each representing a 200-gamma interval with intermediate 100-gamma contours.

The limit of the Southern Province of Guyana is readily evident on the map, and the sediment-filled Takatu graben appears as an elongated magnetic high. The Takatu graben appears to form one arm of a failed triple-spreading junction that did not progress to maturity in the opening of the North Atlantic Ocean.

It has been found in studies of residual magnetic-anomaly maps that a system of regional magnetic anomalies exists having wavelengths up to several hundred kilometers in length that are usually related to the major geologic units and are also closely related to crustal structure. Some of the anomalies also appear to correspond to magnetic units of considerable areal extent that occur in the lower part of the Earth's crust. The bottoms of the units terminate deep within the crust, probably at the Curie-point geotherm. As has been found elsewhere, regional highs appear to correspond with predominantly felsic rocks, such as granite intrusions. The geologic model pertinent to the discussion of the residual magnetic-anomaly map of Guyana is represented in Figure 5, which shows a generalized profile across the central and northern parts of Guyana. The profile consists of a series of longer wavelength anomalies, as shown by the dashed line, upon which are superimposed the much shorter wavelength anomalies from the near-surface geology. The generalized profile also shows differences in am-

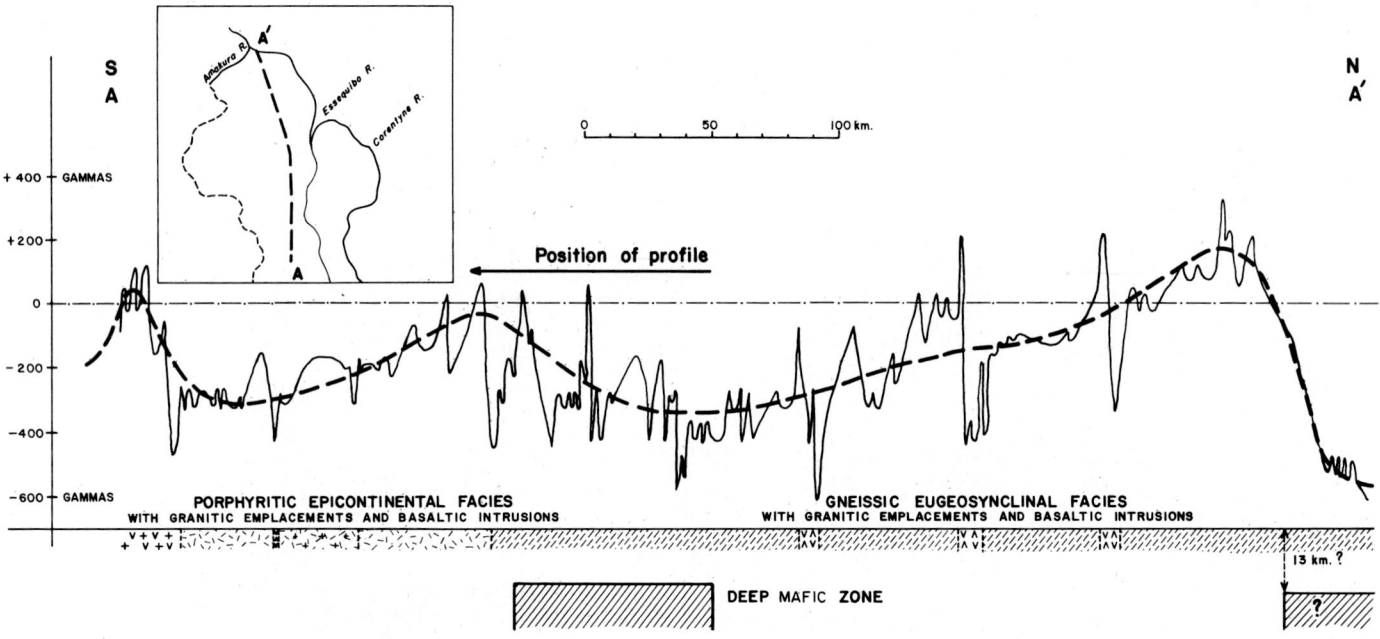

FIG. 5. South–north aeromagnetic profile, with interpreted geology, across central and northern Guyana, South America (Hood and Tyl, 1977).

plitude and wavelengths that appear to be related to the geology of the lower crust. For the latitude of Guyana, where the dip of the Earth's field is about 30 degrees, a large mafic rock formation will produce a magnetic high at its southern extremity with an associated low, which is approximately coincident with its northern edge. On Figure 5, such a formation, labeled "Deep mafic zone," has been inferred from the profile. It is also interesting that there appears to be some correlation of diamond-occurrence localities in northwestern Guyana and the area delineated by the lowest residual magnetic-anomaly value.

Subsequently, in 1981, a digitally compiled colored 1:1 000 000-scale magnetic-anomaly map of Baluchistan in western Pakistan was specified in a CIDA interpretation contract in accordance with the techniques recently developed at the GSC. This map is described in the paper by Spector et al. in this volume. A similar map was specified in the interpretation contract for the Liptako–Gourma contract of West Africa, which was awarded to Paterson, Grant and Watson of Toronto in March 1983.

In conclusion, it is abundantly clear that magnetic-anomaly maps should be made a standard end product of regional aeromagnetic surveys because of their several uses. They serve not only as index maps to the aeromagnetic-survey coverage available on a national basis (as well as to gaps in the coverage) but also stimulate comparison of the regional magnetic features with those appearing on similar-scale geologic and geophysical (e.g., gravity) maps and with Landsat imagery. They also make excellent office wall displays for hand-waving geophysicists.

REFERENCES

Hood, P. J., and Ready, E., 1976, Federal–Provincial aeromagnetic survey program of Canada: a progress report: Pap. 76–1B, Geol. Surv. Can., 267–272.

Hood, P., and Tyl, I., 1977, Residual magnetic anomaly map of Guyana and its regional geological interpretation: Mem. 2nd Latin Am. Congr., Caracas, 3, 2219–2235.

Kornik, L. J., 1969, An aeromagnetic study of the Moak Lake–Setting Lake structure in northern Manitoba: Can. J. Earth Sci., 6, 373–381.

———1971, Magnetic subdivision of Precambrian rocks in Manitoba: Spec. Pap. 9, Geol. Assoc. Can., 51–60.

Kornik, L. J., and MacLaren, A. S., 1966, Aeromagnetic study of the Churchill–Superior boundary in northern Manitoba: Can. J. Earth Sci., 3, 547–557.

MacLaren, A. S., and Charbonneau, B. W., 1968, Characteristics of magnetic data over major subdivisions of the Canadian Shield: Proc., Geol. Assoc. Can., 19, 57–65.

Magnetic Anomaly Map of Canada: 1st ed., Morley, L. W., MacLaren, A. S., and Charbonneau, B. W., 1967; 2d ed., Anonymous; 3d ed., McGrath, P. H., Hood, P. J., and Darnley, A. G., 1977; 4th ed., Dods, S. D., Hood, P. J., Teskey, D. J., and McGrath, P. H., 1984: Map 1255A, Geol. Surv. Can., scale 1:5 000 000.

McGrath, P. H., Haley, E. L., Reveler, D. A., and Letourneau, C. P., 1978, Compilation techniques employed in constructing the Magnetic Anomaly Map of Canada: Pap. 78–1A, Geol. Surv. Can., 509–515.

The new series of 1:1 000 000-scale magnetic anomaly maps of the Geological Survey of Canada: Compilation techniques and interpretation

S. D. Dods,* D. J. Teskey,* and P. J. Hood*

ABSTRACT

In 1977 the Geological Survey of Canada initiated a project to produce colored magnetic-anomaly maps at a scale of 1:1 000 000 for those areas of Canada covered by published aeromagnetic maps at a scale of 1:50 000 or 1 inch:1 mile. An additional objective of this project was to create a magnetic data bank for Canada which could be used at scales from 1:250 000.

One of the more interesting uses of these data has been the generation of magnetic shaded-relief maps in which the magnetic-field values are assigned a vertical scale to represent topography and illuminated from a light source (usually considered to be the sun). This procedure has the effect of enhancing features such as dikes and contacts that are not seen on the original magnetic-anomaly maps because of the coarseness of the color scale. The data can also be used with standard continuation or derivative techniques.

It is intended that the 1:1 000 000-scale magnetic-anomaly maps will be utilized essentially as basic building blocks in compiling future editions of the 1:5 000 000-scale Magnetic Anomaly Map of Canada, which will in turn be used for the Magnetic Anomaly Map of North America being compiled as part of the Geological Society of America's Decade of North American Geology program.

INTRODUCTION

Since 1947 the Geological Survey of Canada has published aeromagnetic total-field maps at scales of 1 inch:1 mile (1:50 000 after metrification) and 1 inch:4 miles (1:250 000). Beginning in 1967, three editions of 1:5 000 000-scale colored Magnetic Anomaly Maps of Canada have been published. It was felt, however, that an intermediate scale of 1:1 000 000 was required to provide the necessary detail for regional studies. The decision to commence this program was given impetus by the development of the Applicon plotter, capable of producing maps with 39 distinct color ranges. An additional objective was the creation of a digital magnetic data bank for Canada which could be used for production of maps at scales from 1:250 000 to 1:5 000 000 and which would be accessible for data enhancement and interpretation. The aeromagnetic and shipboard coverage for Canada to December 1982 is shown in Figure 1a. Figure 1b shows the progress of the 1:1 000 000-scale magnetic-anomaly-map project to date.

COMPILATION

The basic compilation steps in producing a 1:1 000 000-scale magnetic-anomaly map are as follows:

1. *Digitization*—Contour intercepts along flight lines on 1:50 000-scale or 1 inch:1 mile aeromagnetic map sheets are digitized, as this is considered to be the most accurate method of recovering the original total-field data. The digitized flight lines are plotted back for editing and error correction.
2. *Filing*—Each data point is converted to geographical coordinates and filed by survey area. This file constitutes the basic unit for the magnetic data bank.
3. *Gridding*—After subtraction of the Definitive Geomagnetic Reference Field for the average year of survey, the data are converted to a common-origin Lambert conformal conic projection and gridded, using a simple inverse distance-weighted average technique. This technique is considered adequate, as the average flight-line spacing (805 m) is approximately equal to the grid size (812.8 m); thus it is essentially a sampling procedure.
4. *Continuation*—In some cases, private data that satisfy the basic specifications have been included in the original data set. If these data were recorded at a different altitude from the surrounding published sheets, they were continued upward or downward where possible to the standard datum, using fast Fourier transform techniques.
5. *Boundary adjustments*—Discontinuities at survey

*Geological Survey of Canada, 601 Booth Street, Ottawa, Ont. K1A 0E8, Canada.

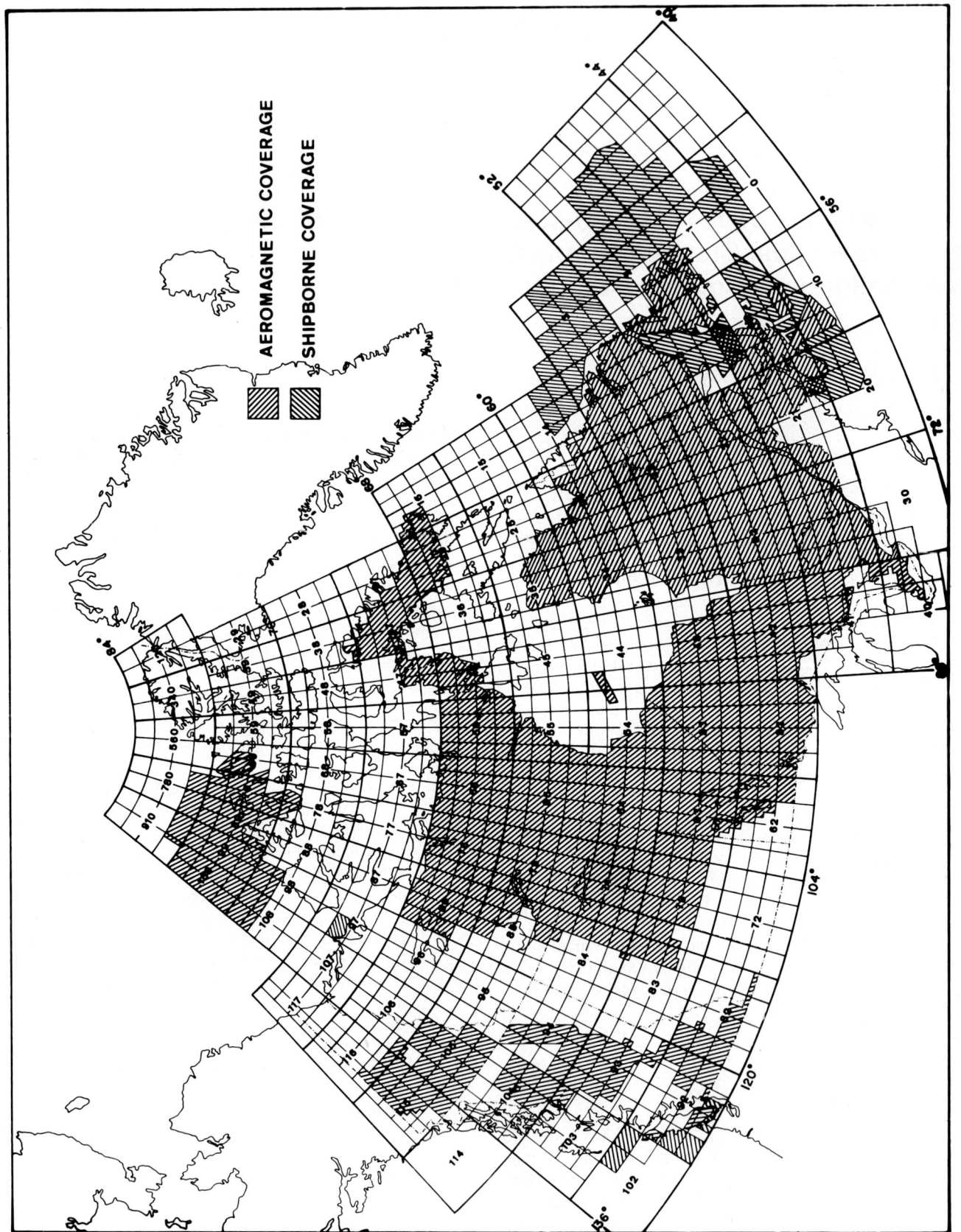

FIG. 1a. Aeromagnetic and shipborne magnetometer map coverage of Canada to December 31, 1982.

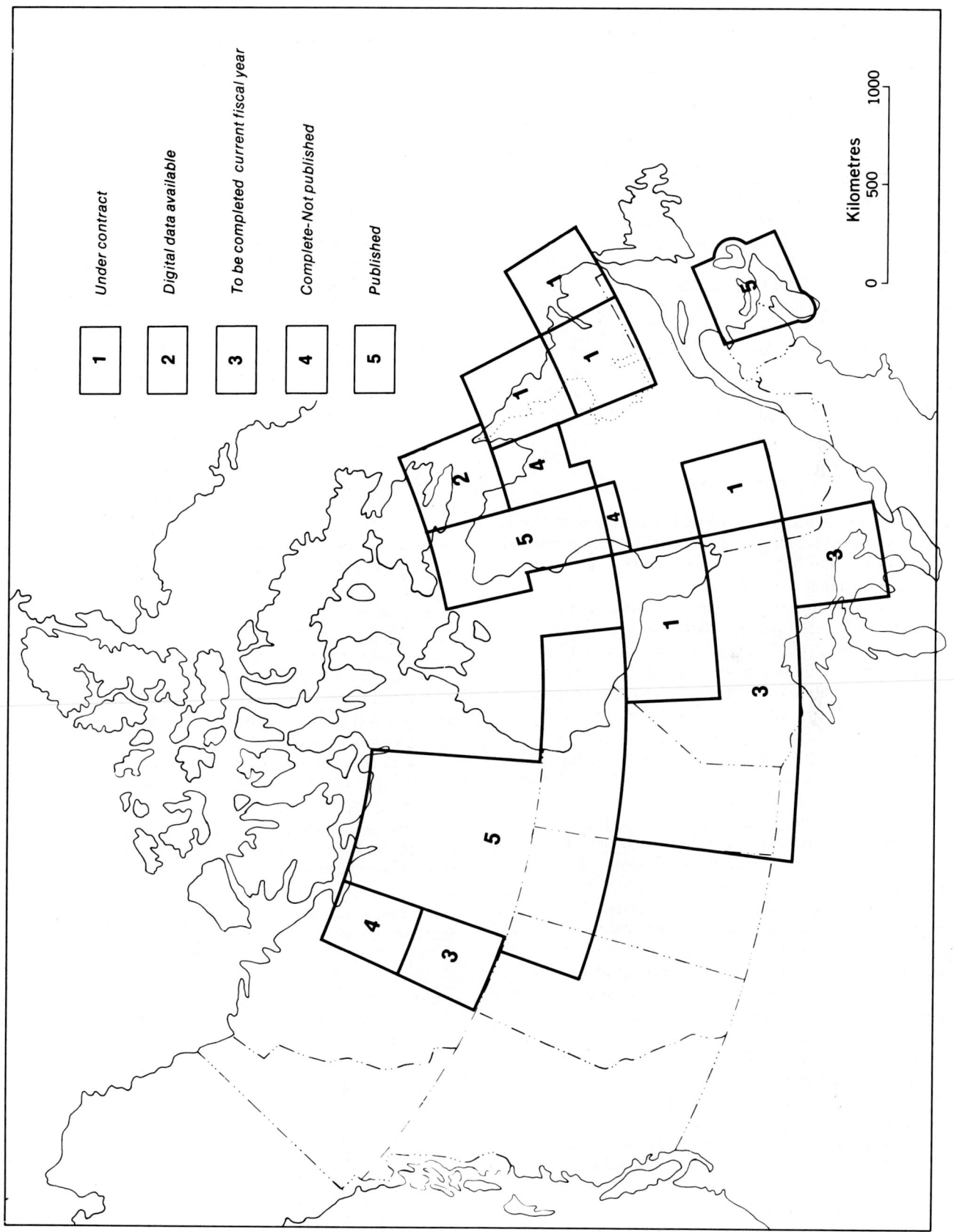

Fig. 1b. Progress of 1:1 000 000-scale magnetic-anomaly-map project to December 31, 1982.

boundaries are removed by first calculating a first-order fit along the boundary of the area to be adjusted and the fixed area. The resulting first-order surface is subtracted. (In all cases to date only a constant term has been required.) Remaining discrepancies are removed by treating the residual errors around the boundary as the boundary values for a simple Laplacian differential equation, which is solved numerically. Thus the smoothest possible surface consistent with the boundary differences is derived and subtracted from the survey to be adjusted. It is anticipated that some situations will occur (i.e., in rugged terrain) in which a complete removal of boundary differences will not be desirable.

6. *Merging*—The adjusted surveys are combined as International Map of the World (IMW) sheets and converted to the final Lambert conformal conic projection.

7. *Printing*—The three primary colors are plotted on the Applicon plotter and used as the color separates for the offset-printing process.

DISCUSSION

A black and white copy of the colored composite 1:1 000 000-scale magnetic-anomaly map for the published area west of Hudson Bay is shown in Figure 2. In Figure 3, known fault systems are indicated by solid lines, whereas faults or contacts suggested by the magnetic-anomaly map, or from the magnetic shaded-relief maps, are shown by dashed lines. Most of this map area is in the Churchill structural province of the Canadian shield, last affected by a major orogeny about 1 735 m.y. ago. This area is separated from the older (2 480-m.y.) Slave province to the northwest by the MacDonald fault, which has a clear magnetic expression, and the Thelon front, a zone of abrupt metamorphic change with a parallel magnetic low and high just to the east. The precise position of the Thelon front is under study, and its final placement may be influenced by the magnetic pattern. On the southeast the Churchill province is separated from the older Superior province by the Thompson fold belt. In the Churchill province the dominant magnetic patterns are northeast, except in the portion west of Lake Athabasca, where they are rotated to a north–south direction, and in Northern Manitoba, where they are primarily east–west. (Note the exception of the northeast-striking Nelson fault.) These trends are parallel to a series of gravity lows from the eastern end of Lake Athabasca to Baker Lake, which has been suggested as the site of a major suture. For example, Cavanaugh and Seyfert (1977), on the basis of radiometric age determinations and apparent polar wander paths, consider that the northwest portion of the Churchill province was attached to the Slave plate and the southeast portion to the Superior plate prior to 1 750 m.y. ago. According to their interpretation, the Hudsonian orogeny was initiated by the collision of these plates. Darnley (1981) suggested that the Athabasca axis is a rift structure and has pointed out the coincidence of some of the gravity lows with granitic rocks anomalously high in radioelement distribution. The Needle Falls shear zone, which separates the Wollaston fold belt and the Rottenstone migmatite zone, and which is clearly defined as a narrow magnetic low bisecting an elongated high, has also been suggested as the site of complex rifting, followed by a closing and subduction process (Ray and Wanless, 1980). Similar patterns are seen in the portion of the Nelson fault extending below Paleozoic cover west of Hudson Bay, along the convoluted magnetic pattern extending northeast from Lake Athabasca and at the Amer Lake shear zone, which is considered to be a possible extension of the MacDonald fault system. One of the more interesting features on the map is the sudden change in magnetic character across a line (A–A' in Figure 3) from the western end of Lake Athabasca to Chantrey Inlet. The area northwest of this line, which is bounded on the northwest by the MacDonald fault and on the west by the Thelon front, is characterized by elongated, fragmented magnetic highs with a "stretched" northeast–southwest appearance, while that on the southeast is relatively featureless except for isolated anomalies. This line appears to define a major tectonic contact.

DIGITAL ENHANCEMENT

As mentioned above, one of the prime objectives of this program has been to acquire a digital data bank for Canadian magnetic data. This data bank makes possible enhancement techniques which emphasize features on the magnetic-anomaly map that do not appear in the original colored anomaly map because of the coarseness of the color scale. Many dikes, for example, produce magnetic anomalies that are clearly defined but have a relatively small amplitude.

One technique that has been used extensively to emphasize topographic landforms, and more recently for such purposes as automatic navigation and analysis of satellite imagery, is that of shaded relief (Horn and Bachman, 1978). This technique consists essentially of calculating the intensity at an observation point of light reflected from a surface element with the light source at a given distant position. For the above applications with topographic relief, the calculations are complicated by such factors as mutual illumination of surface elements, shading by neighboring elements, texture of the reflecting surface, and scattering in the atmosphere (Horn and Bachman, 1978). In our application, since the purpose is simply to emphasize the desired magnetic features, the above factors need not be considered.

The basic technique is described by Horn and Bachman (1978) and is illustrated by Figure 4. Each pixel can be considered as a basic surface element. If x, y, and z represent the two horizontal and the vertical axes, respectively, the gradient

$$p = \frac{\partial z}{\partial x} \quad \text{and} \quad q = \frac{\partial z}{\partial y}$$

can be estimated using first differences. Since the vertical axis corresponds in our case to the magnetic-field value rather than elevation, an appropriate scale factor in nanoteslas/distance unit must be assigned. The normal vector to the surface element has components

$$(-p/\sqrt{p^2 + q^2 + 1}, -q/\sqrt{p^2 + q^2 + 1}, 1/\sqrt{p^2 + q^2 + 1}).$$

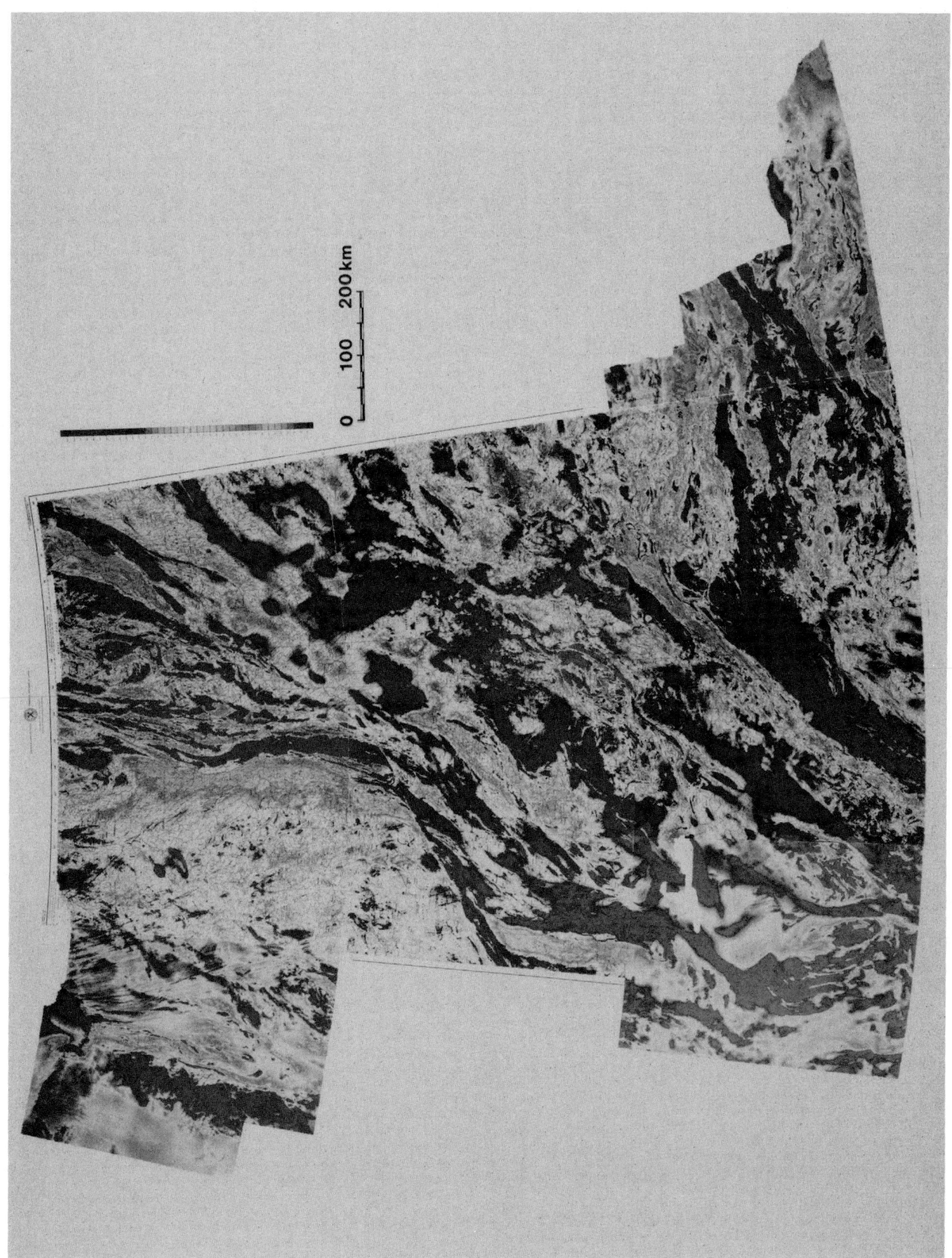

FIG. 2. Black and white reproduction of completed 1:1 000 000-scale magnetic-anomaly maps west of Hudson Bay to December 31, 1982.

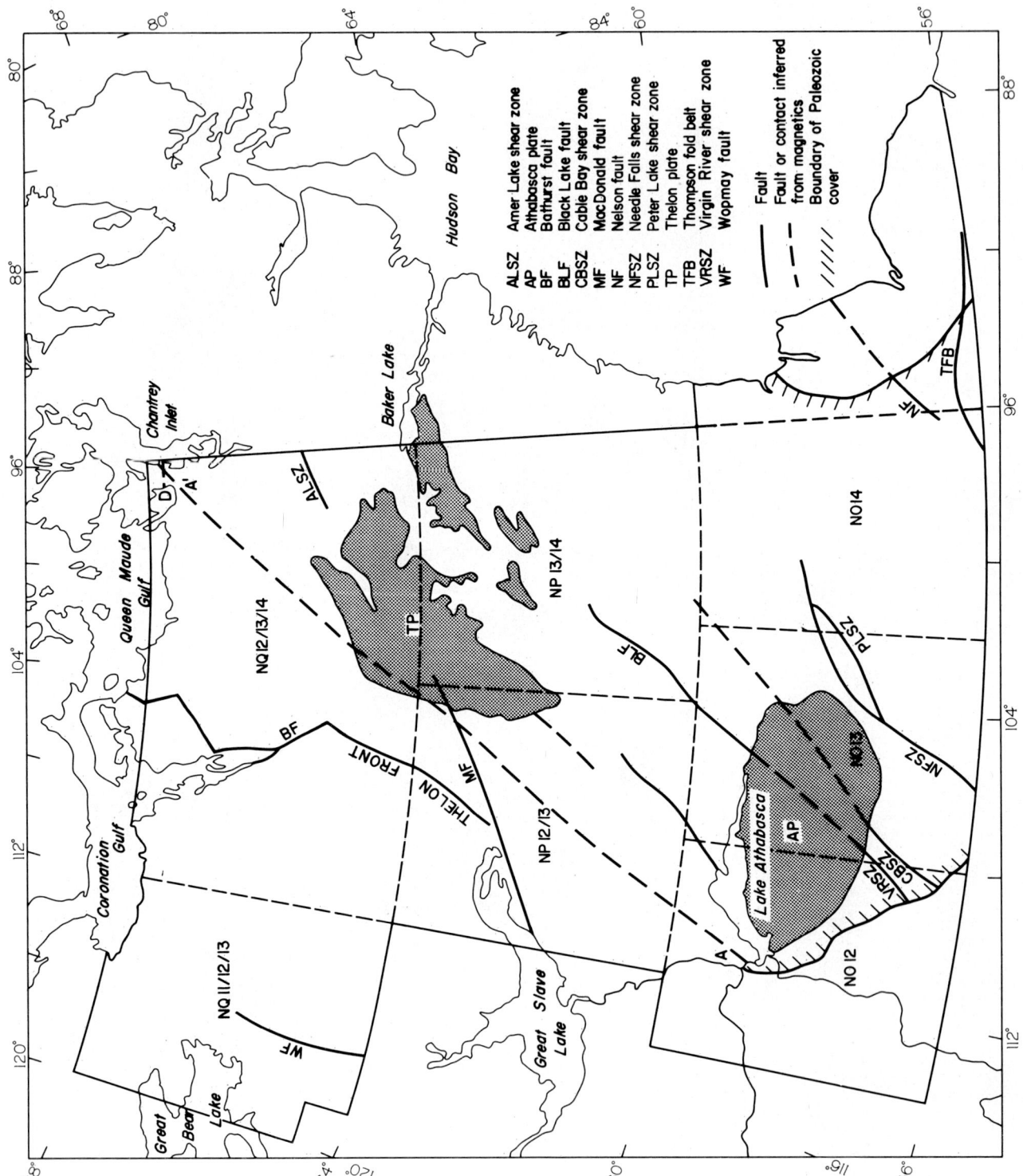

Fig. 3. Sketch map of area corresponding to the magnetic-anomaly map shown in Figure 2, identifying major features referred to in the text.

If the direction between the surface element and the light source is given as an inclination and declination, the vector from the surface element to the source will have direction cosines (cos θ cos φ, sin θ cos φ, sin φ). The cosine of the angle between the two vectors is given by the scalar product

$$\cos \lambda = \frac{-p \cos \theta \cos \phi - q \sin \theta \cos \phi + \sin \phi}{\sqrt{p^2 + q^2 + 1}}. \quad (1)$$

Again, because we are not required to duplicate precisely the effect of a light source on a real surface, we are able to simplify the relationship between λ and the measured light intensity. In particular, we assume that the viewing surface is at a great distance directly above the map surface and that the surface is a perfect diffuser; that is, the measured intensity is independent of the angle between the normal to the surface element and vector from the surface element to the viewing point. In fact, it is found that a simple function in which the brightness is proportional to cos λ is adequate. When a color plotter is being used for the output device, the appearance is affected as well by the choice of colors to represent the brightness levels so that the function used and the colors chosen to produce the desired effect are clearly interdependent. In our case we have chosen yellow to represent the brightest case (normal incidence), increasing magenta (red) content, and finally cyan (blue) as cos λ decreases to a dark purple for λ greater than 90 degrees. The process is generally applied to the digitized map in two orthogonal directions to bring out the desired details. The process is primarily sensitive to the gradients p and q and thus is similar to a calculated horizontal gradient map. (Note that in this paper all original 1:1 000 000-scale magnetic and shaded-relief maps are shown as black and white reproductions of the original colored maps.)

The technique was applied to the Wollaston Lake sheet (IMW NO 13), shown in Figure 5. The resultant shaded-relief maps for 45- and 135-degree light-source declination are shown in Figures 6a and 6b. Northwest-striking features, such as the dikes in the northeast corner (shown by arrows in Figure 6a), are emphasized with the light source at 45 degrees, while northeast-striking features, such as the Needle Falls (NF), Cable Bay (CB), Black Lake (BL), and Virgin River (VR) shear zones (Figures 3, 6b), are brought out clearly with the source at 135 degrees. The Virgin River shear zone can be easily traced under the Athabasca basin to the Black Lake fault. Similarly, the Cable Bay shear zone can be traced under the basin to the northeast corner of the map.

The shaded maps for the 1:1 000 000-scale Lockhart River sheet (IMW NP 12/13) (Figure 7) are shown in Figures 8a and 8b. Here, the dominant feature is the MacDonald fault (MF). Perhaps of greater interest is the previously discussed southwest–northeast contact, which is clearly defined on the shaded-relief map illuminated from 135 degrees (shown as A–A' in Figures 3 and 8b). A number of subtle east–west features shown by arrows in Figure 8b can also be seen north of the fault in the Slave structural province. As before, the northwest-striking features—in particular, the Mackenzie dike swarms—are emphasized when the area is illuminated from 45 degrees (Figure 8a).

The magnetic-anomaly map for the Thelon River sheet

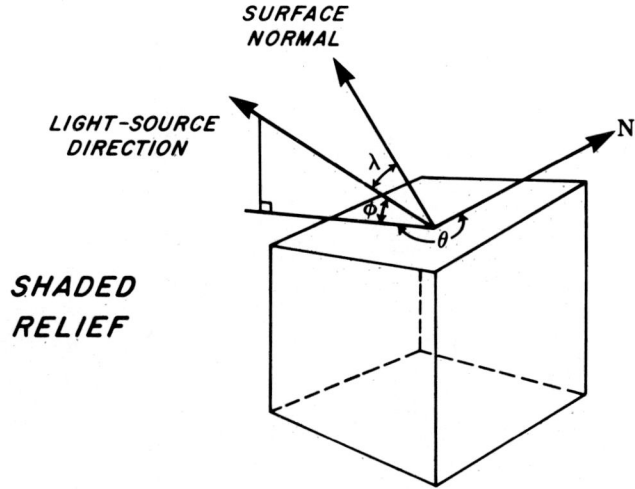

FIG. 4. Principle of the shaded-relief technique, showing the geometry and relevant nomenclature.

(NQ 12/13/14) is shown in Figure 9. The shaded maps were run at 90- and 180-degree declination and are shown in Figures 10a and 10b. Again, the diabase-dike systems are brought out in the 90-degree map, as are the sinistral displacements along the Bathurst fault (BF) and the dextral offset in the northeast corner (D in Figures 3 and 10a) of the map.

Additional processes, such as upward continuation, which can be useful to deemphasize surface effects when modeling deeper sources, or vertical gradient, to emphasize contacts and faults, can be carried out by standard fast Fourier transform techniques. The magnetic-anomaly map continued to 5 000 m for the eastern half of the Thelon sheet is shown in Figure 11, while the calculated vertical gradient is shown in Figure 12.

SUMMARY

The 1:1 000 000-scale magnetic-anomaly maps currently being produced by the Geological Survey of Canada will not only serve as a useful interpretation tool, but the magnetic data bank that is being collected in the process will be valuable for enhancement, processing, and interpretation. This has been demonstrated for the completed area west of Hudson Bay in Figure 2, where known and previously inferred features are clearly shown as well as features that are evident only on the 1:1 000 000-scale magnetic-anomaly map, such as contact A–A' (Figure 3) or on the shaded-relief maps, which are particularly effective for delineating low-amplitude features not clearly visible on the original map. The data bank will make possible production of maps at scales down to 1:250 000 at arbitrary projections. In particular, it will serve as the basis for the succeeding 1:5 000 000-scale magnetic-anomaly maps and for the Magnetic Anomaly Map of North America being compiled as part of the Decade of North American Geology program to celebrate the centenary of the Geological Society of America.

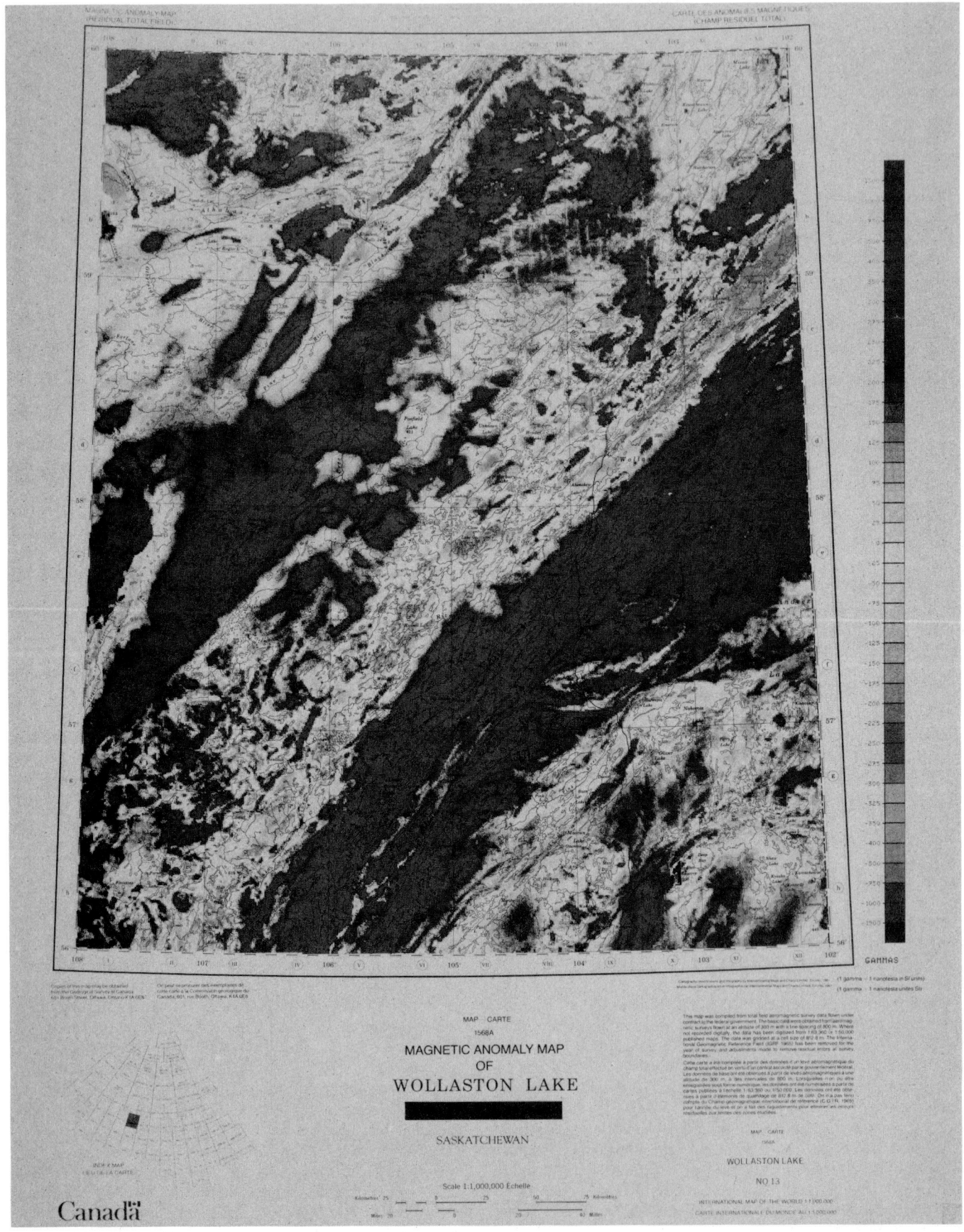

FIG. 5. Reproduction of 1:1 000 000-scale colored magnetic-anomaly map of Wollaston Lake, Saskatchewan.

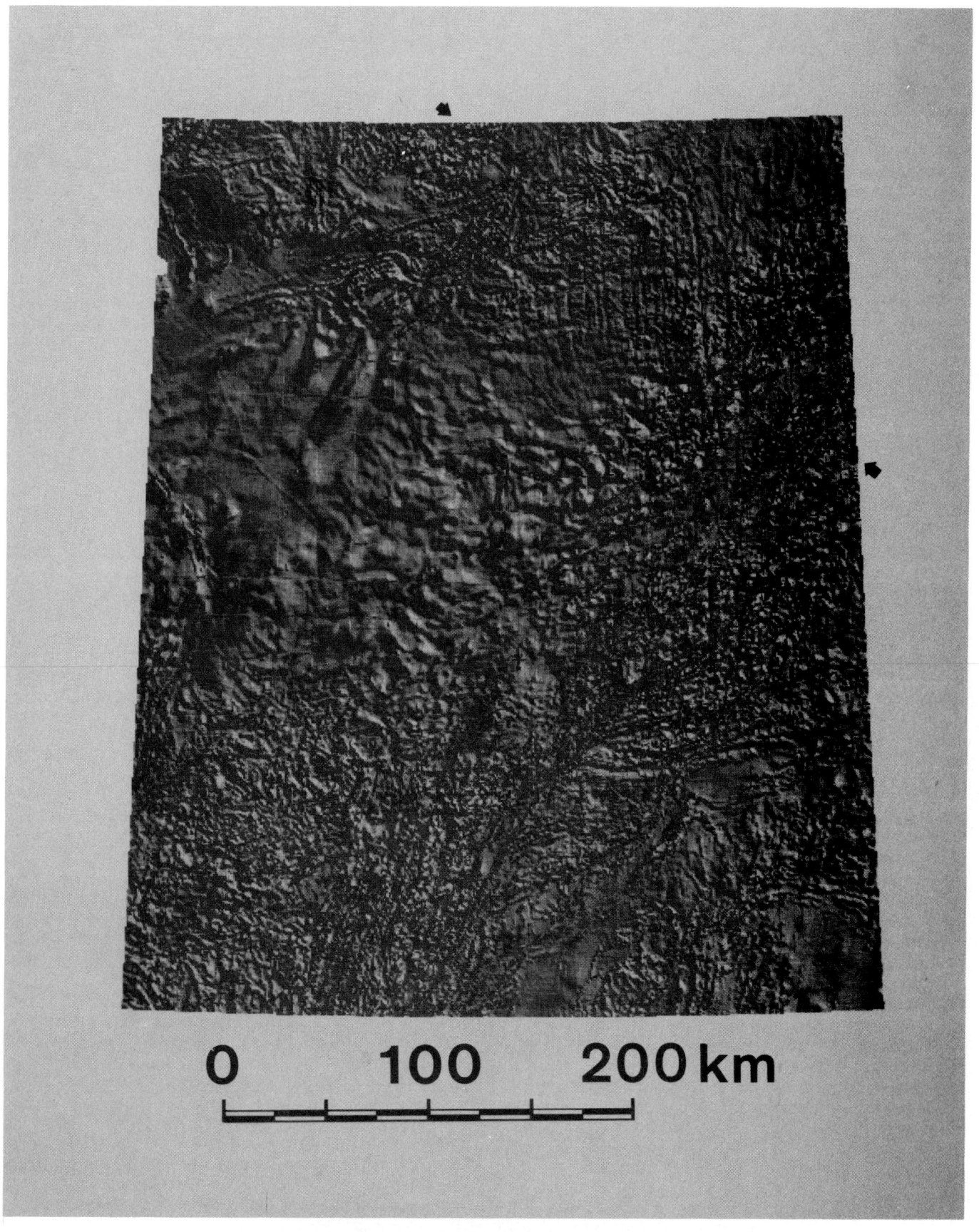

FIG. 6a. Shaded-relief map calculated from data used to prepare Figure 5. Inclination, 30 degrees; declination, 45 degrees. Northwest–southeast structure indicated by arrows.

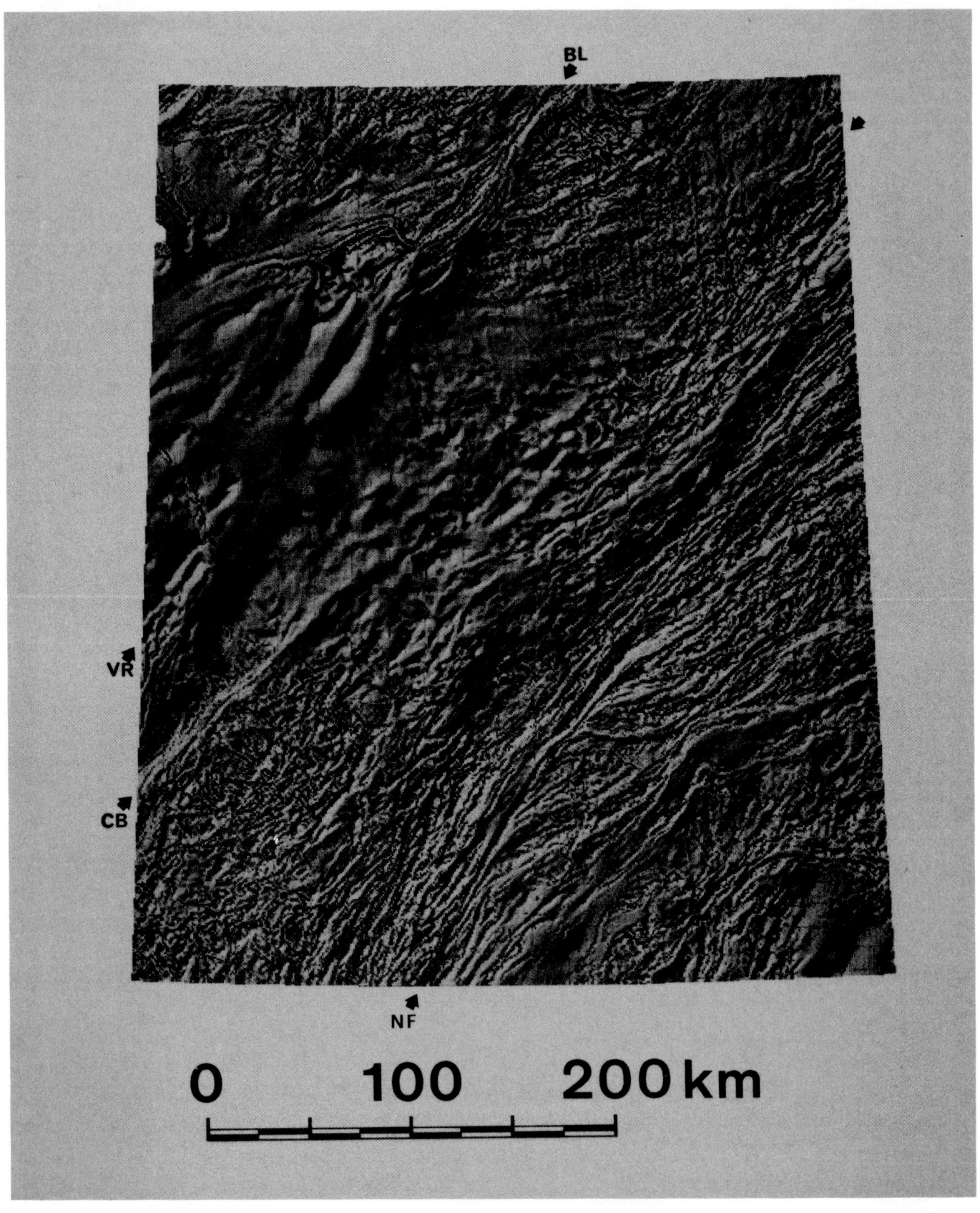

FIG. 6b. Shaded-relief map calculated from data used to prepare Figure 5. Inclination, 30 degrees; declination, 35 degrees. NF = Needle Falls shear zone; CB = Cable Bay shear zone; VR = Virgin River shear zone; BL = Black Lake shear zone.

FIG. 7. Reproduction of 1:1 000 000-scale colored magnetic-anomaly map of Lockhart River, N.W.T.

FIG. 8a. Shaded-relief map calculated from data used to prepare Figure 7. Inclination, 30 degrees; declination, 45 degrees.

FIG. 8b. Shaded-relief map calculated from data used to prepare Figure 7. Inclination, 30 degrees; declination, 135 degrees. MF = MacDonald fault; A–A' is a part of the major contact referred to in the text (shown also in Figure 3). Subtle east–west features are shown by arrows in the northwestern part of the map.

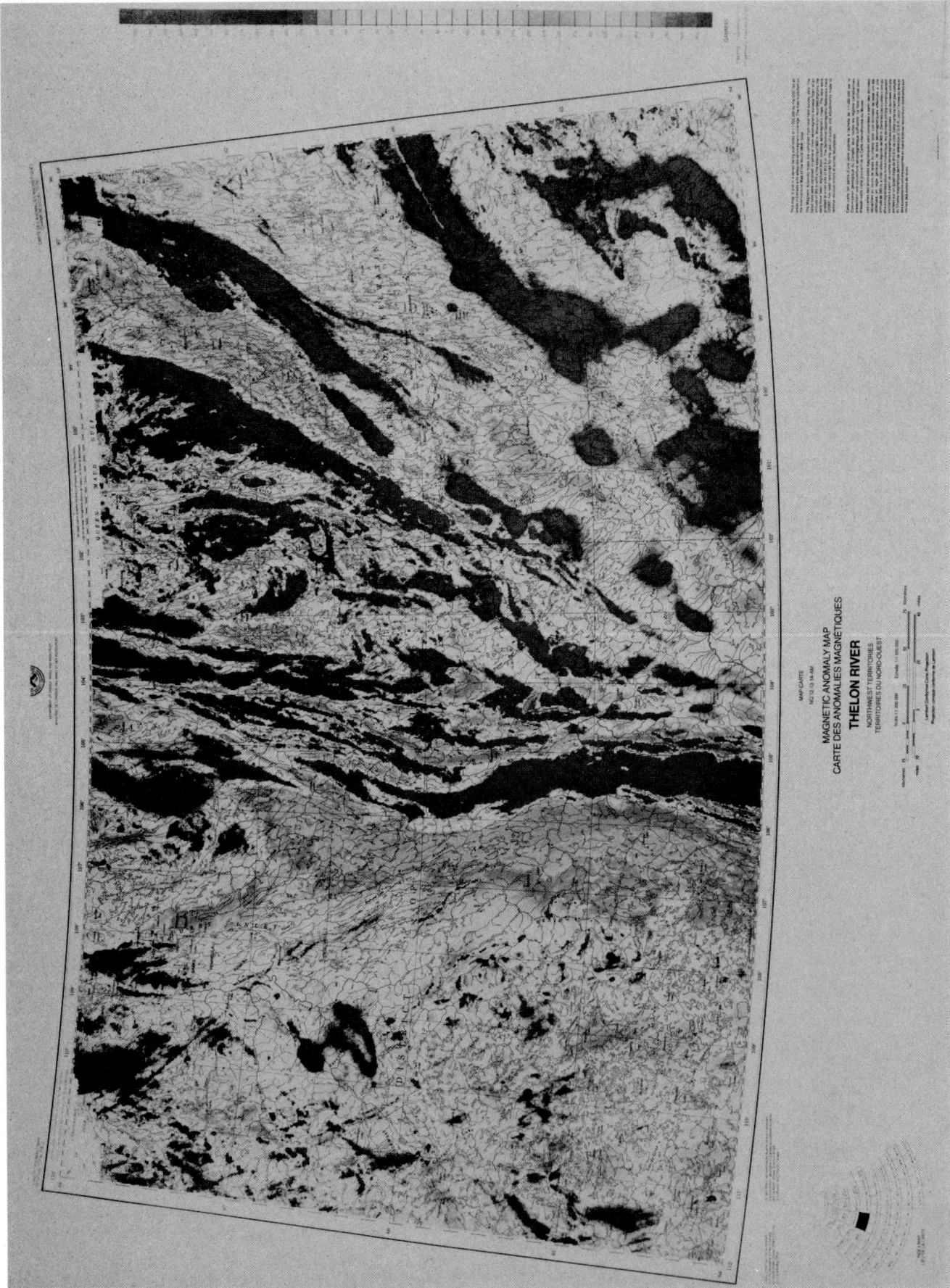

FIG. 9. Reproduction of 1:1 000 000-scale color magnetic-anomaly map of Thelon River, N.W.T.

FIG. 10a. Shaded-relief map calculated from data used to prepare Figure 9. Inclination, 30 degrees; declination, 90 degrees. BF = Bathurst fault; dextral offset indicated by arrows at D.

FIG. 10b. Shaded-relief map calculated from data used to prepare Figure 9. Inclination, 30 degrees; declination, 180 degrees.

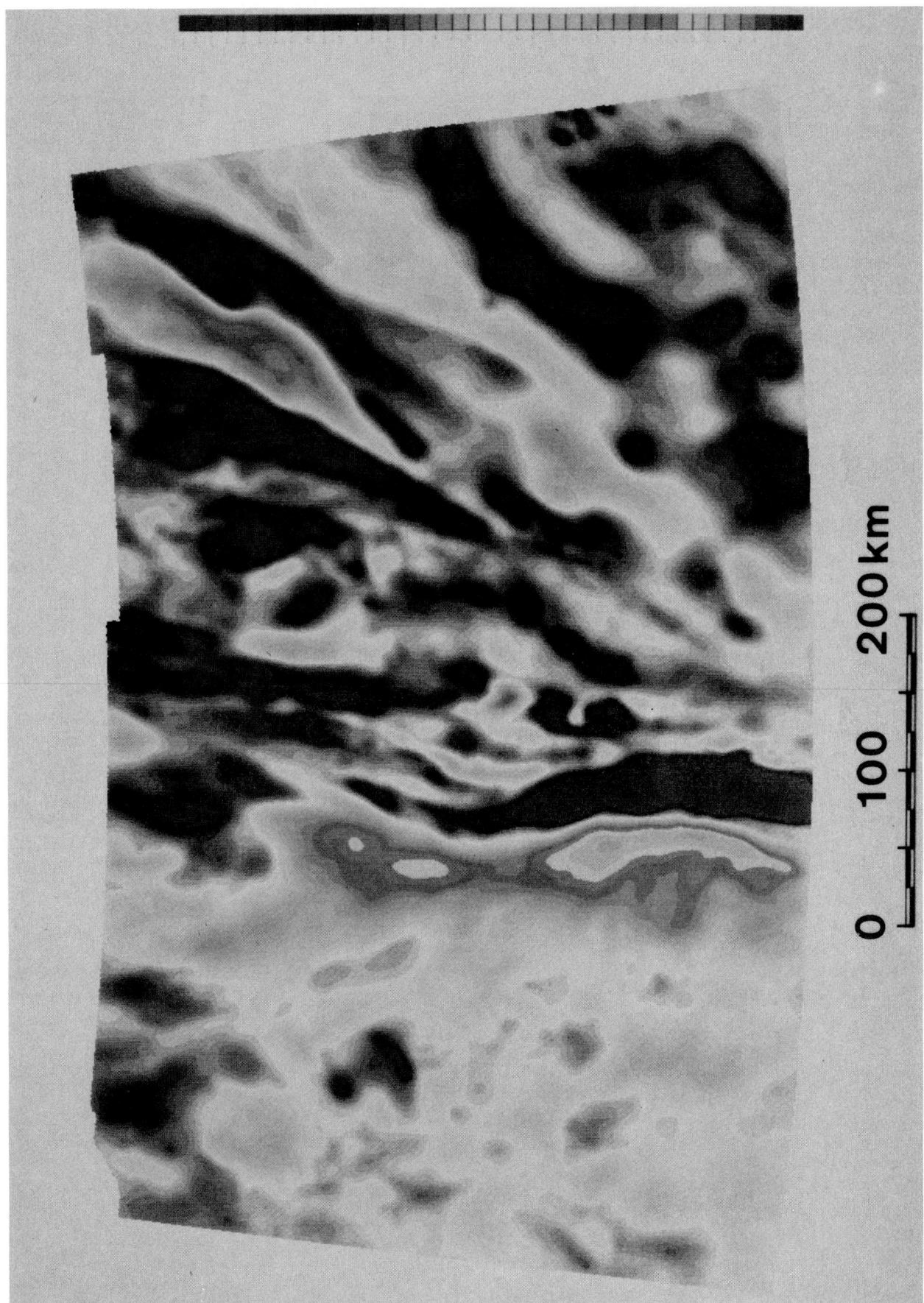

Fig. 11. Data used to prepare Figure 9 (Thelon River), upward continued to 5 000 m.

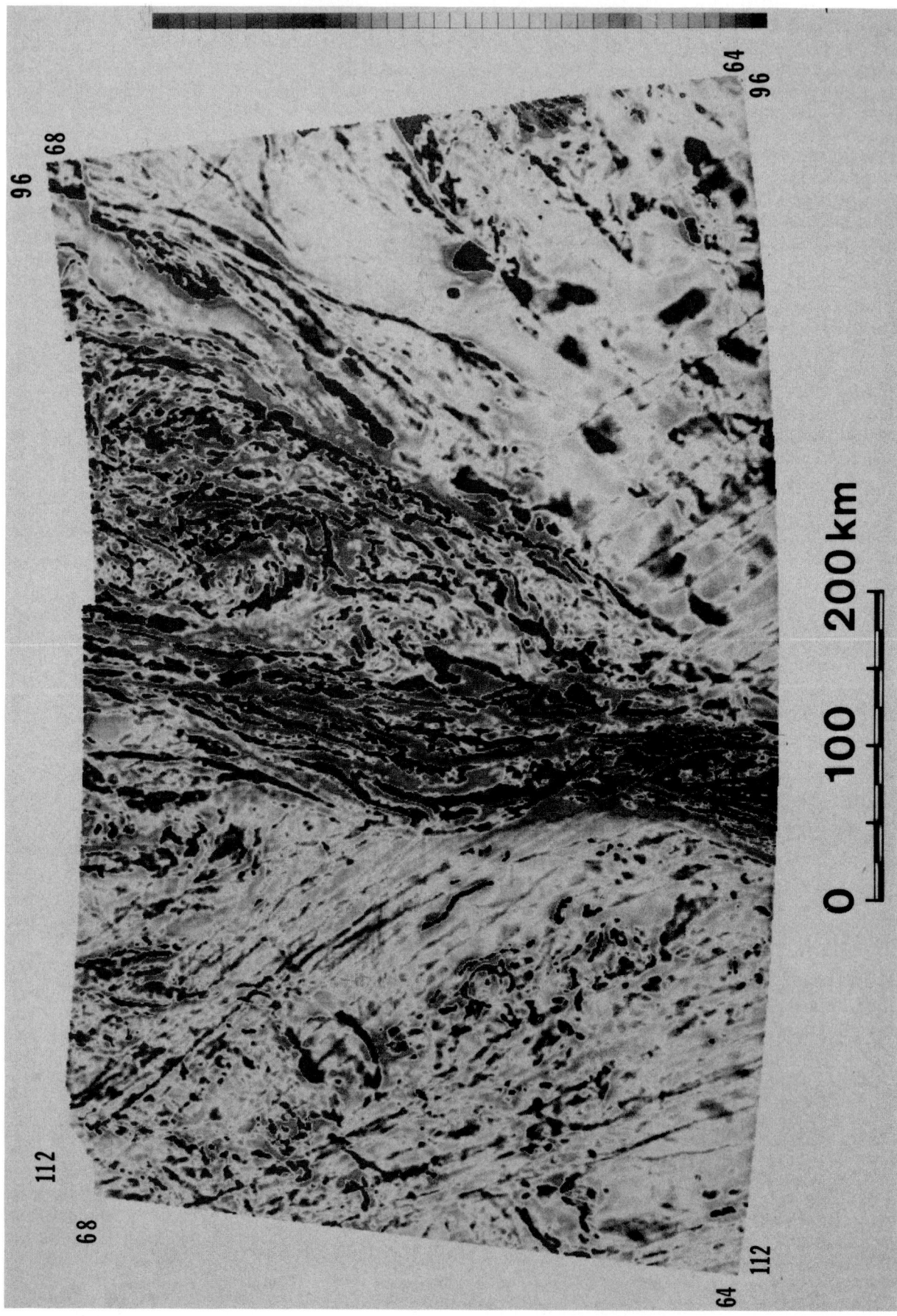

FIG. 12. Calculated vertical gradient of data used to prepare Figure 9 (Thelon River).

ACKNOWLEDGMENT

This program was initiated under the direction of M. T. Holroyd. Companies that have contributed to the digitizing and data processing under contract include Dataplotting, Ltd., Sander Geophysics, Kenting Earth Sciences, and Les Releves Geophysiques.

REFERENCES

Cavanaugh, M. D., and Seyfert, C. K., 1977, Apparent polar wander paths and the joining of the Superior and Slave Provinces during Early Proterozoic time: Geology, **5**, 207–211.

Darnley, A. G., 1981, The relationship between uranium distribution and some major crustal features in Canada: Miner. Mag., **44**, 425–436.

Horn, B.K.P., and Bachman, B. L., 1978, Using synthetic images to register real images with surface models: Commun. Assoc. for Comput. Mach., **21**, 914–924.

Ray, G. E., and Wanless, R. K., 1980, The age and geological history of the Wollaston, Peter Lake and Rottenstone domains in northern Saskatchewan: Can. J. Earth Sci., **17**, 333–347.

Global gravity maps and the structure of the Earth

Carl Bowin*

ABSTRACT

It is now possible to view global maps depicting long-wavelength and short-wavelength features of the Earth's gravity potential field, both as geoid anomalies and as free-air anomalies. This progress has resulted from (1) an increased number of surface gravity measurements, (2) orbiting of artificial satellites, (3) installation of radar altimeters in three spacecraft, and (4) improved computing and display capabilities. The ratio of gravity to geoid anomalies (g/N) has been utilized to suggest a decomposition of the Earth's gravity field. The g/N ratio for each individual spherical harmonic degree is independent of the harmonic coefficient values, being determined only by the degree and values for normal gravity and the Earth's radius. This g/N-ratio value has an associated point-mass depth that is more limiting of source depth than that provided by wavelength considerations. Mass anomalies that may exist in the Earth at great depth will be best represented in the coefficients of low-degree spherical harmonics, not only because of their longer wavelength anomalies at the surface but also to provide a better match of the resulting ratio of gravity to geoid at the center of the anomaly feature.

Shallow (less than 600-km-deep), broad masses that might be considered possible sources of the degree 2* and 3* Sri Lanka geoid low and the New Guinea geoid high are considered geologically unlikely, whereas 1 to 2 km of relief on the core–mantle boundary is plausible and can explain the combined degree 2 and 3 geoid and gravity anomalies. The core–mantle boundary may become warped as a result of stresses developed by hydrodynamic motions in the outer part of the core, possibly related to those that produce the Earth's magnetic field. Combined contributions from harmonic degrees 4 through 10 identify a narrow positive mass anomaly beneath convergent-plate zones. These positive anomalies are not due to the anomaly of a downgoing slab because they are much broader and have larger magnitude. The fact that association of this positive mass anomaly with convergent-plate zones is more sharply identified in the selected packet of harmonic coefficients than in the full field is a strong indication that the preliminary decomposition is proceeding in the correct direction.

Cumulative geoid degree contribution curves for the low harmonic terms of the gravity field of Venus have a different pattern than that for the Earth. Maps of the low harmonic coefficients for Venus indicate correspondence with surface topographic features in marked contrast to the Earth. Venus rotates very slowly and has no internal magnetic field. Perhaps coherent motion within the core is negligible, so that there is no magnetic field and no distortion of the core–mantle boundary. These relations support the conclusion that long-wavelength undulations at the Earth's core–mantle boundary may contribute significantly to the lowest degree harmonic coefficients of the Earth's potential field.

INTRODUCTION

Irregularities in the gravity field of the Earth result from lateral variations in mass relative to a spherical model that varies only with radius. The longest wavelength irregularity in the gravity field is due to the ellipsoid shape of the Earth and its rotation; these two factors can be computed, given the flattening of the shape of the Earth, the average acceleration owing to gravity at the equator, and the sidereal rotation rate. When these effects, and the change in gravity with height above mean sea level (because of increasing distance to the center of mass of the Earth), are corrected for, residual variations (called free-air gravity anomalies) still remain.

One method of displaying such remaining variations is

---

*Woods Hole Oceanographic Institution, Woods Hole, MA 02543. Contribution no. 5549 of Woods Hole Oceanographic Institution.

Global Gravity Maps and the Earth's Structure

with a diagram of free-air gravity anomalies as a function of elevation of the Earth's surface at the observation site. Figure 1 shows that the great majority of the Earth's surface has free-air-anomaly values within ±75 mGal of zero. This indicates that most of the Earth's surface is in isostatic equilibrium. The negative free-air-anomaly values in regions of great water depths are associated with the trenches of island arcs, which are also regions of large negative isostatic gravity anomalies. Variations in frequency of observations along the zero-milligal level approximate the variations of the hypsometric curve of frequency for topographic heights on the Earth. The greater range of free-air anomalies at zero elevation shown by frequency contours presumably results from edge-effect anomalies at continental margins. Although Figure 1 shows that the range of free-air-anomaly values is less than 0.08 percent of the magnitude of the Earth's gravity field (980 000 mGal), they still remain a fundamentally important geophysical parameter. Important long-wavelength anomalies are not apparent from the point measurements until the anomalies are viewed in map form. A free-air gravity-anomaly map of the world (Bowin et al., 1982), at a 25-mGal contour interval, prepared from surface measurements, shows broad regions that have predominantly negative anomaly values (in the Indian Ocean near southern India, northeastern Canada, southeastern to western South Atlantic Ocean, and southeastern Asia) and regions with predominantly positive values (northern North Atlantic Ocean, southeastern Southeast Asia eastward along Melanesia, and in the southern Indian Ocean). These broad regions coincide with anomalies previously observed in global gravity and geoid anomaly maps obtained from satellite-tracking data. Small perturbations in the orbits of satellites indicate irregularities in the Earth's gravity potential field at orbit altitude, and the solutions from these analyses are generally given as sets of spherical harmonic coefficients. Figures 2 and 3 are global maps of gravity and geoid anomalies, respectively, calculated from the GEM-9 set of spherical harmonic coefficients for the Earth's potential field (Lerch et al., 1979) with complete coefficients to degree 20 and some coefficients to harmonic degree 30. The GEM-9 coefficients represent a refined solution obtained exclusively from observations of satellite-orbit perturbation. They do not incorporate surface-gravity or radar-altimeter data. I prefer this independence because one of my objectives is to compare surface observations with the satellite-perturbation solutions.

Although the existence of large long-wavelength geoid anomalies (Figure 3) has been recognized for several decades, their origin has been unknown. In this paper I present evidence that mass anomalies at the core–mantle boundary region contribute significantly to the magnitude of the Earth's largest geoid anomalies, thus supporting the statistical covariance-analysis study of Jordan (1978). When this first estimate of the core–mantle boundary mass-anomaly contribution to the Earth's long-wavelength potential field is subtracted from the Earth's degree-10 field, a group of spherical harmonic coefficients is discovered that define narrow bands of positive geoid anomalies (Figure 12) coinciding almost everywhere with sites of plate convergence. These positive anomalies are not due to the downgoing slab: they are much broader, and have larger magnitude, than that

anomaly. This correlation of large positive mass anomaly beneath nearly all plate-convergence zones led to the speculation (Bowin, 1983a) that the mass anomaly may be an important driving force for plate tectonics and that its existence may localize most of the sites of plate convergence. This paper summarizes and extends the substantiation for this analysis of the deep structure of the Earth.

METHODOLOGY

The gravity (g) and geoid (N) maps (Figures 2, 3) were calculated by summing the harmonic contributions

$$g = \frac{GM}{a_e} \times \frac{1}{R} \sum_{n=2}^{30} \sum_{m=0}^{n} (n-1) \left(\frac{a_e}{R}\right)^n$$
$$\times (C_{nm} \cos m\theta + S_{nm} \sin m\theta) P_{nm}(\sin \phi); \quad (1)$$

$$N = \frac{GM}{a_e} \times \frac{1}{g_o} \sum_{n=2}^{30} \sum_{m=0}^{n} \left(\frac{a_e}{R}\right)^n$$
$$\times (C_{nm} \cos m\theta + S_{nm} \sin m\theta) P_{nm}(\sin \phi); \quad (2)$$

where G is the Universal gravity constant; M the mass of the Earth; a_e its equatorial radius; g_o normal gravity, 980 cm/s^2; ϕ, θ, and R the geocentric latitude, longitude, and radius, respectively, of the location for the calculation; C and S the fully normalized spherical harmonic coefficients for degree, n, and order, m (reference values for $f = 298.247$ have been subtracted); and P_{nm} the Legendre polynomials.

A global harmonic degree-10 map of the ratio values of gravity divided by geoid (g/N) is shown in Figure 4. This is a smoothed representation of the map obtained by dividing the data (up to degree 10) of Figure 2 by that of Figure 3. Although the pattern of Figure 4 at first looks chaotic, note that at the locations for the centers of the major geoid anomalies of Figure 3 the contours have simple patterns, usually local maxima. For example, the great low south of India (herein referred to as the Sri Lanka low) has a g/N-ratio value of 0.44 mGal/m at the center. Explanations for such low ratio values will be developed.

Bowin (1983a) noted that the ratio of the spherical harmonic expressions (1) and (2) defining gravity and geoid produces a fixed value for each individual degree:

$$\frac{g_n}{N_n} = \frac{980(n-1)}{R} \quad (3)$$

where

$$g_n = \frac{GM}{a_e R} \left(\frac{a_e}{R}\right)^n (n-1) \sum_{m=0}^{n} (C_{nm} \cos \theta + S_{nm} \sin \theta) P_{nm}(\sin \phi);$$

$$N_n = \frac{GM}{g_o a_e} \left(\frac{a_e}{R}\right)^n \sum_{m=0}^{n} (C_{nm} \cos \theta + S_{nm} \sin \theta) P_{nm}(\sin \phi).$$

Thus, the ratio g/N for the contribution from each degree is independent of the harmonic-coefficient values, being determined only by the degree and the values for normal gravity and the Earth's radius.

Spherical harmonic analysis of spherical surfaces is analogous to Fourier spectral analysis of plane surfaces. Zero degree is a bias value, and each succeeding degree provides increasingly detailed information on the potential field. To

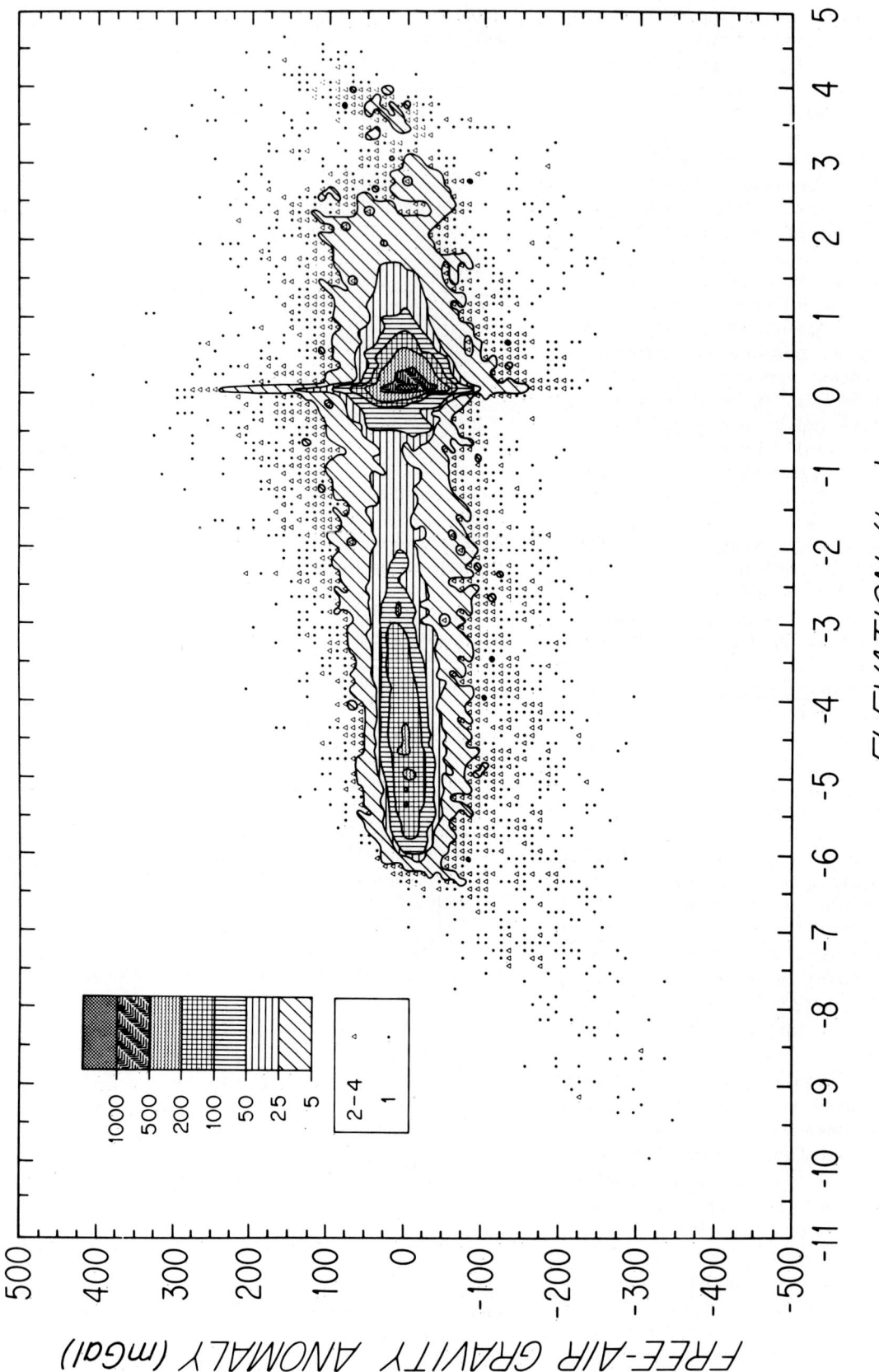

Fig. 1. Frequency diagram of free-air gravity anomalies versus elevation for the Earth. Determined from a representative sample of 91 067 measurements distributed over the globe from a digital data file at Woods Hole Oceanographic Institution abstracted 2 × 2 per degree square.

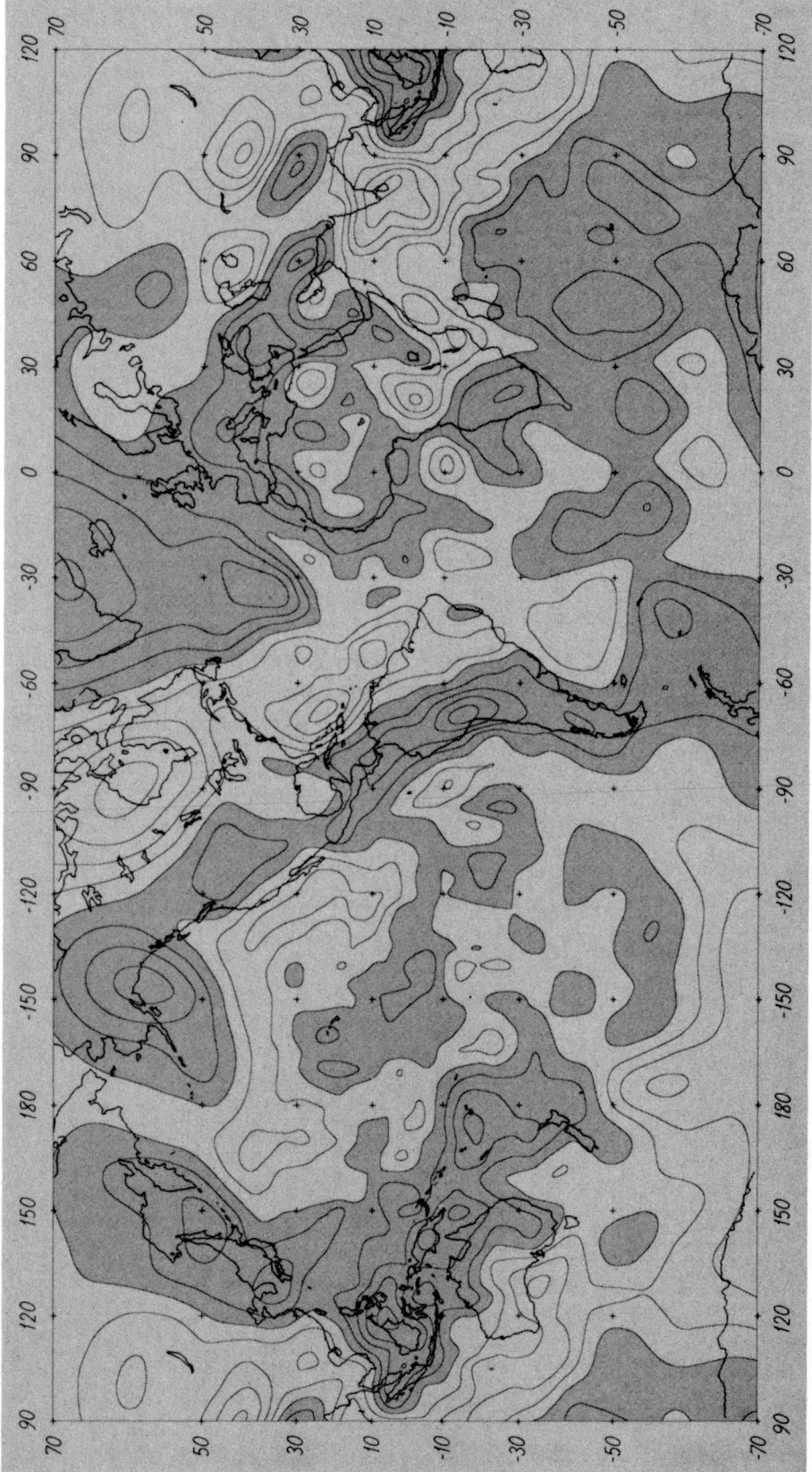

Fig. 2. Free-air gravity-anomaly map, GEM-9 degree 30. Obtained by summing contributions from degrees 2 through 30. Referenced to ellipsoid with reciprocal flattening of 298.247. Positive anomalies are shaded. Contour interval, 10 mGal.

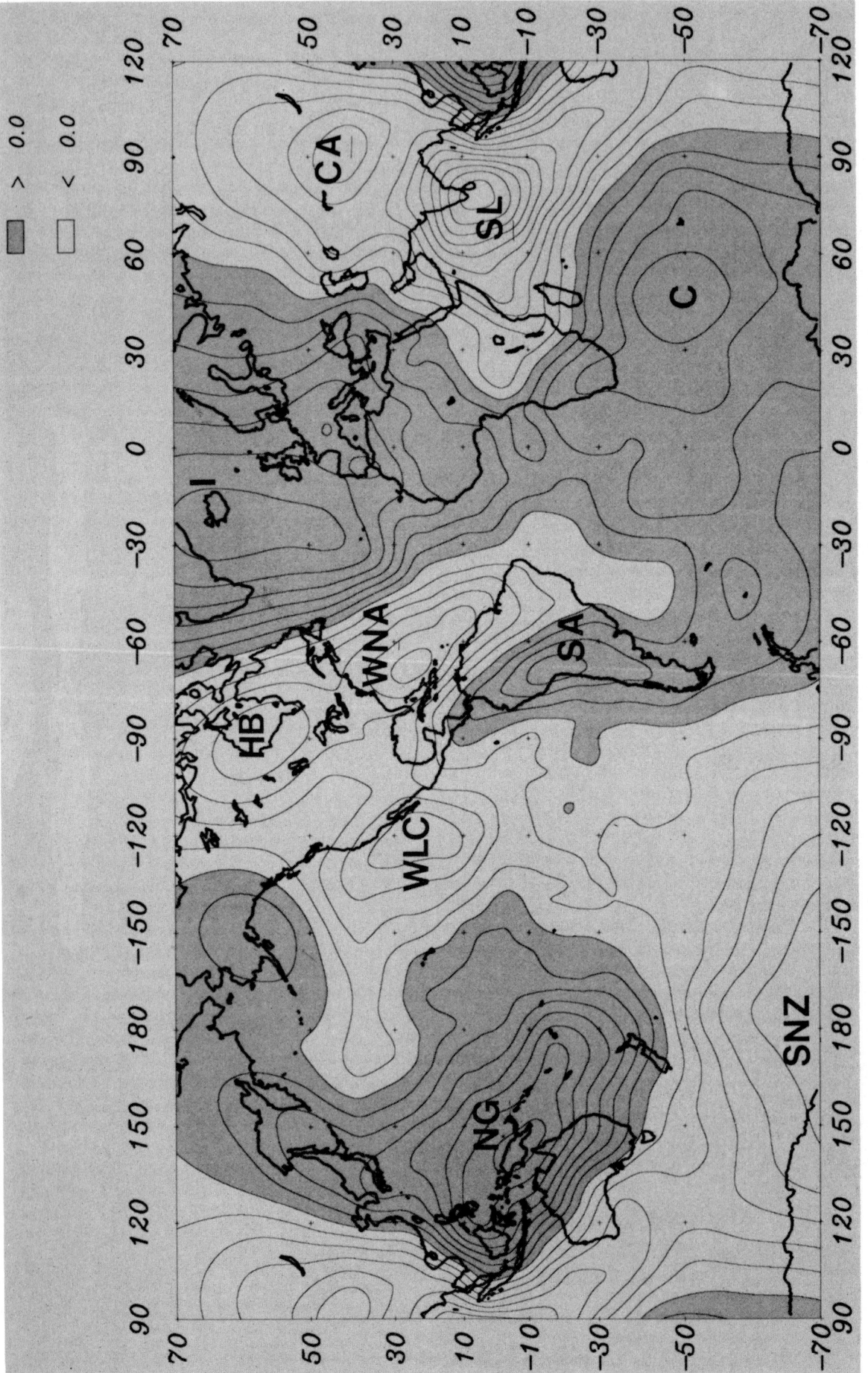

FIG. 3. Geoid-anomaly map, GEM-9 degree 30. Obtained by summing contributions from degrees 2 through 30. Referenced to ellipsoid with reciprocal flattening of 298.247. The 10 major geoid anomalies are labeled. The lows are Sri Lanka (SL), west of Lower California (WLC), Central Asia (CA), western North Atlantic (WNA), Hudson Bay (HB), and south of New Zealand (SNZ). The highs are New Guinea (NG), Iceland (I), Crozet (C), and South America (SA). Positive anomalies are shaded. Contour interval, 10 m.

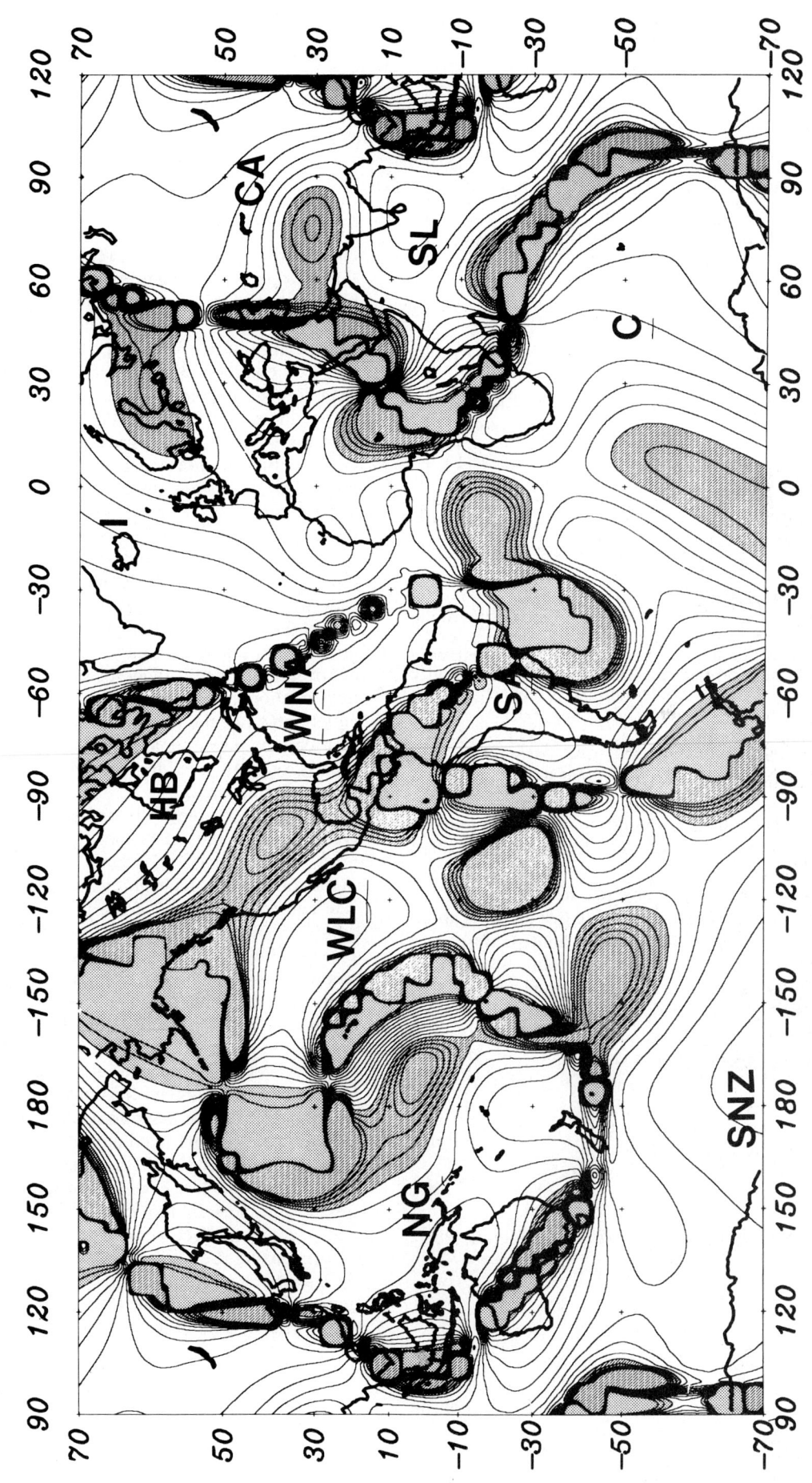

FIG. 4. Map showing g/N ratio. Ratios of gravity to geoid values, GEM-9 degree 10. Obtained by dividing gravity anomalies from degrees 2 through 10 by geoid anomalies from degrees 2 through 10. Degree 10 rather than 30 is used for this map to show more clearly the relations for the major geoid anomalies (see text for further explanation). Contour interval is 0.1 mGal/m. Ruled areas are less than zero; dotted areas are greater than 1.0. See caption of Figure 3 for identification of label abbreviations.

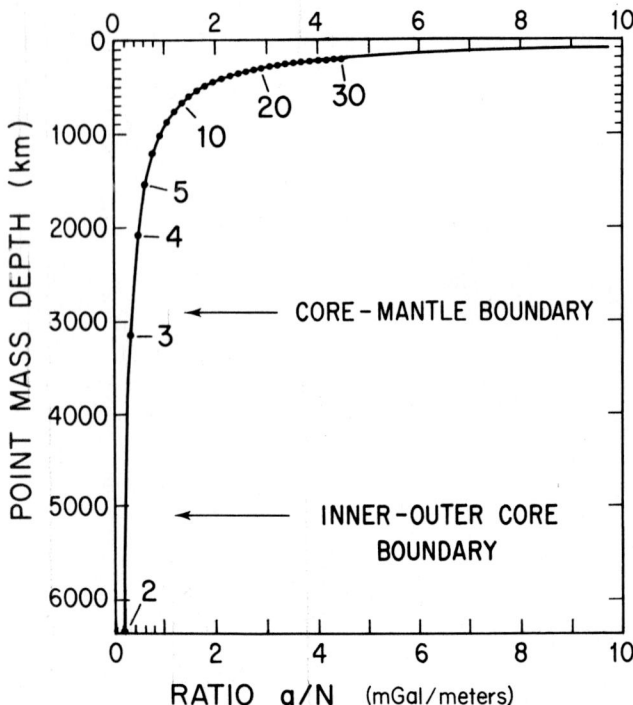

FIG. 5. Ratio of gravity to geoid anomalies (g/N) as a function of depth to a point-mass source. Filled circles along curve indicate ratio for individual harmonic-degree contributions (labeled as to degree). The depth associated with each degree is that of a point mass that has the same gravity to geoid ratio as do the contributions of that harmonic degree. See text for derivation.

distinguish contributions for an individual harmonic degree from an integral contribution from degree 0 to a particular degree, we star (*) the degree number to indicate an individual degree contribution. A wavelength can be associated with each harmonic, which, by convention, takes the approximate form $\lambda = 2\pi R/n$. For the low harmonics, in particular, these wavelengths are so great (for example about 10 000 km at degree 4) that they provide limited constraints on possible source depths. Greater constraint, however, is made available by associating the g/N ratio given by equation (3) with the g/N ratio obtained directly above a point mass at depth, d. Flat-earth and spherical-earth values are identical directly above a point mass:

$$\frac{g}{N} \leftrightarrow \frac{\frac{GM}{d^2}}{\frac{GM}{g_o d}} = \frac{g_o}{d} = \frac{980}{d}. \qquad (4)$$

To obtain an equivalent point-mass depth for each degree:

$$d_n = \frac{R}{n-1} \qquad n = 2, 3, \ldots 30. \qquad (5)$$

Thus the gravity and geoid contributions from each individual degree have a fixed ratio value that can be related to the g/N ratio of point masses at a depth of some fraction of the Earth's radius (Figure 5). The utility of making this association, as expressed in equation (5), follows.

The gravity and geoid anomalies at any location on the Earth are calculated by summing the contributions of the spherical harmonic coefficients from all degrees, thus:

$$\frac{g}{N} = \frac{g_2^* + g_3^* + g_4^* + \cdots g_n^*}{N_2^* + N_3^* + N_4^* + \cdots N_n^*}. \qquad (6)$$

The coefficients determine the maximum magnitude of the contributions from each degree; the actual degree contribution at a particular location also depends on that location's phase between the maximum and null locations of each order within the degree. Except at null locations, the ratio for the sum of contribution from all orders of each degree will be as given by equation (3). The sum of contributions from more than a single degree, as in equation (6), has no fixed ratio value; instead, that value depends on the proportions of contributions for each degree and whether or not the principal coefficients are consistent in sign. For degree contributions of consistent sign, if the contributions of the low-degree harmonics (e.g., seven or less) are minor, from inspection of equation (3) we see that the resulting ratio of the final gravity and geoid sums cannot have a low value (i.e., less than 1.0). Thus, in this case, the lower the final g/N ratio for an anomaly, the more lower harmonic terms must contribute (i.e., in greater proportion) to the final summed gravity and geoid anomaly values. The ratio of sums of contributions from a range of harmonic degrees, except at null locations, have as a *lower limit* the ratio value, from equation (3), of the lowest harmonic degree included. This fact can be used to assign maximum-source depths for groups of harmonic degrees whose degree contributions have the same sign.

From the foregoing it is obvious that mass anomalies at great depth in the Earth have g/N values much less than unity and that mass anomalies at shallow depth, unless of broad dimensions, have g/N values much greater than unity. The curves of Figure 5, for a point mass, and those in the appendix of Bowin (1983a), and also in a study by Carl Bowin, Edward Scheer, and Woollcott Smith (in press), quantify those relations for simple geometric bodies. Therefore, mass anomalies that may exist in the Earth at great depth will be best represented in the coefficients of the low-degree spherical harmonics, not only because of their longer wavelength anomalies at the surface but also to match better the ratio of gravity to geoid values at the center of the anomaly from those mass anomalies. The g/N criteria are a greater constraint on maximum depth than is wavelength.

In situations in which the area of interest is of limited extent so that a "flat-earth" viewpoint is valid, the Fourier transform is a useful tool in the expansion of potential fields and occupies a place similar to that of spherical-harmonic expansions for the whole Earth. In the aforementioned study by Bowin, Scheer, and Smith (in press) we show an analogous relation to equation (5) for two-dimensional Fourier transforms of a point-mass field:

$$d_\lambda = \frac{\lambda}{2\pi} \simeq k^{-1}$$

where d_λ is the depth of a point mass and λ is linear wavelength. We note that wavenumber, k, in this flat-earth relation, plays a similar role as degree, n, in spherical-harmonic expansions.

PRELIMINARY DECOMPOSITION OF EARTH'S GRAVITY FIELD

Figure 5 shows that the contribution from mass anomalies that may lie at the core–mantle boundary region would be primarily contained in the harmonic coefficients for degrees 2* and 3*. Since degrees 2* and 3* together only have 12 coefficients, they provide only limited detail (Figure 11) but constitute our first estimate of the location and relative magnitudes of mass anomalies at the core–mantle region. Thus, unless the actual mass variations at the core–mantle boundary region happened to meet the resolution limitations of those few terms, some residual amount no doubt is included above degree 3*; but the higher degrees will be increasingly less able to match the g/N ratio produced by the actual mass anomalies. The Crozet geoid high is mainly a feature of degrees 3* and 4* (Figures 3, 6, 9) and is inferred to be an example of a core–mantle boundary-region mass anomaly that is unusual because it has an important contribution from a harmonic term higher than degree 3*. Therefore, our second estimate of mass anomalies at the core–mantle region comprises the anomalies of Figure 11 plus the Crozet anomaly of Figure 3.

Another confirmation that mass anomalies at the core–mantle boundary, of radius R_{cmb}, would predominantly be represented in harmonic degrees 2* and 3* is to consider that the core–mantle boundary is the outer surface of the Earth and that spherical-harmonic coefficients were determined for the potential of that smaller Earth with radius R_{cmb}. Assume an extreme case that the spherical-harmonic coefficients for a point mass on that surface have equal magnitude at all degrees from 1 to infinity. This would be analogous to the Fourier transform of an impulse function producing a white spectrum in the frequency domain. However, if those harmonic coefficients are evaluated at an altitude above R_{cmb} corresponding to the real surface of the Earth, R_e, then the computed gravity- and geoid-anomaly contribution at that height for each degree would be reduced by the factor $(R_{cmb}/R_e)^n$ as seen from equations (1) and (2). Thus, point masses at the core–mantle boundary with a white spectrum would have a red spectrum at a height corresponding to the Earth's actual surface (Table 1). Note in Table 1 that coefficients for degrees 2* and 3* contain a minimum of about 79 percent of the energy represented by all harmonic terms from 2 to 10 even for the assumed extreme case for a point mass. Contributions from degrees 10 and above are minuscule. Actual mass distributions will increase the proportion of energy in the low harmonic terms.

Obviously, mass anomalies at more than one depth and location all can contribute significantly to the energy incorporated in a particular coefficient term. Low-density mass anomalies having large vertical extent in the mantle could contain significant energy over an extended range of harmonic degree terms. It is obvious, too, that without the existence of actual mass anomalies, the coefficient values would equal zero. I expect that in the regression calculations

Table 1. Spherical-harmonic upward-continuation factor from core-mantle boundary to Earth's surface as function of harmonic degree, with percentage contribution for each degree. Radius of core-mantle boundary, R_{cmb}, equals 2 900 km. Mean radius of earth, R_e, equals 6 371 km. See text for discussion.

Degree N	$\left(\dfrac{R_{cmb}}{R_e}\right)^n$	Percentage of sum degrees 2 to 10	Approximate cumulative percentage
2	0.207	54.5	54.5
3	0.093	24.7	79.2
4	0.043	11.3	90.5
5	0.020	5.3	95.8
6	0.009	2.4	98.2
7	0.004	1.1	99.3
8	0.002	0.5	99.8
9	0.001	0.3	100.1
10	0.000	0.0	100.1

the harmonic coefficients will incorporate as much energy as possible at degree coefficients most appropriate for the mass-anomaly depths and their g/N ratios.

The 10 major geoid anomalies of the earth are labeled in Figure 3, and "cumulative degree-contribution curves" for each are plotted in Figures 6 and 7. The cumulative degree-contribution curves are constructed stepwise by accumulating the spherical-harmonic contributions for each degree. This is equivalent to displaying the spectral content of an anomaly. The consistency of the sign of the degree contributions is readily observed by consistency of the sign of the slopes for the cumulative degree-contribution curves. I reference the anomalies to those on the best-fitting ellipsoid for the actual Earth rather than to a theoretical hydrostatic figure. Goldreich and Toomre (1969) demonstrated that the "non-hydrostatic bulge" is not a spheroid and that the nonhydrostatic Earth should be considered as a collection of more or less random-density inhomogeneities.

The Sri Lanka low (the greatest geoid anomaly on the Earth) is made significantly lower than all other geoid lows by the source of the large 2* and 3* contributions (Figure 6). The Sri Lanka low is also noteworthy in showing the closest coincidence of locations of minima arising from 2-, 3-, 4- to 6-, and 7- to 10-degree contributions (see Bowin, 1983a, for discussion of possible explanations). Most slopes in Figure 6 do not change signs so that the previous analyses are applicable. Note also that the major change in slope magnitude for the Sri Lanka, south New Zealand, and New Guinea geoid curves occurs between degrees 3 and 4. Most of the major geoid anomalies attain about 95 percent of their 30-degree values by degree 10. Exceptions are the western North Atlantic and Crozet curves, where only about 90 percent of the 30-degree value is reached at degree 10, and the South America curve, where only 76 percent is attained by degree 10. This last anomaly, however, also has the lowest degree-30 magnitude. Most of the major geoid anomalies either flatten or change slope above degree 10. The degree-10 map (Figure 8) does indeed retain the major features of the degree-30 map (Figure 3), as expected from the cumulative degree curves (Figure 6). Thus the degree-10 geoid is a good choice for providing a smoothed representation of the major global geoid anomalies both for studying

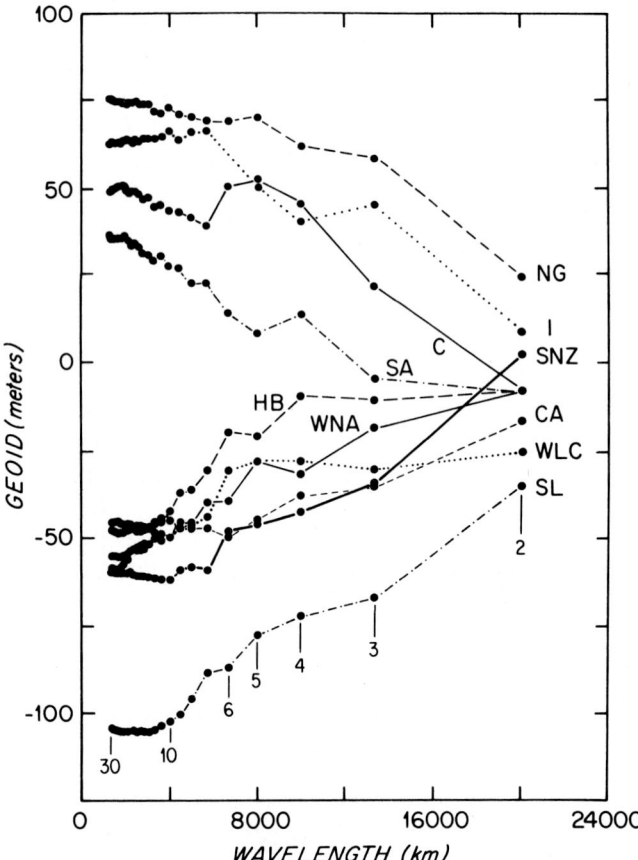

FIG. 6. Cumulative geoid-anomaly curves for the 10 major geoid anomalies computed from GEM-9 spherical-harmonic coefficients. Numbers below lowermost curve indicate harmonic degree. The lows are Sri Lanka (SL), west of Lower California (WLC), Central Asia (CA), western North Atlantic (WNA), Hudson Bay (HB), and south of New Zealand (SNZ). The highs are New Guinea (NG), Iceland (I), Crozet (C), and South America (SA). Their locations are shown in Figure 3.

0.5, indicating dominance of the lowest harmonic degrees 2 and 3.

From the cumulative geoid degree curves of Figure 6 and the individual degree-contribution curves of Figure 9, and an examination of the resulting global maps, I identify three packets of harmonic terms that suggest separate sources: degrees 2 and 3, degrees 4 through 10, and degree 11 and above. These groups of coefficients characterize the preliminary decomposition of the Earth's gravity field, and the spatial distribution of their anomalies further demonstrates the utility of maps for synthesizing information and aiding interpretation. This initial decomposition (Bowin, 1981, 1983a) comprises four main sources for mass anomalies: (1) mass anomalies at the core–mantle boundary region, expressed principally in coefficients of harmonic degrees 2* and 3* (Figure 11); (2) mantle anomalies, and the mass excess of the deeper parts of convergent-plate zones, expressed principally in coefficients of harmonic degrees 4* through 10* (Figure 12); (3) upper mantle mass anomalies in the outer 600 km of the Earth, revealed by subtracting the GEM-9 degree-10 geoid field from radar-altimeter observations; and (4) crustal anomalies, revealed by residual free-air gravity anomalies.

Large mass anomalies at the core–mantle boundary region commonly have been considered unlikely because they normally imply large stress differences—for example, 1 kilobar in the case of a 2-km bump on the boundary. However, the core–mantle boundary may become warped as a result of stresses developed by motions in the outer part of the core (Garland, 1957). An equivalent surface distribution of mass, proportional to the difference in density between core and mantle, will result. Such motions may be related to those that produce the Earth's magnetic field (e.g., Cuong and Busse, 1981; Busse, 1975; Cox and Cain, 1972). The

their nature and for providing a regional field to be removed for studies of local shorter wavelength features.

A different way of displaying the information in the cumulative degree-contribution curves is simply to plot the individual degree contributions. Such a plot is shown in Figure 9. This figure clearly demonstrates that the coefficients for degrees 2, 3, and 4 provide the largest magnitude contributions for individual degrees. Note also in Figure 9 that the contributions from degrees 11 and higher are small. The relatively large degree-4* contribution (Figure 9) to the South America geoid high is inferred to arise from an upper mantle mass anomaly and contrasts with the inferred deep source for the degree-4* contribution to the Crozet geoid high.

Solutions for equation (6) as a function of degree for the 10 major geoid anomalies are plotted in Figure 10. These diagrams present yet another way to view the relative degree contributions making up an anomaly feature and thus aid in the interpretation of Figure 4. Note that the Sri Lanka and the New Guinea curves remain below a g/N ratio value of

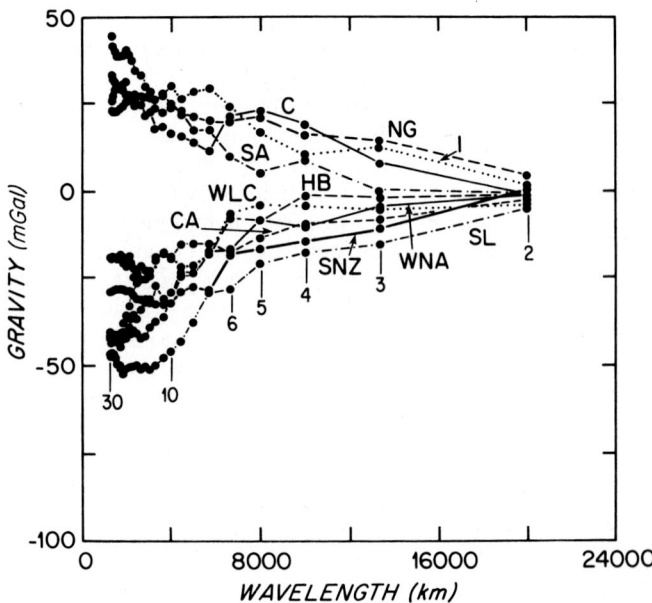

FIG. 7. Cumulative gravity-anomaly curves for 10 major geoid anomalies computed from GEM-9 spherical-harmonic coefficients. See caption of Figure 6 for explanation of labels.

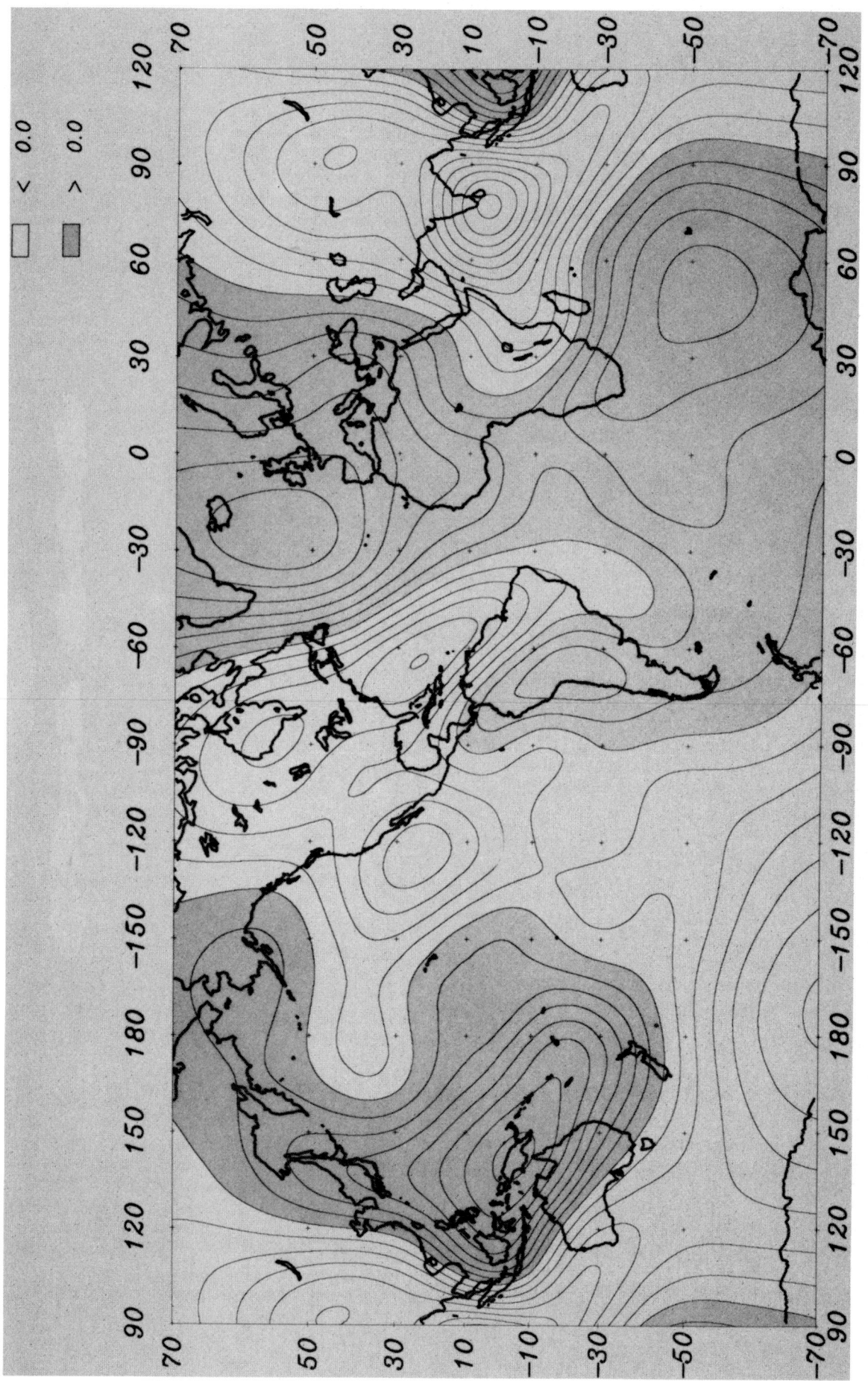

FIG. 8. Geoid-anomaly map, GEM-9 degree 10. Obtained by summing harmonic contributions from degrees 2 through 10. Positive anomalies are shaded. Contour interval, 10 m.

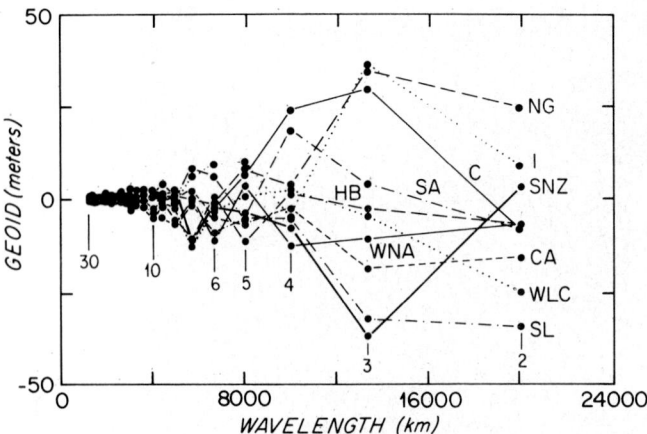

FIG. 9. Individual geoid degree-contribution curves for the 10 major geoid anomalies computed from GEM-9 spherical-harmonic coefficients. See caption of Figure 6 for explanation of labels.

above decomposition is compatible with Jordan's (1978) statistical analysis on the Earth's gravity field by comparing covariance functions with a model having five a priori horizons of density contrast. Because the deepest two sources of our decomposition are estimated from simple truncation of harmonic degrees, they are only approximately defined. Similarly, partial contributions to the lowest 10 harmonic-degree terms arising from shallower sources than those inferred in the initial decomposition have not been resolved.

My preliminary decomposition is based on simple truncation of sets of harmonic coefficients. Harmonic degrees 2 and 3 are the first estimate of mass anomalies at the core–mantle boundary (Figure 11), and harmonic degrees 4 through 10 identify a narrow positive mass anomaly beneath convergent-plate zones (Figure 12). As indicated previously, I infer that the Crozet anomaly (principally a harmonic-degree-4 feature) also originates at the core–mantle boundary region. The fact that the association of positive mass anomaly with convergent-plate zones is more sharply identified in the selected packet of harmonic coefficients than in the full field is a strong indication that the preliminary decomposition is proceeding in the correct direction.

I find further support for that decomposition from (1) my analysis of regional gravity and geoid anomalies in the Philippine Basin (Bowin, 1981); (2) the pattern of residual geoid anomalies (equivalent to contributions from degree 11 and higher), obtained by subtracting the GEM-9 degree-10 field from GEOS-3 and SEASAT-1 radar-altimeter data (Bowin, 1983a; Figure 11); and (3) in my comparison between the Earth's gravity field and that of the similarly sized planet Venus (Bowin, 1983b; Bowin et al., 1985). In a profile eastward across the center of the Philippine Basin, the geoid field is concave downward, whereas the gravity field is concave upward, indicating at least two mass anomalies (one positive and one negative); the positive mass anomaly must be deeper than the negative mass anomaly. Subtraction of a degree-10 field from the geoid anomalies yields a negative residual geoid anomaly that has good spatial correspondence with the negative free-air gravity anomaly. Because the average density of Venus is similar to that of the Earth, Venus presumably also has a core. However, Venus has no internal magnetic field and rotates very slowly. Thus, perhaps motion within the core is negligible, resulting in no magnetic field and no distortion of the core–mantle boundary. The slopes of geoid cumulative degree-contribution curves for Venus (Figure 13) have a different pattern than those shown for Earth (Figure 6). The Venus curves commonly have greater slopes between harmonics 4, 5, 6, and 7 and lesser slopes between harmonics 2, 3, and 4, whereas the opposite relations are true for the Earth. Thus the shapes of the cumulative geoid-contribution curves for

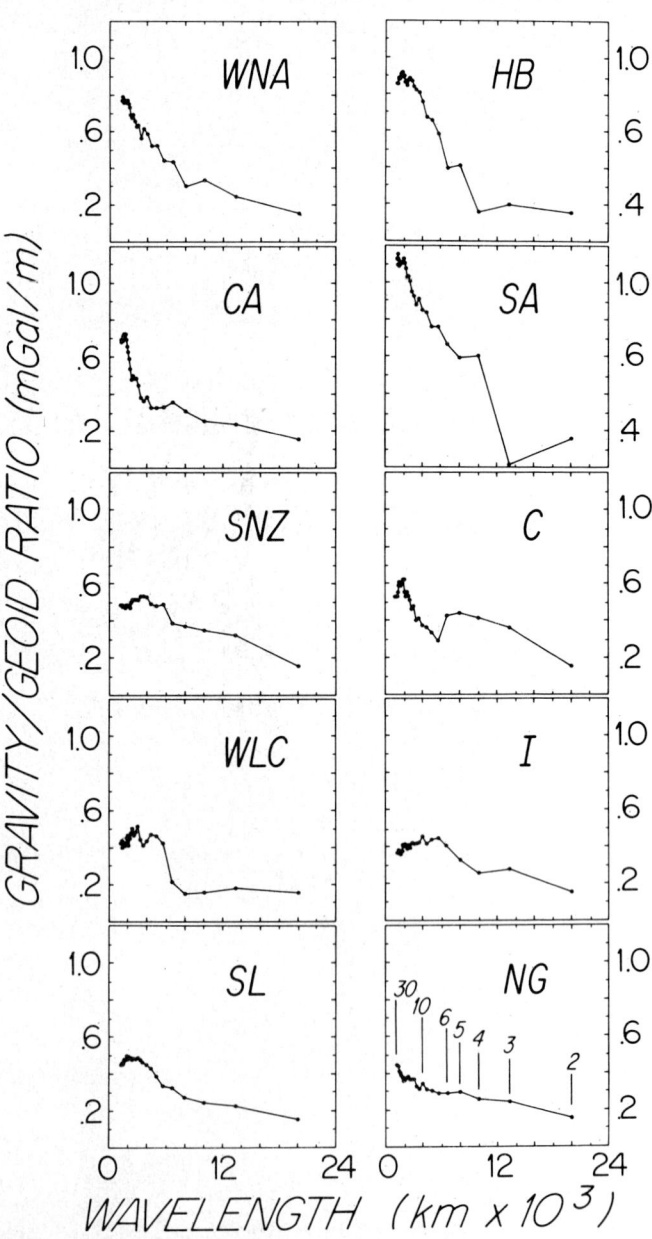

FIG. 10. Gravity to geoid ratio as a function of harmonic degree. Computed by equation (6) from GEM-9 spherical-harmonic coefficients. See caption of Figure 6 for explanation of labels.

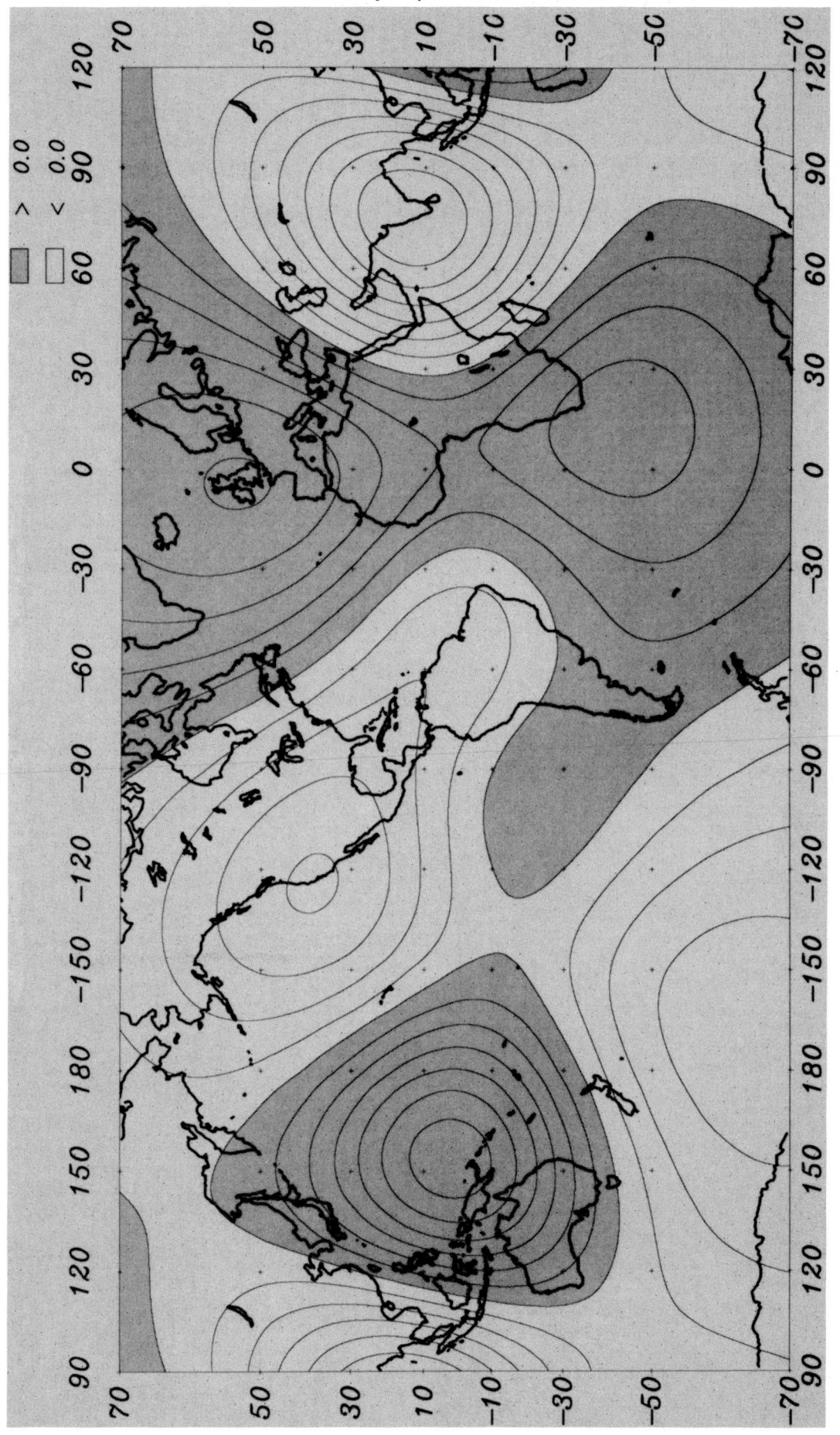

FIG. 11. Geoid-anomaly map, GEM-9 degree 3. Obtained by summing harmonic contributions from degrees 2 and 3 only. Positive anomalies are shaded. Contour interval, 10 m.

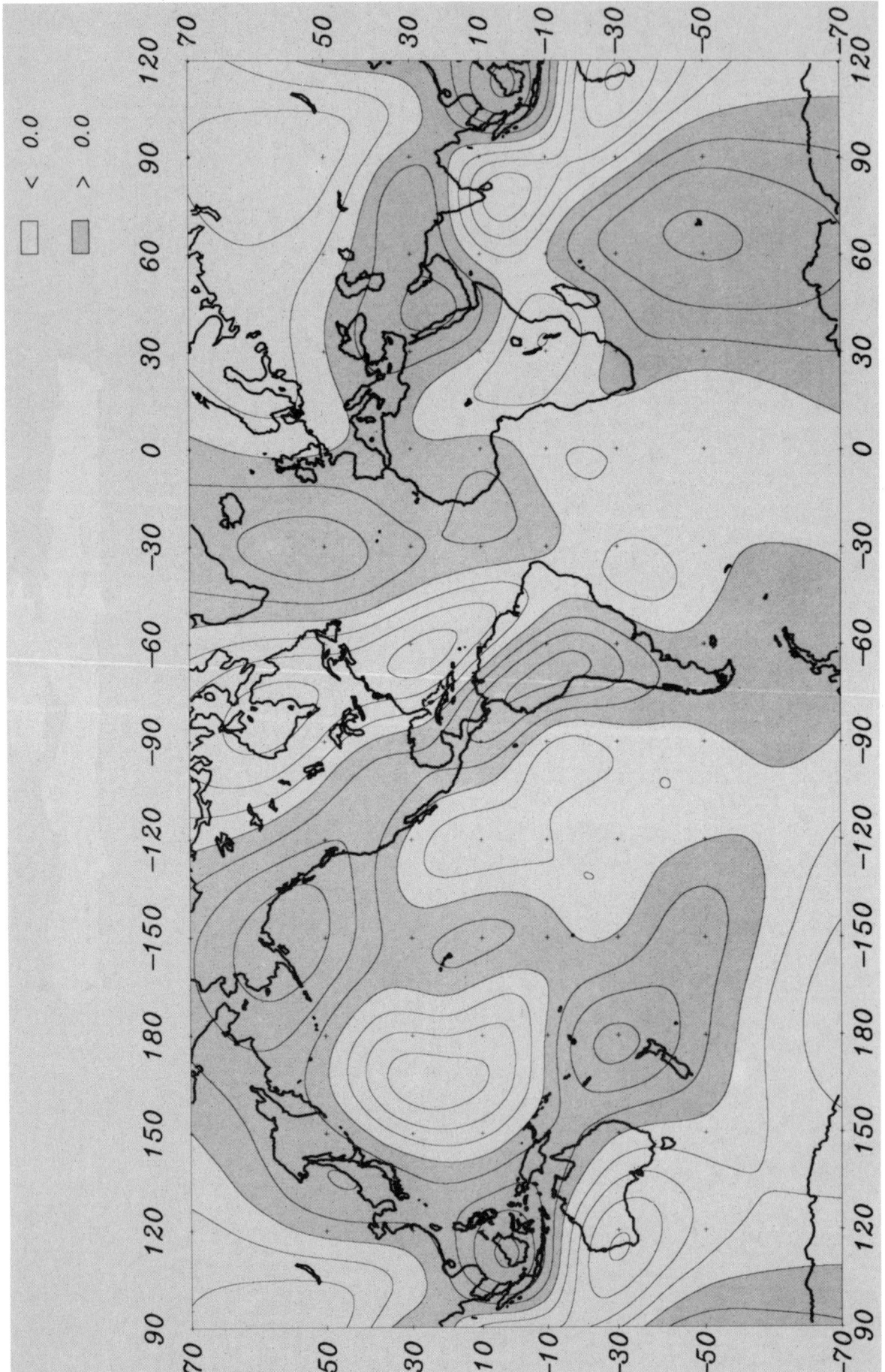

FIG. 12. Geoid-anomaly map from GEM-9 spherical-harmonic coefficients degrees 4 through 10. Equivalent to difference geoid anomaly obtained by subtracting degree-3 anomalies (Figure 9) from degree-10 anomalies (Figure 8). Positive anomalies are shaded. Note that most convergent-plate zones lie within the narrow belts of positive anomalies. Contour interval, 10 m.

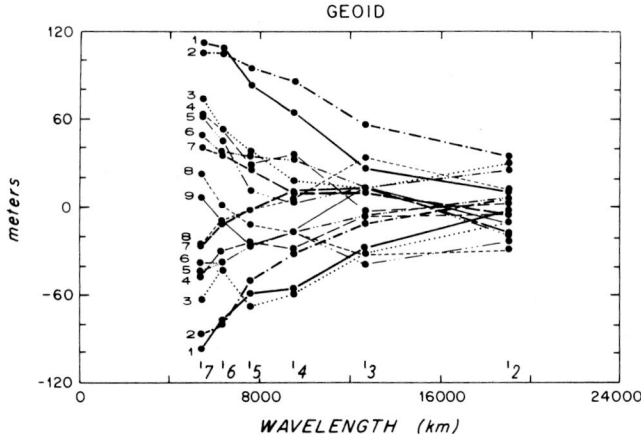

FIG. 13. Cumulative geoid-anomaly curves for the 17 major gravity-anomaly features of Venus, nine positive and eight negative. Computed from the spherical-harmonic coefficients of B. G. Williams (pers. comm., 1981). Numbers beneath lowest curve indicate harmonic degree.

positive-anomaly features are concave upward for Venus and convex upward for Earth, and for negative-anomaly features they are concave downward for Venus and convex downward for Earth. The pattern of geoid and gravity anomalies from Venus' low-order harmonic terms correspond well with surface topography (B. G. Williams, pers. comm., 1981; Bowin, 1983b; Bowin et al., 1985), as is also true for the more detailed line-of-sight tracking-acceleration data (Sjogren, et al., 1980, 1981; Bowin, 1983b; Bowin et al., 1985). The information available indicates a good correspondence of long- and intermediate-wavelength gravity variations with surface topography of Venus in marked contrast to the Earth. These relations further strengthen the conclusion that long-wavelength undulations at the Earth's core-mantle boundary contribute significantly to the lowest degree harmonic coefficients of the Earth's anomalous potential field.

SUMMARY

1. One principal and directly observable *fact* is that the degree-4 through -10 geoid *does* pinpoint convergence zones. This is an important empirical confirmation of the theory.
2. Second empirical fact: Maps of Venus' low-degree potential-field coefficients show correspondence (B. G. Williams, pers. comm., 1981; Bowin, 1983b; Bowin et al., 1985) with the major topographic features, unlike the situation for the Earth. Venus rotates very slowly, has no internal magnetic field, and thus may have an undisturbed core-mantle boundary in contrast to the Earth's.
3. Third fact: If there are significant mass anomalies at the core-mantle boundary, then more than 79 percent of their contribution will be contained in the coefficients for degrees 2 and 3. Large contributions from those two degrees cause the greatest magnitudes for seven of the 10 (Figure 9) major geoid anomalies of the Earth.

ACKNOWLEDGMENTS

I thank E. K. Scheer for discussion and comment during the progress of this research and B. G. Williams for providing spherical-harmonic coefficients for the gravity field of Venus. R. A. Stephen, G. M. Purdy, and four anonymous reviewers provided helpful comments. Support from the Office of Naval Research, Contract N00014-79-C-0071, is gratefully acknowledged.

REFERENCES

Bowin, C. O., 1980, Why the Earth's greatest geoid anomaly is so negative: Eos, **61**, 209.
———1981, Mass and topographic anomalies at the core-mantle boundary region from gravity and geoid observations: Abstr. 21st Gen. Assem., IASPEI, London/Canada (July 21–30), B2.2.
———1983a, Depth of principal mass anomalies contributing to the Earth's geoidal undulations and gravity anomalies: Mar. Geod., **7**, 61–100.
———1983b, Gravity, topography and crustal evolution of Venus: Icarus, **56**, 345–371.
Bowin, C., Abers, G., and Shure, L., 1985, Gravity field of Venus at constant altitude and comparison with Earth: J. Geophys. Res., **90**, Suppl., Proc. 15th Lunar Planet. Sci. Conf., pt. 2, C757–C770.
Bowin, C. O., Scheer, E. K., and Smith, W., in press, Depth estimates from ratios of gravity, geoid, and gravity gradient anomalies: Geophysics.
Bowin, C. O., Warsi, W., and Milligan, Julie, 1982, Free-air gravity anomaly map and atlas of the world: Map Chart Ser. MC-45 and MC-46, Geol. Soc. Am.
Busse, F. H., 1975, Core motions and the geodynamo: Rev. Geophys. Space Phys., **12**, 206–208.
Cox, A., and Cain, J. C., 1972, International conference on the core-mantle interface: Eos, **53**, 591–618.
Cuong, P. G., and Busse, F. H., 1981, Generation of magnetic fields by convection in a rotating sphere. 1: Phys. Earth Planet. Inter., **24**, 272–283.
Garland, G. D., 1957, The figure of the Earth's core and the dipole field: J. Geophys. Res., **62**, 486–487.
Goldreich, P., and Toomre, A., 1969, Some remarks on polar wandering: J. Geophys. Res., **74**, 2555–2567.
Jordan, S. K., 1978, Statistical model for gravity, topography, and density contrasts in the Earth: J. Geophys. Res., **83**, 1816–1824.
Lerch, F. J., Klisko, S. M., Laubscher, L. E., and Wagner, C. A., 1979, Gravity model improvement using Geos 3 (GEM 9 and 10): J. Geophys. Res., **84**, 3897–3916.
Sjogren, W. L., Birkeland, P. W., Espositio, P. B., Wimberly, R. N., and Ritke, S. J., 1981, Venus gravity field: Local Trans., Am. Geophys. Union, **62**, 386.
Sjogren, W. L., Phillips, R. J., Birkeland, P. W., and Wimberly, R. N., 1980, Gravity anomalies on Venus: J. Geophys. Res., **85**, 8295–8302.

Block structure of continental crust derived from gravity and magnetic maps, with Australian examples

Peter Wellman*

ABSTRACT

A continent is considered to consist of a mosaic of crustal blocks each 500 to 1 000 km across, the blocks having different cratonization history. Within each block, gravity anomalies of wavelength 20 to 100 km are generally elongated and subparallel. The relative ages of the gravity trends in the adjacent blocks can be determined because those blocks with trends oblique to the boundary are likely to be older than the boundary, whereas those parallel to the boundary are likely to be younger. These major gravity trends are thought to be caused by large-scale folding initiated at the first major deformation of each block. The boundaries between Precambrian blocks commonly coincide with a gravity gradient between a major high and a major low anomaly, the total amplitude being more than 50 mGal. The dipole anomaly is interpreted as due to a large and abrupt change in mean upper-crustal density, the crust being in regional isostatic equilibrium. Within the crustal blocks, major cross fractures can be mapped using the combined gravity and magnetic data. The cross fractures cause minor gaps or offsets of the primary gravity and primary magnetic trends, and are coincident along part of their length with the axes of magnetic bodies. These cross fractures are commonly subparallel and are generally not at right angles to the major trends. Where basement geology is known, the cross fractures coincide with geologically mapped faults and, rarely, folds.

INTRODUCTION

The upper crust of the continents can be divided into a "basement" of igneous, metamorphic, and deformed sedimentary rocks, and "cover" rocks of relatively undeformed sedimentary and volcanic rocks. The structure of the basement is known from geologic mapping only in areas where cover rocks are largely absent. However, these areas of well-exposed basement make up only a small fraction of the area of continental crust. It is important to determine basement structure over the whole of the continental crust, because basement structure is crucial in determining the origin, deformation, and cratonization of continental crust, as well as the effects of basement on cover rocks, igneous activity, and mineralization.

Because of the importance of basement structure in geology, there have been many studies of it using geologic, magnetic, and gravity maps and satellite imagery, the data sets being studied either alone or in combination. The relative value of these techniques depends critically on the thickness of cover rocks and surface conditions. In areas of little cover, surface geologic mapping can determine detailed basement structure, except where the rocks are masked by deep weathering or thick vegetation. In areas of cover, the basement structure cannot be mapped adequately by drilling into basement, or from the structure of the cover rocks. Aeromagnetic-anomaly maps generally give an extremely detailed anomaly pattern in areas of thin or no cover; however, the correlation between anomaly patterns and geology is obscure in some areas. Aeromagnetic anomalies resulting from the basement have high-amplitude anomalies of short wavelength and low-amplitude anomalies of long wavelength; the short-wavelength anomalies are rapidly attenuated as cover rocks become thicker. Gravity-anomaly maps are dominated by long-wavelength features that reflect the gross basement structure relatively well. These long-wavelength anomalies are only gradually attenuated as the cover thickness increases. Lineaments determined from satellite imagery or air photographs may be useful in areas of outcropping or thinly covered basement, but they cannot be effectively utilized to map basement structure in areas of thick vegetation or thick cover rocks.

This paper discusses the use of gravity anomalies to give gross basement structure for areas of continental size, and the combined interpretation of gravity and magnetic anomalies to delineate basement fractures in areas of thin or absent cover. Such studies can be carried out only in regions that have high-quality gravity- and aeromagnetic-map coverage. Suitable potential-field maps covering large parts of continents have become available for Australia, North

---

*Bureau of Mineral Resources, P.O. Box 378, Canberra City, A.C.T., 2601, Australia.

America, and Europe; they are still not available for most continents.

The gravity and aeromagnetic maps contain very large amounts of information on anomalous bodies. The information can be divided roughly into two types: (1) the magnitude, gradients, and second derivatives of anomalies, giving information on the depth and physical-property contrasts of the bodies; and (2) the shape of the contour lines, giving information on the horizontal position, size, and shape of the anomalous bodies. This paper is based almost completely on the shape of the contour lines as a measure of the trend and extent of anomalous bodies. The interpretation is derived by eye from contour maps, so details of the interpretation are subjective. An assessment of the reliability of the conclusions comes from the self consistency of the derived geophysical model and the consistency of this model with geologically mapped basement structure where exposed. The text has been written so that the argument applies to all continents. Examples are from Australia, with reference to prior work in the USSR.

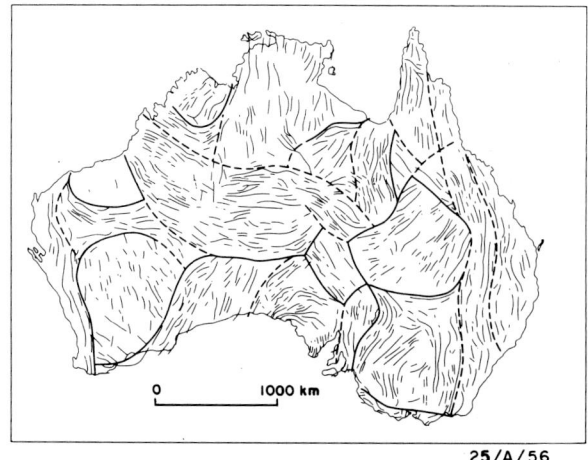

FIG. 1. Australian gravity trends (thin lines) and crustal blocks defined by these trends (thick lines). From Wellman (1976).

MODEL OF BASEMENT GEOLOGY

The primary structure of the continent is considered to be the mosaic of blocks generally 500 to 1 000 km across, each with a distinctive history and structure. The secondary structure of the continent is considered to be the dominant linear or dome structure within each of these blocks. The tertiary structure is considered to be the cross fractures segmenting the secondary structure.

The origin of the primary structure of the continents has been widely discussed. It is thought to be due to one, or both, of the following two processes: (1) the systematic accretion of continental crust onto existing continents, the structure at the time of accretion remaining largely unaltered to the present; (2) part of an existing cratonized crust being completely reworked at a later time to form cross-cutting mobile belts, with a new basement structure.

The origin and control on the formation of the secondary structure has been less widely discussed. It is envisaged here to be controlled by lithospheric strength at the time of the last major deformation that reworks the whole crust. At this time the crust would be relatively hot and weak, so areas of the crust about 50–100 km across would be in almost isostatic equilibrium. This would impose a cross-structural wavelength of 50–100 km on the new crustal structure.

The cross fractures are thought to be of variable origin, some being formed at the time of formation of the main trends and some at a later time.

GRAVITY-TREND PATTERN

Most regions of continental basement can be divided into smaller areas, each about 50 km across, distinguished by being relatively less and relatively more uplifted. In granite/sedimentary rock terranes the relatively uplifted areas have outcropping granites and higher grade sedimentary rocks. The relatively less uplifted areas are low-density younger sedimentary rocks and are gravity lows. In purely sedimentary terranes the relatively more uplifted rocks are older and of higher metamorphic grade. In most areas the structures are elongate and subparallel. Rarely, the structure has no dominant direction but is dominated by approximately randomly spaced, circular granitic domes (e.g., Pilbara craton of Australia); these areas without linear structure are not considered further.

The elongate, subparallel, geologic-basement structures within each crustal block cause major elongate gravity anomalies. In granite/greenstone terranes the relatively less uplifted areas are the greenstone belts, which have higher density than the surrounding granites, and so they are associated with gravity-anomaly highs. In terranes of both uncompacted and compacted sediment the relatively less uplifted areas will be less dense and are associated with gravity lows. Almost no upper crustal gravity anomalies occur in areas where relatively uniform rocks have been metamorphosed and intensely deformed (granulites in the southwestern part of the Yilgarn block of Australia, and graywackes and schists of eastern South Island of New Zealand). In some areas where the cover varies in thickness, the gravity anomalies may solely reflect variation in the thickness of the cover sedimentary rocks.

Elongate and subparallel structures within the basement are theoretically best delineated by their elongate gravity gradients, as these gradients reflect the position of actual density changes within the crust. Only slightly less satisfactory than the use of gradients is the use of axes of elongate high and low anomalies. The continent can be divided into crustal blocks of subparallel trends. The relative ages of the adjacent blocks can be inferred where trends on one side are parallel to the boundary and on the other side are oblique to the boundary. The block containing trends oblique to the boundary is inferred to be older than the boundary, and the block with trends parallel to be younger. Figure 1 shows the pattern of gravity trends and crustal blocks mapped for the Australian continent by Wellman (1976).

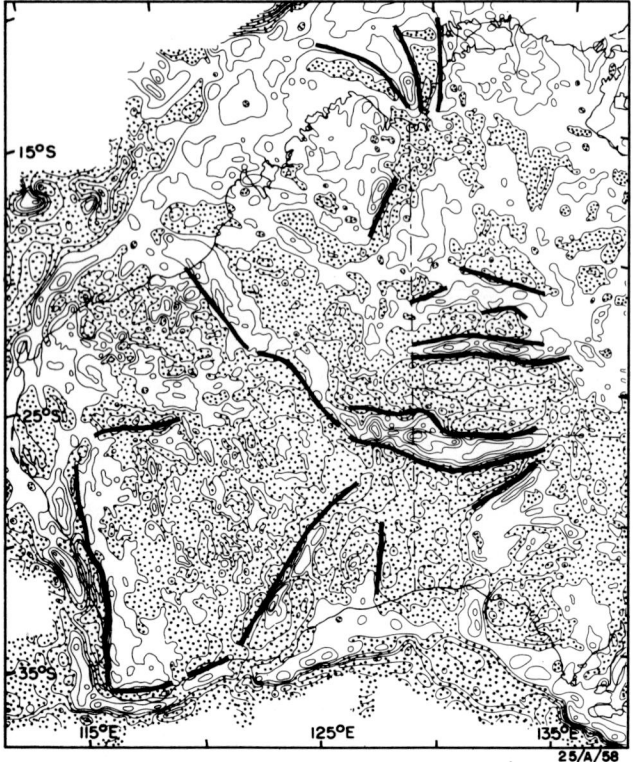

FIG. 2. Location of the steep gradients of the major gravity "dipole" anomalies. Base shows free-air map of central and western Australia with contour interval of 20 mGal; areas of negative free-air anomaly are stippled. After Wellman (1978).

GRAVITY "DIPOLE" ANOMALIES

Figure 2 shows the major gravity "dipole" anomalies. These are thought to overlie crustal-block boundaries. The gravity high of the "dipole" anomaly lies over the younger crust inferred from geology or gravity-trend pattern in six out of eight boundaries. The block boundaries lie near the center of the steep gravity gradient between the high and low of the "dipole," and one of the gravity anomalies overlies the oblique trends on the side of the block boundary with older crust. Figure 3 shows the block boundaries inferred from the "dipole" anomalies superimposed on the trend pattern to show the agreement in the position of the block boundaries inferred by the two methods. The positional agreement is very good if the part of the major "dipole" over the older block (that is due to isostatic compensation) is ignored in analysis of the gravity-trend pattern.

The major gravity "dipole" anomalies are interpreted as due to large and abrupt changes in mean upper crustal density, the area being in regional isostatic compensation (Gibb and Thomas, 1976; Thomas and Gibb, 1977; Wellman, 1978). For some boundaries the anomalies are in addition partly due to overthrusting, lithospheric flexure, and sedimentation (Karner and Watts, 1983; Lambeck, 1983).

CRUSTAL BLOCKS

The geophysically determined relative age of the Australian crustal blocks is roughly consistent with geologically determined block ages. In particular the gravity trend pattern is consistent, with the oldest blocks being those with radiogenic ages showing that they are Archean (Yilgarn, Pilbara, Pine Creek, and Gawler blocks) (Page et al., in press) and with the youngest blocks being those with Phanerozoic basement (Figure 4).

The Pilbara, Yilgarn, and Gawler blocks are of Archean age and are relatively close together, but they differ in gravity-trend pattern and in geologic structure. These Archean blocks are therefore unlikely to have been a single large uniform Archean block, parts of which were subsequently reworked to form the intervening Proterozoic mobile zones. The Archean blocks are likely to have been formed separately. The crust forming the Proterozoic blocks is likely to be either the reworked edges of existing Archean blocks or crust formed subsequent to the formation of the Archean blocks and before the present relative positioning. Paleomagnetic results (Embleton, 1981) show that the Pilbara, Yilgarn, and Gawler blocks have not moved significantly relative to one another since 1.8 b.y. ago. This constrains crustal models to those with central and western Australia having been cratonized by this time.

A "dipole" anomaly is not found at the few boundaries in Antarctica where the younger crust is identified as formed by intraplate reworking of an Archean granulite crust (Wellman

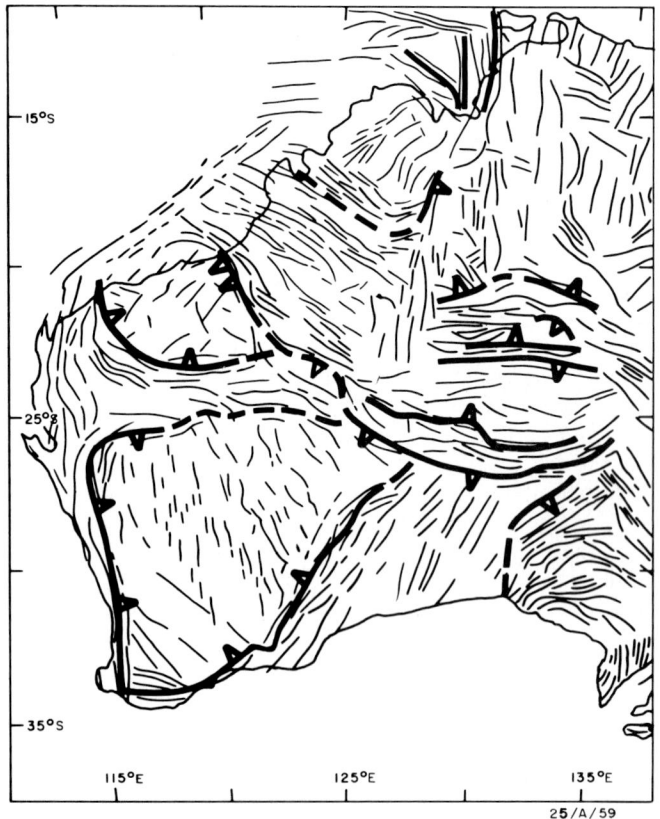

FIG. 3. Crustal-block boundaries inferred from gravity "dipole" anomalies superimposed on the gravity-trend pattern to show that the two techniques give similar results. The "V" marks are on the crust inferred to be older from the gravity "dipole" anomalies. After Wellman (1978).

and Williams, 1982; Wellman, 1983). This is presumably because the crustal density and crustal thickness do not change significantly during intraplate crustal reworking. Other boundaries may abut without collision and without reworking by processes such as faulting. The densities of these abutting crusts may not be determined by the abutting process, so there may be no gravity anomaly or a gravity "dipole" with a high on the older or younger side.

The major basement blocks in the USSR have been mapped by the trend-direction method. The major blocks of European USSR were mapped by Simonenko and Tolstihina (1968) by delineating areas on maps with similar patterns of anomalies; the pattern could be linear or a mosaic. Gravity and magnetic maps were used concurrently. Provodnikov (1975) mapped the trend pattern and block boundaries over most of Siberia, mainly using aeromagnetic-flight-line profiles rather than contour maps. The pattern of blocks and block boundaries found in the European USSR and Siberia is similar to that found in Australia by Wellman (1976), suggesting that the same sort of structure is being mapped and that the trend-pattern technique can be applied to all continents.

CROSS FRACTURES

The gravity and magnetic trends within a basement block, although in a broad sense continuous, are in detail broken into short, linear segments. The features that separate the trend segments displace the geologic structures forming the trends, the movement being vertical and/or horizontal. On geologic maps, the cross-cutting features are faults or folds at an angle between 30 and 90 degrees to the geologic trend. In rare cases they are coincident with an intrusion. In this paper an attempt is made to map the larger of these cross-

FIG. 5. Trends and cross fractures in a part of Western Australia. Contours are of 5-mGal Bouguer gravity anomalies; densely stippled lines are gravity trends drawn along axes of residual gravity highs. Short lines are axes of magnetic bodies. Less densely stippled lines are inferred cross fractures. Location of area is given in Figure 6.

cutting features—called hereafter "cross fractures." The cross fractures are relatively minor features, so their identification and mapping are relatively difficult. Both their direction and location have to be determined. To make the mapping more objective, only those cross fractures that have both gravity and magnetic expressions are mapped.

The major cross fractures are here taken to be those zones that both mark the ends of linear segments of the major gravity trends and are concordant along parts of their lengths with the elongate axes of bodies causing aeromagnetic anomalies. The preferred method of showing the geophysical information is an overlay of the axes of magnetically anomalous bodies over a gravity-anomaly contour map. This is shown in Figure 5. This type of map cannot be reduced to a small scale, however, so the maps of larger areas illustrating the subsequent discussion do not contain the gravity contours.

The following examples (Figure 6) are for areas where cover rocks are thin or absent, gravity trends are strongly developed, and both gravity- and magnetic-anomaly maps are available. In areas of Figures 7–12 the dominant gravity trends are nearly linear, with gravity lows overlying basement granitic intrusives and gravity highs overlying basement sedimentary and volcanic rocks.

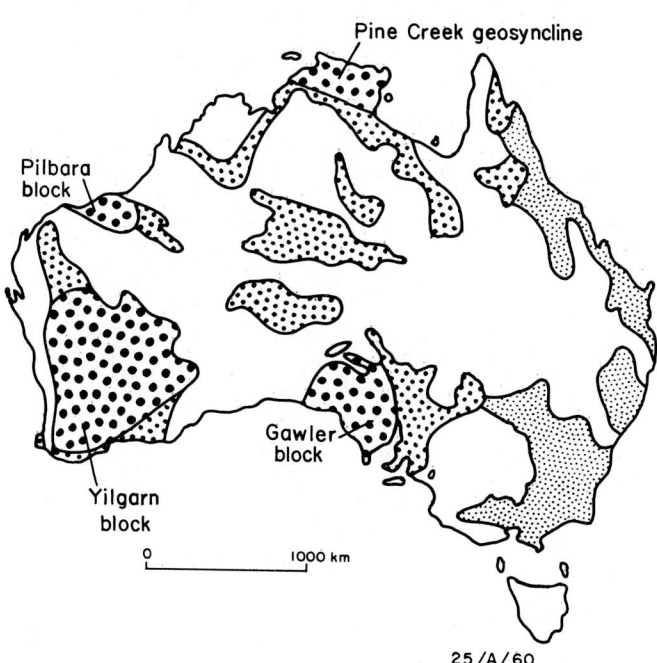

FIG. 4. Basement age of Australia determined from geology. Large stippling, Archean; medium stippling, Proterozoic; fine stippling, Phanerozoic; unstippled, cover rocks.

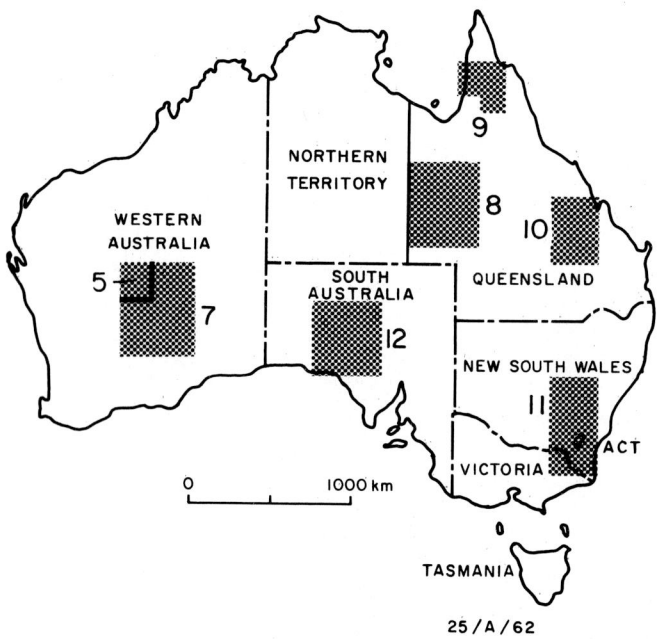

Fig. 6. Map of Australia showing locations of areas covered by Figure 5 and Figures 7–12.

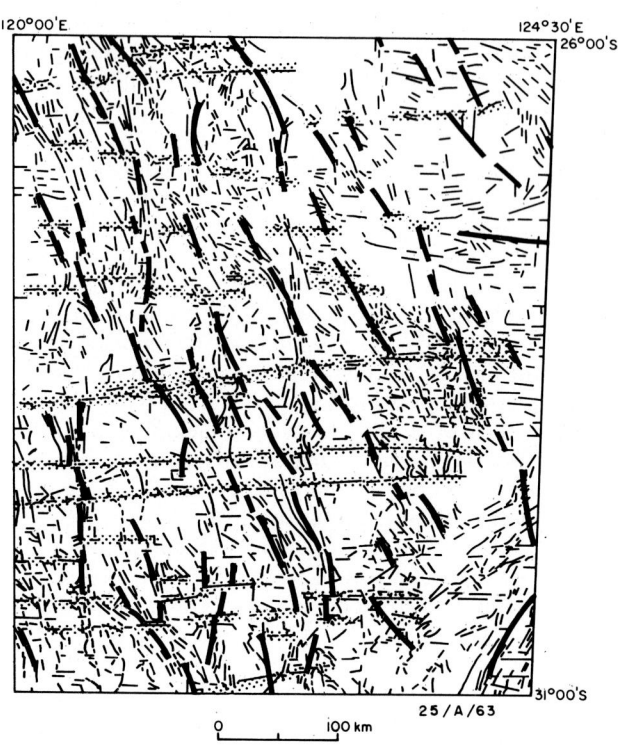

Fig. 7. Eastern Yilgarn shield, Western Australia, showing gravity trends (thick lines), magnetic-body axes (thin lines), and inferred cross fractures (stippled lines). Location shown in Figure 6.

In the eastern Yilgarn shield (Figures 5, 7) the basement is Archean with a north-northwest trend. Cross fractures are east striking and are commonly coincident with long, wide basaltic dikes of 2.4–2.5–b.y. age.

In the southern Mount Isa block (Figure 8), basement rocks are of Proterozoic age. The main basement trends are north-northwest to north, and the cross fractures are mainly short with a northeast or northwest strike. The cross fractures correspond to the set of dominant cross faults of short length.

In the Coen inlier area (Figure 9) the basement consists of metamorphosed Proterozoic rock with Paleozoic granitic intrusives. Gravity trends are northwest to north, and cross fractures are at right angles. The cross fractures appear to represent major folds in the basement.

In the Bowen basin area of the central Tasman geosyncline (Figure 10) the gravity trends strike north-northwest and are composed of short segments. The gravity highs correspond to deformed basement sedimentary and volcanic rocks of Silurian to Devonian age, and the gravity lows to granitic intrusions. The cross fractures strike east.

In the Canberra area of the southern Tasman geosyncline (Figure 11), gravity trends strike north. Gravity highs correspond to Ordovician–Devonian deformed sedimentary and volcanic rocks, and gravity lows to Silurian–Carboniferous granitic intrusions. The cross fractures strike northwest, and at least some correspond to faulting.

In the northern Gawler block (Figure 12), the basement rocks are granitic intrusives and metamorphosed sedimentary, volcanic, and intrusive rocks, all of Precambrian age. The gravity trends are northwest in the northeast quadrant of the map and arcuate from northeast through east to northwest over the remainder of the map. Several sets of cross

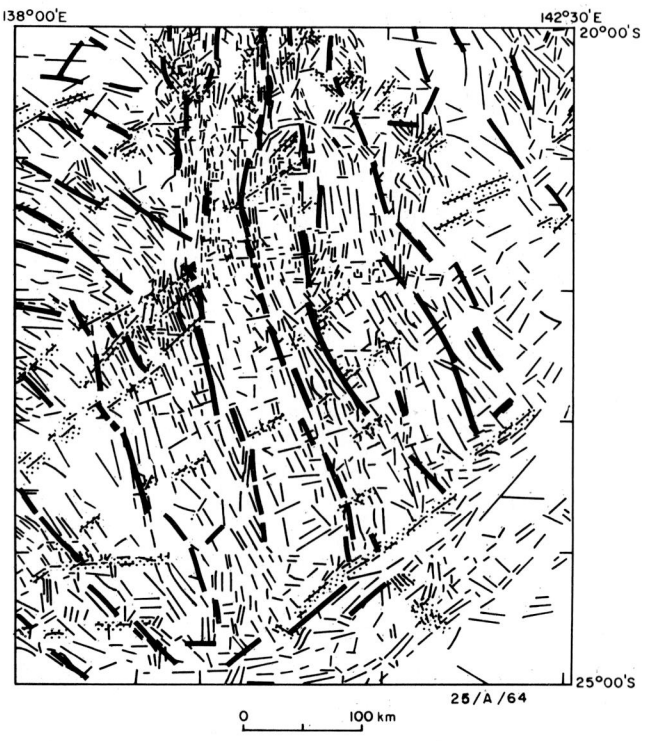

Fig. 8. Southern Mount Isa block, Queensland. See Figure 7 for explanation of symbols. Location shown in Figure 6.

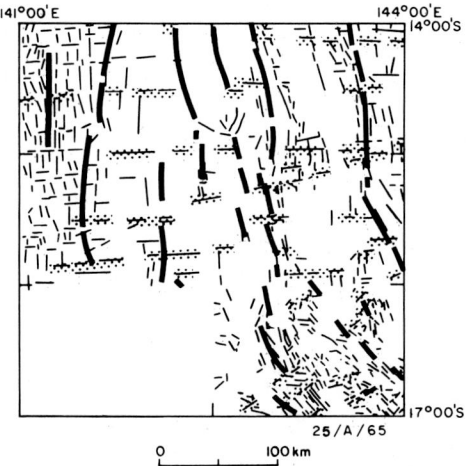

FIG. 9. Part of Coen inlier, northern Queensland. See Figure 7 for explanation of symbols. Location shown in Figure 6.

FIG. 11. Part of southern Tasman geosyncline, southeastern Australia. See Figure 7 for explanation of symbols. Location shown in Figure 6.

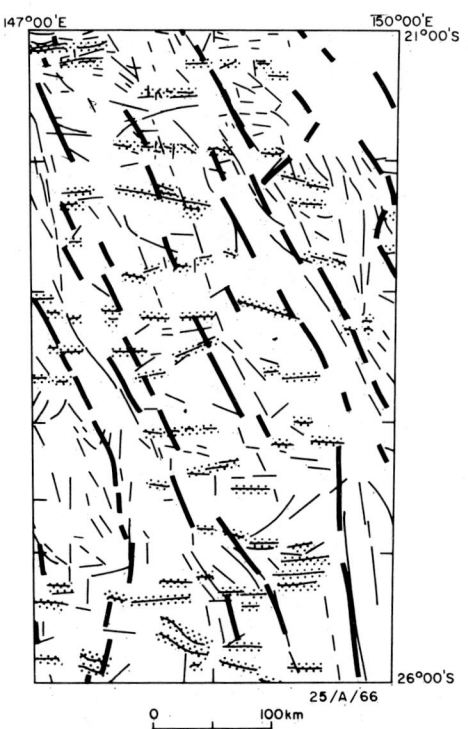

FIG. 10. Part of central Tasman geosyncline, eastern Queensland. See Figure 7 for explanation of symbols. Location shown in Figure 6.

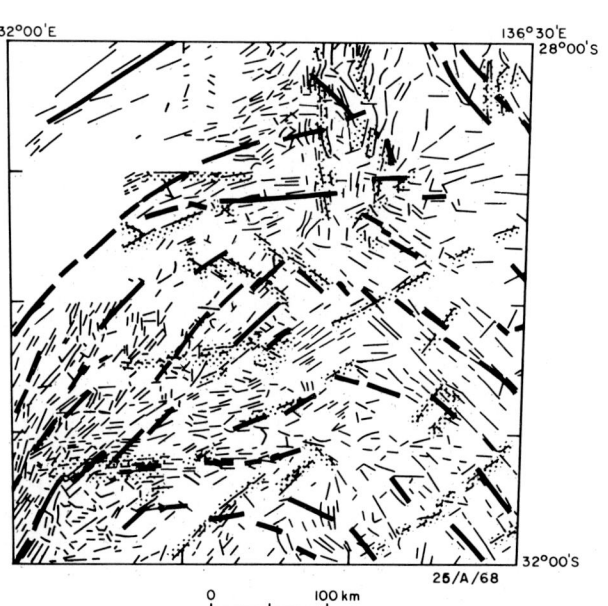

FIG. 12. Part of Gawler block, South Australia. See Figure 7 for explanation of symbols. Location shown in Figure 6.

fractures with different directions are present. The dominant directions of both trends and cross fractures are northeast and northwest.

In the above examples, the subparallel cross fractures have been emphasized, but even so in most areas there seems to be a remarkable lack of magnetic-body axes that could represent sets of fractures in directions other than those of the trends and the main set of cross fractures.

Provodnikov (1975) mapped cross fractures that displace the main-trend pattern of Siberia. Almost all his mapped cross fractures are exactly east–west, however. It is thought that these cross fractures are biased in direction, because the interpretation is based on wide-spaced flight-line profiles flown mainly in an east–west direction. The preferred display is a detailed contour map based on closely spaced flight lines "leveled" by tie lines.

MEANING OF CROSS FRACTURES

Several independent lines of evidence suggest that the nonsedimentary crust is cut by deep faults into blocks 50 to 200 km across. Deep seismic sounding in eastern Europe has consistently been interpreted as showing displacements of the deep crustal layers at irregular intervals by steep faults cutting most of the crust (Kosminskaya and Pavlenkova, 1980). These steep faults are commonly correlated with crustal fractures mapped from gravity and magnetic anomalies (Tyapkin, 1980). The vertical tectonic movement during cratonization cannot be by lithospheric flexure, because the isostatic equilibrium is too local and the wavelength of the tectonic movement too short; the movement is interpreted as mainly on deep faults. Major geologic faults and some lineations on satellite images have also been interpreted as deep faults bounding crustal blocks.

Although the deep-sounding and surface-lineation techniques have been used to map deep faults, the use of gravity and magnetic anomalies has the advantage of both areal coverage and of reflecting the physical properties of the average rocks of the top few kilometers of the basement. Reliable mapping of deep faults is likely to depend on the combined use of maps of gravity, magnetics, remote-sensing lineations, and geology. The real nature of deep faults is poorly understood; of particular importance is the mechanism of their formation as well as their suitability as sites for igneous intrusions or as paths for mineral-bearing solutions during either average stress conditions or abnormal stress conditions.

ACKNOWLEDGMENTS

Figures were drawn by T. J. Kimber. The paper is published with the permission of the Director, Bureau of Mineral Resources, Canberra.

REFERENCES

Embleton, B.J.J., 1981, A review of the palaeomagnetism of Australia and Antarctica, *in* McElhinny, M. W., and Valencio, D. A., Palaeomagnetic reconstruction of the continents: Geodynamic Ser., Am. Geophys. Union and Geol. Soc. Am., **2**, 77–92.

Gibb, R. A., and Thomas, M. D., 1976, Gravity signature of fossil plate boundaries in the Canadian Shield: Nature, **262**, 199–200.

Karner, G. D., and Watts, A. B., 1983, Gravity anomalies and flexure of the lithosphere at mountain ranges: J. Geophys. Res., **88B**, 10449–10477.

Kosminskaya, I. P., and Pavlenkova, N. I., 1980, A generalized seismic model of the continental crust, *in* Seismic models of the lithosphere for the major geostructures on the territory of the USSR: Nauka, Soviet Geophys. Comm., Acad. Sci. USSR, 141–152.

Lambeck, K., 1983, Structure and evolution of the intracratonic basins of central Australia: Geophys. J., **74**, 843–886.

Page, R. W., McCulloch, M. T., and Black, L. P., in press, Isotopic record of major Precambrian events in Australia: Proc. Int. Geol. Congr., Moscow, 4–14 Aug. 1984.

Provodnikov, L. Y., 1975, The basement of platformean regions of Siberia: Nauka, Trans. Inst. Geophys. Siberian Branch Acad. Sci. USSR, **194**.

Simonenko, T. N., and Tolstihina, M. M., 1968, Block structure of the folded basement in the European USSR: Geotectonics 1968, 222–229.

Thomas, M. D., and Gibb, R. A., 1977, Reply to Comment by A. J. Baer on Gravity anomalies and deep structure of the Cape Smith fold belt, northern Ungava, Quebec: Geology, **5**, 651–653.

Tyapkin, K. F., 1980, A study of abyssal fractures in the Ukrainian Shield by geophysical methods: Geophys. Prospect., **28**, 935–944.

Wellman, P., 1976, Gravity trends and the growth of Australia: a tentative correlation: J. Geol. Soc. Austr., **23**, 11–14.

———1978, Gravity evidence for abrupt changes in mean crustal density at the junction of Australian crustal blocks: BMR J. Austr. Geol. Geophys., **3**, 153–162.

———1983, Interpretation of geophysical surveys, longitude 45° to 65° E, Antarctica, *in* Oliver, R. L., James, P. R., and Jogo, J. B., Antarctic earth science: Austr. Acad. Sci., 522–526.

Wellman, P., and Williams, J., 1982, The extent of Archean and Proterozoic rocks under the ice cap of Princess Elizabeth Land, Antarctica, inferred from geophysics: BMR J. Austr. Geol. Geophys., **7**, 213–218.

Gravity studies of the Grenville province: Significance for Precambrian plate collision and the origin of anorthosite

M. D. Thomas*

ABSTRACT

One of the most prominent negative Bouguer gravity anomalies in the Canadian shield follows the northeastern part of the Grenville front for almost 1 200 km. It is attended along its southeastern flank, within the Grenville province, by a belt of positive anomalies that are significantly more positive than the general level of the gravity field along the opposite flank, over the older Superior, Churchill, and Nain provinces. Together, the belts of negative and positive anomalies constitute a typical example of a particular kind of paired gravity anomaly that signifies the presence of collisional sutures produced by convergent plate tectonics. The tectonic significance of the gravity signature first became apparent by way of the Grenville example when it was noted that the steep gradient separating negative and positive components, ascribed to a steep discontinuity penetrating the entire crust, was located within a zone defined on paleomagnetic grounds as one likely to contain a suture; the discontinuity was subsequently equated with a cryptic suture. Studies of other paired gravity anomalies in both Phanerozoic and Precambrian terranes have provided a variety of evidence supporting a collisional origin for the signature.

In the northeastern half of the Grenville province, gabbroic areas of terranes formed of rocks of the anorthositic suite correlate with large positive gravity anomalies. Gravity models suggest that these are related to saucer- or funnel-shaped gabbroic masses extending, in cases, to mid-crustal depths. These masses are interpreted as dense mafic cumulates produced by gravity-assisted differentiation of gabbroic magma, and as such represent the lower levels of former magma chambers. Associated anorthosites are believed to have developed by flotation of plagioclase to higher levels.

INTRODUCTION

The principal utility of geophysics is its capability to provide an insight into the three-dimensional structure of the lithosphere (gravity, magnetic, electromagnetic, and seismic methods) and to trace past relative motions of various elements of the lithosphere (paleomagnetic method). Individually, or in combination, with an appropriate blend of geological (including geochemical, geochronological, etc.) data, geophysical methods can lead to a better understanding of not only the existing static geologic situation but also of the dynamic processes that produced the situation. Here, geophysical evidence, with emphasis on the significance of gravity anomalies, is presented in support of the possible role of two separate processes in developing respective features of the Precambrian Grenville structural province.

One process is plate tectonics, which involved, at some stage, collision between the Grenville province and the remainder of the Canadian shield. The timing of collision is debatable. Certain lines of evidence suggest that it was Grenvillian, whereas others indicate that it was Hudsonian or slightly later. The second process, gravity differentiation (of gabbroic magmas), is considered to have played a major role in the formation of anorthositic intrusions. The material presented draws to a large extent on the results of previous studies and is therefore in the nature of a review.

GEOLOGIC SETTING

The Grenville structural province is an approximately 375-km-wide orogenic belt trending northeasterly along the southeastern margin of the Canadian shield (Figure 1). It is

*Department of Energy, Mines & Resources, Earth Physics Branch, Gravity, Geothermics & Geodynamics Division, 1 Observatory Crescent, Ottawa, Ont. K1A 0Y3, Canada.
Contribution of the Earth Physics Branch No. 1100.

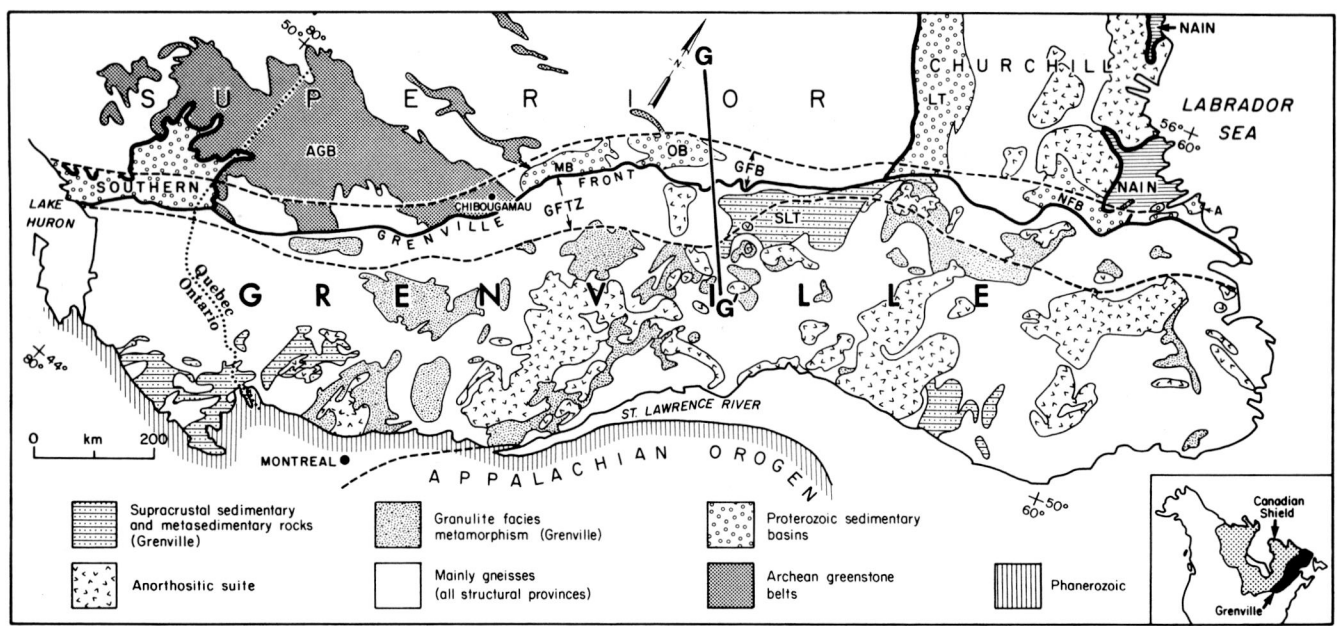

FIG. 1. Simplified geologic map of Grenville province and adjacent marginal areas of Southern, Superior, Churchill, and Nain provinces. AGB = Abitibi greenstone belt; MB = Mistassini basin; OB = Otish Mountains basin; NFB = Naskaupi fold belt; A = Aillik Group; LT = Labrador trough; SLT = Southern Labrador trough; GFB = Grenville foreland belt; GFTZ = Grenville front tectonic zone; G–G' = line of gravity profile and model (Figure 4a, b).

the youngest Precambrian structural province, having experienced its last major deformation during the Grenvillian orogeny, which, according to Stockwell (1982), probably commenced, culminated, and terminated approximately 1 140, 1 090, and 1 000 m.y. ago, respectively. While the characteristics of the Grenville province are primarily a product of the Grenvillian orogeny, geochronological evidence indicates that various parts have been involved in at least five orogenic events, and rocks ranging in age from Archean to Hadrynian are recognized (Stockwell, 1982). Names and ages of time units referred to in this paper are presented in Table 1.

Most of the province represents a reworked basement complex formed largely of quartzofeldspathic gneisses considered to be mainly of Archean age and including some of Aphebian age (Wynne-Edwards, 1972). The complex probably forms the basement for Aphebian and Helikian supracrustal rocks (Wynne-Edwards, 1972; Stockwell, 1982) that occupy relatively small areas scattered throughout the Grenville province. Sedimentary and metasedimentary rocks dominate the supracrustal sequences, but some volcanic and metavolcanic rocks do occur. The Grenville mosaic is completed by large complexes of anorthositic rocks and some granitic intrusions. Although few ages older than Grenvillian regional metamorphism (~1 100 m.y.) have been determined for the anorthositic rocks, circumstantial evidence indicates that they are products of magmatic activity that occurred about 1 450 m.y. ago (Emslie, 1978), an episode of activity termed the Elsonian disturbance by Stockwell (1982).

The greater part of the Grenville province is affected by regional metamorphism in the upper amphibolite or granulite facies grade (Wynne-Edwards, 1972). The general depth of erosion over most of the province ranges from 15 to 25 km, although in a large region identified as the eastern Grenville province it is probably less than 10 km (Wynne-Edwards, 1972). The deepest levels, 25–30 km, are observed in the Grenville front tectonic zone (Figure 1), where metamorphic assemblages are indicative of the highest pressures recorded in the Grenville province.

The Grenvillian orogeny produced structures trending northeasterly, but large-scale structures (i.e., those defined by geologic mapping and aeromagnetic anomalies) with a northeasterly trend do not dominate all parts of the orogen. Earlier orogenies and the influence of large massifs of anorthosite and mangerite during the Grenvillian orogeny have produced trends other than northeasterly. However, at a smaller scale of investigation, fabric analysis indicates that linear elements are systematically oriented in a northeasterly direction within large areas of the province (Wynne-Edwards, 1972), testifying to the pervasiveness of the Grenvillian orogeny. The elements are interpreted as a younger set of northeast-trending and southeast-dipping axial planes intersecting older S-surfaces in a lineation.

The criteria defining the Grenville front are variable from place to place and may be a fault, mylonite or cataclasite zone, a marked change in structural style or metamorphic grade, or a combination of these features. If such criteria are absent, the front is positioned where K–Ar ages decrease from older values typical of adjacent provinces to values of about 1 000 m.y. or slightly less, which are typical of the

Table 1. Time units for Canadian shield (after Stockwell, 1982).

Age (m.y.)		
570 — Hadrynian		
1 000 — Helikian	1 000 — Neohelikian	1 000 — Grenvillian orogeny
		1 140 —
	1 400 — Paleohelikian	1 400 — Elsonian disturbance
		1 500 —
1 750 — Aphebian	1 750 —	1 750 — Hudsonian orogeny Penokean orogeny
2 510 — Archean		1 850 —

Note: Vertical spacings of unit-boundary ages (m.y.) are not to scale.

Grenville province; the change from older to Grenville ages generally takes place within a few kilometers. Several major features within the older provinces terminate abruptly at the front (Figure 1); examples are the Proterozoic Mistassini basin and Naskaupi fold belt, which are both truncated by major thrust faults. Other features undergo a marked increase in metamorphic grade at the front, including the Archean Abitibi greenstone belt and the Aphebian succession of the Labrador trough. The latter is traced southward across the front into the Southern Labrador trough of Dimroth et al. (1970).

The presence of rocks south of the front representing more highly metamorphosed counterparts of rocks in the older provinces is also indicated by radiometric dating (e.g., Stockwell, 1982). Rocks yielding such older ages apparently are more or less confined to the 15–80-km-wide Grenville front tectonic zone (Figure 1) that is typified by a strong northeast-trending foliation and many parallel zones of cataclasis and mylonitization. Archean dates have been obtained from several localities; in Ontario these occur within a 10–15-km-wide zone southeast of the Grenville front, whereas in Quebec, near Chibougamau, they occur up to 50 km from the front (Doig, 1977). Most of the Grenville gneisses southeast of the front in Ontario yield primary ages of about 1 800 m.y. (Doig, 1977), characteristic of the Hudsonian orogeny. Krogh and Davis (1971) believed that these gneisses may occur in a belt, up to 80 km wide, that may extend for at least 600 km into Quebec.

Immediately northwest of the front, Wynne-Edwards (1972) defined a Grenville foreland belt, 30–95 km wide, embracing a number of basins of deformed but largely unmetamorphosed supracrustal rocks of Aphebian or Helikian age that rest unconformably on mainly crystalline basement. The northwestern boundary is located at the limit of northeast-trending, southeast-dipping cleavages, faults, and folds that affect the basins.

TECTONIC DEVELOPMENT

Geosyncline and "millipede" models

The Grenvillian orogeny and the tectonic development of the Grenville orogen have long been and continue to be subjects of intense debate. Most early opinions were influenced by classical geosynclinal theory in which mountains and attendant orogens evolve from geosynclines, and it was generally regarded that the Grenville province represented the root zone of a former mountain system (e.g., Shillibeer and Cumming, 1956), although some workers had expressed doubts about a geosynclinal origin (Osborne and Morin, 1962). One of the later proponents of a geosynclinal origin introduced a variation on a theme; Dietz (1966) suggested that the Grenville province evolved from an ensimatic eugeosynclinal sedimentary prism located on a continental rise. The prism was deformed by continentward underthrusting of a mobile oceanic lithosphere, a mechanism reminiscent of plate subduction, that produced collapse and igneous intrusion of the prism to create an orogen that was compressed (accreted) against the adjacent continental margin.

Strong arguments against a geosynclinal origin were presented by Wynne-Edwards (1972), who, while noting that geosynclinal development may have begun around 1 750 m.y. ago just south of the Grenville front, emphasized that the Paleohelikian record of the province indicated subsiding continental crust in a miogeosynclinal belt and stated further that evidence of Helikian eugeosynclinal deposition was not to be found. He concluded, therefore, that the Grenville province developed on preexisting continental crust. This conclusion, coupled with his interpretation of two sets of pre-Grenvillian foliation planes trending easterly and north-northwesterly, trends that occur in the adjacent Superior and Churchill provinces, respectively, led him to suggest that the Grenville province represented a reworked section of the Canadian shield. The mechanism by which reworking was achieved was termed the millipede model of ductile tectonics, which Wynne-Edwards (1976) believed was the dominant process for continental reworking from about 2 500 to about 600 m.y. ago. A mantle heat cell produces ductile stretching of crust, leading to subsidence, miogeosynclinal-type sedimentation, intrusion of mafic complexes, and crustal melting. Unilateral movement of crust over the cell produces migration of the aforementioned phenomena. Crust that has moved away from the spreading center becomes compressed and undergoes anatexis, deformation, and metamorphism as isotherms rise, and the ductile crust thickens against cooler crust that was remote from the initial region of mantle upwelling and is now recognized as a cratonic foreland.

Plate-tectonic models

The concept of plate tectonics had been applied to Precambrian terranes at least as early as 1970 in a conference on African geology (Burke and Dewey, 1972), but Wynne-Edwards (1972, 1976) felt that inadequate geologic evidence for subduction/collision processes—e.g., ophiolites, blueschist metamorphism, island-arc sequences—ruled out its application to the Grenville province. The idea that the development of the Grenville province could have been governed by modern plate tectonics can be credited to Wilson (1962), who, in what has turned out to be a visionary paper, stated: "The fact that two provinces of the Canadian Shield have been together during post-Cambrian time does not necessarily mean that they were formed close together. . . ." It was not until 1971, however, that the possible role of plate tectonics in developing the Grenville province was stated categorically for the first time when Hess in Vine and Hess (1971) and Schenk (1971), without elaborating, independently suggested that the province was a product of continent–continent collision.

The first hard evidence in support of a collision origin was provided by geophysical data. Irving et al. (1972) interpreted paleomagnetic data to indicate that the Grenville province was at one time located many thousands of kilometers from the remainder of the shield and had subsequently collided and joined with it between 1 100 and 1 200 m.y. ago. The Grenville front was viewed as a strike-slip juncture produced by predominantly dextral strike-slip motion, though a component of motion perpendicular to the front may have been present; zones of cataclasis and mylonitization were explained by these motions. The possibility of a plate-tectonic origin for the province was examined from a geologic standpoint the following year by Dewey and Burke (1973). Like Irving et al. (1972), they too proposed a continent–continent collision at about 1 100 m.y. ago, but in contrast to Irving et al. they did not view the Grenville front as a collisional juncture, preferring instead to locate the suture beneath younger Appalachian cover. Their argument for doing so is based on the contention that the front cannot be a suture because rocks of the Labrador trough have been traced across the front into the Grenville province. According to them, the front simply marks the limit of reactivation associated with a former lithospheric slab that subducted northwestward beneath the Grenville province.

Dewey and Burke (1973) viewed the Grenville province as an analog of the Tibet plateau eroded to a deep level. The overriding Grenville plate, following collision, was subjected to considerable crustal thickening, a consequence of continued convergence following consumption of all oceanic lithosphere. The thickening, together with a higher than normal heat flow, commonly observed in overriding plates (Toksoz and Bird, 1977), induced partial melting of the lower crust, leading to partitioning of the whole crust into a relatively dense, lower refractory crust typified by granulitic rocks, anorthosites and gabbros, and a lighter, granite-rich upper crust. Isostatic uplift and concomitant erosion eventually led to exposure of the lower crust to produce the picture observed in the Grenville province today. A discrepancy in this model with respect to the origin of Grenville anorthosites is that the latter probably have an age of about 1 450 m.y. (Emslie, 1978), whereas Dewey and Burke (1973) suggested that they were formed in response to continental collision about 1 100 m.y. ago.

GRAVITY ANOMALIES AND COLLISION TECTONICS

It is a testimony to the complex nature of the Grenville province and to the difficulty in extrapolating modern-day and geologically recent plate-tectonic models back into the Precambrian that these first two plate-tectonic interpretations of the Grenville province (Irving et al., 1972; Dewey and Burke, 1973) should be diametrically opposed on the questions of where the collisional suture is located and the paleosubduction direction. It is in respect to answering these questions that gravity studies focusing on the most prominent negative gravity anomaly in the shield (Figure 2) are believed to have made a major contribution. That they were able to do so is in no small measure credited to the previous paleomagnetic investigations.

Irving et al. (1974), extending earlier paleomagnetic work, were able to define a broad intra-Grenville zone within which a collisional suture was predicted to lie. The zone lay south of regions of Grenville rocks considered to be metamorphic equivalents of rocks in older provinces, thus effectively ruling out the Grenville front itself as a suture line, and north of sampling sites used to determine Grenville poles (Figure 3). Irving et al. (1974) admitted that apparently there was no geologic evidence to support the suture hypothesis but pointed out that at the deep levels of crust now exposed any evidence of a former intervening ocean may have been removed. However, what is present is a belt of steep

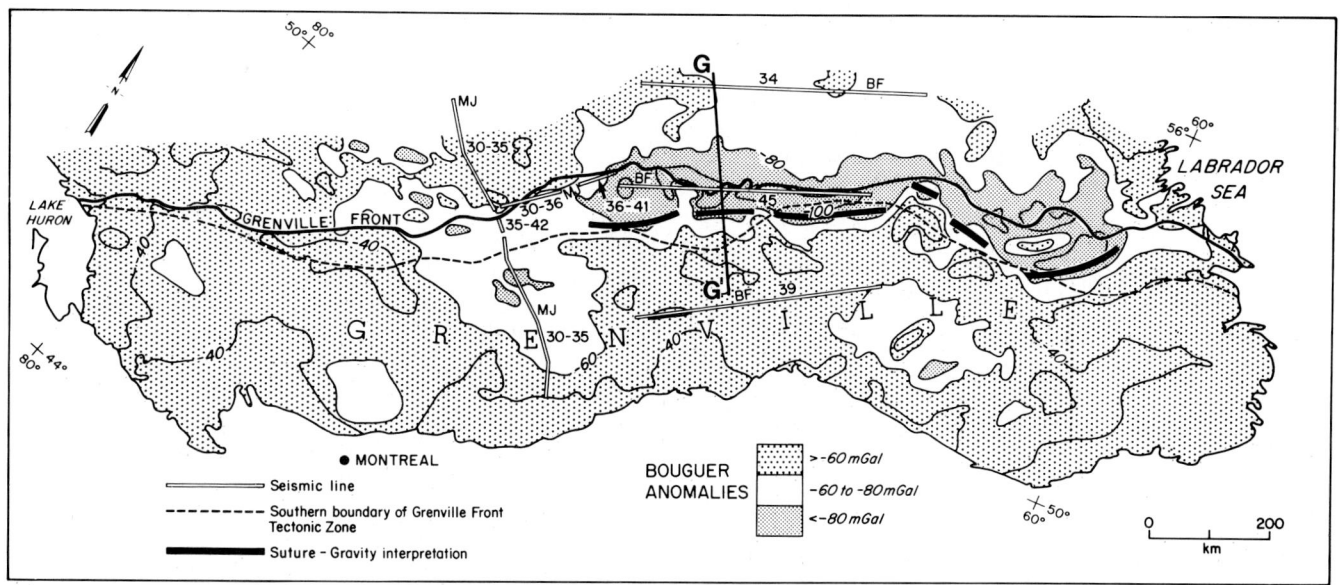

FIG. 2. Simplified Bouguer gravity-anomaly map of Grenville province and marginal areas of neighboring provinces; contour interval, 20 mGal. Seismic-line key: BF = Berry and Fuchs (1973); MJ = Mereu and Jobidon (1971). Crustal thicknesses are in kilometers. G–G' = line of gravity profile and model (Figure 4a, b).

gradients associated with the southeastern flank of a regional negative gravity anomaly that extends for almost 1 200 km from the vicinity of the Mistassini basin to the Labrador coast (Figure 2). The negative anomaly and a flanking complementary belt of relatively high gravity values to the southeast form a typical paired gravity-anomaly signature that has been recognized at other structural boundaries in the shield (Gibb and Thomas, 1976).

Tanner (1969) interpreted the signature to indicate that the upper 20 km of Grenville crust was 0.1 g/cm^3 more dense than that of the Superior, the contact between the contrasting upper crusts dipping about 45 degrees southeastward from the vicinity of the Grenville front; his model, constrained to be in isostatic equilibrium, also suggested that the Grenville crust was about 40 km thick compared to thinner (35 km) crust of the Superior province. Support for the model was provided by the seismic experiments of Mereu and Jobidon (1971) near the southwestern extremity of the anomaly (Figure 2), which indicated considerable crustal thickening south of the Grenville front along the strike of the

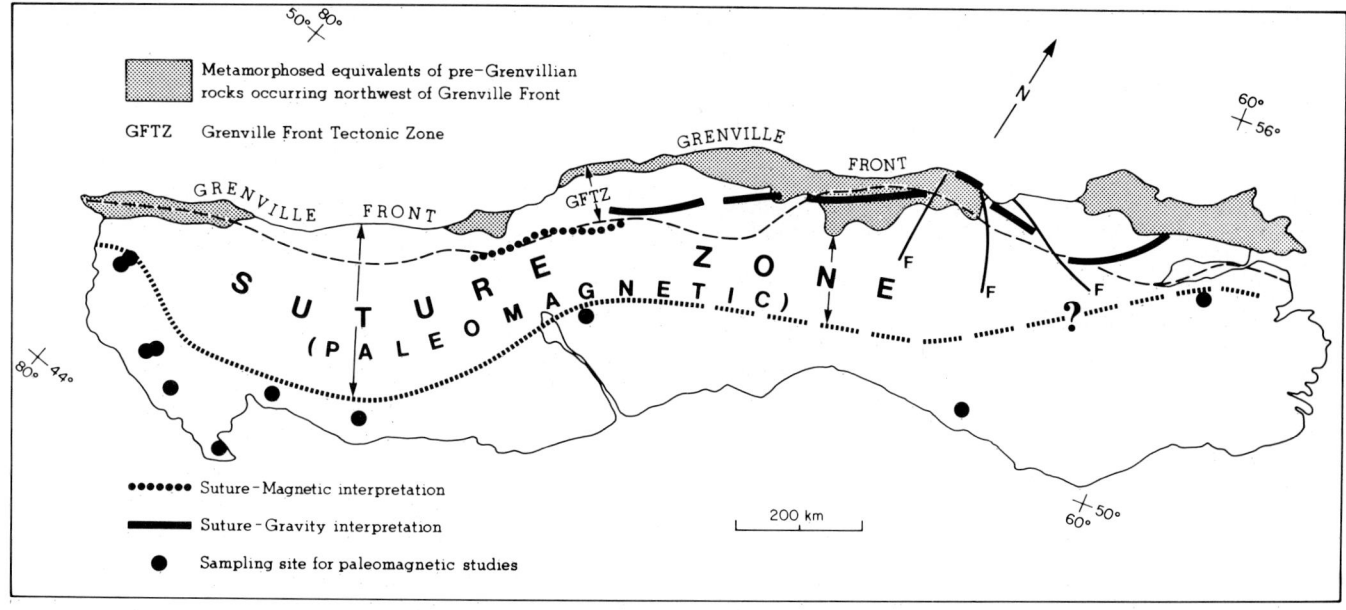

FIG. 3. Suture zone within Grenville province defined by paleomagnetic studies (Irving et al., 1974), and suture position based on gravity and magnetic anomalies (Thomas and Tanner, 1975). F = faults interpreted from gravity anomalies (Thomas, 1974).

gravity anomaly, leading them to conclude that the front is not a superficial structure but is probably a major "fossil" fault penetrating well into the mantle. Further seismic experiments by Berry and Fuchs (1973) confirmed the existence of major changes in crustal thickness across the front and the anomaly (Figure 2) postulated by Tanner (1969).

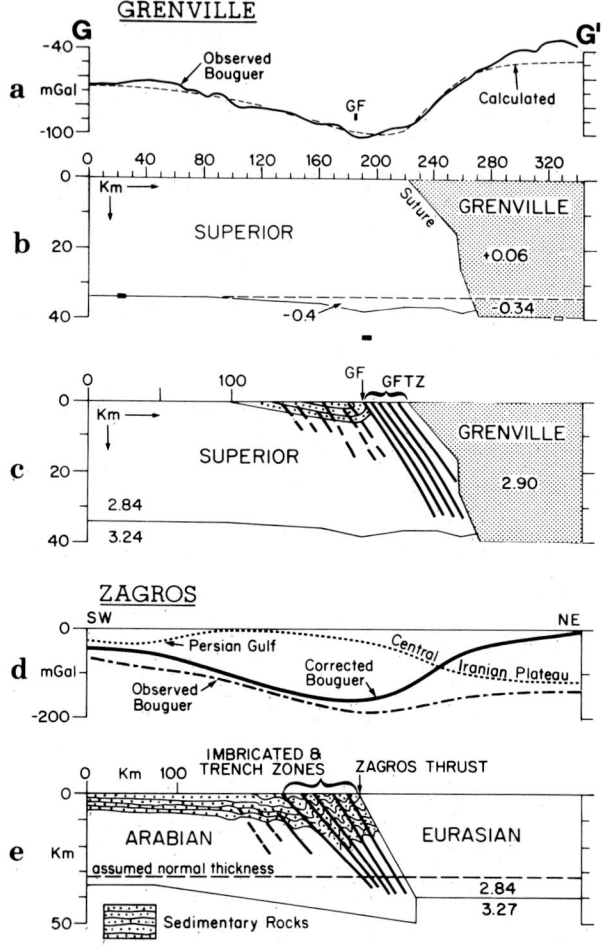

FIG. 4. a. Observed gravity profile G–G' (Figure 2) across Grenville front and gravity profile (calculated) corresponding to crustal model in b. GF = Grenville front.
  b. Crustal model interpreted from observed gravity profile shown in a. Horizontal bars represent positions of Mohorovičić discontinuity as determined by Berry and Fuchs (1973). Solid bars = true horizontal positions; hollow bar = position extrapolated from just outside limit of model (see Figure 2).
  c. Schematic crustal section based on b. Heavy continuous lines represent faults in Grenville front tectonic zone (GFTZ); heavy broken lines represent cleavages and faults in Grenville foreland belt.
  d. Observed Bouguer gravity profile across the Zagros region of Iran. Negative contributions of Persian Gulf and Central Iranian Plateau are also illustrated, as is a Bouguer profile corrected for these effects. After Bird (1978).
  e. Crustal model of Zagros region, based on gravity interpretation (after Bird, 1978). Heavy lines represent faults.

In light of these various indications for major structure and changes in crustal composition within or close to a zone postulated on paleomagnetic grounds to contain a collisional suture, Thomas and Tanner (1975) reinterpreted the gravity anomaly in terms of a cryptic suture. Simple two-dimensional gravity models were calculated for gravity profiles crossing the front, using the following assumptions and constraints: single-layer crusts having mean densities were assumed for both the Superior and Grenville provinces; the density contrast between the Superior crust and the mantle was assumed to be 0.4 g/cm^3; the thicknesses of the Superior crust (34 km) and Grenville crust (39 km) were constrained, using the seismic estimates obtained by Berry and Fuchs (1973); the crusts of both provinces were assumed to be in relative isostatic equilibrium. A typical gravity model is shown in Figure 4b. In this, a thin wedge of nonisostatic crust underlies the southeastern margin of the Superior province, a demand of the model in order to obtain a reasonable fit to the profile in the region of the negative anomaly. The wedge increases the thickness of the Superior crust to about 38 km, a value that falls considerably short of the 45 km predicted from the seismic experiment. The discrepancy results from using average crustal densities and could be eliminated by thickening the crust by an appropriate amount and increasing its density near its base. However, the geologic significance of such a model would differ little from that shown. The interesting point of the model is the density boundary that dips southeastward under the Grenville province from a position near the bottom of the steep gradient that forms the southeastern flank of the anomaly. This boundary is interpreted as a cryptic suture, and on the basis of this kind of model it has been possible to sketch the surface trace of the suture (Figures 2, 3).

Analogs supporting Grenville collision

Irving et al. (1974) noted that, because of the deep crustal levels now exposed, all remnants of any former intervening oceans may have been obliterated, thus destroying evidence that could have been critical to the Grenville collision hypothesis. In the absence of such evidence, the collision argument has been strengthened by the use of analogs.

An analog that demonstrates the collision significance of the paired gravity signature is provided in the geologic evolution and setting of the eastern margin of the West African craton and adjacent Pan African terrane, interpreted as a former passive and active continental margin, respectively, that collided approximately 600 m.y. ago (Black et al., 1979). Three geologic features, all regarded as key indicators of former subduction/collision processes, have been identified: ophiolites and high-pressure–low-temperature metamorphism in the supracrustal rocks of the craton, and a calc-alkaline batholith within the reactivated Pan African terrane. Of relevance to the collision-related origin proposed for the gravity anomalies along the Grenville front is the fact that a section of the West African suture zone coincides with a similar paired gravity anomaly, a collision origin for which seems highly probable in view of the associated geologic evidence for collision. The association of paired gravity anomalies and collision zones was discussed also by Fountain and Salisbury (1981). They noted that several exposed cross-sections of lower continental crust, regarded as being uplifted by obduction during continental collisions, are attended by a paired gravity signature similar to those observed at several other suture zones, among which they cited the Grenville example.

A second analog, the Zagros suture zone of Iran, is used to show that certain geologic features near the Grenville front, if not diagnostic of collision tectonics, do conform to an origin by such a mechanism. Specifically, it is proposed that miogeosynclinal sedimentary basins, the front itself, and the Grenville front tectonic zone fit the picture of a passive margin whose leading edge (represented by the margins of the Superior, Churchill, and Nain provinces) was tectonically deformed following collision with an active margin (Grenville orogen). Gravity modeling (Figure 4c) indicates that the Grenville front tectonic zone, whose southern boundary coincides closely with the proposed suture (Figure 3), has the gross density characteristics of the older Superior province and as such is regarded as an integral part of that province. The zone is characterized by strongly developed northeast-trending foliations and many parallel belts of cataclasis and mylonitization, low magnetic relief, and relatively thick crust. Metamorphic assemblages within it are indicative of the highest pressures recorded in the Grenville province and suggest that a former very deep level has been exposed by overthrusting onto the craton to the north (Wynne-Edwards, 1972). Thrusting has also produced thickening. Wynne-Edwards (1972) suggested that many of the Proterozoic basins lying north of the Grenville front, adjacent to the tectonic zone (Figure 1), represent the remains of foreland sequences related to the Grenville orogenic belt, although he noted that some also have had an earlier history as part of the Circum-Ungava geosyncline. It is speculated that close to the suture, within the Grenville front tectonic zone, upthrusting was most intense. The effect of this was to expose basement and remove most of the passive-margin sedimentary record, although some sedimentary rocks may have been transformed into metamorphic equivalents through deep burial at a stage when limited subduction of the leading edge may have operated. The Grenville front itself would represent the continentward limit of major thrusting.

In the Zagros region of Iran, the structural equivalent of the Grenville front tectonic zone occurs along the leading edge of the Arabian plate and takes the form of a belt with a maximum width of 120 km that includes the Imbricated and Trench zones of Haynes and McQuillan (1974). These zones were developed following late Cretaceous collision with the Eurasian plate (Figure 4e). Possibly the comparison should be restricted to the Imbricated zone alone, as trench mélange has not been identified within the Grenville front tectonic zone. The Imbricated zone lies along the leading edge of the miogeosynclinal wedge (Arabian plate) and has been tightly folded and thrust southwestward toward shelf and miogeosynclinal sedimentary rocks affected by relatively open folding (Simply folded belt). The Grenville front is analogous to the boundary between the Simply folded belt and the Imbricated zone, which is represented by a recumbent isoclinal fold or a fault scarp. Miogeosynclinal basins northwest of the front may be compared with the sedimentary successions of the Simply folded belt. An obvious difference in the regions is the degree of erosion.

A close similarity between the regions is observed also in crustal models interpreted from gravity anomalies by Thomas and Tanner (1975) for the Grenville province and by Bird (1978) for the Zagros (Figure 4c, e). The crusts associated with both passive plates have normal thicknesses that are similar (~34 km), and both thicken toward the suture, the Superior by 4 km and the Arabian crust dramatically by 17 km. The greater thickening of the latter is a product of tectonic shortening and thickening that is still in progress (Bird, 1978), and the crustal root is maintained isostatically by the overlying Zagros Mountains. An interesting feature of the Bouguer gravity profile across the Zagros suture is that if it is modified by applying corrections for the negative effects of the Persian Gulf and Central Iranian Plateau, the resulting signature is not unlike that along the Grenville front, a remarkable feature considering the great difference in ages of the two regions.

Timing of collision: Grenvillian or Hudsonian?

Collectively, paleomagnetic studies, the presence of a paired gravity signature along the Grenville front, and the various described analogs and characteristics of the geology provide a reasonably strong case for southeastward subduction and related collision approximately 1 000 m.y. ago (Irving et al., 1974) near the Grenville front. However, there is considerable uncertainty in regard to the timing of the proposed collision. During the time since the paleomagnetic data that led to the interpretation of the Grenville gravity anomaly as a collisional phenomenon were published by Irving et al. (1974), redating of rocks used to determine poles and new paleomagnetic data have led to the questioning of the validity of the two-plate collisional model. Morris and Roy (1977), and more recently Berger et al. (1979), claimed that the two-plate model is untenable and that if a collision was indeed the cause of the Grenvillian orogeny it probably occurred along a subduction zone associated with a suture that is now hidden beneath the Appalachians, as suggested by Dewey and Burke (1973).

Others who subscribe to a sub-Appalachian suture are Seyfert (1980), who viewed the Grenville province as a product of North America's underthrusting of Gondwana following collision 1 250 to 1 150 m.y. ago effected by southeastward subduction (the Grenville front may have been located beneath the northwestern margin of Gondwanaland), and Greenhouse and Bailey (1981), who, in one possible model to explain conductivity anomalies, proposed a suture some 450 km south of the front that originated in the manner proposed by Dewey and Burke (1973).

In view of these opinions, and because the latest geochronological and paleomagnetic data seemingly rule out a *Grenvillian collision* as an explanation for the gravity signature, it is appropriate to consider that it might be an artifact of an earlier collision. In this respect it is interesting to note that rocks of the anorthositic suite, which are relatively confined within the eastern part of the Canadian shield, are not restricted to the Grenville province but occur also in proximal areas of the Churchill and Nain provinces (Figure

1). If they were all intruded approximately 1 450 m.y. ago, as suggested by Emslie (1978), then it would suggest contiguity of the various provinces at that time. It suggests also that the anorthosites are not a refractory product of crustal segregation influenced by *Grenvillian* collision, as suggested by Dewey and Burke (1973) in their reactivation model. One anorthosite massif, in particular, occurring immediately southeast of the Otish Mountains basin (Figure 1), cuts perpendicularly across the steep gravity gradient related to the suture. If the massif has an age of 1 450 m.y., as predicted for Grenville province anorthosites, and if it bridges the suture, then collision prior to this date is suggested. This argument, of course, is only as good as the assumption that all Grenville anorthosites are 1 450 m.y. old; surprisingly, one major anorthosite body has been dated as 540 m.y. old (Higgins and Doig, 1977). Further, since the Grenville front tectonic zone truncates north-northwesterly structures of the Labrador trough, and the gravity anomaly borders the Superior and Churchill provinces, it is apparent that collision postdates the Hudsonian orogeny in the eastern Churchill province, which terminated about 1 750 m.y. ago (Stockwell, 1982) and which may itself have been a product of continental collision between the Superior and Churchill protocontinents (Kearey, 1976).

Could there have been a collision in the period 1 750 to 1 450 m.y. ago? The existence of a subduction system near the front about the beginning of this period has been suggested (Krogh and Davis, 1971; Baragar and Scoates, 1981), and there is a variety of evidence to suggest that a major tectonic event, possibly collision-related, took place during the same period. Wynne-Edwards (1972) believed that certain basins of the Grenville foreland belt may have had a pre-Grenvillian involvement in the history of the Circum-Ungava geosyncline and drew attention to dates indicative of a Hudsonian episode of intrusion and deformation that affected rocks of the Huronian Supergroup and Aillik Group. Brooks (1967) proposed that Huronian rocks were regionally metamorphosed about 1 700 m.y. ago. Various structural phenomena in the neighborhood of the front have been attributed to a Hudsonian event: Krogh and Davis (1971) related foliation, lineation, and mylonitization to a pre-1 600-m.y.-old event; Brocoum and Dalziel (1974) thought that mylonite foliation and isoclinal folding may have been produced by the Penokean orogeny.

Granite intrusion extending from 1 700 to 1 200 m.y. ago apparently is confined to the northwesternmost 80 km of the Grenville province in Ontario, and this pattern may extend for more than 600 km adjacent to and south of the front (Krogh and Davis, 1971). The longevity and geographic confinement of granite intrusion formed the basis of Krogh and Davis' proposal that the region of the Grenville front was located along an ancient plate boundary associated with a subduction zone. As far as can be determined, it is not known whether the granites are a product related to a subducting slab or are anatectic in origin, formed in response to thrusting associated with crustal thickening and shortening, as are the granites of the Himalayas, for example (Hamet and Allègre, 1976).

Krogh and Davis (1971) suggested that the Superior province had been rifted and developed a continental margin that accumulated sediments that were coeval with Huronian Supergroup rocks of the Southern province (~2 000 m.y. old) and rocks of the Mistassini basin and Labrador trough. This sedimentary wedge was later thrust against the Superior craton and experienced metamorphism and granite intrusion prior to 1 600 m.y. ago, during which period "the Front probably was a zone between a mobile recrystallizing sediment wedge and a stable craton at an ancient subduction zone"; details were not given.

Baragar and Scoates (1981), on the basis of a much wider study of the Circum-Superior belt, linked the sequences of the Mistassini and Otish Mountains basins with rocks of the belt and proposed that the Grenville front was associated with a northwestward-dipping subduction zone of Hudsonian age. Other evidence for large-scale movements of crustal blocks at this time is provided by Clark (1979), who has documented closure of a northeast-trending trough or sea between two cratons as having taken place within the southern terrane of the Nain province, just north of the Grenville front. Clark linked the Nain trough southwestward with the Labrador trough and proposed that both troughs closed in Hudsonian times; closure of the Labrador trough has been attributed to plate tectonics (Kearey, 1976).

Although critical evidence is obviously missing, it is tempting to speculate that the paired gravity-anomaly signature is a product of plate collision that occurred during the Hudsonian orogenic period or shortly thereafter. Under this hypothesis the Grenvillian orogeny might then be accommodated by a model involving northwestward subduction from a location within the Appalachians such as advanced by Dewey and Burke (1973). Certain lines of evidence are consistent with this. Brocoum and Dalziel (1974) concluded that a major orogenic event occurred approximately 1 000 m.y. ago with associated polyphase deformation, high-grade metamorphism, and igneous activity in the southeastern part of the Grenville province, but they noted that in the northwestern part there is evidence only of faulting, relatively brittle deformation, and a thermal event that reset isotopic clocks. In the northeastern part of the Grenville province a 1 457-m.y.-old gabbro within the Grenville front tectonic zone is unmetamorphosed and undeformed, indicating the absence of the Grenvillian orogeny (Stockwell, 1982). Stockwell indicated that the Grenvillian orogeny has not been recognized isotopically in the Grenville front tectonic zone, and because much work has been done on rocks from the zone he believed that it is unlikely that this orogeny occurred there or that it was neither widespread nor strongly developed; probably the Grenvillian orogeny terminated at or near the southeastern boundary of the Grenville front tectonic zone.

GRAVITY ANOMALIES AND THE ORIGIN OF ANORTHOSITE

The Grenville province and adjacent regions of the Churchill and Nain provinces form a belt of terrane that is host to perhaps 75 percent of the world's known anorthositic rocks (Wynne-Edwards, 1972) (Figures 1, 5). The origin of these rocks remains a controversial and problematical issue, so much so in fact that Philpotts (1981) considered that "the origin of large Precambrian anorthosite massifs remains one of the major unsolved problems of petrology." One contro-

FIG. 5. Bouguer gravity-anomaly pattern in the Grenville province and adjacent marginal areas with simplified map of rocks of the anorthositic suite superposed. Contour interval, 10 mGal. + and − signs denote relatively positive and negative gravity anomalies, respectively, that are coincident with or close to anorthositic rocks. "A" indicates areas of anorthositic rocks where relatively pure anorthosite is the dominant rock type.

versy concerns the "granitic" components (including mangerite, adamellite, syenite, and charnockite) of the anorthositic suite that are often found intimately associated with anorthositic and mafic elements. On one hand, these have been explained as a fractional differentiation product of the parent magma that produced the anorthositic components (e.g., Morse, 1969), while on the other hand it has been argued that "granite" and anorthosite are not comagmatic (e.g., Emslie, 1978). A related, and more fundamental, issue concerns the composition of the parent magma itself. A wide range of primary compositions spanning mafic to felsic have been proposed, including gabbroic anorthosite, gabbro, diorite, monzodiorite, quartz diorite, and quartz monzonite (Isachsen, 1969).

Gravity signatures of anorthosites

Although workers may differ radically in opinion concerning the composition of the parent magma, they are generally agreed that anorthosites are accumulations of magmatically differentiated plagioclase crystals (Isachsen, 1969). In this respect, the role of gravity differentiation whereby plagioclase crystals are separated from a fractionating gabbroic magma by sinking (e.g., Bowen, 1917) or floating (e.g., Morse, 1969) has received considerable attention. However, in the case of a gabbroic parent magma, such a differentiation mechanism would inevitably produce also significant quantities of associated mafic and ultramafic differentiates. It has been argued that one of the inconsistencies with this mode of origin for anorthosites is that the presence of large quantities of associated mafic and ultramafic differentiates has not been confirmed, either by field relationships or gravity studies (Isachsen, 1969). Coincidentally, the first anorthosites to be covered by gravity surveys were all associated with negative gravity anomalies: Simmons (1964) outlined negative anomalies over Adirondack anorthosites,

and Philpotts (1969) noted that several anorthosite bodies in the southwestern part of the Grenville province coincided with areas of highly negative Bouguer anomalies (Figure 5). Curiously, Philpotts also stated that a recently compiled but unpublished gravity map of Canada showed that, with one major exception, all of the anorthosite masses are associated with negative Bouguer anomalies. Yet Tanner (1969), using presumably the same database, demonstrated that gravity anomalies over anorthosite intrusions in the Grenville province east of longitude 73° W are nearly all positive and that their sources can be explained as large gabbroic masses with maximum thicknesses ranging from 5 to 15 km; Thomas (1974) presented a similar picture.

Generally, this close relationship between large positive gravity anomalies and rocks of the anorthositic suite within the Grenville province appears to have gone unnoticed. Often, where discussion of gravity anomalies occurs, authors have tended to emphasize the association with negative anomalies (e.g., Middlemost, 1970; Simmons and Hanson, 1978; Wiebe, 1980) and in some cases have not fully apprised themselves of the facts, Simmons and Hanson, for example, stating that anorthosites of the eastern part of the Grenville province are associated with small positive gravity anomalies; anomalies with amplitudes ranging from about 30 to 80 mGal and covering areas measuring tens of kilometers by tens of kilometers (Thomas, 1974) are hardly small. Kearey and Thomas (1979) attempted to show that the large gravity anomalies in the eastern part of the Grenville province can be meaningfully interpreted in terms of gabbroic masses produced comagmatically with anorthosite under the dominating influence of gravity differentiation; much of the remaining part of this paper is based on their publication.

A possible reason for the lack of attention paid to the prominent positive gravity anomalies of the Grenville province and their possible link to anorthosite genesis is the fact that many occur in terrane dominated by gabbroic rocks and where anorthosite is relatively restricted; the anorthosite

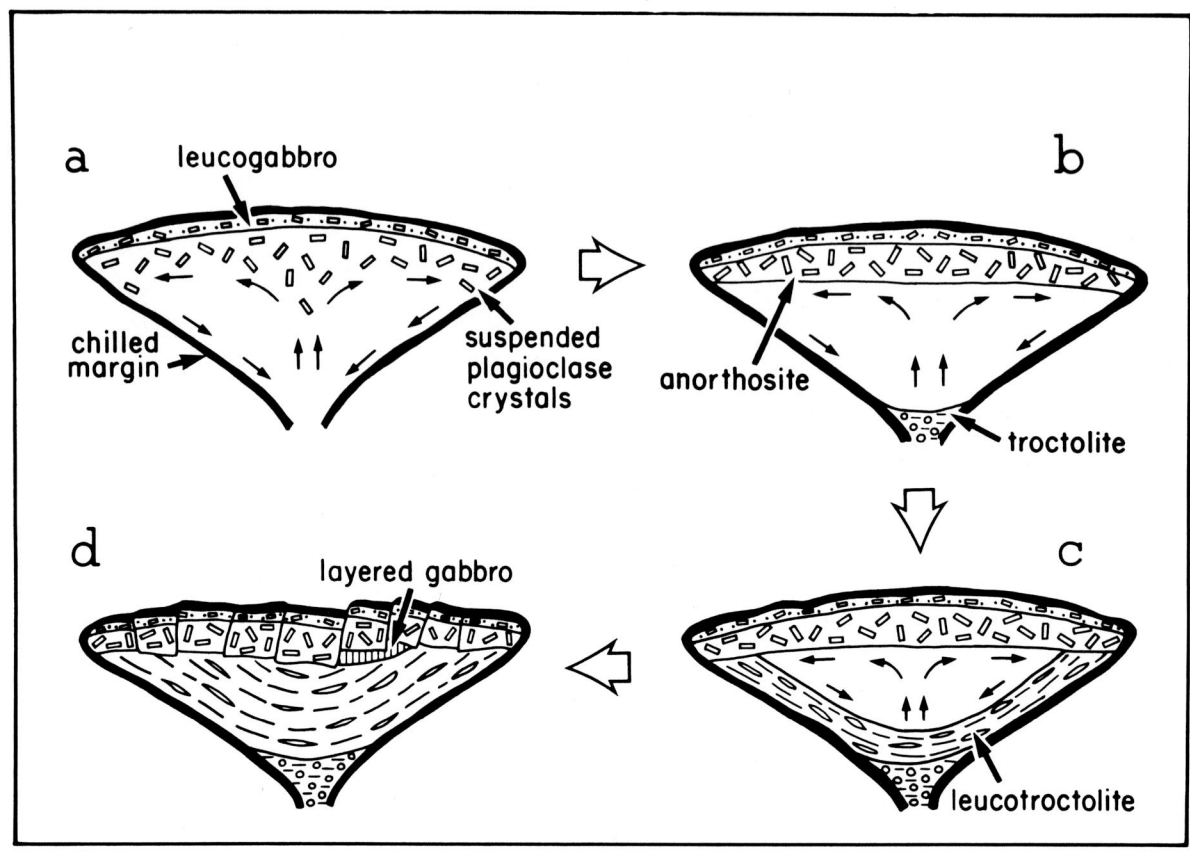

FIG. 6. Model for formation of anorthosite by plagioclase flotation (after Emslie, 1970).
  a. Plagioclase crystals suspended in basaltic liquid rise rapidly initially and trap interstitial liquid near roof of magma chamber, forming leucogabbro.
  b. Crystal accumulation proceeds more slowly, allowing diffusion and adcumulus growth of plagioclase and expulsion of interstitial liquid, leading to formation of anorthosite. Mafic crystals accumulate by gravity settling to form troctolite at bottom of chamber.
  c. Anorthosite formation is terminated, and bottom accumulation of olivine and plagioclase crystals produces leucotroctolite.
  d. Roof collapse results in contact between anorthosite and leucotroctolite. Residual liquid is squeezed to upper part of intrusion and trapped in pockets, forming gabbro.

connection may thus have been overlooked. In addition, geologic mapping has been largely of a reconnaissance nature because access to most of the bodies is by air-transport; conditions have not therefore been conducive to detailed studies that might have led to interest in the gravity anomalies. However, a somewhat unique opportunity to study the association of a large positive gravity anomaly and large areas of anorthositic rocks was provided following the mapping of the Lac Fournier and Romaine River massifs described by Sharma and Franconi (1975). The positive anomaly in the area coincides with anorthositic gabbro, which, according to the hypothesis advanced by Kearey and Thomas (1979), represents the lower levels of a former magma chamber in which anorthosite was produced by flotation at higher levels.

Plagioclase flotation

The hypothesis is supported by observations that anorthosite is found at high levels and overlies more mafic and dense rock types in several anorthositic intrusions. Some examples are the Michikamau intrusion, Labrador, where anorthosite is underlain by leucotroctolite and troctolite (Emslie, 1970); the Kalka intrusion, Australia, in which anorthosite passes downward into norite, followed by pyroxenite (Goode, 1977); the Mineral Lake intrusion, Wisconsin, where anorthosite and gabbroic anorthosite overlie anorthositic gabbro containing zones of pyroxenite and picrite (Olmsted, 1969). Flotation, operating in one form or another, is held to have played a role in the development of all these intrusions. Such stratigraphic sequences suggest that claims for the absence of large quantities of mafic and/or ultramafic rocks in association with anorthositic massifs (Isachsen, 1969; Philpotts, 1969), used as evidence against a gabbroic parent magma, may be somewhat overstated. An example of one flotation model is reproduced in Figure 6. In this, Emslie (1970) appealed to an initial basaltic magma in which sodic plagioclase crystals were suspended. At first, rapid accumulation of the crystals trapped interstitial liquid and developed the upper border group of gabbros and leucogabbros; this was followed by more gradual accumula-

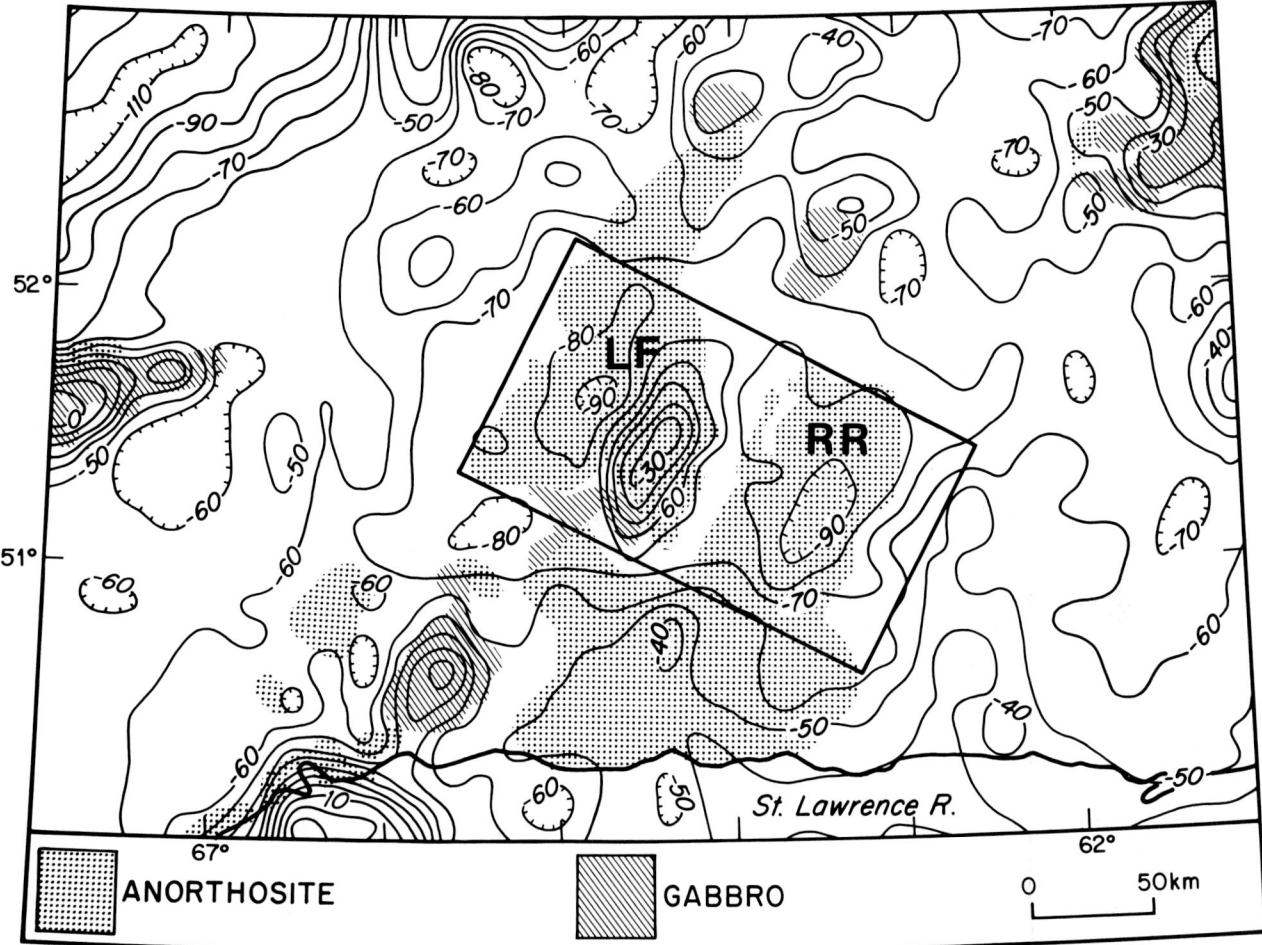

FIG. 7. Setting of Lac Fournier ("LF") and Romaine River ("RR") massifs with respect to regional Bouguer gravity field. Contour interval, 10 mGal. Closed contours with tick marks denote gravity lows. Contours are computer drawn and slightly smoothed.

tion, which allowed adcumulus growth of the plagioclase and anorthosite formation. Finally, bottom accumulation of olivine and more calcic plagioclase produced the layered series of leucotroctolite below the anorthosite, at the very base of which is developed more dense troctolite, which Emslie suggested accumulated in the central, deeper parts of a funnel-shaped chamber. The development of the Lac Fournier and Romaine River anorthosite massifs is considered to be analogous to this model.

Lac Fournier and Romaine River massifs

The setting of the Lac Fournier and Romaine River massifs with respect to the regional gravity field is shown in Figure 7, which illustrates also the common association of mainly gabbroic intrusions with positive gravity anomalies. The geology of the massifs is shown in more detail in Figure 8. The most striking feature of the gravity field associated with the two massifs is a large positive anomaly, amplitude 60 mGal, trending north-northeasterly along the eastern part of the Lac Fournier massif, an area henceforth referred to as the Magpie anorthosite. The latter anorthosite is relatively enriched in mafic minerals compared to the anorthosite in the western part of the Lac Fournier massif and in the Romaine River massif; it is in fact described as quite heterogeneous in compositon and texture and consists largely of gabbroic anorthosite (11–20 percent mafic minerals) and anorthositic gabbro (21–35 percent) with some gabbro (~35 percent) (Sharma and Franconi, 1975). The contact between the western and eastern parts of the massif is gradational. The gravity field over the flanking anorthosites (0–10 percent mafic minerals) takes the form of smaller negative anomalies that attain about −15 mGal and −20 mGal amplitude to the west and east of the Magpie anorthosite, respectively. This pattern of gravity anomalies reflects the pattern of rock densities provided by 1 378 rock samples. Kearey and Thomas (1979) adopted the following mean densities for their modeling of gravity anomalies: background density of rocks surrounding the massifs (mainly a variety of gneisses), 2.74 g/cm^3; anorthosite, 2.67 g/cm^3; gabbro, 3.00 g/cm^3.

The more mafic nature of the Magpie anorthosite and the close correlation of the area with the positive anomaly leave

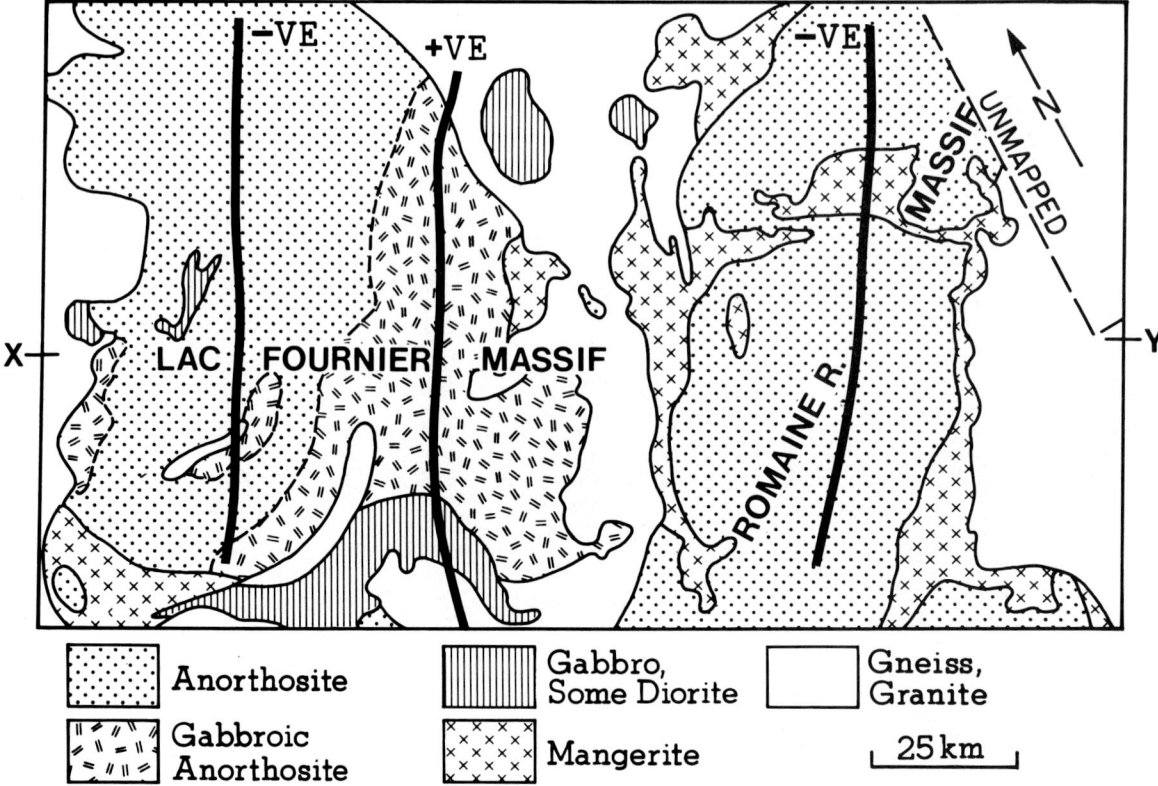

FIG. 8. Simplified geologic map of Lac Fournier and Romaine River massifs (after Sharma and Franconi, 1975). Axes of main negative and positive gravity anomalies are superposed.

little doubt as to the cause of the latter. However, the mean density of the Magpie anorthosite is only 0.10 g/cm^3 more dense than the background density, and this contrast is insufficient to reproduce the form of the anomaly. On the basis of gradient-amplitude ratios (Bott and Smith, 1958), Kearey and Thomas (1979) demonstrated that an equivalent point source of the causative structure is no deeper than 10 km, with the consequence that its density contrast is at least 0.21 g/cm^3; that is, its density is at least 2.95 g/cm^3. This information indicates that the mafic-rich heterogeneous mixture of gabbroic anorthosite, anorthositic gabbro, and gabbro of the Magpie anorthosite grades downward into increasingly more mafic varieties, ultimately probably becoming homogenized as a gabbro. In addition, there is the likelihood of ultramafic phases below the gabbro, but this is a problem that cannot be resolved by gravity studies. That gabbro is probably present at depth is suggested also by its occurrence at the surface within the mélange of rock types that constitute the Magpie anorthosite, and also as separate gabbroic bodies along the southern margin of that anorthosite; other smaller gabbroic bodies occur in the western part of the Lac Fournier massif. Wiebe (1980) had questioned whether the mafic intrusion (gabbro) interpreted by Kearey and Thomas (1979) to underlie the Magpie anorthosite was comagmatic with the anorthosites. The mixed lithologies present in the Magpie anorthosite, the fact that Sharma and Franconi (1975) stated that there is no sharp limit between this mafic portion of the Lac Fournier massif and more pure anorthosite in the western part of the massif, and the gravity anomaly/density requirements for gabbroic rocks at no great depth would seemingly indicate that the rocks in question are comagmatic.

Gravity models

A three-dimensional model of the interpreted gabbro produced by Kearey and Thomas (1979) has the general form of a saucer shape in its upper 5 km, thickening gradually toward its center at gentle dips of 10 to 15 degrees along its northern and eastern flanks and more steeply, up to 50 degrees, along western and southern flanks. At a depth of about 5 km the body increases in thickness abruptly, with the sides dipping at about 60 degrees, and below 8 km it divides into two roots that reach depths of 10.8 and 12.5 km.

A body with inward-dipping contacts was favored over one with outward-dipping contacts for the following reasons. First, indirect evidence for inward-dipping contacts is provided by numerous examples of mafic intrusions in which the geometry has been established; the Kiglapait intrusion (Morse, 1969) and Michikamau intrusion (Emslie, 1970) are just two examples. Further, Petraske et al. (1978), using an approach based on plane-strain plate theory, concluded that mafic intrusions emplaced at depths greater than about 10

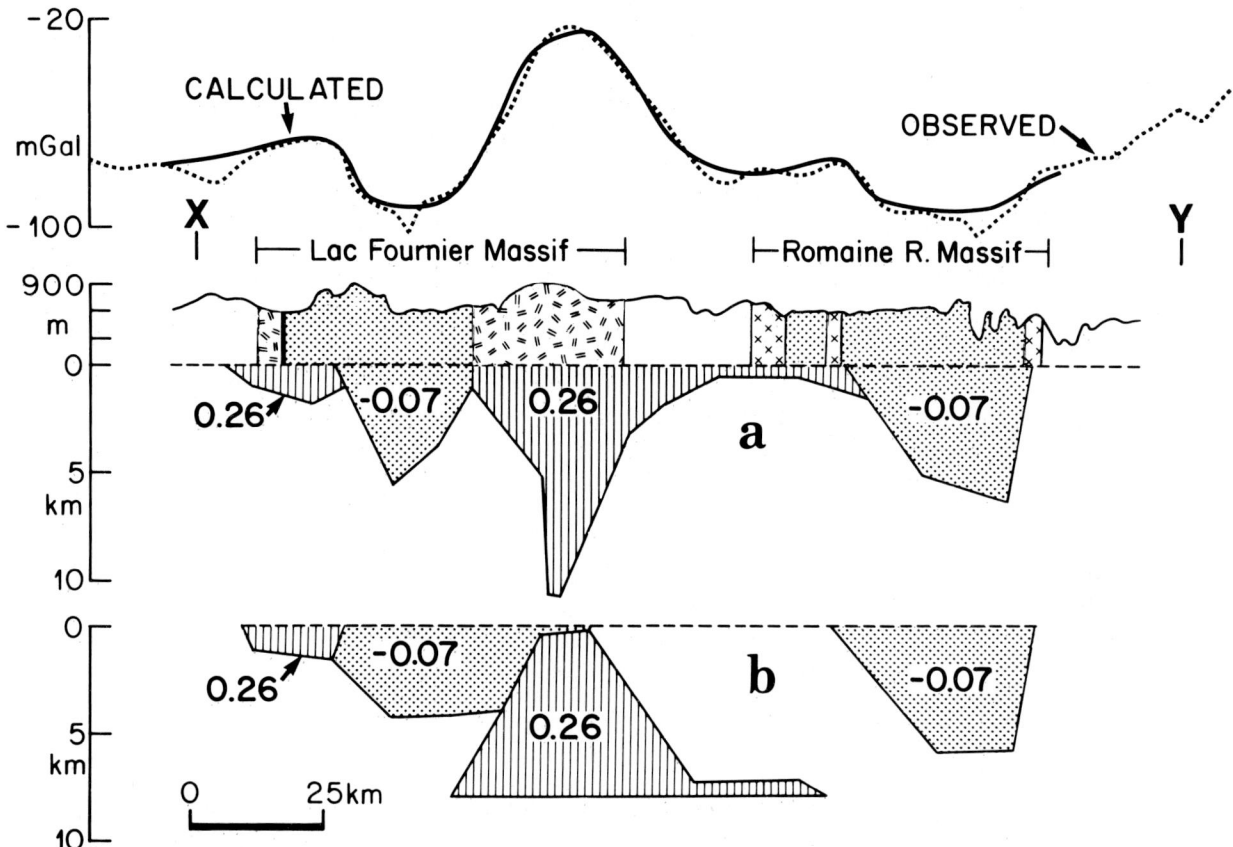

FIG. 9. Two-dimensional gravity models across Lac Fournier and Romaine River massifs; geologic explanation as in Figure 8.
a. Preferred model with inward-sloping contacts for gabbro underlying positive gravity anomaly.
b. Model with outward-sloping contacts.

km tend to be bowl shaped as opposed to laccolithic shaped for those intruded at shallower depths. Berg (1977) suggested that anorthositic rocks are intruded at depths in the range 12 to 22 km, so that on the basis of the results of Petraske et al. (1978) inward-dipping contacts are indicated.

Models based on two-dimensional gravity interpretation across the entire width of the Lac Fournier and Romaine River massifs are shown in Figure 9. The model in Figure 9a features the preferred inward-sloping contacts for the gabbroic body. Adjacent to the latter are two bodies of anorthosite that reach depths of about 5 km and 6 km, respectively, to the west and east. The anorthosites are separated from the gabbro by basement ridges. Stratigraphically, these anorthosites are at the same level as the upper part of the gabbro. If the hypothesis that they evolved by gravity differentiation and flotation is correct, tectonic adjustments resulting in relative uplift of the central gabbro must have occurred subsequent to crystallization. Evidence for tectonic discontinuities that would fit in with this picture is provided by a major topographic and magnetic lineament along the northwestern side of the Magpie anorthosite and another prominent topographic lineament along its southeastern margin. Alternatively, it is possible that the basement ridges in part are actually anorthositic rocks of about the same density as the country rocks and represent a gradation in composition from the more gabbroic mass interpreted to underlie the Magpie anorthosite to the lighter varieties that surround it. The model shown in Figure 9b, outward-dipping contacts for the gabbro, does not seem reasonable for reasons previously cited in support of inward-dipping contacts, and because the gabbroic mass is not connected to the Romaine River massif.

CONCLUSIONS

The ability of the gravity method to probe deeply into the crust has made a significant contribution to the understanding of geologic processes that shaped various features of the Grenville orogen. The presence of a paired gravity anomaly along the Grenville front, analyzed in the context of paleomagnetic, seismic, geochronological, and a variety of geologic data, almost certainly signifies a collision event effected by modern-type plate tectonics. The collisional suture, now cryptic, is interpreted to coincide roughly with the southeastern margin of the Grenville front tectonic zone. Paleosubduction was southeastward beneath the Grenville province. The exact timing of the collision is debatable. The initial set of paleomagnetic data that was instrumental in first drawing attention to the possible collision significance of the gravity signature suggests a Grenvillian date (~1 000 m.y.), but a Hudsonian (~1 750 m.y.) or slightly younger date cannot be ruled out.

A prominent positive gravity anomaly associated with a large region of anorthositic terrane that is relatively enriched in mafic minerals has been attributed to a large underlying body of gabbro having inward-dipping contacts. Large bodies of relatively pure anorthosite occur on either side of the mafic anorthosite. The overall picture that emerges from the surface geology and modeling of the gravity data is regarded as consistent with the hypothesis that the gabbro body represents the lower levels of a former magma chamber and was formed under the dominating influence of gravity differentiation of a parent gabbroic magma. The adjacent anorthosites, which subsequent to crystallization probably were lowered tectonically to the same level as the upper part of the gabbro, are viewed as having formed by flotation of plagioclase in the upper levels of the same magma chamber. The common association of prominent positive anomalies with anorthositic terranes in the Grenville province indicates that simple gravity differentiation of a gabbroic magma involving plagioclase flotation was a major influence in the development of anorthosites in these particular terranes.

Grenville anorthositic intrusions associated with negative gravity anomalies may have originated in similar fashion to those associated with positive anomalies. If they did, their negative gravity signature indicates that they were subsequently separated from their gabbroic roots. Such a possibility was raised by Wynne-Edwards (1972), who suggested that deformation (probably Grenvillian) has detached the more deformed Grenville anorthosites from their roots.

ACKNOWLEDGMENTS

I thank M. R. Dence and R. A. Gibb of the Earth Physics Branch for reviewing the manuscript and two anonymous referees for their comments.

REFERENCES

Baragar, W.R.A., and Scoates, R.F.J., 1981, The Circum-Superior belt: A Proterozoic plate margin?, *in* Kröner, A., Ed., Precambrian plate tectonics: Elsevier, 297–330.

Berg, J. H., 1977, Regional geobarometry in the contact aureoles of the anorthositic Nain Complex, Labrador: J. Petrol., **18**, 399–430.

Berger, G. W., York, D., and Dunlop, D. J., 1979, Calibration of Grenvillian palaeopoles by $^{40}Ar/^{39}Ar$ dating: Nature, **277**, 46–48.

Berry, M. J., and Fuchs, K., 1973, Crustal structure of the Superior and Grenville provinces of the northeastern Canadian Shield: Seismol. Soc. Am. Bull., **63**, 1393–1432.

Bird, P., 1978, Finite element modeling of lithosphere deformation: the Zagros collision orogeny: Tectonophysics, **50**, 307–336.

Black, R., Caby, R., Moussine-Pouchkine, A., Bayer, R., Bertrand, J. M., Boullier, A. M., Fabre, J., and Lesquer, A., 1979, Evidence for late Precambrian plate tectonics in West Africa: Nature, **278**, 223–227.

Bott, M.H.P., and Smith, R. A., 1958, The estimation of the limiting depth of gravitating bodies: Geophys. Prospect., **6**, 1–10.

Bowen, N. L., 1917, The problem of the anorthosites: J. Geol., **25**, 209–243.

Brocoum, S. J., and Dalziel, I.W.D., 1974, The Sudbury Basin, the Southern Province, the Grenville Front, and the Penokean Orogeny: Geol. Soc. Am. Bull., **85**, 1571–1580.

Brooks, E. R., 1967, Multiple metamorphism along the Grenville front, north of Georgian Bay, Ontario: Geol. Soc. Am. Bull., **78**, 1267–1280.

Burke, K. C., and Dewey, J. F., 1972, Orogeny in Africa, *in* Dessauvagie, T.F.J., and Whiteman, A. J., Eds., African geology, Ibadan 1970: Univ. Ibadan, Geol. Dep., 583–608.

Clark, A.M.S., 1979, Proterozoic deformation and igneous intrusions in part of the Makkovik sub-province, Labrador: Precambrian Res., **10**, 95–114.

Dewey, J. F., and Burke, K.C.A., 1973, Tibetan, Variscan, and Precambrian basement reactivation: Products of continental collision: J. Geol., **81**, 683–692.

Dietz, R. S., 1966, Passive continents, spreading sea floors, and collapsing continental rises: Am. J. Sci., **264**, 177–193.

Dimroth, E., Baragar, W.R.A., Bergeron, R., and Jackson, G. D., 1970, The filling of the Circum-Ungava geosyncline, *in* Baer, A. J., Ed., Basins and geosynclines of the Canadian Shield: Pap. 70-40, Geol. Surv. Can., 45–142.

Doig, R., 1977, Rb-Sr geochronology and evolution of the Grenville province in northwestern Quebec, Canada: Geol. Soc. Am. Bull., **88**, 1843–1856.

Emslie, R. F., 1970, The geology of the Michikamau intrusion, Labrador: Pap. 68-57, Geol. Surv. Can.

——— 1978, Anorthosite massifs, rapakivi granites, and late Proterozoic rifting of North America: Precambrian Res., **7**, 61–98.

Fountain, D. M., and Salisbury, M. H., 1981, Exposed cross-sections through the continental crust: implications for crustal structure, petrology, and evolution: Earth Planet. Sci. Lett., **56**, 263–277.

Gibb, R. A., and Thomas, M. D., 1976, Gravity signature of fossil plate boundaries in the Canadian Shield: Nature, **262**, 199–200.

Goode, A.D.T., 1977, Flotation and remelting of plagioclase in the Kalka intrusion, central Australia: Petrological implications for anorthosite genesis: Earth Planet. Sci. Lett., **34**, 375–380.

Greenhouse, J. P., and Bailey, R. C., 1981, A review of geomagnetic variation measurements in the United States: implications for continental tectonics: Can. J. Earth Sci., **18**, 1268–1289.

Hamet, J., and Allègre, C.-J., 1976, Rb-Sr systematics in granite from central Nepal (Manaslu): Significance of the Oligocene age and high $^{87}Sr/^{86}Sr$ ratio in Himalayan orogeny: Geology, **4**, 470–472.

Haynes, S. J., and McQuillan, H., 1974, Evolution of the Zagros suture zone, southern Iran: Geol. Soc. Am. Bull., **85**, 739–744.

Higgins, M. D., and Doig, R., 1977, 540-Myr-old anorthosite complex in the Grenville Province of Quebec, Canada: Nature, **267**, 40–41.

Irving, E., Emslie, R. F., and Ueno, H., 1974, Upper Proterozoic paleomagnetic poles from Laurentia and the history of the Grenville structural province: J. Geophys. Res., **79**, 5491–5502.

Irving, E., Park, J. K., and Roy, J. L., 1972, Paleomagnetism and the origin of the Grenville Front: Nature, **236**, 344–346.

Isachsen, Y. W., 1969, Origin of anorthosite and related rocks—a summarization, *in* Isachsen, Y. W., Ed., Origin of anorthosite and related rocks: Mem. 18, New York State Mus. and Sci. Serv., 435–445.

Kearey, P., 1976, A regional structural model of the Labrador Trough, northern Quebec, from gravity studies, and its relevance to continent collision in the Precambrian: Earth Planet. Sci. Lett., **28**, 371–378.

Kearey, P., and Thomas, M. D., 1979, Interpretation of the gravity field of the Lac Fournier and Romaine River anorthosite massifs, eastern Grenville Province: significance to the origin of anorthosite: J. Geol. Soc. London, **136**, 725–736.

Krogh, T. E., and Davis, G. L., 1971, The Grenville Front interpreted as an ancient plate boundary: Carnegie Inst. Washington Year Book, **70**, 239–240.

Mereu, R. F., and Jobidon, G., 1971, A seismic investigation of the crust and Moho on a line perpendicular to the Grenville front: Can. J. Earth Sci., **8**, 1553–1583.

Middlemost, E.A.K., 1970, Anorthosites: A graduated series: Earth-Sci. Rev., **6**, 257–265.

Morris, W. A., and Roy, J. L., 1977, Discovery of the Hadrynian polar track and further study of the Grenville problem: Nature, **266**, 689–692.

Morse, S. A., 1969, Layered intrusions and anorthosite genesis, *in* Isachsen, Y. W., Ed., Origin of anorthosite and related rocks: Mem. 18, New York State Mus. and Sci. Serv., 175–187.

Olmsted, J. F., 1969, Petrology of the Mineral Lake intrusion, northwestern Wisconsin, *in* Isachsen, Y. W., Ed., Origin of anorthosite and related rocks: Mem. 18, New York State Mus. and Sci. Serv., 149–161.

Osborne, F. F., and Morin, M., 1962, Tectonics of part of the Grenville subprovince in Quebec, *in* Stevenson, J. S., Ed., The tectonics of the Canadian Shield: Spec. Publ. 4, R. Soc. Can., 118–143.

Petraske, A. K., Hodge, D. S., and Shaw, R., 1978, Mechanics of emplacement of basic intrusions: Tectonophysics, **46**, 41–63.

Philpotts, A. R., 1969, Parental magma of the anorthosite–mangerite suite, *in* Isachsen, Y. W., Ed., Origin of anorthosite and related rocks: Mem. 18, New York State Mus. and Sci. Serv., 207–212.

——— 1981, A model for the generation of massif-type anorthosites: Can. Mineral., **19**, 233–253.

Schenk, P. E., 1971, Southeastern Atlantic Canada, northwestern Africa, and continental drift: Can. J. Earth Sci., **8**, 1218–1251.

Seyfert, C. K., 1980, Paleomagnetic evidence in support of a middle Proterozoic (Helikian) collision between North America and Gondwanaland as a cause of the metamorphism and deformation in the Adirondacks: Summary: Geol. Soc. Am. Bull., pt. 1, **91**, 118–120.

Sharma, K.N.M., and Franconi, A., 1975, Magpie, Saint-Jean, Romaine Rivers area (Grenville 1970): Geol. Rep. 163, Quebec Dep. Nat. Resources.

Shillibeer, H. A., and Cumming, G. L., 1956, The bearing of age determination on the relation between the Keewatin and Grenville provinces, in Thomson, J. E., Ed., The Grenville Province: Spec. Publ. 1, R. Soc. Can., 54–73.

Simmons, E. C., and Hanson, G. N., 1978, Geochemistry and origin of massif-type anorthosites: Contrib. Mineral. Petrol., **66**, 119–135.

Simmons, G., 1964, Gravity survey and geological interpretation, northern New York: Geol. Soc. Am. Bull., **75**, 81–98.

Stockwell, C. H., 1982, Proposals for time classification and correlation of Precambrian rocks and events in Canada and adjacent areas of the Canadian Shield. Part I: Time classification of Precambrian rocks and events: Pap. 80–19, Geol. Surv. Can.

Tanner, J. G., 1969, A geophysical interpretation of structural boundaries in the eastern Canadian Shield: Ph.D. thesis, Univ. Durham.

Thomas, M. D., 1974, The correlation of gravity and geology in southeastern Quebec and southern Labrador: Gravity Map Ser. nos. 64–67, 96–98, Can. Dep. Energy, Mines Resources, Earth Phys. Branch.

Thomas, M. D., and Tanner, J. G., 1975, Cryptic suture in the eastern Grenville Province: Nature, **256**, 392–394.

Toksoz, M. N., and Bird, P., 1977, Formation and evolution of marginal basins and continental plateaus, in Talwani, M., and Pitman, W. C., III, Eds., Island arcs, deep sea trenches and back-arc basins: Maurice Ewing Ser. 1, Am. Geophys. Union, 379–393.

Vine, F. J., and Hess, H. H., 1971, Sea-floor spreading, in Maxwell, A. E., Ed., The sea, Vol. 4 (New concepts of sea floor evolution), Part III (Concepts): Wiley Intersci., 587–622.

Wiebe, R. A., 1980, Anorthositic magmas and the origin of Proterozoic anorthosite massifs: Nature, **286**, 564–567.

Wilson, J. T., 1962, The effect of new orogenetic theories upon ideas of the tectonics of the Canadian Shield, in Stevenson, J. S., Ed., The tectonics of the Canadian Shield: Spec. Publ. 4, R. Soc. Can., 174–180.

Wynne-Edwards, H. R., 1972, The Grenville Province, in Price, R. A., and Douglas, R.J.W., Eds., Variations in tectonic styles in Canada: Spec. Pap. 11, Geol. Assoc. Can., 263–334.

———1976, Proterozoic ensialic orogenesis: The millipede model of ductile plate tectonics: Am. J. Sci., **276**, 927–953.

Interpretation of gravity and magnetic data, central and eastern Brazil

Nicolau L.E. Haralyi* and Yociteru Hasui‡

ABSTRACT

The Bouguer gravity-anomaly and total-intensity magnetic-anomaly maps covering large areas of central and eastern Brazil define many major structures of the Precambrian basement. Archean granite–gneiss terranes with intervening greenstone belts and high-grade metamorphic zones occur in a series of crustal blocks separated by major discontinuities. This structural pattern essentially was not modified by the Proterozoic polycyclic thermo-tectonic processes and Upper Jurassic–Tertiary reactivation of the area. Rather, the Archean block structure greatly influenced the deformation involved in these later events. The magnetic-anomaly data are especially useful in mapping regional fault systems with lengths measured in hundreds of kilometers and mapping of mafic–ultramafic and alkaline intrusions. The distribution of many major mineral deposits is closely related to the major regional crustal structure interpreted from the geophysical data, and thus the regional geophysical data can be used to delimit regional mineralized domains.

INTRODUCTION

Geophysical information over a large part of central and eastern Brazil has been acquired since 1950 by many institutions and companies. Gravity surveys have been carried out mainly by the Brazilian Institute of Geography and Statistics (FIBGE) since 1950; the National Observatory of the National Council for Technological and Scientific Development (CNPq/ON) since 1971; the National Department for Mineral Production of the Ministry of the Mines and Energy/Mineral Resources Research Co. Agreement (DNPM/CPRM), 1979–80; the Brazilian petroleum enterprise (Petrobras) since 1950; the National Council for Technological and Scientific Development/Office de la Recherche Scientifique et Technique d'Outre Mer/Institute of Geosciences of the University of São Paulo/National Department for Mineral Production Agreement (CNPq/ORSTOM/IGUSP/DNPM), 1978–79; and the Institute of Astronomy and Geophysics, University of São Paulo (IAGUSP), 1977–78.

Aeromagnetic surveys have been carried out by joint missions of the Brazilian and West German (CGBA) governments in the States of Minas Gerais and Espirito Santo, 1971–75; and the Brazilian and the Canadian governments (PGBC) in the States of Goiás, Pará, and Mato Grosso, 1978–80. DNPM/CPRM carried out surveys in southeastern Goiás areas, and DNPM and Consulting and Aerial Surveys (ENCAL) in the State of Rio de Janeiro, 1978.

The purpose of this paper is to investigate regional structural features as indicated by regional gravity and magnetic anomalies and to relate the distribution of mineral deposits to these anomalies. The geological information used in this analysis is portrayed on a number of 1:1 000 000-scale maps and a compilation map of Brazil at a scale of 1:2 500 000 (Schobbenhaus Filho et al., 1981).

REGIONAL GEOLOGY

Archean granite–gneiss terranes in central and eastern Brazil are made up chiefly of migmatites, gneisses, and granites with scattered inliers of schists, quartzites, itabirites, marbles, calc-silicate rocks, amphibolites, and other rocks (Inda and Barbosa, 1978; Berbert, 1980; Cordani and Neves, 1982; Hasui, 1982; Mascarenhas and Sá, 1982). Archean greenstone belts, which are important sources of gold and maganese, have been described in the States of Goiás, Minas Gerais and Bahia (Mascarenhas, 1979; Saboia, 1979; Teixeira and Danni, 1978; Danni et al., 1981; Barreira and Dardenne, 1981; Montalvão et al., 1982; Schorscher et al., 1982). The granite–gneiss terranes are separated by belts of high-grade felsic to ultramafic granulites (Chodhuri et al., 1978; Wernick and Almeida, 1979; Oliveira, 1980; Lindenmayer, 1981; Sighinolfi et al., 1981; Oliveira et al., 1982; Costa and Mascarenhas, 1982; Hasui et al., 1982). These belts generally exhibit effects of repeated Proterozoic reactivation as shown by migmatization and retrograde metamor-

---

*Formerly CNPq/Observatorio Nacional, Rua General Bruce 586, CEP 20091, Rio de Janeiro, RJ, Brazil; presently P.O. Box 19026, CEP 04599, São Paulo, SP, Brazil.
‡Instituto de Pesquisas Tecnológicas do Estado de São Paulo, P.O. Box 7141, CEP 01000, São Paulo, SP, Brazil.

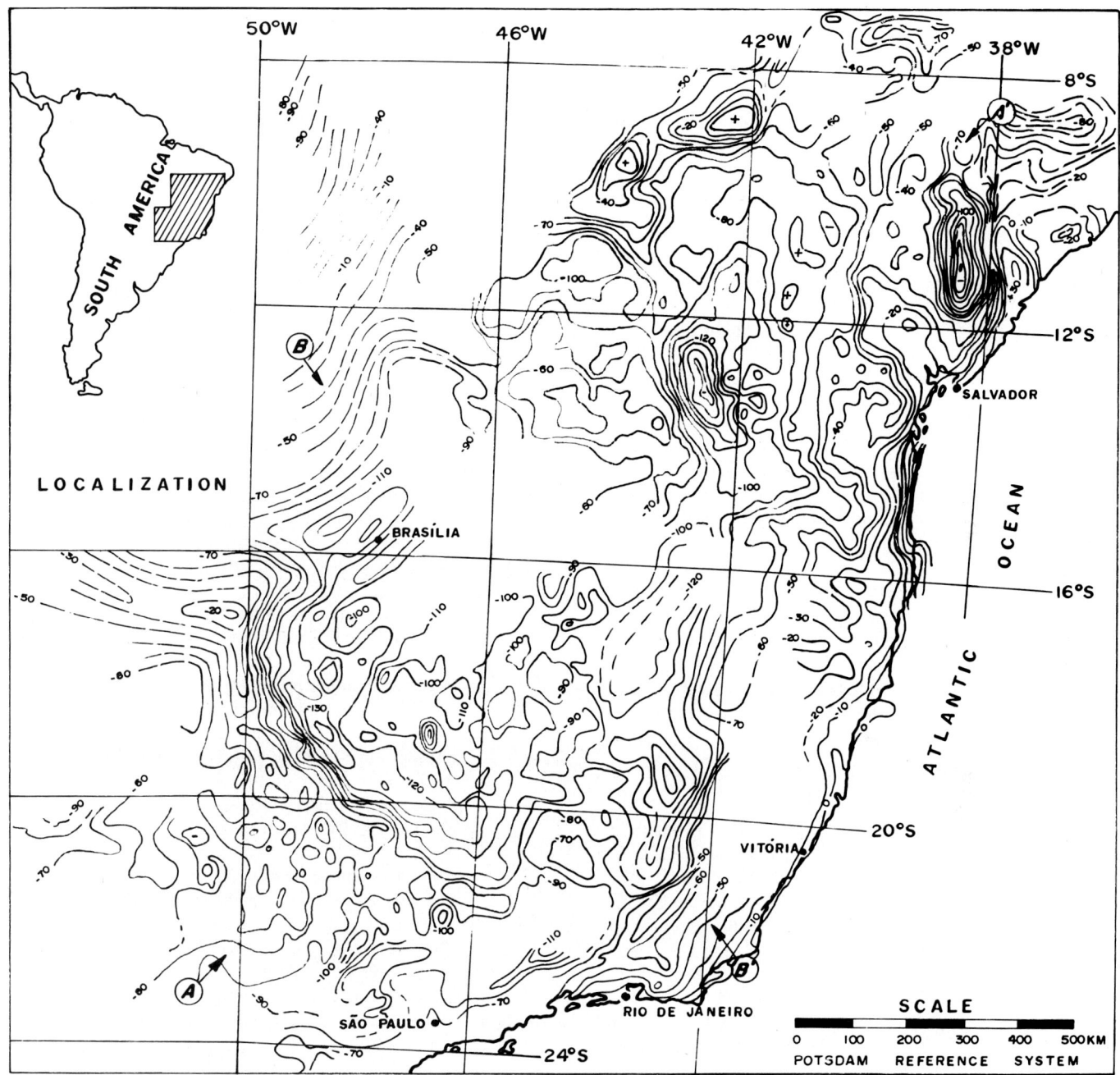

FIG. 1. Bouguer gravity-anomaly map of parts of central and eastern Brazil. Contour interval, 10 mGal. Density = 2.67 g/cm³. (Modified from Haralyi and Hasui, 1982.)

phism to amphibolite and greenschist facies. In addition, the belts are faulted and cut by felsic intrusives.

Radiometric ages from rocks in the granite–gneiss terranes are as old as 3.3–3.0 b.y., but most are related to the younger Jequié event (2.7 ± 0.2 b.y.) (Sá et al., 1976; Cordani and Iyer, 1979; Cordani and Teixeira, 1979; Tassinari and Montalvão, 1980; Hasui et al., 1980; Neves et al., 1980; Wernick et al., 1981; Teixeira, 1982). The distribution of granite–gneiss terranes, with intervening greenstone belts and belts of high-grade metamorphic rocks, clearly defines a blocklike structure. Individual blocks are bounded by major geophysical anomalies; this structural pattern is due to the Late Archean Jequié event (Almeida et al., 1980; Haralyi and Hasui, 1981, 1982).

During the Proterozoic Era, supracrustal sequences were laid down on the Archean rocks, granite intrusions were emplaced, and migmatization and retrograde metamorphism were locally overprinted. In addition, regional imprints, strike-slip faults and shear zones, and many types of mineral deposits are related to the progressively younger Transamazonian (2.0 ± 0.2 b.y.), Uruassuan (1 050 ± 150 m.y.), and Brasilian (550 ± 150 m.y.) events (Cordani and Teixeira,

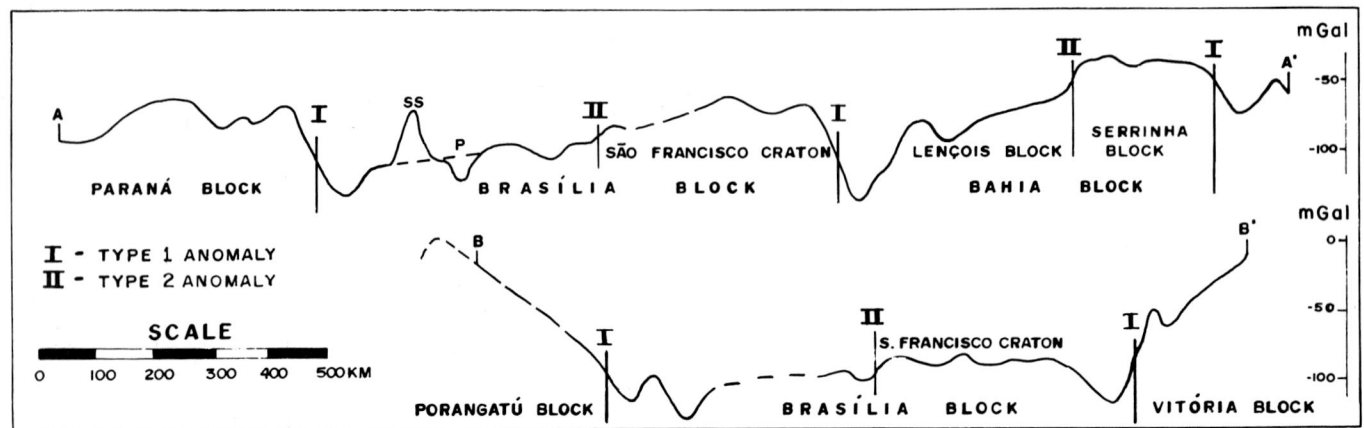

FIG. 2. Gravity-anomaly profiles A–A' and B–B'. For location, see Figure 1.

1979; Hasui et al., 1980; Neves et al., 1980; Teixeira, 1982). Mobile belts developed around the Brasilian-age São Francisco craton (Almeida et al., 1976; Almeida, 1977; Marini et al., 1977; Hasui et al., 1978; Hasui, 1982).

Extensive Phanerozoic sedimentary rocks cover many of the Precambrian units. During Late Jurassic and Cretaceous time, the region was again reactivated with concurrent faulting, basaltic volcanism, alkaline intrusions, kimberlitic-affinity diatremes, taphrogenic-basin development, and the formation of mineral deposits (Almeida, 1967; Almeida et al., 1980).

GRAVITY DATA

Systematic gravity surveys presently cover Bahia State and partially extend over Rio de Janeiro, Minas Gerais, São Paulo, and Goiás States (E. P. Hoyler, unpublished information; Gama, 1971, 1972; Haralyi, 1978; Gomes and Motta, 1978; Blitzkow et al., 1980; Escobar, 1980; Escobar and Santos, 1980; Lesquer et al., 1981).

The Bouguer gravity-anomaly map (Figure 1) prepared by Haralyi and Hasui (1982) was computed using a uniform rock density of 2.67 g/cm^3. Systematic errors are introduced in this map by charnockites and granulites of the high-grade metamorphic rock terranes that have relatively greater densities ranging from 2.69 to 2.86 g/cm^3 (Michale, 1982). For example, at a station 500 m above sea level in a high-grade terrane with an average density of 2.8 g/cm^3, the anomaly values shown in Figure 1 will be 2.73 mGal too negative. Terrain corrections were carried out for most of the gravity stations; they average less than 0.5 mGal but reach 5 mGal in areas of rugged topography. The map was machine contoured at a 10-mGal interval with a 5- × 5-km moving-average smoothing filter. Dashed isogal contours are used where less than one station per 900 km^2 is available.

The Bouguer gravity-anomaly field of central and eastern Brazil is characterized by the following features (Figures 1 and 2):

1. Elongated zones of negative anomalies, mostly 100 to 200 km wide with values less than −120 mGal. Their areal distribution is shown in Figure 3. They are interpreted to be zones of crustal thickening.

2. Narrow belts of high gradients that are interpreted as representing major lateral crustal discontinuities. The belts are classified as:

   a. Type 1 anomalies, which have gradients of 1.0 to 1.5 mGal/km with total amplitudes greater than 40 mGal and are adjacent to elongated zones of negative anomalies. This type of anomaly indicates major crustal discontinuities of low-angle dip, which separate high-grade terranes overthrusted onto granite–gneiss regions. Figure 3 shows the Archean crustal units, termed the Brasilia, Bahia, Araguacema, Paraná, São Paulo, and Vitória blocks (Haralyi and Hasui, 1982).

   b. Type 2 anomalies, which are similar to type 1 but have gradients of 0.5 to 0.6 mGal/km and amplitudes of less than 30 mGal. This type of anomaly indicates major discontinuities that subdivide the above-mentioned Archean blocks.

3. Regional gravity gradients of about 0.06 mGal/km, which become progressively more negative toward the southwest and occur between type 1 anomalies (Figure 2). They are interpreted as being due to the tilting of the Archean blocks.

4. A broad zone along the coast with seaward gradients close to 1.0 mGal/km, which reflect the offshore transition from continental to oceanic crust.

5. Dislocations of the narrow belts of high gradients. The dislocations are interpreted as originating with major strike-slip faults hundreds of kilometers long with northwest, west-northwest, and northeast trends. Horizontal displacements may exceed 100 km in places.

6. Circular gravity lows and highs, mostly less than 60 km in diameter. These anomalies are associated with Mesozoic alkaline intrusions, such as the one at Serra Negra–Salitre (46°50′ W, 19° S; SS on profile A–A' in Figure 2) and Possos de Caldas (46°30′ W, 21°50′ S), or tectonic depressions as at Caldas Novas (48°40′ W, 17°45′ S), or to small sedimentary basins such as the one at Patos (46°20′ W, 17°30′ S; P on profile A–A' in Figure 2).

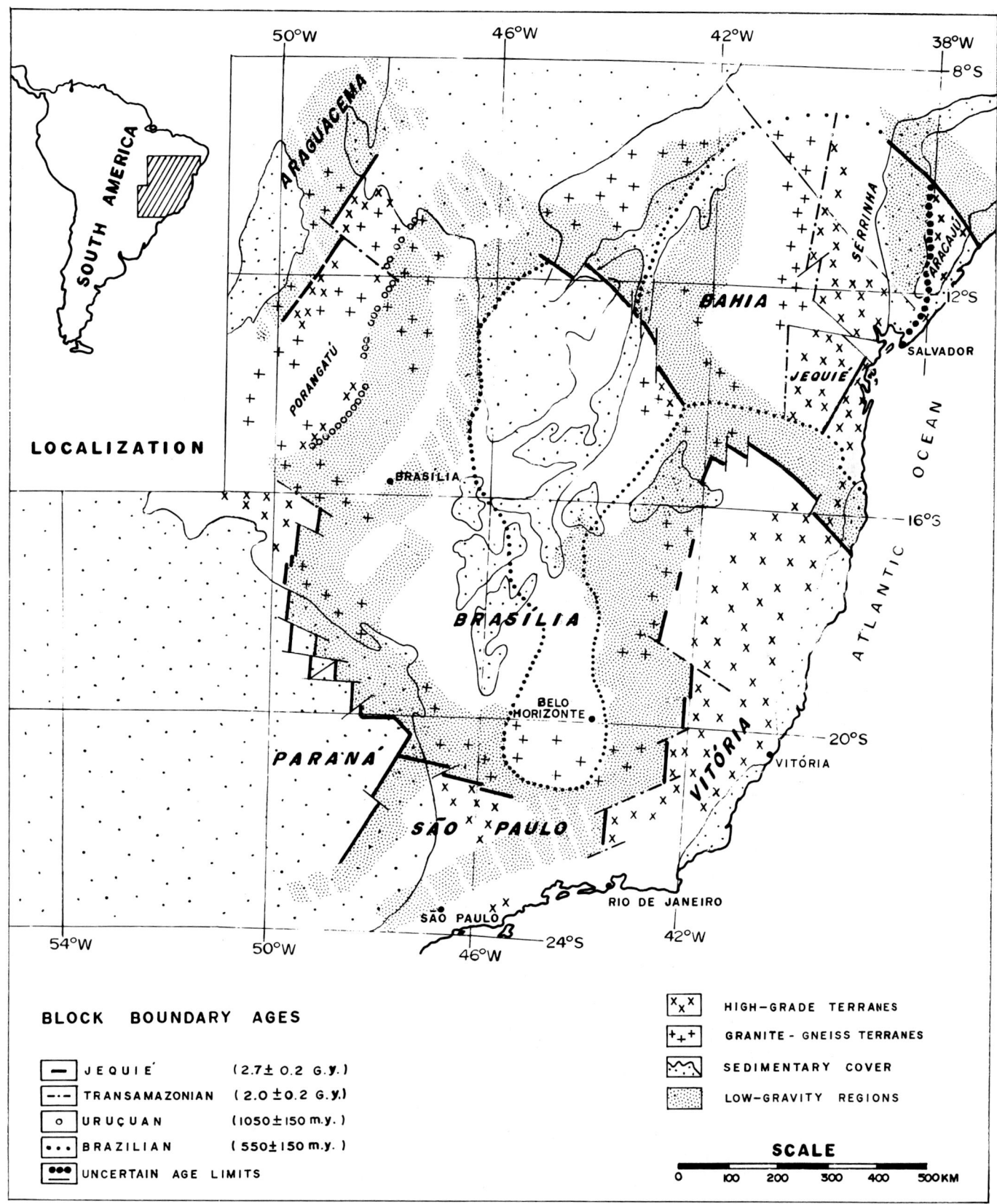

Fig. 3. Map showing block structure of central and eastern Brazil.

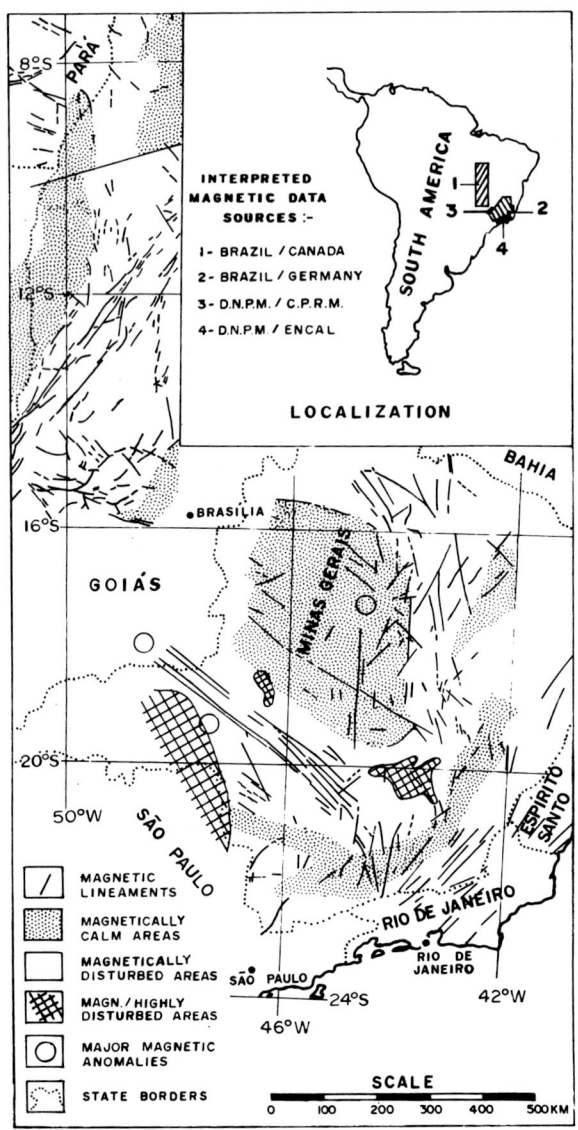

Fig. 4. Interpretive aeromagnetic map of parts of central and eastern Brazil.

MAGNETIC DATA

Total-magnetic-intensity anomaly maps at scales of 1:100 000, 1:250 000, and 1:500 000 (CGBA, 1971–73, PGBC, 1979–80) and total-intensity magnetic-anomaly profiles at a scale of 1:50 000 (ENCAL, 1978) were subjected to filtering to isolate anomalies with wavelengths less than 4 km from the larger regional anomalies.

Five major magnetic features are recognized:

1. Lineaments (Figure 4), which are expressed as linear dipole anomalies of short wavelengths (usually less than 2.5 km) with normal or reverse polarity and peak-to-peak amplitudes from 50 to more than 500 nT. Second-vertical-derivative anomaly maps enhance the lineaments owing to their short-wavelength character. The lineaments are interpreted to originate from fault zones. Some of the lineaments can be followed for more than 400 km into the west Minas Gerais lineament system (Bosum, 1973) and into the Transbrasiliano lineament zone in the State of Goiás (Schobbenhaus Filho, 1975; Schobbenhaus Filho et al., 1981). Lineament trends differ from block to block. In the Brasilia block, the most prominent magnetic lineaments strike N 55 degrees W, parallel to the orientation of mafic dikes and faults and related alkaline stocks and kimberlite diatremes. Other lineaments trend N 5 degrees E and N 25 degrees E in the same block. In the Vitória block, the lineament system strikes N 35 degrees E, whereas in the Porangatu block it strikes N 40 degrees E, with weaker trends at N 10 degrees E. At some places, lineaments correspond to major strike-slip faults interpreted from gravity data.

2. Magnetically calm areas, which are extensive and exhibit magnetic relief of 50 nT or less. One magnetically calm area constitutes a belt 30–60 km wide and 800 km long near the southern border of the State of Minas Gerais, where it coincides approximately with the edge of the Brasilia block as defined by the gravity anomalies.

3. Magnetically disturbed areas, which are areas of accentuated magnetic relief with lineaments and anomalies having amplitudes of 100 to 500 nT. These areas lie along the edge of the gravity-low belts at the borders of the tectonic blocks.

4. Magnetically highly disturbed areas, with anomaly amplitudes of more than 500 nT and characterized by short wavelengths. In part, as in western Minas Gerais, these areas coincide with basaltic lava flows that disturb the magnetic signature of the underlying basement rocks. Gravity information is particularly important in these areas for the identification of the basement structures. Similar anomalies are found over the sedimentary basin of the Patos region and over the iron formations near Belo Horizonte.

5. Three major anomalies derived from reversely polarized sources, which have been mapped (Figure 4). One of these, near the town of Pirapora (44° 40′ W, 17° 20′ S), about 365 km southeast of Brasilia, has an amplitude greater than 1 300 nT. Another, the Perdizes anomaly (Haralyi, 1978), with an amplitude greater than 500 nT, is near Araxá (47° 15′ W, 19° 20′ S), 370 km north-northwest of Belo Horizonte. A third anomaly is found near Caldas Novas (48° 40′ W, 17° 35′ S), 250 km south-southwest of Brasilia (Haralyi, 1978, 1980). All three anomalies are closely associated with major magnetic lineaments and gravity lows.

MINERAL DEPOSITS

Chromium, nickel, cobalt, asbestos, and talc deposits in mafic–ultramafic rocks (Suszczynski, 1975; Damasceno, 1982) are closely associated with major lateral crustal discontinuities (types 1 and 2) defined by gravity anomalies.

Alkaline stocks mineralized with titanium, platinum, and niobium (Barbosa et al., 1970) and diatremes of kimberlitic affinity define the alkaline-kimberlitic province of Alto Paranaiba (Hasui et al., 1976; Svisero et al., 1982). This province is related to major magnetic lineaments in west Minas Gerais State, notably around the Perdizes magnetic anomaly. The eastern pegmatite province, largely restricted to the western edge of the Vitória block, includes pegmatites mineralized with tantalum, niobium, beryllium, lithium, and tin (Abreu, 1973; Suszczynski, 1975). Deposits of gold are dispersed, but the most important ones are related to the

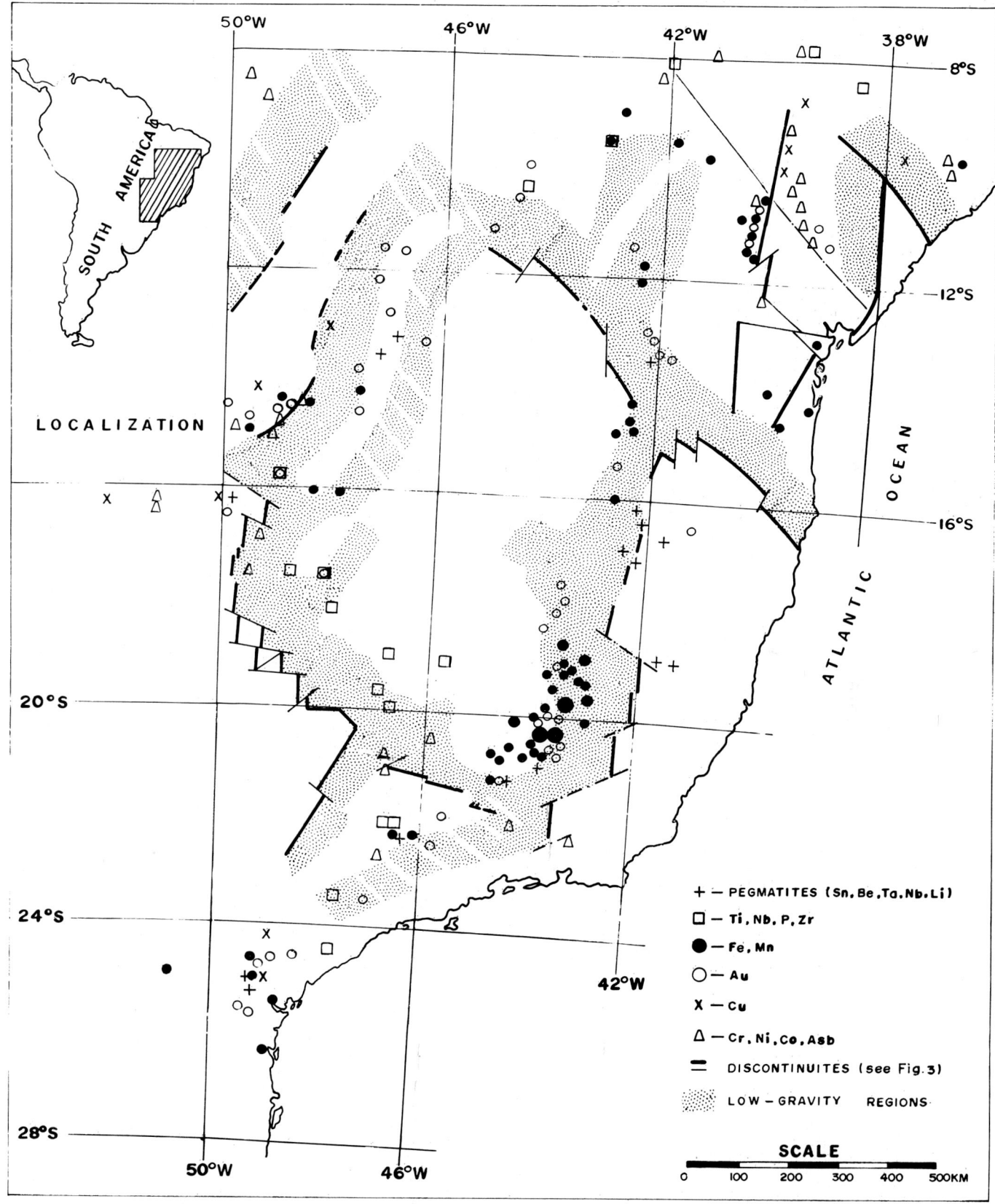

FIG. 5. Distribution of mineral deposits in eastern Brazil. The size of the symbols is proportional to the economic importance of the deposits. (After Schobbenhaus Filho et al., 1981.)

greenstone belts that are now in the granite–gneiss terranes, such as Santa Luz, south Goiás, and Rio das Velhas (Dorr, 1969; Berbert et al., 1980; Inda and Duarte, 1980; Teixeira and Kishida, 1980; Schorscher et al., 1982). Iron and manganese occur in deposits of different types and ages. Their distribution coincides with the inner parts of the gravity lows that border the major discontinuities. The relationship between the distribution of these examples of mineral deposits and geophysical anomalies is illustrated in Figure 5.

DISCUSSION

Figure 2 presents the block-structure framework of the region under study, defined mainly by type 1 gravity anomalies. Overthrusting of the blocks along major discontinuities caused the local superposition of one block over another, resulting in a zone of crustal thickening and a tilted-block pattern. Gravity anomalies over the Vitória block are influenced by the complex crustal structure of the adjacent Atlantic-type continental margin.

The fragmentation of the Archean Brasilia and Bahia blocks is interpreted as being related to the evolution of mobile belts during the Proterozoic. The Serrinha and Jequié blocks are fragments of the Bahia block, and the Porangatu block is a fragment of the Brasilia block. The profile B–B′ (Figure 2) suggests that in a northwest–southeast direction the Porangatu block is sharply tilted to the southeast, whereas the Brasilia block is subhorizontal.

Despite the block displacements and strong deformation in these mobile belts, the Archean structural framework can be recognized and represents an important constraint to the application of the Wilson cycle tectonics to Proterozoic geodynamics. Other constraints also arise, such as the parallelism of Proterozoic belts to Archean discontinuities, the arch described by the Arassuai Brasilian fold belt crosscutting the north–south–trending features of the Archean rocks, and the absence of evidence for collisional tectonics. Therefore, the general picture is one of a continental mass with block structure, subjected to Proterozoic deformation and metamorphism along belts whose orientations were controlled by Archean structures (Haralyi and Hasui, 1982) mostly in an ensialic environment.

Faults and shear zones from a variety of ages are readily recognized in the magnetic surveys by virtue of magnetic lineaments.

Important chromium, nickel, copper, cobalt, asbestos, and talc deposits in the mafic–ultramafic rocks are closely associated with major crustal discontinuities indicated by type 1 gravity anomalies.

CONCLUSION

Geophysical data aid in a better definition of the regional structural framework in central and eastern Brazil and provide some insight to crustal structure. Several important features, not clearly identified by geologic mapping alone, are recognized. These features include many major faults and most structural features buried under sedimentary cover. In the study area, geophysics has shed new light on the interpretation of geological data, on the spatial distribution of mineral deposits, and on the geologic evolution of a major cratonic area of the Earth's crust.

REFERENCES

Abreu, S. F. de, 1973, Recursos minerais do Brasil: São Paulo, Ed. Edgard Blücher.

Almeida, F. F. M. de, 1967, Origem e Evolução da Plataforma Brasileira: Bull. 241, Div. Geol./Dep. Nac. Prod. Min.

Almeida, F. F. M., 1977, O craton do São Francisco: Rev. Bras. Geocienc., **7**, 349–364.

Almeida, F. F. M., Hasui, Y., Davino, A., and Haralyi, N. L. E., 1980, Informações geofísicas sobre o oeste mineiro e seu significado tectônico: Ann. Acad. Bras. Cienc., **52**, 49–60.

Almeida, F. F. M. de, Hasui, Y., and Neves, B. B. B., 1976, The Upper Precambrian of South America: Bull. IG, **7**, 45–80.

Barbosa, O., Braun, O. P. G., Dyer, R. C., and Cunha, C. A. B. R., 1970, Geologia da região do Triangulo Mineiro: Bull. 136, Div. Geol./Dep. Nac. Prod. Min.

Barreira, C. F., and Dardenne, M. A., 1981, A sequencia vulcano-sedimentar de Rio do Coco: Ata Simp. Geol. Centro-Oeste, 241–264.

Berbert, C. O., 1980, Complexo basal Goiano: Ann. 31 Congr. Bras. Geol., **5**, 2837–2849.

Berbert, C. O., Olivatti, O., Correia Filho, F. C. L, and Oliveira, C. C., 1980, Controles da mineralização aurifera no Centro-Oeste Brasileiro: Ann. 31 Congr. Bras. Geol., **3**, 1388–1401.

Blitzkow, D., Gasparini, P., Mantovani, M. S. M., and Sá, M. C., 1980, Crustal structures of southeastern Minas Gerais, Brazil, deduced from gravity measurements: Rev. Bras. Geocienc., **9**, 39–43.

Bosum, W., 1973, O levantamento aeromagnético de Minas Gerais e Espirito Santo e a sua sequencia quanto a estrutura geológica: Rev. Bras. Geocienc., **3**, 149–259.

CGBA-Convenio Geofísica Brasil-Alemanha, 1971–73, Mapas de anomalias magnéticas: Brasilia, Dep. Nac. Prod. Min.

Chodhuri, A., Fiori, A. P., and Bettencourt, J. S., 1978, Charnockitic gneisses and granulites of the Botelhos region, southern Minas Gerais: Ann. 30 Congr. Bras. Geol., **3**, 1236–1249.

Cordani, U. G., and Iyer, S. S., 1979, Geochronological investigation on the Precambrian granulitic terrain of Bahia, Brazil: Prec. Res., **9**, 255–274.

Cordani, U. G., and Neves, B. B. B., 1982, The geologic evolution of South America during the Archean and Early Proterozoic: Rev. Bras. Geocienc., **12**, 78–88.

Cordani, U. G., and Teixeira, W., 1979, Comentários sobre as determinações geocronológicas existentes para as regiões das folhas Rio de Janeiro, Vitória e Iguape: Texto explicativo das folhas Rio de Janeiro SF.23, Vitória SF.24 e Iguape SG.24: Brasilia, Dep. Nac. Prod. Min.

Costa, L. A. M. da, and Mascarenhas, J. de F., 1982, The high-grade metamorphic terrains in the interval Matuipe-Jequié: Archean and Lower Proterozoic of east-central Bahia: Int. Symp. on Archean and Early Prot. geol. evol. and metall., Abstr. and Excursions, 19–37.

Damasceno, E. C., 1982, Archean and the Early Proterozoic mineral deposits in Brazil: Rev. Bras. Geocienc., **12**, 426–436.

Danni, J. M. C., Dardenne, M. A, and Fuck, R. A., 1981, Geologia de região de Goiás, Go: o greenstone belt Serra de Santa Rita e sequencia Serra Cantagalo: Ata Simp. Geol. Centro-Oeste, 265–280.

Dorr, J. V. N., II, 1969, Physiographic, stratigraphic, and structural development of the Quadrilatero Ferrifero, Minas Gerais, Brasil: Prof. Pap. 641-A, U.S. Geol. Surv.

ENCAL S.A., Consultoria e Levantamentos Aéreos, 1978, Perfis rebatidos de anomalia magnética: Brasilia, Dep. Nac. Prod. Min.

Escobar, I. P., 1980, Métodos de levantamento e ajustamento de observações gravimétricas visando a implantação da rede gravimetrica fundamental Brasileira: Rio de Janeiro, Obs. Nac.

Escobar, I. P., and Santos, J. P., 1980, Ajustamento da rede gravimétrica do Observatorio Nacional: Rio de Janeiro, Obs. Nac.

Gama, L. I., 1971, Valores da gravidade no nordeste e região centro-leste do Brasil: Rio de Janeiro, Obs. Nac.

——— 1972, Valores da gravidade nas regiões centro e sul do Brasil: Rio de Janeiro, Obs. Nac.

Gomes, R. A. A. D., and Motta, A. C., 1978, Projeto levantamento gravimétrico no estado da Bahia: Salvador, Conv. DNPM/CPRM.

Haralyi, N. L. E., 1978, Carta gravimétrica Bouguer do oeste de Minas Gerais, sudeste de Goiás e norte de São Paulo: Ph.D. thesis, Univ. São Paulo.

———1980, Depressões tectonicas no cráton do Paramirim: Ann. 31 Congr. Bras. Geol., **5**, 2634–2638.
Haralyi, N. L. E., and Hasui, Y., 1981, Anomalias gravimétricas e estruturas maiores do sul de Goiás: Ata I Simp. Geol. Centro-Oeste, 73–92.
———1982, The gravimetric information and the Archean–Proterozoic structural framework of Eastern Brazil: Rev. Bras. Geocienc., **12**, 160–166.
Hasui, Y., 1982, The Mantiqueira Province: Archean structure and Proterozoic evolution: Rev. Bras. Geocienc., **12**, 167–171.
Hasui, Y., Almeida, F. F. M. de, Haralyi, N. L. E., Davino, A., and Svisero, D. P., 1976, Contexto tectônico dos carbonatitos do oeste de Minas Gerais: Int. Symp. on Carbonatites, unpubl. abstr.
Hasui, Y., Almeida, F. F. M., and Neves, B. B., 1978, As estruturas brasilianas: Ann. 30 Congr. Bras. Geol., **6**, 2423–2437.
Hasui, Y., Del'Rey, L. J. H., Silva, F. J. L, Mandetta, P., Moraes, J. A. C., Oliveira, J. G., and Miola, W., 1982, Geology and copper mineralization of Curacá river valley, Bahia: Rev. Bras. Geocienc., **12**, 463–474.
Hasui, Y., Tassinari, C. G. C., Siga Jr, O., Teixeira, W., Almeida, F. F. M., and Kawashita, K., 1980, Datacões Rb–Sr e K–Ar do centro-norte do Brasil e seu significado geologico–geotectônico: Ann. 31 Congr. Bras. Geol., **5**, 2659–2676.
Inda, H. A. V, and Barbosa, J. F., 1978, Texto explicativo para o mapa geológico da Bahia, escala 1:1 000 000: Salvador, Secret. Minas Energia.
Inda, H. A. V., and Duarte, F. B., 1980, Geologia e recursos minerais do estado da Bahia: Salvador, Secret. Minas Energia.
Lesquer, A., Almeida, F. F. M. de, Davino, Aa., Lachaud, J. C., and Mallard, P., 1981, Importance structurale des anomalies gravimetriques de la partie sud du craton de São Francisco et de sa bordure occidentalle (Bresil): Tectonophysics, **76**, 273–293.
Lindenmayer, Z. G., 1981, Evolucão geológico do vale do Curacá e dos corpos máfico–ultramáficos mineralizados a cobre: M.S. thesis, Salvador, Inst. Geocienc. Univ. Fed. Bahia.
Marini, O. J., Fuck, R. A., Dardenne, M. A., and Teixeira, N. A., 1977, Dobramentos na borda oeste do cráton do São Francisco: Bol. Esp. Soc. Bras. Geol., **3**, 155–204.
Mascarenhas, J. F., 1979, Estruturas tipo "Greenstone Belts" no leste da Bahia: Textos básicos, Geol. Recursos Miner. do Estado da Bahia, **2**, 25–26.
Mascarenhas, J. F., and Sá, J. H. S., 1982, Geological and metallogenic patterns in the Archean and Proterozoic of Bahia State, Eastern Brazil: Rev. Bras. Geocienc., **12**, 193–214.
Michale, M., 1982, Densidades de rochas do embasamento Pré-Cambriano do Brasil: Rep. CNPq/ON, Rio de Janeiro (unpubl.).
Montalvão, R. M. G., Hildred, P. D., Bezerra, P. E. L., Prado, P., and Silva, S.deJ., 1982, Petrographic and chemical aspects of the mafic–ultramafic rocks of the Crixás, Guarinos, Pilar de Goiás-Hidrolina and Goiás greenstone belts, Central Brasil: Rev. Bras. Geocienc., **12**, 331–347.
Neves, B. B. B., Cordani, U. G., and Torquato, J. R., 1980, Evolucão geocronológica do Precambriano do Estado da Bahia: Textos básicos, Geol. Recursos Miner. do Estado da Bahia., **3**, 1–101.
Oliveira, E. P., Lima, M. I. C., Carmo, U. F., and Wernick, E., 1982, The Archean granulite terrain from east Bahia, Brazil: Rev. Bras. Geocienc., **12**, 356–368.
Oliveira, M. A. F., 1980, Petrologia das rochas granuliticas da faixa Paraiba do Sul, estados do Rio de Janeiro e Minas Gerais: Thesis, Inst. Geocienc. e Cienc. Ex. Univ. Est. Paulista.
PGBC-Projeto Geofísico Brasil–Canada, 1979–80, Mapas de intensidade magnética total: Brasilia, Dep. Nac. Prod. Min.
Sá, E. F. J., McReath, I., Neves, B. B. B., and Bartels, R. L., 1976, Novo dados geocronológicos sobre o Cráton do São Francisco no Estado da Bahia: Congr. Bras. Geol., 29, Proc. SBG, **4**, 185–204.
Saboia, L. A., 1979, Os greenstone belts de Crixás e Goiás: Rev. Nucleo Centro-Oeste, **9**, 44–72.
Schobbenhaus Filho, C., Ed., 1975, Folha Goiás, SD 22. Carta Geol. Brasil Milionésimo: Brasilia, Dep. Nac. Prod. Min.
Schobbenhaus Filho, C., Campos, D. A., Derze, G. R., and Asmus, H. E., Eds., 1981, Geologic map of Brazil and adjoining ocean floor including mineral deposits, 1:2 500 00: Brasilia, Dep. Nac. Prod. Min.
Schorscher, H. D., Santana, F. C., Polonia, J. C., and Moreira, J. M. P., 1982, Quadrilátero ferrifero—Minas Gerais State: Rio das Velhas greenstone belt and Proterozoic rocks: Int. Symp. on Archean and Early Proterozoic Geol. Evol. and Metall., Excursions Annex.
Sighinolfi, G. P., Figueiredo, M. C. H., Fyfe, W. S., Kronberg, B. I., and Oliveira, M. A. F. T., 1981, Geochemistry and petrology of the Jequié granulitic complex (Brazil): an Archean granitic basement complex: Contrib. Mineral. Petrol., **78**, 263–271.
Suszczynski, E. F., 1975, Os recursos minerais reais e potenciais do Brasil e sua metalogenia: Rio de Janeiro, Liv. Intercienc.
Svisero, D. P., Meyer, H. O. A., Haralyi, N. L. E., and Hasui, Y., 1982, Geology of Brazilian kimberlites: Terra Cognita, **2**, 246–247.
Tassinari, C. C. C., and Montalvão, R. M. G., 1980, Estudo geocronológico do "Greenstone Belt" Crixás: Ann. 31 Congr. Bras. Geol., **5**, 2752–2759.
Teixeira, W., 1982, Geochronology of the southern part of the São Francisco craton: Rev. Bras. Geocienc., **12**, 268–277.
Teixeira, W., and Danni, J. C. M., 1978, Geologia da raiz de um greenstone belt na região de Fortaleza de Minas, Minas Gerais: Rev. Bras. Geocienc., **9**, 17–26.
Teixeira, J. B. G., and Kishida, A., 1980, Geologia das mineralizacões auriferas estratiformes da faixa Weber, Araci (BA): Ann. 31 Congr. Bras. Geol., **3**, 1802–1811.
Wernick, E., and Almeida, F. F. M., 1979, The geotectonic environments of Early Precambrian granulites in Brazil: Precambrian Res., **8**, 1–17.
Wernick, E., Artur, A. C., and Fiori, A. P., 1981, Reavaliacão de dados geocronológicos da região nordeste do Estado de São Paulo e unidades equivalentes nos estados de Minas Gerais e Rio de Janeiro: Atas Simp. Reg. Geol. **1**, 328–342.

Applications of aeromagnetic data to mineral-resources exploration—Baluchistan, Pakistan

Allan Spector,* Peter J. Hood,‡ Abul Farah,§ and Waheeduddin Ahmed§

ABSTRACT

An interpretation of regional aeromagnetic-survey data for Baluchistan Province, Pakistan, demonstrates important implications of magnetic data with regard to plate tectonics as well as to mining and hydrocarbon and ground-water exploration.

Magnetic anomalies in the western part of the survey show the locations of intrusive magmatic belts that have been uplifted and displaced through thrust faulting as a consequence of oceanic-crust subduction. Observed copper and iron deposits are spatially related to the intrusive bodies. Fifty-two sites that involve magnetic features prospective for porphyry-type and metasomatic-type mineralization have been identified from the survey data and are recommended for mining exploration. For the eastern part of the survey, magnetic anomalies show the presence of Precambrian basement rocks developed in a north-trending structure at depths varying from 5 000 to 14 000 m. Structures in sedimentary rocks overlying this structure may have hydrocarbon potential. In addition, magnetic depth determinations in the arid western part of the survey show the positions of 14 sediment-filled depressions which are targets for ground-water exploration.

INTRODUCTION

As a result of a request in 1975 from the Geological Survey of Pakistan to the Canadian International Development Agency (CIDA), provision was made for an aeromagnetic survey and its interpretation of "promising mineral belts in Baluchistan covering an area of nearly 108 000 km². " The designated area is outlined in Figure 1.

*Allan Spector & Associates, Ltd., 24 Strathallan Boulevard, Toronto, Ont. M5N 1S7, Canada.
‡Geological Survey of Canada, Ottawa, Ont., Canada.
§Geological Survey of Pakistan, Quetta, Pakistan.

The responsibility for the specifications and monitoring of the aeromagnetic contract was given to the Geological Survey of Canada.

Because the area in question is in large part mountainous, the techniques of aeromagnetic surveying that were developed to survey the Cordillera of western Canada were employed (Hood and Holroyd, 1973). Consequently, the project area was divided into 10 overlapping subareas. In each subarea a specific survey altitude, line spacing, and direction were adopted. Altitudes ranged from 1 200 to 3 300 m above sea level. Line spacing and direction were 800 m and north–south in the western part, and 1 600 m and east–west in the eastern and southern parts.

The successful bidder for the execution of the aeromagnetic survey, which amounted to approximately 125 000 line-km, was Photosur, Inc., of Montreal, Quebec. The survey contractor utilized an Aero Commander 500B aircraft equipped with a Geometrics Model G803 proton precession magnetometer mounted in a tail-stinger installation. Throughout the survey, which extended over the period January 1976 to April 1977, a magnetometer was operated at the base of operations as a diurnal ground monitor. The aeromagnetic data were digitally compiled, and 1:50 000-scale maps were contoured at a 5-gamma interval.

In January 1981, the firm of Allan Spector & Associates, Ltd., was contracted by CIDA to perform a geologic interpretation of the aeromagnetic-survey data with the collaboration of officers of the Geological Survey of Pakistan. The results of the interpretation were presented in the form of maps at a scale of 1:250 000 and a report by Spector and Pichette (1981).

ANALYSIS OF THE MAGNETIC DATA

Calcomp profiles of the magnetic and radar altimeter data at a scale of 1:50 000 were produced for the analysis which emphasized a thorough assessment of the survey data, particularly in profile form. Figure 2 shows an example of the analysis of the magnetic-profile data. The magnetic data have been corrected for geomagnetic gradient, which is about 4.3 gammas/km in Pakistan. The profile analysis involves definition of anomalies, magnetic gradients, measurement of gradients using a maximum slope technique to

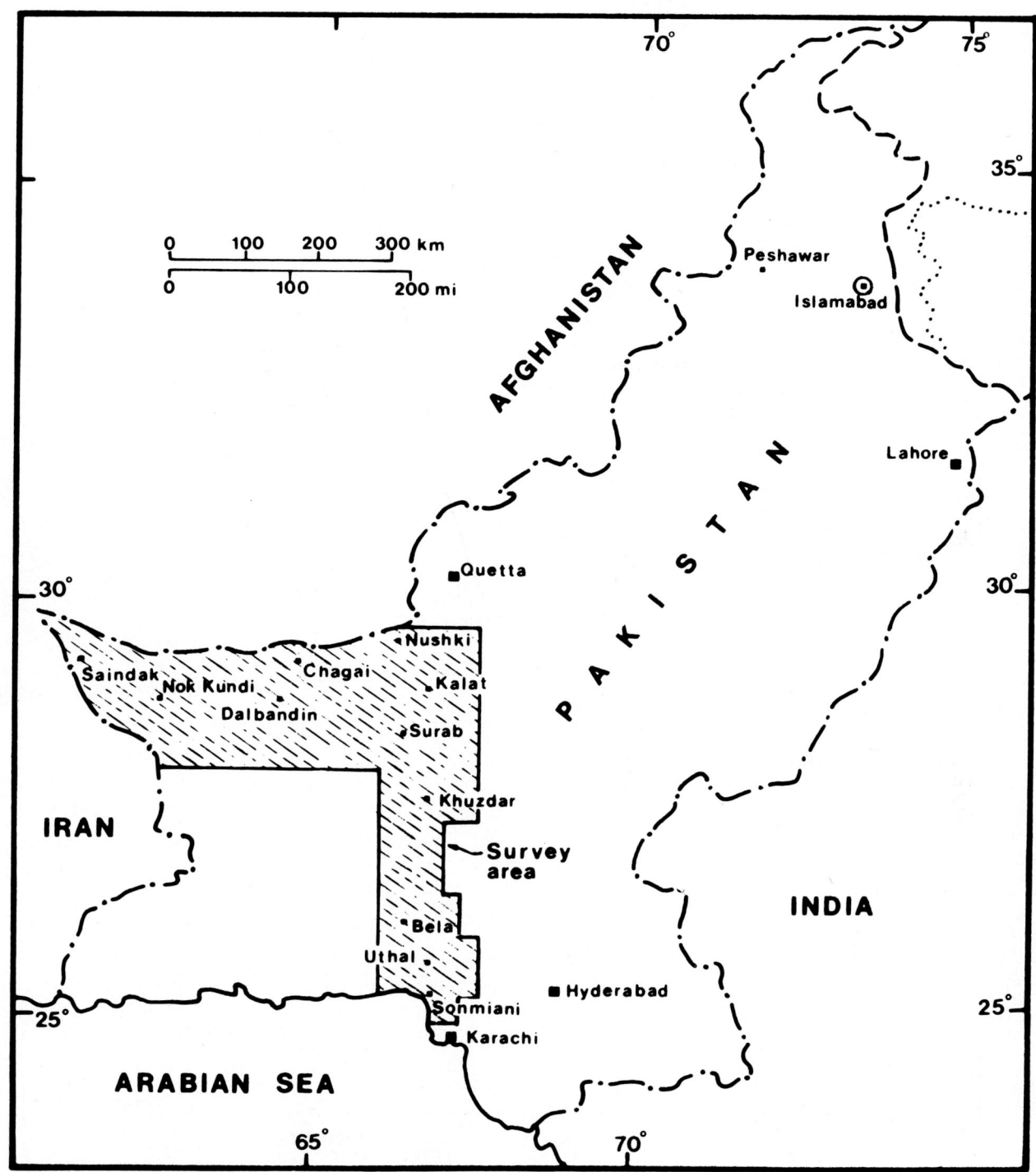

Fig. 1. Map of Pakistan showing survey area.

determine depth to magnetic basement, location of magnetic contacts, measurement of anomaly amplitude to determine rock magnetization, and examination of anomaly-shape characteristics to interpret magnetic-bedding attitude.

A basis for profile-form analysis is provided from model curves. Figure 3 describes the shape of computed magnetic anomalies from prismatic models in Pakistan (field inclination, 40 degrees; declination, 0 degrees). These charts help to define the outer contacts of intrusive bodies from magnetic contour maps. Figure 4 is an example of computed magnetic anomalies in profile form, illustrating how magnetic contacts are definable and how magnetic-anomaly gradients are related to depth to magnetic zone. The computed susceptibility factor, K_f, is used to convert magnetic-anomaly amplitude to susceptibility contrast once the appropriate model is selected. From the profile analysis, overlays were compiled showing magnetic contacts, anomaly amplitude, bedding-dip determinations, and basement-depth de-

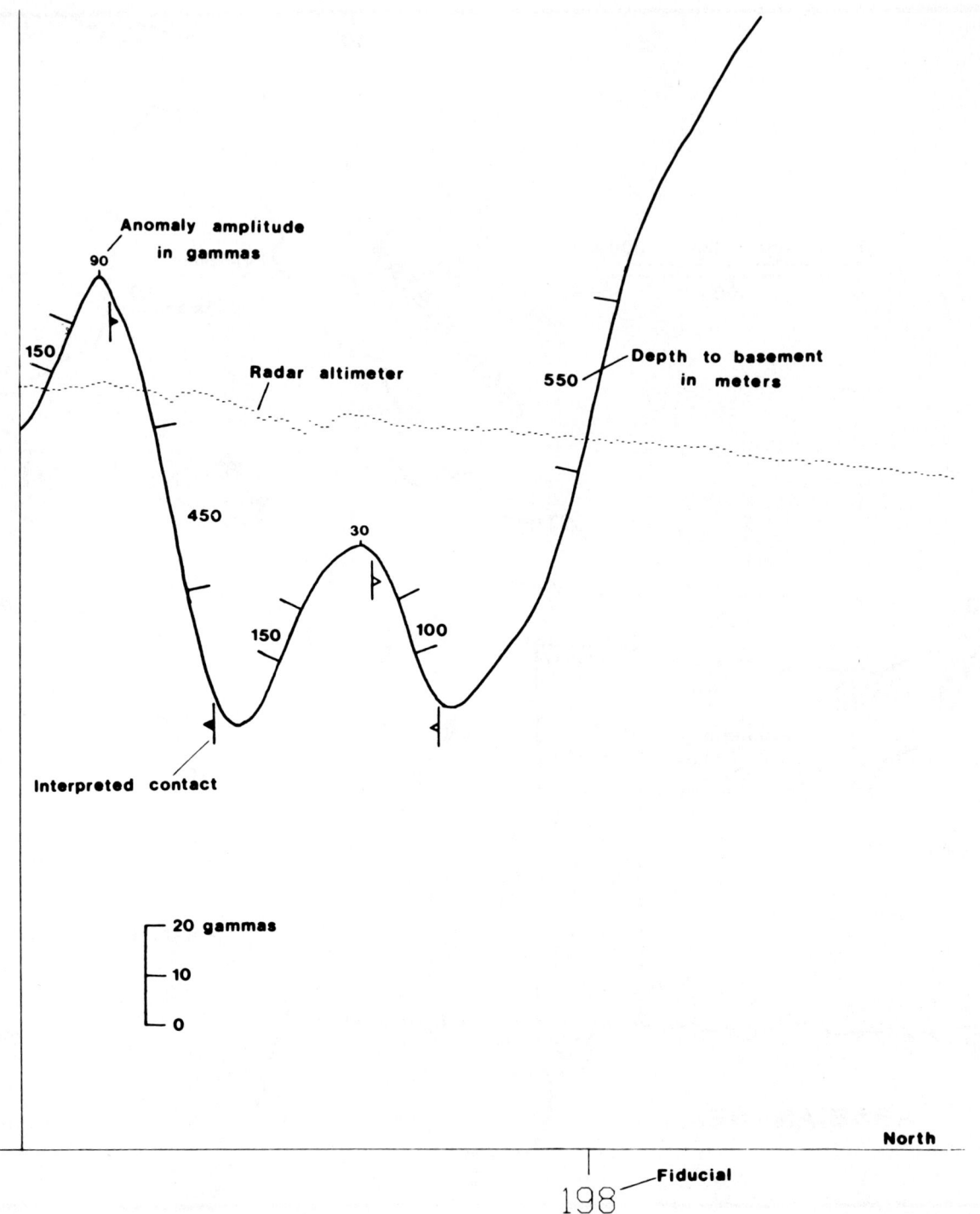

FIG. 2. Profile illustrating example of computer-plotted magnetic intensity and survey altimeter data.

terminations. Geologic interpretation involved the use of these overlays, superimposed on the magnetic contour maps (see Figure 5), together with all available geologic control that was furnished by the Geological Survey of Pakistan, mainly at 1:50 000 scale (covering 13 percent of the project area). Photogeologic-interpretation maps, which cover the project area, were found to be invaluable for structural interpretation.

A priority target was the magnetic definition of intrusive bodies that were prospective for porphyry copper–type mineralization. To assist in the resolution of these features, the magnetic data were filtered and also reduced to the magnetic pole. The filtering was done with the objective of suppressing magnetic effects associated with recent volcanic rocks which overlie the intrusive bodies. The computer processing was done using techniques described by Spector

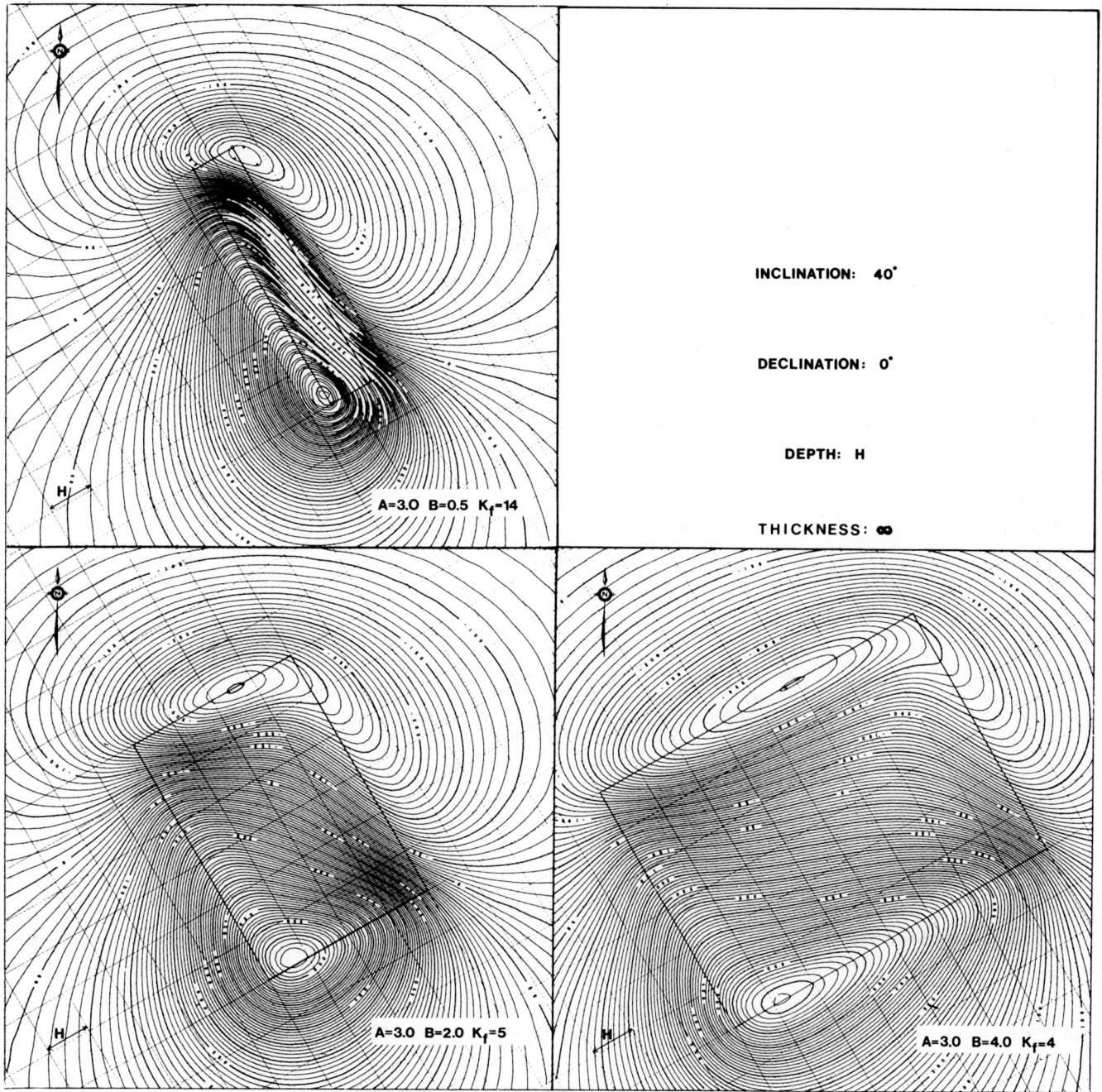

FIG. 3. Prism model: total-magnetic-intensity maps. Anomaly amplitude in each case is normalized to 1 000 gammas. Contour interval, 10 gammas. Parameter K_f is the susceptibility factor described in the text. Parameters A and B are prism half-widths.

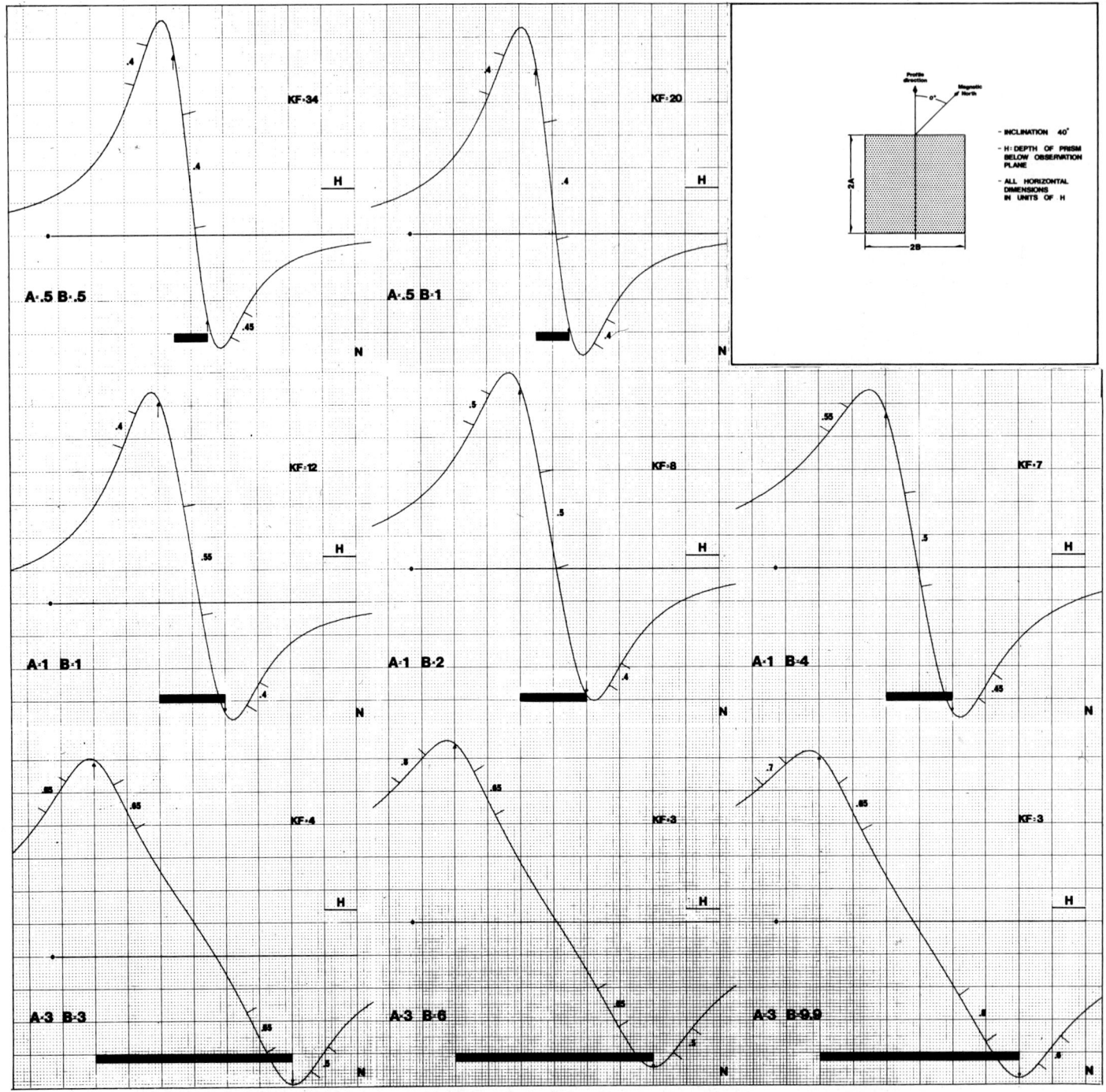

FIG. 4. Prism model: total-magnetic-intensity profiles. Anomaly amplitude in each case is normalized to 1 000 gammas. Parameter K_f is the susceptibility factor described in the text. Parameters A and B are prism half-widths, given in units of $H = 1$.

(1968). First, the power spectrum of the digitized aeromagnetic map was computed to ascertain the spectral characteristics of the shallow volcanics as opposed to the deeper intrusives. These characteristics were used to specify a matched-filter operator. Filtering was done by multiplication of the Fourier transform of the aeromagnetic data with this operator. Magnetic-pole reduction was accomplished, again in the frequency domain, by multiplication with an operator, which is a function of the Earth's field inclination and declination.

Figure 6 shows a comparison of aeromagnetic data with the data after they have undergone matched filtering and magnetic-pole reduction. The example is taken from the western part of the project area, which includes the porphyry-type Saindak copper–gold deposit (Ahmed, 1981).

As a further aid in the resolution of regional structure, the

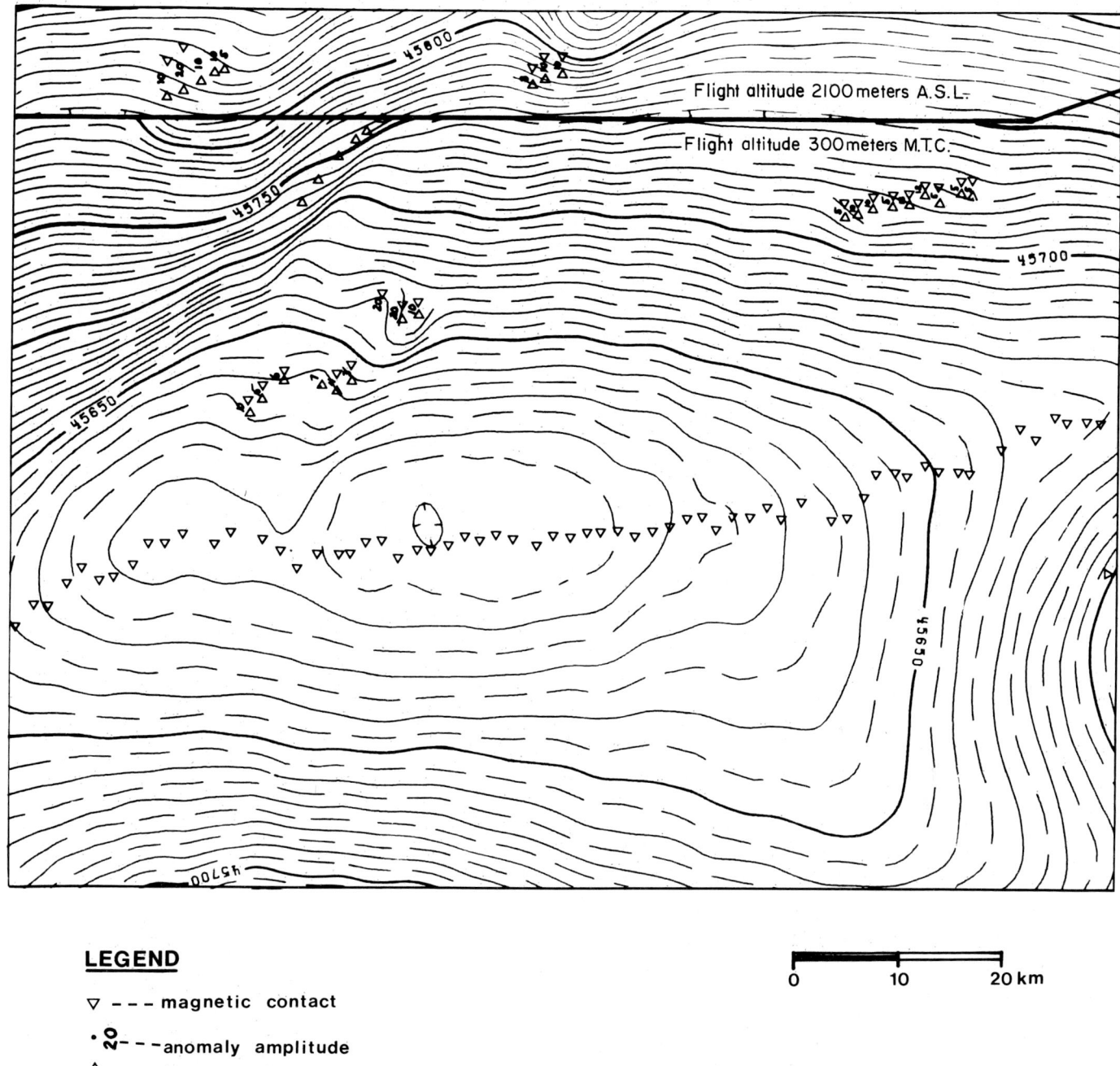

FIG. 5. Overlay of magnetic contacts superimposed on the magnetic contour map. Contour interval, 5 gammas.

digital data (after correction for geomagnetic gradient) were used to produce a colored map of the aeromagnetic data at 1:1 000 000 scale. Forty different color variations, representing different levels of magnetic intensity, were produced by combinations of three major colors (yellow, cyan, and magenta), using an Applicon color jet plotter. (Copies of this magnetic-anomaly map can be obtained by writing to the Director General, Geological Survey of Pakistan, Quetta, Pakistan.)

GEOLOGY OF THE PROJECT AREA

As seen in Figure 7, the project area lies in an area of intense deformation that resulted from collision of continental plates: the Indian plate to the east, the Eurasian plate to the north, and the oceanic portion of the Arabian plate to the west. Subduction of oceanic-crustal material is postulated in the Makran region in the western part of the project area.

Figure 8 shows the locations of igneous rocks, ophiolites,

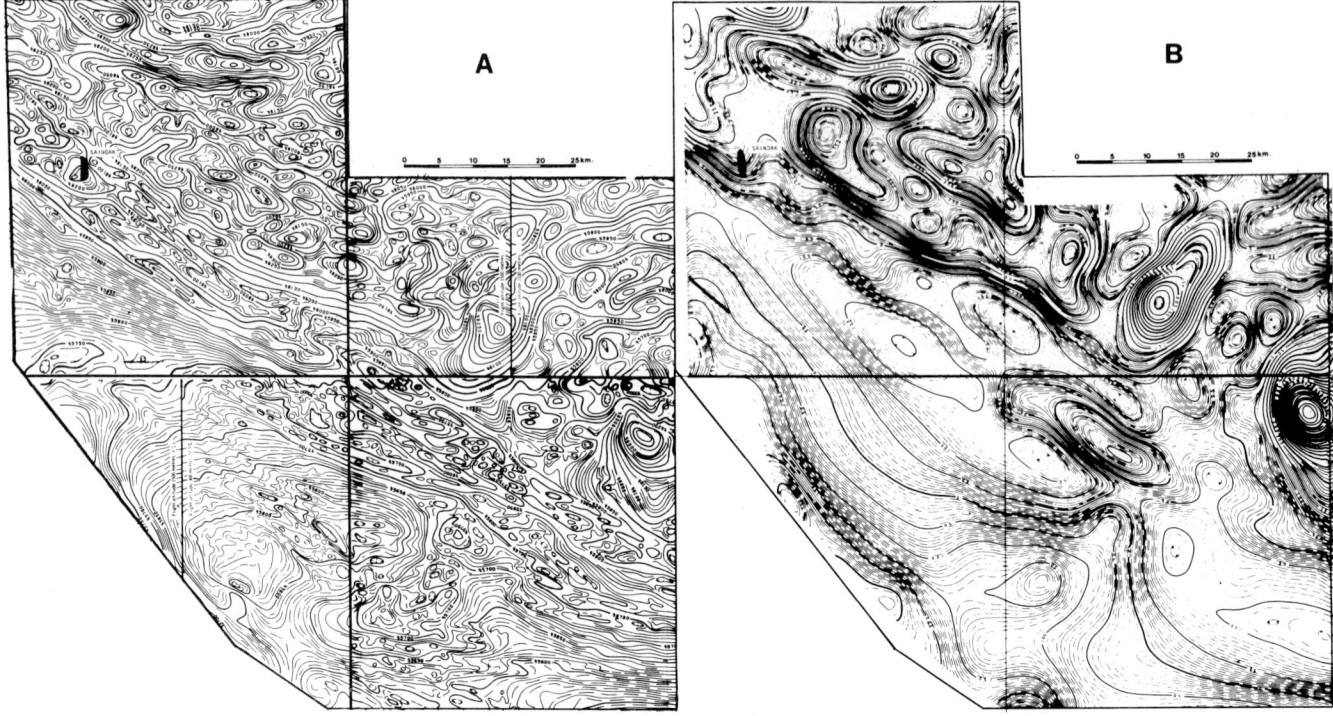

FIG. 6. Comparison of contoured magnetic-intensity data before (A) and after (B) filtering/magnetic-pole reduction. Contour interval, 5 gammas. Example is taken from extreme western part of project area.

major fault zones, and known mineral occurrences within the project area. The Chagai belt includes outcropping volcanic and calc-alkaline intrusive rocks that are Cretaceous and younger in age. The evolution of the Chagai belt began with an extensive, thick sequence of volcanics in Late Cretaceous time, folding and batholithic intrusion (plate collision) in middle Tertiary time, and finally Quaternary volcanism. Iron and copper mineralization are noted in the older Cretaceous volcanics and in the intrusive rocks. The most notable example is the copper–molybdenum–gold deposit at Saindak, which is currently under development. This porphyry deposit is included in a series of deposits that extend into eastern Iran.

The Ras Koh belt, in which occur ophiolites, Paleogene graywacke, and volcaniclastics, is tightly folded and intruded by calc-alkaline plutons. It is an allochthonous structure emplaced by thrust faulting.

Thick deposits of Tertiary sediments flank the Chagai and Ras Koh belts to the south, e.g., in the Kharan basin. To the east, the Axial belt involves older sedimentary rocks that have been folded into north-trending anticlines in the Jurassic–Cretaceous interval. Ophiolitic rocks were emplaced from the west by thrust faulting in a narrow, wedge-like structure. The Ophiolite belt is believed to have been obducted in a major suture zone created by movement of the Indian plate against the Eurasian plate. Chromite, manganese, and copper mineralization are associated with the ophiolites. At Gunga, Mississippi Valley–type lead–zinc–silver mineralization is currently under development in Jurassic carbonates that flank the Ophiolite belt.

DISTRIBUTION OF MAGNETIC ROCKS AND REGIONAL STRUCTURE

Magnetic anomalies ranging from 3 to 3 000 gammas were observed in the data. This relief is attributed to differences in magnetite mineral content between various rock units and to variations in the depth to magnetic rocks. Rock-magnetization measurements were made in Pakistan by R. Pichette and the Geological Survey of Pakistan and are summarized in Table 1. Intrusive rocks are seen to be highly magnetized, i.e., up to $6\,000 \times 10^{-6}$ cgs units; however, the mild form of magnetization exhibited by volcanic rocks, i.e., 100 to 1 000 $\times 10^{-6}$ cgs units, would make their detection difficult where they overlie intrusive rocks.

Figure 9 is extracted from the aeromagnetic-interpretation results. It illustrates in plan and in section the following three major types of geologic information: (a) intrusive zones distinguished by high rock magnetization, (b) regional faults, and (c) magnetic depressions with indications of sedimentary thickness in kilometers.

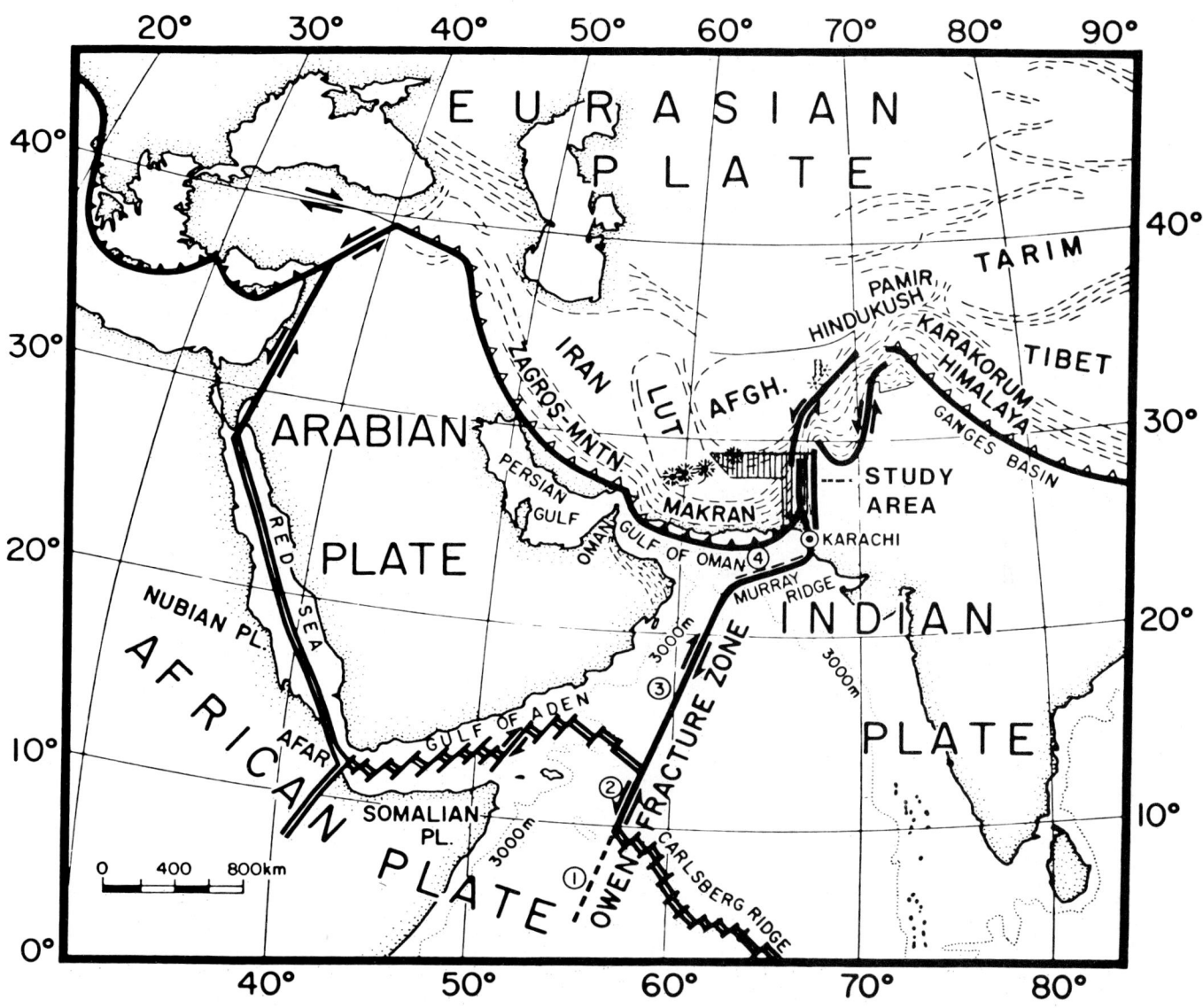

Fig. 7. Regional plate-tectonics map, after Jacob and Quittmeyer (1979).

As is evident in Figure 9, the surveyed area is divided into three parts according to contrasting magnetic and tectonic signatures: (1) the Chagai–Ras Koh–Kharan region, which covers the western and northern parts of the area; (2) the Brahui belt, located east of 67° E longitude; and (3) the Ophiolite belt, in the southern part of the area, extending to Karachi.

Chagai–Ras Koh–Kharan region

A number of magmatic belts are associated with a series of east-trending highly magnetic zones in this region. The belts have been uplifted and horizontally displaced, mostly by thrust faults, in response to forces associated with subduction on the margin of the Eurasian plate to the north.

The Chagai belt is the most extensive of the magmatic belts. Calc-alkaline intrusive rocks are reflected by relief of 1 000 to 2 000 gammas. This relief is indicative of magnetic susceptibility contrasts of 2 000 to 20 000 × 10^{-6} cgs (1 to 10 percent Fe_3O_4). Features that exhibit magnetic zoning, or elliptically shaped zones of high magnetization, were interpreted as intrusives. Magnetic-pole reduction of the magnetic data improved the definition of these features. The south boundary of the Chagai belt can be continuously traced for a distance of 400 km as the Great Chhappar fault zone ("GCFZ" in Figure 9).

Sites of observed copper sulfide mineralization, as well as skarn and volcanogenic iron deposits, are found to be spatially related to the intrusives. The Saindak porphyry copper deposit is adjacent to an interpreted batholithic intrusion. It is also linked with a major fault intersection.

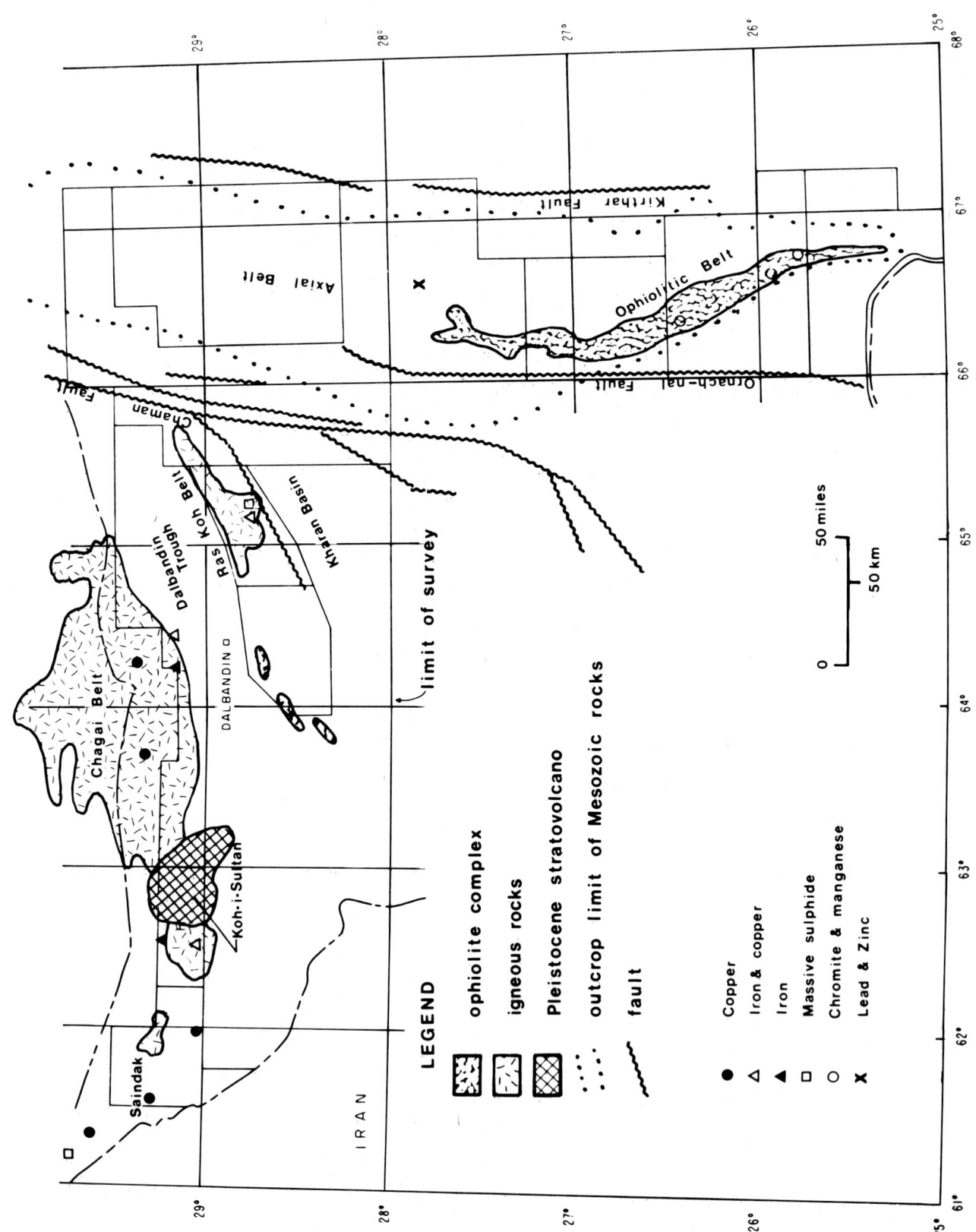

FIG. 8. Map of study area showing regional geology and mineral occurrences.

Table 1. Magnetic-susceptibility measurements in western Pakistan.

Rock type	Area	No. of samples	Magnetic susceptibility in 10^{-6} cgs units
A. Intrusive rocks			
Diorite	Ras Koh	7	0–1000
Diorite	Quetta	3	3400
Syenodiorite	Ras Koh	4	1200–3800
Granodiorite	Chagai	5	0–5200
Norite	Ras Koh	1	500
Serpentinized ultramafic	Khuzdar	8	200–3000
Gabbro	Khuzdar	2	600–6000
B. Volcanic rocks			
Basalt	Ras Koh	1	100
Trachyte	Koh-i-Sultan	1	800
Andesite	Chagai	20	0–1500
Tuff, agglomerate	Chagai	2	200
C. Sedimentary rocks			
Limestone, shale	Quetta–Khuzdar	40	nil

Magnetic anomalies that reflect either localized zones of increased iron mineral content or, conversely, zones of reduced magnetization indicative of alteration were isolated as targets for mineral exploration. Fifty-two sites in all were identified. Most of these are within the Chagai belt.

Several arcuate magmatic zones are evident south of the Chagai belt. For the most part, they are concealed by deposits of younger sedimentary rocks. These, together with numerous narrow sill- and dike-like bodies, are grouped into the Central structural complex (Figure 9). The imbricated fault complex is about 40 km in width and can be traced 400 km in an "S"-shaped arc that terminates to the east against the Chaman transform fault zone. Most of the complex is made up of plate-like structures that have been displaced from the north by thrust faulting. This is evident from magnetic anomalies in profile form, which display an asymmetry indicative of a northerly dip. The Ras Koh belt is such an example and is associated with the most intense anomaly in the survey, 3 000 gammas.

Farther to the south, an island-arc system, buried by as much as 6 000 m of Tertiary sediments, is named the Amiri belt. One could speculate that the Chagai belt, the Central structural complex, and the Amiri belt each represents a formerly separate island-arc system.

Brahui belt

Along the east border of the survey a north-trending linear zone of high magnetization is observed, the Brahui belt. The magnetized rocks are buried by 6 000 to 10 000 m of Mesozoic and Paleozoic(?) rocks. The magnetized rocks may represent Precambrian basement. The linearity of the west contact of the magnetized belt suggests a major fault. Structures in the thick, sedimentary succession overlying this structure may have hydrocarbon potential.

Ophiolite belt

Highly magnetized Cretaceous ophiolitic rocks are included in a narrow, 300-km-long wedge-shaped belt that extends south from Khuzdar. From the magnetic data, the northern part of the belt appears to have been sheared off and rotated to the northeast, apparently in response to north–south compressive forces. Sites of Mississippi Valley-type lead–zinc–fluorite mineralization are developed in Mesozoic carbonates near Gunga. Tectonics that caused rotation of the Ophiolite belt may have had a controlling influence in the movement of the mineral-rich fluids.

Other anticlinal structures involving Cretaceous volcanic material are traced by 10- to 50-gamma anomalies in the Pab and Central belts.

Chaman fault zone

The Chaman fault zone, which is a major, left-lateral transform fault, divides the survey area into two distinct structural regions. It is correlated with a series of north-northeast-trending magnetic contacts at depth. The structure is seen to comprise a number of faults, which are distributed within a 100-km-wide interval.

DEPTH TO MAGNETIC BASEMENT

From analysis of the magnetic data, areas can be outlined where the magnetic rocks are within 200 m of ground level, as opposed to areas where they are buried in places by as much as 10 000 m of sedimentary rocks and alluvium. Magnetic basement takes on different meanings in various parts of the study area. Over the Chagai belt, it consists of Cretaceous volcanics. To the south, however, it consists of Paleogene sills and dikes, whereas on the east side of the Axial belt it is possibly made up of Precambrian rocks in the Brahui belt.

Fourteen sediment-filled magnetic-basement depressions were outlined from the magnetic depth determinations south of the Chagai belt. Their distribution is shown in Figure 9, together with estimates of sedimentary thicknesses, which range from 400 to more than 4 000 m. Most of the depres-

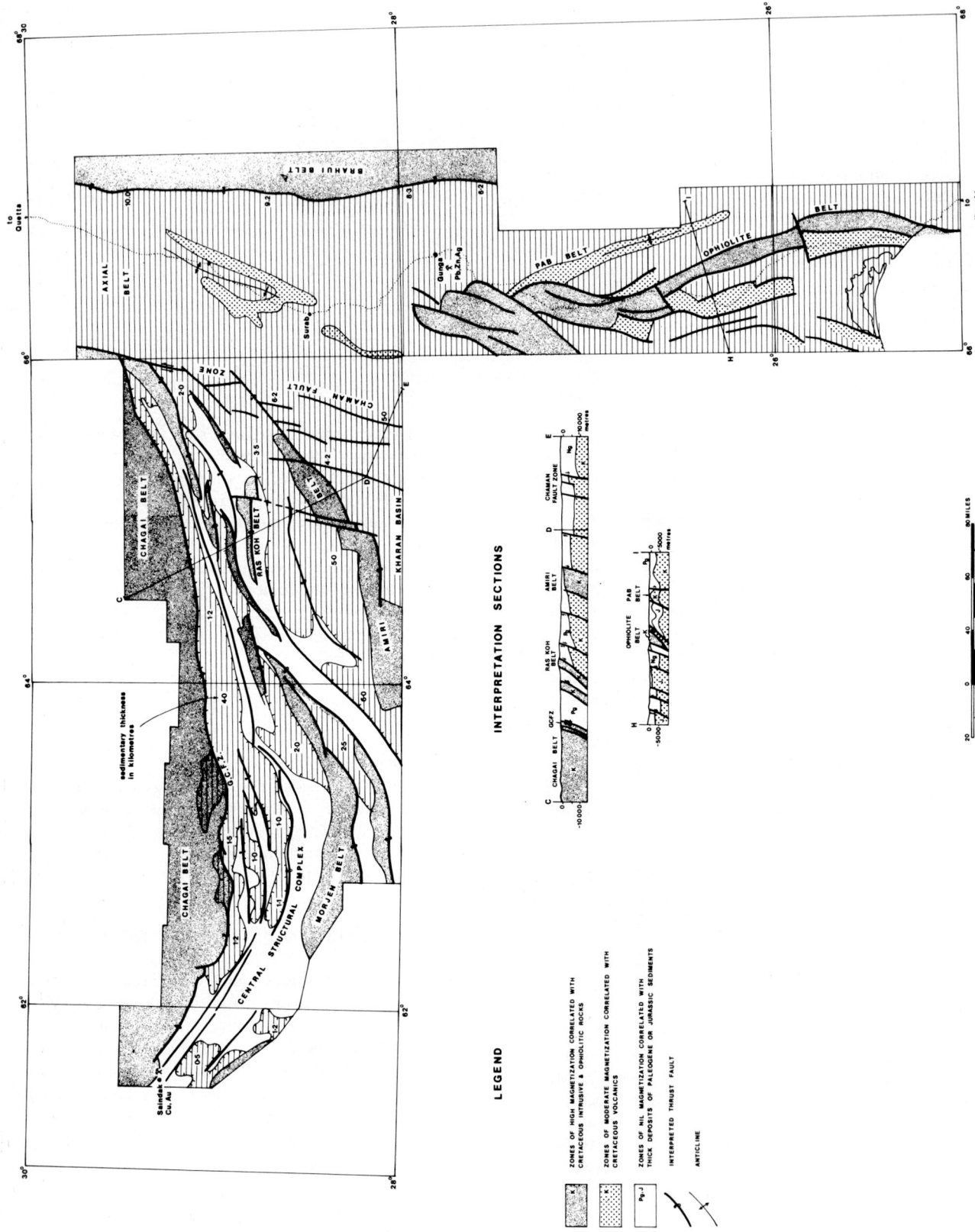

FIG. 9. Map of study area showing basement depth, structure, and lithology. GCFZ = Great Chhappar fault zone.

sions appear to be structurally controlled in the form of graben-like structures.

ACKNOWLEDGMENTS

The authors ackn/owledge with thanks the assistance and cooperation of a number of individuals who were directly involved in this project: Messrs. Joe G. Foster and Arif Turkeli of the Canadian International Development Agency; Mr. B. E. Manistre of the Geological Survey of Canada; and Messrs. Nayyer A. Zaigham, Anwaruddin Ahwed, and Mohammed Ali Mirza of the Geological Survey of Pakistan.

REFERENCES

Ahmed, W., 1981, Metallogenic framework and mineral resources of Pakistan: Rep. 261, Geol. Surv. Japan, 47–76.

Hood, P., and Holroyd, M. T., 1973, Results of GSC experiments indicate high resolution aeromagnetic surveys have potential for more development: North. Miner, **59**, 42–44.

Jacob, K. H., and Quittmeyer, R. L., 1979, The Makran region of Pakistan and Iran: trench-arc system, with active plate subduction, in Farah, A., and Dejong, K. A., Eds., Geodynamics of Pakistan: Geol. Surv. Pakistan, 305–317.

Spector, A., 1968, Spectral analysis of aeromagnetic data: Ph.D. thesis, Univ. Toronto, Dep. Phys.

Spector, A., and Pichette, R., 1981, Report on interpretation of aeromagnetic survey data, Baluchistan Province, Pakistan: Rep. to CIDA on behalf of Geol. Surv. Pakistan.

The Kalahari Desert, central southern Africa:
A case history of regional gravity and magnetic exploration

Colin V. Reeves*

ABSTRACT

Botswana, in central southern Africa, is an area of high mineral potential, where conventional mapping of the basement geology is thwarted by the ubiquitous but generally thin cover of younger rocks, principally the sands of the Kalahari Desert. Some examples are taken from the regional gravity and aeromagnetic coverage of the country to demonstrate the exploration role of potential-field surveys in such large areas of concealed Precambrian metamorphic terrane. These examples include (a) the detection of unseen Archean greenstone belts by their gravity anomalies, (b) the delineation of a mineralized Proterozoic fold belt from its gravity and magnetic characteristics, (c) the discovery of a remarkable Mesozoic dike swarm from magnetic anomalies, and (d) the interpretation of a major linear feature some 600 km in length (evident in both gravity and magnetic data) as a suture between two crustal provinces, one of Archean age and the other of Proterozoic. Such interpretations of potential-field data are playing an important part in the strategy for mineral exploration in the country.

INTRODUCTION

Botswana, a country the size of Texas, is located centrally on the mineral-rich Precambrian shield of southern Africa (units 4–7, Figure 1; Baldock et al., 1976). Within most of Botswana, however, basement rocks are totally concealed by a veneer of Tertiary to Holocene Kalahari sand (unit 1) and Karoo deposits (Paleozoic–Mesozoic sedimentary rocks and lavas, unit 2), which occupy the Kalahari inland drainage basin (Green, 1966). As an initial phase of exploration of this large area, regional geophysical programs, funded by the British and Canadian governments, were initiated by the Geological Survey of Botswana about 10 years ago. A 2 100-station regional gravity survey was carried out during 1972 and 1973 and gave the first indication of major structures in the concealed geology. Reconnaissance aeromagnetic surveys of the sand-covered areas followed in 1975 and 1976.

Results of these programs have been published in some detail (Reeves and Hutchins, 1982; Hutchins and Reeves, 1980). The purpose of the present paper is simply to highlight a few examples of the contribution these studies have made to a logically developed program of mineral exploration in the Kalahari.

GREENSTONE BELTS

In Precambrian metamorphic terranes around the world, explorationists have achieved the greatest success in the location of economic base-metal and precious-metal deposits in the areas of lightly metamorphosed volcano-sedimentary rocks or "greenstones," which occur locally amongst the granitic or gneissic rocks that usually make up 70–80 percent of such terranes (Pretorius and Maske, 1976). The location by geophysics of greenstone belts in metamorphic terranes veiled by more recent geologic units is already, therefore, a significant step forward in mineral exploration.

Figure 2a is taken from the 1:1 000 000-scale geologic map of Botswana and shows (at "A") the Tati greenstone belt, where gold was first discovered in southern Africa more than 100 years ago (Baldock et al., 1976) and where promising showings of Cu and Ni mineralization are known (ibid.). The gravity-anomaly map at the same scale (Figure 2b) indicates that the high proportion of mafic minerals in this belt produces a sufficient positive density contrast with the surrounding granitic rocks to generate a sizable gravity anomaly. The anomaly is not closely defined by the widely spaced regional gravity survey, but the presence of a significant anomaly is indisputable.

Farther west (at "B") the areally more extensive Matsitama greenstone belt (greenschist to lower amphibolite facies) is seen to be progressively obscured to the west by the cover of Kalahari and Karoo sediments. The marked positive gravity anomaly over this feature indicates that dense rocks extend west, well beyond the area of outcrop, then curve in an arcuate fashion to the north and northeast. This interpretation was confirmed unintentionally at one locality,

---

*International Institute for Aerospace Survey and Earth Sciences (ITC), 3 Kanaalweg, 2628 EB Delft, The Netherlands.

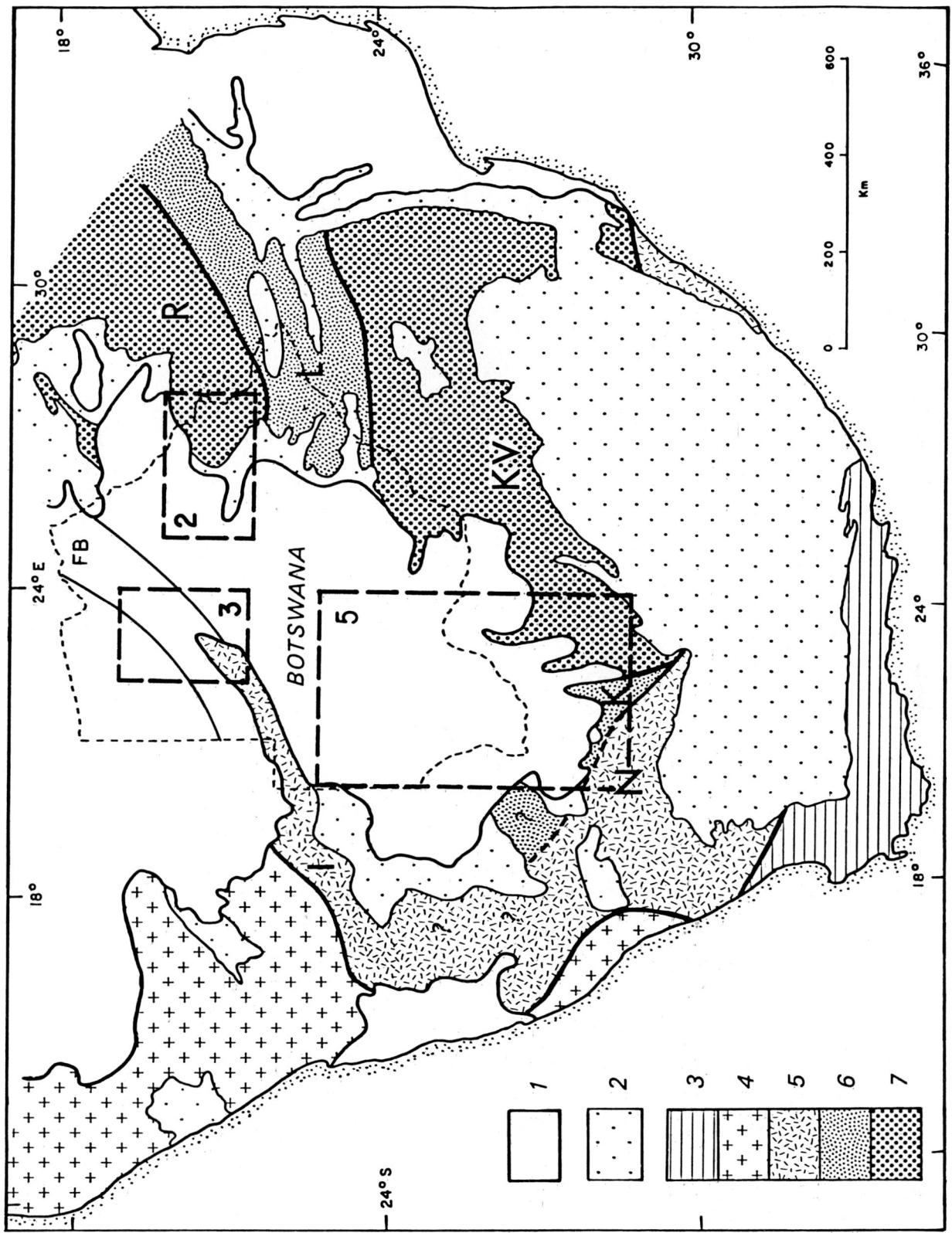

FIG. 1. Botswana (short-dashed lines) in relation to the main geologic units of southern Africa: Kalahari cover (1), Karoo cover (2), and basement rocks (3–7). KV = Kaapvaal craton; L = Limpopo mobile belt; R = Rhodesian (Zimbabwean) craton; N = Namaqualand mobile belt; K = Kheis Group; I = Irumide belt; FB = fold belt (Ghanzi-Chobe fold belt), shown partly in Figure 3. The areas covered by Figures 2, 3, and 5 are indicated by heavy broken lines.

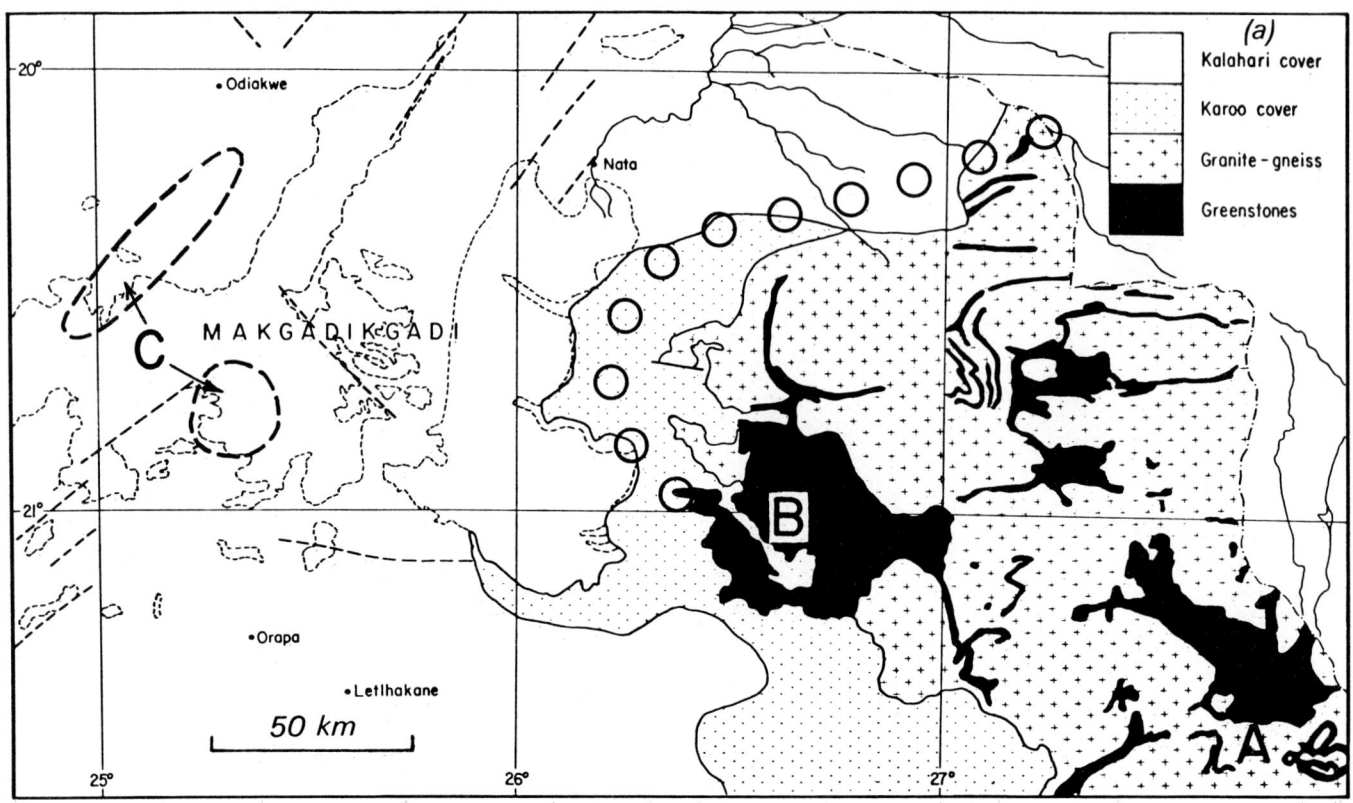

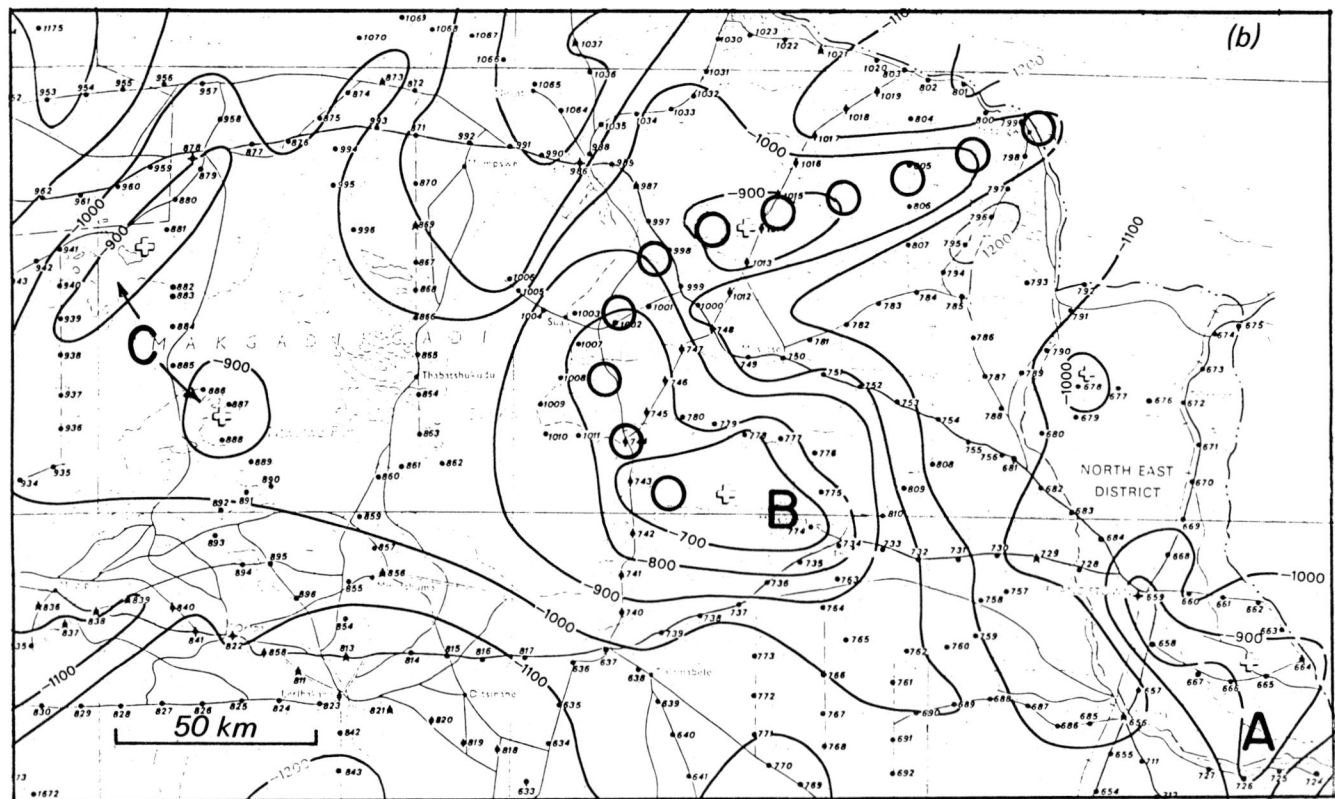

FIG. 2. a. Simplified geologic map of northeast Botswana. b. Bouguer gravity-anomaly map of the same area. Contour interval, 10 mGal. A = Tati greenstone belt; B = Matsitama greenstone belt and its interpreted extension (circles) below younger cover rocks; C = interpreted, unseen greenstone belts.

where a shallow drill hole for coal exploration in the Karoo cover entered Precambrian ultramafic rocks.

Farther west again (at "C") other positive gravity anomalies indicate the possible presence of unseen greenstone belts, remote from any Precambrian outcrop, but in an area where interpretation of the areomagnetic data indicates that the crystalline basement is within 100 m of the ground surface. No drill holes in this area penetrate more deeply than a few tens of meters.

It is clear that more detailed gravity surveys would be invaluable in defining features "B" and "C" in Figure 2 more accurately. The magnetic expression of all three greenstone belts unfortunately is almost totally obscured by that of much younger dolerite dikes (discussed in a later section), which invade this area extensively.

REGIONAL STRUCTURE IN NORTHWEST BOTSWANA

Figure 3a shows another area of the 1:1 000 000-scale geologic map, south of the Okavango Delta in northwest Botswana. Limited outcrops here consist of isoclinally folded sedimentary rocks containing strata-bound mineralization in Cu and Ag that has been the subject of economic investigations over several years (Baldock et al., 1976). Intrusive and/or extrusive igneous rocks, dated at 980 m.y. (Key and Rundle, 1981), occur in the same region, but their relation to the (probably younger) sedimentary rocks is unclear on field evidence (Thomas, 1973).

Amongst the igneous rocks, two geophysical signatures are highly significant. First, an erosion-resistant quartz-feldspar porphyry, which forms isolated inselberg exposures, is highly magnetic and gives rise to the largest magnetic anomalies of the region. Second, an epidotized diabase, which is known from drilling to contain blebs of native copper locally, is very dense (2.95 g/cm^3) and gives local positive gravity anomalies, elongate along the northeast–southwest direction of the folding. A third signature, resulting from the high magnetite content of some units of the folded sedimentary sequence, enables the outcrop traces of these rocks to be determined from the aeromagnetic maps, and hence the fold structures can be plotted.

The interpretation of the aeromagnetic data for the area of Figure 3a is shown in Figure 3b. Here it is seen that these geophysical characteristics, identified over the known outcrop areas, occupy a belt about 100 km in width which can be traced beyond the area of the figure to the international border with Namibia in the southwest and to the border with Zambia in the north ("FB," Figure 1). The area of outcrop is thus only a very small part of this belt. However, depth determinations for magnetic anomalies indicate that the cover of Kalahari sediments over much of the belt is generally less than 100 m, indicating possibilities for exploration in new areas.

POST-KAROO DIKE SWARM

The aeromagnetic interpretation in Figure 3b indicates a large number of linear features (D–D′) cutting almost at right angles to the northeast–southwest grain of the basement geology described above. There is no evidence of such a feature on the geologic map in Figure 3a, but the aeromagnetic data enable the anomalies to be traced with certainty some 400 km to the east, where they are seen to occur over known dolerite dikes cutting the youngest of the Karoo formations in areas of better exposure (Crockett, 1967). In fact, the aeromagnetic data may be used to delineate this band of intense post-Karoo dike injection across the entire 800-km width of Botswana. When this is done, the dike swarm is seen to narrow from a width of more than 100 km in the east to less than 60 km in the west.

The significance of this new, rather geometrical, feature in the tectonic history of the region has to be sought outside the confines of Botswana (Figure 4). It is interpreted (Reeves, 1978) as the failed third arm of a triple junction, of which the other two arms are represented by the Lebombo and Sabi monoclines, which were demonstrably active as major faults during the fragmentation of Gondwanaland in Late Jurassic or Early Cretaceous time (Cox, 1970).

The significance of this may seem rather academic, but dike injection was probably contemporaneous with the emplacement of the diamondiferous kimberlite pipes found immediately south of the Makgadikgadi (Baldock et al., 1976), which include the Orapa pipe, one of the largest known.

KALAHARI LINE—SOUTHWEST BOTSWANA

Turning to the area southwest of Botswana, both the gravity and the magnetic data here indicate a major north–south feature in the subsurface. This feature falls approximately along meridian 22° E and has become known as the Kalahari line. The area is the remotest from any basement outcrop, and, once again, data from outside Botswana must be referred to in any attempt to interpret the geophysical results in terms of known geology. In this case the published gravity (Smit et al., 1962) and aeromagnetic surveys (South African Geological Survey, 1971–75) for the adjacent areas of South Africa have been used in order to trace the geophysical features seen in Botswana into the area of basement outcrop in the Upington area, some 300 km farther south (Figure 5a).

Even here the geologic history of the exposed rocks is, as yet, only poorly understood (Truswell, 1977, p. 81) despite the increased level of interest following the important base-metal discoveries at Copperton and Aggeneys in the region of Prieska, immediately south of Figure 5. The Archean Kaapvaal craton with its Lower Proterozoic cover rocks lies east of 22° E, while Namaqualand granites and gneisses of Middle Proterozoic age (900–1 200 m.y.) are in evidence southwest of a line approximately from Upington to Prieska (unit "N" in Figure 1). These rocks occur only in the extreme southwest part of Figure 5a and b. The area between the Kaapvaal craton and the Namaqualand belt (i.e., the western half of Figure 5) is virtually devoid of basement exposure, except for somewhat enigmatic rocks of the Kheis Group—metamorphosed sediments and volcanics—which occur in the south part of the area, striking in a north–south direction. Here they are infolded with Matsap rocks, which consist of quartzites, shales, and volcanic rocks (Truswell, 1977, p. 57), and are the time-equivalent (c.2 000 m.y.) of the Waterberg supracrustal rocks of the Kaapvaal craton.

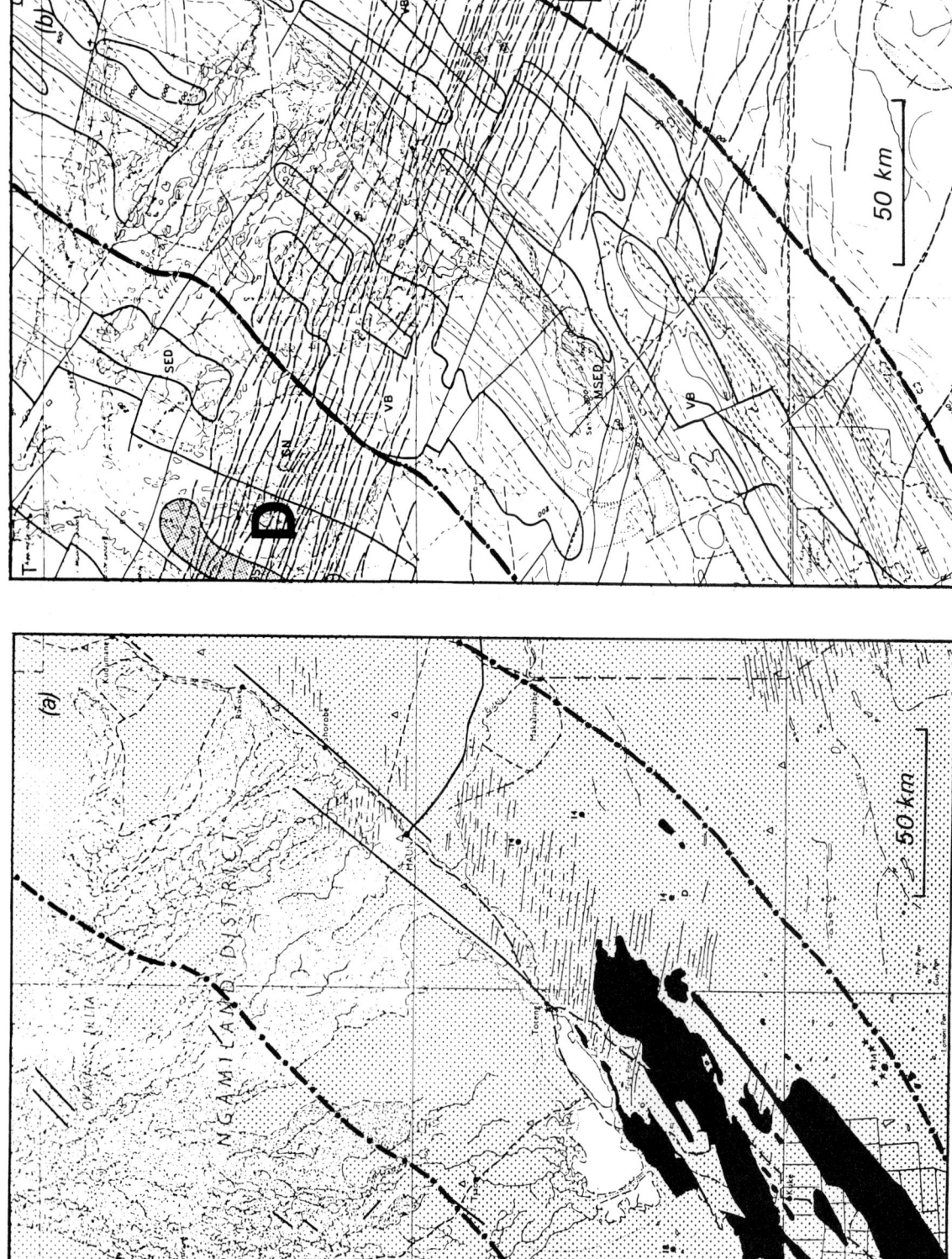

FIG. 3. a. Geologic map of part of northwest Botswana. Outcrops of pre-Karoo rocks have solid shading; dot pattern indicates cover of Kalahari sand. b. The same area of the aeromagnetic-interpretation map of Botswana. The heaviest dot-dash lines on both maps indicate the limits of a fold belt interpreted from aeromagnetic and gravity data. Within the fold belt broken lines indicate the axes of magnetic anomalies; solid lines (both light and heavy), the outlines of magnetic rock units. Heavier broken lines trending west-northwest (D–D') are interpreted dolerite dikes, part of the swarm shown in Figure 4. One such interpreted dike was proven by drilling through the sand cover in the southeast corner of the area of Figure 3b.

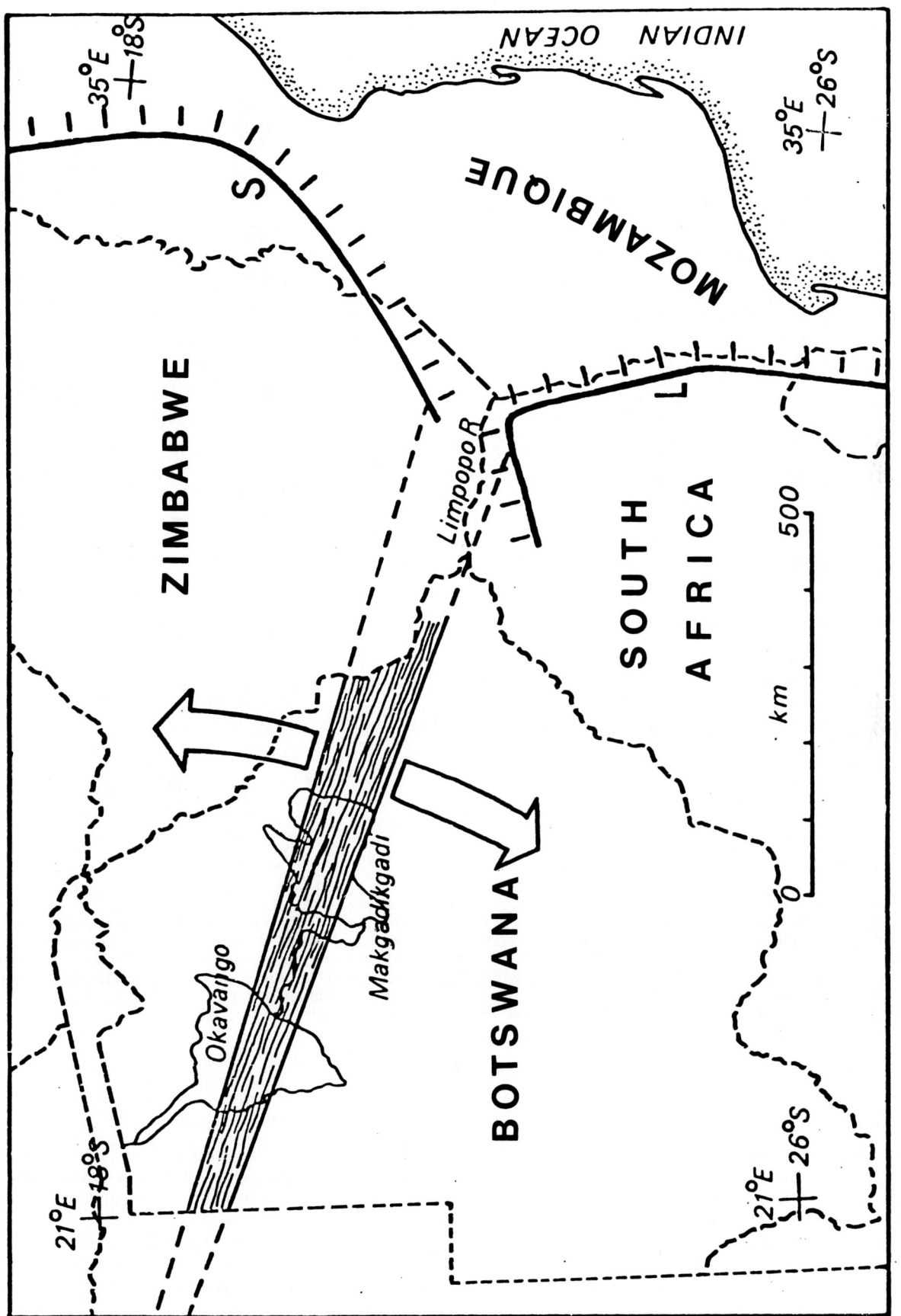

FIG. 4. The geometrical relationship of the post-Karoo dike swarm to the Lebombo monocline ("L") and Sabi monocline ("S"). About 2 degrees of relative rotation between the two crustal plates separated by the dike swarm is interpreted in the direction shown by the arrows. Field observations in the eastern part of the swarm in Botswana (Key, 1976, p. 43) indicate dikes occupying about 5 percent of the surface area.

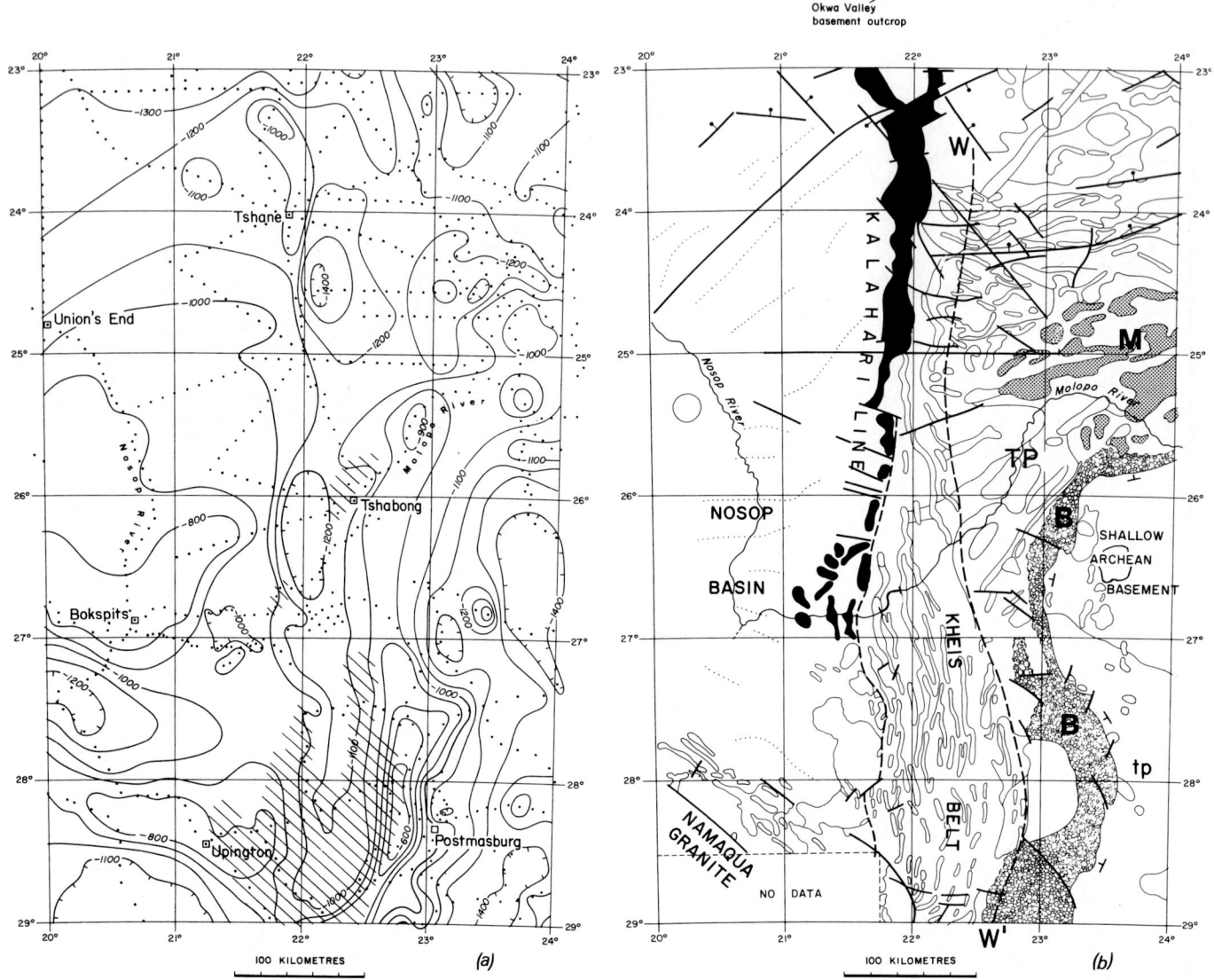

FIG. 5. a. Bouguer gravity-anomaly contours in the vicinity of the Kalahari line, which falls along the 22° E meridian. Contour interval, 10 mGal. Areas of tightly folded Kheis and Matsap rocks, which crop out sporadically between Upington and Postmasburg, are shown by shading. b. Outline interpretation of aeromagnetic data for the same area. Highly magnetic rocks of the Tshane complex, which marks the Kalahari line in Botswana, are shown solid black. W–W' = interpreted western limit of flat-lying Proterozoic cover rocks on the Archean basement; B = banded ironstones in the cover rocks; tp = thin Proterozoic cover; TP = thick Proterozoic cover; M = Molopo Farms ultramafic complex in southern Botswana. Outlines of other interpreted rock units are indicated by thin solid lines. Axes of broad magnetic anomalies originating below the Nosop basin are indicated by dotted lines. The international boundary between Botswana in the north and South Africa in the south is formed by the Nosop and Molopo rivers.

KALAHARI LINE—GRAVITY DATA

A re-contouring of the published gravity-station data for the region (Smit et al., 1962; Reeves and Hutchins, 1976) is shown in Figure 5a. It is seen here that high Bouguer anomaly values occur west of the Kalahari line, toward the Nosop River. To the east, low gravity values are underlain by the near-surface (locally outcropping) Archean basement of the Kaapvaal craton in the northern Cape Province and southern Botswana. Along the Kalahari line itself, an area of extremely low topographic relief, a narrow gravity low is evident. At places this is flanked by an equally narrow gravity high to the east.

The Kalahari line was initially equated (Reeves, 1976) with the unseen western margin of the Archean cratonic area of South Africa, Botswana, and Zimbabwe (unit 7, Figure 1). Much younger rocks are known in northeast Botswana, in most parts of Namibia, and in the southern parts of South

Africa (units 4 and 5). However, an enigmatic triangular region remains to the west of the Kalahari line, coincident with the area of high regional gravity.

The gravity high would lead one to expect the presence of dense rocks in the region. Paradoxically, only Kalahari sand and isolated outcrops of Karoo sediments are seen at the surface, and even in ground-water boreholes drilled to a depth of up to 500 m. Densities of these formations are known from published data in South Africa to fall largely in the range 2.4 to 2.6 g/cm^3 (Smit and Maree, 1966). One deeper hole entered older rock—Late Proterozoic Nama sedimentary rocks—at a depth of almost 900 m. No crystalline basement is exposed or penetrated by any drill hole within the area.

KALAHARI LINE—MAGNETIC DATA

The distinct change in geophysical character from east to west across the Kalahari line is further emphasized by the aeromagnetic data, an outline interpretation of which is shown in Figure 5b.

From east to west, three major zones are identified. The first is the area of shallow Archean basement of the Kaapvaal craton in the east. This is progressively overlain by Proterozoic sedimentary rocks, such as the highly magnetic banded ironstones labeled "B" that dip gently to the west. Dips of all exposed rock units become steep to vertical in the second or central zone where tightly folded Matsap and Kheis rocks are found in the south part of the figure area. This zone, labeled "Kheis belt" (Figure 5b), is 50–100 km wide and is interpreted to extend 400 km north into Botswana from the Kheis and Matsap exposures of the Northwest Cape. The narrow, well-defined gravity low mentioned earlier lies over these rocks, suggesting a density deficiency extending to considerable depth. Within Botswana this belt of tightly folded metasedimentary rocks is flanked immediately to the west by the highly magnetic rocks of the Tshane complex. Interpretation of their magnetic anomalies suggests they lie at depths of 500 to 1 000 m.

The general magnetic-anomaly level is higher west of the line, indicating the presence of rocks with a high susceptibility, probably of mafic character. The anomalies here strike southwest to northeast, and their long wavelength indicates a considerable depth of burial; depth estimates for two-dimensional sources give values as high as 10 km. However, any mafic material giving rise to the anomalies must be of sufficient volume to more than offset the density deficiency attributable to the overlying Nama, Karoo, and Kalahari sediments of the Nosop basin, and hence give rise to the overall positive gravity anomalies observed.

KALAHARI LINE AS A SUTURE

The picture then emerges of a deeply buried, relatively mafic crust in the west having been thrust against the sialic Kaapvaal craton and perhaps partially obducted. In the process, the Proterozoic cover rocks of the adjacent Kaapvaal craton were tightly folded into what we have called the Kheis belt. Recent studies of previously unmapped exposures of massive gray to red Matsap quartzites in the Molopo River valley within the Kheis belt have shown low-temperature–high-pressure metamorphism and a marked increase in metamorphic grade and tectonic intensity westward toward the Kalahari line, which would support this model (R. M. Key, pers. comm.).

Key and Rundle (1981) recently published a Rb/Sr whole-rock age of 1 813 ± 68 m.y. for the granitic gneisses in the Okwa Valley (shown in the north part of Figure 5b, on strike with the Kalahari line), which form one of the few basement outcrops within the Kalahari 450 km farther north. An age of about 1 800 m.y. has also been obtained for volcanic rocks within the Kheis belt in an area illustrated in the south part of Figure 5 (Cornell, 1977).

Integration of the regional geophysical information with what little is known of the geology of the area thus leads to the interpretation that the Kalahari line is a suture, the age of which is equated with the 1 800-m.y. date for the Okwa Valley granitic rocks and the volcanic rocks of the Kheis belt itself. This is an intermediate age between the Archean rocks of the eastern part of southern Africa and the Middle and Upper Proterozoic rocks of the western regions and indicates a period of extensive tectogenesis in central Botswana, which had not previously been recognized. Such an event would postdate the igneous activity of the Bushveld igneous complex within the Kaapvaal craton (Truswell, 1977) by about 150 m.y. and be contemporaneous with the final stages of Eburnean tectonism in West Africa (Tagini, 1972).

GRAVITY AND MAGNETIC MODELING

Some support for the interpretation of the Kalahari line as a suture comes from the modeling of gravity and magnetic profiles across the Kalahari line using interactive computer programs written for the IBM Personal Computer.

Figure 6a shows the average gravity profile over five distinct province boundaries or "cryptic sutures" in the Canadian shield and the interpretive model developed by Gibb and Thomas (1976). Their studies indicate that the typical gravity anomaly over such sutures, consisting of a "high" over the marginal regions of the younger province and a "low" on the older margin, may be accounted for by a thicker, denser crust in the younger province in contact with a thinner, less dense crust in the older province, and appropriate isostatic compensation.

This model was adapted to the present situation, where the rocks covering the crystalline basement remain in place and contribute to the anomaly. For an east–west profile at 24°30′ S, the outcrop positions of three anomalous density units, The Tshane complex, the Kheis belt, and the Proterozoic platform cover, were fixed from interpretation of their aeromagnetic expression (Figure 5b). The densities and subsurface geometry of these units were then adjusted to account for the difference between the gravity profile observed and that from the adapted Gibb–Thomas model. The theoretical gravity profile for the entire model is shown in comparison with the observed profile in Figure 6b. The resulting model appears reasonable in geological terms; the steep westerly dip of the highly magnetic Tshane complex is confirmed by interpretation of its magnetic anomaly.

The mineral potential of such a crustal feature will clearly need to be evaluated. The recurrence farther north of base-metal deposits (Cu–Pb–Zn) of the type found near

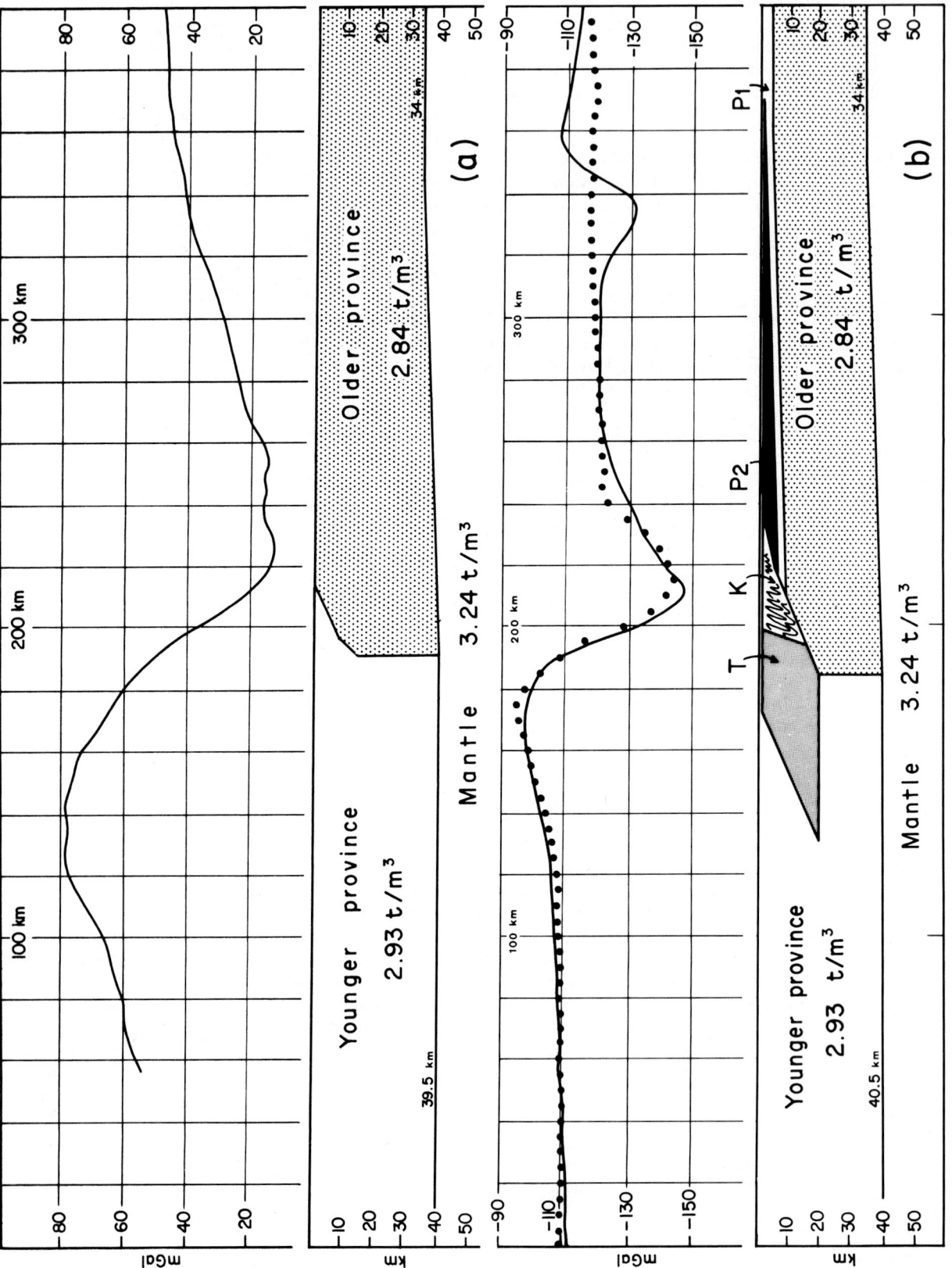

FIG 6. a. Average gravity profile over five province boundaries in the Canadian shield and its interpretive model (after Gibb and Thomas, 1976). b. Gravity profile over the Kalahari line (solid line) and the theoretical anomaly (large dots) computed for the crustal model shown. T = Tshane complex (2.98 g/cm³); K = Kheis belt (2.93 g/cm³); P1 and P2 = Proterozoic cover rocks (2.93 and 3.08 g/cm³, respectively).

Prieska is just one possibility that could be pursued. Kimberlite emplacement controlled by reactivation of this tectonic lineament, and the occurrence of coal within the thick Karoo succession lying west of the Kalahari line, are other possibilities that have already attracted some attention.

CONCLUSION

It has been seen that the interpretation of regional geophysical data often starts with the identification of anomalies over known geologic features which, in most of the above examples, are quite remote from the (exposureless) areas of interest. The geophysical data then provide a means whereby the known geology may be extrapolated into these exposureless areas. A rationale for selecting interesting areas for more detailed investigation may thus be developed.

Stratigraphic drilling at widely spaced sites across Botswana has been carried out (1979–83) under Canadian government funding as a control for the geophysical interpretation. Publication of drilling results was scheduled for 1984 and will undoubtedly lend new direction to mineral exploration in the Kalahari.

The above examples show only the application of gravity and magnetic methods at a highly regional scale, but these same methods will certainly find renewed application in the phases of more detailed exploration that now lie ahead.

ACKNOWLEDGMENTS

I thank Dr. R. M. Key for bringing recent geological and geochronological data to my attention and Dr. M. D. Thomas for providing data for Figure 6. Permission from the Geological Survey of Botswana to publish this paper is gratefully acknowledged.

REFERENCES

Baldock, J. W., Hepworth, J. V., and Marengwa, B. S., 1976, Gold, base metals and diamonds in Botswana: Econ. Geology, 71, 139–156.
Cornell, D. H., 1977, A post-Transvaal age for the Marydale Formation, Kheis Group, Southern Africa: Earth Planet. Sci. Lett., 37, 117–123.
Cox, K. G., 1970, in Clifford, T. N., and Gass, I. G., Eds., African magmatism and tectonics: Oliver and Boyd (Edinburgh).
Crockett, R. N., 1967, Shashi, Post-Karoo System dykes: Geol. Map Sheet 2127A, Geol. Surv. Lobatse, scale 1:125 000.
Gibb, R. A., and Thomas, M. D., 1976, Gravity signature of fossil plate boundaries in the Canadian Shield: Nature, 262, 199–200.
Green, D., 1966, The Karoo System in Bechuanaland: Bull. 2, Geol. Surv. Botswana.
Hutchins, D. G., and Reeves, C. V., 1980, Regional geophysical exploration of the Kalahari in Botswana—a review: Tectonophysics, 69, 201–220.
Key, R. M., 1976, The geology of the area around Francistown and Phikwe, Northeast and Central Districts, Botswana: Dist. Mem. 3, Geol. Surv. Botswana.
Key, R. M., and Rundle, C. C., 1981, The regional significance of new isotopic ages from Precambrian windows through the Kalahari Desert in northwestern Botswana: Trans. Geol. Soc. South Africa, 84, 51–60.
Pretorius, D. A., and Maske, S., 1976, An issue devoted to mineral deposits in southern Africa: Preface: Econ. Geol., 71, 1.
Reeves, C. V., 1976, The delineation of crustal provinces in southern Africa from a compilation of gravity data: 20th Annu. Rep., Res. Inst. African Geol. (Leeds).
——1978, Failed Gondwana spreading axis in Southern Africa: Nature, 273, 222–223.
Reeves, C. V., and Hutchins, D. G., 1976, The National Gravity Survey of Botswana, 1972–73: Bull. 5, Geol. Surv. Botswana.
——1982, A progress report on the geophysical exploration of the Kalahari in Botswana: Geoexploration, 20, 209–224.
Smit, P. J., Hales, A. L., and Gough, D. I., 1962, Gravity Survey of the Republic of South Africa: Handb. 3, Geol. Surv. South Africa.
Smit, P. J., and Maree, B. D., 1966, Densities of South African rocks for the interpretation of gravity anomalies: Bull. 48, Geol. Surv. South Africa.
Tagini, B., 1972, Notice explicative à la carte géologique de la Côte d'Ivoire au 1/2 000 000: SODEMI (Abidjan).
Thomas, C. M., 1973, South Ngamiland: Geol. Map Sheet 2022D with pts. of 2022C and 2023C, Geol. Surv. Lobatse, scale 1:125 000.
Truswell, J. F., 1977, The geological evolution of South Africa: Purnell (Cape Town).

Geologic mapping of the basement of the Paris basin (France) by gravity- and magnetic-data interpretation

N. Debeglia* and C. Weber*

ABSTRACT

Simultaneous interpretation of gravity and magnetic data, taking into account all available information (geology, borehole, and nonconfidential seismic surveys), has been carried out with the help of potential-field transformation, modeling, and inversion software. Geophysical synthesis led to the delineation of a geologic map of the pre-Triassic basement of the Paris basin and to the delineation of the most important structural features of the Hercynian and Caledonian Ranges in France between outcropping Paleozoic massifs. Recent boreholes, drilled at the end of this synthesis, largely confirm the proposed interpretation. Thanks to simultaneous gravity and magnetic inversion, models of the Magnetic Anomaly of the Paris Basin can be proposed, and the origin of this anomaly can be related to a scheme of the structural evolution of the basin.

INTRODUCTION

The Paris basin is an intracratonic basin of approximately 180 000 km² lying between areas of outcropping Paleozoic rocks (Armorican massif, Central massif, Vosges, Ardenno-rhenan massif). The depth of the Permian basement is greater than 3 000 m in the center of the basin (Figure 1).

Knowledge of the basement has been vastly improved from petroleum-exploration studies, particularly seismic prospecting and borehole investigations (Heritier and Villemin, 1971). Interpretation of gravity data (Goguel, 1954; Gerard, 1971) has led to hypotheses on the intrabasement origin of some of the most important anomalies. The study of aeromagnetic maps gives complementary information (Gerard and Weber, 1971).

The present work resulted from an integrated investigation using gravity and magnetic data, and which took into account all other available information, such as geology, borehole logs, and nonconfidential seismic surveys. These interpretations were carried out with the help of transformations, modeling, and inversion software.[1]

The nonconfidential petroleum seismic data and information from the deep boreholes were used to map the depth of the pre-Triassic basement. The interpretation of the gravity and magnetic data enabled us to sketch a geologic map of the basement under the sedimentary cover.

BASIC DATA

The gravity map of France, surveyed for the most part and published by the Bureau de Recherches Géologiques et Minières (BRGM) includes, for the Paris basin, a measurement density ranging from 1 to 0.2 station per square kilometer. The error of the measured anomaly is about 0.2 mGal, so the mean accuracy of the mapping is generally better than 1 mGal.

The basic aeromagnetic data used for this synthesis were taken from the aeromagnetic map of France published by BRGM at a scale of 1:1 000 000 and issued from the survey carried out by the Centre National de la Recherche Scientifique in 1964 (barometric flight altitude, 3 000 m; spacing of lines and traverses, respectively, 10 and 100 km). More detailed surveys for petroleum prospecting are available locally. The most important anomaly of this map is termed the Magnetic Anomaly of the Paris Basin (MAPB), which extends from the English Channel to the northern boundary of the Central massif (Figure 1), about 400 km in length. The intensity of this anomaly reaches 250 nT in its southern part.

GEOLOGIC MAPPING OF THE BASEMENT

The synthetic interpretation of the gravity and magnetic data in the central part of the Paris basin was preceded by study of the geophysical behavior of the surrounding Hercynian massifs (Weber, 1967, 1968). Some well-logging results were also examined.

Figure 2 summarizes the density properties of the Hercynian basement of the Armorican massif and of the northern

---

*Bureau de Recherches Géologiques et Minières, Orléans, France.

[1] Most of this software was supported by a joint research program of TOTAL Compagnie française des Pétroles et Bureau de Recherches Géologiques et Minières.

Central massif. The Brioverian and Armorican Paleozoic sequences show densities varying largely with age and lithology; the Brioverian has the largest density (2.69 g/cm^3 on the average), and the Carboniferous the smallest (2.57 g/cm^3 on the average).

In the eastern part of the Paris basin, well data indicate strong density contrasts within Paleozoic strata, particularly between the Devonian and the Carboniferous formations. The metamorphic formations range from a low density of 2.55 g/cm^3 to a high density of 2.70 g/cm^3 where the metamorphic grade decreases. Amphibolite densities range from 2.7 to 3.0 g/cm^3. Likewise among the plutonic rocks the most mafic formations have the highest densities (2.90 g/cm^3). Granites show a wide variety of density: 2.6 g/cm^3 on the average for leucogranites, 2.65 g/cm^3 for biotite granites, and up to 2.75 g/cm^3 for granodiorites.

With respect to magnetic properties, except for the south Armorican iron-ore bodies, the highest values of magnetic susceptibility (up to 0.025 SI) are found in some granitoids (granodiorite type) on the northern border of the Central massif. Most granites have a very weak magnetic susceptibility, as do the Brioverian and Paleozoic series, in the absence of contact metamorphism. The susceptibility of metamorphic rocks rarely exceeds 0.0025 (SI). The volcano-sedimentary formations can reach values of 0.01 (SI).

Preliminary interpretation of the gravity and magnetic data, carried out on the Hercynian basement and its border (Weber, 1972), confirm the characteristic behavior of Paleozoic units whose limits can be extrapolated beneath the sedimentary cover. Localized control of these interpretations is possible from the data obtained from deep boreholes.

Structural interpretations using transformed maps

The goal of these transformations is to put potential-field data into a form that facilitates structural interpretation.

Maps of the vertical gravity gradient (Figure 3a) and of the

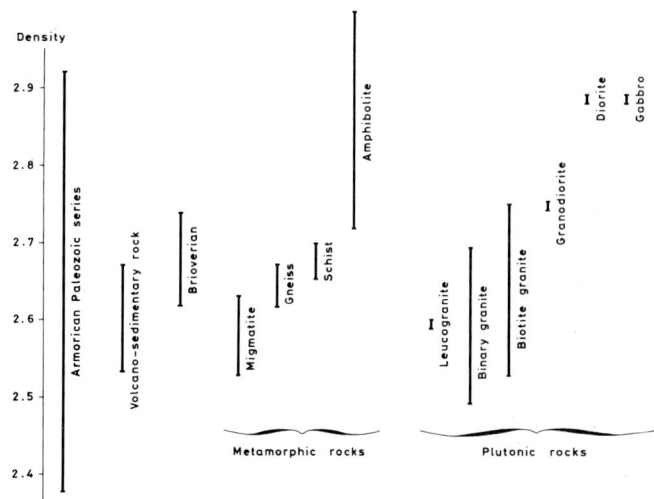

FIG. 2. Density (g/cm^3) of outcropping basement.

magnetic field reduced to the pole (Figure 3b) have thus been calculated. Gravity has been, in this way, brought to a degree of derivation comparable to that of the magnetic field. In fact, if the susceptibility of the body considered for interpretation were proportional to the density, and if the Earth's magnetic field were vertical, the vertical gradient of gravity would be proportional to the vertical component of the magnetic field (Goguel, 1972). The preceding transformations were carried out by applying, in the frequency domain, the corresponding operator (Gerard and Griveau, 1972).

Assuming that the gravity effect of the sedimentary sequence is negligible, the downward continuation of gravity data to the top of the Paleozoic basement can be calculated by transforming and filtering in the frequency domain (Debeglia, 1979). In the central part of the Paris basin, the Bouguer anomaly (Figure 4a) has been continued downward to a surface approaching the top of the Paleozoic basement (Figure 4b). The vertical gradient of this continuation (Figure 4c) allows preparation of a map of the schematic distribution of basement densities (Figure 4d; here, on the hypothesis of infinite downward structures). The effect of the resulting model is compatible with the initial field, as profile modeling shows (Figure 5). With the aid of well data, a table can be proposed (Table 1) relating the structures thus defined and the possible lithologies.

The transformations have greatly contributed to geologic mapping of the top of the basement (Figure 6). This map also includes some intrabasement units that have a strong gravity or magnetic effect (in particular, the magnetic and dense formations associated with MAPB).

Control by recent boreholes

It is possible that the basement mapping here presented will be superseded when direct knowledge of pre-Triassic terrain is obtained. In fact, the low resolution of potential-field data does not allow us to point out the complexity of the

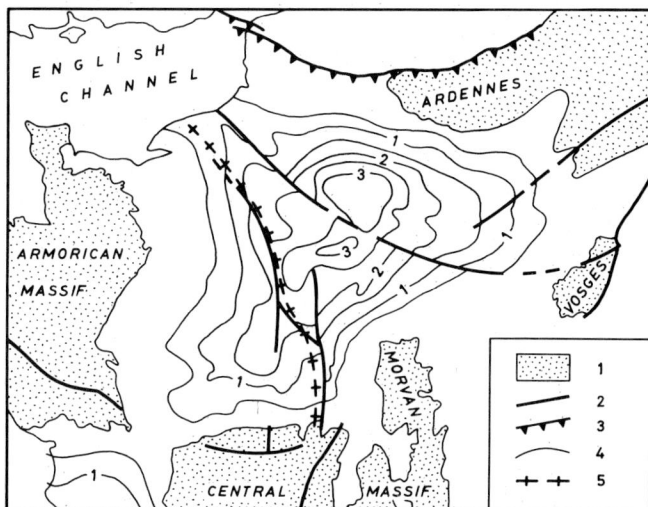

FIG. 1. Geologic setting of Paris basin. Explanation: 1 = outcropping basement; 2 = fault; 3 = overthrust fault; 4 = depth of basement, in thousands of meters; 5 = axis of Magnetic Anomaly of Paris Basin (MAPB).

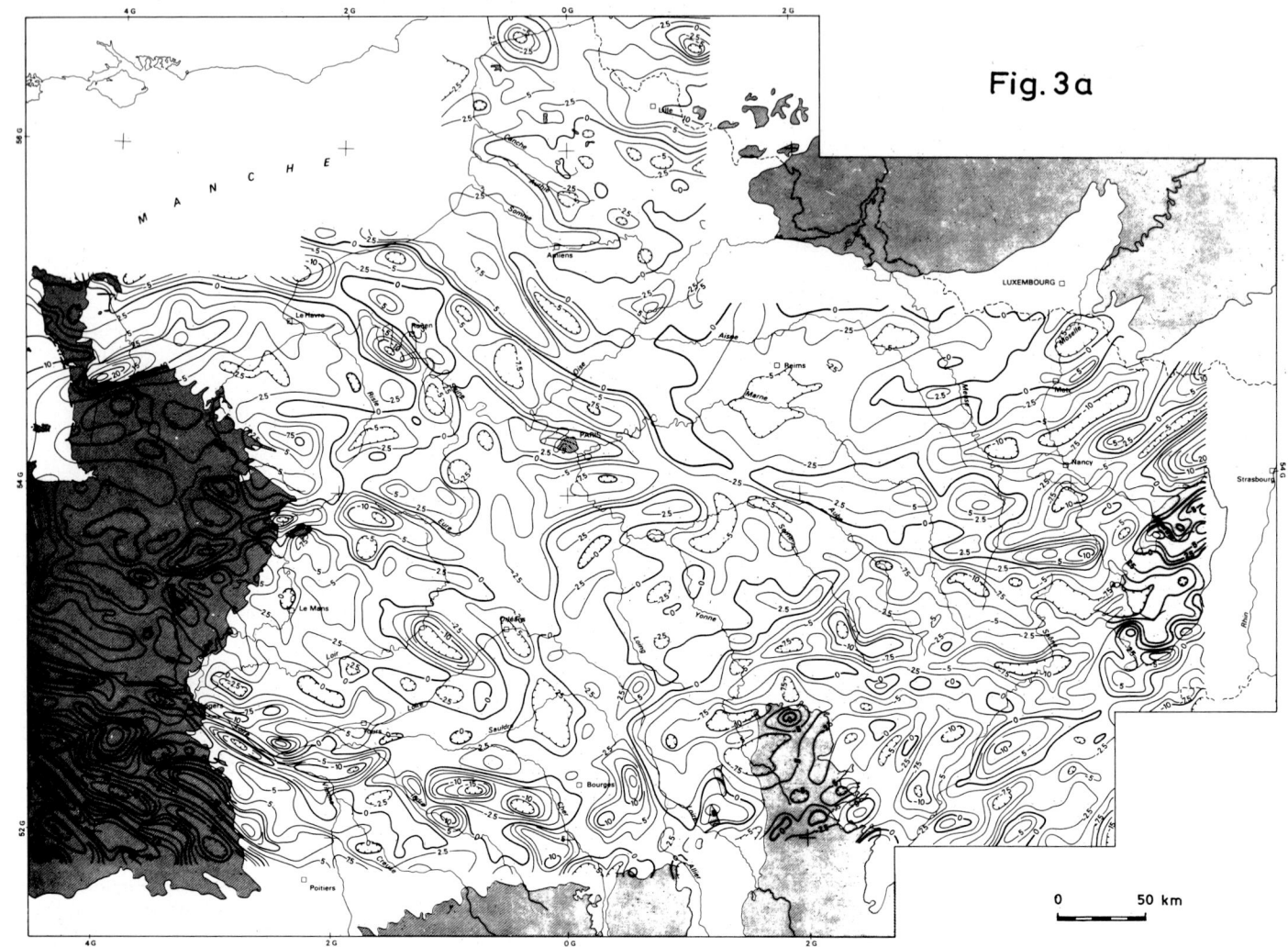

FIG. 3a. Paris basin region. Vertical gradient of gravity field (in 0.1 mGal/km).

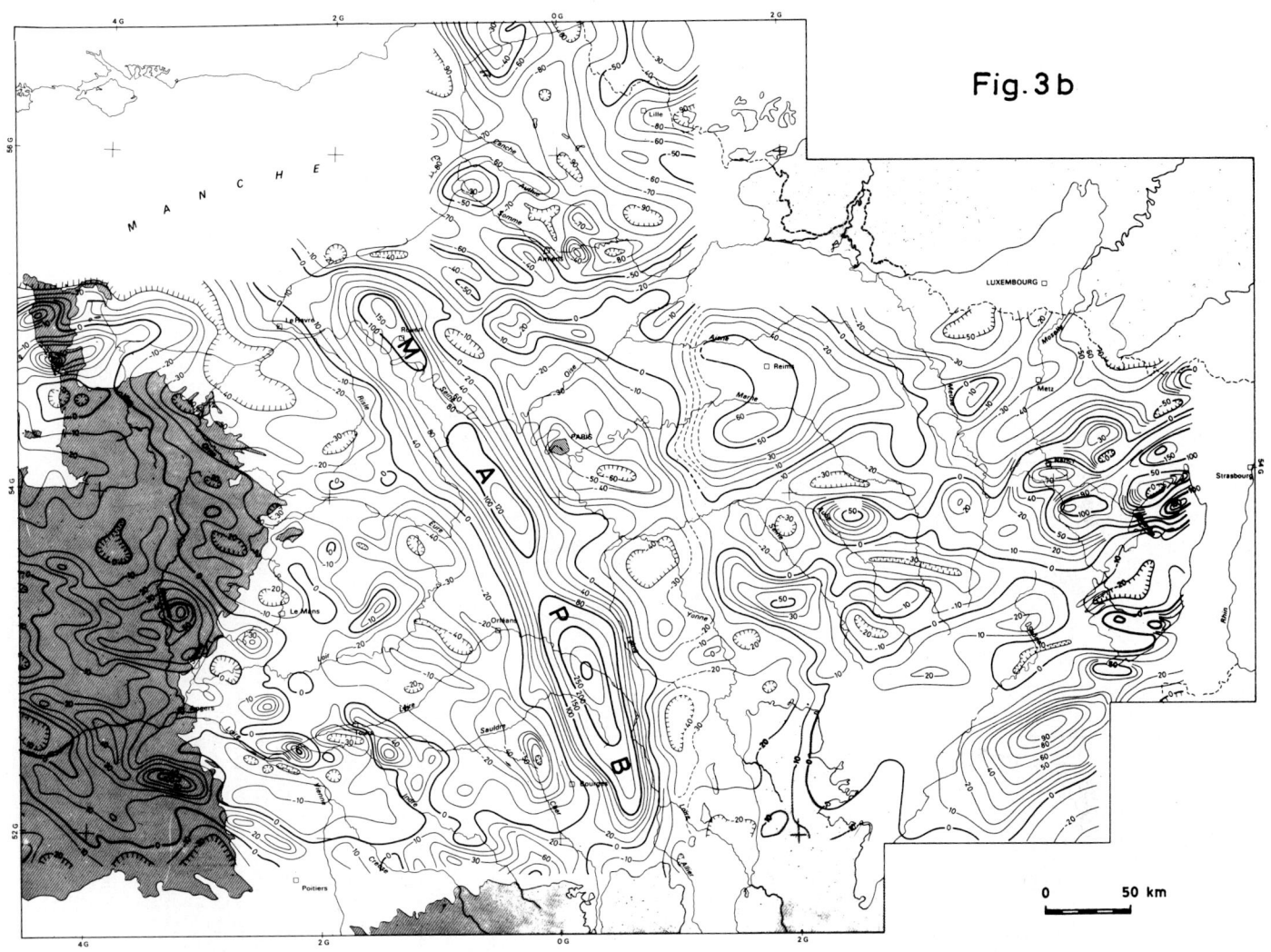

FIG. 3b. Paris basin region. Magnetic field reduced to the pole, elevation 3 000 (in nT).

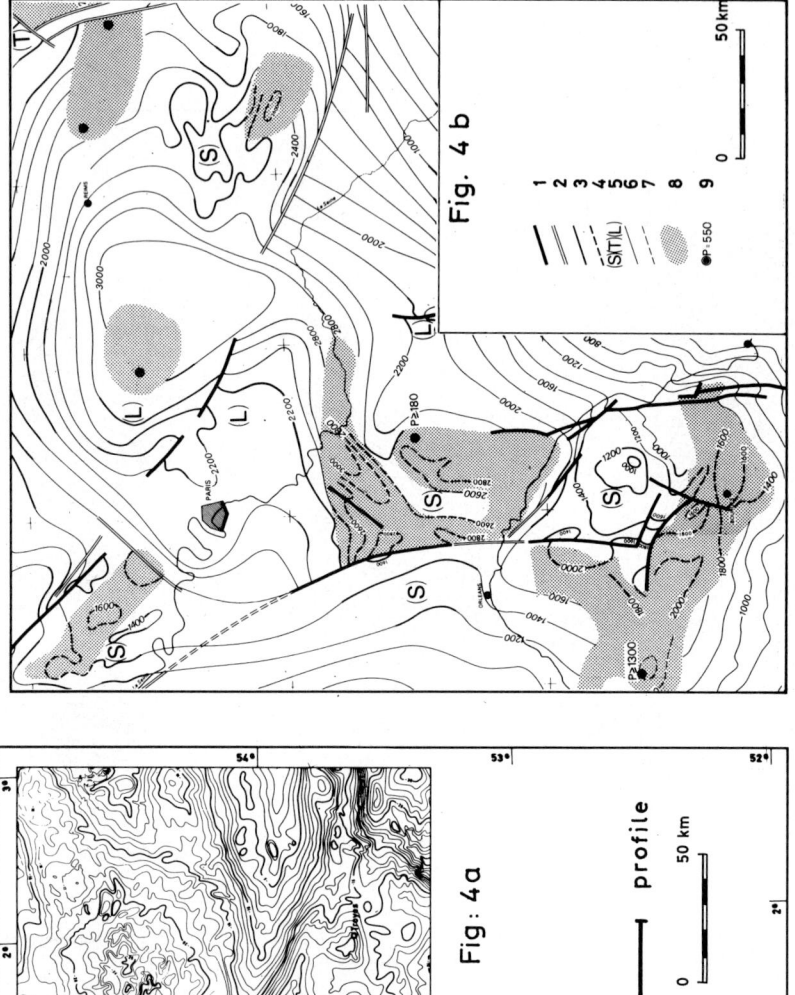

FIG. 4. Examples of the use of downward continuation to the top of basement, Paris basin region.
4a: Bouguer anomaly (in mGal).
4b: Isobath map of the top of basement:
  1 = discontinuities according to seismic data.
  2 = discontinuities according to gravity and aeromagnetic data.
  3 = isobaths according to seismic data on Paleozoic basement.
  4 = isobaths according to seismic data on pre-Permian basement.
  5 = levels mapped by seismics: basement (S), Trias (T), or Lias (L).
  6 = isobaths according to well data on Paleozoic basement.
  7 = isobaths according to well data on pre-Permian basement.
  8 = Permian basins.
  9 = boreholes drilled to the Permian, showing Permian thickness (in meters).

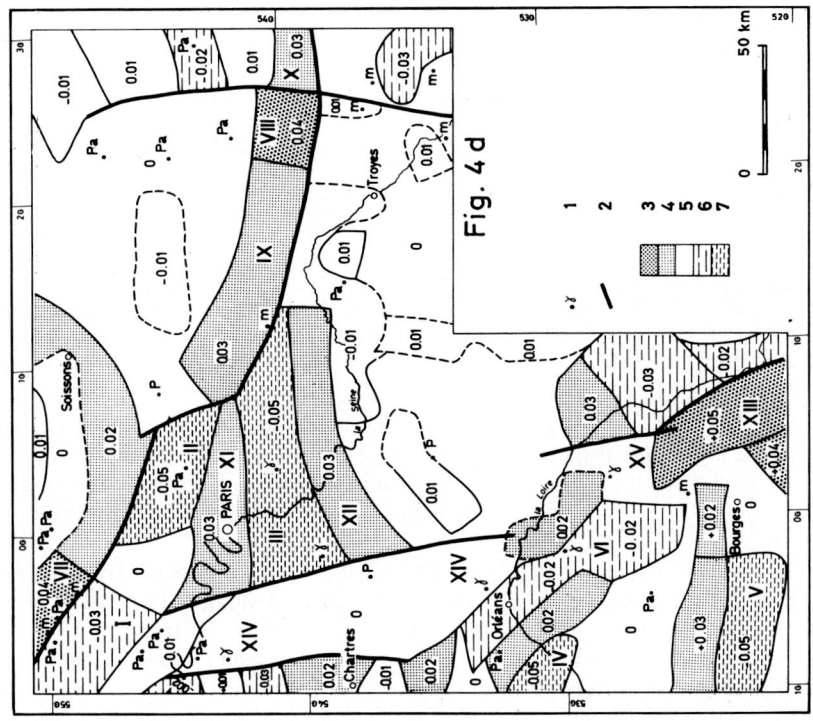
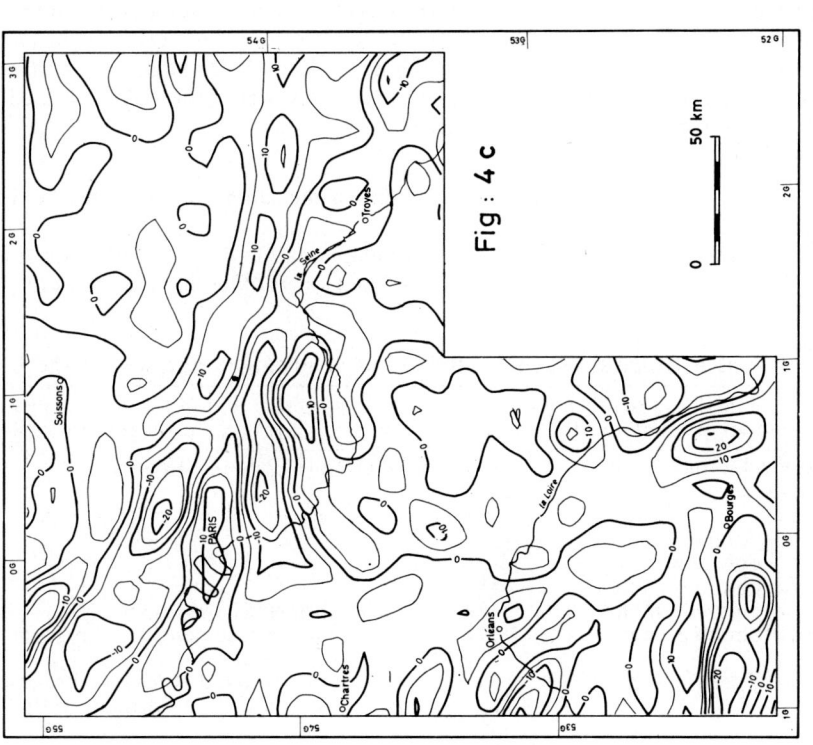

4c: Vertical gravity gradient at the top of basement (in 0.1 mGal/km).
4d: Density sketch map of the top of basement:
1 = principal boreholes showing the nature of basement (γ = granite; m = metamorphic; Pa = Paleozoic; P = Permian).
2 = gravity faults.
3 = gravity structures with highly positive contrasts (mafic rocks).
4 = gravity structures with positive contrasts (metamorphic rocks).
5 = gravity structures with near zero contrasts.
6 = gravity structures with negative contrasts (granites, Permian basins).
7 = gravity structures with highly negative contrasts (leucogranites).

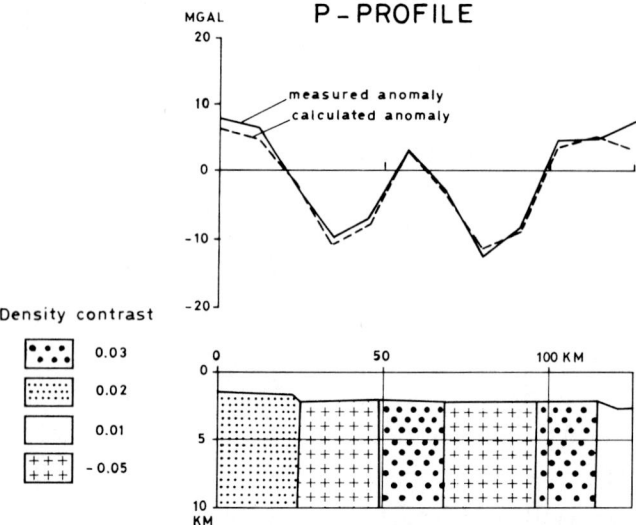

FIG. 5. Gravity modeling of a profile, P′, of density distribution taken from Figure 4d, with depth of basement taken from Figure 4b.

Table 1. Proposed correlation between structures of Figure 4d and lithologies.

Structures	Mean density (for the hypothesis of infinite downward structures)	Proposed lithology
II, III, IV, V	−0.05 g/cm^3	Leucogranite
I	−0.03 g/cm^3	Permian
VI, XIV, XV	0 to −0.02 g/cm^3	Granite and Paleozoic
VII, VIII, IX, X, XI, XII	0.02 to 0.04 g/cm^3	Metamorphic rocks
XIII	0.05 g/cm^3	Mafic rocks

geological reality. Thus, a single geophysical discontinuity can correspond to a complex grouping of faults or of contacts. Moreover, only the large geologic units that have strong density or magnetic-susceptibility contrasts can be detected in this manner. Finally, a thin cover with a density approaching that of overlying sedimentary rocks (Permian, Paleozoic, schists) cannot be detected.

This is the reason why it is important to verify the interpretations established by geophysical synthesis, using recent well data. Since completion of the geophysical interpretation, nine boreholes have reached the Paleozoic basement. For the most part, they confirm the given interpretation (Table 2).

Structural results

Two major geophysical features are evident in the geophysical synthesis.

The first, the Bray–Vittel fault (BVF in Figure 6), extends from Pays de Bray to the Vosges, striking N 130 degrees to east–west. North of this discontinuity a metamorphic unit can be considered as the western extension of the crystalline Hercynian zone that crops out in Germany. Northward, a thick Paleozoic sequence (Rhino–Hercynian zone) occurs, into which magnetic and highly dense igneous bodies have intruded. In the north of France this unit overthrusts autochthonous Paleozoic strata.

The second feature is a grouping of nearly north–south faults extending along the MAPB. The MAPB axis marks an obvious discontinuity of structural trends, averaging N 110 degrees W (Armorican direction) and N 70 degrees E (Morvano–Vosgian direction).

The intensity of the magnetic effect observed in the MAPB has led us to assume that extensive mafic units occur in the Paleozoic basement. However, none has been reached by boreholes. Along the MAPB, only granitic formations have been shown by well data and confirmed by local negative gravity anomalies.

Modeling of the southern part of the MAPB

Simultaneous gravity and magnetic inversion have been carried out on profiles in the southern part of the MAPB

Table 2. Results of principal petroleum boreholes drilled after geologic mapping of the basement.

Borehole	Level reached and depth below sea level (in meters)		Lithology and depth (below sea level, in meters) predicted by geophysical synthesis	
La Houssay en Brie	Metamorphic basement	2390	Granite–Paleozoic contact	2350
Lhuitre 1	Permo–Triassic	2322		
	Permian	2593	Permian	2500
	Lower Autunian / Upper Stephanian	3304		
	Metamorphic basement	3555		
Crecy en Brie 1	Metamorphic basement	2469	Metamorphic basement	2300
Heurtebise	Metamorphic basement	2594	Paleozoic basement, not far from contact with granite	2700
Donnemarie 1	Triassic to Permo–Triassic	2387 to 3489	Undifferentiated Paleozoic basement, less than 5 km from Permian basin	2900
Corfelix	Permian	3250	Permian	3200
Marsangis 1	Permian	2765	Undifferentiated Paleozoic	2800
Tousson 101	Permo–Triassic	2300	Permian	undefined (2700: pre-Permian basement)
Videlles 1	Undifferentiated basement	2300	Undifferentiated Paleozoic	2100

Geologic Mapping of Basement, Paris Basin

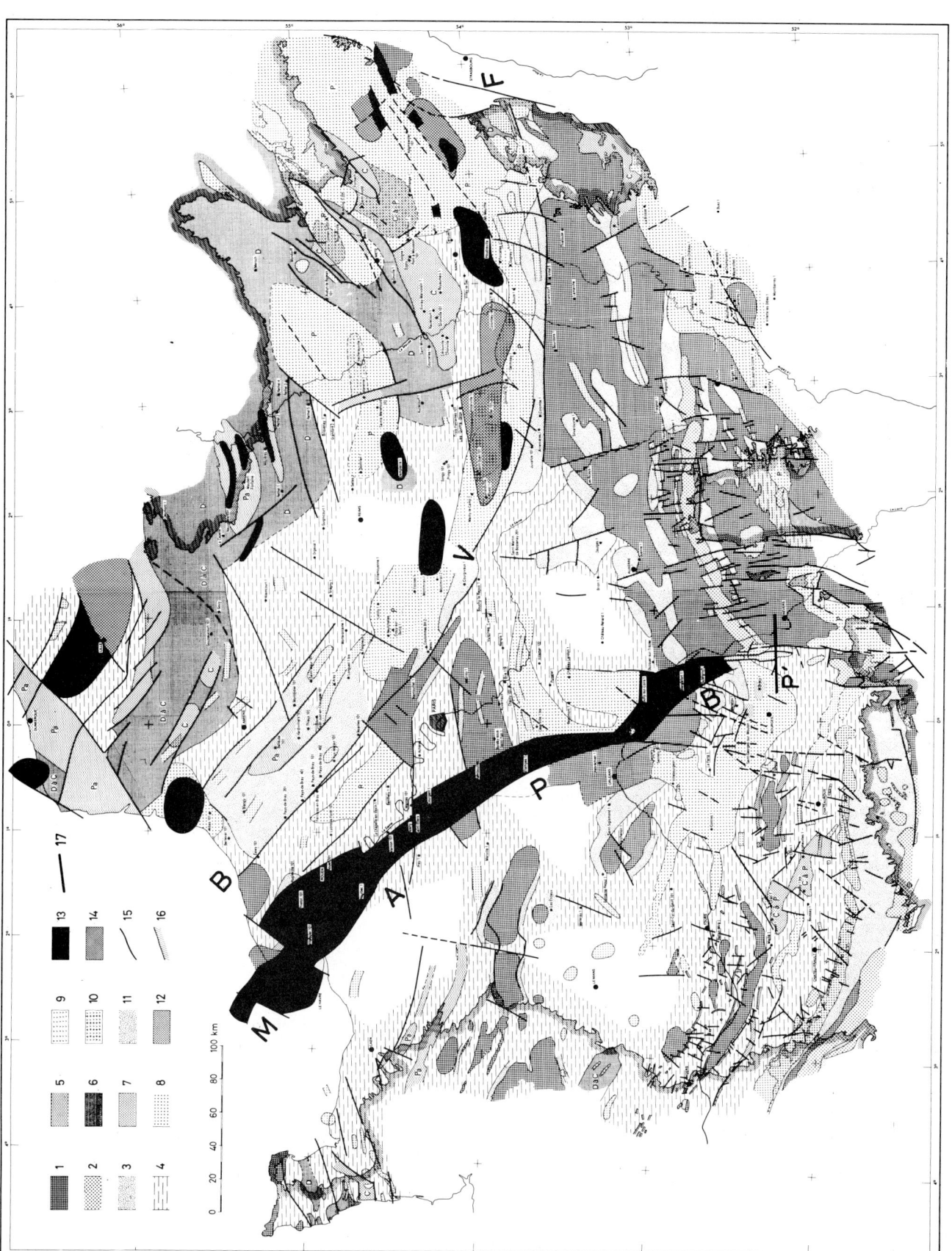

FIG. 6. Geologic map of the Paleozoic basement. BVF: Bray-Vittel fault. MAPB: Magnetic Anomaly of Paris Basin. Low-density and weakly magnetic formations: 1 = granite (sensu lato); 2 = leucogranites; 3 = metamorphic rocks; 4 = undifferentiated Brioverian or Paleozoic; 5 = lower Paleozoic; 6 = Devonian; 7 = Carboniferous; 8 = Permian. Dense and magnetic formations: 9 = mafic rocks; 10 = granodiorites; 11 = metamorphic rocks and contact metamorphism; 12 = iron-bearing formations; 13 = deep magnetic masses; 14 = deep heavy masses; 15 = fault; 16 = limit of outcropping basement; 17 = location of profile P' of Figure 7.

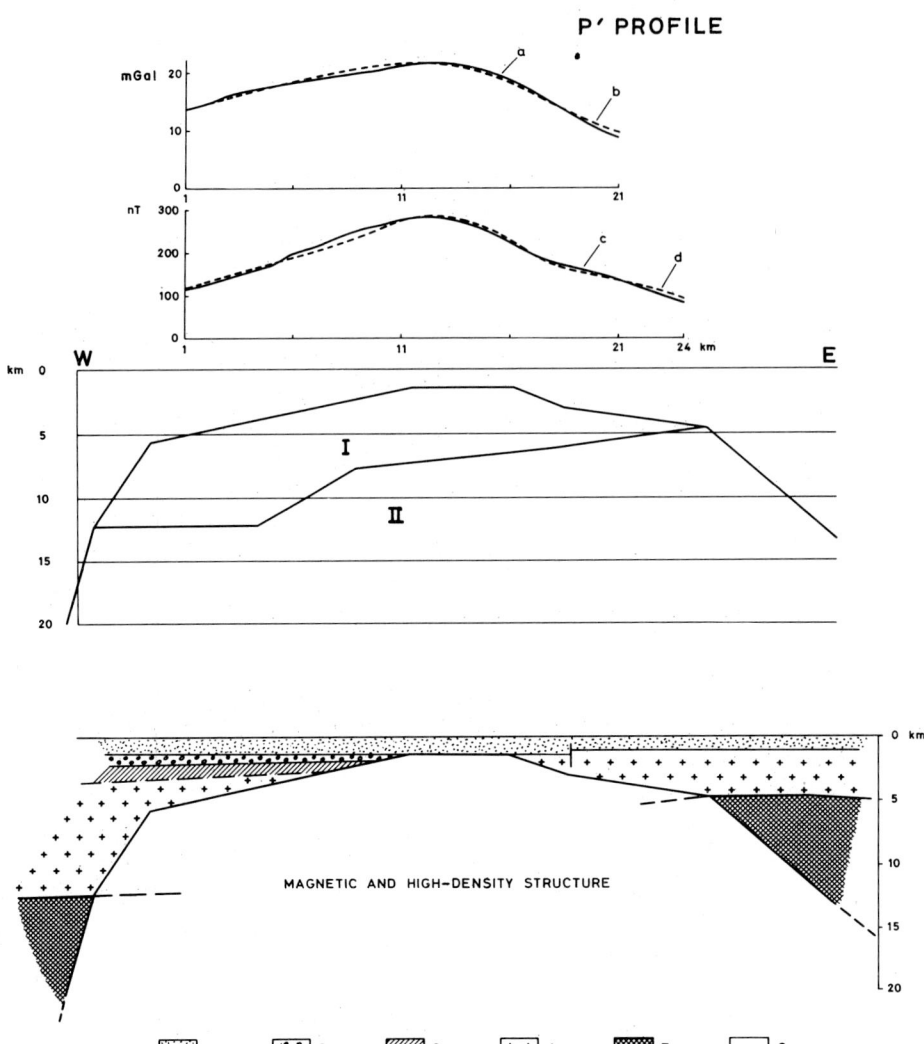

FIG. 7. P' profile: modeling example of MAPB (located on Figure 6). a = Bouguer anomaly; b = gravity effect of modeling; c = aeromagnetic data; d = magnetic effect of modeling; I: $\Delta X = 0.0016$ SI units, $\Delta \rho = 0.12$ g/cm^3; II: $\Delta \rho = 0$, $\Delta X = 0.0016$ SI units.
Explanation of lithologies:
1 = sedimentary rocks.
2 = Permian.
3 = undifferentiated Brioverian and Paleozoic.
4 = granite.
5 = basement (nonmagnetic and dense).
6 = magnetic and high-density structure; from Menichetti and Guillen (1983).

(Menichetti and Guillen, 1983; Menichetti, 1984). As shown in Figure 7, a positive gravity anomaly is related to the magnetic anomaly. Inversion has been performed for different pairs of density and susceptibility contrasts using an initial model obtained by direct interactive modeling. It has been assumed that the structure is an intrabasement intrusion with homogeneous magnetization from its upper surface to the Curie depth (about 20 km). We assume also that its density contrast decreases with depth. This last assumption probably involves an increase with depth of the density of surrounding formations (granitic units overlying a denser series of unknown nature).

To arrive at a suitable solution for this inversion, the density contrast (at the top of the structure) would vary from 0.1 to 0.25 g/cm^3, and the susceptibility contrast between 0.016 and 0.032 (SI). The top of the structure is between 1.8 km (nearly outcropping at the top of the basement) and 4 km deep, according to the chosen density-susceptibility pair. One of these solutions is shown in Figure 7; for this solution, the top of the structure crops out at the top of the basement.

In the center of the Paris basin, the gravity anomaly related to the MAPB is much smaller and partly masked by various intrabasement heterogeneities (granite batholiths, for example). The maximum depth of the magnetic structure has been found to be great, up to 8 km (Le Mouel, 1969). So the whole magnetic body can probably lie inside the lower dense unit (unit 5 in Figure 6), and thus no gravity effect is observed.

CONCLUSION

Geophysical synthesis allowed preparation of a map of the lithology and structure of the pre-Triassic basement of one of the most important French sedimentary basins, the Paris basin. It is now possible to sketch the hidden continuation of the principal Caledonian and Variscan units under the sedimentary cover (Autran, 1980).

A preliminary hypothesis for the structural evolution of the Paris basin and of the Magnetic Anomaly of the Paris Basin (MAPB) can be suggested. Prior to emplacement of leucogranitic eohercynian units extending from the Vosges to the Armorican massif (dated at 340 m.y. in its outcropping part), a tensional phase (rifting) allowed intrusion of mafic rocks, which explains the MAPB. Later, a compressional phase with sinistral strike-slip faults, well known in the Central massif (Sillon Houiller fault, dated at 280 m.y.), occurred with a major overthrust mainly in the northern part of the Paris basin. At the same time the intrusion of granitic bodies marked a new magmatic event along the MAPB axis. Later vertical displacements controlled sedimentation from Permian to Tertiary time. These displacements can be observed on the faults bordering the MAPB (Debeglia and Debrand–Passard, 1980). Subsidence was the dominant tectonic movement up to Mio–Pliocene time.

REFERENCES

Autran, A., 1980, Place du socle du bassin dans le cadre structural varisque, *in* Synthèse géologique du bassin de Paris (ouvrage collectif publié sous la direction de C. Megnien): Mém. 101, BRGM Fr.

Debeglia, N., 1979, L'interprétation des champs dérivants d'un potentiel par modèles équivalents: Thèse 3ème cycle, Univ. Strasbourg.

Debeglia, N., and Debrand–Passard, S., 1980, Principaux accidents tectoniques issus des corrélations entre les données de géophysique et les données de terrain (au sens large), dans le Sud-Ouest du bassin de Paris: Bull. Soc. Geol. Fr., **22**, 639–646.

Gerard, A., 1971, Apports de la gravimétrie à la connaissance de la tectonique profonde du bassin de Paris: Bull. 2, BRGM Fr., sect. 1, (2), 75–88.

Gerard, A., and Griveau, P., 1972, Interprétation quantitative en gravimétrie et magnétisme à partir des cartes transformées de gradient vertical: Geophys. Prospect., **20**, 459–481.

Gerard, A., and Weber, C., 1971, L'anomalie magnétique du bassin de Paris interprété comme élément structural majeur dans l'histoire géologique de la France (C.R. Acad. Sci. Fr., **272**, 921–923).

Goguel, J., 1954, Levé gravimétrique détaillé du bassin parisien: Publ. 15, BRGM Fr.

——— 1972, Tendances modernes dans l'interprétation géologique des données gravimétriques: Bull. 5, BRGM Fr., 2d ser.

Heritier, F., and Villemin, J., 1971, Mise en évidence de la tectonique profonde du bassin de Paris par l'exploration pétrolière: Bull. 2, BRGM Fr., sect. 1, (2), 11–30.

Le Mouel, J., 1969, Sur la distribution des éléments magnétiques en France: Thèse, Fac. Sci. Paris.

Menichetti, V., 1984, Techniques inverses en gravimétrie et magnétisme. Application à la terminaison sud de l'anomalie magnétique du bassin de Paris: Thèse 3ème cycle, U.S.T.L. Montpellier.

Menichetti, V., and Guillen, A., 1983, Simultaneous interactive magnetic and gravity inversion: Geophys. Prospect., **31**, 929–944.

Weber, C., 1967, Le prolongement oriental des granites du Lanvaux d'après la gravimétrie et l'aéromagnétisme: Mém. 52, BRGM Fr., 83–90.

——— 1968, Données géophysiques sur le prolongement du socle cristallin du Morvan sous ses bordures sédimentaires: Bull. Soc. Géol. Fr., **10**, 263–272.

——— 1972, Le socle antétriasique de la partie sud du bassin de Paris d'après les données géophysiques: Bull. 3 et 4, BRGM Fr., 2d ser., sect. 2, 219–343.

Crustal geology of Arizona as interpreted from magnetic, gravity, and geologic data

John S. Sumner*

ABSTRACT

Regional aeromagnetic and gravity maps of Arizona show a variable pattern of basement-fabric trend directions, commonly correlating with physiographic and structural geologic features in the state. The most prominent correlation is seen in the two major physiographic provinces, the Basin and Range and the Colorado Plateau, where the tectonism appears to be controlled by the underlying trend directions of the basement fabric.

At the southern edge of the Colorado Plateau province an ancient buried intrusive paleorift is identified from its distinctive magnetic signature. Also in that province and to the east near the state border there is a major 30-mGal low. This gravity low is on the Datil volcanic pile and measures about 100 km in diameter. The Mesa Butte magnetic anomaly and gravity gradient is a major linear feature in the state, which extends from southeastern Utah to west-central Arizona.

In the Basin and Range province, geophysical patterns are obvious for a northwesterly basement-fabric trend, whereas the more recent normal faulting typical of this Cenozoic tectonic province is generally north-northwesterly in direction.

Several of the gneiss-dome "metamorphic core complexes" exist in the Basin and Range portion of the state. Evidence is offered that these bodies are structural remnants of northeast-trending Precambrian basement blocks now being assimilated into the Basin and Range structural regime.

Some of the larger porphyry copper districts are related to arcuate magnetic lows, probably caused by the destruction of magnetite in the deep intersecting crustal fractures that controlled the circulating hydrothermal fluids.

The Phoenix arc, identified from magnetic data, is at least 400 km long and concave to the south. It is interpreted as a late Precambrian transform fault at the south end of the White Mountains intrusive body, the 50- × 100-km paleorift feature mentioned above. Most of Arizona's porphyry copper deposits are found in the lithospheric dilatational stress regime along and south of the Phoenix arc.

Gravity modeling of all alluvial basins within the state has been carried out in a program that integrated all available subsurface data. The resulting depth-to-bedrock map infers the Basin and Range fault strike and throw, and also quantitatively assesses the ground-water resources in this region.

Other regional geophysical techniques, including deep reflection seismic, electrical and electromagnetic, and heat-flow studies, have assisted in understanding the crustal geology of the state.

INTRODUCTION

The surface geology of Arizona is dominated by the Basin and Range province to the southwest and the Colorado Plateau province to the northeast. These physiographic features are controlled by vertical and horizontal basement geologic structures and are complicated by concealing alluvial, volcanic, and sedimentary deposits. Although soil and vegetative cover is minimal, there is much less than 10 percent of exposed basement rock over the state. Surface rock generally is weathered to depths of 10 to 100 m, and the water table is seldom shallow. In this environment, regional geophysical mapping, such as discussed in this report, can be useful in determining basement geologic structure and lithology between exposures and also at depth. The generally shorter wavelength anomalies of the magnetic map

---

*Laboratory of Geophysics, Department of Geosciences, University of Arizona, Tucson, AZ 85721.
A modified version of this paper will appear in the Arizona Geological Society Digest Volume on the *Geology of Arizona*.

(Figure 1, after Sauck and Sumner, 1970) resolve basement features more clearly than the anomalies of the residual-gravity map (Figure 2, after Lysonski et al., 1980), but interpretation of the magnetic data is more complex.

An important simplifying assumption made in the analysis of continental aeromagnetic anomalies (Reford and Sumner, 1964) concludes that induced rather than remanent magnetization is largely responsible for most anomalies. The larger grain size of continental rocks enhances the induced magnetization component. At depth with higher temperatures the induced magnetization parameter can be supplemented by colinear viscous remanent magnetization. Usually the validity of this directional-magnetization assumption can be seen in the anomaly signature.

From the magnetic map of Arizona, it is possible to observe distinctive trends underlying the Basin and Range and Colorado Plateau provinces. Many of these trends may indicate structural features that predate exposed geology and that probably have controlled the tectonic expression of these provinces.

DATA BASES

The industry-sponsored regional aeromagnetic data were obtained in 1968 by University of Arizona personnel using a digitally recording proton-precession magnetometer with 1-gamma (1-nT) precision with photo positioning of flight lines. North–south flight lines were spaced 5 km apart at 9 000- and 11 000-ft (2 743- and 3 353-m) sea-level altitudes. The higher flight elevation is in the eastern fifth of the state, east of 110° W longitude. The minor noise was manually smoothed, and the International Geomagnetic Reference Field (IGRF), tie-line level changes, and diurnal effects were removed before compilation. All basic data from the magnetic survey are available from the University of Arizona Geophysical Society. This SEG student chapter also markets the regional geophysical maps that are reviewed herein.

The Arizona Gravity Data Base (Schmidt et al., 1973) has been very useful as an integrated data filing system. The database is being continually modified and updated, most recently by Lysonski (1980) and Aiken et al. (1981). Since 1980, all of the gravity data have been compiled on the International Gravity Standardization Network of 1971 (IGSN71), referenced to the Geodetic Reference System of 1967, are terrain corrected using a density of $2.67 g/cm^3$, and are corrected for geodetic curvature.

The gravity data for this report have been gathered over the years by several groups and many individuals and include data from the University of Arizona, Arizona State University, Northern Arizona University, the U.S. Geological Survey, the U.S. Defense Mapping Agency, and industry-supported research programs. A comprehensive gravity base-station network and gravimeter-calibration course was established in 1964 (Sumner, 1965). The gravity station interval has been at most a 5-mile (8-km) interval in northern Arizona and at best a 1-mile (1.6-km) interval in some of the urbanized valleys of southern Arizona. Data from approximately 44 000 gravity stations are presently compiled in the University of Arizona Gravity Data Base. The displayed gravity residual map is effectively filtered by removal of the very long-wavelength, deep-seated Bouguer gravity-anomaly features, almost all of which are related to isostasy. The elevation-correlation method of constructing the 23 Complete Residual Bouguer Anomaly Maps of Arizona (Lysonski et al., 1982), which are similar to but more detailed than Figure 2, is described by Aiken et al. (1981).

METHODS OF ANALYSIS

Many of the geophysical anomalies discussed in this paper have been previously modeled by others so as to determine, by comparison, interpreted depth, size, shape, and physical-property contrast. This type of analysis has been carried out by most of the governmental and academic groups that originally gathered the geophysical data. In particular, Sauck (1972), West (1972), Klein (1982), Oppenheimer (1980), and MacInnes (1982) have made major contributions to the interpretation of specific anomalies.

The figures accompanying this report are admittedly inadequate to portray fully every feature being discussed. The larger and more detailed maps being referred to may be needed for the comprehensive review of specific anomaly relationships.

PROVINCES

The State of Arizona can be divided into two provinces (Figure 3) by study of mapped magnetic and gravity signatures. In this context, definition of the two provinces may be different than that defined by physiographic and surface structural-geological relationships. The conventional "transition zone" is difficult to define precisely using geophysical data alone, although a poorly resolved zone of lateral change is implied from the regional data.

Transition zone

The transition zone primarily is a region of physiographic change between the Basin and Range and the Colorado Plateau provinces. This region sometimes appears to be tectonically active as an extension of the intermountain seismic belt (Sumner, 1976), with generally north–south high-angle faults in a horst-and-graben structural mode. Transition-zone geologic structures can be identified by locally detailed geophysical surveys, and surveying is currently being conducted by Northern Arizona University researchers D. Best and D. Brumbaugh (pers. comm., 1983) on some of these features in northern and central Arizona.

The concept of a transition zone is in some ways linked to the idea that the Colorado Plateau might have been differentially uplifted during Basin and Range tectonism, subsequent to the Laramide orogeny about 70 m.y. ago. This particular kind of "uplift" is contested by geologic field observations (Peirce et al., 1979) to the extent that, at least in Arizona, the Colorado Plateau should be viewed as the still relatively undisturbed erosional remnant of strata that originally extended into the Basin and Range province. Geophysically, a minimum-depth wavelength analysis of the Fourier components of the digitized aeromagnetic data indicates that the basement rock in the Colorado Plateau region

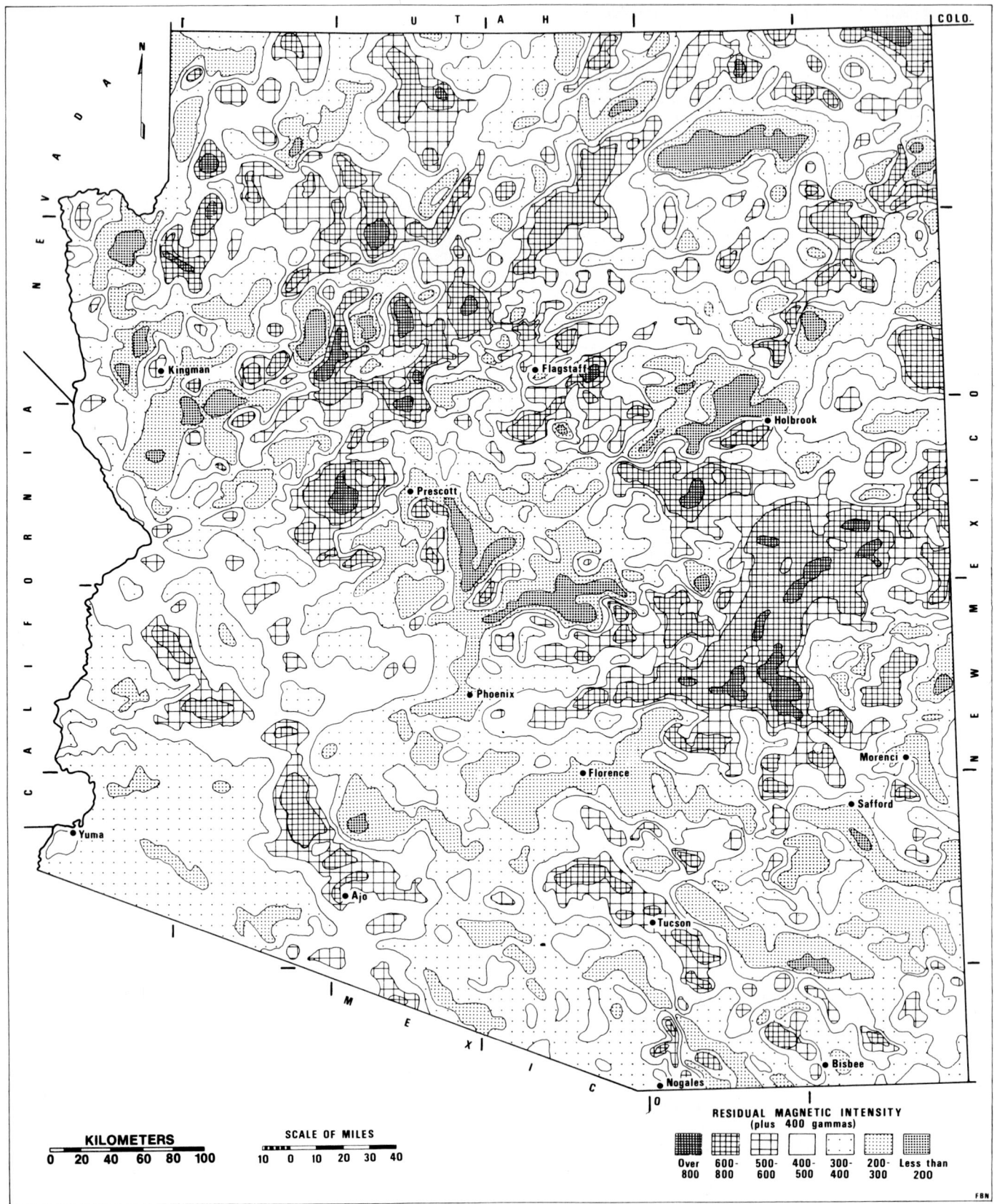

FIG. 1. Residual aeromagnetic map of Arizona. After Sauck and Sumner (1970).

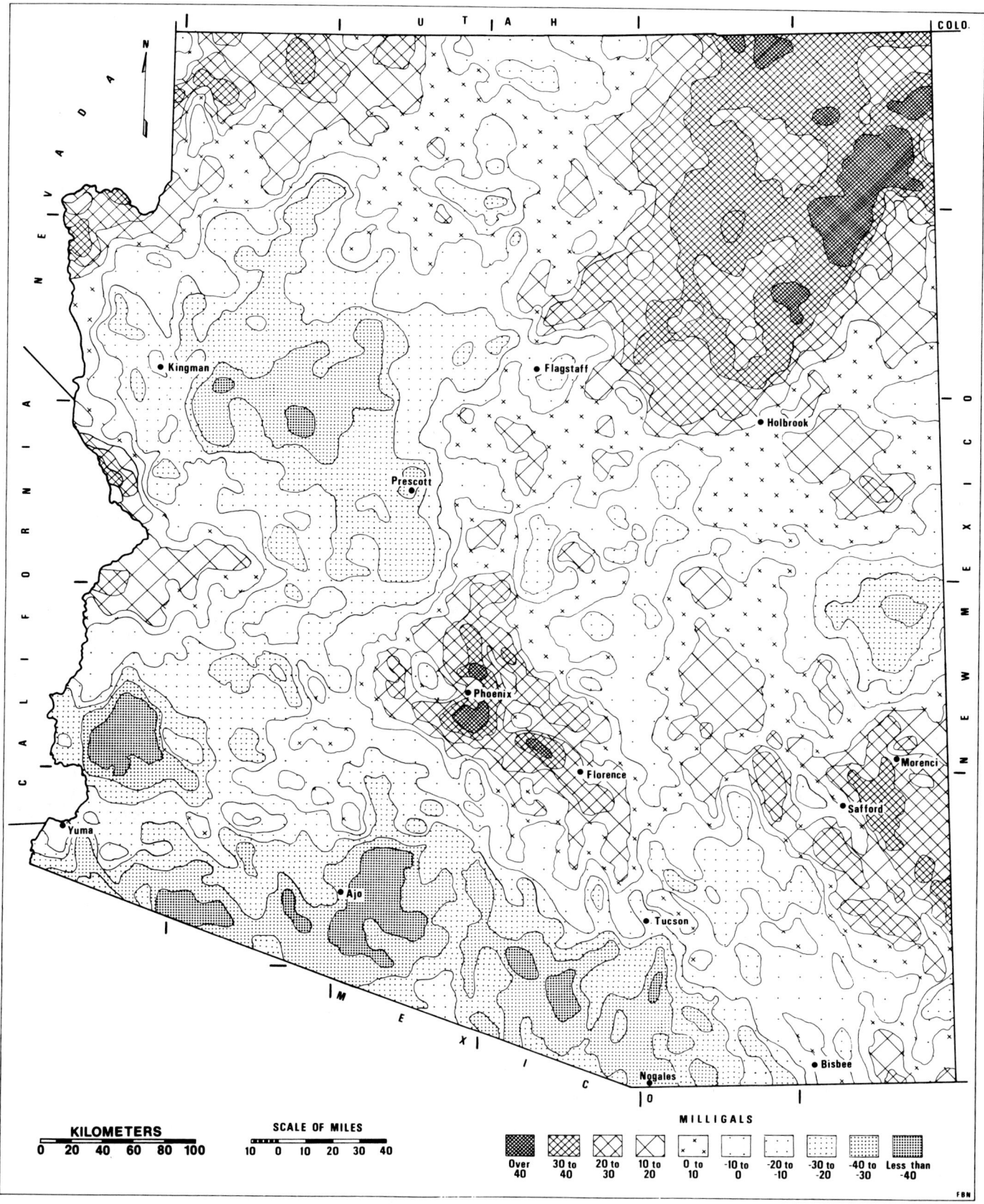

Fig. 2. Residual Bouguer gravity-anomaly map of Arizona. After Lysonski et al. (1980).

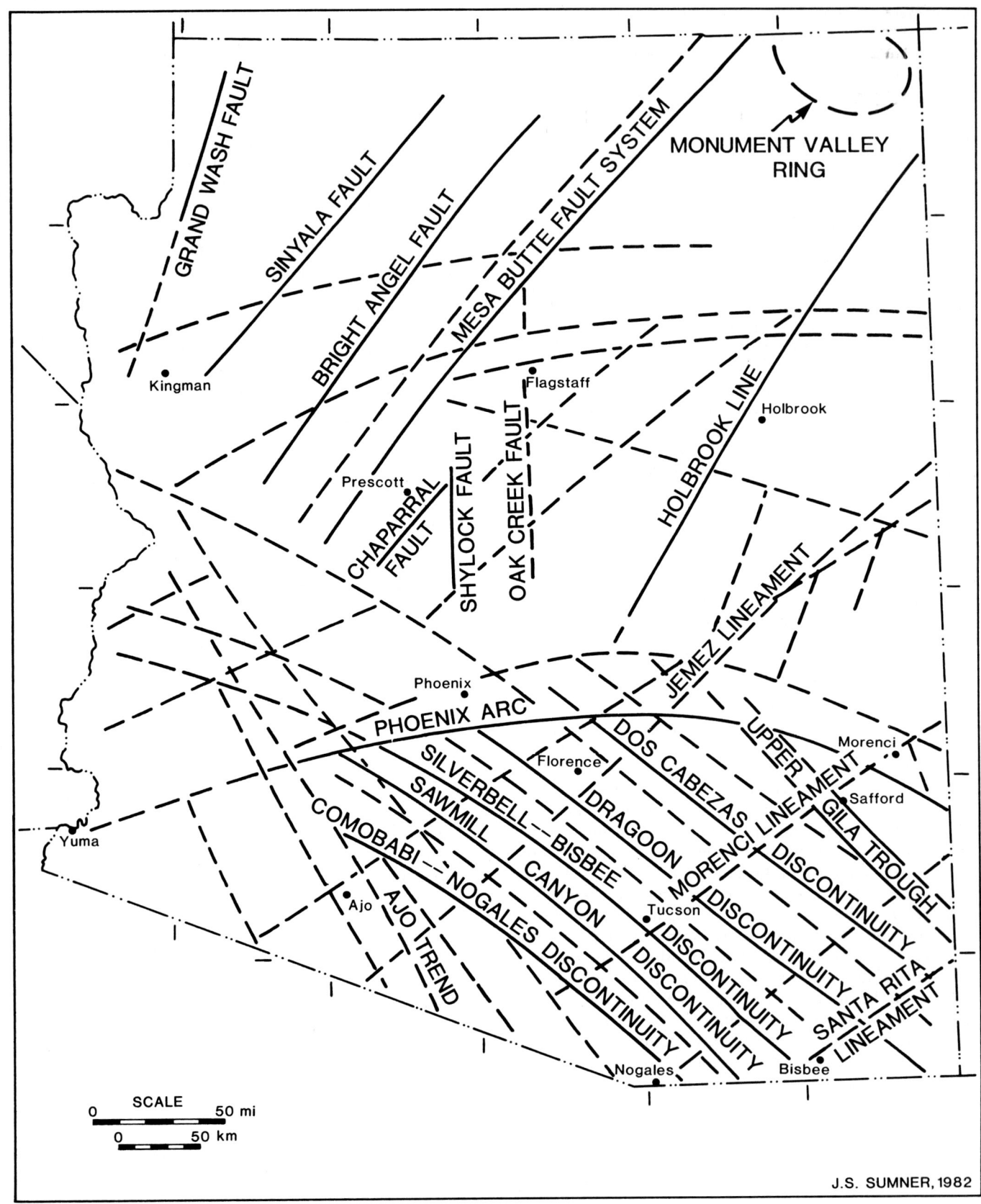

FIG. 4. Map of Arizona showing regional magnetic trends interpreted from the colored 1:1 000 000-scale aeromagnetic map by Sauck and Sumner (1970). Dashed lines indicate less certain trends.

intrusive body, it probably is an ancient tectonic rift and spreading center. Such failed rifts are postulated from regional geophysical mapping to exist in several parts of the North American craton (Carmichael and Black, 1984). Magnetic-anomaly "stripes" can be identified that are bilaterally symmetrical with and parallel to the long axis of the rectangular center, and the extended serifed ends could be related transform faults extending east and west. The Phoenix arc forms the southern edge of the body. The original age of this intrusive body seems to have been late Precambrian—older than the overlying Troy quartzite and younger or contemporaneous with deposition of the associated Apache Group. Unlike some other ancient rifts, like the Midcontinent geophysical anomaly (Woollard and Joesting, 1964), the correlative gravity expression of this paleorift is more subdued, with only an irregularly shaped 30-mGal positive anomaly near the south end. The time span for creating this rift width would likely be almost 1 million years, assuming average rifting rates of about 5 cm per year.

Tectonic activity in major paleorift zones usually recurs over geologic time, in part owing to isostatic readjustment. More recent extrusive events probably associated with the White Mountains intrusive anomaly are to be found in the White Mountains volcanic field on the northeast corner of the feature, and along the southern edge of the body.

Datil volcanics gravity anomaly.—A circular, 30-mGal Bouguer gravity-anomaly low with a 100-km diameter is centered about 25 km east of Baldy Peak in the White Mountains about 95 km north of Morenci (Figure 3). This gravity anomaly is associated with an andesitic volcanic pile that appears to have been sporadically deposited from Tertiary time to the present (Ratte et al., 1969). The Holocene volcanism that has created spatter cones and flows in the area is on the northwest edge of the anomaly and probably contributes only slightly to the anomaly amplitude.

It is not unusual for the bulk density of andesitic volcanic rocks to be lower than average values for crustal rocks because of extensive fractures, voids, and tuffaceous strata present. The bulk-density contrast can be calculated from the gravity anomaly and the probable depth of the body, assumed as being 3 km, giving a density value of about 2.50 g/cm^3. On comparison with the estimated particle density of about 2.70 g/cm^3 of the body, the porosity is seen to be large, on the order of 12 percent. As a result of the large size of this volcanic body, its high porosity, and its fracture-controlled permeability, a major ground-water resource probably exists within the Datil volcanics anomalous body. Other similar large, porous volcanic-rock masses also exist elsewhere in the southwestern United States, such as the San Francisco volcanics near Flagstaff, the Aquarius Mountains east of Kingman, and the Castle Dome Mountains northeast of Yuma.

San Francisco volcanic field.—Several of the details in mapped magnetic patterns of the volcanic area immediately north of Flagstaff (Sauck and Sumner, 1970) disclose an east–west trend to the short-wavelength anomalies, superimposed on northeast-trending medium-wavelength anomalies. This observation suggests that the surficial volcanic features may have been partly controlled by east–west fractures in nonmagnetic sedimentary rocks but that deeper crustal fractures in the area have a northeast trend in keeping with the fabric of the basement rocks in the region. U.S. Geological Survey geomagnetic-variation studies confirm the existence of a crustal conductive zone along the Mesa Butte fault system near Flagstaff but provide no information about east–west crustal conduits. Similar short-wavelength, east–west–trending aeromagnetic features can be seen on the more detailed map (Sauck and Sumner, 1970); these features are associated with the White Mountains volcanic field at the northeast edge of the White Mountains intrusions (Figure 3) and the extensive volcanism in the Mount Trumbull area of northwestern Arizona.

Canyon de Chelly and Monument Valley ring.—Some of the strongest residual gravity and magnetic anomalies in the state are related to intrusive basaltic plugs near Canyon de Chelly and near Monument Valley in northeastern Arizona (Figures 1, 2, 4). Of course, the high background values from the dense, magnetic basement rock in the area contribute to the intensity level. It is noteworthy that the only petroleum production in the state occurs from a volcanic sill near Teec Nos Pos in the extreme northeastern corner of Arizona (see Figure 3). Some of the volcanic diapirs in Arizona and in southeastern Utah appear to have mantle-derived petrologic materials contained in them. These kimberlite plugs are highly magnetic, so here there is question about the rock-type change involved in mantle magnetism (Wasilewski et al., 1979). It is interesting to see that the Monument Valley ring aeromagnetic anomaly has a central high surrounded in turn by a circular low followed by a circular high, reminiscent of the signature of major impact and extrusive ring structures. Inasmuch as this probably is a buried intrusive ring structure, the anomaly shape may be evidence of behavioral stresses under internal or external load conditions.

Basin and Range province

In the southwestern part of the state, Basin and Range tectonism is superimposed on Mesozoic and early Cenozoic mineralization and intrusion, and an underlying Precambrian basement. The geophysical patterns are not simple because of shorter wavelength topographic noise and longer wavelength regional effects. In general, Basin and Range faulting in Arizona trends about N 20 degrees W (Oppenheimer, 1980), whereas the older Laramide basement fabric and many of the prominent aeromagnetic anomalies trend N 45 degrees W (Sauck, 1972; Titley, 1981).

Tucson–Ajo–Kofa magnetic belt.—A dominant west-northwest-trending magnetic positive feature that may indicate a deep-seated intrusive(?) belt (Figure 3) is strongly expressed near Ajo and Tucson. The aeromagnetic indication of this inferred intrusive belt averages 40 km in width and is at some places offset, perhaps by the northeast-trending Jemez lineament, for example, crossing between Tucson and Ajo as suggested by Klein (1982). The gravity expression of this feature is minimal, so the density correlates more with a granodiorite rather than a granite. Trend direction and position could be related to crustal tensional stresses resulting from Mesozoic tectonism (Sauck, 1972). In the Tucson area the aeromagnetic anomaly occupies the central part of the Tucson basin, but source rocks suggestive of an intrusive belt are not obviously exposed in outcrop. The existence of more magnetic and probably denser rocks in the Tucson basin basement would cause errors in gravity modeling if a simple two-dimensional assumption of a laterally homogeneous basement is used in keeping with the classical normal-faulted layer-cake concept of Basin and Range structures.

Southeast of Tucson, crossing the magnetic belt at the lowest part of the gravity anomaly in the Tucson basin, Eberly and Stanley (1978) relate deep-drilling data with seismic profiles and probable geology. The drill hole in this anomaly encountered a mineralized intrusive at about a 3 600-m depth.

Castle Dome Mountains and Ajo negative gravity anomalies.—From information on the nature and extent of low-density volcanic rocks near these anomalies, a 3-km thickness of this material could easily be present in these areas, located about 60 km northeast of Yuma and 30 km east of Ajo. Similar 40-mGal gravity lows are found near the Aquarius Mountains between Kingman and Prescott, and on the United States–Mexico border near the Pinacate Mountains between Ajo and Yuma. The Aquarius Mountains (Figure 3) are also near the center of a broad negative gravity anomaly (Figure 2), which has been ascribed as being caused by a shallow asthenosphere in the area (Stone and Witcher, 1982). If this be so, the heat flow would be anomalously high, and there would be a chance for geothermal prospects in this area.

Gila Valley trough gravity low near Safford.—This long, narrow 20-mGal anomaly has almost a rift-like appearance, and modeling (Oppenheimer and Sumner, 1980) indicates that the alluvium must be at least 4 km thick in this northwest-trending 120-km-long trough. The strike of Basin and Range faulting along this trough closely parallels the older Laramide structural grain, which enhances its structural prominence.

The Gila Valley trough is the easternmost of several alluvium-filled valleys (Figure 5) that appear to radiate from the south end of the White Mountains intrusive anomaly on the south edge of the Colorado Plateau (Figure 3). There may well be a stress-related structural as well as an erosional/depositional relationship of these valleys with the White Mountains intrusives. Structurally, the White Mountains body could have acted as a deep crustal bulwark constraining the Basin and Range tectonism in southern Arizona, and the Gila trough would be a vestige of Basin and Range tectonism superimposed on the underlying rock fabric.

Alluvial valleys in southern Arizona.—In a project supported by the Water Resources Division of the U.S. Geological Survey, Oppenheimer and Sumner (1981) modeled the depth to bedrock in Basin and Range valleys in Arizona (Figure 5). All available geological and geophysical data were merged into the interpretation, providing an unusual amount of control for the analysis. Also, borehole gravity data (Tucci et al., 1982) provided density information from several localities, improving the accuracy of interpretation.

Except as relatively short-wavelength and small-amplitude linear anomalies on the gravity map, most of the Basin and Range structures are not very obvious on the statewide geophysical maps (Figures 1, 2). Indeed, but for a rough correspondence in trend direction, the Basin and Range structures appear to be superimposed on the older northwesterly terrane, and mainly only the deeper, structurally parallel trends are seen except on locally detailed maps. Where the older fabric is normal to the Basin and Range trend, as in the Harcuvar and Harquehala Mountains (75 km southeast of Parker in western Arizona) and at South Mountain (15 km south of Phoenix), some of the older structures persist.

Although primarily an interpretation of regional gravity data, Figure 5 is also a Basin and Range tectonic map, because structural directions are indicated by the trends of the alluvium-filled valleys. And the amplitude of fault throw is measured by the depth of alluvium in the valleys.

The Colorado River canyon north of Parker has the physiographic appearance of a contemporary rift valley, and the region exhibits some seismic activity (Sumner, 1976). North of Kingman and at about 40-km intervals from the Colorado River are parallel, alluvium-filled valleys (Figure 5) that may have been ancient courses of the Colorado River.

Much of southern Arizona depends on ground water for irrigated agriculture and domestic requirements, so an assessment of subsurface-water reserves is desirable for planning purposes. The depth-to-bedrock map (Figure 5) has proved useful for water exploration and quantity-evaluation purposes. Also, alluvium in some of the deeper valleys may yield low-temperature geothermal water. Land subsidence and earth fissuring owing to ground-water withdrawal, which is an increasing problem in southern Arizona, sometimes can also be predicted from strong gravity gradients over basement highs and above subsurface scarps near pediment edges.

Metamorphic core complexes.—The core-complex bodies of the West have been described by Armstrong (1982) and Coney (1979). Several Cordilleran metamorphic core complexes occur in the Basin and Range province of Arizona. These bodies seem to vary in age and date from Paleozoic to Cenozoic at the time of metamorphism. The rock type also varies somewhat but mainly is a granite gneiss, sheared with detachment or décollement denudation(?) faults. In Arizona, core-complex structures form the heart of most of the east-northeast-trending mountain ranges lying within the Basin and Range province.

Concerning regional-strike direction, the fact that most of the Arizona core complexes lie contrary to both the Basin and Range and the older Laramide northwesterly structural trends is a clue that they may be remnants of the older Precambrian northeast-trending terrane such as is seen in the basement rocks of the Colorado Plateau. And here it is important to note that individual core complexes in the Basin and Range province can be correlated with specific Colorado Plateau anomaly extensions.

Except for their unusual orientation and anomaly-trend association, none of these bodies in Arizona has a particularly outstanding regional geophysical signature, perhaps in part because the associated low-angle fault structures are not clearly observed in the regional gravity and magnetic data. An exception to this statement occurs at South Mountain, about 15 km south of Phoenix, where high-density amphibolites (Reynolds and Rehrig, 1980) cause a prominent 20-mGal high near this structure (Figure 3). The Bouguer gravity map

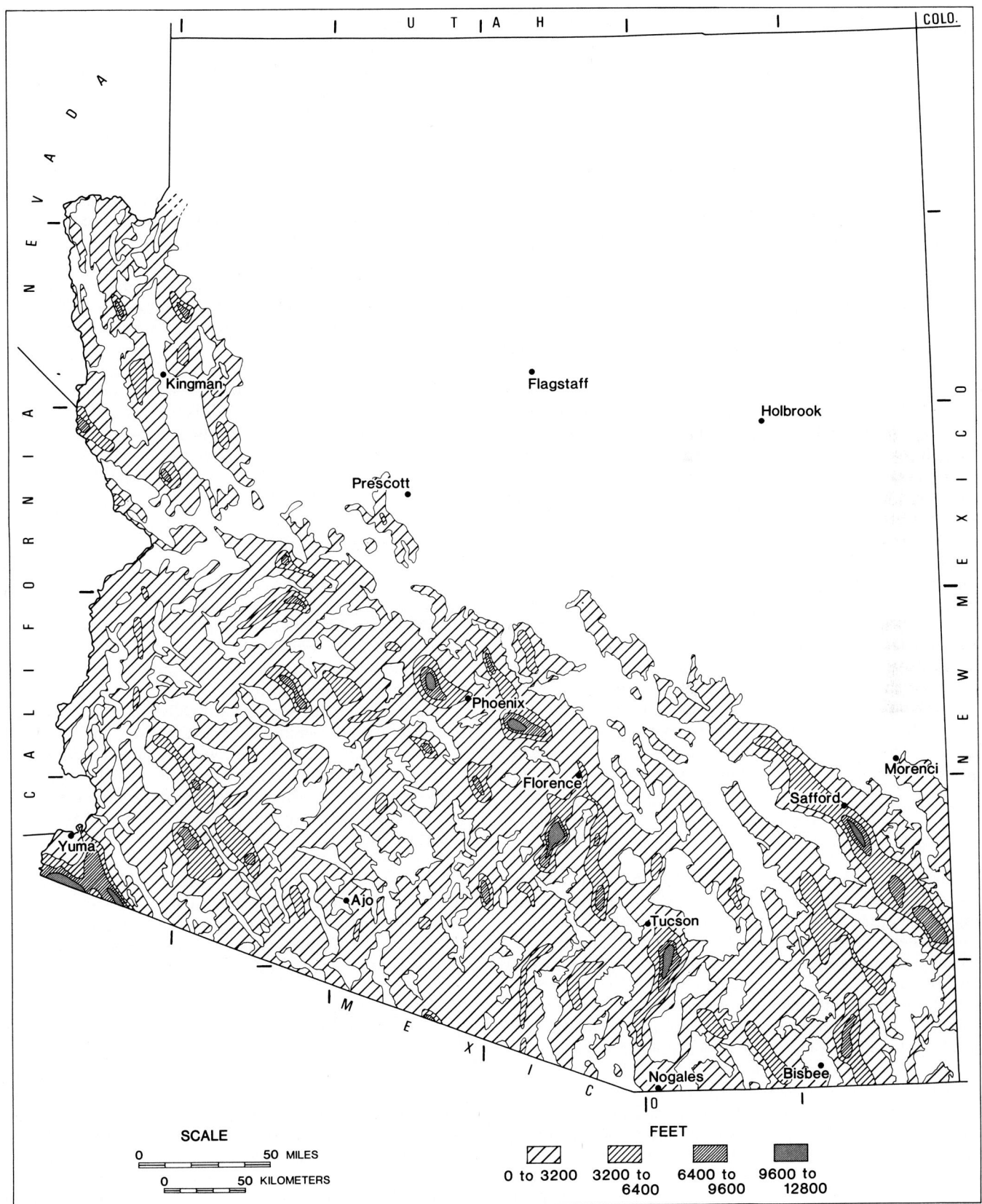

FIG. 5. Map showing generalized depth to bedrock, Basin and Range province, Arizona. After Oppenheimer and Sumner (1981).

also shows a high-density basement to be rather extensive nearby in central Arizona (Figure 2).

South Mountain is aligned with the Holbrook line gravity gradient and has a similar density and susceptibility contrast. The strike of the South Mountain foliation is rotated clockwise about 15 degrees from the strike of the Holbrook line trend.

Some of these gneiss-dome bodies are magnetic lows, such as the Tortilita–Catalina–Rincon complex about 20 km east of Tucson, which appears to be almost devoid of magnetite. The amphibolitic South Mountain area also is low in magnetite content, but the Harcuvar and Harquehala Ranges about 75 km southeast of Parker are moderately magnetic. Where intrusive granite stocks are present, such as in the Catalina Mountains, there is an associated gravity low of 5 to 10 mGal. Granite stocks elsewhere also produce gravity lows, as in the Santa Rita Mountains about 25 km south of Tucson.

One of the distinctive structural features of core complexes is the presence of low-angle listric, or curved, normal faulting. This type of structure is difficult to detect and trace using the regional gravity and aeromagnetic methods of this report. Nevertheless, where a susceptibility and density contrast exists, modeling of these structures is a useful supplement to other data.

Over the eons, erosional unloading of the Basin and Range province will isostatically cause the ranges to rise relative to the basins, predictably producing detachment fault structures that could be mistaken for major tectonic features.

None of the core complexes is extensive enough in horizontal dimension to be locally supported by a sizable, low-density root, a feature that would be indicated by the Bouguer gravity anomaly. One must conclude that these gneissic blisters are primarily rejuvenated intra- or supra-crustal features, supported by the deeper and surrounding flexural rigidity of the crust and lithosphere. This is not to say that regional isostatic compensation does not hold in Arizona.

LINEAMENTS

Over any past or present tectonically active area, surface structural alignments are often visible as expressions on topographic, geologic, geochemical, and geophysical maps. These lineaments become less distinct with age but are still meaningful in interpreting past stresses, displacements, and disturbance patterns. In Arizona, Mayo (1958), Lepley (1977), and many others have observed linear features from structural and photographic data. In recent years it has been fashionable to promote the use of satellite imagery over the spectral band from infrared to ultraviolet.

In the attempt to measure and quantify observations, computational methods have been devised that employ a directionally sensitive filter so as to remove any personal bias in evaluating lineaments. However, the features shown in Figures 3 and 4 were visually selected from the aeromagnetic-anomaly maps using the following criteria: (1) obvious relationship to a linear geophysical anomaly; (2) recognition from correlation with mappable geologic alignments; and (3) relationship with known mineral deposits or structural features such as discontinuities.

Texas zone

Perhaps historically the most discussed geologic alignment in the state has been the Texas zone or lineament (Mayo, 1958), defined as a belt up to 200 km wide extending northwestward across the southern part of the state. This structurally disturbed zone appears to be the left-lateral offset expression of an ancient (Jurassic?) lithospheric displacement with northwest-trending en-echelon fault blocks about 40–50 km wide. Titley (1976) discussed the tectonic relationship of these blocks at length for the area of southeastern Arizona. Although the southeast-trending features described by Titley (1976) are primarily geologic discontinuities and are labeled as such in Figure 4, they are also identifiable magnetic lineaments trending at a slight angle from the strike of the Basin and Range fault blocks that are implied in Figure 5.

Farther southwest a transcontinental "megashear," also with a left-lateral displacement, has been described as a trend crossing Arizona from Lukeville on the southern border to near Blythe, California (Anderson and Silver, 1979), crossing the Texas zone at a low angle. This megashear would lie roughly between the lineaments of Figure 4 southwest of the Ajo trend.

Magnetic and gravity alignments

Figures 3 and 4 indicate the positions of several linear trends noteworthy from their geologic exposures, such as the Oak Creek, Shylock, Chaparral, and Grand Wash faults. Detailed on-site geophysical surveys over these structures undoubtedly would give more exact information about location, dip, and perhaps throw.

It is interesting that some geologically important faults and structures (not shown in Figures 1–6), such as the Hurricane and Kaibab and the north-northwest-trending Colorado Plateau monoclines, do not have a more discernible expression at this regional geophysical-mapping scale discussed here. This observation points out the fact that there has to be a consistent contrast in physical properties for a feature to be traceable within sizable limits in regional geophysical data.

A well-developed horst and a graben are displayed in the gravity data, coincidentally along the lower and the upper Gila River Valley (Figure 3). The lower Gila positive anomaly is also outlined by satellite imagery (Lepley, 1977).

Phoenix arc.—An intriguing arcuate magnetic lineament, concave to the south and about 400 km long, is designated the Phoenix arc in Figure 4. No geologically mapped fault crops out along this feature, but there is an abrupt change in rock type across it. Alluvial and volcanic cover conceals the nature of this arcuate lineament along its entire length. This curved lineation may well be an ancient failed transform-fault trace, extending east and west from the southern end of the White Mountains intrusive anomaly (Figure 3). If this relationship is correct, the Phoenix arc probably is late Precambrian in age. It seems possible that the structural conditions that imposed the sharp curvature also contributed to the failure of the associated 100-km-long and 50-km-wide White Mountains intrusive rift.

The Phoenix arc is close to the northern boundary outlining the distribution pattern for most of the porphyry copper deposits in the state. And several major mining districts are within 15 km of its trace.

North of the Phoenix arc the lithospheric-stress regime would have been compressional, whereas south of this structure the stresses would have been dilational because of the stress field created by the White Mountains intrusives. If the White Mountains intrusives and Phoenix arc were reactivated during Laramide time, lithospheric stresses would have been favorable for the passive intrusion of porphyry copper bodies and convective hydrothermal circulation. The Laramide-stress conditions studied and reported by Heidrick and Titley (1982) and Titley (1982) are not contrary to this hypothesis.

The Phoenix arc is, in terms of terrestrial spherical trigonometry, a small circle, and as such its pole of rotation can be calculated. The present position of this Euler pole is N 29°21′, 111°40′ W, which lies about 65 km northwest of Hermosillo, Sonora.

ECONOMIC APPLICATIONS OF REGIONAL GEOPHYSICS

The economic and population growth in Arizona is largely related to the change in the bulk value of the mineral commodities being exploited. From an initial emphasis on the precious metals, mining evolved into high-grade and then low-grade copper. Next, energy sources, including uranium, coal, and geothermal deposits, have been sought. Currently there is a serious interest in water, sand and gravel, and industrial minerals. Recently, interest is returning to precious metals, and another cycle may be starting. Regional geophysics can be helpful in the exploration and development of all these materials, and even in determining favorable environments for relatively small deposits.

Concepts of interpretive settings evolve with developing theories of ore formation and economic occurrence. Regional geophysical mapping is always at least an indirect guide to economic deposits. Observable, known relationships can be extended into the unknown, inferred situations. The successful interpreter must be versed in geophysics, geology, economics, and their practical interrelationship.

Copper deposits

Regional geophysics is closely akin to regional geologic mapping. In this context, major structure and lithologic contrasts can be identified and traced. The regional environment of the resource is an important consideration in geologic and geophysical interpretation.

Massive sulfide deposits.—It is now well accepted that older and contemporary marine massive sulfide deposits are volcanogenic in origin. In Precambrian time, banded iron formation (usually later to become magnetic iron formation) often was stratigraphically related, in shallow seas, to ocean-ridge volcanogenic thermal springs in the depositional environment that existed at that time. Later geological events often complicated this favorable exploration setting through structural disturbances and through metamorphism. In Arizona, iron formations and associated Precambrian massive sulfide deposits are present. Anomalies related to magnetic iron formation are to be found south and east of Prescott, not far from the sizable massive sulfide deposits at Jerome, and at the Bradshaw Mountains.

Porphyry copper deposits.—Partly because of the nature of desert weathering processes and the resulting supergene enrichment of bulk low-grade disseminated copper deposits, Arizona is a leading copper producer in North America. These huge, low-grade, usually mineralogically zoned deposits (Lowell and Guilbert, 1970) appear to have formed in the roots of volcanic complexes where fracture conduits allowed hydrothermal plumbing systems (Norton, 1982) to concentrate the ore-forming fluids.

In Arizona, where most of the porphyry copper deposits happen to be quartz-monzonite bodies of the Lowell and Guilbert (1970) type, circulating fluids in the deep fracture zones below the porphyry copper deposits have often altered and destroyed the nearby magnetite present, with the iron having been redeposited as sulfides, skarn, or silicate minerals in the porphyry deposits. Types of chemical reactions, the nature of hydrothermal fluids, and alteration mineralogy are discussed in detail by Bean (1982). Regional magnetic surveying of the type described in this paper is a higher altitude method and thus "sees" longer wavelength anomalies from deep sources such as those related to the plumbing systems of the deposits themselves. Near-surface, high-susceptibility volcanic intrusives and flows usually have too short a wavelength to be resolved, and the dipole character of shallow features further reduces their magnetic expression.

Several of the major Arizona porphyry copper districts display an association with arcuate magnetic lows that are probably related to deep intersecting fracture systems. Examples are indicated in Figure 3 as Twin Buttes, Morenci, and Mineral Park. The Twin Buttes magnetic low is also correlated with the Sawmill Canyon discontinuity (Figure 4), and several copper deposits are to be found along its length extending into northern Sonora. With its sharp magnetic gradient, the Sawmill Canyon fault must be the surface expression of a deep-crustal hydrothermal system.

Certainly not all porphyry copper deposits are to be found on regional magnetic lows. The New Cornelia deposit at Ajo is high in magnetite and lies along an arcuate magnetic high. The mineralogy of this particular deposit is more like the dioritic-porphyry bodies outside of Arizona in that it is somewhat more mafic, similar to the diorite model of Hollister (1978) in which magnetite readily forms. The Silver Bell and Lakeshore deposits 50 and 80 km west-northwest of Tucson are also on or near aeromagnetic highs. Nevertheless, most of Arizona's copper production comes from districts that are associated either with the Phoenix arc, regional lows, or arcuate magnetic anomalies.

Salt (halite) and evaporite deposits

The Luke salt mass (Figure 3) is the source of a distinct, circular 25-mGal gravity low 25 km west-northwest of Phoenix, just north of the Phoenix arc. The body measures 8 × 12 km in horizontal extent and is up to 4 000 m thick (Eaton et al., 1972; Eberly and Stanley, 1978).

Salt beds are also known in the Red Lake area 55 km north

of Kingman, where there is a less distinct 15-mGal gravity low.

Another major evaporite deposit with a 2 000-m thickness of anhydrite occurs about halfway between Tucson and Phoenix, in the Pichaco basin (Figure 3). The gravity anomaly there is not large because of the high density of anhydrite; in fact, the 5-mGal low is mainly due to the thickness of overlying alluvium.

Other evaporite deposits are known to exist in the alluvial valleys and playa-lake beds of the state. The favorable environment can often be seen in gravity data, but extent and thickness will not be positively known until there is development drilling.

Ground water

As has been previously mentioned, large amounts of ground water exist in the alluvial valleys and volcanic rocks of Arizona. West and Sumner (1972) discussed methods of estimating ground-water volumes in basins, using gravity-survey data. Oppenheimer and Sumner (1981) estimated that approximately 2 200 km^3 of water are to be found in alluvial basins and that a similar quantity probably is available in the volcanic rocks. The depth of alluvial valleys in Arizona is shown in Figure 5. Although the water in volcanic rocks is a major resource in the southwest, research into this matter has not yet gained support.

Oil and gas

Hydrocarbon shows occur in the southeast and northeast parts of Arizona, but so far production is not significant. The search continues, however, and exploration for energy sources, including geothermal occurrences, will be guided by regional geophysical and geologic maps. Interpretation of geophysical data in southeast Arizona by Aiken (1978) shows what can be done to help localize the exploration areas. Many exploration crews have been active in Arizona, but findings are seldom revealed.

Breccia pipes

The Orphan Mine is a uranium producer in a breccia pipe located on the south rim of the Grand Canyon about 1 km west of the Bright Angel fault (Figure 4), and the mineralization appears to be related to the fault structure. Breccia pipes in igneous rocks invariably produce magnetic lows, but the thickness of nonmagnetic sedimentary rocks at the Orphan Mine attenuates the local anomaly. Detailing magnetic surveys on mineralized faults, and elsewhere, can readily locate mineralized breccia-pipe deposits if the host rock is somewhat magnetic.

FREE-AIR GRAVITY SURVEYING

The main feature of the free-air gravity-anomaly map (Figure 6) is the northwest-trending gradient extending across the central part of the state, correlating with the southwest edge of the Colorado Plateau. This gradient trend is largely due to an abrupt elevation change but also to a mass increase under the Colorado Plateau corresponding to, perhaps, a thinning asthenosphere. This gradient trend also coincides with the intermountain seismic belt (Sumner, 1976). Other smaller, subparallel negative and positive gradient trends occur in both the Basin and Range and Colorado Plateau provinces.

In parts of Arizona it has been noted (Sumner et al., 1976) that the zero free-air gravity contour tends to follow the pediment edge in the alluvial basins. This feature of the free-air map makes it of some use for exploration purposes. Otherwise, the Bouguer anomaly maps best show near-surface density changes of more interest to most geologists.

Arizona is largely in isostatic equilibrium, judging from the lack of long-wavelength free-air anomaly patterns in the region.

Regional gravity and isostasy

Over the southwest-to-northeast span of the entire state the long-wavelength crustal-density structure is an important matter, as brought out by Aiken et al. (1981). After removing the elevation-related regional trend, the residual Bouguer gravity-anomaly values can be closely related to the underlying shallow-crustal causative geology. This is a considerable improvement over the older, typically untreated Bouguer gravity-anomaly maps, such as the one by West and Sumner (1973), and any wavelength filtering method not based on a -34 mGal per 1 000 ft (-112 mGal per km) regional gravity change with elevation, derived from the Bouguer formula $g = -2\pi G\sigma h$. A wavelength-filtered residual without elevation control makes no assumption about isostasy so that elevation-related anomalies cannot be interpreted. Also, negative artifacts will be inserted next to positive anomalies, and the regional datum is left "floating." Although gravity-anomaly wavelength filtering may appear to enhance the data by emphasizing certain features, one usually is left with a residual map without knowing exactly what was removed from the original data. If one does not know specific regional trends, major geologic features cannot be adequately modeled (except visually) from an uncontrolled wavelength-filtered map.

REFLECTION SEISMOLOGY

A report on reflection-seismic exploration by Exxon Corporation by Eberly and Stanley (1978) over the Luke salt mass and the Tucson basin has been most helpful in understanding the basin-fill, evaporite, and basin alluvial-fill environment in Arizona. The seismic surveys and the subsequent drilling were initially laid out from interpretation of gravity data. Lack of velocity information in the Tucson basin caused initial depth estimates to be much too shallow, by at least 50 percent, for the subsequent drill budget. The semi-indurated Tucson basin-fill sediments finally proved to be more than 3 600 m thick.

Phillips Petroleum Company (Reif and Robinson, 1981) and Anshutz Corporation (Keith, 1980) have released some of their reflection-seismic data that were obtained during searches for an illusive overthrust belt in central Arizona. Although a phantom reflecting horizon was found and inter-

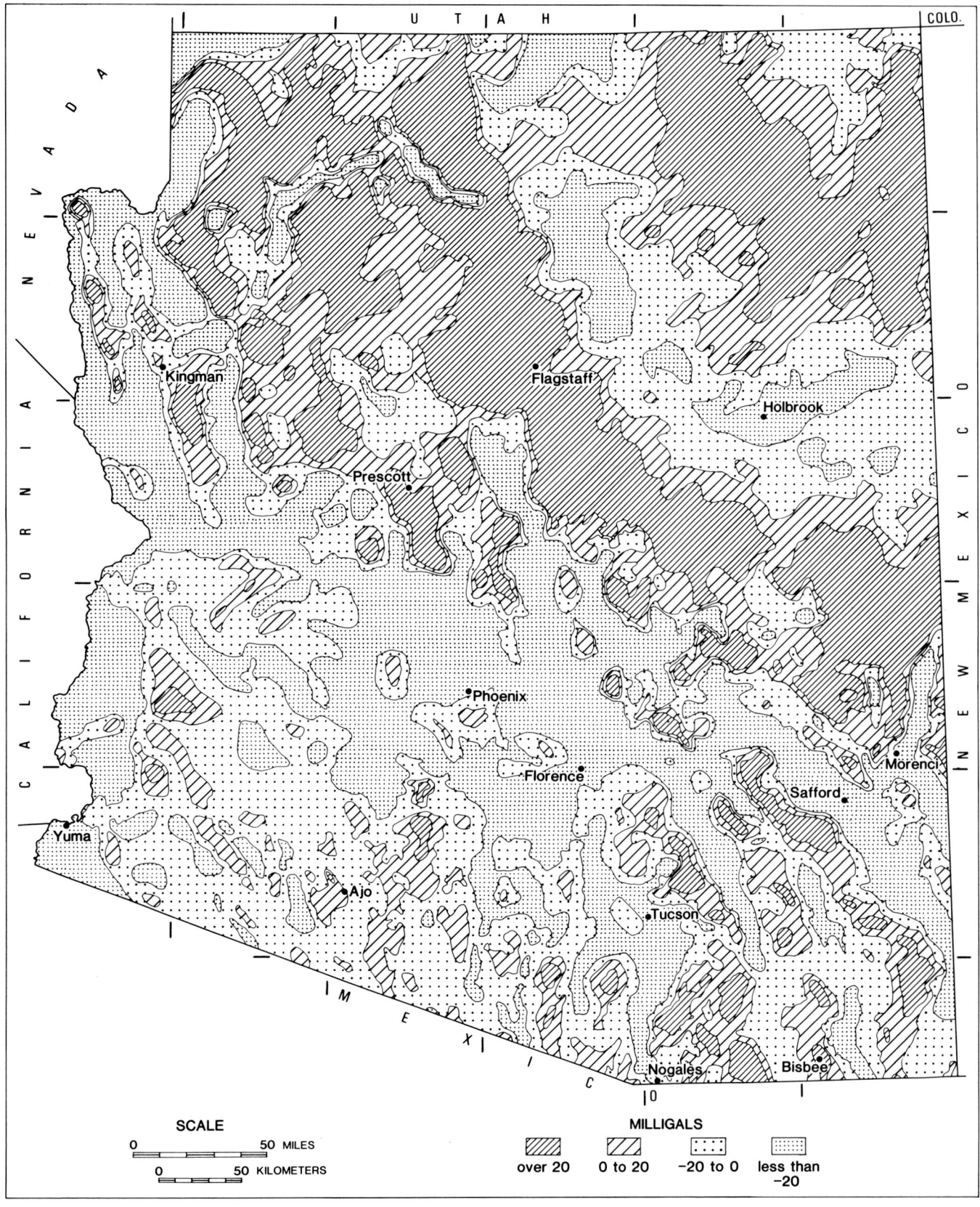

FIG. 6. Free-air gravity-anomaly map of Arizona, uncorrected for elevation effects. After Lysonski and Sumner (1981).

preted to correlate with thrusted sediments, the follow-up deep drilling near Florence failed to confirm the inference. Other reflection-seismic surveys from petroleum exploration in the state have not been made available.

DEEP ELECTRICAL AND ELECTROMAGNETIC SOUNDINGS

Over the past 30 years, several deep electrical soundings have been made in Arizona. Most of these determinations have used natural electrical sources, such as the low-frequency micropulsations originating from solar disturbances. Deep magnetotelluric surveys were made by Swift (1967) and Wojniak (1979), and geomagnetic-array studies were made by Schmucker (1964) and Porath (1971). Warren et al. (1969) discuss some of this work relative to heat-flow data.

The older, indurated sediments in Arizona's alluvial basins are found to have a high electrical conductivity, ranging from 0.2 to 0.01 mhos/m because of the ion-charged pore water and the clay minerals present. Basement rocks have the higher resistivity of about 100 to 1 000 ohm-meters. At the base of the Earth's crust the conductivity rises back again, perhaps because of electronic conduction in the minerals, to a level of about 0.1 to 0.5 mhos/m, decreasing again at the base of the lithosphere, interpreted to be 80 km in depth. The conductivities appear to be lower under the Colorado Plateau and have been shown to have positive correlation with heat-flow values in the southwestern United States (Warren et al., 1969).

HEAT FLOW IN ARIZONA

Numerous temperature measurements have been made in drill holes and have been compiled (Calvo, 1982) as a national contribution for promoting geothermal development in the state. A geothermal map, including the lineament analysis by Lepley (1977), has been published (Hahman et al., 1978), and data have been interpreted (Stone and Witcher, 1982).

The Basin and Range province has an anomalously high heat flow, averaging about 2.7 observed heat-flow units (HFU), almost twice the average value over the Earth (White and Williams, 1975). In terms of reduced HFU, which removes the contribution of upper-crust radioactivity, the Basin and Range area in Arizona averages 1.4 HFU. The Colorado Plateau averages about 0.8 in reduced HFU, although few measurements have been made in this part of Arizona.

By modeling the bottom depth of magnetized prismatic blocks, Byerly and Stolt (1977) produced a contour map of the Curie-point isotherm in northern and central Arizona. In this regard it should be pointed out that some researchers (Wasilewski et al., 1979) regard the Mohorovičić seismic discontinuity (Moho) as a magnetic boundary, with the logic that most of the recognized samples of mantle rock are relatively nonmagnetic—that is, they lack magnetite. If this is so, there may be a linkage between thermal, seismic, and long-wavelength magnetic characteristics of the Earth's crust. However, Haggerty (1978) has opposing views as to correlation of the Moho with magnetic properties.

SUMMARY AND CONCLUSIONS

Interpreted regional geophysical surveys in Arizona have assisted in providing a three-dimensional view of the complex basement geology of the state. It is hoped that these results will be of beneficial use both to residents of the region and to scientists and engineers who may be able to make contributions to the understanding of this and similar areas. If there is a shortcoming with respect to regional aeromagnetic and gravity surveying, it is that there has been insufficient support for adequately continuing these worthwhile programs.

In particular, aeromagnetic data (Sauck and Sumner, 1970) define the Mesa Butte fault system anomaly as being related to a major basement feature, which continues, by simple extension, to some of the enigmatic metamorphic core complexes in west-central Arizona. The South Mountain core complex, from its gravity and magnetic characteristics, is correlated as being an extension of the Holbrook-line anomaly. By inference, the other core complexes may well be similar remnants of Precambrian basement fabric.

The White Mountains intrusive body is geologically unknown except for its distinctive aeromagnetic anomaly, which identifies it as a buried paleorift. This intrusive rift is associated with an arcuate transform fault at its southern terminus, here named the Phoenix arc, which also is unrecognized except for its aeromagnetic anomaly. Most of the porphyry copper districts of Arizona lie near or south of the Phoenix arc. By interpretation of the stress fields from the White Mountains intrusion, a dilatant regime existed south of the Phoenix arc, allowing the passive intrusion of most of Arizona's porphyry copper bodies.

Several of the larger porphyry copper districts occur in broad aeromagnetic lows, implying magnetite alteration in associated deep fracture zones. Circulating ore fluids within the fracture zones concentrated the copper in the porphyry intrusives.

Interpretation and modeling of the regional gravity data clearly show details of the alluvial basins in the southwestern part of the state. These alluvial basins also allow preservation of Basin and Range tectonism, indicating fault strike and throw. The resulting basin patterns can then be perceived as vestiges of the Cenozoic stress fields in southwestern Arizona. The alluvial basins are Arizona's primary source of water, and a volume estimate made from the gravity data indicates about 2 200 km^3 of ground water recoverable from storage (Oppenheimer and Sumner, 1981).

Large bodies of low-density, porous and permeable volcanic rock are noted from comparison of geological and gravity data, and a considerable but as yet unknown groundwater resource probably exists in these bodies.

Finally, regional aeromagnetic and gravity data assist in locating energy resources, evaporite deposits, and land-subsidence hazards. Information is also provided for land-planning purposes and for comprehensive geologic mapping.

ACKNOWLEDGMENTS

I am grateful to the many field researchers of the past who have provided data for this review. The names and projects are numerous, but many of them are contained in the citations of the references in this report. I particularly appreciate the encouragement of Bill Hinze in compiling this report. A manuscript review by Doug Klein also was most helpful in organizing material for the reader.

I gratefully acknowledge the typing support from the Department of Geosciences at the University of Arizona.

REFERENCES

Aiken, C.L.V., 1978, Gravity and aeromagnetic anomalies of southeastern Arizona: Guideb., 29th Field Conf., N.Mex. Geol. Soc., 301–314.

Aiken, C.L.V., Lysonski, J., Sumner, J., and Hahman, W. R., 1981, A series of 1:250 000 complete residual Bouguer gravity anomaly maps of Arizona: Ariz. Geol. Soc. Dig., 13, 31–39.

Anderson, T. H., and Silver, L. T., 1979, The role of the Mojave–Sonora megashear in the tectonic evolution of northern Sonora, in Geology of northern Sonora: Guideb. 27, Geol. Soc. Am., 59–68.

Armstrong, R. L., 1982, Cordilleran metamorphic core complexes: Earth Planet. Sci. Lett. Annu. Rev., 10, 129–154.

Bean, R. E., 1982, Hydrothermal alteration in silicate rocks, in Titley, S. R., Ed., Advances in geology of the porphyry copper deposits: Univ. Ariz. Press, 117–137.

Byerly, P. E., and Stolt, R. H., 1977, An attempt to define the Curie point isotherm in northern and central Arizona: Geophysics, 42, 1394–1400.

Calvo, S. S., 1982, Geothermal resources in Arizona—A bibliography: Circ. 23, Ariz. Bur. Geol. Miner. Tech.

Carmichael, R. S., and Black, R. A., 1984, Analysis and use of Magsat magnetic data for interpretation of the crustal structure and character in the U.S. midcontinent [abstr.]: Trans. Am. Geophys. Union, 65, 202.

Coney, P. J., 1979, Tertiary evolution of Cordilleran metamorphic core complexes, in Cenozoic paleography of the western United States: SEPM, Pac. Sec., 15–28.

Eaton, G. P., Peterson, D. L., and Shumann, H. H., 1972, Geophysical, geohydrological and geochemical reconnaissance of the Luke salt body, central Arizona: Prof. Pap. 753, U.S. Geol. Surv.

Eberhart-Phillips, D., Richardson, R. M., Sbar, M. L., and Herrman, R. B., 1981, Analysis of the 4 February 1976 Chino Valley, Arizona earthquake: Seismol. Soc. Am. Bull., 71, 787–801.

Eberly, L. D., and Stanley, T. B., 1978, Cenozoic stratigraphy and geologic history of southwestern Arizona: Geol. Soc. Am. Bull., 86, 921–940.

Haggerty, S. E., 1978, Mineralogical constraints on Curie isotherms in deep crustal magnetic anomalies: Geophys. Res. Lett., 5, 105–108.

Hahman, W. R., Sr., Stone, C., and Witcher, J. C., 1978, Preliminary map of geothermal energy resources of Arizona: Ariz. Bur. Geol. Miner. Tech.

Heidrick, T. L., and Titley, S. R., 1982, Fracture and dike patterns in Laramide plutons and their structure and tectonic implications, in Titley, S. R., Ed., Advances in geology of the porphyry copper deposits: Univ. Ariz. Press, 73–91.

Hollister, V. F., 1978, Geology of the porphyry copper deposits of the western hemisphere: Am. Inst. Min. Eng.

Keith, S. B., 1980, The great southwestern Arizona overthrust oil and gas play: Ariz. Bur. Geol. Miner. Tech., 10, 1–8.

Klein, D. P., 1982, Regional gravity and magnetic evidence for a tectonic weakness across southwestern Arizona, in Frost, E. G., and Martin, D. L., Eds., Mesozoic–Cenozoic tectonic evolution of the Colorado River region, California, Arizona, and Nevada: Cordilleran Publ. (San Diego), Anderson–Hamilton v., 61–67.

Lepley, L. K., 1977, Landsat lineament map of Arizona: Open-File Rep., U.S. Dep. Energy.

Lowell, D., and Guilbert, J., 1970, Lateral and vertical alteration-mineralization zoning in porphyry ore deposits: Econ. Geol., 65, 373–408.

Lucchitta, I., 1978, Basement control of structural features near the boundaries of the Colorado plateau in Arizona: Contrib. 46, 3d Int. Conf. on Basement Tectonics, Durango, Colo.

Lysonski, J., 1980, the IGSN 71 residual gravity anomaly map of Arizona: M.S. thesis, Univ. Ariz.

Lysonski, J., Aiken, C., and Sumner, J., 1982, The complete residual Bouguer gravity anomaly map of Arizona: Ariz. Bur. Geol. Miner. Tech., scale 1:250 000, 23 sheets.

Lysonski, J., Schmidt, J., Sumner, J., and Aiken, C., 1980, IGSN 71 residual Bouguer gravity map of Arizona: Univ. Ariz., Lab. Geophys., Dep. Geosci.

Lysonski, J., and Sumner, J., 1981, Free-air gravity map of Arizona: Univ. Ariz., Lab. Geophys., Dep. Geosci.

MacInnes, S., 1982, The inversion of gravity data into three-dimensional polyhedral models: Prepubl. M.S. manus., Univ. Ariz.

Mayo, E. B., 1958, Lineament tectonics and some ore districts of the Southwest: Min. Eng., 10, 1169–1175.

Norton, D., 1982, Fluid and heat transport phenomena typical of copper-bearing pluton environments, in Titley, S. R., Ed., Advances in geology of the porphyry copper deposits: Univ. Ariz. Press, 59–72.

Oppenheimer, J. M., 1980, Gravity modeling of the alluvial basins, southern Arizona: M.S. thesis, Univ. Ariz.

Oppenheimer, J. M., and Sumner, J. S., 1980, Depth-to-bedrock map, Basin and Range Province, Arizona: Univ. Ariz., Lab. Geophys., Dep. Geosci.

——— 1981, Gravity modeling of the basins in the Basin and Range province, Arizona: Ariz. Geol. Soc. Dig., 13, 111–116.

Peirce, H. W., 1970, Coal, oil, natural gas, helium, and uranium in Arizona: Bull. 182, Ariz. Bur. Mines.

Peirce, H. W., Damon, P. E., and Shafiqullah, M., 1979, An Oligocene (?) Colorado Plateau edge in Arizona: Tectonophysics, 61, 1–24.

Porath, H., 1971, Magnetic variation anomalies and seismic low velocity zone in the western United States: J. Geophys. Res., 76, 2643.

Ratte, J. C., Landis, E. R., Gaskill, D. L., and Raabe, R. G., 1969, Mineral resources of the Blue Range primitive area, Greenlee County, Arizona, and Catron County, New Mexico: Bull. 1261-E, U.S. Geol. Surv.

Reford, M. S., and Sumner, J. S., 1964, Aeromagnetics: Geophysics, 29, 482–516.

Reif, D. M., and Robinson, J. P., 1981, Geophysical, geochemical and petrographic data and regional correlation from the Arizona State A-1 well, Pinal County, Arizona: Ariz. Geol. Soc. Dig., 13, 99–109.

Reynolds, S., and Rehrig, W., 1980, Mid-Tertiary plutonism and mylonitization, South Mountains, central Arizona: Mem. 153, Geol. Soc. Am., 159–175.

Sauck, W. A., 1972, Compilation and preliminary interpretation of the Arizona aeromagnetic map: Ph.D. diss., Univ. Ariz.

Sauk, W. A., and Sumner, J. S., 1970, Residual aeromagnetic map of Arizona: Univ. Ariz., Lab. Geophys., Dep. Geosci.

Schmidt, J. S., Aiken, C.L.V., and Sumner, J. S., 1973, An integrated data base management and analysis system for gravity: Proc., 11th Annu. Symp. Comput. Appl. Miner. Ind., Univ. Ariz., H1–H19.

Schmucker, U., 1964, Anomalies of geomagnetic variation in the southwestern United States: J. Geomagn. Geoelect., 15, 193.

Sears, J. D., 1973, Structural geology of the Precambrian Grand Canyon series: M.S. diss., Univ. Wyo.

Shoemaker, E. M., Squires, R. L., and Abrams, M. J., 1978, Bright Angel and Mesa Butte fault system of northern Arizona, in Smith, R. B., and Eaton, G. P., Eds., Cenozoic tectonics and regional geophysics of the western Cordillera: Mem. 152, Geol. Soc. Am., 341–367.

Shride, A. F., and Wrucke, C. T., 1969, Unpubl. U.S. Geol. Surv. rep., in Wilson, E. D., Moore, R. T., and Cooper, J. R., Geologic map of Arizona: Ariz. Bur. Mines.

Stone, C., and Witcher, J. C., 1982, Geothermal energy in Arizona: Final DOE report: Ariz. Bur. Geol. Miner. Tech.

Sumner, J. S., 1965, Gravity measurements in Arizona: Trans. Am. Geophys. Union, 46, 560–563.

——— 1976, Earthquakes in Arizona: Ariz. Bur. Mines Fieldnotes, 6, 1–5.

Sumner, J. S., Aiken, C.L.V., and Schmidt, J., 1976, The free-air gravity of Arizona: Ariz. Geol. Soc. Dig., 10, 7–12.

Swift, C. M., 1967, A magnetotelluric investigation of an electrical conductivity anomaly in the southwestern U.S.: Ph.D. thesis, MIT.

Titley, S. R., 1976, Evidence for a Mesozoic linear tectonic pattern in southeastern Arizona: Ariz. Geol. Soc. Dig., 10, 71–102.

——— 1981, Geologic and geotectonic setting of porphyry copper deposits in southern Arizona: Ariz. Geol. Soc. Dig., 14, 79–97.

——— 1982, Geologic setting of porphyry copper deposits, in Titley, S. R., Ed., Advances in geology of the porphyry copper deposits: Univ. Ariz. Press, 37–58.

Tucci, P., Schmoker, J., and Robbins, S., 1982, Borehole-gravity surveys in basin fill deposits of central and southern Arizona: Open-File Rep. 82–473, U.S. Geol. Surv.

Warren, R. E., Sclater, J. G., Vacquier, V., and Roy, R., 1969, A comparison of terrestrial heat flow and transient geomagnetic fluctuations in the southwestern United States: Geophysics, **34**, 463–478.

Wasilewski, P. J., Thomas, H., and Mayhew, M., 1979, The Moho as a magnetic boundary: Geophys. Res. Lett., **6**, 541–544.

West, R. E., 1972, A regional Bouguer gravity anomaly map of Arizona: Ph.D. diss., Univ. Ariz.

West, R. E., and Sumner, J. S., 1972, Ground-water volumes from anomalous mass determination for alluvial basins: Ground Water, **10**, 24–32.

———1973, Regional Bouguer gravity map of Arizona: Univ. Ariz., Lab. Geophys., Dep. Geosci.

White, D. F., and Williams, D. L., 1975, Eds., Assessment of geothermal resources of the United States—1975: Circ. 726, U.S. Geol. Surv.

Wilson, E. D., Moore, R. T., and Cooper, J. R., 1969, Geologic map of Arizona: Ariz. Bur. Mines.

Wojniak, S. J., 1979, A magnetotelluric sounding at the Tucson magnetic and seismological observatory: M.S. thesis, Univ. Ariz.

Woollard, G. P., and Joesting, H., 1964, Bouguer gravity anomaly map of the United States: Am. Geophys. Union and U.S. Geol. Surv., scale 1:2 500 000.

Mapping basement magnetization zones from aeromagnetic data in the San Juan basin, New Mexico

Lindrith Cordell* and V.J.S. Grauch*

ABSTRACT

Two new techniques are employed in analysis of aeromagnetic data from the San Juan basin, New Mexico, to enhance the expression of buried basement structure and lithology: (1) the data are analytically continued downward onto the irregular basement surface in order to reduce the effect of variable depth to basement, and (2) magnetization boundaries are delineated by a linear filter based on the gradient of pseudogravity. Reduction to the basement involves three procedures that, together, provide a practical method for continuation of potential fields between general surfaces. Data are continued from the nonlevel flight-elevation surface onto a level surface (drape-to-level continuation) by means of a system of successive approximations based on expansion of a Taylor series. Data are continued from the new level surface downward to another level surface at the mean elevation of the basement (level-to-level continuation) by analytical downward continuation based on the fast Fourier transform, incorporating a high-cut filter. Data are continued from this level onto the nonlevel basement surface (level-to-drape continuation) by direct evaluation of a truncated Taylor series expansion, in the vertical dimension, of the field represented on the level surface.

Magnetization boundaries are determined by evaluating the magnitude of the horizontal gradient of the pseudogravity transform of the magnetic data. Lines drawn along ridges of the horizontal gradient mark inferred basement-magnetization boundaries.

After application of these techniques, the aeromagnetic data from the San Juan basin, New Mexico, revealed features of the Precambrian basement not evident, or only vaguely so, in the original data. Together with scanty drill data, the aeromagnetic data indicate a 70-km-wide belt of predominantly metasupracrustal rocks trending east-northeast across the center of the basin. Predominantly granitic terranes border this belt on the northwest and southeast. Numerous magnetization boundaries tens of kilometers in length are identified. Structural grain is strongly east-northeast; an area of prominent northerly trends occurs in the north-central part of the basin and locally elsewhere. Cenozoic intrusive rocks are unquestionably aligned with basement structural grain. Laramide and Neogene structures in general are not so aligned. There is no evidence of large strike-slip displacement of basement structures within the area surveyed.

INTRODUCTION

The "crystalline basement complex" served nineteenth-century geologists as a catch-all for the uninteresting fundament upon which the Earth's structural architecture and stratigraphic record rested. Yet most of the geological record is within the basement rocks, and basement structure inevitably influences the evolution of the structure, facies, and fluid migrations within the overlying sedimentary-rock cover. To some twentieth-century geophysicists, therefore, the problem has come full cycle: the basement is the object of study, and the cover gets in the way.

Aeromagnetic data provide a means of seeing through effectively nonmagnetic sedimentary-rock cover to reveal patterns of magnetization in crystalline basement rocks. By inference, magnetization shows something about lithology, although the inference is thinly drawn. Magnetization is primarily a function of magnetite content, domain size, age, and metamorphism (including structural and hydrothermal alteration)—none of which is reliably diagnostic of lithology. Even so, magnetization patterns reflected in aeromagnetic

---

*U.S. Geological Survey, Box 25046, MS 964, Denver Federal Center, Denver, CO 80225.

data can reveal general lithologic and structural grain that can, at times, be referred to specific rock type if independent data are available.

In this spirit we seek to map basement magnetization in a deep, bowl-shaped sedimentary basin and to present our treatment of the following associated problems: (1) variation both in survey elevation and depth-to-basement give rise to anomaly variation owing to changing depth-to-source, unrelated to the nature of the source itself; and (2) the aeromagnetic data are displayed in the form of continuous contours of intensity of magnetic field strength, whereas what we are primarily interested in is the locus of magnetization boundaries delineating geologic contacts. We shall make use of a pragmatic approach to continue potential-field data between arbitrary surfaces, together with a study of the gradient of the pseudogravity anomaly. The overall objective is to make, as best we can, a geologic-like map from aeromagnetic data.

SAN JUAN BASIN, NEW MEXICO

The location, basement-structure contours, and generalized geologic structure of the San Juan basin are shown in Figure 1. The region is a dissected tableland typically at 2 000 m elevation, rising as high as 3 400 m in structural uplifts flanking the basin's margins. The basement forms a bowl-shaped surface that crops out on the south and east sides of the basin and is nearly 5 km below the surface in the center. The basin is filled with unmetamorphosed Cambrian through Holocene shallow-marine and continental strata.

In places, the borders of the basin are delineated by faults or monoclines that lie over faults in the subsurface. These faults are generally considered to be Late Cretaceous to Paleocene (Laramide) in age; many are likely to be Neogene and Quaternary, however, and related to extension of the Rio Grande rift. Other than various epeirogenic and depositional structures, the Laramide and younger(?) high-angle normal faults account for most of the structures observed in Phanerozoic rocks of the basin. An exception is the north–south–trending Nacimiento fault marking the east side of the basin. The Nacimiento fault has a component of right-lateral strike-slip displacement, according to Baltz (1967). This would be compatible with Hamilton's (1981) suggestion of Laramide movement of the Colorado Plateau and San Juan basin northward, relative to the High Plains, as defined by an Euler pole in Texas. Significant right-lateral north–south displacements, therefore, might be anticipated in the aeromagnetic data. Recent aeromagnetic studies (Cordell and Keller, 1984) indicate unsuspected major north–south–trending right-lateral strike-slip faults immediately east of the San Juan area, in the Rio Grande rift.

At the level of sensitivity of the aeromagnetic data, effectively the only rocks in the San Juan basin sufficiently magnetic to cause aeromagnetic anomalies are either Precambrian basement or a scattering of Laramide through Quaternary-age intrusives and lava fields. The younger magmatic rocks are exposed primarily around the south and west sides of the basin (Figure 1), and more are present in the subsurface within the basement and within the overlying sedimentary rock cover. We have identified some of these, using the aeromagnetic data, as discussed below. In general, the Cenozoic intrusive and associated extrusive rocks, where present, mask the aeromagnetic expression of basement rocks, which is our primary objective here.

Precambrian basement in the San Juan basin is known from a few outcrops around the basin margins and a few drill holes within it (Figure 1), and from regional studies. Condie and Budding (1979) characterized the basement as generally 30 percent low- to medium-grade metasupracrustal rocks foundered in regional (70 percent) "granite." Both are in the range 1.65–1.73 b.y. old; an east–west–trending boundary with rocks younger than 1.65 b.y. occurs somewhere in central New Mexico, possibly within the area of our study, but more likely to the south (see Condie and Budding, 1979; Condie, 1981).

Basement rocks are exposed in the Nacimiento and Zuni Mountains (Figure 1) and in the Tusas area about 50 km beyond the area of Figure 1, to the east. The Nacimiento Mountains are composed predominantly of granitic and related rocks (Woodward et al., 1974). The Tusas and Zuni areas expose metasupracrustal rocks intruded by "granites," and planar elements there trend northwest. Apparent bedding, including prominent metaquartzite units in the Tusas area, seem to follow this trend. Recent mapping by J. Callender (oral commun., 1982) and students, however, shows Precambrian rocks in New Mexico generally to be so severely deformed penetratively that true bedding often cannot be discerned.

Bottom-hole lithology of Precambrian rocks from drill data (Figure 1) was obtained primarily from unpublished data from the New Mexico Bureau of Mines and Mineral Resources and also from R. E. Denison, Dallas, Texas. All deep holes to Precambrian known to us are shown. We attempted to group these data into a relatively few units on the basis of possible tectonic affinities, as suggested in Figure 1. The quartzite localities suggest possible continuity with extensive quartzite and related metasedimentary units of the Tusas area. Little else can be said of the Precambrian terrane underlying the San Juan basin.

AEROMAGNETIC DATA

Aeromagnetic surveys of an extensive region in New Mexico and Arizona were flown by the U.S. Geological Survey in 1979 in support of uranium and other resource studies on Indian lands and in the San Juan basin and the Grants uranium district. These surveys were flown along north–south lines spaced 1.6 km (1 mile) apart, at an elevation 305 m (1 000 ft) above ground surface. Contour maps of the data are available at 1:250 000 scale from the U.S. Geological Survey (U.S. Geological Survey, 1980). Data from these surveys covering the San Juan basin were used to compile the anomaly map shown in Figure 2. For purposes of digital analysis, data of Figure 2 were interpolated from the original survey data tapes onto a 1-km grid (Webring, 1981) and contoured by computer. In the process, the International Geomagnetic Reference Field (IGRF) was removed from the data (R. Sweeney, U.S. Geological Survey, unpubl. computer program, 1982).

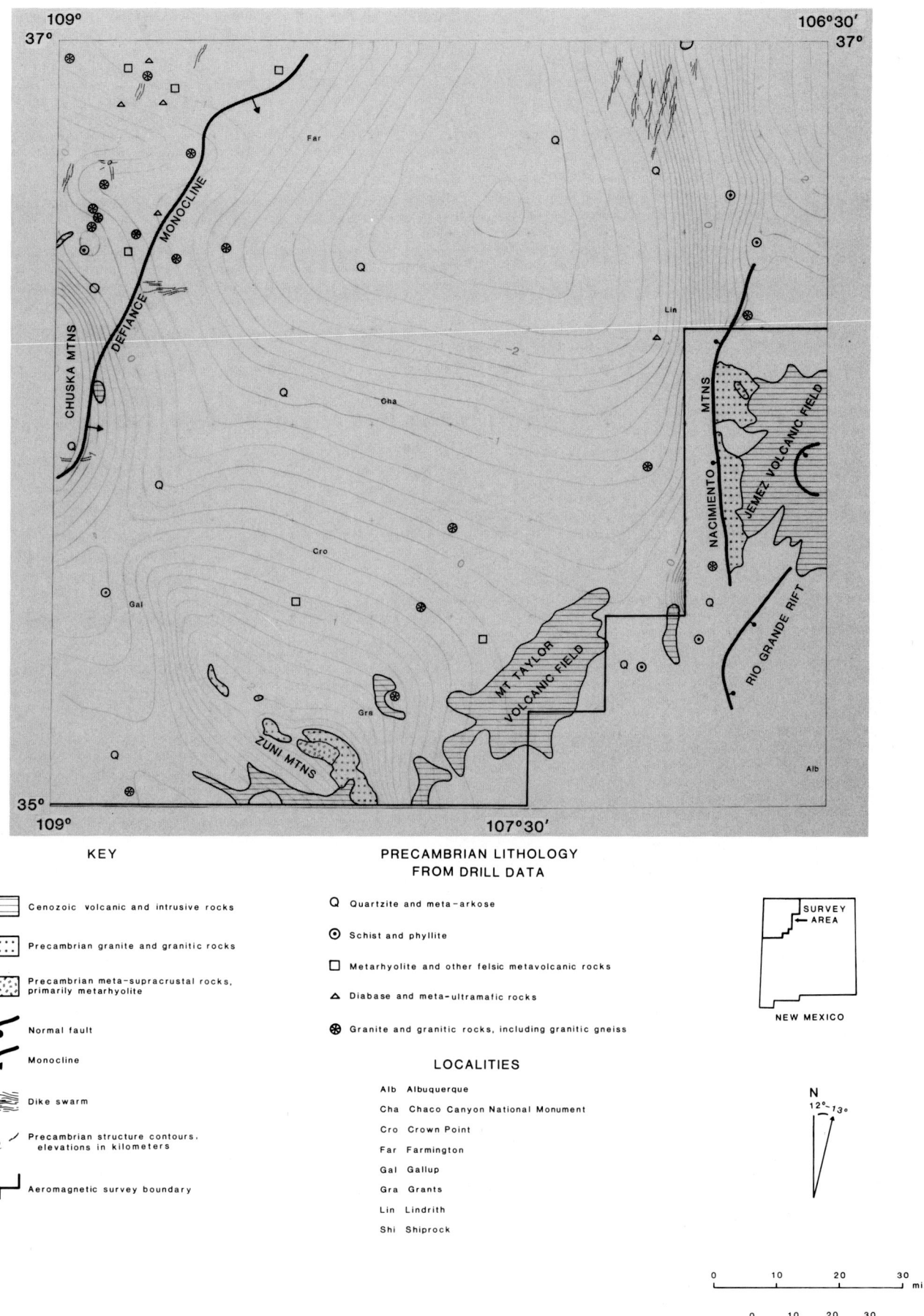

FIG. 1. Location, basement-structure contours, and generalized geologic structure of San Juan basin area. Basement-structure contours (screened) are based on drill-hole data and stratigraphic projection.

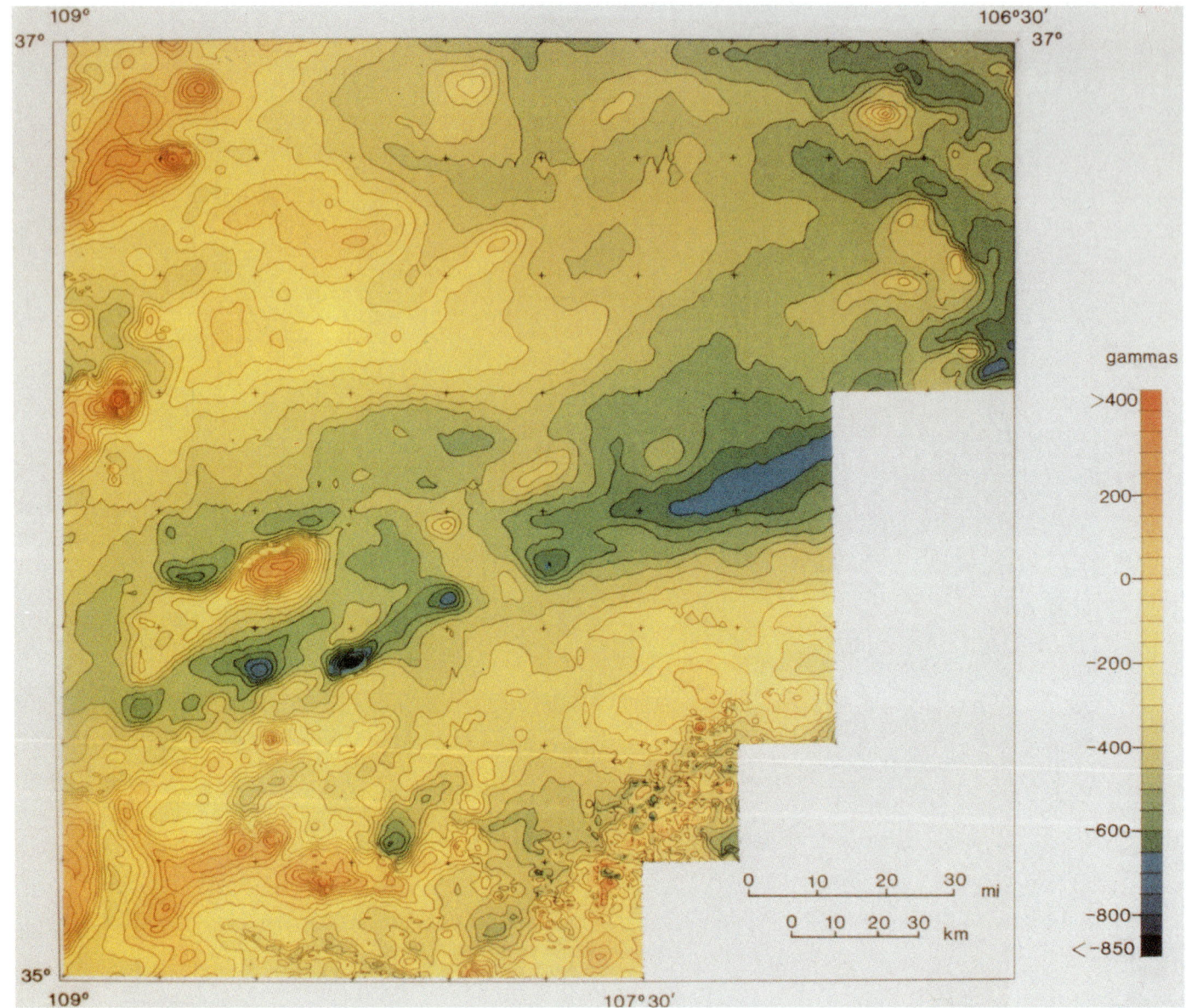

FIG. 2. Residual total-intensity-magnetic-field data, gridded at a 1-km interval and corrected for the International Geomagnetic Reference Field (IGRF). Data are draped 305 m (1 000 ft) above ground surface.

REDUCTION OF AEROMAGNETIC DATA TO BASEMENT

Besides a few prominent anomalies over scattered Cenozoic intrusive and extrusive rocks, the aeromagnetic patterns in Figure 2 reflect variation in magnetization in the Precambrian basement. However, major variation in the aeromagnetic signature is caused by the severe attenuation of the field with depth in the central part of the basin, where basement is as much as 2 700 m below sea level and is 5 000 m distant from the sensor (cf. Figures 1 and 2). If the nonmagnetic sedimentary-rock cover could be stripped away and we could fly immediately above the basement everywhere, the variable attenuation would not be observed. Knowing both the flight elevation and at least the approximate elevation of the basement surface, we attempted to simulate the field on the basement surface by analytical continuation.

Derivations

Continuation of a potential from one level onto a higher level is made possible by the famous integral solution of the Dirichlet problem, which can be validly applied to the total-intensity-anomaly field (Henderson, 1970), provided the anomaly field is small relative to the Earth's field and the latter is effectively constant. Such level-to-level continuation can be effected numerically either by convolution (e.g. Henderson, 1960) or, as is currently more popular, by use of the fast Fourier transform (FFT). We use the FFT routine of Hildenbrand (1983).

Continuation between general surfaces can be treated by means of a generalization of the Dirichlet-problem integral (Syberg, 1972) or by inversion of a matrix in the space (Dampney, 1969) or frequency (Henderson and Cordell, 1971) domains. Error is an especially important consideration in general-surface continuation (Huestis and Parker, 1979) as is, in practical terms, the inevitably large computer cost.

As a technique that can be rapidly applied to very large data sets defined on regular grids, we use the following approximate solution to the general continuation problem. Let $t[x,y,z,(x,y)]$ represent the total magnetic-intensity-anomaly field on the surface $z(x,y)$ in a right-handed x,y,z Cartesian system with the z direction positive down. In terms of discrete digital data on a regular grid, let the equivalent discrete function $t_{ij}(z_{ij})$ represent the magnetic field on the surface z_{ij} at the i,jth grid point. Following Evjen (1936), we approximate the field on the surface z_{ij} by means of a Taylor series expansion of the field about the level $z = 0$:

$$t_{ij}(z_{ij}) = t_{ij}(0) + z_{ij} \frac{\partial}{\partial z} t_{ij}(0) + \frac{1}{2} z_{ij}^2 \frac{\partial^2}{\partial z^2} t_{ij}(0) + \cdots. \quad (1)$$

It is advantageous in terms of convergence of the series in equation (1) to place the $z = 0$ origin of the coordinate system near the mean of $\{z_{ij}\}$, so that the range of $\{|z_{ij}|\}$ will be minimum. In practice, we truncate the series at two or three terms; we continue the data grid from its original level onto the $z = 0$ level, and we obtain the first- and second-vertical-derivative grids of the data on that level using the discrete Fourier transform (Hildenbrand, 1983). These three grids, together with the grid defining the surface z_{ij}, make up the ingredients of the right side of (1), from which we obtain the data draped on the z_{ij} surface, $t_{ij}(z_{ij})$. We refer to this as the level-to-drape continuation to distinguish it from the more complicated drape-to-level continuation problem to be considered next.

Rearranging (1) suggests a possible recursive formula for the drape-to-level problem, i.e., the problem of continuing data observed on the surface z_{ij} onto the level $z = 0$. From (1),

$$t_{ij}(0) = t_{ij}(z_{ij}) - z_{ij} \frac{\partial}{\partial z} t_{ij}(0)$$
$$- \frac{1}{2} z_{ij}^2 \frac{\partial^2}{\partial z^2} t_{ij}(0) - \cdots. \quad (2)$$

Both the surface z_{ij} and the field on that surface, i.e., $t_{ij}(z_{ij})$, are known. We intend to evaluate the $n + 1$ th estimate of $t_{ij}(0)$ on the level surface $z = 0$, using derivatives of the nth estimates of $t_{ij}(0)$. Thus, indicating successive estimates with subscript n, the recursive formula would have the form

$$t_{ij_{n+1}}(0) = t_{ij}(z_{ij}) - z_{ij} \frac{\partial}{\partial z} t_{ij_n}(0)$$
$$- \frac{1}{2} z_{ij}^2 \frac{\partial^2}{\partial z^2} t_{ij_n}(0) - \cdots. \quad (3)$$

We need a first approximation to get started. We obtained fairly rapid convergence using the draped data as the first approximation, i.e.,

$$t_{ij_1}(0) = t_{ij}(z_{ij}).$$

We obtained better results, however, by treating $t_{ij}(z_{ij})$ as though it were on a level and projecting it onto $-z_{ij}$ by the level-to-drape equation, equation (1). Thus, instead we define

$$t_{ij_1}(0) = t_{ij}(z_{ij}) - z_{ij} \frac{\partial}{\partial z} t_{ij}(z_{ij})$$
$$+ \frac{1}{2} z_{ij}^2 \frac{\partial^2}{\partial z^2} t_{ij}(z_{ij}) - \cdots. \quad (4)$$

In practice, we approximate the vertical-derivative terms for the first approximation by treating the grid $t_{ij}(z_{ij})$ as though it were on a level surface.

Defining an error function for each grid point at the nth iteration to be

$$\varepsilon_{ij_n} = t_{ij}(z_{ij}) - \left[t_{ij_n}(0) + z_{ij} \frac{\partial}{\partial z} t_{ij_n}(0) \right.$$
$$\left. + \frac{1}{2} z_{ij}^2 \frac{\partial^2}{\partial z^2} t_{ij_n}(0) \cdots \right], \quad (5)$$

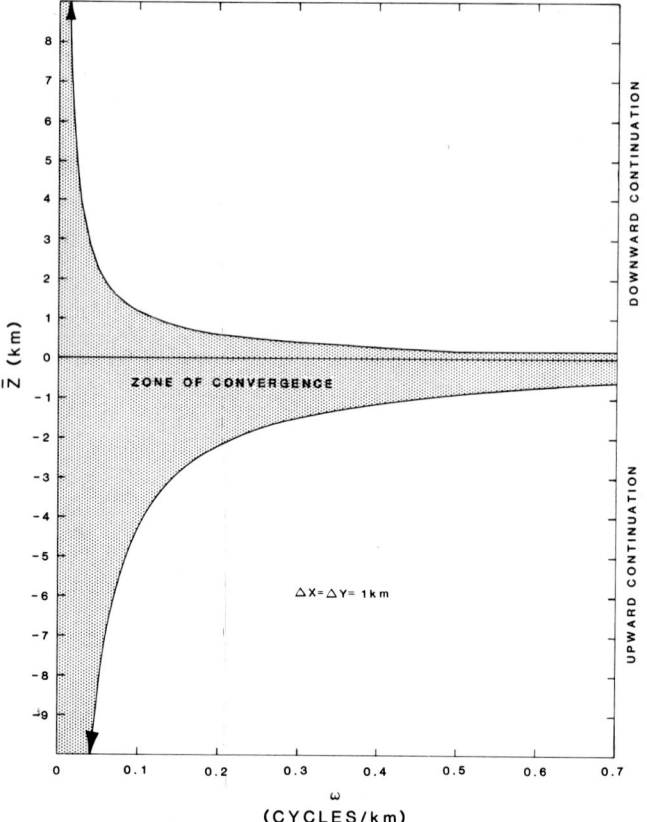

FIG. 3. Range over which the error in successive iterations of the drape-to-level approximation converges to zero (equation 8), given the simple case of level-to-level continuation. The zone of convergence is shown in the frequency domain, where $\omega = \sqrt{v_1^2 + v_2^2}$ and v_1 and v_2 (in units of cycles/km) correspond to independent variables x and y in the space domain. In this case the grid intervals $\Delta x = \Delta y = 1$ km so that radial Nyquist frequency is $\sqrt{2}/2$. The original "draped" surface is on the level $z = 0$; the level to which we are approximating the data is $z = \bar{z}$, where $\bar{z}$ constitutes upward continuation if it is negative and downward continuation if it is positive.

and combining (5) and (3), we have

$$t_{ij_{n+1}}(0) = t_{ij_n}(0) + \varepsilon_{ij_n}. \tag{6}$$

The term in brackets in (5) represents the same computation needed for the level-to-drape problem (right side of equation 1), and ε_n provides an estimate of error at successive approximations. Once ε_n is evaluated, the $n + 1$ th estimate is obtained by the simple addition of grids indicated in (6). Equations (1), (5), and (6), together with standard methods for obtaining vertical derivatives and continuing between level surfaces, form the basis for our method of continuation between arbitrary surfaces. An approximate method, we emphasize—designed for rapid application to large data grids.

The error inherent in applying (1), (5), and (6) will depend on the roughness of both the surface z_{ij} and the data function t_{ij}. Convergence of (6) cannot be proved, inasmuch as a divergent counter-example can easily be devised. We are more concerned in testing whether (6) converges under *any* circumstances. To this end, consider that $z_{ij} = \bar{z} = $ constant, and equation (5) in the frequency domain (truncated beyond the second order) becomes

$$E_n(\omega) \simeq e^{2\pi\omega\bar{z}}T(\omega) - \left[T_n(\omega) + \bar{z}(2\pi\omega)T_n(\omega) \right. $$
$$\left. + \frac{\bar{z}^2}{2}(2\pi\omega)^2 T_n(\omega) \right], \tag{7}$$

where we denote corresponding Fourier transforms by capital letters and the independent radial-frequency variable by $\omega = \sqrt{v_1^2 + v_2^2}$, where v_1, v_2 (in units of cycles/km) correspond to independent variables x, y in the space domain. Continuation onto level $\bar{z}$ is effected by multiplication by $e^{2\pi\omega\bar{z}}$; vertical differentiation is effected by multiplication by $2\pi\omega$ (see, for example, Bhattacharyya, 1967). $T(\omega)$ denotes the true value of the transform of the function on $z = 0$ (to be determined), and $T_n(\omega)$ denotes the nth estimate of $T(\omega)$.

From (6),

$$T_{n+1}(\omega) = T_n(\omega) + E_n(\omega);$$

evaluating (7) for E_{n+1}, substituting the above expression for $T_{n+1}(\omega)$, and using the definition of $E_n(\omega)$ gives

$$E_{n+1}(\omega) \simeq -(2\pi\omega\bar{z})[1 + \pi\omega\bar{z}]E_n(\omega). \tag{8}$$

Equation (8) shows that (1) successive estimates alternately overshoot and undershoot the solution; (2) convergence depends on frequency (the lower the frequency, the better the convergence) and requires, implicitly, a cutoff frequency; and (3) for a given frequency the sequence can be shown to converge if z is sufficiently small, and that convergence is more rapid for upward than for downward continuation. To be specific, we can show the sequence to converge absolutely, provided

$$\frac{-\sqrt{3}-1}{2\pi\omega} < \bar{z} < \frac{\sqrt{3}-1}{2\pi\omega}. \tag{9}$$

In digital analysis with discrete-data grids and the FFT (see Brigham, 1974), z is scaled to the grid interval, and we only consider ω in the range from 0 to Nyquist. For a grid having equal grid interval $\Delta x = \Delta y$, the radial Nyquist frequency is $\sqrt{2}/2\Delta x$ and $\bar{z}$ becomes $\bar{z}/\Delta x$ in (9). For $\Delta x = 1$, the range over which vertical continuation is possible in this example can then be shown as sketched in Figure 3.

Test example

As a test, we analytically continued the vertical component of gravitational force from a homogeneous sphere from a level onto an irregular surface (level-to-drape, by equation 1) and vice versa (drape-to-level, by equations 5 and 6). The surface is shown in plan view in Figure 4 and in profile across the center of the sphere in Figure 5c.

The level-to-drape results are shown in Figure 5a. The gravity field on the level $z = 0$ was draped on the irregular

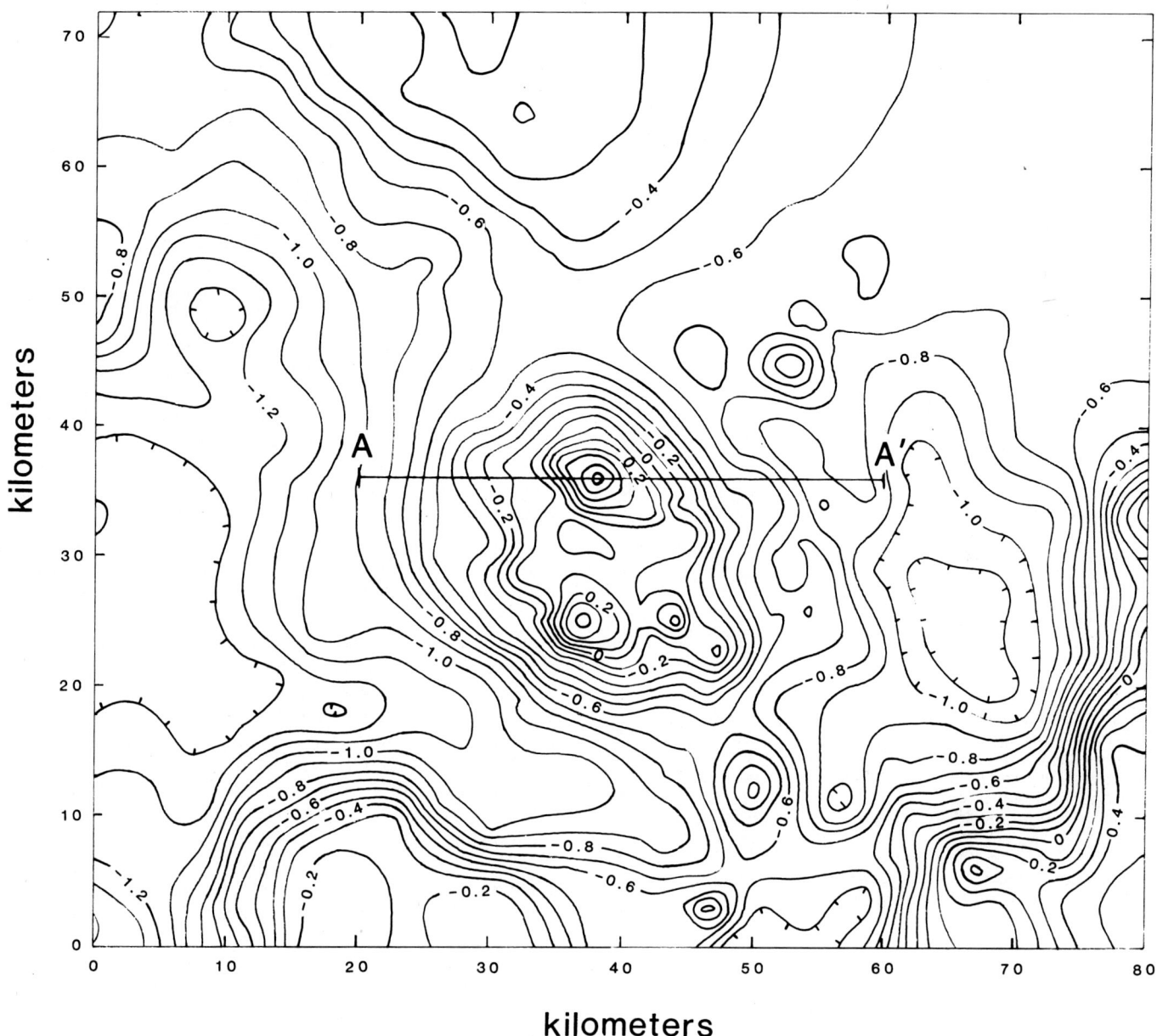

FIG. 4. Plan view of irregular draped surface used in the test example. Profiles for Figure 5 were taken along A–A'. Contour interval is 0.1 km.

surface z_{ij} (Figure 5c) by means of equation (1), using the expansion only to and including the second-order (z^2) term. Agreement between the actual draped field and the draped field calculated by equation (1) is good.

Drape-to-level results for the same configuration are shown in Figure 5b. This time we calculated the field from the irregular surface z_{ij} onto the $z = 0$ level, using two iterations of equations (5) and (6). Guided by the convergence diagram of Figure 3, we placed the $z = 0$ datum somewhat above the mean of z_{ij}.

Application to San Juan basin data

Both the drape-to-level and level-to-drape operations were performed on the San Juan basin aeromagnetic data. We first calculated the originally drape-flown data onto a level at approximately the mean flight elevation by means of equations (5) and (6). We continued this down to approximately the mean of the basement surface by standard level-to-level FFT continuation (Hildenbrand, 1983), and then draped the data onto the basement surface, using equation (1).

The approximate flight elevation of the San Juan basin survey (shown in Figure 6) was obtained by adding 305 m (1 000 ft) to smoothed digital topography. We used a smoothed elevation surface, partly because some smoothing of the draped surface is inevitable with fixed-wing aircraft but primarily to reduce possible error at high frequencies. The flight elevation ranged from a minimum of 1 760 m to a maximum of 3 400 m, with a mean value of 2 350 m sea-level

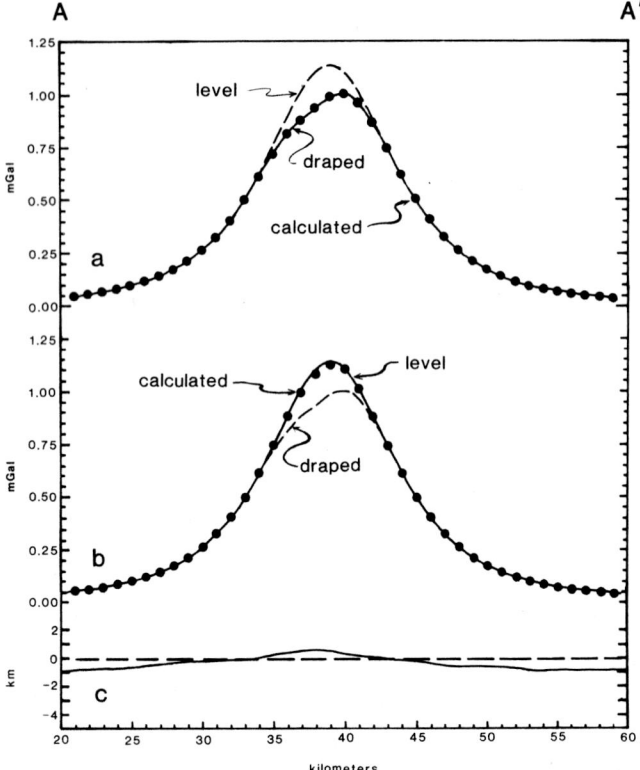

FIG. 5. Results of the test example along the profile A–A' of Figure 4. The gravity effect of a sphere at a horizontal location of 39 km and a depth of 7 km was calculated on the irregular surface of Figure 4 to simulate draped data over a hill, and on level 0 to simulate level data (c). The level-to-drape approximation was calculated using equation (1) (truncated after the second-order term) and is shown in a. The drape-to-level approximation is shown in b after two iterations of equations (5) and (6) (also truncated after the second order term), using a first approximation as given in equation (4).

elevation. Despite an overall range of 1 640 m, local relief of the flight surface, both in actuality and as represented in Figure 6, was generally small.

Aeromagnetic data on the irregular flight-elevation surface shown in Figure 6 were projected onto a level surface at 2 400 m elevation by the drape-to-level procedure of equations (5) and (6), using two iterations; wavelengths less than 4 km (4 times the grid interval) were filtered in the derivatives (Hildenbrand, 1983).

From the mean flight level at 2 400 m the data were continued upward 800 m to simulate the field observed on a level at 3 200 m elevation in Figure 7 (the highest topography reaches 3 100 m). The draped original survey and the calculated level field show little difference (compare Figures 2 and 7). We show the results, however, in order to make the point that had the survey been originally obtained on a constant level it would have been flown, of necessity, at 3 200 m elevation, and much of the high-frequency signal (hidden, perhaps, in Figure 2 but present in the raw data and profiles) would have been lost. Having the data on the lower, draped surface, however, we can easily recover the data on a higher, level surface, as might be advantageous for regional compilation and analysis. By contrast, it is only with difficulty and possible introduction of error that we can project data from a higher level downward. With respect to design of reconnaissance, fixed-wing aeromagnetic surveys, therefore, we would recommend flying drape rather than constant-level mode (see also Grauch and Campbell, 1984). We should perhaps refer to these as smooth-drape surveys to distinguish them from helicopter-draped surveys, which can generate extreme magnetic-field gradients and warrant special consideration.

Having data on a level, we are ready to continue the magnetic field downward onto the irregular basement surface. To simplify interpretation of these data by inspection, we also reduced the data to the pole (Baranov, 1957; Cordell and Taylor, 1971) to simulate the effect of vertical magnetization. The basement surface has a maximum elevation of 2 800 m above sea level, a minimum of 2 600 m below sea level, and a mean elevation of approximately sea level. In view of the range in basement elevation, and recognizing that little high-frequency content existed in the data over the deep part of the basin in the first place, we regridded the data at a 3-km interval. We continued the data from 3 200 m elevation to sea level by FFT methods (Hildenbrand, 1983), then we draped the data onto the basement surface by means of equation (1) (Figure 8). The lines in Figure 8 show magnetization boundaries and are discussed below. Before making the downward continuation to sea level and the vertical-derivative calculations, we employed a tapered high-cut filter similar in effect to a filter we developed to deal with high-frequency distortion inherent in the discrete Fourier transform (Cordell and Grauch, 1982a).

We recognize that, strictly speaking, continuation immediately onto the basement surface is only barely possible theoretically and impossible in practice. We only purport to do so, of course, by filtering out distortion inevitably occurring in the higher frequencies of the data. The filtering does not make our procedure equivalent to continuation onto a surface some arbitrary distance above the basement. Most of the significant part of the data in the lower frequencies was unaffected by the filter (except that the 3-km grid interval has doubtless introduced minor aliasing into the lower frequencies) and genuinely does represent the broad features of the data when reduced to the basement surface.

The San Juan basin reduced-to-basement map (Figure 8) shows fairly uniform frequency content throughout the basin and considerably enhanced expression of basement anomalies over the deep, central part of the basin, as was intended (compare Figures 2, 6, and 8). The seeming preference in the data for many little waves and circles about 4 km wide is an artifact of the high-cut filter. Otherwise, the generally east-northeast grain evident in Figure 2 persists in Figure 8. However, the reduced-to-basement data bring out individual anomalies, typically 25 km in length, with some being aligned with the general grain and others showing more northerly and northeasterly trends.

The reduction-to-basement procedure predictably produces erratic results where magnetic sources occur at shallow levels above the basement, such as the Mount Taylor

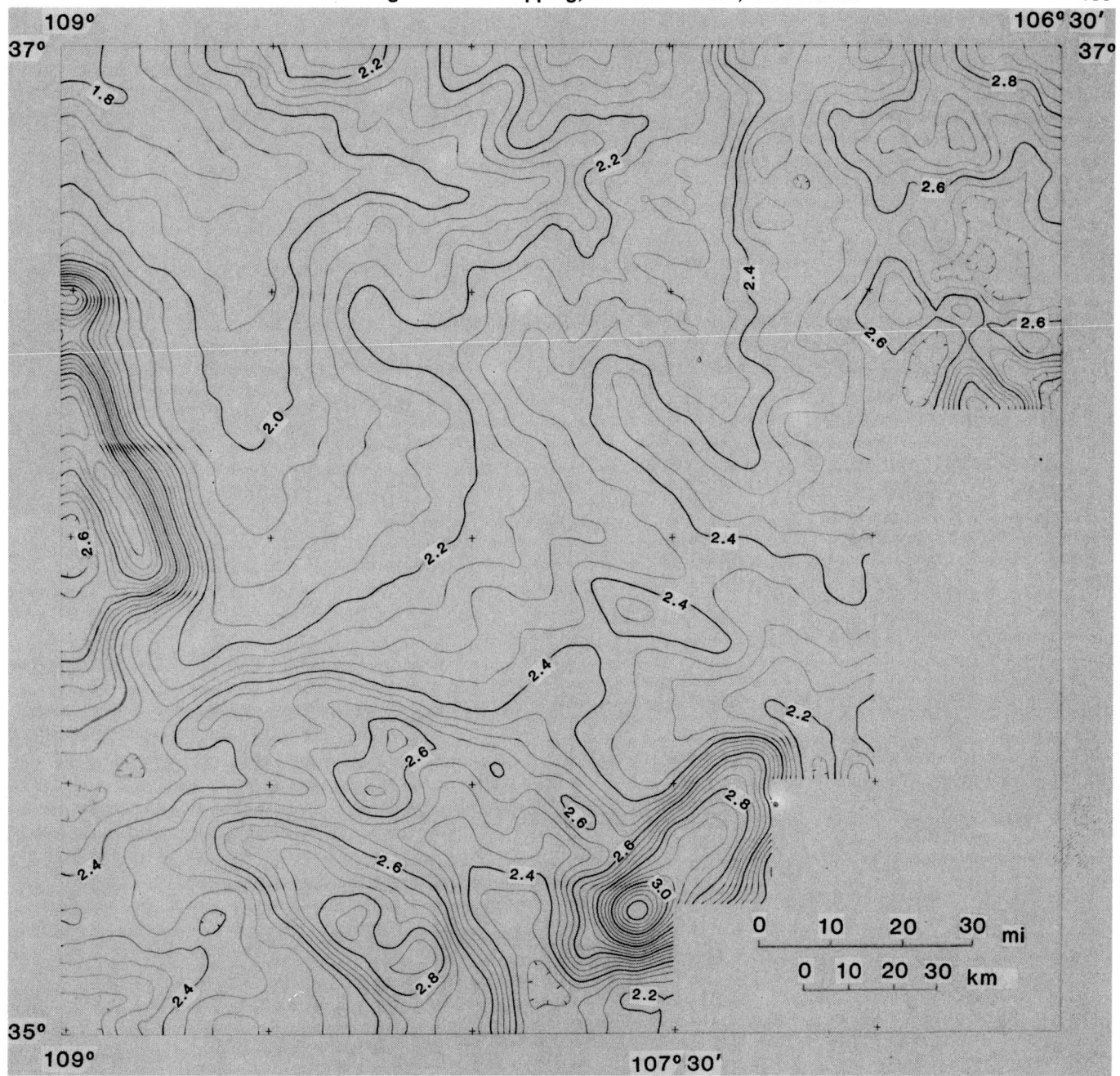

FIG. 6. Approximate flight elevation of the San Juan basin aeromagnetic survey, in kilometers. Surface was determined by adding draping constant (305 m or 1 000 ft) to smoothed digital topography. Contour interval is 0.05 km.

volcanic field and local intrusive bodies (Figure 1). We contoured the data in these areas anyway, however, hoping that broad, and thereby possibly deep, basement effects might show through. Most of these areas of suprabasement sources are small enough so they do not mask the regional picture. The extensive Mount Taylor volcanic field in the southeast corner of the survey area, however, is an exception. Here, a northeast-trending magnetic gradient occurs along the north edge of the Mount Taylor field, but extends beyond it to the northeast, and therefore may reflect a basement feature along which the volcanic field is causally aligned.

MAGNETIZATION BOUNDARIES

Although by reduction to basement we have to some extent removed the effects of variable depth to basement and magnetic polarization, we still have a map showing a

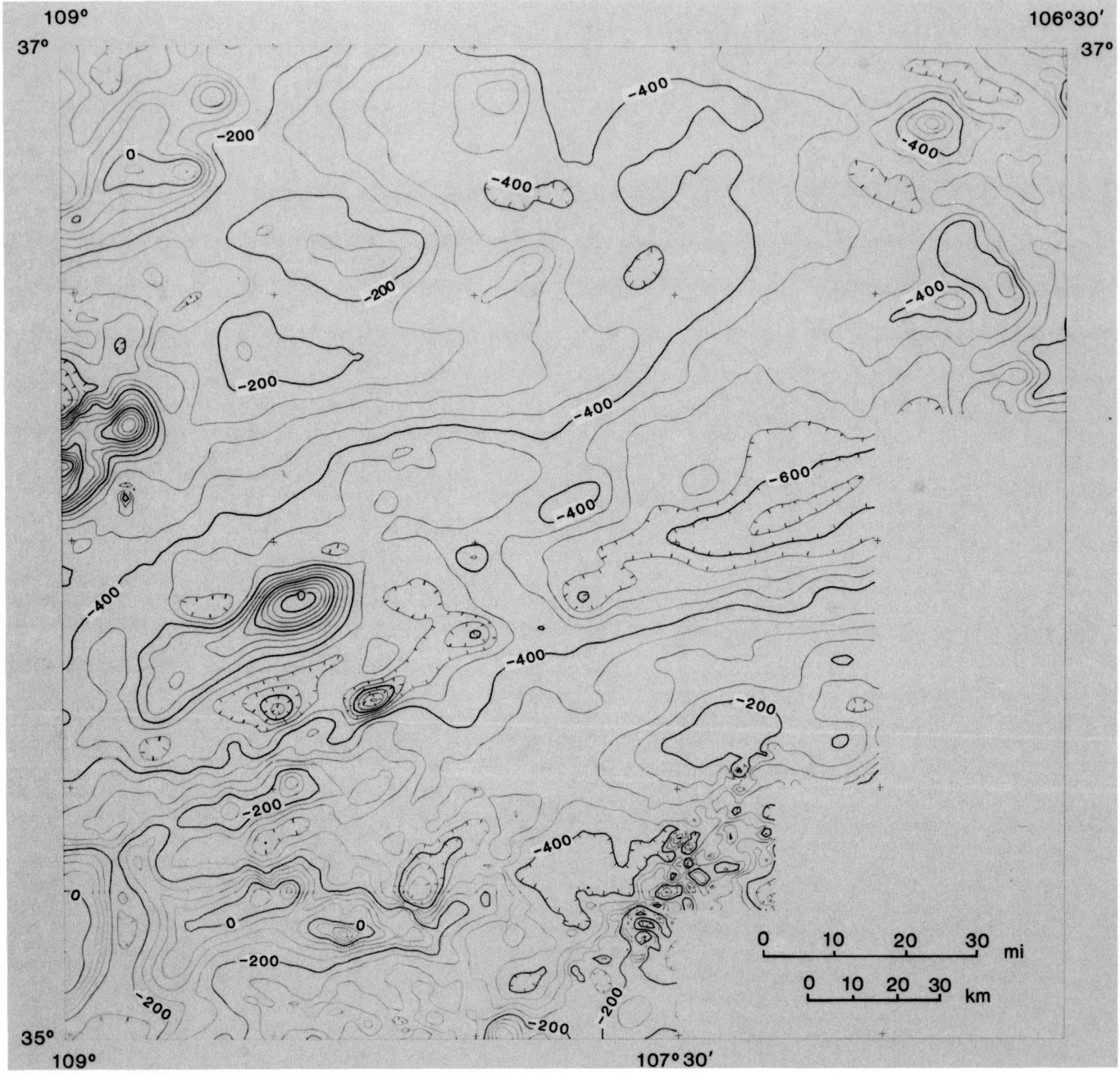

FIG. 7. Residual total-magnetic-field data of Figure 2 put on the level 3 200 m above sea level, using drape-to-level equations (5) and (6).

smoothly varying potential field (Figure 8), which reflects basement lithologic boundaries in only a diffuse and out-of-focus way. Experience with rock samples and with the aeromagnetic fields themselves, in general, indicates that, although rock magnetization in a sense is a smoothly varying three-dimensional continuum, it can generally be best characterized in terms of magnetization units delimited by fairly narrow boundaries. So, without necessarily equating magnetization with geological formation boundaries, we will proceed here from the premise that magnetic-field effects of sharp boundaries between regions of fairly uniform magnetization are common in the data and will seek to delineate these boundaries in an objective and reproducible way.

Horizontal-gradient method

Cordell (1979) used the maxima of gravity gradients associated with low-density graben fill to map the position of suspected graben-bounding faults. Making use of the so-called pseudogravity anomaly (Baranov, 1957), we extend basically the same analysis to magnetic data. Dole and

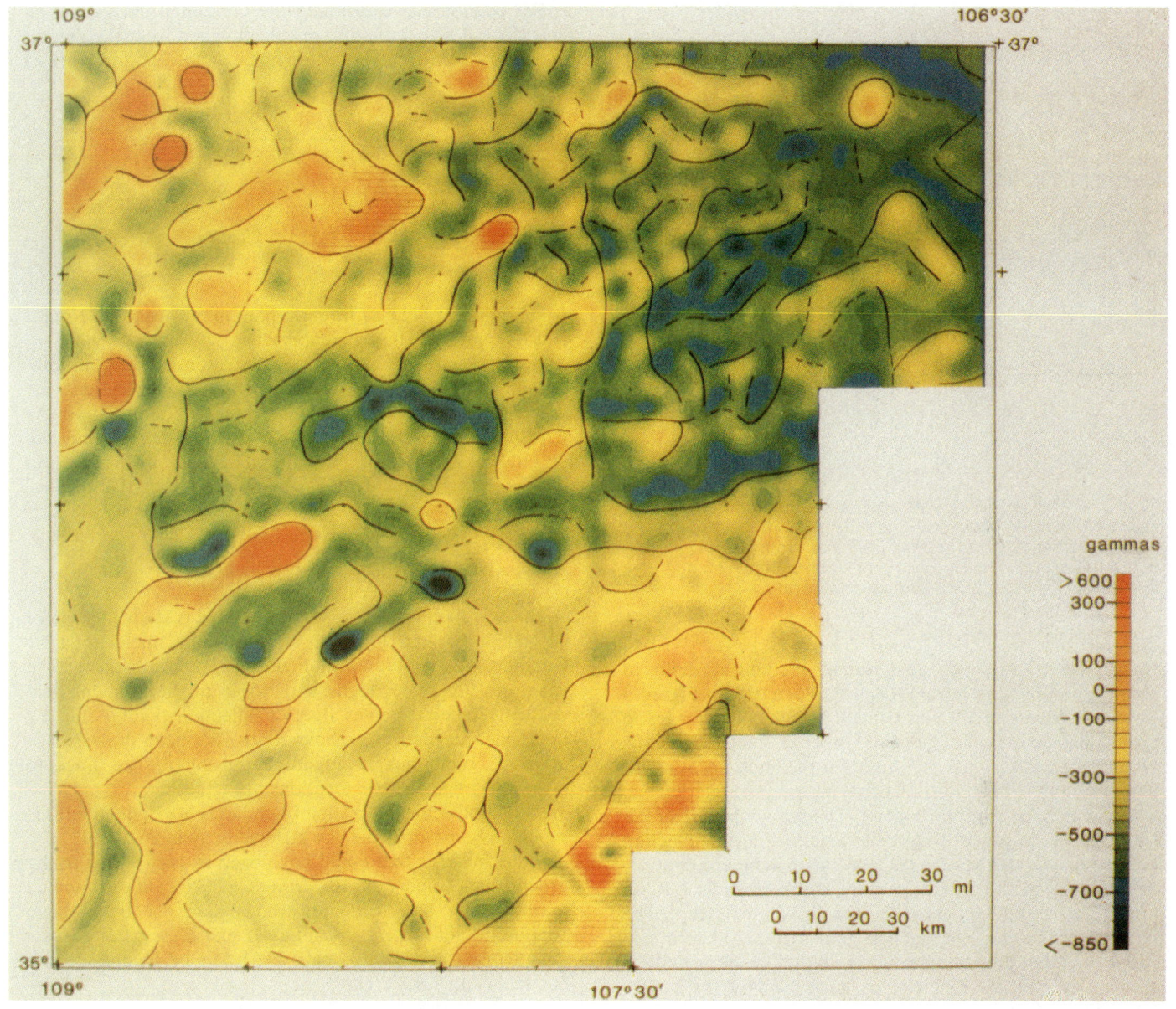

FIG. 8. San Juan basin aeromagnetic data continued to the basement surface. Black lines were obtained through the horizontal-gradient method and show magnetization boundaries (see text).

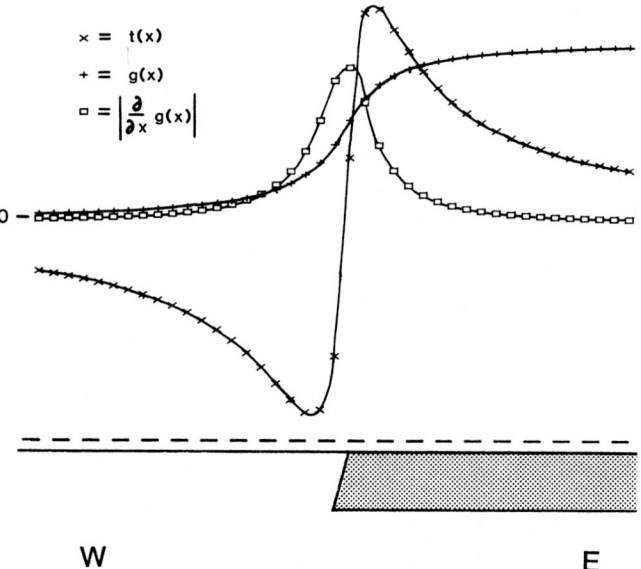

FIG. 9. Hypothetical two-dimensional steep contact, showing the observed magnetic anomaly $t(x)$, its pseudogravity counterpart $g(x)$, and the magnitude of the horizontal gradient of the pseudogravity $|\partial/\partial x\, g(x)|$. Inclination = 60 degrees; declination = 0 degrees. Units are arbitrary.

Jordan (1978) applied the gradient method directly to aeromagnetic data. Our approach is illustrated in Figure 9 (modified from Cordell and Grauch, 1982b). The magnetic profile (labeled $t(x)$ in Figure 9) over a steep contact is typically a high–low pair, with neither the high, the low, nor any of the three inflection points having a simple relationship to the surface trace of the contact. The gravity field over the same model, however, is a simple ramp, having an inflection point immediately over the contact. (With a dipping contact, the position of the inflection point shifts downdip but remains near the contact as long as the dip is fairly steep). The position of the gravity inflection point is easily determined by locating the culmination in the amplitude of the horizontal gradient of the gravity field, as suggested by the profile labeled $|\partial g/\partial x|$ in Figure 9.

In applying this technique to aeromagnetic-map data, we first obtain the pseudogravity field $g(x,y)$ from the aeromagnetic data, and from the pseudogravity data determine the gradient function,

$$f(x, y) = \sqrt{\left(\frac{\partial g}{\partial x}\right)^2 + \left(\frac{\partial g}{\partial y}\right)^2}. \quad (10)$$

We refer to f loosely as the "horizontal gradient." Strictly speaking, it represents the amplitude of the horizontal component of the pseudogravity gradient. Normally, use of the pseudogravity technique occasions some justification of assumed consanguinity of the magnetic and the hypothetical density distributions. No such assumption is required here, and the analysis could proceed directly from a convolution of the magnetic field with the appropriate operator, without reference to gravity. The development in terms of "gravity" simply takes advantage of an intuitively evident rationale and already existing computer programs.

For economy, we estimate the horizontal derivatives in equation (10) by finite differences obtained from the digital-data grids. For example, let $g(x,y)$ be defined at grid point i,j as $g_{i,j}$. Then

$$\frac{\partial g}{\partial x} \simeq \frac{g_{i+1,j} - g_{i-1,j}}{2\Delta x}$$

and

$$\frac{\partial g}{\partial y} \simeq \frac{g_{i,j+1} - g_{i,j-1}}{2\Delta y}, \quad (11)$$

which is equivalent to fitting a parabola to each grid point and its two immediate neighbors.

Note that the pseudogravity calculation involves an integration of the field along a given direction, whereas the gradient operation requires a subsequent differentiation. As a result, the frequency content of the gradient and the original magnetic fields remains essentially unchanged. In the frequency domain it is easy to show that the spectral phase, but not its average amplitude, is changed.

Application to San Juan basin

The San Juan basin pseudogravity field is shown in Figure 10; the horizontal gradient of these data, as defined by equation (10), is shown in Figure 11. As with the profile example (Figure 9), highs and ridges in the gradient field delineate inflection lines (loci of inflection points) and, by inference, magnetization boundaries. We have drawn these inferred boundaries in Figure 11, using dashed lines to designate ridges whose locations are uncertain.

Many major magnetization boundaries are defined by this process. Some of these, as at locality A in Figure 11, are obvious features that could have been as easily seen in the original data (Figures 8, 10). Others, as at localities B and C, follow longer and straighter trends than one might have supposed. Still others (localities D and E, e.g.) appear as complete surprises, although these can be identified with confidence in Figure 11.

BASEMENT MAGNETIZATION ZONES IN SAN JUAN BASIN

Correction for variation in source depth and delineation of principal magnetization boundaries are both incorporated in Figure 8. These techniques represent objective aids in the inherently subjective process of translating magnetic-contour lines into information about basement geology. Procedures leading to and including Figure 8 can be exactly prescribed and the results would be essentially reproducible by anyone. On the other hand, the following discussion focuses on our subjective opinion, with which other interpreters might differ. We believe, however, that the objective treatment developed above serves to make the subjective interpretation more straightforward and possibly serves also to constrain the margin of reasonable disagreement.

Up to a point, zonation of the anomaly patterns in Figure 8 is obvious by inspection. There are broad zones of differing magnetic-intensity levels, and also broad zones having in

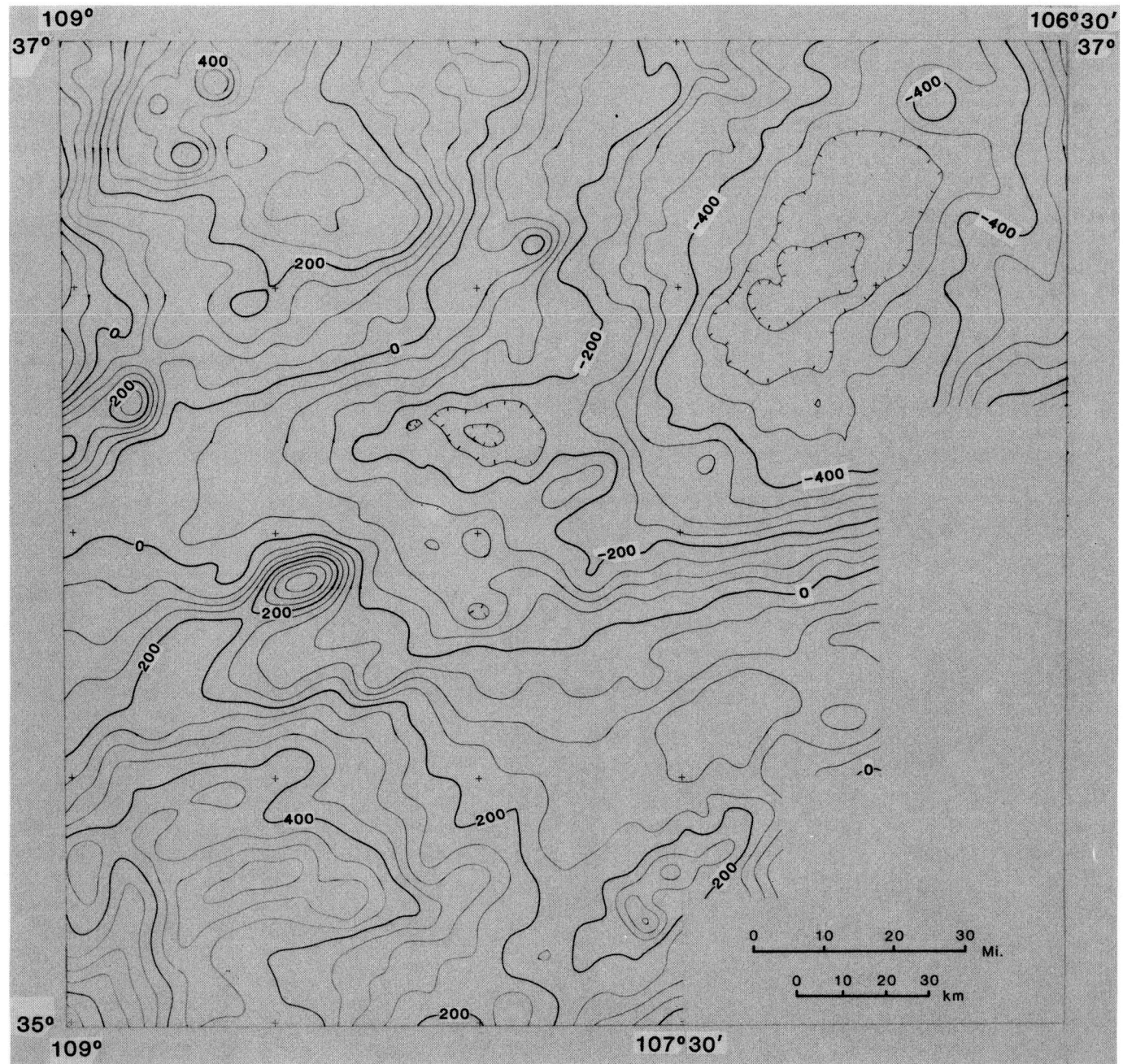

FIG. 10. Pseudogravity of the reduced-to-basement San Juan basin aeromagnetic data. Magnetization assumed induced with inclination of 63 degrees and declination of 12 degrees E. Magnetization/density ratio chosen arbitrarily.

common a characteristic gradient-line trend. By combining magnetic intensity and gradient-line trends, the latter both as linear-pattern guides and as zone boundaries, we identified four regional magnetization zones (Figure 12). Sharp magnetization boundaries as delineated by horizontal-gradient lines are traced directly from Figure 11. Interpreted zone boundaries are indicated by ragged lines, a technique borrowed from geologic maps of basement terrane in shield areas under extensive cover. The zones are colored according to relative magnetic-intensity level: blue for low intensity, green and yellow for intermediate, and red for high.

The most prominent regional feature of the data is the east-northeast-trending belt of generally low magnetic intensity and strong east-northeast gradient trends, extending from the center of the west border of the map to the northeast corner (blue zone in Figure 12). The south side of this zone is clearly defined by a line of gradient-line segments. The north side of the zone is fairly clear except from the center of the map eastward; we draw it along the longest of the northeast-trending gradient lines in this area. South of this zone is a region of intermediate magnetic-field intensity and pervasive northeast gradient-line trend. A similar zone occupies the northwest corner of the map. These zones are indicated in yellow in Figure 12, with red for the very high

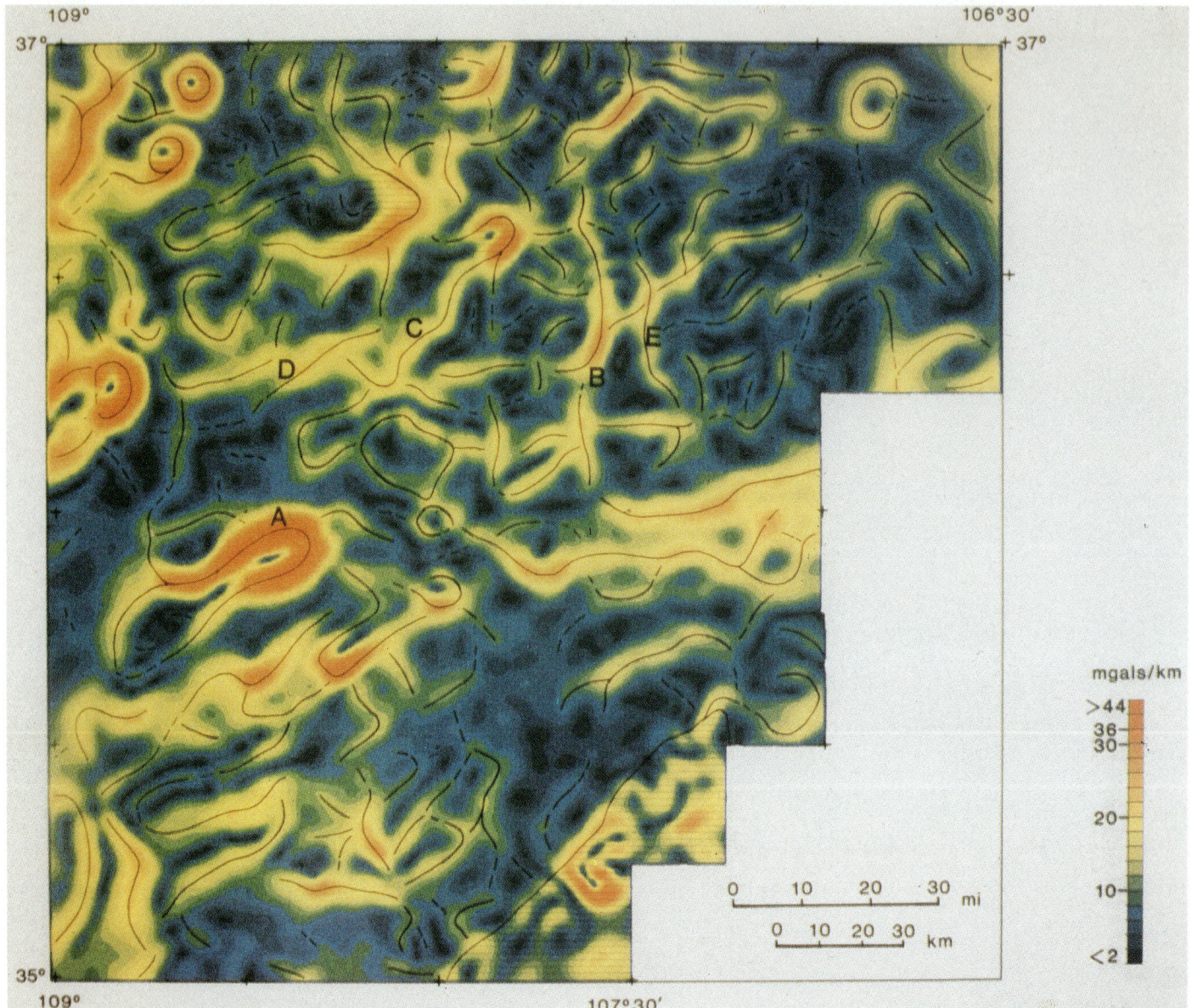

FIG. 11. Magnitude of the horizontal gradient of the pseudogravity. Lines drawn along the ridges represent magnetization boundaries, as shown in Figure 8 and discussed in text; dashed lines designate ridges whose locations are uncertain.

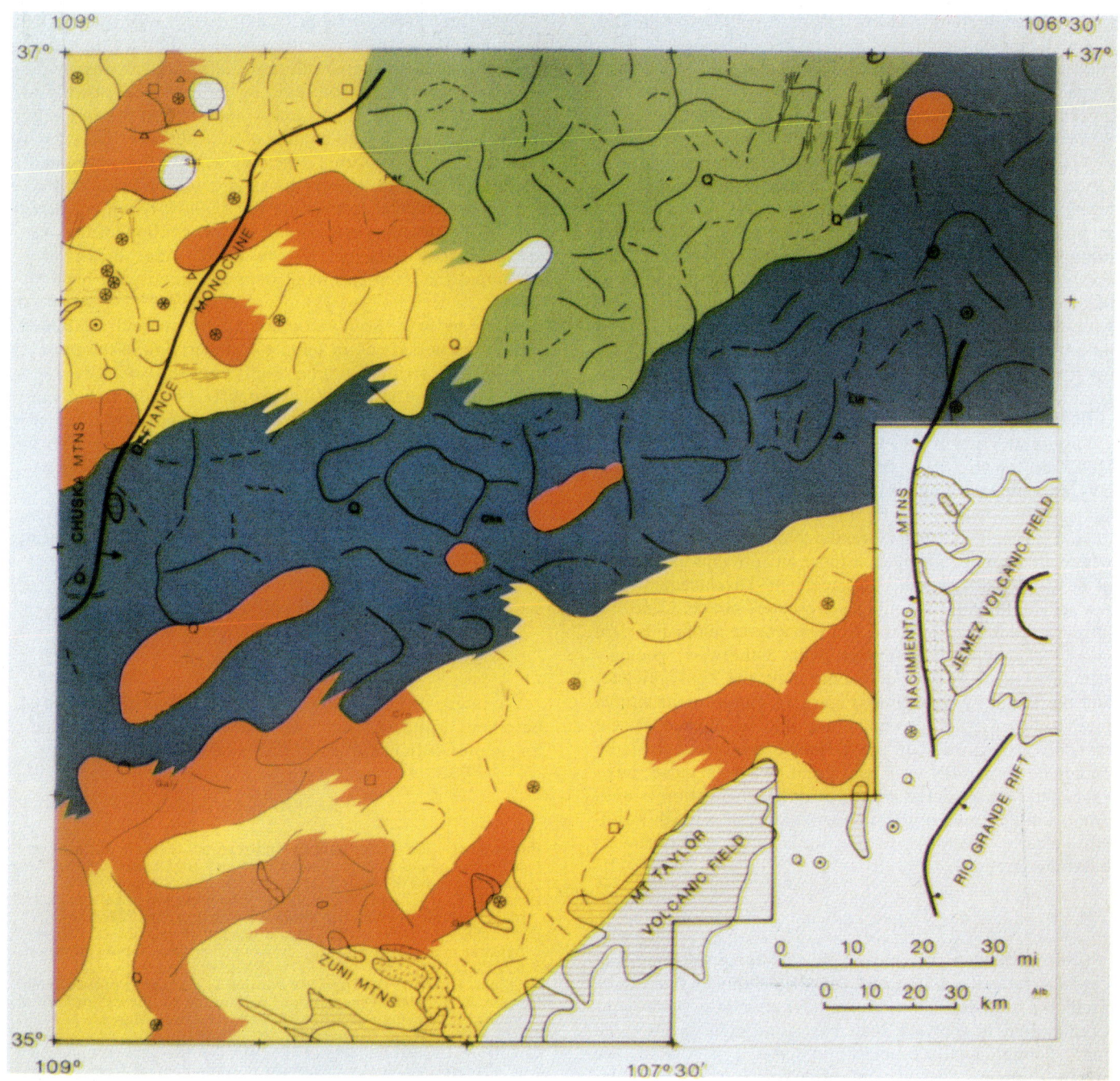

FIG. 12. Basement magnetization zones derived from Figure 8, showing the gradient lines of Figure 11 and the drill-hole and basement-outcrop geology of Figure 1. See text for explanation.

magnetic-intensity areas. A region in the north-central part of the map, shown in green in Figure 12, contains prominent north and northwest gradient trends as well as the ubiquitous northeast trends.

Strong magnetic anomalies over circular features 7–10+ km in diameter in the western part of the map probably indicate plutons intruded into sedimentary cover and therefore Phanerozoic (and probably Paleocene or younger) in age. A lined pattern with no color on such anomalies in Figure 12 indicates where depth analysis on magnetic profiles shows that source depth is significantly shallower than Precambrian basement. Other such pluton anomalies may be present but cannot be determined for certain.

Precambrian drill-hole and outcrop data are generalized in Figures 1 and 12 to distinguish possible sedimentary, volcanic, and plutonic protoliths, with such nondiscriminatory terms as "schist," "gneiss," and "phyllite" included in the grouping, as seemed most appropriate. Although other interpretations would be possible, the limited data on hand, in combination with the aeromagnetic data, show the basin to be traversed by an east-northeast-trending zone, 40 to 70 km wide and at least 250 km long, of quartzite and related metasupracrustal rocks. The zone is bounded on the south by predominantly granitic terrane (as is exposed in the Nacimiento Mountains) and on the north by granites that perhaps lessen eastward relative to supracrustal rocks.

Structural grain is strongly northeast in all four zones. However, north trends are also important in the green-colored zone, the southern apex of which merges, near the center of the map, with the inferred metasupracrustal belt having strongly northeast trends. Neither here, nor elsewhere on the map, however, is there any obvious expression of unsuspected strike-slip displacement. Displacement of a few tens of kilometers could have escaped notice. Cenozoic intrusive rocks (as inferred from magnetic anomalies) are clearly aligned with basement grain. Otherwise, geologic structures, such as the Nacimiento fault, the Zuni anticline, and the Defiance monocline (Figure 1), and local structures (see Kelley, 1955), generally are not so aligned. Local en-echelon segments of the Defiance monocline may be perceived to parallel basement grain. None of these, however, can be associated with specific magnetization boundaries, and the trend of the monocline overall is north-northeast, clearly transverse to the northeast to east-northeast basement grain.

DISCUSSION

Consonant with the general philosophy of applying successive "reductions" to geophysical data until all that can be predicted has been removed, we have treated two problems encountered in the San Juan basin and also encountered fairly commonly in aeromagnetic surveys in general. We sought to reduce the effect of variable attenuation of anomalies related to variation in source depth and to delineate the sharp physical-property boundaries that are implicit, but vaguely determined, in geophysical contour maps. In a nutshell: If the rocks of the two areas are the same, we want the geophysical maps to look the same, and we want to draw our initial interpretive lines on geophysical maps by a method that is objective and deterministic.

These goals have been essentially realized in the San Juan basin study. Basement features are unquestionably more finely resolved than in the original data; magnetization boundaries and, by inference, structural or lithologic trends, are established with much greater confidence than would be possible in qualitative interpretation by inspection. Furthermore, the necessary calculations are rapid and simple, and accurate enough for the purpose at hand.

There are some problems. In the case of reduction to basement, these are primarily related to the fact that high-frequency and low-amplitude anomalies are missing in the deep-basement part of the data in the first place, owing to diminished signal-to-noise ratio, and therefore cannot be recovered by downward continuation. It follows that, even if "the rocks were the same," the only way the deep-basement data and the shallow-basement data could be made to look the same would be to degrade the shallow-basement data. This would obtain whether one upward-continued the shallow-basement data or, as we have done, downward-continued the deep-basement data and imposed a high-cut filter.

Problems with the horizontal-gradient method are minor as long as one keeps in mind the underlying assumptions. The model strictly applies only to an isolated, two-dimensional vertical physical-property boundary that may be either sharp or gradational. Dip, superposition of anomalies, and trends having small radius of curvature all can cause mislocation of the horizontal-gradient lines relative to magnetization boundaries to some extent.

Our limited experience with these techniques at this time suggests that a horizontal-gradient map is worth making in nearly every case, whereas the reduction-to-basement operation is only warranted where the basement-relief problem significantly hinders interpretation of the data.

ACKNOWLEDGMENTS

We thank F. E. Kottlowski and R. E. Denison for use of basement drill data, and T. G. Hildenbrand and J. W. Cady for helpful criticism of early drafts of the manuscript.

REFERENCES

Baltz, E. H., 1967, Stratigraphy and regional tectonic implications of part of upper Cretaceous and Tertiary rocks, east-central San Juan basin, New Mexico: Prof. Pap. 552, U.S. Geol. Surv.

Baranov, V., 1957, A new method for interpretation of aeromagnetic maps: pseudo-gravimetric anomalies: Geophysics, **22**, 359–383.

Bhattacharyya, B. K., 1967, Some general properties of potential fields in space and frequency domains: a review: Geoexploration, **5**, 127–143.

Brigham, E. O., 1974, The Fast Fourier Transform: Prentice-Hall.

Condie, K. C., 1981, Precambrian rocks of the southwestern United States and adjacent areas of Mexico: Resource Map 13, N.Mex. Bur. Mines Miner. Resources.

Condie, K. C., and Budding, A. J., 1979, Geology and geochemistry of Precambrian rocks, central and south-central New Mexico: Mem. 35, N.Mex. Bur. Mines Miner. Resources.

Cordell, L., 1979, Gravimetric expression of graben faulting in Santa Fe country and the Espanola basin, New Mexico: Guideb., 30th Field Conf., Santa Fe Country, N.Mex. Geol. Soc., 59–64.

Cordell, L., and Grauch, V.J.S., 1982a, Reconciliation of the discrete and integral Fourier transforms: Geophysics, **47**, 237–243.

——— 1982b, Mapping basement magnetization zones from aeromagnetic data in the San Juan basin, New Mexico [abstr.]: Abstr. Program, SEG, 1982 Annu. Meet., 246–247.

Cordell, L., and Keller, G. R., 1984, Regional structural trends inferred from gravity and aeromagnetic data in the New Mexico–Colorado border region: Guideb., 35th Field Conf., Rio Grande rift, N.Mex. Geol. Soc., 21–23, color pls. x, xi.

Cordell, Lindrith, and Taylor P. T., 1971, Investigation of magnetization and density of a North Atlantic seamount using Poisson's theorem: Geophysics, **36**, 919–937.

Dampney, C.N.G., 1969, The equivalent source technique: Geophysics, **34**, 39–53.

Dole, W. E., and Jordan, N. F., 1978, Slope mapping: AAPG Bull., **62**, 2427–2440.

Evjen, H. M., 1936, The place of the vertical gradient in gravitational interpretations: Geophysics, **1**, 127–136.

Grauch, V.J.S., and Campbell, D. L., 1984, Does draping aeromagnetic data reduce terrain-induced effects?: Geophysics, **49**, 75–80.

Hamilton, W., 1981, Plate-tectonic mechanism of Laramide deformation: Contrib. Geol. 19, Univ. Wyo., 87–92.

Henderson, R. G., 1960, A comprehensive system of automatic computation in magnetic and gravity interpretation: Geophysics, **25**, 569–585.

——— 1970, On the validity of the use of the upward continuation integral for total magnetic intensity data: Geophysics, **35**, 916–919.

Henderson, R. G., and Cordell, L., 1971, Reduction of unevenly spaced potential field data to a horizontal plane by means of finite harmonic series: Geophysics, **36**, 856–866.

Hildenbrand, T. G., 1983, FFTFIL: A filtering program based on two-dimensional Fourier analysis of geophysical data: Open-File Rep. 83–237, U.S. Geol. Surv.

Huestis, S. P., and Parker, R. L., 1979, Upward and downward continuation as inverse problems: Geophys. J. R. Astron. Soc., **57**, 171–188.

Kelley, V. C., 1955, Regional tectonics of the Colorado Plateau and relationship to the origin and distribution of uranium: Publ. Geol. 5, Univ. N.Mex.

Syberg, F.J.R., 1972, Potential field continuation between general surfaces: Geophys. Prospect., **20**, 267–282.

U.S. Geological Survey, 1980: Aeromagnetic map of northeast Arizona and northwest New Mexico: Open-File Rep. 80–614, U.S. Geol. Surv.

Webring, M., 1981, MINC, a Fortran gridding program based on minimum curvature: Open-File Rep. 81–1224, U.S. Geol. Surv.

Woodward, L. A., McLelland, D., and Kaufman, W. H., 1974, Geologic map and sections of Nacimiento Peak quadrangle, New Mexico: Geol. Map 32, N.Mex. Bur. Mines Miner. Resources.

Regional gravity and magnetic study of west Texas

G. R. Keller,* R. A. Smith,‡ W. J. Hinze,‡ and C.L.V. Aiken§

ABSTRACT

West Texas is well suited to a regional gravity and magnetic study, because such data have potential for resolving existing questions regarding the structural relations in this area as well as its tectonic history. In this study, gravity data have been compiled from many sources and analyzed along with aeromagnetic data from the National Uranium Resource Evaluation (NURE) program of the U.S. Department of Energy in an integrated regional study of west Texas. Complete Bouguer gravity-anomaly and total-magnetic-intensity maps and a variety of filtered maps were employed in this analysis. Several basins, such as the Hueco bolson, Salt basin graben, Marfa basin, Presidio graben, and Valentine basin, can be delineated from the maps, as can interesting positive features associated with the Diablo Plateau, Davis Mountains, and Chalk Draw fault. Deep-seated anomalies were found to be associated with the Delaware basin, Central basin platform, and Ouachita orogenic belt.

INTRODUCTION AND TECTONIC SETTING

West Texas (Figure 1) is well suited for a regional gravity and magnetic study, because many tectonic features in the area are poorly known in terms of their origin, subsurface geometry, and structural relations with deeper seated and/or adjacent features. This area has experienced a complex tectonic history, with the first documented activity having occurred in late Precambrian time in the Van Horn area (Figure 1), where outcrops reveal that rocks of the Carrizo Mountain Group were deformed and thrust over younger rocks to the north (King and Flawn, 1953; Davidson, 1980). Other late Precambrian rocks crop out on the Diablo Plateau and in the Franklin and Hueco Mountains. The complex nature of these outcrops (e.g., Thomann, 1981) and the variety of Precambrian rocks encountered in wells drilled into the basement (Flawn, 1956; Denison and Hetherington, 1969) suggest a complicated Precambrian history for the region. A major continental breakup at the close of the Precambrian affected the area (Keller and Cebull, 1973) and may have led to the development of an aulacogen in the Delaware basin–Central basin platform area (Walper, 1977; Shurbet and Cebull, 1980; Keller et al., 1980). This breakup led to the development of a deep ocean basin south and southeast of the study area.

Early Paleozoic tectonism was dominated by the formation of the Tobosa basin (Galley, 1958), a broad feature that covered most of west Texas. The relative tectonic quiescence of the early and middle Paleozoic was terminated by the Ouachita orogeny (Flawn et al., 1961). The major orogenic activity in the Marathon segment of the Ouachita system affected most of west Texas and began in the Late Mississippian and culminated in the Early Permian (e.g., King, 1978). Although the relationships are not completely clear, the fundamental mechanism for the origin of the Marfa basin and the uplift and deformation of the Central basin platform are probably related to the Ouachita orogenic belt. The Permian was a major period of deposition in the area, which led to the formation of several large basins (e.g., Delaware, Midland, and Val Verde).

The Late Jurassic–Cretaceous was a time of major deposition and marine transgression in the Chihuahua trough (DeFord, 1969; DeFord and Haenggi, 1971). The Laramide orogeny began in the Late Cretaceous and formed a thrust belt (Chihuahua tectonic belt) that extends into west Texas (Gries and Haenggi, 1971; Reaser and Underwood, 1980). The subsurface extent of this thrusting has been of particular interest recently because of its petroleum potential.

The Laramide orogeny may have extended into the Eocene and was followed by widespread volcanic activity and block faulting typical of the Basin and Range province. The voluminous volcanic eruptions and alkalic intrusions (38 to 30 m.y. ago) were the result of subduction along the Pacific coast (Barker, 1979). These eruptions formed several calderas (McAnulty, 1976) that have been associated with mineral deposits. Volcanism terminated 16 m.y. ago (McDowell, 1979), and radiometric dating of dikes along the Sierra Vieja indicates that normal faulting began about 20 m.y. ago (Dasch et al., 1969). This faulting preserved signi-

*Department of Geological Sciences, University of Texas at El Paso, El Paso, TX 79968.
‡Department of Geosciences, Purdue University, West Lafayette, IN 47907.
§Program in Geosciences, University of Texas at Dallas, Richardson, TX 75080.

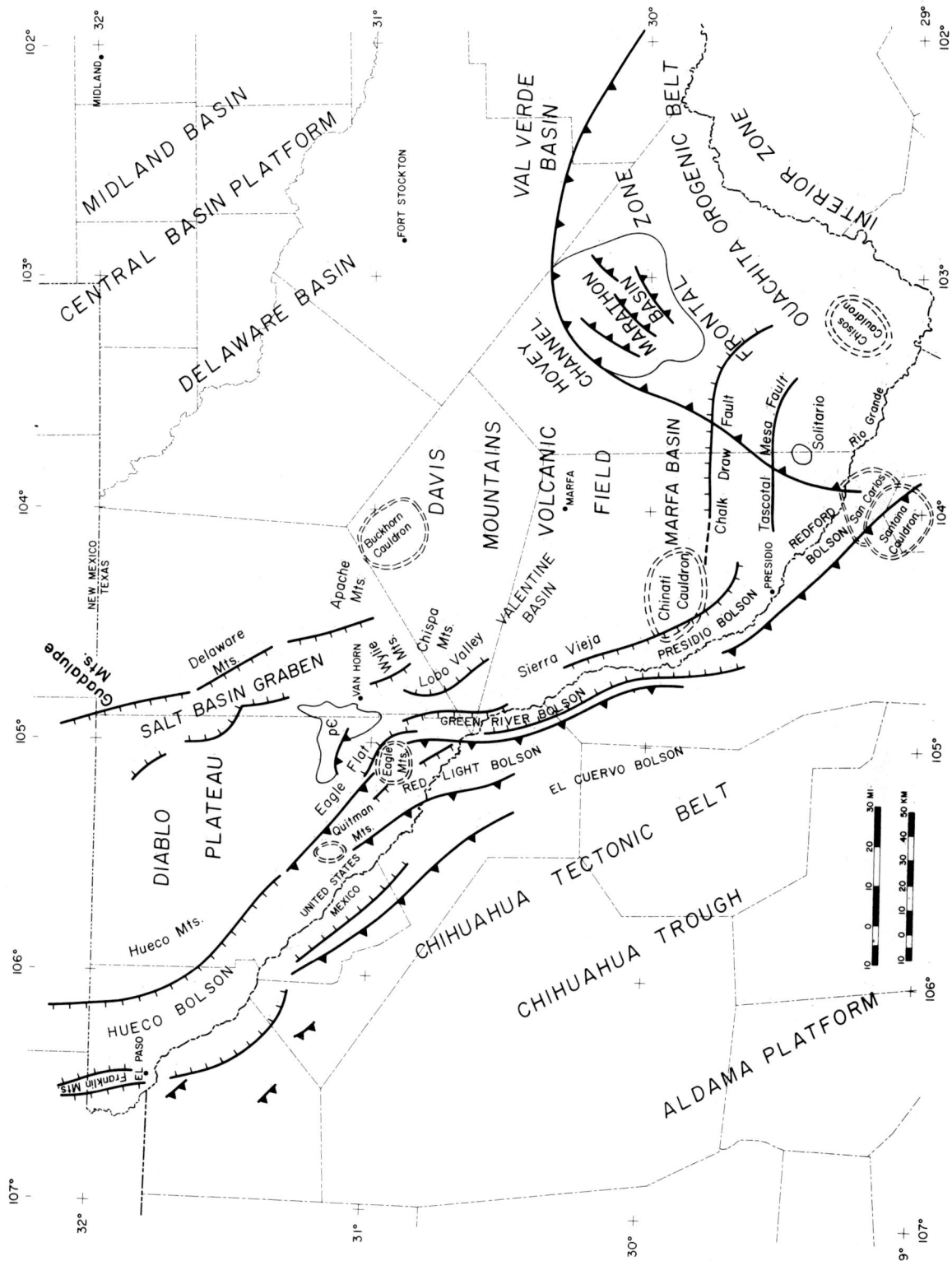

Fig. 1. Index map of major tectonic features in west Texas. Circular features indicated by double-dashed lines are calderas.

ficant volumes of older rocks in areas such as the Salt basin graben (Veldhuis and Keller, 1980) and Presidio bolson (Mraz and Keller, 1980). The possible extension of the Rio Grande rift into the area remains an open question (Seager and Morgan, 1979), but recent fault scarps (Goetz, 1980), elevated heat-flow values (Taylor and Roy, 1980), and the Valentine earthquake of 1931 (Dumas et al., 1980) attest to continued tectonic activity.

GRAVITY DATA

We have been compiling gravity data in the area for several years. Major external sources of data include Mr. Hart Brown, the U.S. Defense Mapping Agency, Texas Tech University, Covert (1976), Wiley (1970), Rice University, and industry. These data have been tied together and adjusted to the IGSN71 base-station network (Morelli, 1976) and reduced using the 1967 International Gravity Formula (Morelli, 1976). A sea-level datum and a reduction density of 2.67 g/cm^3 were employed. A significant improvement over previous gravity databases was the calculation of terrain corrections for the distance range of 0.9 to 167.7 km, using digital terrain data and the technique of Plouff (1977). The resulting data file was gridded, using the minimum-curvature technique (Briggs, 1974), and contoured by computer to produce the complete Bouguer anomaly map shown as Figure 2. The principal facts for the gravity data and a description of the reduction procedure are on open file at the University of Texas at El Paso. A series of gravity maps of the area at a scale of 1:250 000 is in preparation for publication by the Bureau of Economic Geology of the University of Texas.

MAGNETIC DATA

The total-field aeromagnetic surveys were performed by GeoMetrics, Inc., and LKB Resources, Inc., for the Grand Junction office of Bendix Field Engineering Corporation under subcontract nos. 76-033-L and 76-032-S. These surveys were performed as part of the U.S. Department of Energy's National Uranium Resource Evaluation (NURE) program. The results of these surveys are presented in open-file reports GJBX-2 (1977) and GJBX-88 (1979) and can be obtained from Bendix Field Engineering Corporation, Technical Library, Grand Junction, Colorado.

The two companies contracted to perform the survey used different survey specifications and instrumentation. The most significant difference is the type of magnetometer used to observe the total-intensity magnetic-anomaly data. GeoMetrics, Inc., used a proton precession magnetometer, which measures the absolute magnetic field, while LKB Resources, Inc., used a fluxgate magnetometer, which measures the relative magnetic field. Because of this difference, data recorded by the fluxgate magnetometer were adjusted to the absolute datum by subtraction of a constant value of 2 400 gammas. This value was obtained by comparing the differences in the overlapping profiles at the borders of the two surveyed areas. A second-vertical-derivative map of the total-magnetic-intensity anomaly data shows no spurious values at the borders between the two surveyed areas, indicating that the datum level was properly adjusted.

Both surveys were conducted along east–west flight lines at an average spacing of 5 km at an elevation of roughly 130 m above the ground surface. North–south tie lines were flown at an average spacing of 25 km. Both fixed- and rotary-wing aircraft were employed. A 195 km/hr ground speed was maintained for the fixed-wing aircraft, and about 175 km/hr for the rotary-wing aircraft. Observations were made at 0.5- and 1.0-s intervals, respectively, which translates to an average horizontal distance of approximately 27 to 50 m between samples.

The reduction of the magnetic data was performed by the contractor, using different geomagnetic reference fields. The Van Horn and Pecos quadrangles had the 1975 International Geomagnetic Reference Field (IGRF) removed, which was updated to the month of the survey. The magnetic data were also corrected for diurnal variations by using data collected by a ground-based magnetometer and tie-line adjustment. The remaining four quadrangles had the 1965 IGRF removed, which was updated to 1977 with temporal variations corrected by tie-line adjustments.

The magnetic-anomaly data used in this study were gridded, using the same minimum-curvature method (Briggs, 1974) used in gridding the gravity data. The gridded values were calculated from the data set for each NTMS quadrangle provided by Oak Ridge National Laboratory. These data sets were processed by a critical-point selection procedure from high-cut-filtered, contractor-supplied anomaly data, using an error tolerance of 15 gammas (nT) and a maximum separation of 3 km (Tinnel and Hinze, 1981). The gravity- and magnetic-anomaly data sets, which include 202 columns (N–S) and 170 rows (E–W) were registered to a common origin located at 28°59.31′ N latitude and 106°2.88′ W longitude. Both data sets were gridded at a 2-km interval of UTM coordinates, using a central meridian of 105° W. The regional total-magnetic-intensity map is shown in Figure 3.

FILTERING OF DATA SETS

A variety of filtering techniques were employed in order to enhance the data sets used in this study. These techniques included reduction to the pole, upward continuation, vertical derivatives, wave-number filtering, and strike-sensitive filtering. It was not feasible to include all of these maps in this paper, but the reader is referred to Smith (1981) and Keller et al. (1981) for a complete set of maps. A description of the computer codes that were used to create the digitally processed maps can be found in Reed (1980).

All filters were ultimately applied in the wave-number domain. A window was applied so that the edges of the data set were tapered to zero. This window was applied to the 5 percent of the data along each edge. An ideal filter was designed in the frequency domain and transformed to the space domain, where it was smoothed utilizing a Hamming-Tukey window, which passed 80 percent of the rows and columns. This smoothed filter was transformed back to the wave-number domain, where it was multiplied with the data. The results were then transformed to the space domain and contoured.

The total-magnetic-intensity map shown in Figure 3 is complicated by effects of the many Cenozoic volcanic rocks present. Thus, a major concern was to produce a magnetic

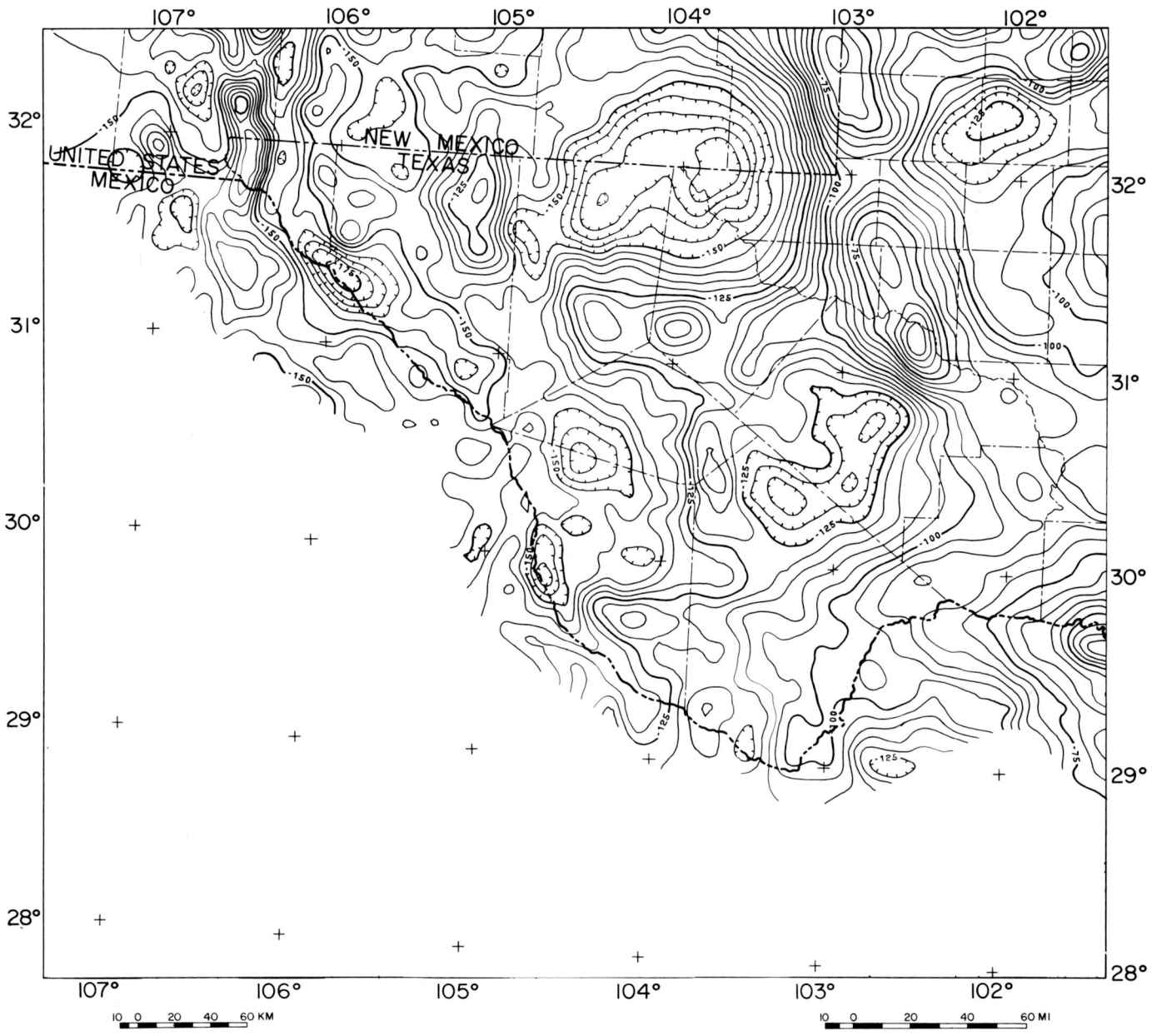

FIG. 2. Complete Bouguer gravity-anomaly map of west Texas region. Contour interval, 5 mGal. Datum, sea level. Reduction density, 2.67 g/cm³.

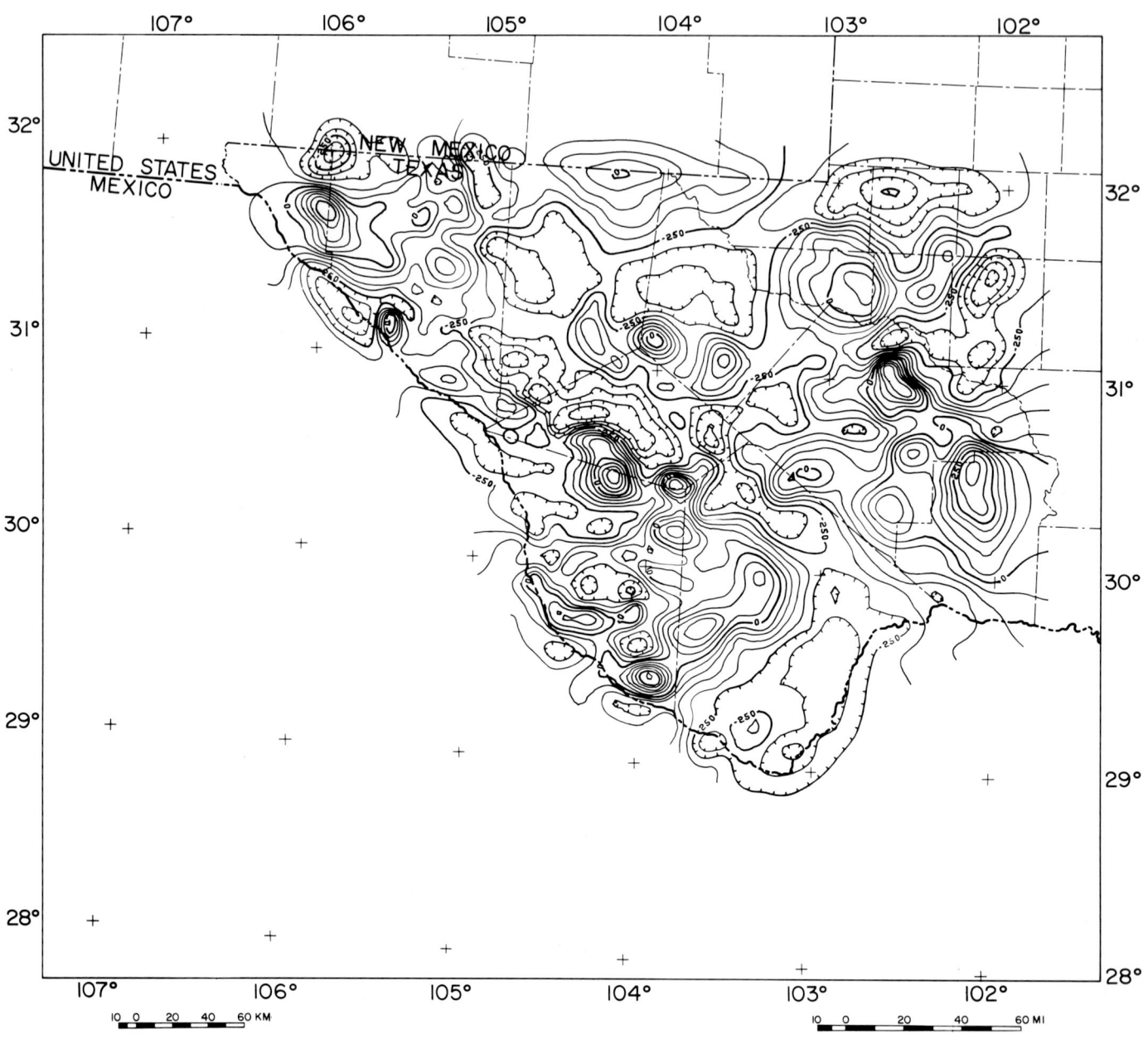

Fig. 3. Total-magnetic-intensity-anomaly map of west Texas region. Contour interval, 50 gammas.

map that could be more readily compared with the Bouguer anomaly map of Figure 2. The first step taken was to reduce the magnetic data to the pole. In this process, we assumed that the observed magnetic anomalies are induced by a magnetic field whose inclination is 59 degrees and whose declination is 11 degrees E. These are average geomagnetic values for the survey area. Although remanent magnetism may locally pose a problem, the resulting magnetic anomalies should in large part be located over the causative bodies.

These reduced-to-the-pole values were then upward continued to 2 km to reduce the effects of near-surface sources. The resulting map is shown as Figure 4.

Other filtered maps are shown as Figures 5–10. Figures 5 and 6 have been upward continued to 2 km and then high-pass filtered to pass wavelengths between 80 and 4 km (wavelengths less than 4 km would be aliased). These maps are, in effect, band-pass filtered with the goal of enhancing major upper crustal features. The maps shown in Figures 7

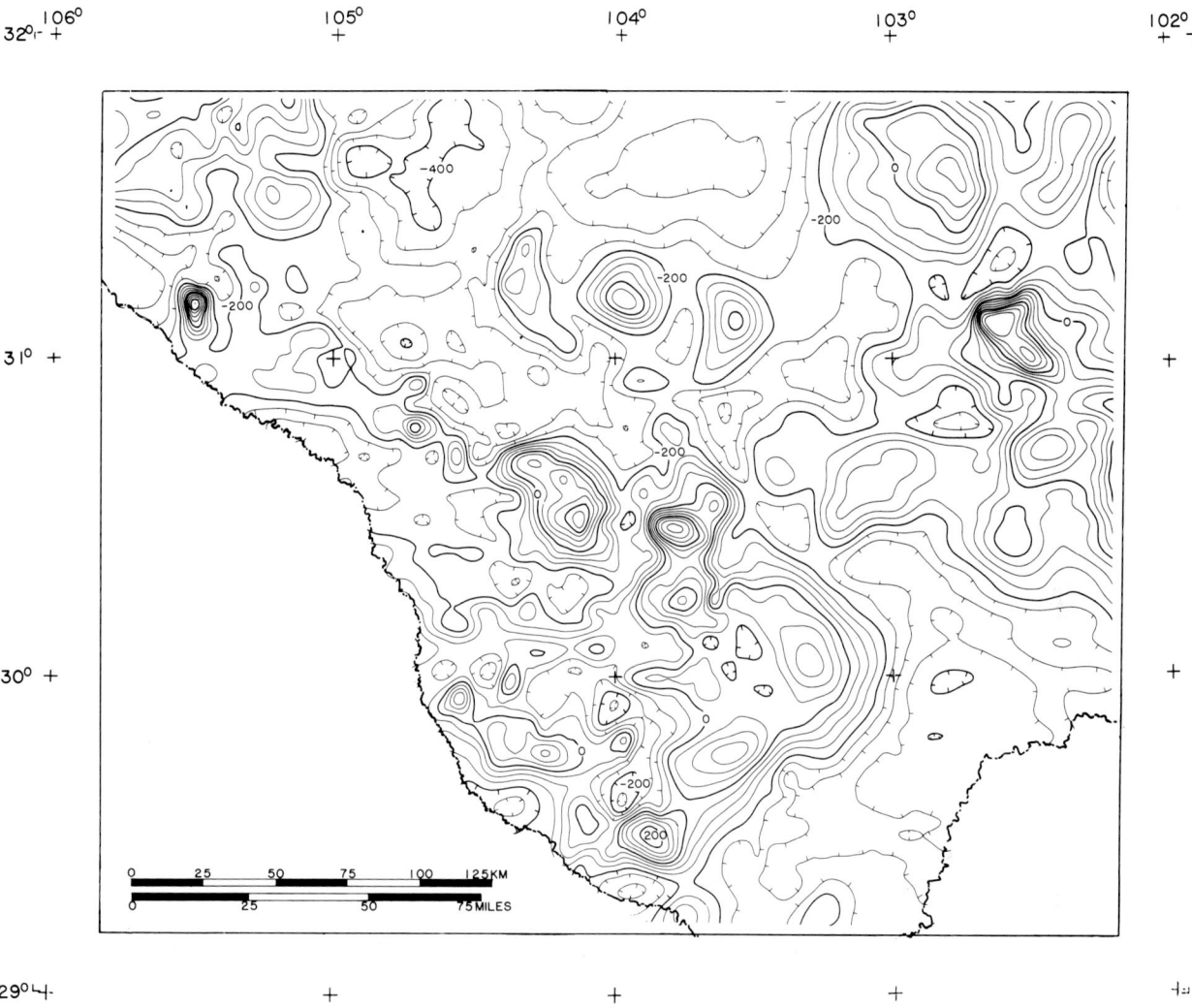

FIG. 4. Upward-continued to 2 km, reduced-to-the-pole, total-magnetic-intensity-anomaly map of a part of west Texas. Contour interval, 50 gammas.

and 8 have been upward continued to 5 and 10 km, respectively, with the goal of enhancing deep-seated anomalies. The use of two different continuation datums is based on the subjective choice of which maps of a suite of upward-continued maps showed the best anomaly correlations. Finally, the maps shown as Figures 9 and 10 have been low-pass filtered to pass wavelengths from ∞ to 80 km. These maps are useful for comparison with Figures 7 and 8 and enhance even deeper seated features.

REGIONAL GRAVITY AND MAGNETIC ANOMALIES

There are many correlations between gravity and aeromagnetic anomalies throughout the study area, and these correlative anomalies provide significant information about the structural units with which they are associated. However, the many surface volcanic rocks present provide additional complications to the patterns of magnetic anomalies. The following discussions refer not only to the gravity- and magnetic-anomaly maps shown as Figures 2 and 3 but also to the filtered maps shown as Figures 4–10. The geologic features discussed are shown in Figure 1.

In the western part of the study area, the Hueco bolson is responsible for the strong northwest-trending minima and gradients (31.1° N, 105.7° W) along the Rio Grande River. This feature is a major graben that contains at least 3 000 m of bolson fill (Mattick, 1967; Uphoff, 1978; Covert, 1976) and is associated with negative magnetic and gravity anomalies that outline areas where these sedimentary rocks attain significant thicknesses.

The nearby northern Quitman Mountains are a resurgent cauldron complex that contains approximately 35 km^3 of welded ash-flow tuffs (Hobbs and Hoffer, 1980). This feature produces a prominent magnetic high. The associated gravity high is poorly defined, because only a limited number of readings were made available for this area of rugged terrain.

A strong northwest-trending gravity and magnetic high (31.8° N, 105.9° W) is associated with the Hueco Mountains. A few scattered outcrops of Precambrian rocks are present

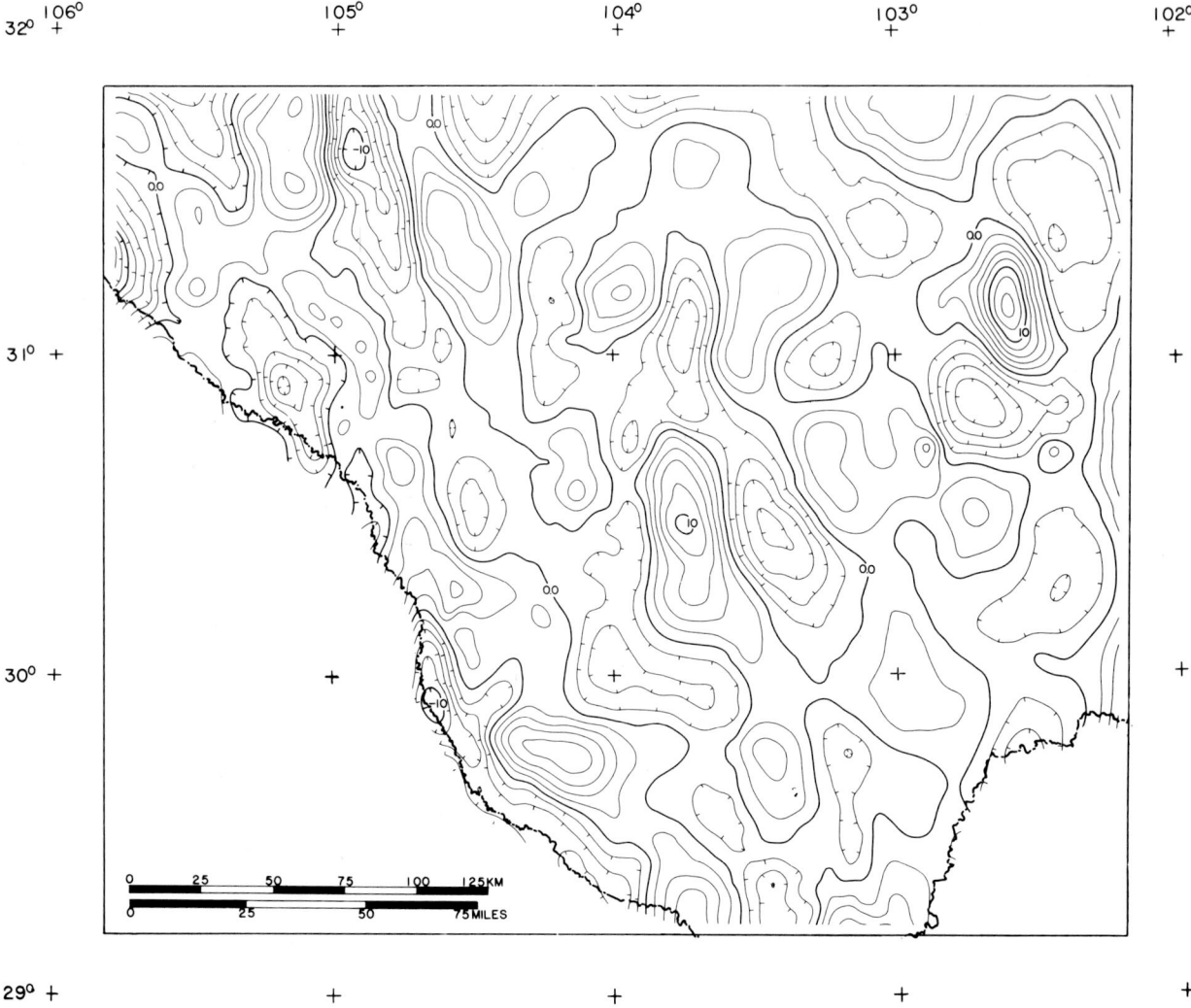

FIG. 5. High-pass-filtered, upward-continued to 2 km, Bouguer gravity-anomaly map of a part of west Texas. Wavelengths from 80 to 4 km are passed. Contour interval, 2 mGal.

in this area, and the geophysical anomalies probably outline an area where Precambrian rocks are near the surface. This area is significantly larger than the outcrop area.

Gravity and magnetic lows east of the Hueco Mountains likely represent an increase in the depth to Precambrian basement. Farther east, the Diablo Plateau is marked by a prominent north–south–trending gravity high. The associated magnetic high is more complicated because there are numerous Tertiary intrusive bodies in the area (Barker, 1977). Modeling of the gravity and magnetic data (Figure 11) indicates that a mafic upper crustal mass is required, which may be the source for these intrusives.

The Salt basin graben is delineated by linear gravity and magnetic lows (31.5° N, 105° W), and computer modeling constrained by well data indicates that this feature is associated with a thick section of Paleozoic rocks in addition to Cenozoic fill (Veldhuis and Keller, 1980). The numerous Quaternary fault scarps in this graben (Goetz, 1980) suggest that it is associated with the Rio Grande rift. However, the thickness of Cenozoic fill in this graben is generally less than 0.5 km (Gates et al., 1978), and Seager and Morgan (1979) suggest that this thickness of fill is too small to be associated with the Rio Grande rift, where the grabens generally contain well over 1 km of fill. Computer modeling (Figure 11) shows that this graben is at least associated with a major change in crustal structure, but the question of the presence of the Rio Grande rift cannot be answered without more data on deep crustal structure in the area.

Precambrian rocks crop out in the Van Horn area (31.1° N, 105° W), and more detailed maps of this area (Keller et al., 1981) show a complicated pattern of gravity and magnetic anomalies. However, a northwest trend is shown in Figures 2 and 3. The magnetic low just north of Van Horn may be related to a Precambrian basin into which late Precambrian and early Paleozoic sediments were deposited.

The Permian basin (Delaware basin, Central basin platform, Midland basin) constitutes the northeast part of the study area. Geophysical anomalies in this area are relatively smooth and uncomplicated because of the thick sedimentary cover. The deepest area of the Delaware basin (31.5° N,

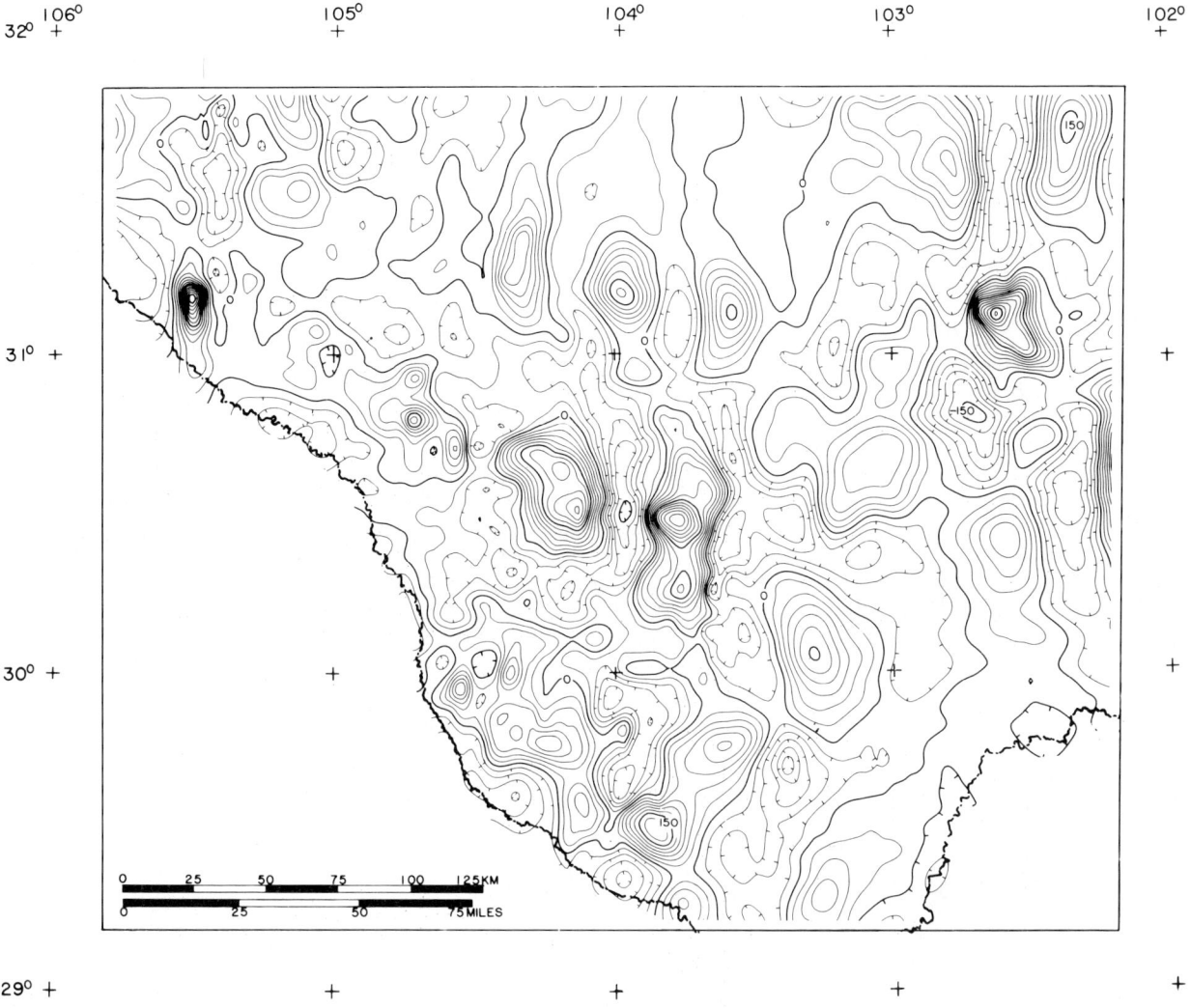

FIG. 6. High-pass-filtered, upward-continued to 2 km, reduced-to-the-pole, total-magnetic-intensity-anomaly map of a part of west Texas. Wavelengths from 80 to 4 km are passed. Contour interval, 25 gammas.

103.5° W) corresponds with large gravity and magnetic lows. To the southwest (31.2° N, 104.4° W), correlative east–west–trending gravity and magnetic highs are attributed to an intra-basement mass (intrusive?) by Veldhuis and Keller (1980). Farther south (30.5° N, 103.5° W), roughly correlative east–west–trending gravity and magnetic lows delineate the southern part of the Delaware basin. A strong north-northwest trend of gravity and magnetic maxima correlate well with the Central basin platform, a major basement horst. As pointed out by Keller et al. (1980), the gravity relief between this platform and the Delaware basin is too great (~ 100 mGal) to be explained by the sedimentary rocks present, because, as shown in Figure 12, the large amount of well control available made it possible to model the gravitational effects of the Phanerozoic rocks accurately. The most plausible explanation for this discrepancy is that there is a mafic, intra-basement mass beneath the Central basin platform. This mass can be interpreted as being associated with an aulacogen that extended through the area in late Precambrian time and was deformed during the late Paleozoic. The east–west–trending highs mentioned above can be interpreted as the signature of an arm of this aulacogen. The Midland basin lies east of the Central basin platform and is associated with gravity and magnetic lows.

The large gravity low centered near Valentine (30.5° N, 104.5° W) delineates the Valentine basin, which has been interpreted to be a Cenozoic graben (Covert, 1976). This anomaly is elongate in a northwest direction parallel to the strike of Basin and Range structures in the area. To the north on more detailed maps (Keller et al., 1981), this basin merges with Lobo Valley, which also correlates with a northwest-trending, linear gravity low. On more detailed maps (Keller et al., 1981), magnetic anomalies in both of these areas are complicated because of near-surface volcanic rocks of the Davis Mountains volcanic field, but in Figure 3 a magnetic high flanked by a smaller low to the north is associated with the Davis Mountains. Although the magnetic data do not outline the Valentine basin, anomalies in the southern part of the basin are subdued and reduced in value, indicating an increase in basement depth. An east–west–trending gravity

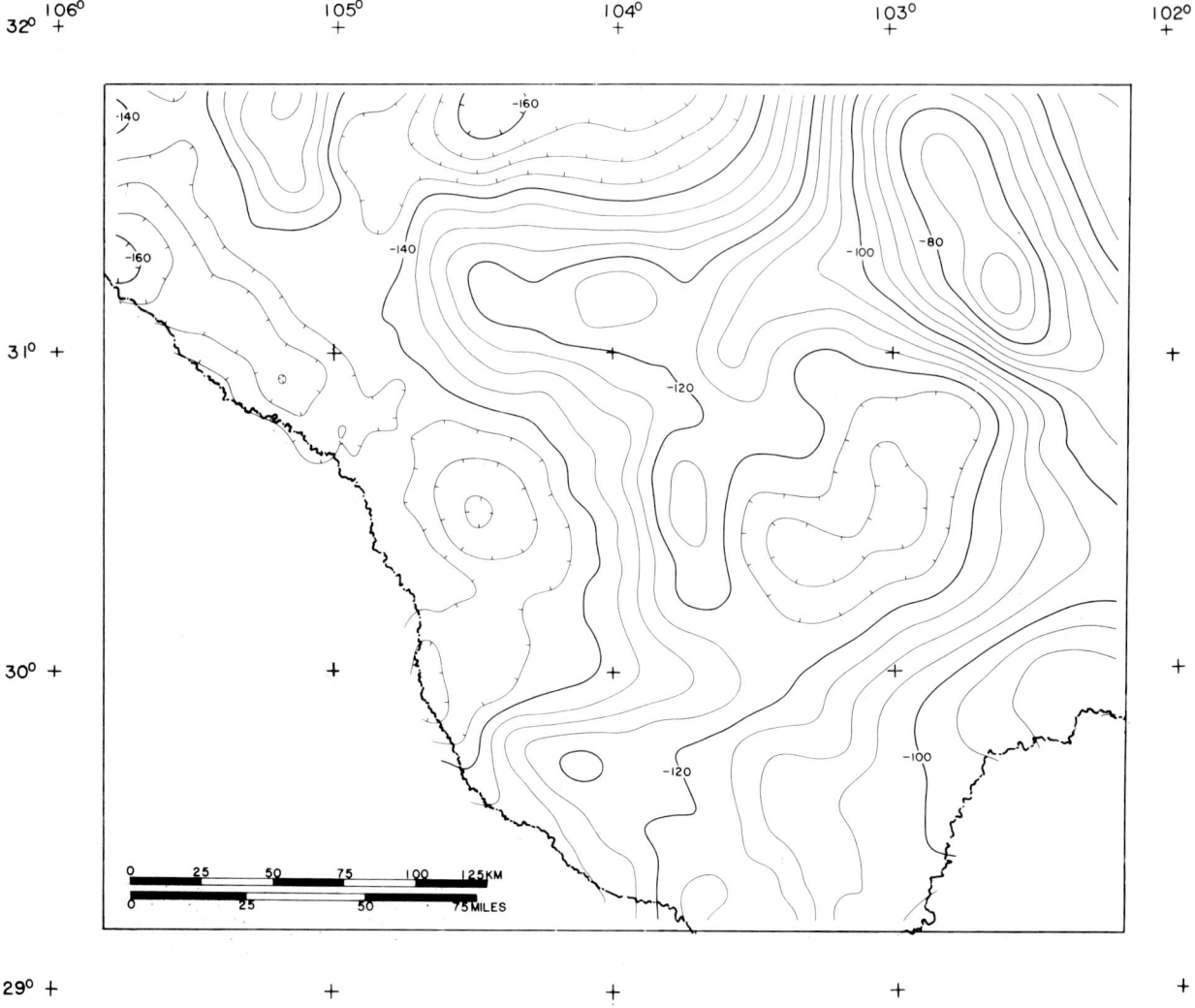

FIG. 7. Upward-continued to 5 km, Bouguer gravity-anomaly map of a part of west Texas. Contour interval, 5 mGal.

and magnetic maximum (30° N, 104° W) separates the Valentine and Marfa basins.

The Marfa basin formed in the late Paleozoic and was filled with more than 3 000 m of Lower Permian sedimentary rocks (Luff, 1981). This basin produces an east–west–trending gravity low (30° N, 104° W), which probably delineates the area in which these sedimentary rocks are thickest. Magnetic anomalies in the area are too complicated from the effect of shallow volcanic rocks to provide much information about this feature.

The Presidio bolson is associated with a north-northwest-trending, major gravity low (29.8° N, 104.5° W), which indicates it is a Basin and Range structure. Mraz and Keller (1980) conducted a gravity study of this feature, and their computer models show that it contains more than 1 km of fill. This amount of fill, plus the Quaternary fault scarps and hot springs (Henry, 1979) in the area, suggests that the Presidio bolson may be part of the Rio Grande rift (Seager and Morgan, 1979). The Redford bolson lies south of the town of Presidio along the course of the Rio Grande. This bolson, along with the Presidio bolson, can be interpreted to be contained within a single major feature, the Presidio graben (Mraz and Keller, 1980).

In the area of the Sierra Vieja (30.5° N, 104.8° W), edge-of-the-data effects and volcanic rocks produce irregular magnetic anomalies along the Rio Grande River. North–south gravity trends follow the trend of faulting in the Sierra Vieja area. To the northwest, the Green River bolson, the Red Light bolson, and Eagle Flat are delineated by a gravity low, but the nearby Eagle Mountains caldera is associated with a prominent positive magnetic anomaly. As shown by Henry (1979), the structural relations in this area (Quitman Mountains to Presidio graben) are highly complex. Although relatively sparse gravity data and flight-line terminations in this remote area make refined interpretations difficult, available gravity data generally outline the grabens associated with the bolsons, and magnetic data delineate the intrusive centers. For example, Eagle Flat is a large bolson,

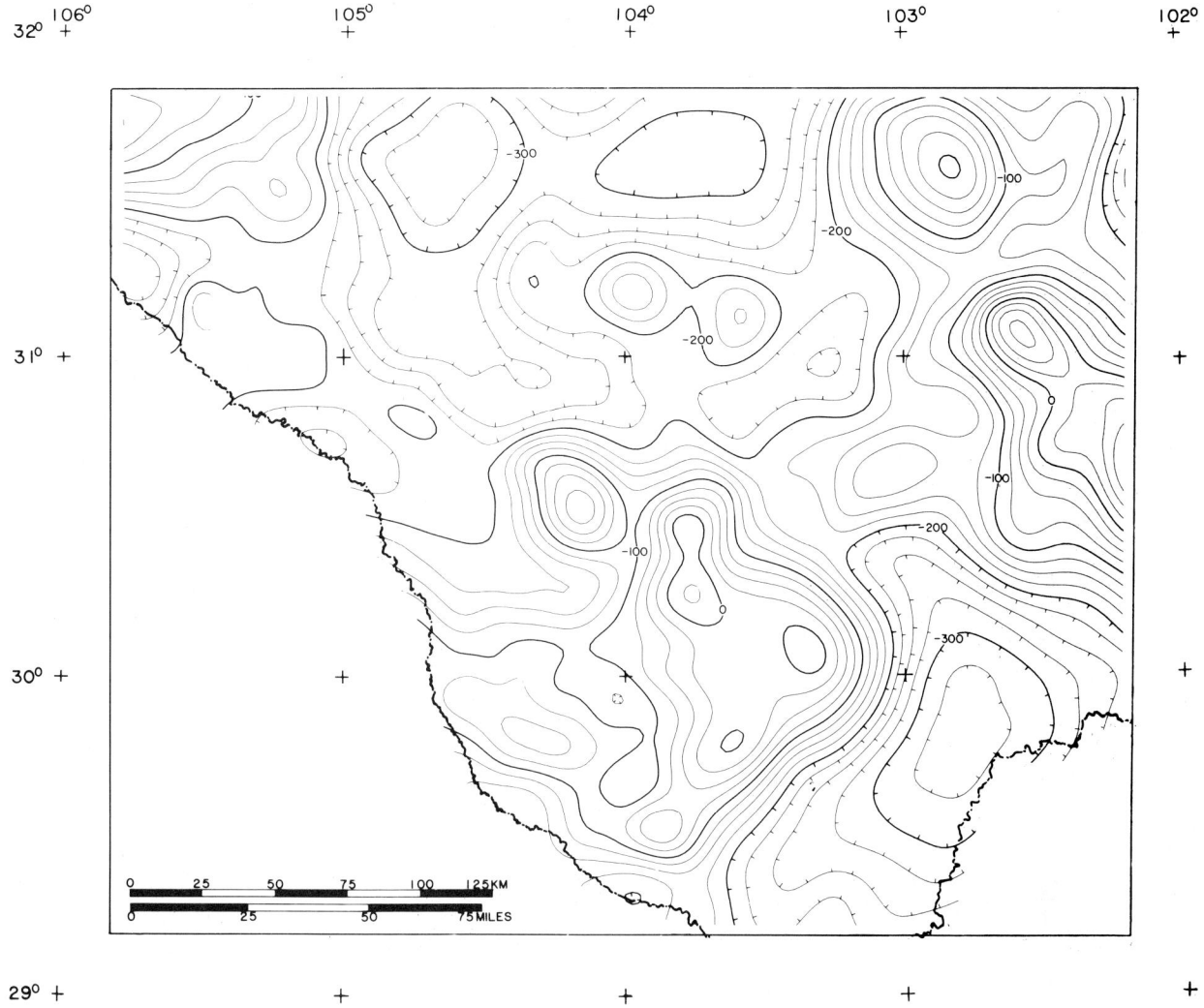

FIG. 8. Upward-continued to 10 km, reduced-to-the-pole, total-magnetic-intensity-anomaly map of a part of west Texas. Contour interval, 25 gammas.

but gravity data (Keller et al., 1981) indicate that it is associated with a relatively narrow northwest-trending graben just north of the Eagle Mountains.

The Chihuahua tectonic belt west of the Presidio graben is represented by a northwest-trending gravity maximum, which is surprising in light of the great thickness of sediments in the area (Gries and Haenggi, 1971). The area west of the Chihuahua tectonic belt in Mexico (not shown here) is dominated by north–south–trending gravity lows that delineate Basin and Range grabens in the area (Aiken et al., 1981; Tovar et al., 1978).

The southeastern part of the study area is dominated by the Ouachita orogenic belt, which is exposed only in the Marathon area and the Solitario uplift area. This orogenic belt is associated with an arcuate gravity high (29.8° N, 102.5° W) that marks its metmorphosed interior zone. Large volumes of sediment were deposited in the Marathon basin during the Ouachita orogeny, and subdued magnetic anomalies in the area indicate that the basement is deep. The gravity high that persists in spite of the thick sedimentary cover suggests a transition in crustal structure. Such a transition is expected, because there is a major fossil plate boundary (more oceanic lithosphere to the southeast) in the area (Keller and Cebull, 1973). Well control in this area indicates that the Ouachita-facies rocks are allocthonous and overlie Paleozoic shelf rocks.

A northwest-trending gravity high extends from the northern Big Bend area (29.6° N, 103.3° N) across the area north of the Solitario and is associated with the Chalk Draw fault and the southern margin of the Marfa basin. The Solitario is associated with a gravity and magnetic high. Aeromagnetic anomalies are surprisingly smooth in the Big Bend area (29.2° N, 103.2° W), which contains several major volcanic centers. A northeast-trending gravity low (30° N, 103.5° W) probably delineates the Hovey channel, which connects the Marfa and Delaware basins.

The Davis Mountains volcanic field is a diverse pile of extrusive and intrusive volcanic rocks that are associated with a complicated pattern of magnetic anomalies. The eastern part of the Davis Mountains area contains pro-

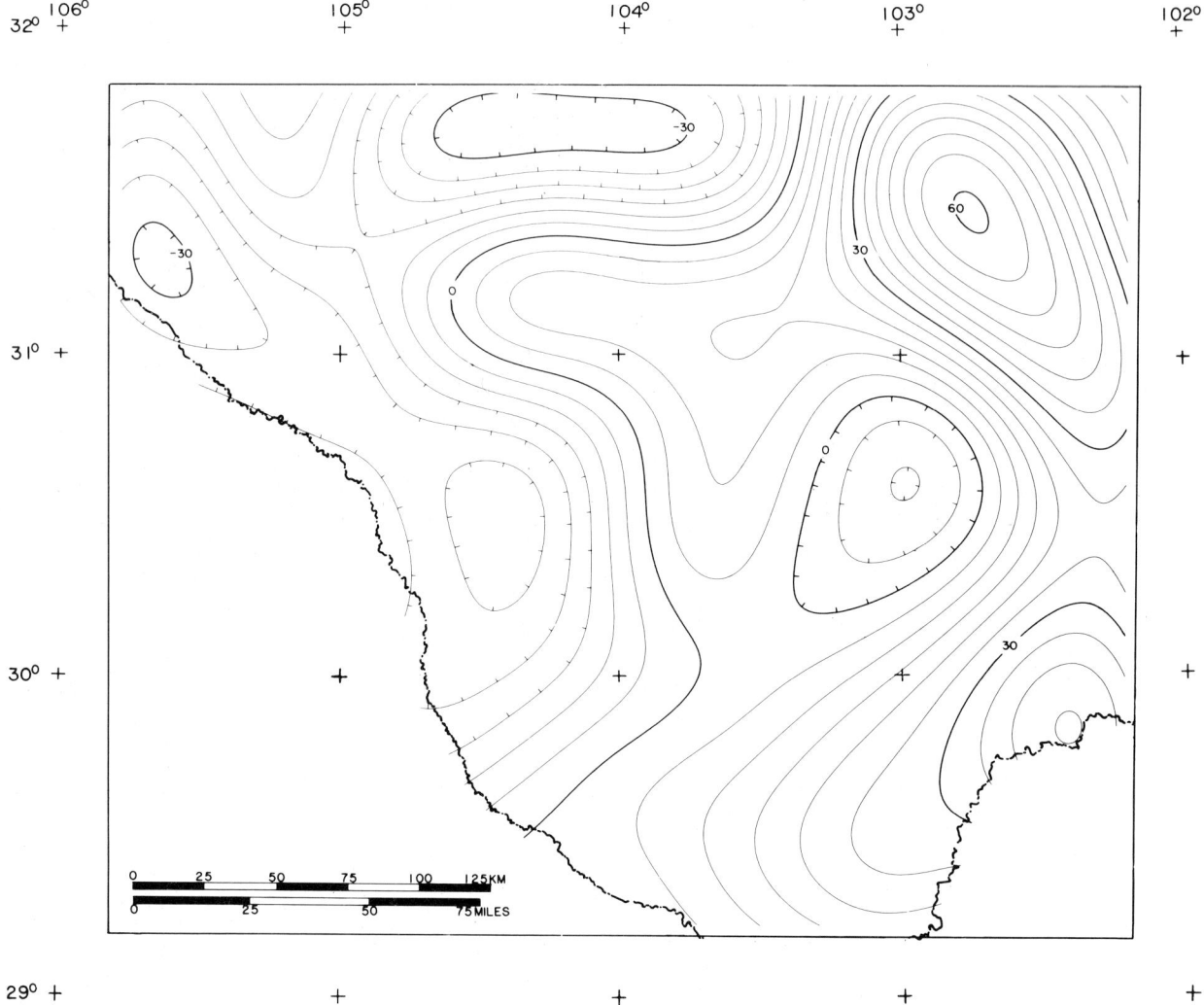

FIG. 9. Low-pass-filtered, upward-continued to 2 km, Bouguer gravity-anomaly map of a part of west Texas. Wavelengths from ∞ to 80 km are passed. Contour interval, 5 mGal.

nounced, north–south–trending, linear gravity and magnetic highs (30.5° N, 103.8° W) that do not correlate well with any known Cenozoic geologic features. This region is covered with large thicknesses of primarily extrusive volcanic rocks, and deep drilling in this area is almost nonexistent. Because this area is in the transition zone between the Basin and Range (Rio Grande rift?) and the Delaware basin, this trend is important from both a tectonic and economic standpoint. These linear anomalies might reflect Basin and Range (Rio Grande rift?) structures buried by the Davis Mountains, but the Davis Mountains volcanic field is approximately 35–40 m.y. old (McDowell, 1979) and clearly predates Basin and Range faulting in this area (Dasch et al., 1969). Although geologic mapping in the Davis Mountains is in large part reconnaissance in nature, it seems unlikely that such a major structure could remain unmapped.

Another possibility is that these anomalies reflect sources for the Davis Mountains volcanic field. The San Juan and Mogollon–Datil volcanic fields of southwestern Colorado and New Mexico, respectively, are of similar age and origin (subduction-zone related; Barker, 1979). Circular gravity lows are associated with these features, suggesting the presence of granitic batholiths that have negative density contrasts with the upper crustal rocks which they intrude (Plouff and Pakiser, 1972; Krohn, 1976). Thus, the positive anomalies in the Davis Mountains area would require a major change in the character of either the batholithic or the upper crustal rocks. The similarities in age and origin and the geographic proximity to the Mogollon–Datil volcanic field suggest that such a variation is unlikely, and the observed anomalies are linear, not circular.

It seems most plausible to suggest that these anomalies are the result of a basement uplift of similar age and origin as the Central basin platform. They could, of course, be of Laramide or much older origin, but their configuration and proximity to the Marathon salient in the Ouachita orogenic

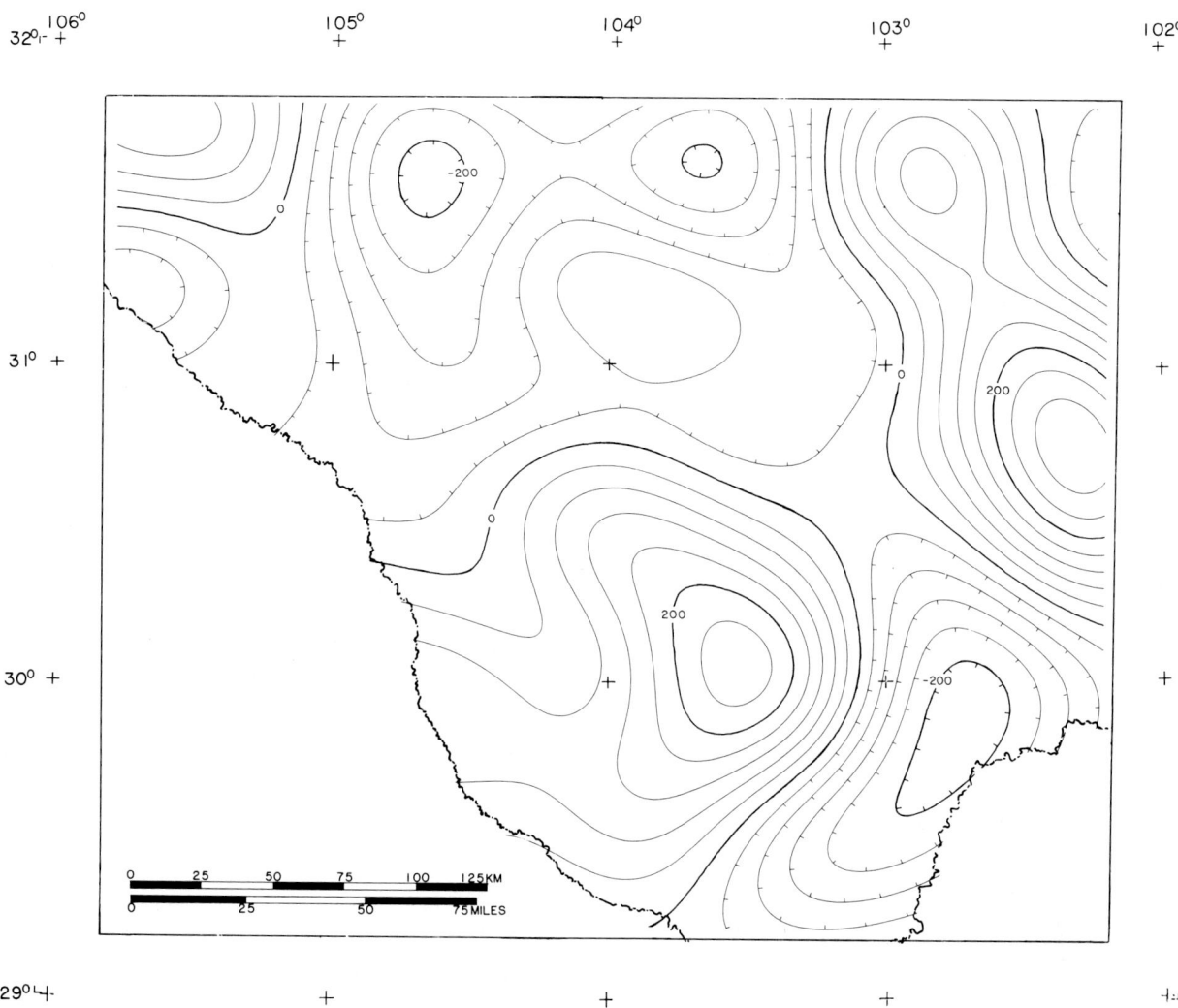

FIG. 10. Low-pass-filtered, upward-continued to 2 km, reduced-to-the-pole, total-magnetic-intensity-anomaly map of a part of west Texas. Wavelengths from ∞ to 80 km are passed. Contour interval, 40 gammas.

belt suggest a genetic relation. Although the geologic cross-sections of this area presented by Luff (1981) permit such an interpretation, additional deep drilling will be required to test this hypothesis.

CONCLUSIONS

The regional gravity and magnetic data analyzed in this study provide considerable insight into the complex structural relations of west Texas. Basins such as the Hueco bolson, Presidio graben, Valentine basin, Marfa basin, and Salt basin graben, whose subsurface geometry is only generally known, can be readily delineated in Figures 2–6. In these same figures, relatively short-wavelength, positive anomalies, such as those beneath the Diablo Plateau, beneath the Davis Mountains, and south of the Chalk Draw fault, delineate structures that are tectonically significant and that warrant further study.

More deeply seated relations are depicted in Figures 7–10. A positive gravity anomaly and a negative magnetic anomaly associated with the Ouachita system indicate thick sediments and a major transition in crustal structure. Long-wavelength highs extend from the Central basin platform into the interior zone of the Ouachita system in a manner analogous to relations where the Southern Oklahoma aulacogen intersects the Ouachita system, suggesting a similar origin for these features. Large gravity and magnetic highs extend westward from the Central basin platform area and would be of similar magnitude to the anomalies associated with the Central basin platform if the Delaware basin did not cover them. It is therefore reasonable to assume that these westward-extending anomalies are related to a deep-seated feature that was a likely target for reactivation by Phanerozoic geologic events. Such reactivations could form targets for deep exploration in the Delaware basin. The page-size maps presented in this study limit the analysis to only general considerations, while larger scale maps (1:250 000) presented by Keller et al. (1981) provide more detailed information.

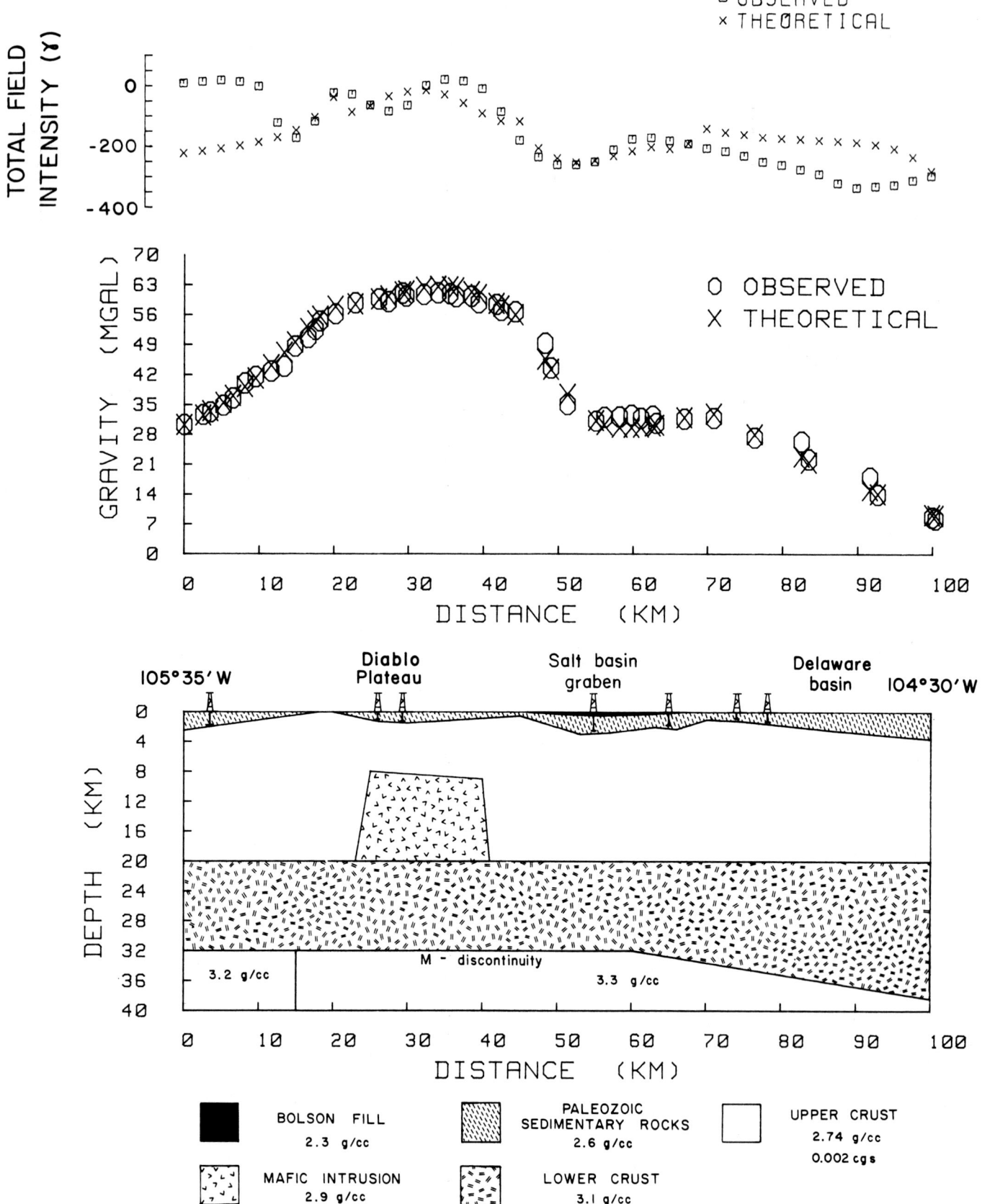

FIG. 11. Computer model for a regional west–east profile across the Salt basin graben at 31°45′ N. Modified from Veldhuis and Keller (1980).

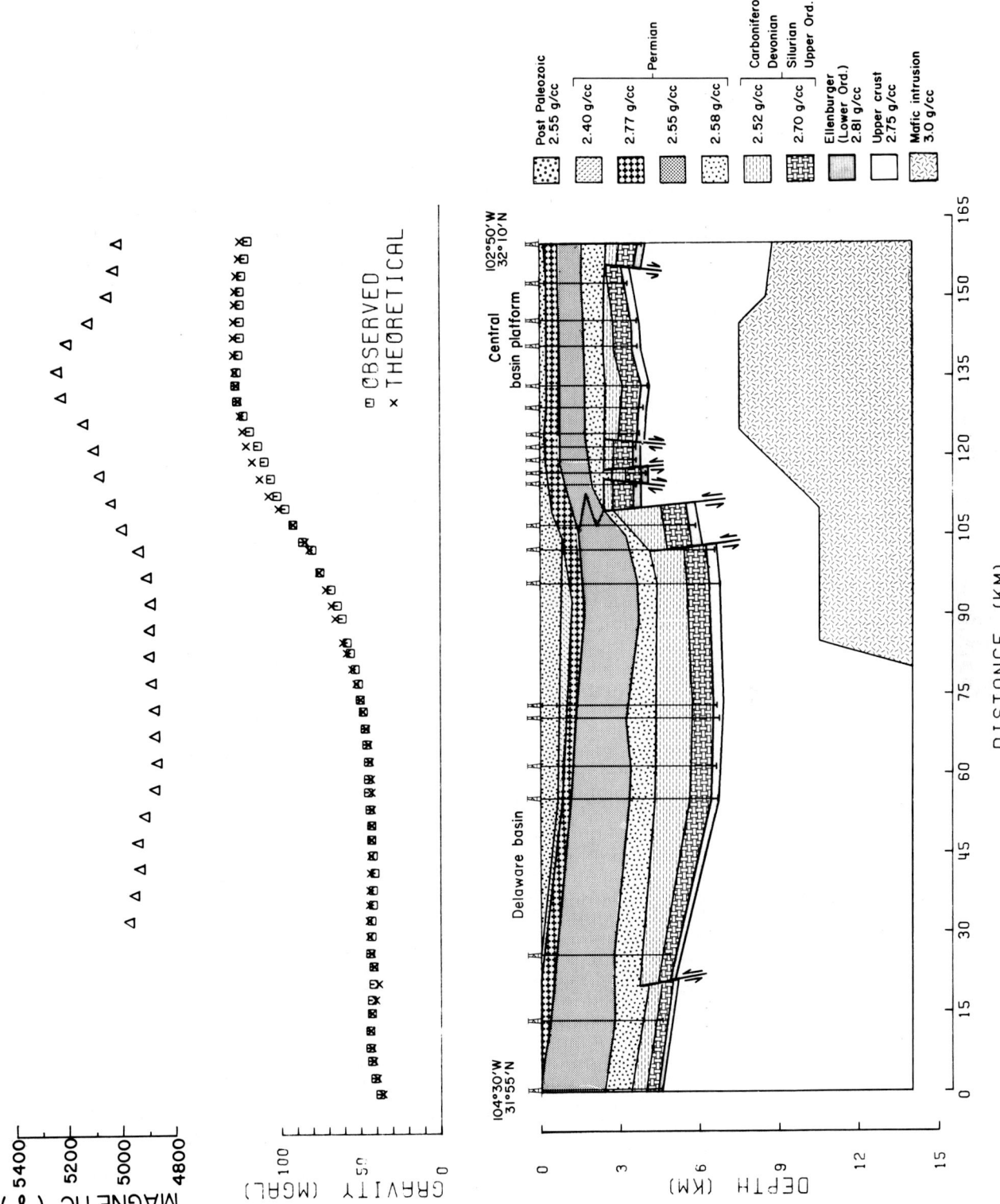

FIG. 12. Computer model for a regional profile across the Delaware basin and Central basin platform. Modified from Keller et al. (1980).

ACKNOWLEDGMENTS

This research was supported in part by the Bendix Field Engineering Corporation (Contract 78-240-E) and the Bureau of Economic Geology of the University of Texas at Austin. The following oil companies also provided financial support and valuable information: Exxon, ARCO, POGO Producing, Getty, Amoco, Chevron, Cities Service, Conoco, Sun, and Sohio.

The efforts of many students made this project possible. We particularly thank J. Reed, J. Mraz, L. Covert, J. Veldhuis, R. Sarno, W. Costello, R. Djeddi, R. Coultrip, and J. Lance. We would also like to acknowledge the insight we have gained through discussions with professional colleagues who are too numerous to mention individually.

REFERENCES

Aiken, C.L.V., Garvey, D. L., Keller, G. R., Goodell, P. C., and Duch, M. de la F., 1981, A regional geophysical study of the Chihuahua City area, Mexico: Stud. Geol. 13, AAPG, 311–328.

Barker, D. S., 1977, Northern Trans-Pecos magmatic province: Introduction and comparison with the Kenya rift: Geol. Soc. Am. Bull., **88**, 1421–1427.

——— 1979, Cenozoic magmatism in the Trans-Pecos province: Relation to the Rio Grande rift, in Rio Grande rift: Tectonics and magmatism: Am. Geophys. Union, 382–392.

Briggs, I. C., 1974, Machine contouring using minimum curvature: Geophysics, **39**, 39–48.

Covert, L. L., 1976, A gravity and tectonic study of Trans-Pecos Texas: M.S. thesis, Univ. Ky.

Dasch, E. J., Armstrong, R. L., and Clabaugh, S. E., 1969, Age of Rim Rock dike swarm, Trans-Pecos Texas: Geol. Soc. Am. Bull., **80**, 1819–1824.

Davidson, D. M., Jr., 1980, Precambrian geology of the Van Horn area, Texas: Guideb., 31st Field Conf., N.Mex. Geol. Soc., 151–154.

DeFord, R. K., 1969, Some keys to the geology of northern Chihuahua: Guideb., 20th Field Conf., N.Mex. Geol. Soc., 61–65.

DeFord, R. K., and Haenggi, W. G., 1971, Stratigraphic nomenclature of Cretaceous rocks in northeastern Chihuahua: Publ. 71–59, W. Tex. Geol. Soc., 173–196.

Denison, R. E., and Hetherington, E. A., Jr., 1969, Basement rocks in far West Texas and south-central New Mexico, in Border stratigraphy symposium: Circ. 104, N.Mex. Bur. Mines Miner. Res., 1–16.

Dumas, D. B., Dorman, H. J., and Lathman, G. V., 1980, A reevaluation of the August 16, 1931 Texas earthquake: Seismol. Soc. Am. Bull., **70**, 1171–1180.

Flawn, P. T., 1956, Basement rocks of Texas and southeast New Mexico: Publ. 5605, Univ. Tex. Bur. Econ. Geol.

Flawn, P. T., Goldstein, A., Jr., King, P. B., and Weaver, C. E., 1961, The Ouachita system: Publ. 6120, Univ. Tex. Bur. Econ. Geol.

Galley, J. E., 1958, Oil and geology in the Permian Basin of Texas and New Mexico, in Habitat of oil: Spec. Publ., AAPG, 395–446.

Gates, J. S., White, D. E., Stanley, W. D., and Ackermann, H. D., 1978, Availability of fresh and slightly saline ground water in the basins, Texas: Open-File Rep. 78–663, U.S. Geol. Surv.

Goetz, L. K., 1980, Quaternary faulting in Salt Basin graben, West Texas: Guideb., 31st Field Conf., N.Mex. Geol. Soc., 83–92.

Gries, J. C., and Haenggi, W. T., 1971, Structural evolution of the eastern Chihuahua tectonic belt: Publ. 71–59, W. Tex. Geol. Soc., 119–139.

Henry, C. D., 1979, Geology setting and geochemistry of thermal water and geothermal assessment, Trans-Pecos Texas: Rep. Invest. 96, Univ. Tex. Bur. Econ. Geol.

Hobbs, T.M.C., and Hoffer, J. M., 1980, The Square Peak Volcanic Series, Northern Quitman Mountains, Hudspeth County, Texas: Guideb., 31st Field Conf., N.Mex. Geol. Soc., 231–236.

Keller, G. R., and Cebull, S. E., 1973, Plate tectonics and the Ouachita system in Texas, Oklahoma, and Arkansas: Geol. Soc. Am. Bull., **83**, 1659–1666.

Keller, G. R., Hills, J. M., and Djeddi, R., 1980, A regional geological and geophysical study of the Delaware Basin, New Mexico and West Texas: Guideb., 31st Field Conf., N.Mex. Geol. Soc., 105–111.

Keller, G. R., Hinze, W. J., Aiken, C.L.V., Goodell, P. C., Roy, R. F., and Pingitore, N. E., 1981, Evaluation and combined geosciences data in the Van Horn, Pecos, Marfa, Fort Stockton, Presidio, and Emory Peak Quadrangles, Texas: Rep. GJBX-365 (81), U.S. Dep. Energy 2 v.

King, P. B., 1978, Tectonics and sedimentation of the Paleozoic rocks in the Marathon region, West Texas: Publ. 78–17, SEPM, Permian Basin Sec., 5–37.

King, P. B., and Flawn, P. T., 1953, Geology and mineral deposits of Precambrian rocks of the Van Horn area, Texas: Publ. 5301, Univ. Tex. Bur. Econ. Geol.

Krohn, D. H., 1976, Gravity survey of the Mogollon plateau volcanic province, southwestern New Mexico: Spec. Publ. 5, N.Mex. Geol. Soc., 113–117.

Luff, G. C., 1981, A brief overview (and oil gas potential) of the Marfa Basin: Publ. 81–20, SEPM, Permian Basin Sec., 110–130.

Mattick, R. E., 1967, A seismic and gravity profile across the Hueco Bolson, Texas: Prof. Pap. 575-D, U.S. Geol. Surv., 85–91.

McAnulty, W. N., Jr., 1976, Resurgent cauldrons and associated mineralization, Trans-Pecos Texas: Spec. Publ. 6, N.Mex. Geol. Soc., 180–186.

McDowell, F. W., 1979, Potassium–Argon dating in the Trans-Pecos Texas volcanic field: Guideb. 19, Univ. Tex. Bur. Econ. Geol., 10–18.

Morelli, C., 1976, Modern standards for gravity surveys: Geophysics, **41**, 1051.

Mraz, J. R., and Keller, G. R., 1980, Structure of the Presidio Bolson area, Texas, interpreted from gravity data: Circ. 80–13, Univ. Tex. Bur. Econ. Geol.

Plouff, D., 1977, Preliminary documentation for a Fortran program to compute gravity terrain corrections based on topography digitized on a geographic grid: Open-File Rep. 77–535, U.S. Geol. Surv.

Plouff, D., and Pakiser, L. C., 1972, Gravity study of the San Juan Mountains, Colorado: Prof. Pap. 800-B, U.S. Geol. Surv., B183–B190.

Reaser, D. F., and Underwood, J. R., Jr., 1980, Tectonic style and deformational environment in the Eagle–Southern Quitman Mountains, western Trans-Pecos Texas: Guideb., 31st Field Conf., N.Mex. Geol. Soc., 123–130.

Reed, J. E., 1980, Enhancement/isolation wavenumber filtering of potential field data: M.S. thesis, Purdue Univ.

Seager, W. R., and Morgan, P., 1979, Rio Grande rift in Southern New Mexico, West Texas, and northern Chihuahua, in Rio Grande rift: Tectonics and magmatism: Am. Geophys. Union, 87–106.

Shurbet, D. H., and Cebull, S. E., 1980, Tabosa–Delaware basin as an aulacogen: Tex. J. Sci., **32**, 17–22.

Smith, R. A., 1981, Gravity and magnetic anomaly studies in West Texas: M.S. thesis, Purdue Univ.

Taylor, B., and Roy, R. F., 1980, A preliminary heat flow map of west Texas: Guideb., 31st Field Conf., N.Mex., Geol. Soc., 137–139.

Thomann, W. F., 1981, Ignimbrites, trachytes, and sedimentary rocks of the Precambrian Thunderbird group, Franklin Mountains, El Paso, Texas: Geol. Soc. Am. Bull., **92**, 94–100.

Tinnel, E. P., and Hinze, W. J., 1981, Preparation of magnetic anomaly profile and contour maps from DOE–NURE aerial survey data, Vol. 1: Processing procedures: Rep. ORNL/CSD/TM-155, Oak Ridge Nat. Lab.

Tovar, J., Vazques, M., and Lozano, S., 1978, Interpretation of the geology and geophysics of the northern portion of Chihuahua: Mex. Assoc. Pet. Geol. Bull., **30**, 57–134.

Uphoff, T. L., 1978, Subsurface stratigraphy and structure of the Mesilla and Hueco Bolson, El Paso, Texas and New Mexico: M.S. thesis, Univ. Tex. at El Paso.

Veldhuis, J. H., and Keller, G. R., 1980, An integrated geological and geophysical study of the Salt Basin graben, West Texas: Guideb., 31st Field Conf., N.Mex. Geol. Soc., 141–150.

Walper, J. L., 1977, Paleozoic tectonics of the southern margin of North America: Trans. Gulf Coast Assoc. Geol. Soc., **27**, 230–241.

Wiley, M. A., 1970, Correlation of geology with gravity and magnetic anomalies, Van Horn–Sierra Blanca region, Trans-Pecos Texas: Ph.D. diss., Univ. Tex. at Austin.

Kansas basement study using spectrally filtered aeromagnetic data

Harold L. Yarger*

ABSTRACT

A recently compiled 80 000-line-kilometer aeromagnetic survey of the State of Kansas, flown under uniform specifications, provides an excellent database for spectral-enhancement techniques. A suite of spectrally filtered maps of Kansas is proving useful in the regional study of the Precambrian basement. The pole-correction map (i.e., the reduced-to-the-pole map minus the original map) enhances an east–west–trending boundary between Precambrian terranes of different age. The second-vertical-derivative map reveals extensive basement faults trending southwest through central Kansas. The combination of high-frequency-pass and trend-pass filters reveals the bounding faults of the Central North American rift system (CNARS). The upward and downward continuation filters are also useful in delineating basement terranes. These maps reveal parallelism between the Humboldt fault, which bounds the eastern side of the Nemaha uplift, and the CNARS. This suggests that the Humboldt fault probably developed as one of the easternmost faults of the CNARS in late Precambrian time and was reactivated in late Paleozoic time. Recent seismicity indicates that some of the CNARS faults are active today.

INTRODUCTION

The Kansas Geological Survey (KGS) recently completed an 80 000-km aeromagnetic survey of the State of Kansas. This study compiles and extends earlier preliminary reports (Yarger, 1980, 1981; Yarger et al., 1978a, 1976) on the regional interpretation of Kansas aeromagnetic data.

The KGS airborne proton precession magnetometer system, which consists primarily of Geometrics equipment with digital recording, was flown at ±1-nT (1 nT = 1 gamma) sensitivity and a 2-s sampling rate. The magnetic sensor ("bird") was trailed 30 m from the Twin Beech D-18 aircraft used in the survey. An Automax G-2 35-mm camera, electronically triggered by the magnetometer, was used for flight-path recovery.

The survey was conducted separately in eastern and western Kansas with ~340-km east–west flight lines and ~345-km north–south tie lines. The flight lines were spaced 3.2 km apart, and the tie lines were spaced approximately 32 km apart. Navigation was accomplished by visual sighting along section-line roads. In eastern Kansas the airplane was flown at a fixed barometric elevation of 760 m above sea level. In western Kansas the flight elevations were 915 m above sea level in the eastern part and 1 370 m above sea level in the westernmost quarter of the state. The ground clearance, which averaged ~365 m, was measured by radar altimeter and digitally recorded along with the magnetic measurements.

Details of the flight-path-recovery procedure appear elsewhere (Yarger, 1981, 1982). After assignment of longitude and latitude to all magnetic measurements, a time-extrapolated IGRF75 value was computed at each measurement location and subtracted from the total-intensity magnetic field. The temporal variations in the magnetic field were removed by analysis of the mismatches of magnetic-field values at the tie-line–flight-line intersections. This procedure, which does not require a recording base station, assumes that diurnal drift during flight is a smoothly varying, low-order polynomial in time. The polynomial coefficients were determined by minimizing magnetic-field residuals at flight-line–tie-line intersections (for further detail, see Yarger et al., 1978b, and Yarger, 1982). The two western altitude blocks within the western flight block were automatically tied together, because the altitude transition was made in flight. For the IGRF calculation within the transition zone, the instantaneous elevation was estimated, using linear interpolation. The eastern and western flight blocks were reduced separately, then tied together by minimizing residuals between overlapping flight lines in central Kansas.

A master grid of total-intensity magnetic-field values for the state was prepared with 0.16-km (0.1-mile) east–west spacing and 3.2-km (2-mile) north–south spacing, which is nearly equivalent to the original measurement spacing. The grid was determined by shifting magnetic-field values from nearly straight flight lines to nearby grid lines. The overall grid location was determined by minimizing (in the least-squares sense) the difference between grid-line and flight-line

*Kansas Geological Survey, University of Kansas, West Campus, Lawrence, KS 66044.

coordinates. The gridding procedure, which takes advantage of a regularly spaced grid-like flight-line pattern, avoids smoothing that normally occurs when gridding by the available computer algorithms written primarily for arbitrarily spaced data. The master grid, which represents the total-intensity magnetic-anomaly field, is useful for machine contouring over a wide range of scales and contouring intervals. It is also useful for most quantitative analyses, because the gridding procedure has preserved the original integrity of the data. A color map, machine contoured from this grid, has been published at a scale of 1:500 000 and a contour interval of 50 nT (Yarger et al., 1981). Figure 1 presents a photoreduction of a black and white version of this map.

FILTER APPLICATION

Several spectrally filtered versions of the original contour map (Figure 1) are used in this regional interpretation. The purpose of filtering a map is to remove certain unwanted characteristics and to enhance desirable characteristics that are diagnostic of the geology. Because of the simple mathematical form of most potential-field filters in the spectral domain, it is advisable to transform the original unfiltered map to the spectral domain, apply the filters, then transform the filtered map back to the spatial domain for use in interpretation (Gunn, 1975). A suite of spectrally filtered magnetic maps has proved useful in this study of Kansas basement composition, paleotectonics, and age terranes. Certain filtered maps reveal magnetic patterns, not readily apparent in the original unfiltered map, that are related to basement geology.

Kansas was divided into two equal 216 by 216 grids for input to the FFT program (Singleton, 1969). The grid cell size is approximately 1.6 km on a side. The eastern and western grids overlapped by 15 cells. The filtered maps were compiled using a variable black-and-white density scale instead of the more traditional contour lines. This format was chosen for several reasons. Density maps are considerably less expensive and faster to produce. Some of the filtered maps have a large dynamic range or may have sharp gradients that are difficult to contour. Finally, the reader can more easily discern relative magnitudes on density maps when restricted to a black-and-white format.

Figures 2 through 9 are a suite of eight filtered maps, which are useful in enhancing basement features. These figures are a subset of maps that were derived from a combination of two or more filters.

Figure 2 is a map reduced to the pole and downward continued, in part, to 760 m above sea level. Reduction to the pole removes the distortion caused by the Earth's inclined magnetic field (approximately 65 degrees from the horizontal in Kansas). This results in a slight migration to the north of anomaly maxima and an increase, in some cases, of anomaly amplitudes. All maps appearing in Figures 2 through 9 have been reduced to the pole. The data of the western half of Figure 2 were downward continued to 760 m above sea level so as to be comparable to the data of the eastern half. All maps that have been downward continued to the eastern flight elevation of 760 m above sea level are referred to as "leveled" maps.

Figure 3 presents a pole-correction map, which is the reduced-to-the-pole map minus the unreduced-to-the-pole map. This map was originally compiled to check the reduction-to-the-pole filter but has proved useful itself as diagnostic of basement terrane. The pole-correction map represents distortion in the original map caused by the inclined magnetic field. The distortion is maximum for high horizontal north–south gradients, so this filter, in effect, emphasizes areas with high north–south gradients.

Figure 4 presents an aeromagnetic map upward continued to 9 km above sea level and emphasizes deep-seated, long-wavelength magnetic sources within the crust. Figure 5 presents an aeromagnetic map downward continued to 850 m below sea level, which corresponds roughly to the average elevation of the Precambrian surface and emphasizes magnetic sources at or near the Precambrian surface.

Figure 6 presents an aeromagnetic map, leveled, and high-frequency-pass filtered. This filter passed all radial frequencies above 0.4 cycles/km and attenuated lower frequencies. We used a Gaussian attenuation function with a half-width of 0.13 cycles/km. This map emphasizes anomalies caused by magnetic sources at or near the Precambrian surface to a greater extent than does the downward-continued map (Figure 5). Figure 7 presents an aeromagnetic map, leveled, high-pass filtered, and trend-pass filtered. This map is the same as Figure 6 except for the addition of a trend-pass filter. This map emphasizes anomalies caused by magnetic sources at or near the Precambrian surface and also trending northeast ±45 degrees.

Figure 8 complements Figure 7 and emphasizes anomalies caused by sources at or near the Precambrian surface and trending northwest ±45 degrees. Note that the simple addition of Figure 8 to Figure 7 would yield the original high-pass-filtered map in Figure 6.

Figure 9 presents the aeromagnetic map leveled, with the second vertical derivative calculated. This map emphasizes contacts between contrasting magnetization at or near the Precambrian surface.

BASEMENT GEOLOGY OF KANSAS

The Precambrian basement complex in Kansas is part of the Midcontinent craton, which is the concealed southern extension of the Canadian shield. A relatively thin mantle of Phanerozoic sedimentary rocks 150 to 3 000 m (500–10 000 ft) thick covers the basement.

Bickford et al. (1981) recently compiled a basement-rock-type map of Kansas and adjacent Midcontinent states, based on available basement-well samples (Figure 10). The Kansas portion of this map is based on a study of more than 800 thin sections from basement-well samples. The basement terrane in northern Kansas is characterized by granitic to quartz-monzonitic intrusive rock, estimated to have been emplaced at depths of 6.5 to 13 km. These mesozonal rocks commonly have cataclastic to extensively sheared textures, particularly along the Nemaha ridge. Zircon dates (U/Pb) in northeastern Kansas and northwestern Missouri indicate an emplacement age of 1 625 m.y. for this terrane. In contrast, basement wells in southern Kansas reveal silicic volcanic rocks and associated shallowly emplaced granite. These volcanic and epizonal rocks are not cataclastically deformed and have a nominal age of 1 400 m.y.

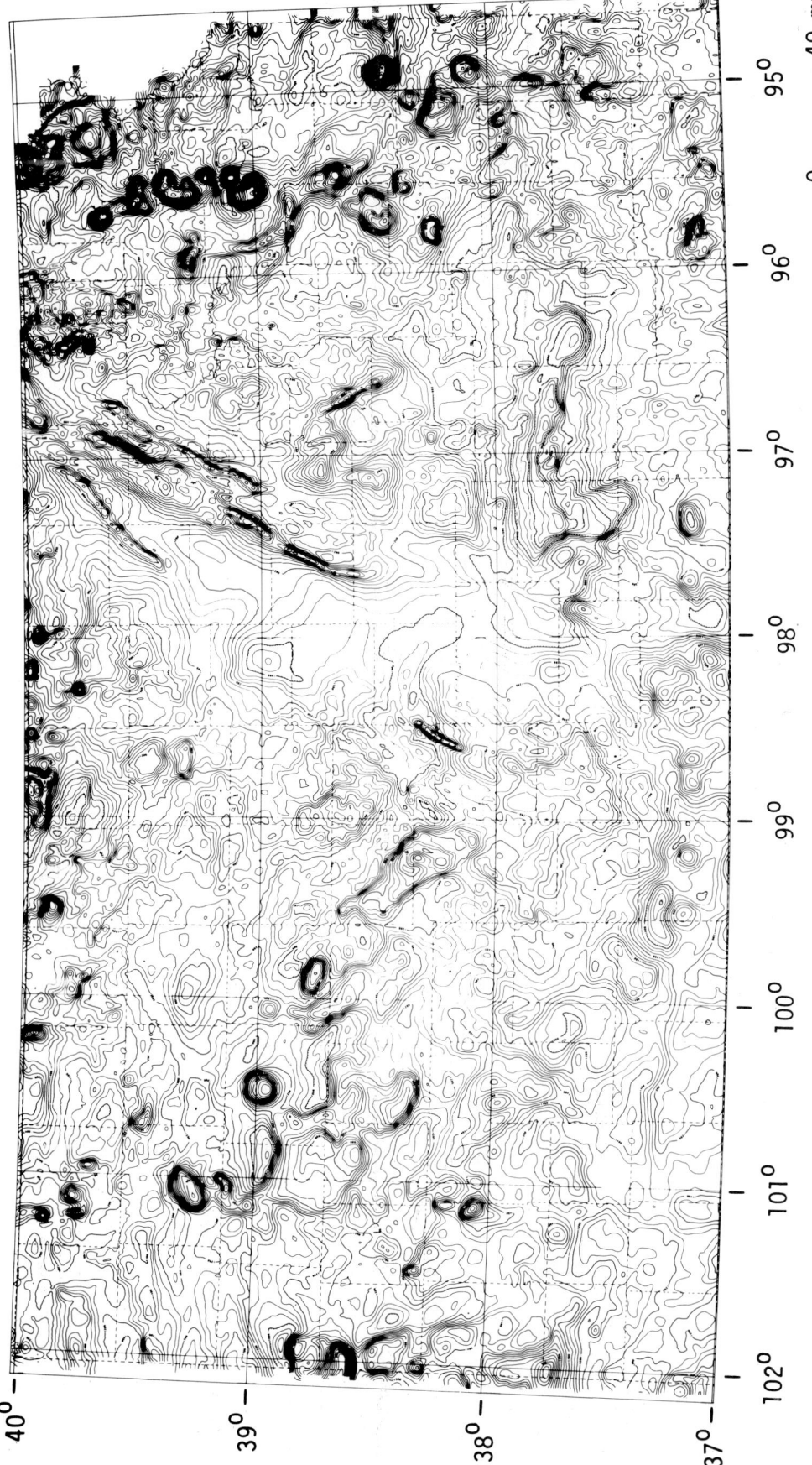

FIG. 1. Aeromagnetic-contour map of Kansas. Contour interval, 50 nT. (Photo-reduction of map compiled at a scale of 1:500 000.)

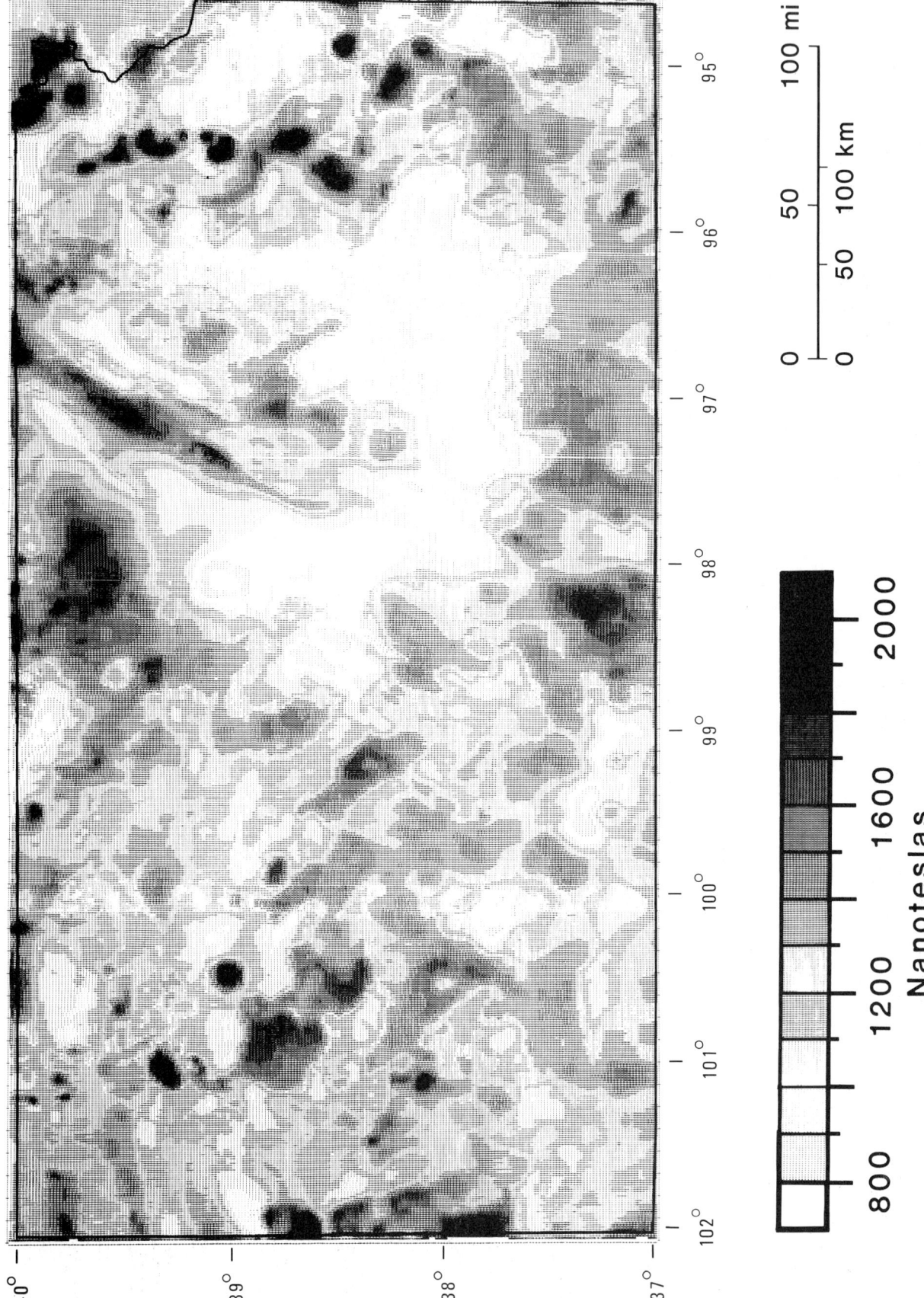

FIG. 2. Aeromagnetic map of Kansas reduced to the pole and downward continued to 760 m above sea level.

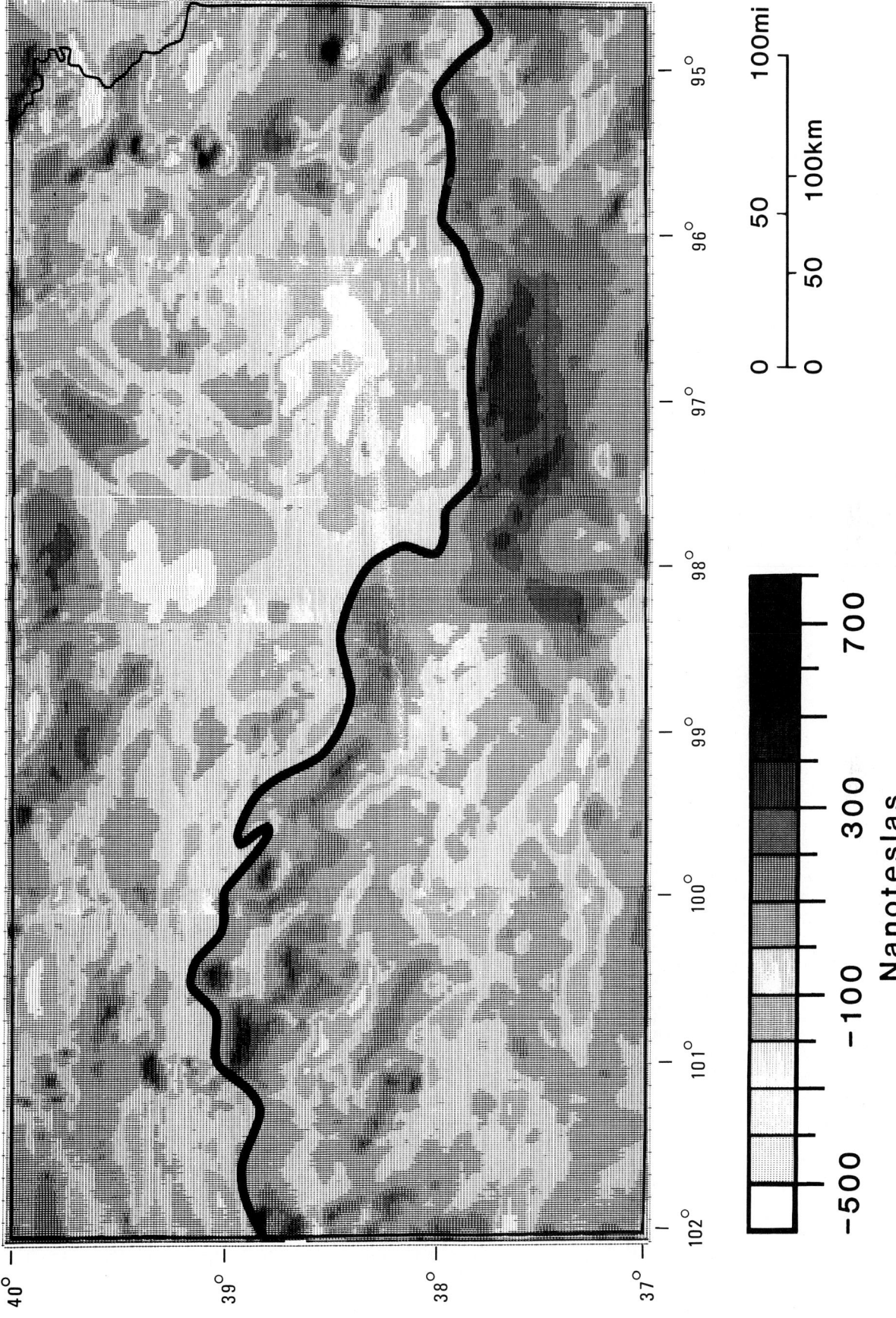

FIG. 3. Pole-correction map of Kansas. This map represents the difference between Figure 2 and the unreduced-to-the-pole version of Figure 2 (not shown). The continuous line delineates a north–south horizontal magnetic gradient trending east–west.

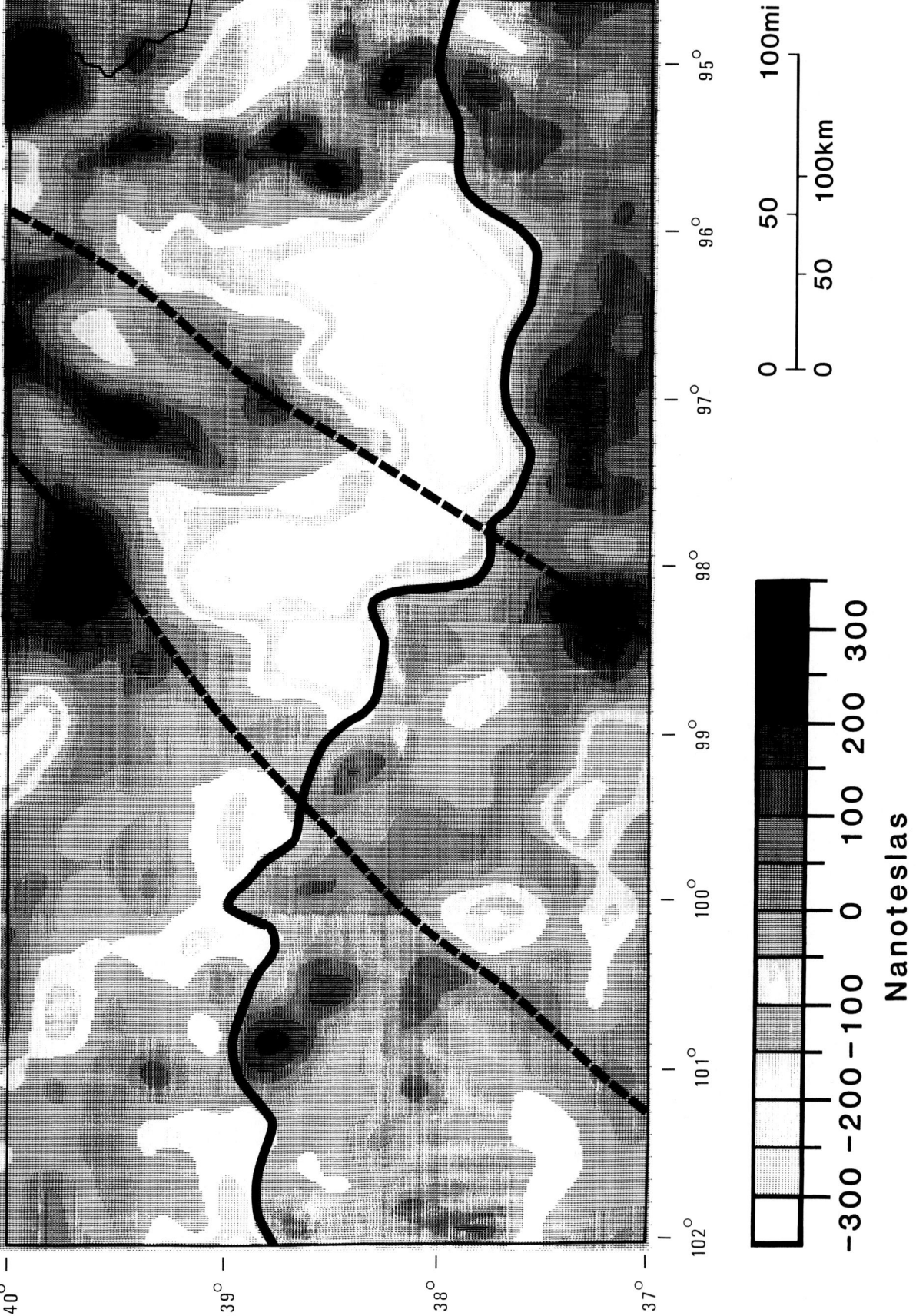

FIG. 4. Aeromagnetic map of Kansas reduced to the pole and upward continued to 9 km above sea level. The continuous east-west-trending line delineates a possible deep-seated paleoplate boundary within the crust. The dashed lines outline the possible deep-seated boundaries of the CNARS.

Kansas Basement Study from Aeromagnetic Data 219

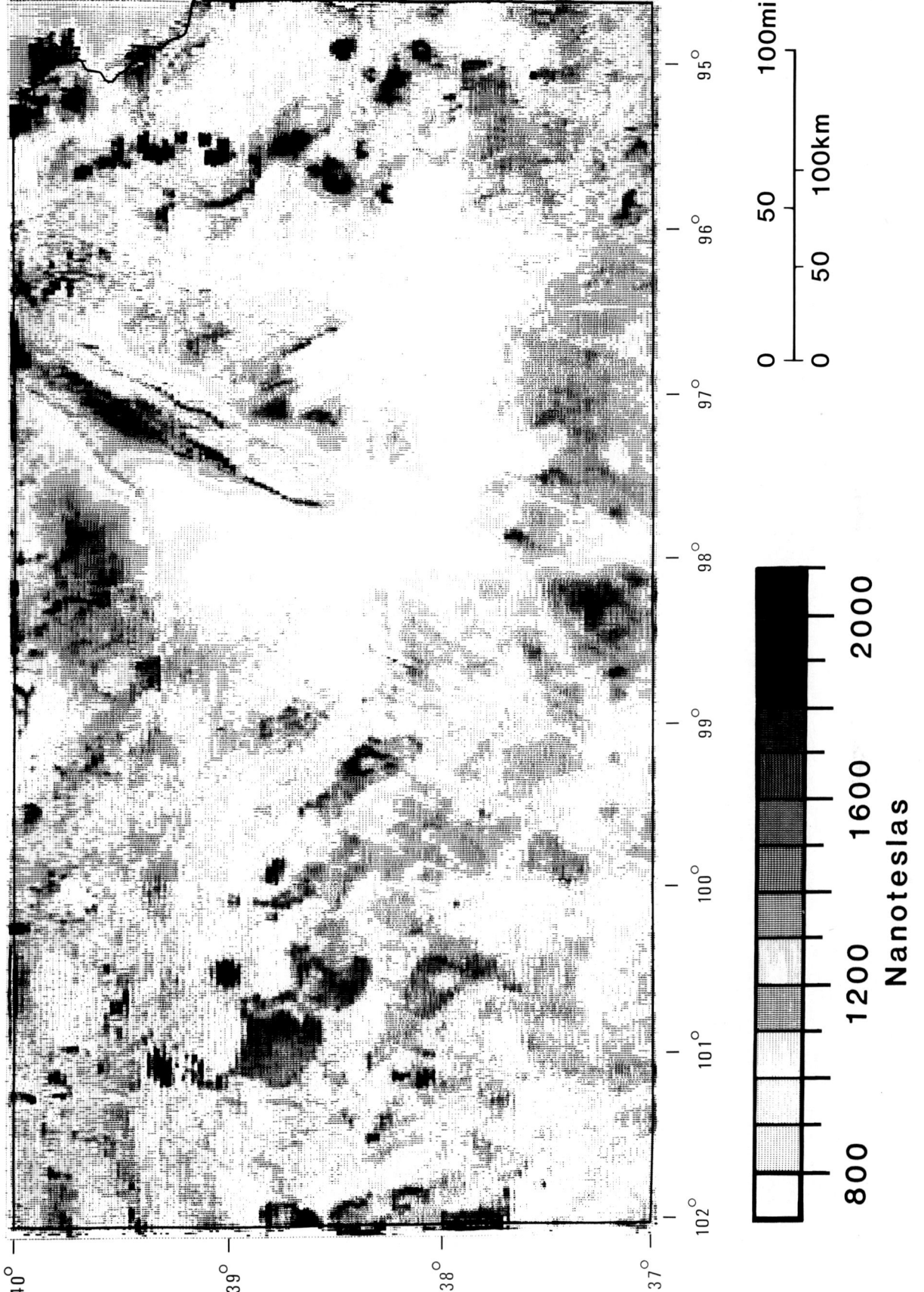

FIG. 5. Aeromagnetic map of Kansas reduced to the pole and downward continued to 850 m below sea level.

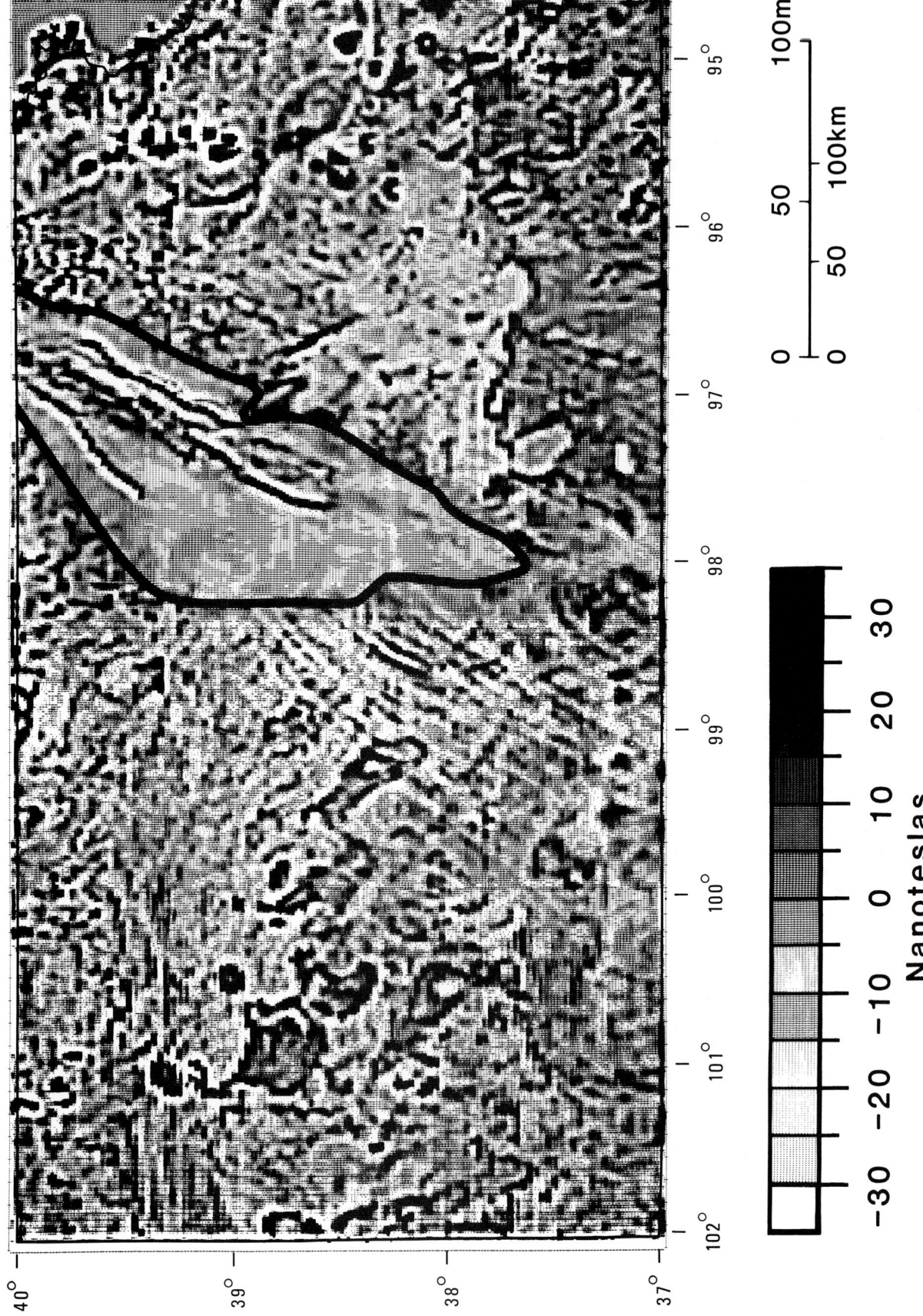

FIG. 6. Aeromagnetic map of Kansas reduced to the pole, leveled, and high-pass filtered. Radial frequencies below 0.4 cycles/km were attenuated by a Gaussian function with a half-width of 0.13 cycles/km. The continuous line outlines the extent of rift sedimentary rocks (Rice Formation).

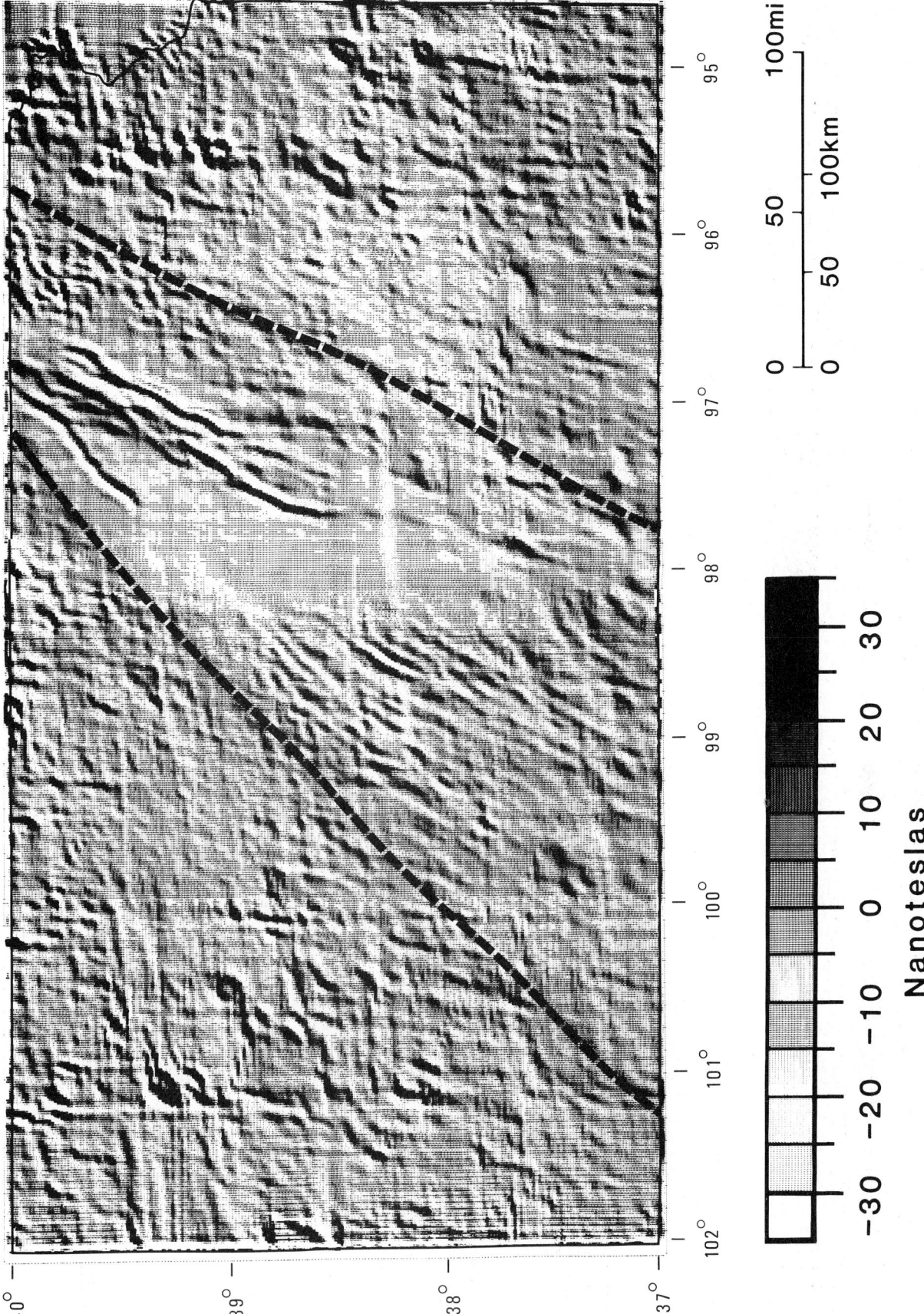

FIG. 7. Aeromagnetic map of Kansas reduced to the pole, leveled, high-pass filtered, and trend-pass filtered northeast ±45 degrees. The dashed lines outline the suggested boundaries of the main part of the CNARS.

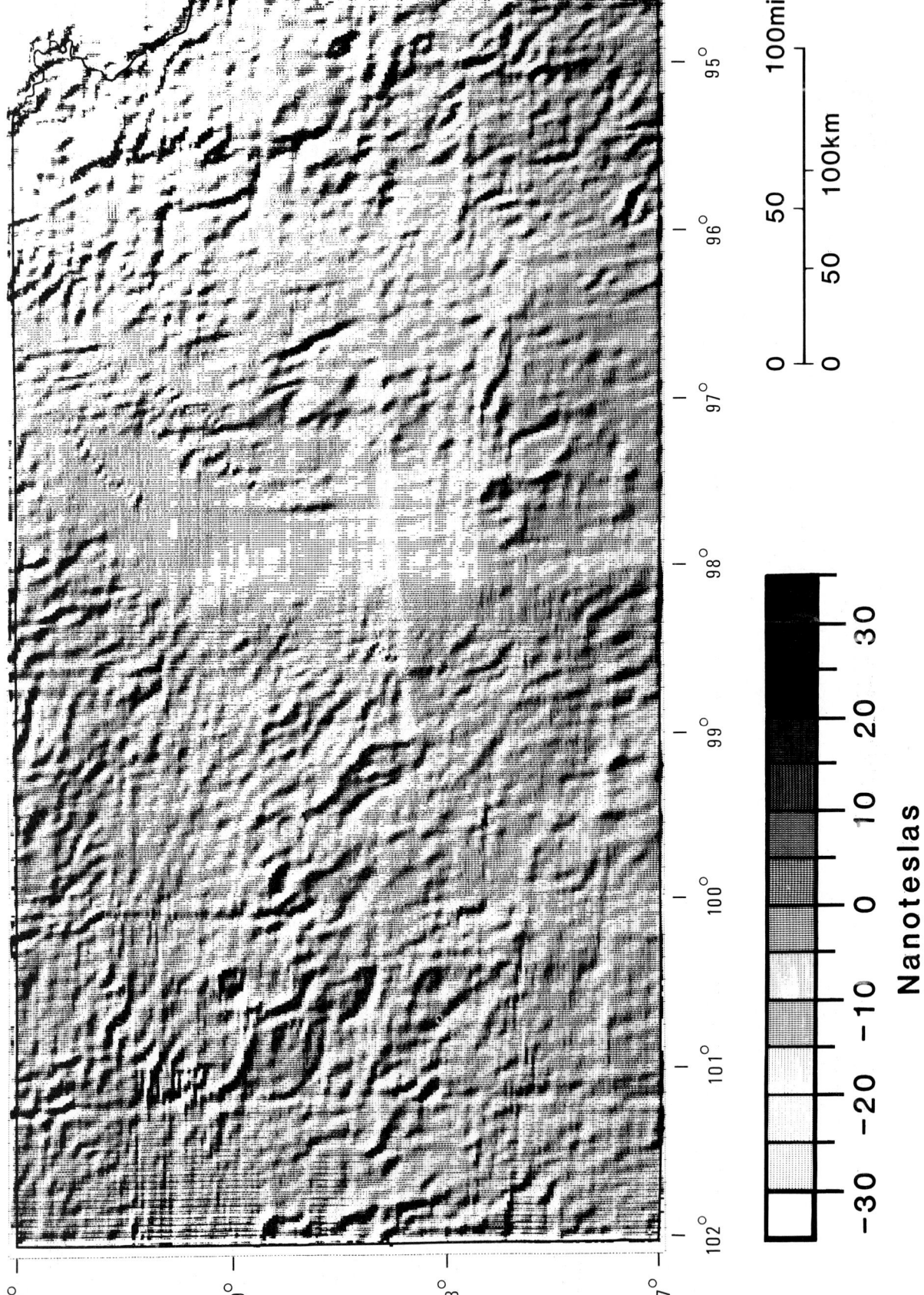

FIG. 8. Aeromagnetic map of Kansas reduced to the pole, leveled, high-pass filtered, and trend-pass filtered northwest ±45 degrees.

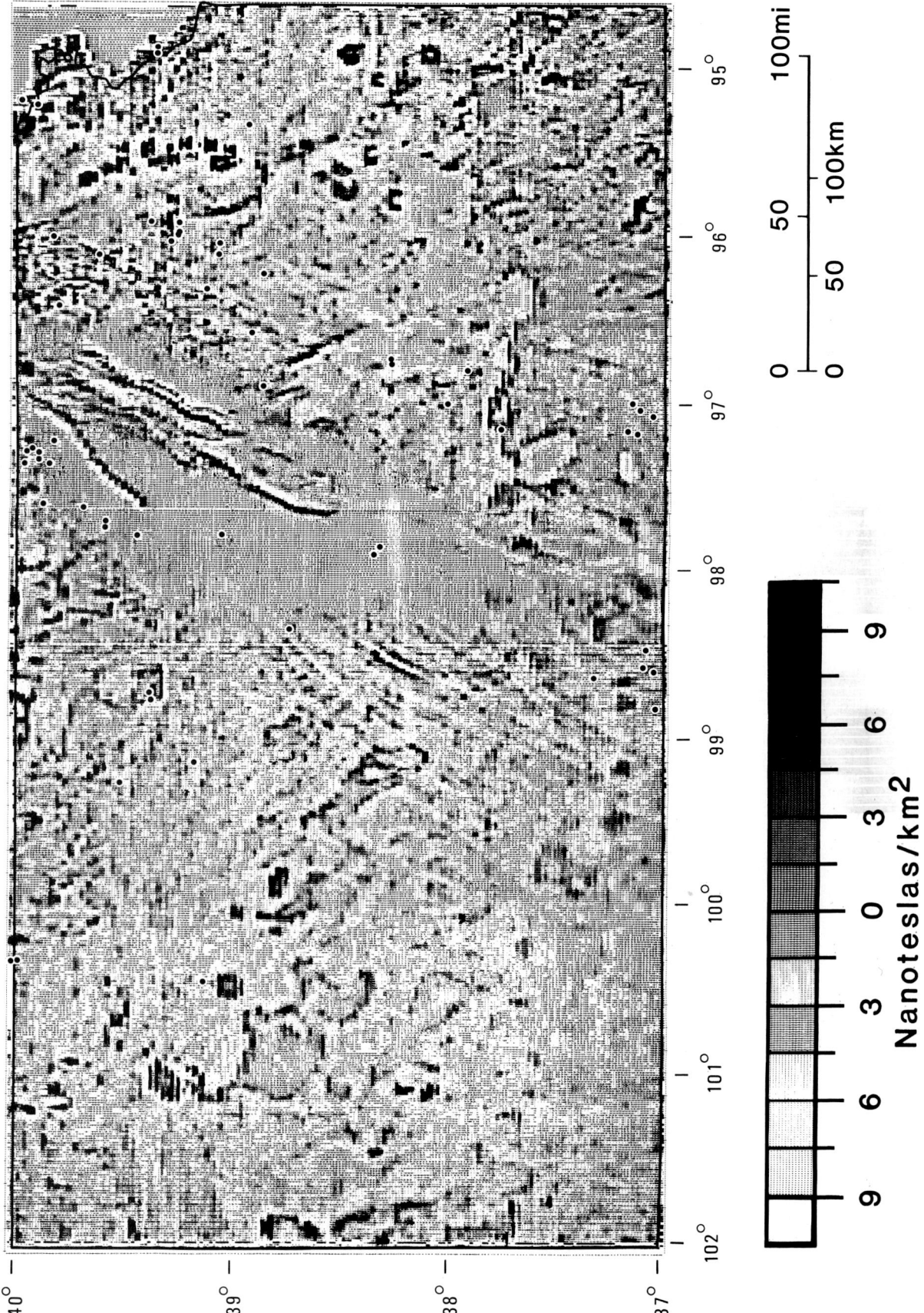

FIG. 9. Aeromagnetic map of Kansas reduced to the pole, leveled, with second vertical derivative calculated. The black dots represent microearthquakes from 1977 to 1982, after Steeples (1980) and Steeples (1982, personal communication).

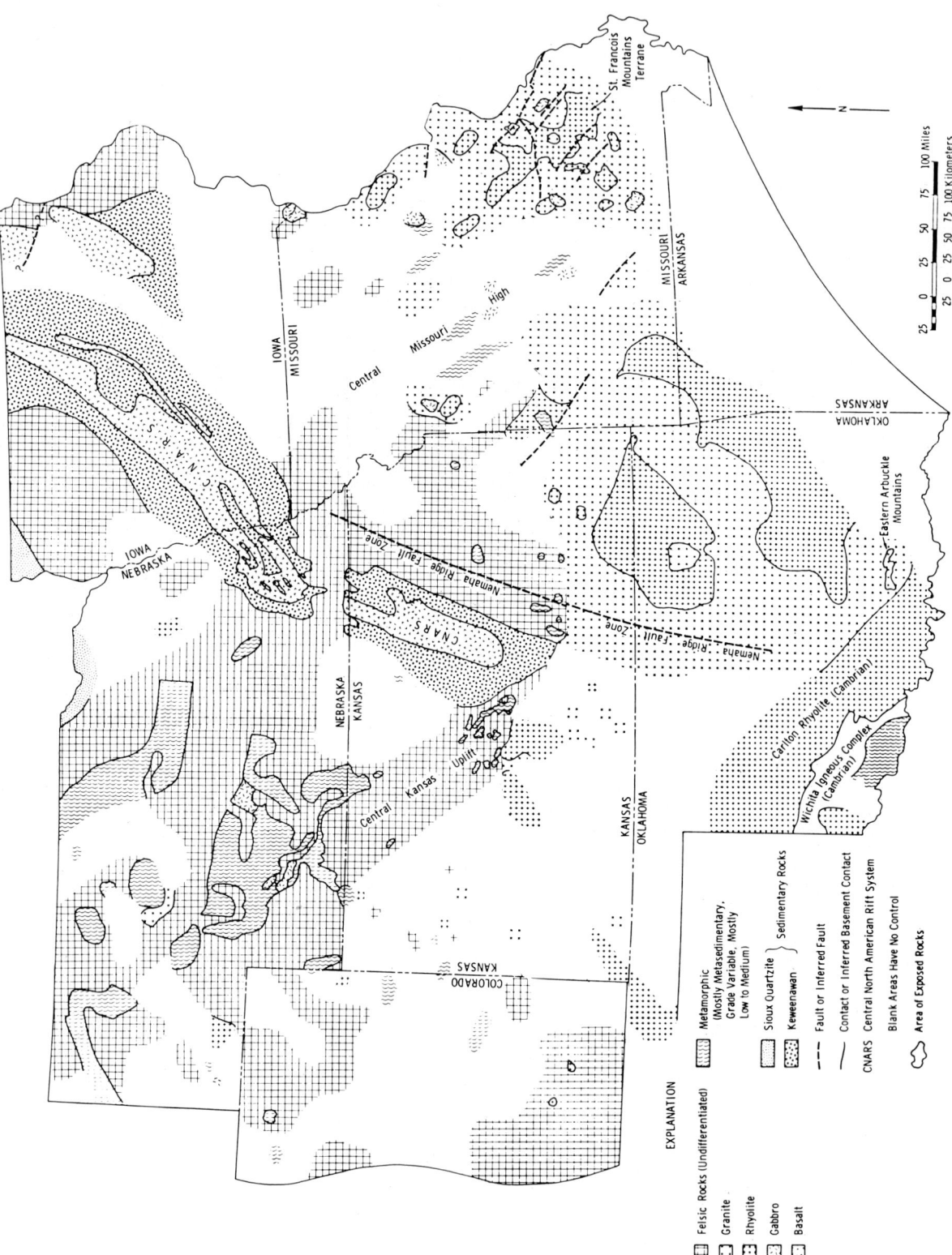

Fig. 10. Map of basement-rock types in the central Midcontinent region. After Bickford et al. (1981).

The belt of northeast-trending gabbroic rock in north-central Kansas is the southern extension of the Central North American rift system (CNARS) (Ocola and Meyer, 1973). This aborted rift system can be traced to the Lake Superior region, where the outcropping rocks are of Keweenawan age (about 1 100 m.y.). A basin of arkosic sandstone to siltstone, designated the Rice Formation of Precambrian age (Scott, 1966) in Kansas, flanks and extends south of the trough of mafic rift intrusives.

The present structural framework, as depicted in Figures 11 and 12, was formed largely during the late Paleozoic Era and has not materially changed since then. The Nemaha ridge is a major linear feature, which crosses eastern Kansas from Nemaha County to Sumner County and extends into Nebraska and Oklahoma. The Nemaha ridge is faulted along the eastern side, where the crystalline basement rocks on the west side of the Humboldt fault are upthrown more than 800 m in some areas. The configuration of the Precambrian surface (Figure 11) exhibits a northeast-trending grain in a swath through Kansas approximately 145 km wide and bounded on the east by the Humboldt fault. Outside this zone, the prevailing grain trends northwest.

The Forest City basin lies mostly in Iowa, Missouri, and Nebraska. The southwestern corner, which is the deepest part of the basin, lies in northeastern Kansas and is bounded on the west by the Nemaha ridge. The Cherokee basin in southeastern Kansas and northeastern Oklahoma could be considered a shallow southern extension of the Forest City basin. However, a mildly positive feature, the Bourbon arch, separates the two basins.

The southern end of the Salina basin, its deepest part, lies in north-central Kansas. In Kansas it is bounded on the east by the Nemaha ridge and on the west by the Central Kansas uplift. It terminates to the south at an unnamed saddle. Before post-Mississippian deformation, it formed part of the larger ancestral North Kansas basin. The maximum thickness of sedimentary rocks found in the basin is 1 400 m (Merriam, 1963). To the south lies the Sedgwick basin, which is a northern-shelf extension of the large Anadarko basin in Oklahoma. The Hugoton embayment, which covers much of western Kansas, is also an extension of the much deeper Anadarko basin in Oklahoma.

REGIONAL INTERPRETATION OF AEROMAGNETIC DATA

The relative total-intensity magnetic-anomaly map (Figure 2) depicts a rather complex crustal-magnetization pattern in eastern Kansas. The dynamic range of the total magnetic field is almost 2 000 nT. The contour levels range from 450 to 2 350 nT, with an average value of approximately 1 200 nT. The base level of the map was established arbitrarily by adding 1 500 nT after subtraction of the IGRF from the absolute total intensity. Relative to the IGRF, the magnetic field in Kansas is anomalously low by some 300 nT. If the IGRF accurately describes the large-scale magnetic-field intensity in the Midcontinent, then the net magnetization of the crust in Kansas is slightly below average.

Basement terranes

Comparison of the magnetic patterns (Figures 1, 2) with the Precambrian relief map (Figure 11) reveals little obvious correlation. The Humboldt fault, which has more than 800 m of vertical displacement at some places, is only weakly discernible. The correlation can be seen in the alignment of magnetic contours along the fault, as in southern Nemaha, northern Pottawatomie, and Wabaunsee counties (see Figures 11 and 12 for location of counties). The influence of vertical displacement can also be seen in southern Morris County, where the fault trace splits the north-northwest-trending anomaly. The horizontal gradient of the anomaly on the high (west) side of the fault is steeper than the gradient on the low side. Thus, the source rock for the anomaly was probably emplaced before vertical displacement occurred along the fault. The presence of this anomaly across the fault also suggests no major (less than 500 m) strike-slip movement along the fault. The difference of 50 nT on either side of the Humboldt fault, which represents maximum basement relief in Kansas, suggests that most other basement-relief features in Kansas will have less than a 50-nT influence on the total magnetic field. Careful evaluation of the low amplitude–maximum gradients along individual magnetic profiles may yield quantitative depth-to-basement information (Steenland, 1965). Most of the magnetic relief, however, is related to magnetic contrasts within the basement.

The total-intensity aeromagnetic maps (Figures 1, 2) exhibit a northwest-trending grain consistent with the Precambrian-surface terrane (Figure 11). This grain is interrupted in northeastern Kansas by the Midcontinent geophysical anomaly (MGA), a northeast-trending linear, positive anomaly flanked by adjacent linear, negative anomalies. (The positive anomaly, caused by CNARS rocks of gabbroic composition, is discussed in the section on the "Central North American rift system.") The MGA appears to terminate in central Kansas against a roughly linear trend of negative anomalies. This negative trend consists of a nearly continuous band across the state defined by total-intensity amplitude <1 200 nT. The most prominent low is centered over Wichita (Sedgwick County); negative anomalies of lesser magnitude continue westward to the Kansas–Colorado border. This trend can also be traced into Missouri (Missouri Geological Survey, 1943). The southern boundary of this band is clearly visible in the pole-correction map (Figure 3). This boundary corresponds to a sharp magnetization contrast, suggesting contrasting rock types. This apparent contrast may correspond to a distinct boundary between the older, mesozonal granitic terrane to the north and the younger, epizonal granitic terrane to the south. This boundary is also evident in the high-frequency-pass-filtered map (Figure 6). This "spectral" boundary is caused by an abrupt attenuation of high-frequency signal, from south to north, and presumably corresponds to the onset of the southern edge of the mesozonal granitic terrane or to some kind of transition zone between the two basement terranes. The northern boundary of the transition zone(?) is not so sharply defined. Outside the CNARS the width of the transition zone varies from 25 to 50 km. The zone may correspond to foundered granitic rock overlain by a Precambrian sedimentary or metasedimentary wedge. Alternatively, the band could be caused by

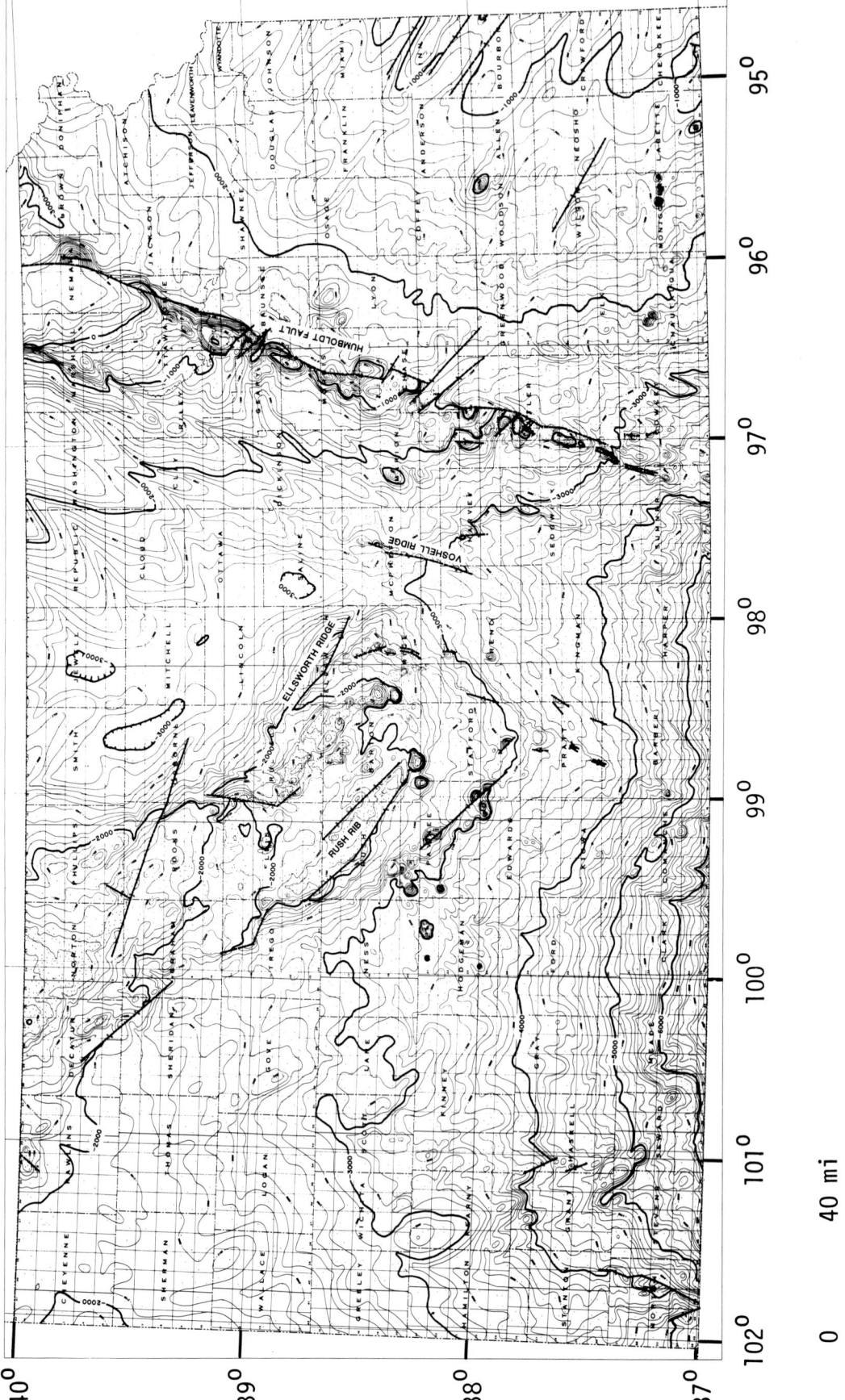

FIG. 11. Configuration of the top of Precambrian rocks in Kansas. After Cole (1976).

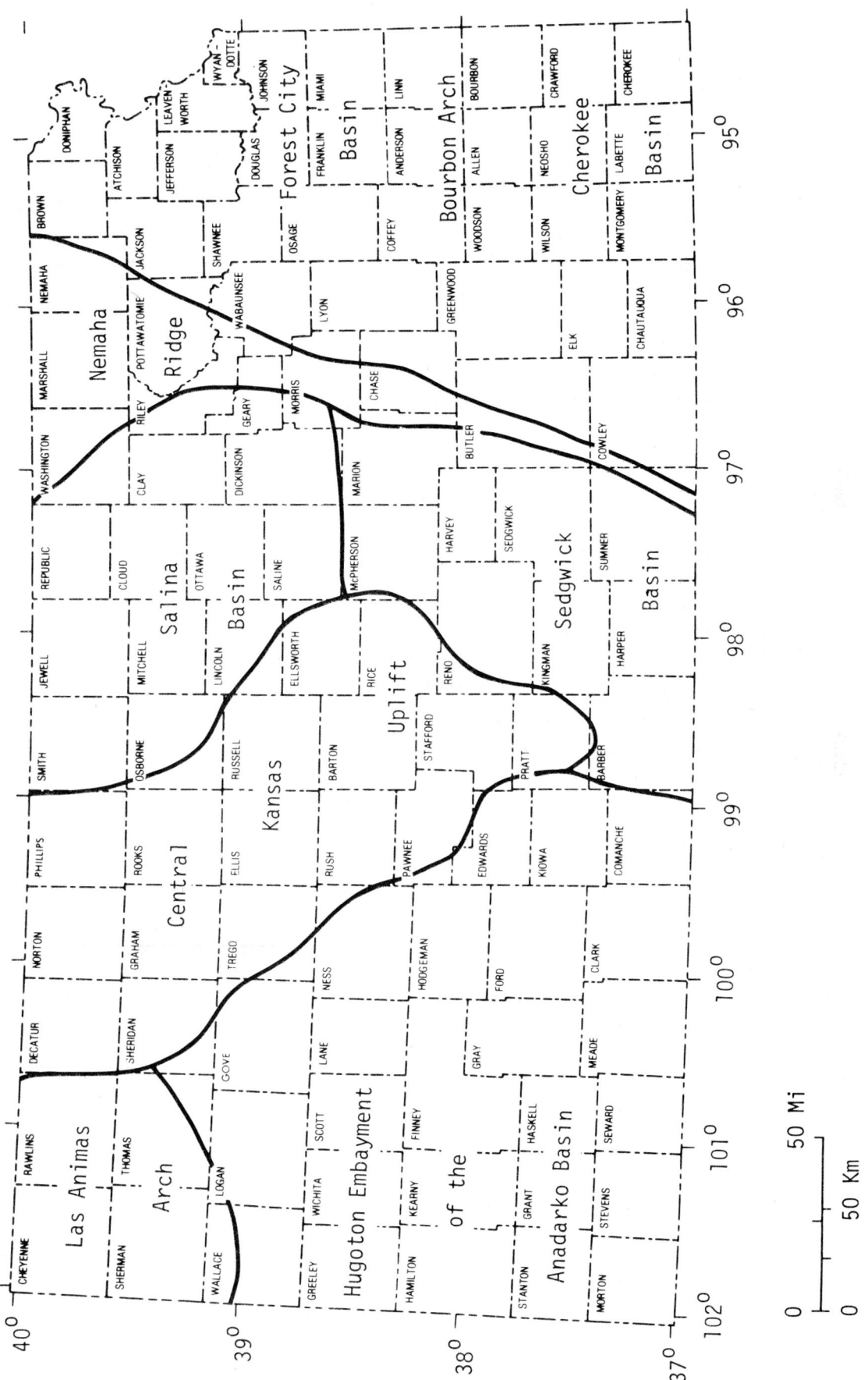

Fig. 12. Generalized structural map of the Precambrian in Kansas. After Merriam (1963).

nonmagnetic granitic basement. The apparent east–west–trending basement-terrane boundary is also evident on the second-vertical-derivative map (Figure 9). The boundary is also discernible on the upward-continued map (Figure 4), which emphasizes the deep-seated magnetic character of the crust.

Intrusives

The extreme northeastern part of the aeromagnetic map (Figures 1, 2) shows a series of strong, positive, roughly circular anomalies with diameters of approximately 15 km. Basement cores from two of these anomalies, in Douglas and Miami counties, yielded U/Pb zircon ages of about 1 350 m.y. (Steeples and Bickford, 1981). Although these rocks are more coarse grained than rocks in the southern terrane, they are clearly epizonal granite and are related to them (Bickford et al., 1981). The Miami County core contains about 2-percent magnetite by weight (Steeples and Bickford, 1981), which may account for the positive magnetic anomalies. The close similarity of the circular magnetic anomalies suggests that most, if not all, of them probably are caused by intrusive bodies similar to those drilled in Douglas and Miami counties. Thus, the 1 625-m.y.-old terrane in northeastern Kansas and northwestern Missouri may be peppered with isolated, shallow granite plutons, possibly related to the 1 400-m.y.-old granitic terrane to the south, but younger by some 50 m.y. This series of circular magnetic highs, along with similar highs in Missouri (Missouri Geological Survey, 1943), roughly tracks the boundary of the southern half of the Forest City basin. These mid-Proterozoic granitic plutons may have been intruded along a zone of weakness that later influenced formation of the late Paleozoic Forest City basin. The plutons themselves do not seem to have any basement relief.

At least six major near-circular anomalies occur in western Kansas, five of which are closely associated with the granitic-terrane boundary discussed previously. Over the entire state there are at least five circular intrusive bodies just north of the terrane boundary (Figure 1). These could possibly be the remnants of volcanic centers associated with Precambrian plate convergence along this boundary.

The Central North American rift system

The pervasive influence of the Central North American rift system on the Kansas crust is best seen in Figures 6, 7, and 9. Figure 7, which emphasizes high-frequency, northeast-trending anomalies, indicates that the rift system extends southward across the entire state.

A magnetic quiet zone (± 10 nT, Figure 6) surrounds the magnetic high along the mafic belt and corresponds to a basin filled with clastic rocks of (presumably) Keweenawan age. The magnetic quiet zone is caused by the extreme depth to magnetic basement, presumably granite, which foundered during the extensional phase of the CNARS. Both the eastern and western boundaries of this Keweenawan basin are fairly sharply defined, suggesting that they may be fault bounded.

The magnetic lineations southwest of the magnetic quiet zone, which are most prominent in Figures 7 and 9, probably correspond to block faulting and possibly to mafic dike intrusion that occurred during the initial stage of rifting.

Apparently the rift in southern Kansas did not progress beyond the stage of block faulting and dike intrusion, whereas the rift in northern Kansas developed to a more mature stage of volcanic flows, followed by foundered crust and clastic deposits. The fracture system in southern Kansas probably is representative of the crust throughout the CNARS during the early stages of rifting, before upwelling of Keweenawan volcanics occurred. This southwest-trending system indicates that the CNARS extends to the Kansas–Oklahoma border.

The Humboldt fault (Figures 10, 11), which borders the eastern side of the Nemaha ridge, closely parallels the CNARS across the state, suggesting that this apparent post-Mississippian fault may be a reactivated CNARS fault. This Proterozoic forerunner of the Humboldt fault may have developed within the easternmost part of the CNARS crust, which was not involved in subsequent foundering. Alternatively, the Humboldt fault may be a zone of pre-Keweenawan crustal weakness that has been reactivated, or it may not have formed until Paleozoic time.

Possible basement faults

The trend-pass-filtered maps (Figures 7, 8) reveal significant magnetic lineations in both the northeast and northwest directions. The most prominent northeast linear trends correspond to the rift system through central Kansas, but a number of northwest trends lie outside the rift, particularly in northwestern Kansas. Figure 8 exhibits northwest-trending magnetic lineations extending over large parts of the state, most of which are interrupted within the rift zone. This, of course, implies that the northwest grain is older than the rift.

For comparison with known basement structure, the magnetic lineations apparent in Figures 4 through 9 are compiled in Figure 13. The magnetic lineations mapped are at least 50 km long or belong to a trend of shorter segments at least 50 km long. In addition to suites of northeast- and northwest-trending lineations across the state (Figure 13), a third suite of east-northeast lineations is evident in southern Kansas.

The number of magnetic lineations far outnumbers the previously mapped faults in the Precambrian. However, a number of one-to-one correlations are evident. The northern half of the Humboldt fault, which borders the eastern sides of the Nemaha ridge, shows up clearly as a continuous magnetic lineation. The northwest-trending magnetic lineation in Rush County matches the southwest-bounding fault of the Rush rib (Figure 11). This fault (Merriam, 1963) lies along the postulated Precambrian-age boundary (Figure 3) discussed earlier. Good correlation with the northeast-trending fault segments in western Reno and Pratt counties also exists. The fault through Ellsworth County, which trends northwest along the southwest side of the Ellsworth ridge (Figure 11), apparently serves as the northern boundary for several of the southwest-trending magnetic linears within the rift zone.

The northwest-trending fault in Pawnee County coincides

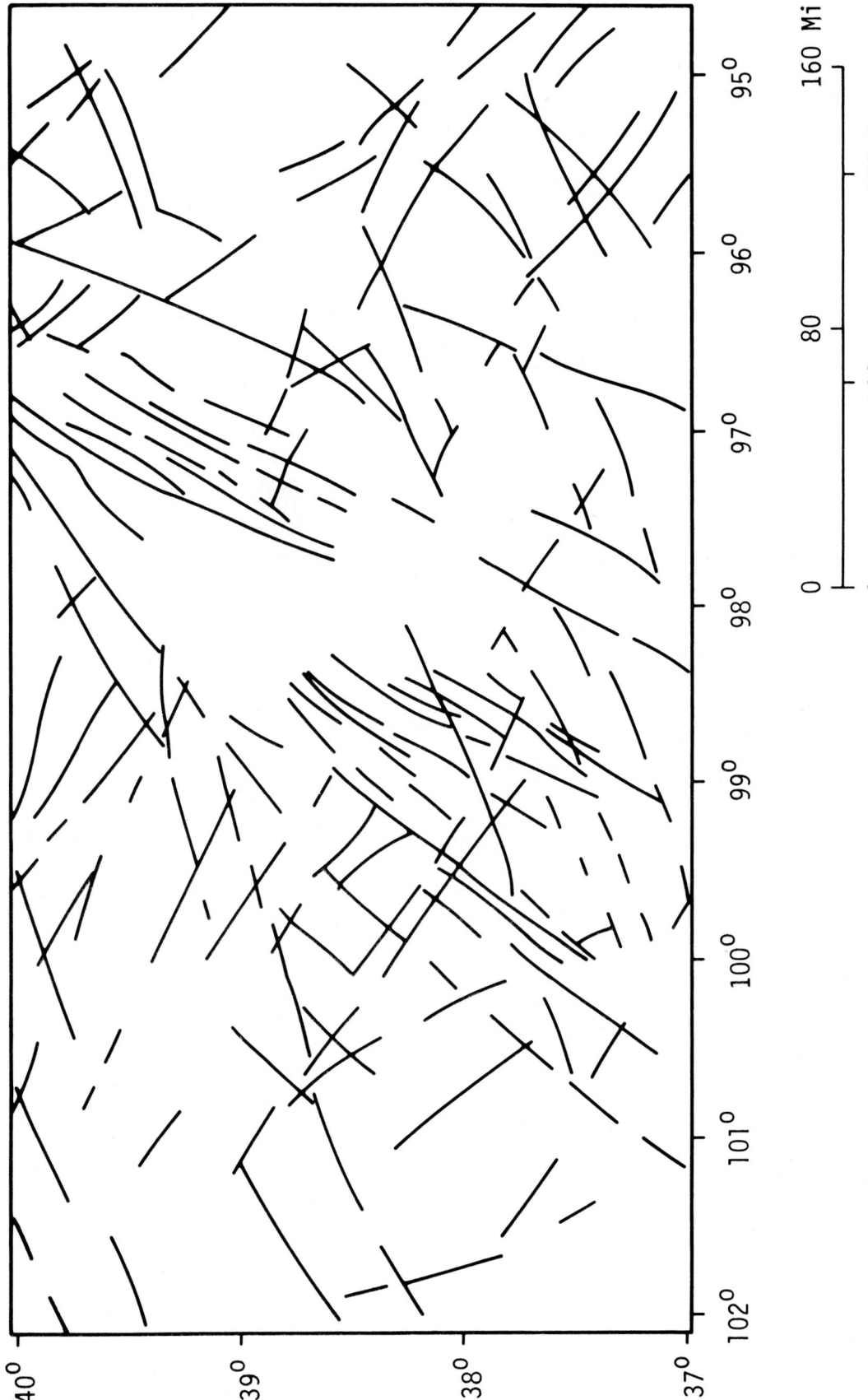

FIG. 13. Magnetic lineations in Kansas. The lineations were derived from Figures 4 through 9.

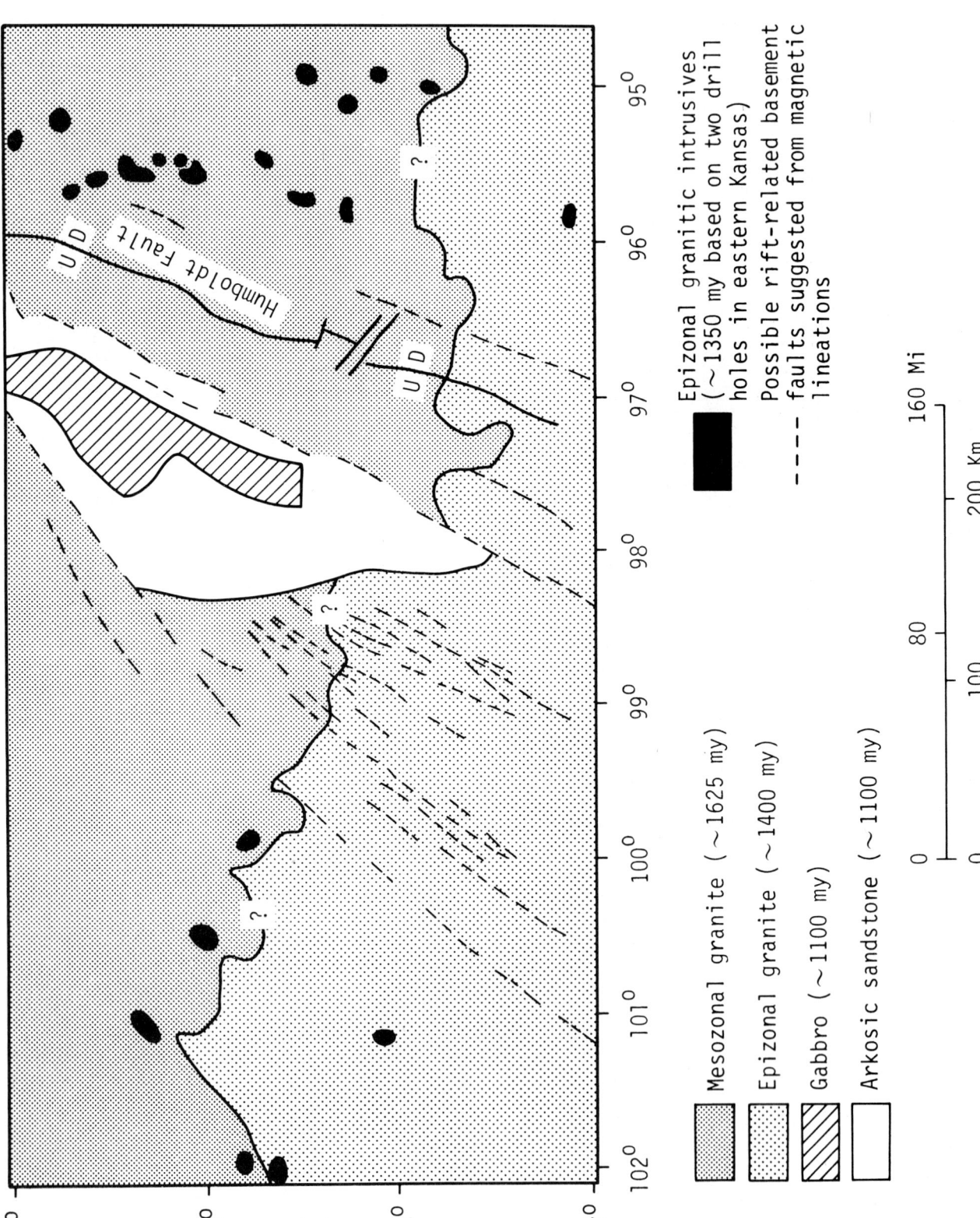

FIG. 14. Precambrian terranes in Kansas. Inferred from magnetic maps and Precambrian age data of Bickford et al. (1981).

with a magnetic trend that continues to the northwest into Rush County. Part of the southwest-trending fault bounding the northwest side of the Voshell ridge (Figure 11) coincides with a short magnetic lineation at the intersection of McPherson, Reno, and Harvey counties.

Although the northwest-trending magnetic grain in eastern Kansas parallels the Precambrian-surface grain, little one-to-one correlation with mapped faults exists. One exception is the fault through Bourbon and Linn counties, which coincides with a magnetic trend that continues northwestward into Anderson and Osage counties.

Whether all magnetic lineations correspond to basement faults is an open question. The trend-pass-filter operation (Figures 7, 8) tends to emphasize lineations by elongating them beyond their actual geographic limit. That some of them correspond to known faults strongly implies that at least some of the remaining lineations must also correspond to faults (so far undetected by boreholes). The northeast-trending magnetic lineations within the rift zone must surely correspond to faulting in Keweenawan time.

Microearthquake results for the last five years indicate that the Humboldt fault is still active. The northeastern magnetic trend through Washington, Republic, and Cloud counties, which corresponds to the postulated basement boundary between the rift sedimentary rocks (Rice Formation) and the older granitic terrane, is seismically active (Figure 9). Several events indicate that the eastern boundary between the rift sedimentary rocks and granitic terrane is also seismically active. The only other significant recent seismic activity recorded within the state is in Barber County and may be related to the southwest magnetic trend through Harper and Barber counties.

CONCLUSIONS

The recently compiled aeromagnetic map of Kansas (Yarger et al., 1981) is highly useful in studying the composition and paleotectonics of the Precambrian crust. Examination of the magnitudes and gradients of the total-intensity magnetic-anomaly map and of a suite of spectrally filtered maps, in light of existing geologic information, has yielded the following regional interpretation of the Precambrian crust in Kansas.

The postulated Precambrian terranes are summarized in Figure 14. A rather distinct boundary appears to exist between the northern 1 625-m.y.-old mesozonal granitic terrane and the southern 1 400-m.y.-old epizonal granitic and rhyolitic terrane, whose magnetic signature is a series of nearly contiguous lows trending west across the state. The southern boundary of this band of lows is sharply defined by steep horizontal gradients and short wavelengths. The magnetic source of this band is not clear. Gravity measurements in this region are being taken by the Kansas Geological Survey; along with potential-field modeling, these measurements should help to clarify the origin of these magnetic minima.

Drilling results from two of the 14 circular magnetic highs in northeastern Kansas suggest that the older 1 625-m.y.-old crust in northeastern Kansas is pockmarked with younger, 1 350-m.y.-old granitic plutons similar in composition to the southern, 1 400-m.y.-old terrane of epizonal granite and rhyolite. Magnetite in amounts of 1 to 2 percent (by weight) found in the two basement cores may account for the positive magnetic anomalies, whose magnitudes range from 500 to more than 1 000 nT.

The CNARS extends through Kansas and probably into Oklahoma. Although large volumes of mafic volcanics clearly did not reach the Proterozoic surface in southern Kansas, magnetic evidence strongly indicates that block faulting and possibly dike intrusion accompanied the initial stages of continental rifting. The southern part of the 1 100-m.y.-old rift that extends into the 1 400-m.y.-old crust in Kansas did not evolve into the more mature stages of deep rift-valley formation, accompanied by voluminous volcanics and clastics, as did the main part of the rift to the north.

Three main suites of magnetic lineations of basement origin are present. Predominantly northwest-trending lineations are found in southeastern and northwestern Kansas, whereas central Kansas is dominated by north-northeast-trending lineations. Both of these trends are present in south-central Kansas, resulting in a system of roughly orthogonal intersecting lineations. A third suite of east-northeast-trending lineations is present in southern Kansas. Several of the magnetic lineations correspond to previously mapped basement faults, suggesting that at least a small fraction of the remaining lineations correspond to previously unknown faults. These remaining lineations must be examined by other geophysical methods to establish which ones are faults.

The Humboldt fault, bounding the eastern side of the Nemaha ridge, parallels the CNARS, suggesting that it may have developed in Keweenawan time and been reactivated in late Paleozoic time. Recent microearthquake results in Kansas indicate that the Humboldt fault is still active. Some activity along two other magnetic lineations within the CNARS has occurred.

ACKNOWLEDGMENTS

I thank William Hinze and Lyle McGinnis for their review of an early version of this manuscript.

This work was supported principally by the Kansas Geological Survey, and in part by special appropriation of the Kansas legislature for an Automated Resource Evaluation System; by the U.S. Geological Survey, under grant 14-08-0001-G-137; by the U.S. Nuclear Regulatory Commission, under contract AT(49-24)-0256; and by the U.S. Department of Energy, under grant DE-AS07-19ET27204.

The following students did substantial work in one or more of the areas of data acquisition, reduction, and interpretation: Roubik Avanessians, Alan Martin, King Ng, Robert Robertson, Rita Sooby, Robert Wentland, and Mike Wolfe. Expert piloting was provided by Dennis Sooby and Stewart Giesick.

REFERENCES

Bickford, M. E., Harrower, K. L., Nussbaum, R. L., Thomas, J. J., Nelson, B. K., and Hoppe, W. J., 1981, Rb–Sr and U–Pb and geochronology and distribution of rock types in the Precambrian basement of Missouri and Kansas: Geol. Soc. Am. Bull., pt. 1, **92**, 323–341.

Cole, V., 1976, Configuration of the top of Precambrian rocks in Kansas: Map M-7, Kans. Geol. Surv.

Gunn, P. J., 1975, Linear transformation of gravity and magnetic fields: Geophys. Prospect., **23**, 300–312.

Merriam, D. F., 1963, The geologic history of Kansas: Bull. 162, Kans. Geol. Surv.

Missouri Geological Survey, 1943, Magnetic map of Missouri: Scale 1:500 000 (repr. 1958).

Ocola, L. C., and Meyer, R. P., 1973, Central North American Rift System 1, structure of the axial zone from seismic and gravimetric data: J. Geophys. Res., **78**, 5173–5194.

Scott, R. W., 1966, New Precambrian formation in Kansas: AAPG Bull., **50**, 380–384.

Singleton, R. C., 1969, An algorithm for computing the mixed radix fast Fourier transform: Inst. Elect. Electron. Eng. Trans. on Audio and Electroacoust., **AU-17**, 93–100.

Steenland, N. C., 1965, Oil fields and aeromagnetic anomalies: Geophysics, **30**, 706–739.

Steeples, D. W., 1980, Microearthquakes recorded by the Kansas Geological Survey: Kans. Geol. Surv. J., **2**, no. 3, 14.

Steeples, D. W., and Bickford, M. E., 1981, Piggyback drilling in Kansas: An example for the Continental Scientific Drilling Program: Trans. Am. Geophys. Union, **62**, 473–476.

Yarger, H. L., 1980, Aeromagnetic analysis of the Keweenawan rift in Kansas [abstr.]: Trans. Am. Geophys. Union, **61**, 1192.

——— 1981, Aeromagnetic survey of Kansas: Trans. Am. Geophys. Union, **62**, 173–178.

——— 1983, Aeromagnetic processing techniques at the Kansas Geological Survey: Proc. Aeromagn. Data Workshop, Nov. 16–18, 1982, Nat. Geophys. Data Cent., NOAA, Boulder, Colo., 51–59.

Yarger, H. L., Robertson, R., Martin, J., Ng, K., Sooby, R., and Wentland, R., 1981, Aeromagnetic map of Kansas: Map M-16, Kans. Geol. Surv., scale 1:500 000.

Yarger, H. L., Robertson, R. R., and Wentland, R. L., 1978a, Aeromagnetic anomalies in eastern Kansas [abstr.]: Presented at Am. Geophys. Union Midwest Meet., St. Louis, Mo., Meet. program and abstr. (AGU Doc. E79-002).

——— 1978b, Diurnal drift removal from aeromagnetic data using least-squares: Geophysics, **46**, 1148–1156.

Yarger, H. L., Robertson, R., Wentland, R., and Zietz, I., 1976, Recent aeromagnetic and gravity data in northeastern Kansas [abstr.]: Trans. Am. Geophys. Union, **57**, 752.

Model of the geothermal system in southwestern South Dakota from gravity and aeromagnetic studies

T. G. Hildenbrand* and R. P. Kucks*

ABSTRACT

The southern flank of the Black Hills uplift is characterized by high geothermal gradients, which exceed 40°C/km at many places and attain a maximum estimated value of 84°C/km. Aeromagnetic, gravity, geologic, geothermal, and Na–K–Ca–Mg geothermometer data are correlated to attempt to understand better the geothermal system of the upper crust of this region. Analyses of the gravity and magnetic fields include compilation of derivative maps to enhance the expressions of lithologic and structural boundaries.

In our model of the geothermal system of the southern Black Hills, ground waters in the topographically high areas descend to a depth of roughly 1.6 km along southward- and southeastward-trending fault zones, migrate upward primarily along faults after becoming heated, and spread outward through permeable sedimentary units. Several interpreted gravity and magnetic features appear to be related to the structural control of the thermal convection. For example, fault-formed lithologic boundaries expressed in the gravity and magnetic data may act as conduits for the southeastward-trending ground waters. These boundaries are abruptly truncated at their southeastern edges by northeast-trending features that presumably represent fault zones. Thermal gradients are significantly higher southeast of the northeast-trending fault zones. Of particular interest are two broad structures, an antiform and a basin, that appear to influence the channeling of warm water.

INTRODUCTION

The Black Hills uplift is thought to have relatively low geothermal gradients. Hydrological and geothermal data (Head et al., 1978; Schoon and McGregor, 1974) indicate that the Black Hills are a cool-water recharge area for aquifers trending outward to the east and west. Thus, the regions flanking the Black Hills might also be expected to have low to intermediate geothermal gradients; low gradients are indeed present except within a region in southwestern South Dakota. Geothermal gradients in southwestern South Dakota range from 5° to 84°C/km and define broad areas where values exceed 40°C/km. Near the town of Hot Springs, South Dakota, springs discharge at a temperature of approximately 29°C, about 22°C higher than the mean annual air temperature. This region is considered to have a potential geothermal resource that could be utilized, for example, in space and agricultural heating.

Upper-crustal structures and lithologies in a 14 000-km^2 area (Figure 1) were investigated, principally through analysis of gravity and magnetic data, to determine the cause of geothermal anomalies. Derivative magnetic and gravity fields were calculated to enhance the expressions of lithologic and structural boundaries. In addition, the Bouguer gravity map was analytically converted to an elevation map, showing estimated elevations of the Triassic surface across which density increases by about 0.3 g/cm^3. The resulting interpretations were correlated with geologic, geothermal-gradient, and Na–K–Ca–Mg geothermometer data to derive a conceptual model describing the geothermal system in southwestern South Dakota.

GENERAL GEOLOGY

The Black Hills uplift is the easternmost and least deformed of the Laramide uplifts associated with Rocky Mountain tectonism (Lisenbee, 1978). It consists of an arcuate north- to northwest-trending dome-shaped anticline that exposes a Precambrian core of metamorphic and igneous rocks (Figure 2). Lithologies and structures in the study area are described briefly below, although the reader is referred to Redden (1975), Kleinkopf and Redden (1975), Darton and O'Harra (1925), Gott et al. (1974), and Lisenbee (1975, 1978) for more detailed descriptions.

Sedimentary units

In Table 1 the Phanerozoic sedimentary-rock formations of the area are divided into two basic groups on the basis of

---

*U.S. Geological Survey, Box 25046, MS 964, Federal Center, Denver, CO 80225.

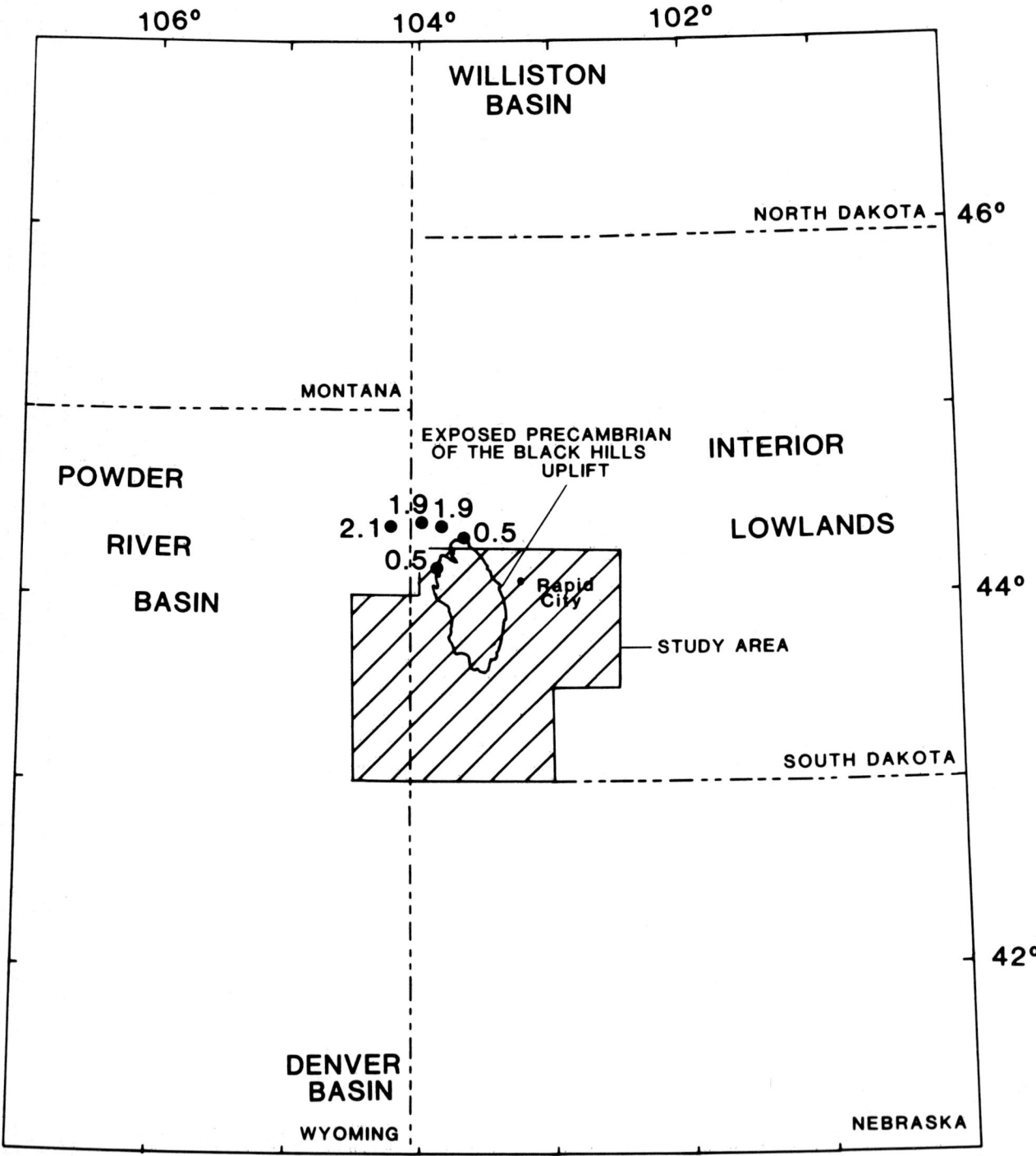

FIG. 1. Location of the study area (hachures) in north-central United States. Numbers represent previously determined heat-flow values (10^{-6} cal/cm^2/s) in the Black Hills region (from Blackwell, 1969; Roy et al., 1968; Sass et al., 1971; Decker et al., 1980).

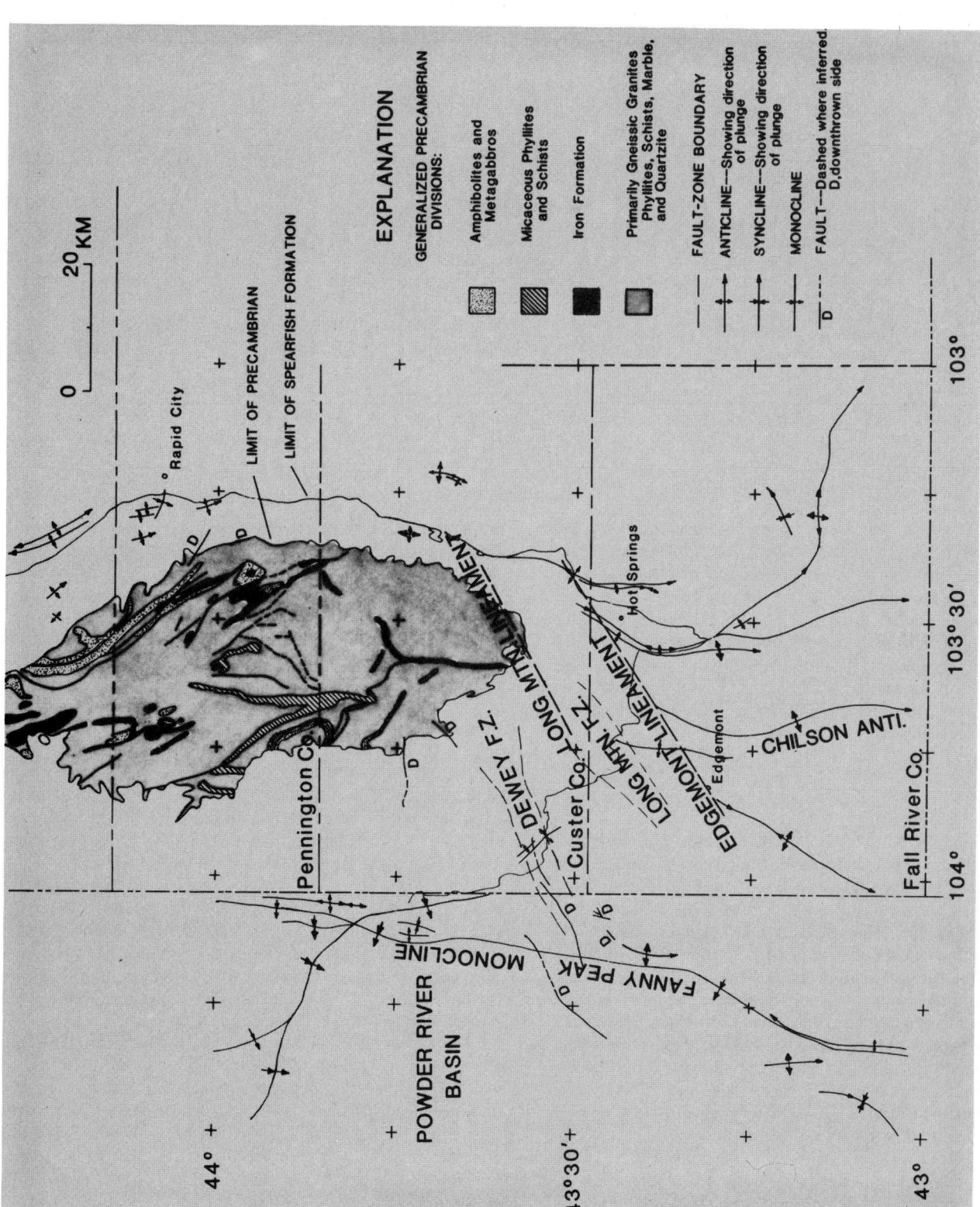

FIG. 2. Tectonic map of the southern region of the Black Hills uplift. Geologic structures and Precambrian units are from Kleinkopf and Redden (1975) and Gott et al. (1974).

Table 1. Mean gamma-gamma densities of Phanerozoic sedimentary rocks encountered in drill holes in the Black Hills area of South Dakota.

Geologic period	Lithologic units	Predominant rock type	State Gary 1 (16-9S-7E)	State 3-16 (16-12S-1E)	Federal Indian Creek 1-31 (31-11S-2E)	Federal Indian Creek 1-3 (3-12S-4E)
Cretaceous	Pierre Shale	Shale	2.35	—	2.4	2.35
	Niobrara Formation	Siltstone	2.35	—	—	—
	Carlile Shale	Shale	2.45	—	—	—
	Greenhorn Formation	Limestone	2.35	—	2.4	2.35
	Belle Fourche and Mowry Shales[1]	Shale	2.45	—	2.45	2.5
	Newcastle Sandstone	Sandstone	2.5	—	2.45	2.4
	Skull Creek Shale	Shale	2.4	—	2.5	2.5
	Inyan Kara Group	Sandstone, siltstone	2.5	2.4	2.35	2.35
Jurassic	Morrison Formation	Shale	—	2.5	2.3	—
	Unkapa Formation	Sandstone	2.4	—	—	—
	Sundance Formation	Shale, sandstone	2.45	2.55	—	—
Triassic–Permian	Spearfish Formation	Sandstone, shale	2.72	2.7	—	—
Permian	Minnekahta Limestone	Limestone	2.75	2.65	—	—
	Opeche Formation	Sandstone	2.7	—	—	—
Permian–Pennsylvanian	Minnelusa Formation	Sandstone, shale, limestone, dolomite	2.75	2.9	—	—

[1] The Bell Fourche and Mowry Shales are equivalent to the Graneros Shale of Colorado.
[Well information from Petrowell Libraries well-log file and subsurface-data microfilm file, Denver, Colo. Densities in g/cm^3. Dash indicates that density was not measured.]

density. These are (1) dense upper Paleozoic and Triassic shales, carbonates, and sandstones; and (2) less dense Jurassic and Cretaceous rocks (primarily shales and sandstones). The elevation of the top of the Triassic Spearfish Formation is shown in Figure 3. The 49 elevation values used to construct this map were obtained from well logs and geologic maps showing outcrops of the Spearfish Formation. The depth of burial of the Spearfish Formation increases outward from the Black Hills and is greatest at the southern and eastern boundaries of the study area.

Precambrian

Exposed Precambrian rocks of the Black Hills are predominantly phyllites, schists, and quartzites. Metabasalts and metagabbros are less common. The Precambrian rock units can be grouped into four divisions (Figure 2), based on their physical properties (density and magnetic susceptibility). Kleinkopf and Redden (1975) describe the 12 lithologic units within these divisions and indicate that their probable densities range from 2.65 to 3.05 g/cm^3. The first division contains dark-green amphibole schists and amphibolite (Figure 2), derived both from intrusive diabase–gabbro rocks and from basalt flows. Amphibolite in the study area is predominantly hornblende and plagioclase. The second division is represented by interbedded schist, impure quartzite, iron formation, streaked quartzite, metaconglomerate, and amphibolite. The iron formation contains interlayered metachert and grunerite strata. Rocks in these first two divisions have moderate to high densities and susceptibilities. A micaceous phyllite or schist, belonging to the third generalized Precambrian division, contains moderate amounts of magnetite. Eight lithologic units having low to moderate susceptibilities and densities are grouped into the last division. These units primarily include gneissic granite, phyllite, schist, marble, and quartzite.

Structures

The Black Hills uplift consists of two nearly flat-topped blocks separated by the Fanny Peak monocline, a north-trending fold lying approximately along the Wyoming–South Dakota border (Noble, 1952). The southeastern block, which lies within the study area (Figure 1), trends northward, is topographically higher than the other block, and has an exposed core of Precambrian rocks. Complex Precambrian structures and lithologies here have resulted from two metamorphic events and possibly six different episodes of deformation (Redden, 1975). Structures in the southeastern block suggest three major periods of Precambrian deformation (Redden, 1968): (1) formation of north-northwest-trending folds and parallel faults, (2) shear deformation along northwest trends forming nearly vertical foliation in the metamorphosed rocks, and (3) intrusion of granite and pegmatite masses that domed the rocks.

During the Mesozoic Era and the Laramide orogeny, the Black Hills were deformed along northeast trends (Gott et al., 1974). Two fault zones, the Dewey and Long Mountain, in southwestern Custer and northwestern Fall River counties, were formed during this period of deformation and are characterized by steeply dipping normal faults, generally upthrown to the north. Having observed abrupt terminations of southeast-trending magnetic and gravity features, Gott et al. (1974) inferred the presence of a concealed Precambrian fault that could be a northeastward extension of the Long Mountain fault zone. They called this feature the Long

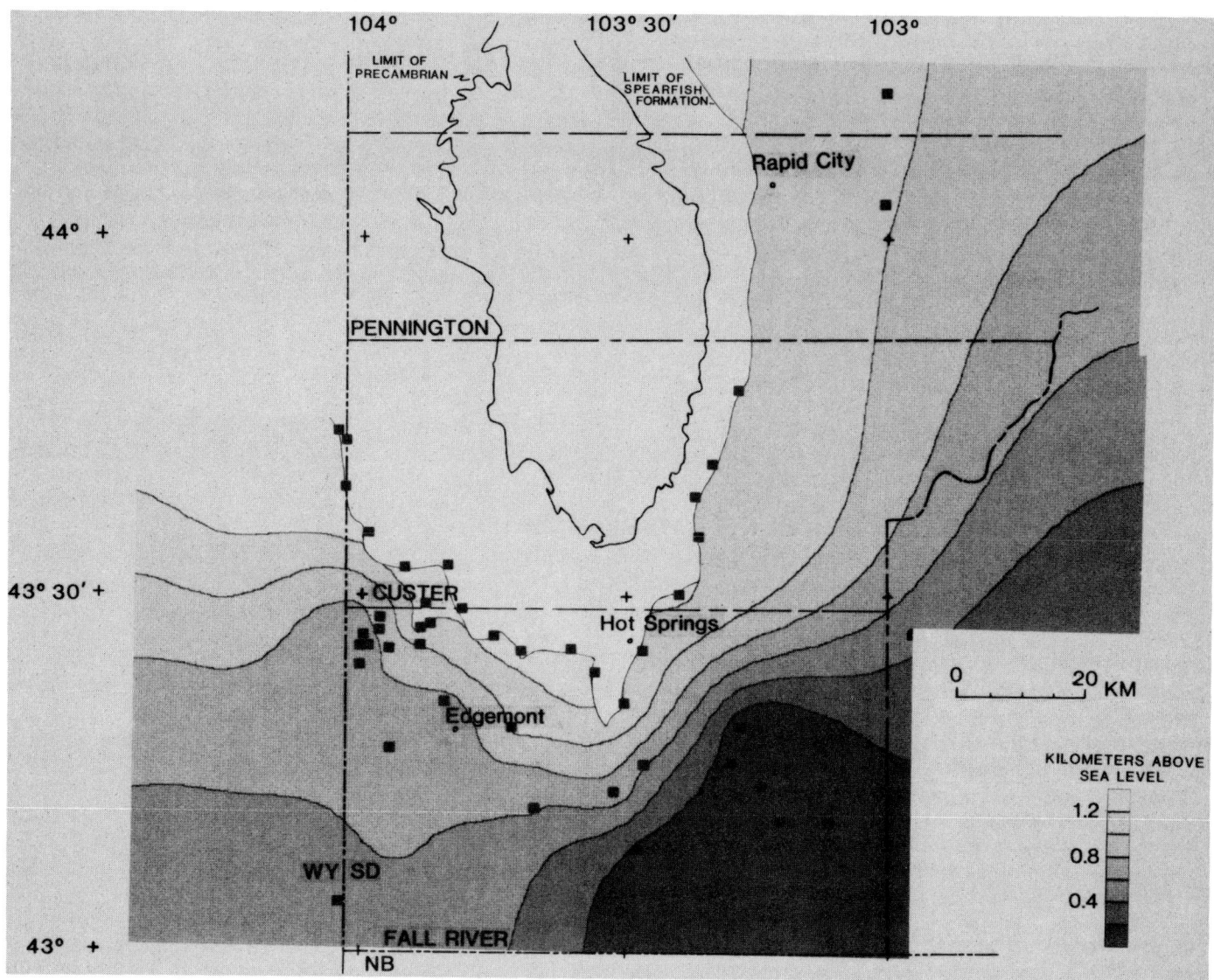

FIG. 3. Elevations above sea level of the top of the Spearfish Formation in the Black Hills area. Small squares represent drill-hole and outcrop locations where elevations are known. Contour interval, 0.2 km.

Mountain lineament. Lying to the south and paralleling this inferred fault, another concealed Precambrian structure, the Edgemont lineament, was interpreted from analysis of magnetic data by Gott et al. (1974). They suggest that this structure was reactivated during the Laramide orogeny, because it apparently influenced the northward terminations of several Laramide anticlines in Fall River County (Figure 2).

ANALYSES OF GEOPHYSICAL DATA

Reduction of magnetic data

The map of the residual aeromagnetic field (Figure 4) was compiled by merging digital data acquired from three individual surveys (Hildenbrand and Kucks, 1981b). The magnetic data were either collected at or analytically adjusted to an elevation of 152 m (500 ft) above terrain. Flight direction was east–west along lines spaced 0.8, 1, and 1.6 km apart. The residual magnetic fields were determined by removing the appropriate International Geomagnetic Reference Field (1965 and 1975) or the GSFC1266 field (Cain et al., 1967), after updating to the years in which the surveys were flown.

A second-vertical-derivative magnetic-anomaly map (Figure 5) resolves or sharpens anomalies of small areal extent. Zero contours of second-vertical-derivative anomalies may be used to outline the approximate boundaries of magnetic sources (Vacquier et al., 1951). Interpreted lithologic and structural boundaries are shown in both Figures 4 and 5.

Reduction of gravity data

A complete Bouguer gravity map (Figure 6) was compiled from a data set of 5 326 gravity stations (Hildenbrand and Kucks, 1981a). Gravity coverage is fairly uniform; average

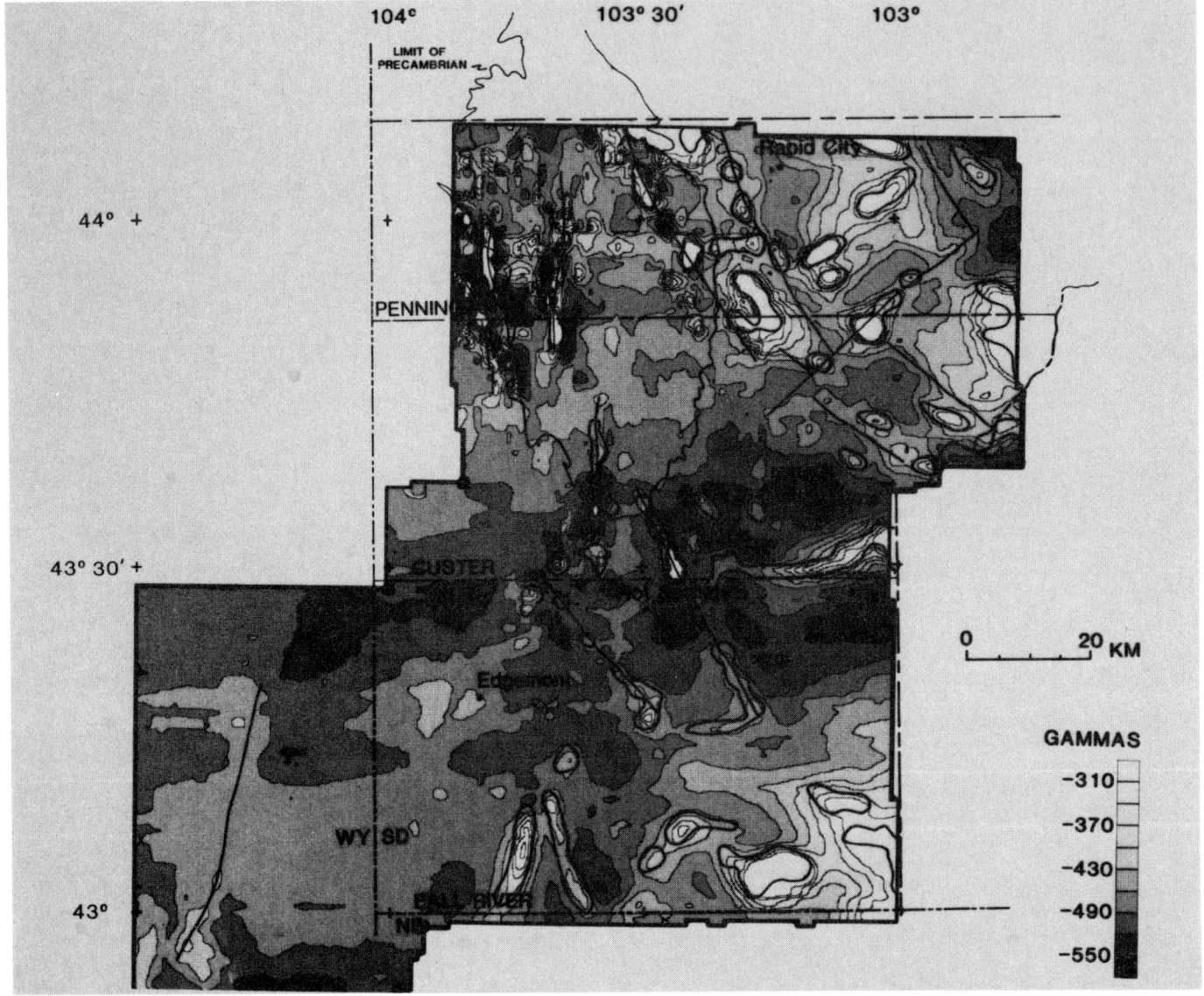

FIG. 4. Aeromagnetic-anomaly map of southwestern South Dakota. Wide lines represent interpreted lithologic and structural boundaries based on the analysis of second-vertical-derivative anomalies (Figure 5). Contour interval, 30 gammas.

spacing between stations is 3 km. Bouguer values were computed employing a reduction density of 2.45 g/cm³ and the 1967 gravity formula (International Association of Geodesy, 1967). Terrain corrections were made for the region extending radially from 0.895 km to 167 km for each gravity station.

The selected Bouguer reduction density of 2.45 g/cm³ is assumed to represent the density of the sedimentary rocks that cover about two-thirds of the study area (Table 1). These low-density sedimentary units include Jurassic and Cretaceous rocks, mainly shales and sandstones. Precambrian crystalline rocks, lower and middle Paleozoic sedimentary rocks (primarily carbonates), and uppermost Paleozoic and Triassic shales and sandstones are characterized by higher densities, ranging from 2.65 to 3.2 g/cm³. Because these rock densities are higher than the reduction density, calculated Bouguer gravity values will be too high over regions where these dense rocks are exposed or lie near the surface. Areas that have high geothermal gradients, however, are blanketed by low-density sedimentary rocks (2.3–2.55 g/cm³), and thus the selected reduction density (2.45 g/cm³) is considered appropriate for the present study.

A new gravity method developed by Cordell and Grauch (this volume) was used to locate lateral mass inhomogeneities by calculating the magnitude of the horizontal-gradient field (Figure 7). Maximum gradient magnitudes delineate the positions of lithologic or structural boundaries. Interpreted boundaries are shown in Figures 6 and 7.

As shown in Table 1, there is a major density contrast at the top of the Spearfish Formation. Assuming an average density of 2.45 g/cm³ for post-Triassic sedimentary units and 2.75 g/cm³ for pre-Jurassic sedimentary and crystalline rocks, the top of the Spearfish Formation represents a boundary across which density increases by approximately

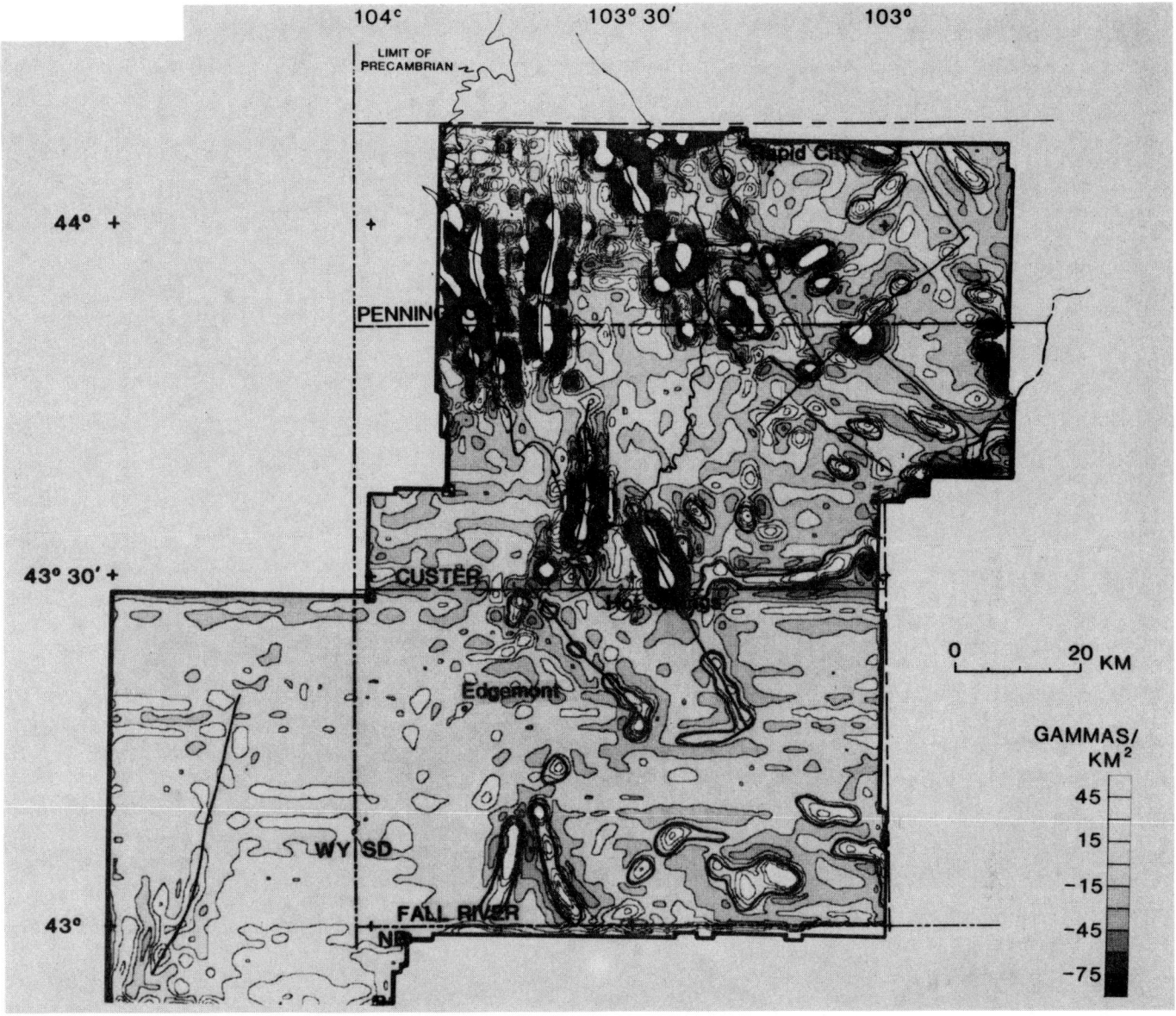

FIG. 5. Second-vertical-derivative magnetic field of southwestern South Dakota. Wide lines denote interpreted lithologic and structural boundaries. Before the derivative-anomaly map was compiled, the data were reduced to the North Pole, assuming the magnetization vector had an inclination of 70.5 degrees N and a declination of 11.8 degrees W. Contour interval, 15 gammas/km^2.

0.3 g/cm^3. If other density contrasts are negligible, the Bouguer gravity map can be converted to a map depicting estimated elevations of the top of the Spearfish Formation. Utilizing a computer algorithm described by Cordell and Henderson (1968), an approximate solution to the inverse gravity problem was determined from a 4-km rectangular grid of gravity values. The resulting elevation estimates of the Spearfish Formation were, however, smaller than the actual elevations (Figure 3) recorded in well logs. These discrepancies in elevations may be largely due to regional gravity sources in the lower crust and upper mantle. To correct the elevation estimates and eliminate gravity effects of deep regional sources, the differences between computed and actual elevations of the Spearfish Formation at well locations were determined and used to create a grid of elevation corrections. These corrections were subtracted from the grid of computed elevation estimates, resulting in Spearfish Formation elevations (Figure 8) that are compatible with well data and that supply additional information between well locations.

The accuracy of computed Spearfish Formation elevations depends on the assumptions made in reducing the gravity data and in generating the interpretational model (Figure 8). The assumed Bouguer reduction density of 2.45 g/cm^3 produces elevation estimates that are too high within regions where dense Precambrian rocks are present above sea level. Where the Spearfish Formation is absent (Figure 3), the calculated elevations are meaningless. Lithologic variations

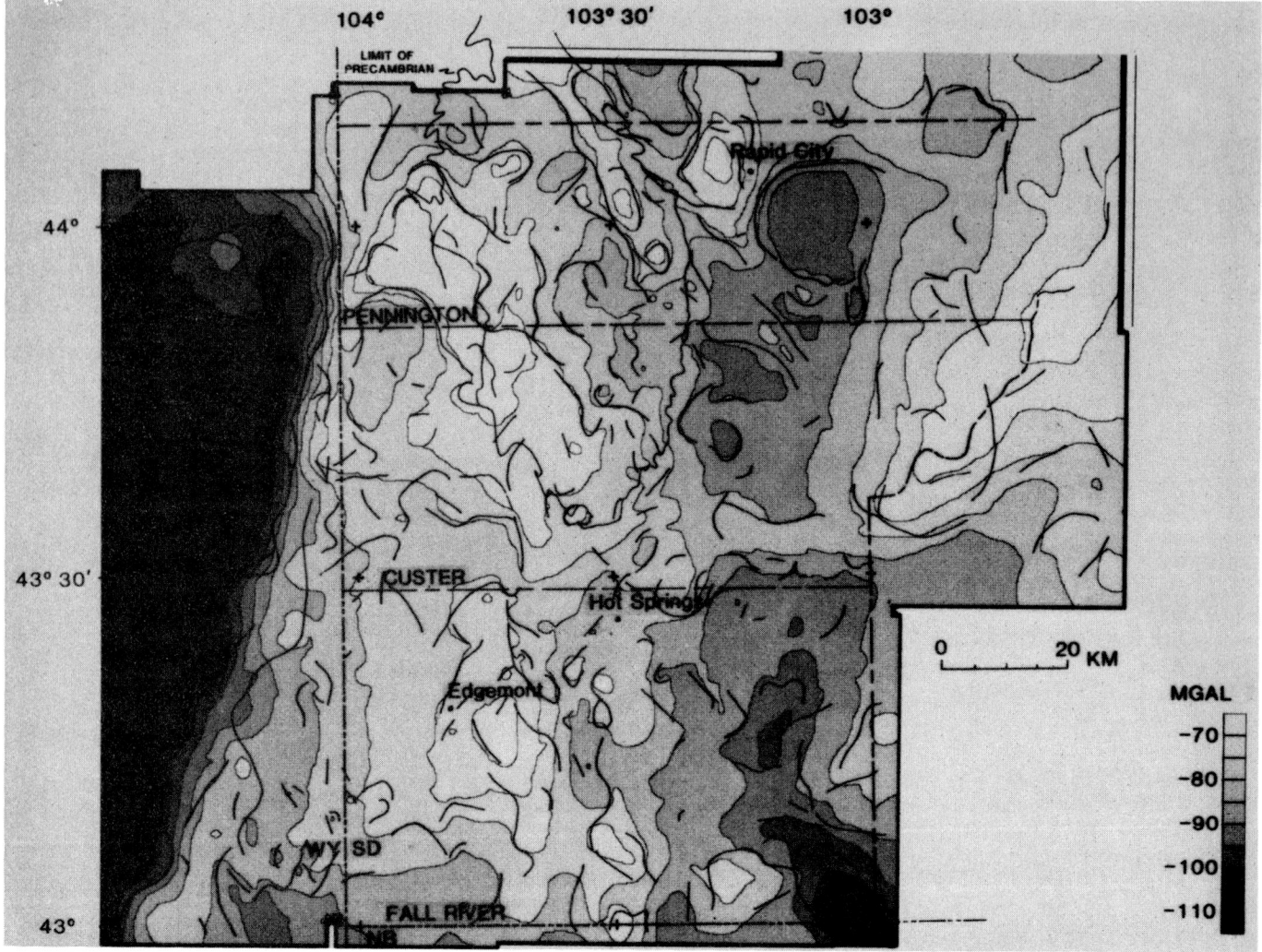

FIG. 6. Complete Bouguer gravity field of southwestern South Dakota. Wide lines represent lithologic and structural boundaries based on the analyses of magnitudes of horizontal gradients (Figure 7). Contour interval, 5 mGal.

within the Precambrian basement may also produce erroneous Spearfish Formation elevations. Regions where calculated elevations may be inaccurate for any of these reasons are indicated by a diagonal-line pattern in Figure 8.

Geophysical features

Analysis of gravity and magnetic data over areas of exposed Precambrian rocks in the study area (Hildenbrand and Kucks, 1981c) allows inferences about the types of probable basement lithologies that they delineate. Exposures of the amphibolite and metagabbro unit and the iron formation (Figure 2) correlate well with prominent gravity and magnetic highs. Similar earlier interpretations were made by Kleinkopf and Redden (1975). An example of this correlation occurs along an elongate magnetic and gravity high (Figures 4, 6) that trends northwest and coincides with outcrops of amphibolites and iron formation at the northeastern edge of the exposed Precambrian region. These rock units are also expressed as magnetic and gravity highs near latitude 44°05′ N and longitude 103°42′ W. The narrow outcrop of iron formation trending primarily south to north in central Custer County coincides with a prominent magnetic high.

The western edge of exposed Precambrian rocks coincides with the axis of a broad gravity high that Gott et al. (1974) called the gravity axis of the Black Hills uplift. Intense, short-wavelength magnetic anomalies also overlie this region. Apparently a thick section of the amphibolite and metagabbro unit and the iron formation is present at depth along the western edge of outcropping Precambrian rocks and extends westward beneath the Paleozoic sedimentary rocks.

Micaceous phyllites and schists (Figure 2), which contain moderate amounts of magnetite, also contribute to the magnetic fields along the western edge of exposed Precambrian rocks. A "V"-shaped outcrop of this unit in south-

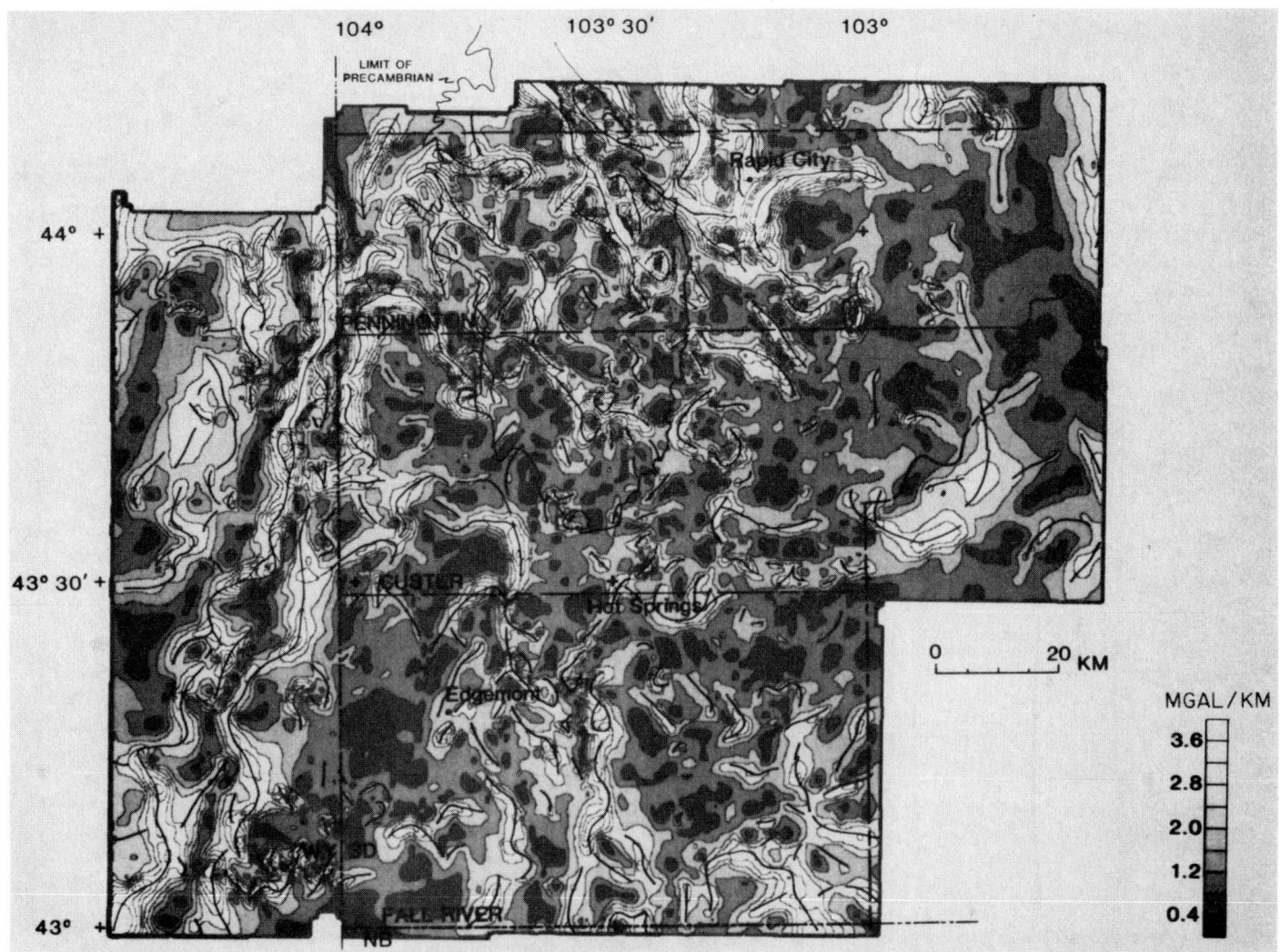

FIG. 7. Horizontal-gravity-gradient magnitudes in southwestern South Dakota. Wide lines represent interpreted lithologic and structural boundaries. Contour interval, 0.4 mGal/km.

western Pennington County coincides with a pronounced magnetic high (Figure 5). The eastern limb of this outcrop is better defined in the magnetics, suggesting that a thicker or more magnetic section of micaceous phyllites and schists is present there.

Magnetic and gravity lows generally overlie exposed granites, schists, phyllites, and quartzites that have low to moderate susceptibilities and densities (Kleinkopf and Redden, 1975). Northwest-trending gravity and magnetic lows in the central region of exposed Precambrian rocks suggest a thick section there of low-density, low-susceptibility rocks of this type.

Magnetic and gravity data from along the eastern flank of the Black Hills uplift also reflect a variety of Precambrian basement lithologies. Northwest of Rapid City (near latitude 44°08′ N, longitude 103°21′ W), the nose of a southeast-trending gravity low coincides with a pronounced magnetic high. The source is presumably a low-density, high-susceptibility phyllite or schist. A gravity high and an associated magnetic high immediately west of Rapid City may reflect underlying amphibolites and metagabbros or iron formation. The eastern extent of these mafic rock types is marked by a steep gravity gradient southeast of Rapid City.

A broad gravity low trending north-northeast from southern Custer County (about latitude 43°35′ N and longitude 103°15′ W) to the northern boundary of the map may represent thick sections of underlying granites, schists, phyllites, or quartzites. The low is transected by northwest-trending gravity highs that have coincident magnetic highs. One linear magnetic feature extends northwestward into the region of exposed Precambrian rocks, where it coincides with surface faults that flank outcrops of amphibolites and metagabbros and iron formation. Apparently the northwest-trending lineaments reflect fault-formed lithologic boundaries associated with these mafic rock types.

A northeast-trending magnetic feature is present in the northeastern corner of the study area (Figure 4). If extended farther southwestward, the magnetic feature coincides with

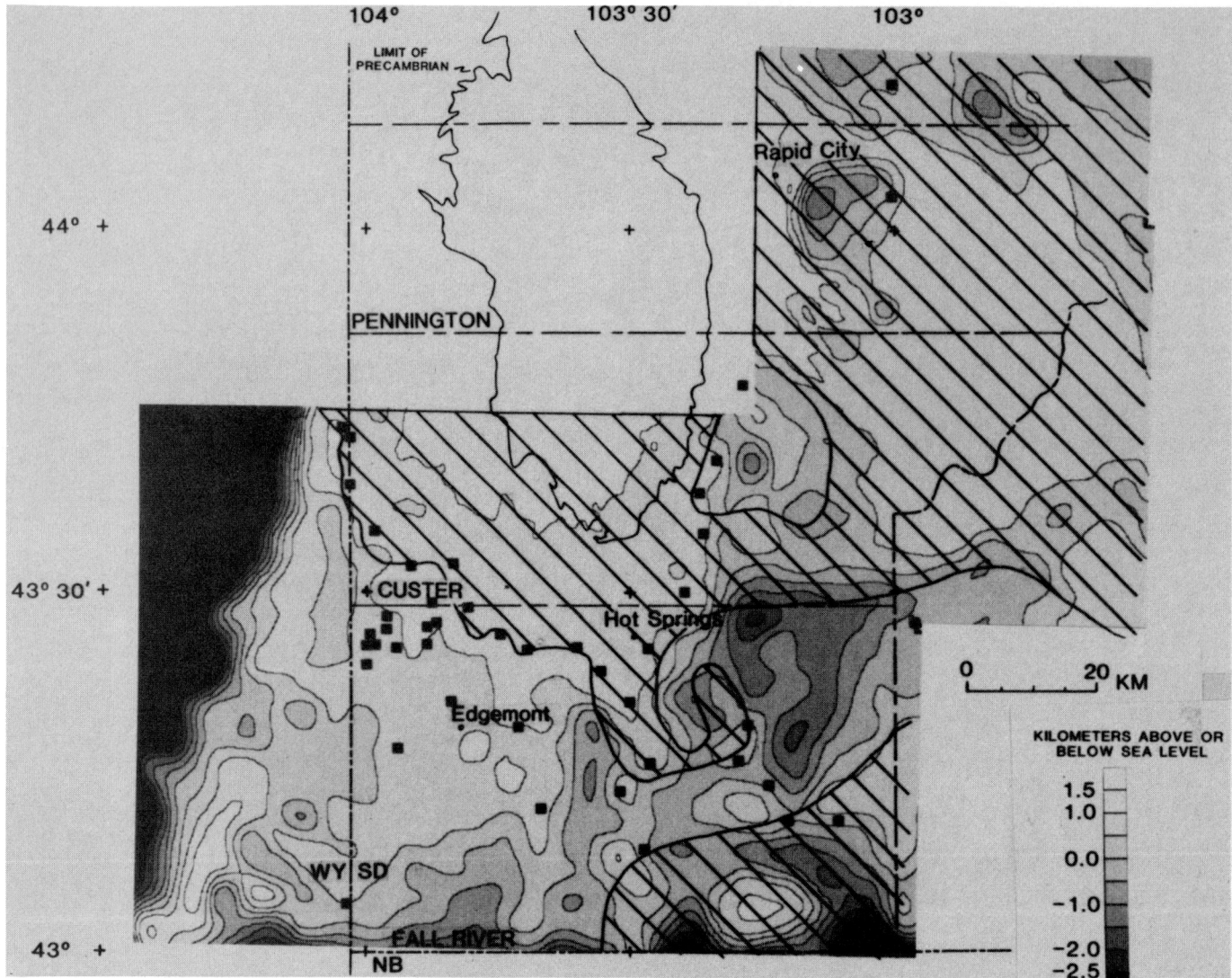

FIG. 8. Elevations of the top of the Spearfish Formation, relative to sea level, based on gravity interpretations. Diagonal-line pattern defines regions where computed elevations may be inaccurate for reasons discussed in text. Locations where elevations are compatible with drill-hole data are represented by small squares. Contour interval, 0.5 km.

the proposed Long Mountain lineament (Gott et al., 1974) representing the northeastward extension of the Long Mountain structural and fault zone. The northeast-trending magnetic lineament may also reflect the continuation of the Long Mountain fault zone. The interpreted northeastward extension of the Long Mountain fault zone abruptly truncates gravity features at their southeastern termini along the eastern flank of the Black Hills uplift.

The broad gravity low is paralleled by an equally pronounced gravity high to the east in the northeast portion of the map (Figure 6). Corresponding highs at the edge of the magnetic-anomaly map (Figure 4) suggest that the causative bodies are massive volumes of amphibolite and metagabbro or iron formation.

The general pattern of the gravity field along the southern flank of the Black Hills uplift includes a broad, north-trending high and low in western and eastern Fall River County, respectively. Comparison with the Spearfish elevation model (Figure 8) shows that the gravity high occurs over a large uplifted region that apparently represents the southward extension of the Black Hills uplift. The structural high begins approximately at the level of the top of the Black Hills uplift in southern Custer County (latitude 43°30′ N), maintains a relatively flat crest for about 35 km southward to latitude 43°12′ N, and then gently dips (5 degrees) southward into Nebraska. Its geometrical configuration resembles an asymmetric antiform facing east. Consequently, we call this structural feature the southern Black Hills antiform.

The steeply dipping eastern limb of the southern Black Hill antiform appears to coincide with a limb of a parallel syncline (Figure 8). The steep dip of the syncline's eastern limb may be due to folding of an adjacent anticline (Figures 2, 8). Farther east a depression or basin probably is delineated by the broad gravity low observed in eastern Fall River County. The implied relief of this basin relative to the crest of the southern Black Hills antiform is about 2 km. Litholog-

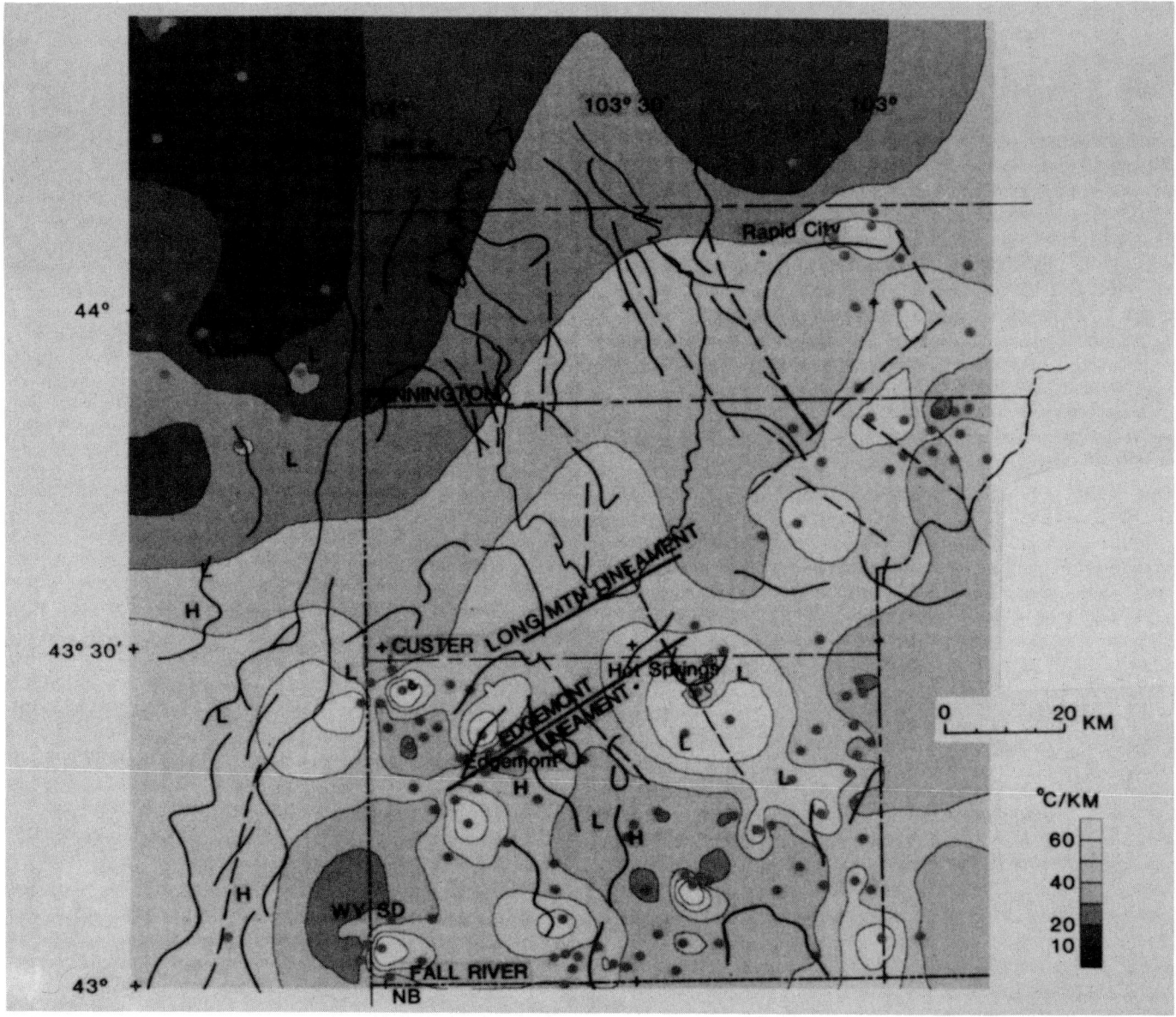

FIG. 9. Geothermal gradients in the southern regions of the Black Hills uplift. Small circles represent drill-hole locations where geothermal gradients were determined. Dashed lines and solid lines denote major magnetic and gravity features, respectively (Figures 4, 6), that may structurally control the convective transport of heat and fluids. Symbols "H" and "L" represent generalized structural highs and lows, respectively, on the surface of the Spearfish Formation (Figure 8). Contour interval, 10°C/km.

ic variations in Precambrian basement may occur in southeastern Fall River County, and thus the approximate boundaries of the basin are difficult to determine.

Well data indicate that Precambrian basement in Fall River County consists primarily of gneiss and schist (Steece, 1975). Although these rock types characteristically have low to moderate densities and magnetic susceptibilities in this region, the magnetic- and gravity-anomaly maps reveal several prominent highs with amplitudes indicative of more mafic sources at depth in central Fall River County. Amphibolites and metagabbros or iron formation, similar to those observed in outcrops of the Black Hills region, may produce most of the magnetic highs along the southern flank of the uplift. Local areas where these Precambrian rock types may be present are indicated in Figure 4.

GEOTHERMAL GRADIENT AND CHEMICAL GEOTHERMOMETER DATA

Variations of geothermal gradients supply near-surface evidence of the nature of the geothermal system in the study area. The thermal-gradient map shown in Figure 9 was compiled using 115 values previously estimated by Schoon and McGregor (1974) and 20 values estimated from well data from the water-quality data bank maintained by the National Water Data Exchange, U.S. Geological Survey. All gradients were estimated by first determining the difference between the formation temperature and the mean annual air temperature (7.2°C) and then dividing by the formation depth. Well depths range from 122 to 2 097 m and have a mean of 753 m and a standard deviation of 284 m. For a formation depth of 753 m, a variation of 5°C in mean annual ground temperature (Sass et al., 1971, Figure 7; Blackwell, 1969, Figure 2) will result in a maximum gradient fluctuation of 7°C/km. Because of thermal convection by ground-water flow, the estimated thermal gradients should be used cautiously in evaluating geothermal prospects in the area. Temperature measurements made in wells encountering aquifers close to cool-water recharge will give gradients above the aquifer much lower than wells in the same sedimentary column without flow. Thermal-gradient anomalies produced by a single measurement are not, therefore, regarded as reliable thermal indicators, although clustering or alignment of a number of anomalously high gradients may reflect principal thermal-discharge areas. Gosnold (1982) studied geothermal gradients in Nebraska and found that gradients determined by bottom-hole temperatures may be too low.

Of particular interest are two regions of very high thermal gradients. In a large area in northeastern Fall River County, values commonly exceed 50°C/km, and thermal anomalies also extend northeast and southeast from this area. The second area that may have high thermal gradients is in western Fall River County, roughly within the interpreted boundaries of the southern Black Hills antiform. Geothermal gradients in northwestern Nebraska (Gosnold, 1982) are compatible with the gradients shown in Figure 9. By contrast, gradients are characteristically lower (less than 30°C/km) along the western flank of the Black Hills uplift, where they probably reflect the westward migration of surface runoff from the Black Hills region through aquifers into the Powder River basin (Head et al., 1978).

Although geothermal-gradient data aid in delineating thermal-discharge areas and in describing the nature of the geothermal system, they generally supply little information about the temperature in deeper positions of the geothermal system. For example, deep waters circulating near molten rock may be substantially cooled in their ascent to near-surface regions. However, geothermal-reservoir temperatures or the temperatures at which water has last interreacted with rock can be estimated, using an empirical method (Fournier and Potter, 1978) based on molar Na, K, Ca, and Mg concentrations in natural waters. This method, called geothermometry, assumes that aqueous Na–K–Ca–Mg relationships are related entirely to temperature-dependent feldspar reactions. Major assumptions are that feldspar is freely available in the reservoir rock and that the water has remained in contact with the rocks long enough to reach chemical equilibrium and that fluid ascent is rapid enough to prevent re-equilibration. Figure 10 shows 48 estimated reservoir temperatures calculated using the Na–K–Ca–Mg geothermometer. Chemical compositions of natural waters were taken from the water-quality data bank, National Water Data Exchange, U.S. Geological Survey. Estimated temperatures range from 14° to 125°C and have a mean of 64°C and a standard deviation of 30°C.

SOURCES OF GEOTHERMAL WATERS

Although estimated thermal gradients (Figure 9) commonly exceed 50°C/km along the southern flank of the Black Hills, geothermal investigations by Adolphson and LeRoux (1968) and Schoon and McGregor (1974) indicate that relatively low temperatures and thermal gradients characterize the Black Hills area. From 42 wells, Adolphson and LeRoux (1968) estimated an average geothermal gradient of 29°C/km and concluded that low thermal gradients are due to the rapid descent of recharging cold water in permeable Paleozoic formations.

Five heat-flow measurements made in the Black Hills region (Figure 1) fail to provide any consistent heat-flow pattern. Two measurements of 0.5 HFU (heat-flow units; 1 HFU = 10^{-6} cal/cm^2/s) indicate low flux (Sass et al., 1971), but three other heat-flow measurements, having an average value of about 2 HFU, suggest high flux (Roy et al., 1968; Blackwell, 1969; Sass et al., 1971; Decker et al., 1980). Perhaps the low heat-flow measurements were affected by circulating ground waters, and high flux actually characterizes this region. More data are needed to resolve the regional heat-flow pattern.

The following discussions address the possible causes of high heat flow in the Black Hills area and, more importantly, the particular geothermal system responsible for the area of greatest heat production along the southern part of the Black Hills. The latter discussions are included because the high geothermal gradients of this area persist in spite of an influx of cool ground waters derived from surface runoff. These ground waters generally reduce geothermal gradients in other parts of the Black Hills area.

Heat flow

Lachenbruch and Sass (1977) show that the Black Hills lie within a broad region extending from New Mexico north to Montana and North Dakota where heat-flow measurements exceed 1.5 HFU. However, the number of measurements used to define this region is small. It is difficult, therefore, to suggest from the available data whether the Black Hills uplift is a localized high-heat-flow area or only part of a much broader high-heat-flow region. In the following discussions, however, we take the view that the Black Hills uplift lies within a localized region characterized by high heat flow.

Geologic and geophysical evidence indicates that near-melting temperatures have persisted in recent geologic time in the lower crust and upper mantle under the Black Hills uplift. A K–Ar date of rhyolitic flows near Lead, South Dakota (Kirchner, 1977), suggests that an igneous event occurred in late Miocene time (10.5 m.y. ago). On the basis of calculated high surface flux and high equilibrium temperatures near the crust–mantle boundary, Decker et al. (1980) proposed that near-melting temperatures prevailed in "younger parts of the Cenozoic." These observations are consistent with high electrical conductivity in the mantle delineated by geomagnetic studies (Camfield et al., 1971). The frequency of historical earthquakes (Schoon and McGregor, 1974) suggests that structural adjustments are occurring there today. Thus, high mantle temperatures may have increased the heat flux in the crust in the Cenozoic, which

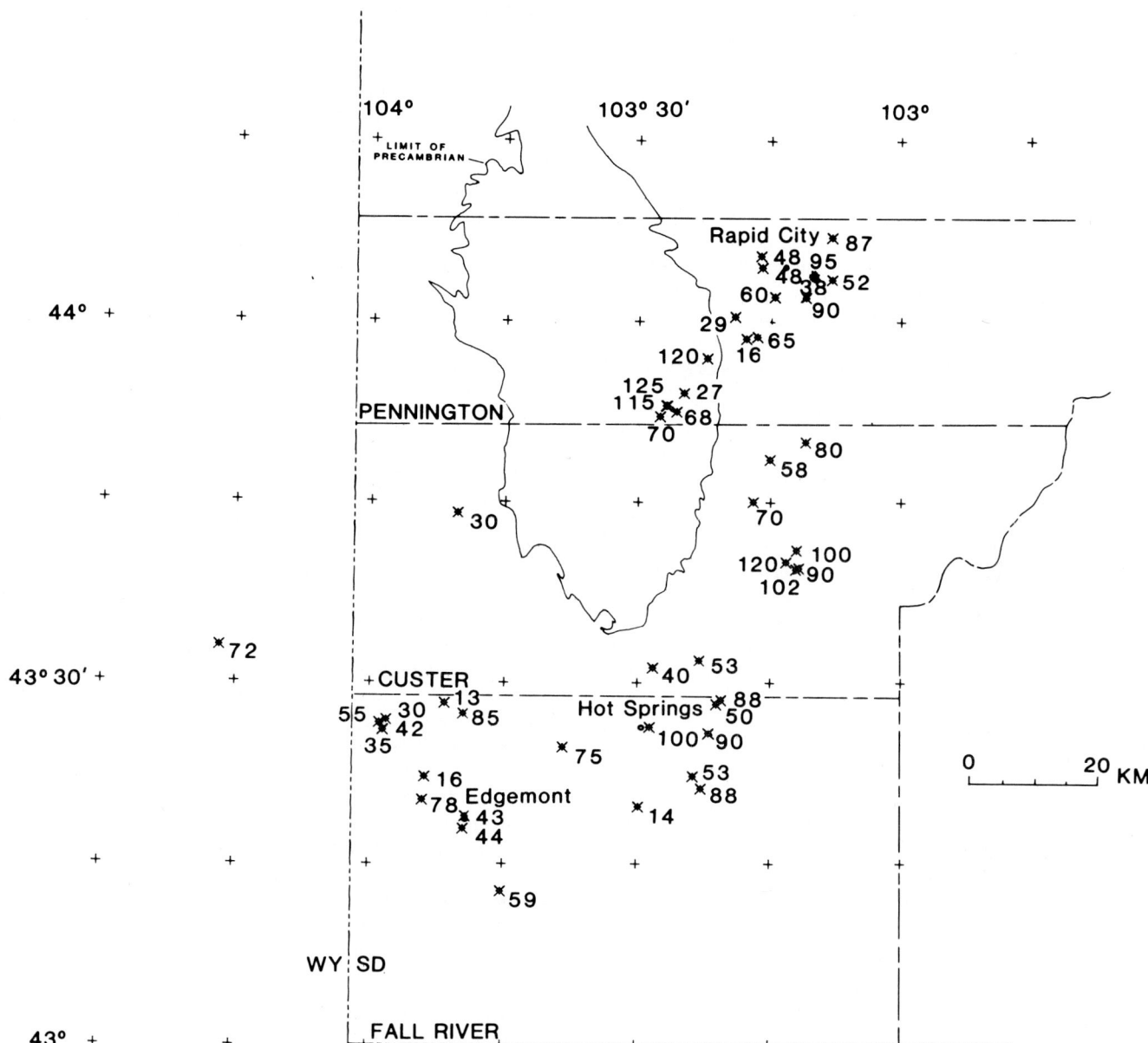

FIG. 10. Geothermal-reservoir temperatures (°C) in southwestern South Dakota, calculated using a Na–K–Ca–Mg geothermometer. "X'ed" boxes represent drill holes where the chemical compositions of ground water were obtained to calculate the indicated temperatures.

may contribute to high estimated geothermal gradients along the southern flanks of the Black Hills.

Geothermal gradients

Common causes of high geothermal gradients include (1) partially cooled magma at shallow crustal depths, (2) exothermic reactions from oxidation and hydration weathering, (3) generation of heat from radiogenic sources, and (4) convective transport of heat and fluids from greater depths. None of the first three causes is thought to be a major contributor of near-surface heat concentrations in the study area. Tertiary intrusions in the northern Black Hills area are expressed as intense, circular magnetic highs of short wavelengths. This magnetic pattern is not present along the southern flank of the Black Hills. In addition, the sources of the magnetic and gravity features in the study area can be related to lithologic variations and structures in the Precambrian basement and in the Phanerozoic sedimentary cover. There is no geologic or geophysical evidence of very young igneous activity having resulted in partially cooled magma in the upper crustal regions in the study area.

Chemical weathering, such as the conversion of anhydrite to gypsum, is also eliminated, because thermal springs near Hot Springs are 22°C warmer than the mean annual temperature—too warm to be explained by exothermic reactions. Heat production by radioactive decay may be improbable, because the thermal springs are not particularly radioactive. Although chemical weathering and radioactive decay may contribute to the heat content of the geothermal system, deep circulation of meteoric water appears to be the most likely cause of the high estimated thermal gradients along the southern flank of the Black Hills.

In our model of the geothermal system, ground waters in the topographically high areas of the Black Hills percolate down along the many south- and southeast-trending faults and fracture zones that were formed during periods of deformation in Precambrian and Laramide time. Assuming a gradient of 40°C/km, the dense, cool water need only descend to a depth of about 1.6 km (1 mile) before attaining a temperature of 64°C, the mean temperature (estimated by geothermometry) at which the water last equilibrated with rocks. The meteoric water then migrates upward along major faults. This vertical flow component along a fault zone is either associated with increased permeability owing to fracturing or is caused by subsurface barriers to flow formed by fault movements (Domenico and Palciauskas, 1973). Most of the water does not reach the surface but spreads outward through permeable zones. Geothermal waters can also mix with local meteoric water, resulting in variations in ground-water temperatures. Some of the heated waters migrate to the surface, forming thermal springs like those observed near the town of Hot Springs. The analysis of the geophysical data in the study area helps to identify structures influencing the distribution of geothermal waters.

DISCUSSION

It is proposed that ground waters, driven by hydraulic gradients, descend to depths of about 1.6 km, become heated, and migrate upward along faults. Regions characterized by high estimated thermal gradients, therefore, presumably lie near fracture zones along which geothermal waters ascend and then migrate outward in aquifers. Similar models to explain geothermal anomalies elsewhere have been proposed. For example, Morgan et al. (1981) suggested that a forced ground-water-convection mechanism is the cause of many geothermal anomalies along the Rio Grande rift.

Some of the geophysically inferred structures in the study area may directly affect the channeling of geothermal waters. Interpreted elevations of the Spearfish Formation (Figure 8) will be assumed roughly to describe undulations associated with major aquifers of the Fall River (of the Inyan Kara Group), Morrison, Sundance, Minnelusa, and Pahasapa (Mississippian) Formations.

The most pronounced thermal-gradient anomaly (Figure 9) encompasses a large area east of Hot Springs. Three geophysical features located in the vicinity of this anomaly may represent faults that permit vertical flow of heated water. Two of these features, the Long Mountain and Edgemont lineaments, are northwest of the thermal-gradient anomaly. The general flow pattern of ground water in this region (Rahn and Gries, 1973) suggests that hot waters emerging from inferred faults associated with these lineaments may conceivably migrate southeastward within aquifers and produce the observed geothermal-gradient anomaly. The northwest-trending magnetic lineament near Hot Springs (crossing latitude 43°24′ N and longitude 103°21′ W) has been interpreted as a fault-formed lithologic boundary (Figure 4). This lineament bisects the thermal-gradient anomaly, suggesting that the corresponding fault influences the distribution of hot waters.

Thermal-gradient anomalies extend to the northeast and southeast from the thermal-gradient anomaly east of Hot Springs. The southeastern extension coincides with the interpreted basin in eastern Fall River County. The basin may collect hot waters flowing southeastward from the Hot Springs area. The northeast-trending high thermal gradient, along the eastern flank of the Black Hills uplift, appears to be related to the inferred extension of the Long Mountain lineament. Ground waters descending along northwest-trending faults in the topographically high eastern block of the Black Hills apparently encounter the interpreted Long Mountain fault zone, ascend to the near-surface regions, and migrate southeastward within aquifers. Thermal gradients are higher southeast of the interpreted extension of the Long Mountain fault and structural zone.

Another region possibly typified by high thermal gradients coincides with the southern Black Hills antiform in western Fall River County. In particular, there may be thermal-gradient highs near Edgemont along a north–south trend approximately paralleling the Chilson anticline (Figure 2). These gradient highs may reflect geothermal waters migrating southward from the Long Mountain fault zone and from the suggested fault represented by the Edgemont lineament. In addition, fracture zones underlying either the Chilson anticline or the southern Black Hills antiform may help distribute geothermal waters. If drape folding of sedimentary units over faulted basement blocks is commonplace in the southern Black Hills region, as suggested by Lisenbee (1978), then either of these positive elements may be associated with deep-seated structures representing channelways for hot waters. For example, the eastern limb of the southern Black Hills antiform may reflect faulting at depth because it is impressively steep and correlates well with the north-trending thermal-gradient anomaly.

CONCLUSION

The analysis of gravity and magnetic data has provided information on subsurface structures in the Black Hills region. Many of the interpreted structures appear to affect the convective transport of heat and fluids. The geothermal-gradient and geothermometer data have aided in delineating thermal-discharge areas and in describing the general nature of the geothermal system. Moreover, the interpretation of the gravity and magnetic data has supplied details of the geothermal system that may be useful in future utilization of the geothermal resource.

ACKNOWLEDGMENTS

We wish to thank reviewers John W. Cady and David L. Campbell for their constructive criticisms. This study was supported by the U.S. Department of Energy through an interagency agreement (ET-78-I-01-3278) with the U.S. Geological Survey.

REFERENCES

Adolphson, D. G., and LeRoux, E. F., 1968, Temperature variations of deep flowing wells in South Dakota: Prof. Pap. 600-D, U.S. Geol. Surv., D60–D62.

Blackwell, D. D., 1969, Heat-flow determinations in the northwestern United States: J. Geophys. Res., 74, 992–1007.

Cain, J. C., Hendricks, S. J., Langel, R. A., and Hudson, W. V., 1967, A proposal model for the International Geomagnetic Reference Field—1965: J. Geomagn. Geoelect., 19, 335–355.

Camfield, P. A., Gough, D. I., and Porath, H., 1971, Magnetometer array studies of northwestern United States and southwestern Canada: Geophys. J. R. Astron. Soc., 22, 201–221.

Cordell, L., and Grauch, V.J.S., Mapping basement magnetization zones from aeromagnetic data in the San Juan basin, New Mexico, this volume.

Cordell, L., and Henderson, R. G., 1968, Iterative three-dimensional solution of gravity anomaly data using a digital computer: Geophysics, 33, 596–601.

Darton, N. H., and O'Harra, C. C., 1925, Description of the central Black Hills, South Dakota: Geol. Atlas, Folio 219, U.S. Geol. Surv.

Decker, E. R., Baker, K. R., Bucher, G. J., and Heasler, H. P., 1980, Preliminary heat flow and radioactivity studies in Wyoming: J. Geophys. Res., 45, 311–321.

Domenico, P. A., and Palciauskas, V. V., 1973, Theoretical analysis of forced convection heat transfer in regional ground-water flow: Geol. Soc. Am. Bull., 84, 3803–3814.

Fournier, R. O., and Potter, R. W., II, 1978, A magnesium correction for the Na–K–Ca chemical geothermometer: Open-File Rep. 78–986, U.S. Geol. Surv.

Gosnold, W. D., Jr., 1982, Geothermal resource maps and bottom hole temperature surveys: Trans. Geotherm. Resources Council, 6, 23–26.

Gott, G. B., Wolcott, D. E., and Bowles, C. G., 1974, Stratigraphy of the Inyan Kara Group and localization of uranium deposits, southern Black Hills, South Dakota and Wyoming: Prof. Pap. 763, U.S. Geol. Surv.

Head, W. J., Kilty, K. T., and Knottek, R. K., 1978, Maps showing temperatures and configurations of the tops of the Minnelusa Formation and the Madison Limestone, Powder River Basin, Wyoming, Montana, and adjacent areas: Open-File Rep. 78–905, U.S. Geol. Surv.

Hildenbrand, T. G., and Kucks, R. P., 1981a, Complete Bouguer gravity map of the southern Black Hills, parts of southwestern South Dakota and eastern Wyoming: Open-File Rep. 81–760, U.S. Geol. Surv.

———1981b, Aeromagnetic map of the southern Black Hills, parts of southwestern South Dakota and eastern Wyoming: Open-File Rep. 81–759, U.S. Geol. Surv.

———1981c, Gravity and magnetic features and their relationship to the geothermal system in southwestern South Dakota: Open-File Rep. 81–1345, U.S. Geol. Surv.

International Association of Geodesy, 1967, Système géodesique de référence 1967: Spec. Publ. 3, Int. Assoc. Geod.

Kirchner, J. G., 1977, Evidence for late Tertiary volcanic activity in the northern Black Hills, South Dakota: Science, 196, 977.

Kleinkopf, M. D., and Redden, J. A., 1975, Bouguer gravity, aeromagnetic, and generalized geologic maps of part of the Black Hills of South Dakota and Wyoming: Geophys. Invest. Map GP-903, U.S. Geol. Surv., scale 1:250 000.

Lachenbruch, A. H., and Sass, J. H., 1977, Heat flow in the United States, in Heacock, J. G., Ed., The Earth's crust: Geophys. Monogr. 20, Am. Geophys. Union, 626–675.

Lisenbee, A. L., 1975, Structural geology—Black Hills, in Mineral and water resources of South Dakota: Rep. to Comm. on Int. Insular Affairs, U.S. Sen., U.S. Gov. Print. Off., 52–56.

———1978, Laramide structure of the Black Hills uplift, South Dakota–Wyoming–Montana, in Matthews, V., Ed., Laramide folding associated with basement block faulting in the western United States: Mem. 151, Geol. Soc. Am., 165–196.

Morgan, P., Harder, V., Swanberg, C. A., and Daggett, P. H., 1981, A groundwater convection model for Rio Grande rift geothermal resources: Trans. Geotherm. Resources Council, 5, 193–196.

Noble, J. A., 1952, Structural features of the Black Hills and adjacent areas developed since Precambrian time: Guideb., 3d Annu. Field Conf., Billings Geol. Soc., 31–37.

Rahn, P. H., and Gries, J. P., 1973, Large springs in the Black Hills, South Dakota and Wyoming: Rep. Invest. 107, S.Dak. Geol. Surv.

Redden, J. A., 1968, Geology of the Berne quadrangle, Black Hills, South Dakota: Prof. Pap. 297-F, U.S. Geol. Surv., 343–408.

———1975, Precambrian geology of the Black Hills, in Mineral and water resources of South Dakota: Rep. to Comm. on Int. Insular Affairs, U.S. Sen., U.S. Gov. Print. Off., 21–28.

Roy, R. F., Decker, E. R., Blackwell, D. D., and Birch, F., 1968, Heat flow in the United States: J. Geophys. Res., 73, 5207–5221.

Sass, J. H., Lachenbruch, A. H., Monroe, R. J., Greene, G. W., and Moses, T. H., 1971, Heat flow in western United States: J. Geophys. Res., 76, 6376–6413.

Schoon, R. A., and McGregor, D. J., 1974, Geothermal potentials in South Dakota: Rep. Invest. 110, S.Dak. Geol. Surv.

Steece, F. V., 1975, Precambrian rocks outside the Black Hills, in Minerals and water resources of South Dakota: Rep. to Comm. on Int. Insular Affairs, U.S. Sen., U.S. Gov. Print. Off., 28–29.

Vacquier, V., Steenland, N. C., Henderson, R. G., and Zietz, I., 1951, Interpretation of aeromagnetic maps: Mem. 47, Geol. Soc. Am.

Magnetic terranes in the central United States determined from the interpretation of digital data

Thomas G. Hildenbrand*

ABSTRACT

A color magnetic-anomaly map compiled from digital data provides a synoptic view of major magnetic anomalies and corresponding geologic features of the central United States. The availability of the data in digital form allowed application of a variety of analytical techniques to enhance the anomalies and provide new interpretive information. Derivative and directional filters were applied to the data to help identify lithologic and structural boundaries. A magnetization-density ratio map was used to delineate major geophysical provinces. In addition, the data were draped over the Precambrian surface to remove the effects of basement relief.

The amplitudes and wavelengths of magnetic anomalies exhibited on the compiled maps vary considerably within the Midcontinent. Magnetic anomalies correlate well with major Precambrian and Paleozoic tectonic features and aid in delineating the features' lateral extent and associated structures. Coherent patterns of anomalies observed on the maps prove useful in delineating basement domains. Of particular interest are several linear magnetic features that may reflect continental-scale geologic features. Two northwest-trending lineaments that show remarkable linearity—the south-central and Great Lakes magnetic lineaments—coincide with several structural and lithologic boundaries in the Midcontinent and apparently express major crustal discontinuities.

INTRODUCTION

Regional geologic investigations of major tectonic features are often aided by complementary studies of their associated magnetic anomalies. The approximate distribution of upper crustal, magnetized rock units interpreted from the anomalies may convey new information about known rock units or may reveal unsuspected new structures. A magnetic-anomaly map of the central United States was compiled from digital data to provide a synoptic view of major magnetic anomalies and to explore what information it might reveal about the tectonic development of the Midcontinent region. Although the ground and airborne surveys used to construct the map were made at different times, flight-line spacing, and elevations, attempts were made to compile a consistent data set by removal of an appropriate geomagnetic reference field and by analytical continuation to a common surface of 1 000 ft (305 m) above mean terrain elevation. The anomaly field values over the map are expected to be generally within 100 gammas of their true values; the greatest source of error is caused by merging several surveys referenced originally to an arbitrary datum level.

The availability of digital data allowed application of a variety of analytical techniques to enhance the anomalies and provide new interpretive information. The application of these techniques, and therefore the results of this study, would have been impossible had the data not been in digital form. Data analyses involved generation of (1) filtered-anomaly maps and shaded-relief maps to enhance lithologic and structural boundaries, (2) anomaly data draped onto the Precambrian basement to remove the magnetic effects of relief on the Precambrian surface, and (3) a magnetization-density ratio map to identify terranes having similar physical properties. The compiled maps reveal a wide variety of magnetic patterns characterized by distinct amplitudes and wavelengths. Magnetic anomalies correlate well with major Precambrian and Paleozoic tectonic features and aid in delineating the features' lateral extent and associated structures. Although the nature of the source of many anomalies is not completely understood, patterns and correlations are observed that seem worthwhile to point out.

MAGNETIC SURVEYS AND DATA REDUCTION

Nearly 70 individual surveys were used to construct the magnetic-anomaly map. In Figure 1 the surveys have been grouped into generalized areas for which data are available in map form. The reader is referred to Hildenbrand et al. (1983) for a more detailed description of the magnetic surveys and data-reduction process.

---

*U.S. Geological Survey, Box 25046, MS 964, Denver Federal Center, Denver, CO 80225.

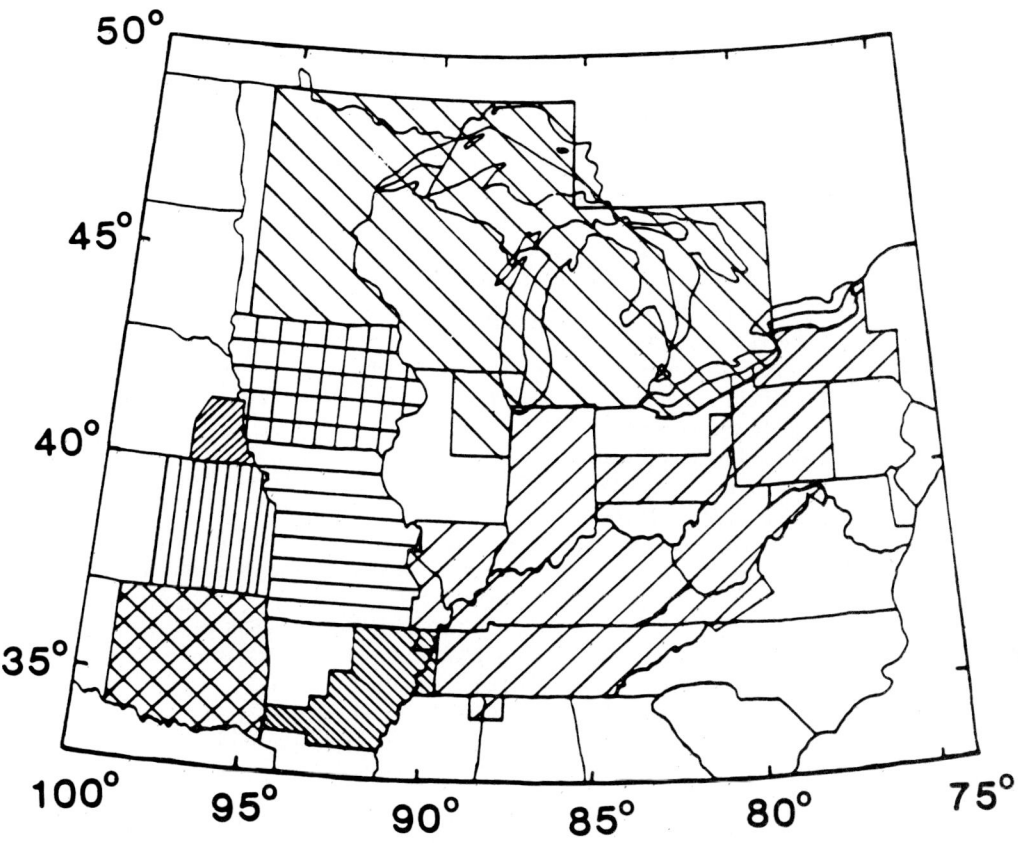

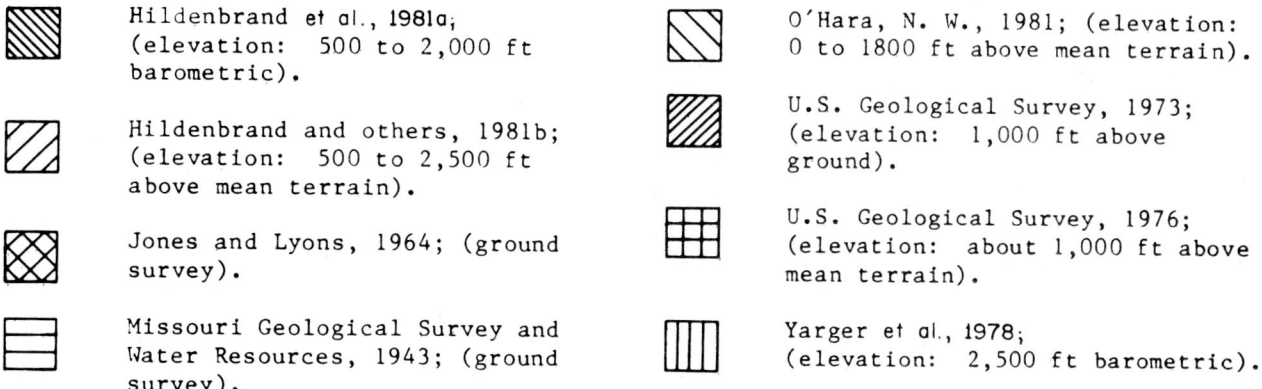

FIG. 1. Index map showing sources of data used in compiling the magnetic-anomaly map (Figure 2).

Approximately 80 percent of the magnetic coverage is airborne data. These surveys were flown at elevations ranging from 500 ft (0.15 km) to 2 500 ft (0.76 km) above mean terrain and with flight-line spacings ranging from 0.5 miles (0.80 km) to 6 miles (9.66 km). Aeromagnetic surveys with flight-line spacings of 4 miles (6.44 km) or greater cover Wisconsin, Lake Superior, Lake Michigan, Lake Huron, and central Pennsylvania. In addition to aeromagnetic surveys, magnetic coverage includes ground surveys of Oklahoma and Missouri (excluding the southeastern part of Missouri), in which only vertical magnetic intensities were measured, and a shipborne total-intensity survey of Lake Erie.

The magnetic-anomaly field was obtained by removing an appropriate reference field [International Geomagnetic Reference Field and Goddard Space Flight Center (Cain et al., 1967)] after updating to the year in which the surveys were flown. An elevation of 1 000 ft (305 m) above the mean terrain elevation of each survey was selected as the datum level. This procedure should not produce significant error because surface relief over the area is generally low. Ground and marine surveys were analytically continued upward to

1 000 ft (305 m) above the mean terrain elevation; surveys flown in a draped mode (constant elevation above terrain) were continued upward or downward as necessary. For surveys flown at a constant barometric altitude, the data were continued an amount equal to the difference between the mean terrain clearance and the selected datum level.

The data digitized from the map compiled by O'Hara (1981) for integration into the present map requires a separate discussion. O'Hara's map was compiled from several surveys that encompass all of Minnesota, Wisconsin, Michigan, the Great Lakes region, and northeastern Illinois, and parts of Canada (Figure 1). These surveys were manually merged by O'Hara and were not separated in this study for individual processing. Some error in the field values of the present map may exist because the differences between survey elevations (0 to 1 800 ft or 549 m above mean terrain) and the selected reduction datum level range from $-1\,000$ to 800 ft (-305 to 244 m). It is not believed, however, that these errors are important in studying anomalies at the present scale.

DESCRIPTION OF MAJOR FEATURES

The resulting magnetic-anomaly map is shown in Figure 2. Although the initial compilation involved a 4-km (2.5-mile) grid of values (Hildenbrand et al., 1983; scale 1:2 500 000), the data were regridded to 8 km (5 miles) to reduce the computer costs of applying the many analytical techniques described below. The magnetic-anomaly map shown in Figure 2 was compiled from the 8-km grid of values. Eight kilometers is less than 1 mm (0.04 inches) at the scale of Figure 2, and utilization of the 8-km grid should not affect the following interpretations of regional-scale anomalies.

The amplitudes and wavelengths of anomalies exhibited on the compiled map vary considerably within the Midcontinent. Lithologic variation accounts for some of the wide variety of magnetic patterns. For example, magnetic anomalies over parts of Minnesota have different amplitudes and wavelengths than those observed in Wisconsin, although Precambrian basement is exposed in both regions. On the other hand, regions separated by large distances, but characterized by magnetic fields that are similar in appearance, are also identifiable. This is evident in comparing some anomalies of Missouri and Iowa with those of northwestern New York and Lake Huron. Similarity in anomaly characteristics does not necessarily indicate basements of similar lithologies but does suggest sources with similar magnetic properties and dimensions. Variation in depth to magnetic basement also produces a wide variety of magnetic patterns. The apparent long-wavelength field over the Michigan basin is caused, in part, by the deepening of Precambrian basement under the basin. In the Midcontinent, magnetic basement is generally assumed to coincide with the surface of Precambrian crystalline rocks or igneous rocks of younger age. Phanerozoic sedimentary rocks here are assumed to be nonmagnetic and to produce little or no change in the magnetic field.

Of particular interest are several magnetic megalineaments observed on the map that may reflect continental-scale geologic features (Figure 3). For the purpose of this study, a magnetic megalineament is defined as a straight or curvilinear alignment of anomalies or gradients that extends horizontally for distances greater than 500 km (311 miles); it may have substantial width, expressing a zone rather than a narrow feature. Examples are the New York–Alabama lineament (King and Zietz, 1978), essentially a boundary between contrasting magnetic terranes, and the Midcontinent magnetic (and gravity) anomaly system (King and Zietz, 1971), a 65-km-wide complex pattern of anomalies in an elongate en-echelon pattern extending for more than 900 km (559 miles). The magnetic expressions of these features are clearly apparent on the east and west sides of the map area and are discussed in the following paragraphs; two other megalineaments and several less prominent magnetic lineaments are also described below.

Midcontinent anomaly and flanking magnetic features

The Midcontinent system of magnetic and gravity anomalies is considered the most striking geophysical feature on the map and represents a major crustal discontinuity. The anomalies delineate a belt of igneous rocks of the Keweenawan Supergroup that extends from the general vicinity of Lake Superior to Kansas. This long semicontinuous zone, averaging 65 km (40 miles) in width, is underlain primarily by layered mafic volcanic flows. Magnetic lows are associated generally with clastic rocks that occur in basins formed by faults along the margins of the belt. King and Zietz (1971) concluded that owing to its appreciable length, igneous composition, and en-echelon pattern, the mafic belt marks a continental rift. More detailed descriptions, rock associations, and discussions are given by Woollard (1951), Thiel (1956), White (1966), Sims and Zietz (1967), and King and Zietz (1971).

Magnetic terrane flanking the Midcontinent rift system is characterized by patterns indicative of a variety of lithologies and structures. To the northwest in central Minnesota, broad, northeast-trending highs generally coincide with mapped Archean metavolcanics, granites, and gneisses that include granulite to greenschist-facies rock (Sims, 1976; Sims and Peterman, 1981). Intervening magnetic lows generally correlate with Archean and Early Proterozoic metasedimentary rocks. The magnetic field apparently delineates the lateral extent of these Precambrian terranes in Minnesota. Similar Archean rock types occur in Wisconsin and in the northern peninsula of Michigan. Because the broad, northeast-trending magnetic highs in Minnesota terminate at or near the Midcontinent anomaly, magnetic near-surface Archean rocks here may be less extensive than those in Minnesota.

In northern and central Wisconsin, isolated short-wavelength highs probably reflect Early to Middle Proterozoic volcanic and mafic intrusive rocks (Van Schmus et al., 1975b; Sims, 1976; Sims and Peterman, 1980). Metasedimentary rocks and low-metamorphic-grade felsic rocks may produce the intervening magnetic lows (O'Hara and Hinze, 1972). Yaghubpur (1979) and Coates et al. (1983) suggested that the broad magnetic highs in southeastern Wisconsin, northern Illinois, and southeastern Iowa are caused by 1.4-b.y. granitic plutons.

Magnetic features trending northwest in Missouri also

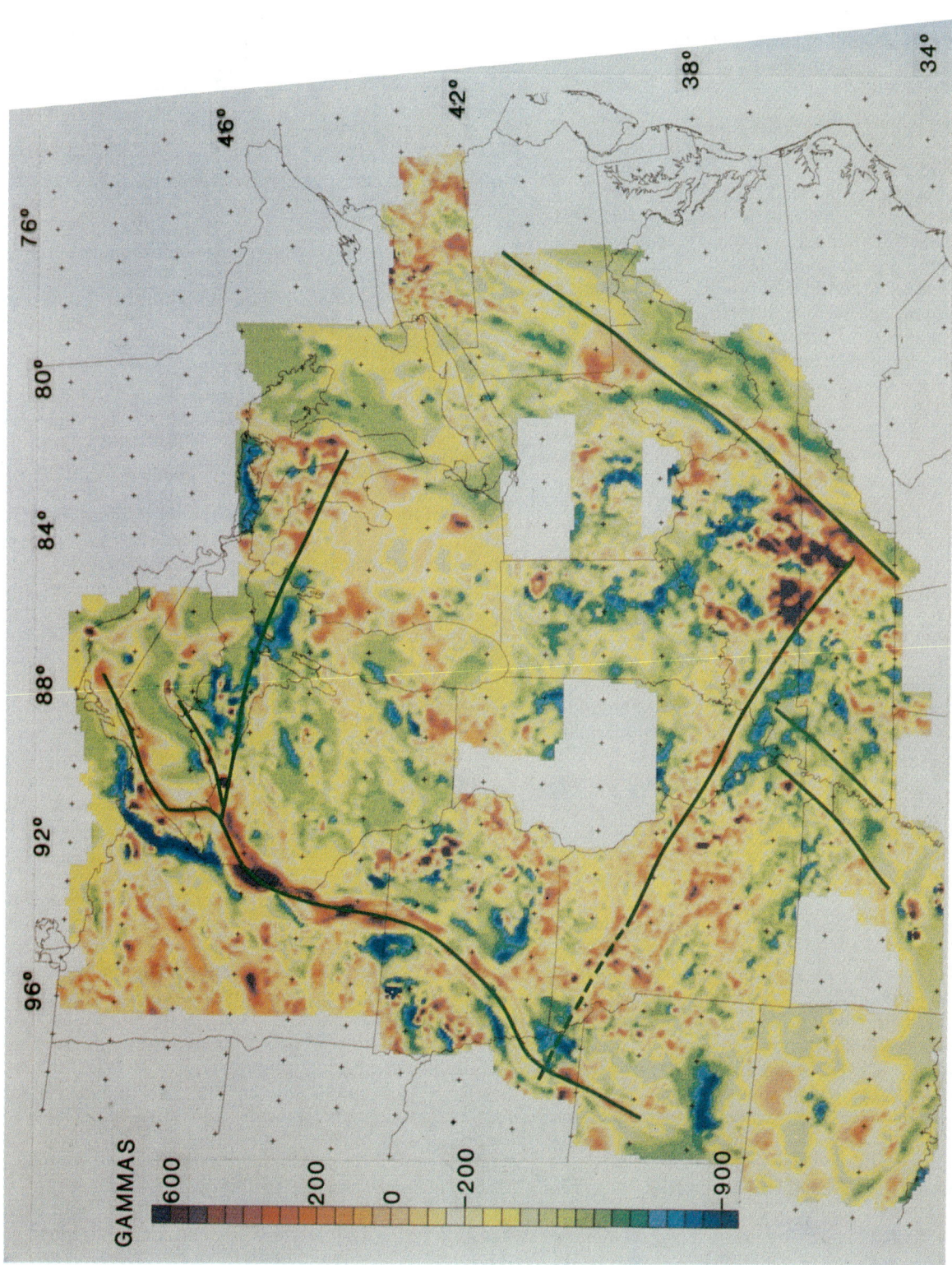

FIG. 2. Magnetic-anomaly map of a portion of central United States. Reference elevation is 1 000 ft (305 m) above ground. Color interval is 50 gammas. Heavy lines indicate major magnetic lineaments discussed in text (dashed where less certain).

follow Precambrian structures (Hays, 1962; McCracken, 1971). The source of the northwest-trending magnetic highs in Missouri may be Precambrian metavolcanic rocks and layered mafic intrusions (Kisvarsanyi, 1974, 1979). Granitic or low-grade metamorphic rocks may produce the northwest-trending magnetic lows in Missouri (Lidiak and Zietz, 1976).

New York–Alabama lineament

A 1 000-km-long segment of the 1 600-km-long New York–Alabama lineament is marked in Figure 2 by a series of prominent linear magnetic gradients trending northeast from southeastern Tennessee (latitude 35° N and longitude 85°10′ W) to central Pennsylvania (latitude 41° N and longitude 78°30′ W). The lineament reflects a profound discontinuity in the magnetic basement. King and Zietz (1978) suggested that the remarkable linearity of this feature is associated with substantial strike-slip displacements. They pointed out that the lineament may represent the southeastern edge of a stable block that acted as a buttress during strong deformation of the Appalachian fold belt. An alternative explanation given by them is that the lineament corresponds to a suture or boundary between two basement terranes with contrasting lithologic and geophysical characteristics. Whatever the cause, the geophysical feature is surely associated with a major crustal discontinuity that is expressed as a magnetic megalineament.

South-central and Great Lakes magnetic lineaments

Two other features are identifiable on the magnetic-anomaly map of the Midcontinent that comply with the above definition of a megalineament, although their magnetic expressions are less distinctive than those discussed previously. One of these, named herein the south-central magnetic lineament (Figures 2, 3), is the most pronounced northwest-trending anomaly on the map, as it extends at least to the New York–Alabama lineament on the southeast and perhaps to the Midcontinent magnetic high on the northwest.

In eastern and central Kentucky and Tennessee, complex high-amplitude anomalies with large horizontal dimensions are present. These highs are abruptly truncated on their southwest edge by the south-central magnetic lineament and on their southeast edge by the New York–Alabama lineament. Basement encountered in drill holes in the region of these intense highs consists of a variety of rock types; among these are mafic and felsic volcanics, amphibolites, and mafic granulites (Hinze et al., 1977). The mafic volcanics are similar in petrologic character to the Keweenawan rocks of the Lake Superior region (Keller et al., 1975; Lidiak and Zietz, 1976). Radiometric ages of basement rocks in Tennessee (Lidiak et al., 1981; Keller et al., 1982) suggest that an igneous event occurred 1.0–1.3 b.y. ago, roughly coinciding with the age (Keweenawan) of rift-related volcanics in the Lake Superior region. When coupled with the geometries of the associated geophysical features, these observations may indicate that the intersection of the south-central magnetic lineament and the New York–Alabama lineament marks an area where the upper crust was extended and invaded by magma during Keweenawan time. The megalineaments may represent deep-seated structures that provided channelways for the ascending magma. In this regard, Lidiak and Zietz (1976) suggested that gravity anomalies in Tennessee, coinciding with a segment of the south-central magnetic lineament, delineate a basement fault.

From western Kentucky (latitude 36°50′ N and longitude 86°50′ W) to eastern Missouri (latitude 38°25′ N and longitude 90°20′ W), a 40-km-wide band of intricate magnetic highs (Lidiak and Zietz, 1976) represents the south-central magnetic lineament. Near the Kentucky–Illinois boundary, syenite to peridotite dikes and sills, together with explosion breccias (Clegg and Bradbury, 1956; Koenig, 1956; Heyl et al., 1965), have been encountered in the vicinity of the highs; micas from these rocks have Rb–Sr and K–Ar ages of approximately 270 m.y. (Zartman, 1977, reconverted to the decay constants of Steiger and Jager, 1978). Hicks dome, a structural feature of the Kentucky–Illinois fluorspar mineral district near latitude 37°30′ N and longitude 88°15′ W, is on these northwest-trending magnetic highs. The dome is an alkalic cryptovolcanic structure that consists of mineralized breccia and kimberlite dikes (Heyl et al., 1965). According to these data, the portion of the megalineament from central Kentucky to eastern Missouri appears to delineate a region intruded by mafic igneous material (of Permian age), which contrasts with the surrounding basement. The south-central magnetic lineament in this region also parallels the Ste. Genevieve fault zone, a high-angle thrust-fault zone that has associated normal faults (Heyl et al., 1965), and its inferred extension into Tennessee (Lidiak and Zietz, 1976; Hildenbrand et al., 1982). Several other fault systems appear to terminate or bend abruptly in the vicinity of the south-central magnetic lineament (Lidiak and Zietz, 1976; Hildenbrand and Keller, 1983). They include the Rough Creek, Wabash, Western Kentucky, and Pennyrile fault systems.

In Missouri, geologic evidence indicates a predominant northwest-trending Precambrian structural grain (Hayes, 1962; McCracken, 1971), a trend that is also reflected in the magnetic data. Although obscured at least in part by a more complex background field, the south-central magnetic lineament is projected northwestward as a boundary separating a northern region, generally characterized by long-wavelength anomalies of variable trend, from a southern region, typified by higher gradients and amplitudes with a northwest trend. Numerous northwest-trending anticlines, synclines, and arches (McCracken, 1971) lie along or near this boundary.

The northwestern continuation of the south-central magnetic lineament beyond north-central Missouri (about latitude 39°20′ N and longitude 93°10′ W) is not clearly evident except as it might coincide with any one of a group of northwest-trending anomalies. If projected to the northwest, however, the lineament intersects the Midcontinent magnetic high in southeastern Nebraska (latitude 40°30′ N and longitude 96°50′ W). In this region, the Midcontinent anomaly is offset in a left-lateral sense and changes trend to a more southerly direction (Gilliland, 1962; Muehlberger et al., 1967). If the apparent displacement of the Midcontinent system is caused by the source of the south-central magnetic lineament, then the source must have had a component of left-lateral movement in post-Keweenawan time.

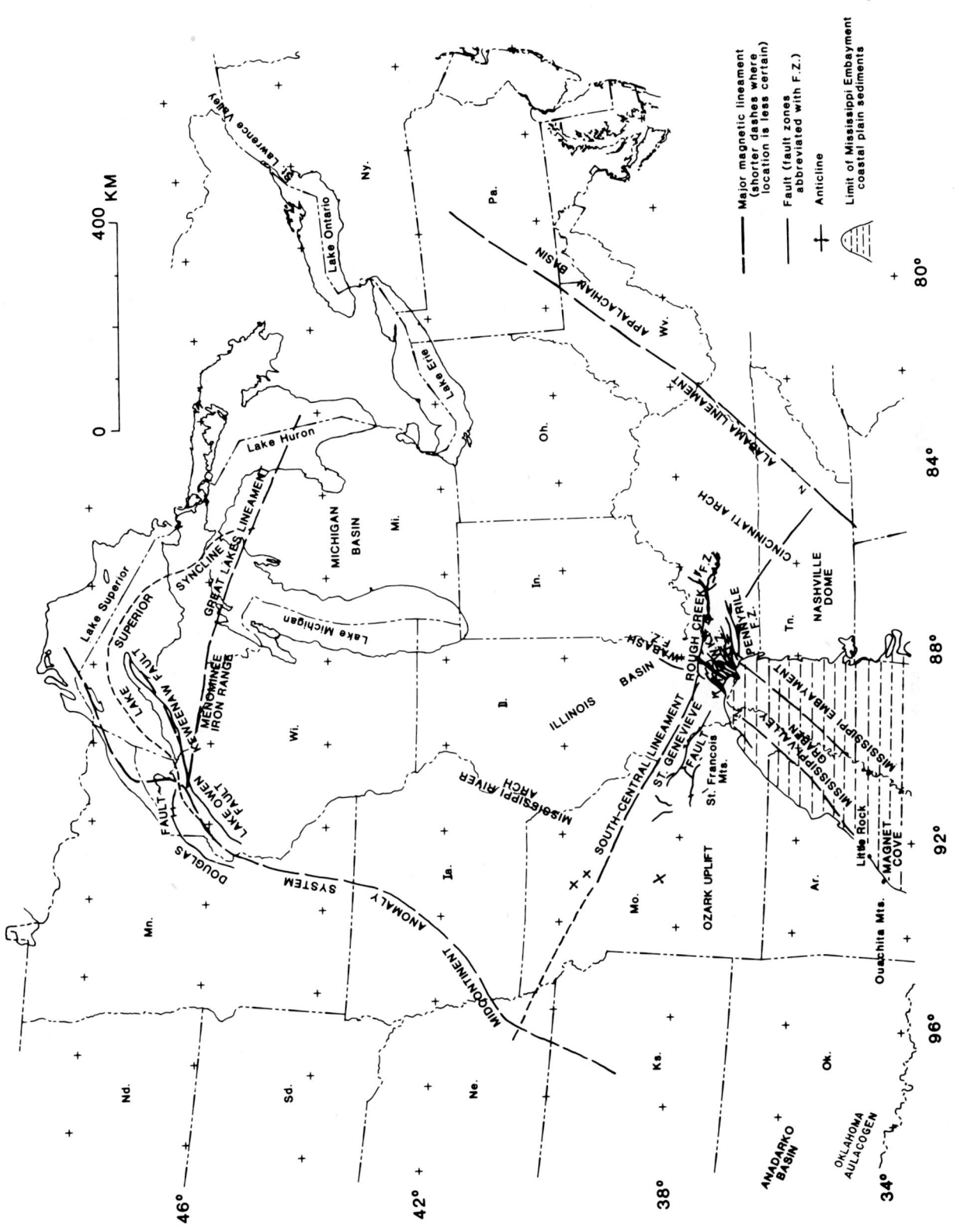

FIG. 3. Tectonic features and major magnetic lineaments in central United States. Tectonic features are from *Tectonic Map of the United States* (American Association of Petroleum Geologists and U.S. Geological Survey, 1962).

The south-central magnetic lineament, therefore, consists of three major anomaly segments along which different geologic processes have occurred. In east-central Tennessee, the lineament is a boundary separating contrasting magnetic terranes and may reflect a feature that acted as a channelway for magma in Keweenawan time (1.0–1.3 b.y.). In western Kentucky and southern Illinois, the lineament is represented as a 40-km-wide zone of magnetic highs that are associated with local Permian (270-m.y.) igneous bodies. In Missouri, the lineament is again expressed as a boundary separating contrasting magnetic terranes, but if extended to southeastern Nebraska, it intersects the Midcontinent rift system, which may have been offset in a left-lateral sense during post-Keweenawan time. From these observations, the south-central magnetic lineament may delineate a zone of crustal weakness that initally developed during or prior to Keweenawan time. Subsequently, segments of this zone of weakness became active as a result of local stress fields and geologic processes that varied along the zone.

The other megalineament, the Great Lakes magnetic lineament, is a west-northwest-trending boundary that separates regions characterized by anomalies of different wavelengths and that correlates with some major geologic boundaries. Near Lake Huron (latitude 44° N and longitude 82° W), the lineament separates moderate-amplitude, long-wavelength anomalies to the south from anomalies with high amplitudes and short wavelengths to the north. The increase in anomaly wavelength and decrease in amplitude to the south over the Michigan basin is due, in part, to a corresponding substantial increase in depth to Precambrian basement. The sharp change in anomaly character over Lake Huron and northern Michigan, however, suggests that the lineament represents a lithologic boundary with rocks of generally higher magnetization lying to the north. The observed magnetic highs immediately north of the Great Lakes magnetic lineament indicate the presence of mafic rocks, such as 1.5-b.y. porphyritic quartz monzonite plutons (Van Schmus et al., 1975a) and Keweenawan igneous rocks (Oray et al., 1973). O'Hara and Hinze (1980) suggested that these highs primarily delineate the Proterozoic (1.6–1.9–b.y.) orogenic terrane intruded by plutons. The amplitudes and wavelengths of these magnetic highs are similar to those reflecting the belt of 1.35–1.5–b.y. anorogenic plutons trending northeast from southeastern Iowa to Lake Michigan (Van Schmus and Bickford, 1981). A suite of these Proterozoic plutons may be present north of the Great Lakes magnetic lineament in northern Michigan and Lake Huron. A Proterozoic (1.6–1.9–b.y.) orogenic terrane, intruded by 1.35–1.5–b.y. plutons, may trend across Wisconsin to northern and central Michigan (Van Schmus and Bickford, 1981; Hoppe et al., 1983), although a 65-km-wide belt of Keweenawan-age rocks transects the southern peninsula of Michigan along a southerly and southeasterly trend (Hinze, 1963).

The eastern terminus of the Lake Superior syncline (Figure 3) impinges on the Great Lakes magnetic lineament in northern Michigan (latitude 45°12′ N and longitude 84°45′ W). The lineament, expressed as an elongate magnetic low, extends across Lake Michigan and the northern peninsula of Michigan to latitude 46°15′ N and longitude 89°30′ W. Farther northwest, the Menominee Iron Range and accompanying faults coincide with the megalineament.

The Great Lakes magnetic lineament transects the Midcontinent anomaly in northwestern Wisconsin (latitude 46°25′ N and longitude 91° W), a tectonically complex region. As the Lake Superior syncline crosses the magnetic lineament in this region, the syncline changes in trend from west-southwest to southwest (Figure 3). This change in trend may be related to right-lateral offsets along faults delineated by linear magnetic features of the Great Lakes magnetic lineament. The notion of a fault of this nature at this location was inferred earlier by Chase and Gilmer (1973) and Sims (1976). Sims et al. (1980) suggested that the fault was reactivated as a transform fault during Keweenawan rifting. South of the Great Lakes lineament, the Midcontinent anomaly is uniaxial, whereas to the north it separates into two magnetic features that coincide approximately with the limbs of the Lake Superior syncline. The Douglas, Lake Owen, and Keweenaw faults also terminate or bend where they intersect the Great Lakes magnetic lineament. Presumably, movement along these faults may have been accommodated along a structural boundary underlying the Great Lakes magnetic lineament.

Mississippi Valley graben

The magnetic expression of the Mississippi Valley graben is an example of a less extensive but important feature. The graben, which probably developed in association with late Precambrian or early Paleozoic rifting (Hildenbrand et al., 1977, 1982; Kane et al., 1981; Hildenbrand, 1982), contains the principal seismicity in the upper Mississippi embayment region and is considered one of the most seismically active regions of the eastern United States. The 1811–12 New Madrid earthquake series, which resulted in widespread damage, occurred within the horizontal limits of the graben. Owing to this intimate relation with seismicity, the graben's lateral extent becomes important to earthquake-hazard and -prediction studies, for which the magnetic-anomaly map can be of use.

The 70-km-wide graben is defined as a region of subdued magnetic expression apparently extending from eastern Arkansas (about latitude 34°30′ N and longitude 91°30′ W) to western Kentucky (about latitude 36°55′ N and longitude 88°45′ W). To the southwest the anomaly associated with the graben is terminated abruptly by a prominent northwest-trending magnetic feature (related to the Ouachita trend) in eastern Arkansas. Four intense magnetic highs lie near the northwest-trending feature and reflect mafic or ultramafic igneous bodies, probably similar to the exposed Cretaceous Magnet Cove ring-dike complex (Erickson and Blade, 1963) and the syenite bodies near Little Rock. An east-northeast-trending zone of magnetic highs, associated with structures of the Ouachita Mountains, is terminated along the southwest extension of the northwest graben boundary. Although the southern terminus of the graben probably lies within this structurally complex region, its ancestral counterpart may have extended farther to the southwest, but thrusting associated with the formation of the Ouachita Mountains may have masked evidence of its existence.

The location of the northern terminus of the Mississippi Valley graben is not clearly defined, but it appears to end southwest of the south-central magnetic lineament. Ahbe (1978) and Braile et al. (1982), however, analyzed magnetic, gravity, and seismicity data of southern Illinois and suggested that northeast-trending features observed north of the south-central magnetic lineament represent an extension of the rift zone. In this regard, Woollard (1958) suggested that a major structural break extends from the St. Lawrence Valley to the head of the Mississippi embayment because of a general alignment of earthquakes.

DATA ANALYSES

Analytical techniques applied to the magnetic data provide new interpretive information by enhancing particular trends or wavelengths of anomalies. Magnetic expressions of lithologic variations within the basement were enhanced by draping the data onto the Precambrian surface. Horizontal-gradient and shaded-relief maps were helpful in identifying and locating lithologic and structural boundaries. A first-vertical-derivative filter was used to resolve or sharpen anomalies of small areal extent. A magnetization-density-ratio map was compiled to delineate quantitatively major geophysical terranes within the Midcontinent region.

Draping onto the Precambrian surface

Generalized depths of Precambrian basement below the ground surface shown in Figure 4 were taken from the *Basement Rock Map of the United States* by Bayley and Muehlberger (1968). Using a technique developed by Cordell and Grauch (this volume), the magnetic data were draped directly onto the Precambrian surface. The downward-continued data represent the magnetic field that would be measured by walking on the Precambrian surface with a magnetometer. The greatest changes in the magnetic field are anticipated to occur in regions where Precambrian basement descends to great depths, such as within the Michigan, Appalachian, Illinois, and Anadarko basins (Figure 3). Compared to the unfiltered magnetic field (Figure 2), the magnetic field draped onto the Precambrian surface (Figure 5) exhibits many more short-wavelength anomalies of higher intensity along the southern and eastern regions of the map area, where basement lies at substantial depths. Where basement is at a relatively shallow depth (for example, the Great Lakes region), the anomaly patterns on the two maps are similar.

Of particular interest are anomaly patterns along the New York–Alabama lineament. Wavelengths and intensities of the anomalies north of the lineament are comparable to those south of it. This was not the case with the magnetic-anomaly map (Figure 2), which was uncorrected for depth to basement. Anomaly patterns in Figure 5 suggest that the lineament reflects a strike-slip fault with substantial displacements, as suggested by King and Zietz (1978). If the low (A' in Figure 5), trending approximately north–south in eastern West Virginia, was once aligned with the low (A) of similar trend in western West Virginia, then about 200 km (124 miles) of left-lateral displacement has occurred along structures associated with the New York–Alabama lineament. This amount of offset is also obtained by aligning magnetic highs (C' and B') in southern West Virginia and western Virginia with highs (B and C) in eastern Tennessee and Kentucky.

Shaded-relief maps

The magnetic-anomaly data, like topographic data, can be artificially illuminated from any azimuth and elevation angle using digital methods to prepare shaded-relief images that enhance features trending normal to the direction of illumination. An illumination azimuth of S 20 degrees W and an elevation angle of 45 degrees were used to produce the shaded-relief map shown in Figure 6. Northwest-trending features, such as the south-central and Great Lakes magnetic lineaments, the Oklahoma aulacogen anomaly, and those in Missouri, are all enhanced on this map. A circular feature, having a diameter of about 100 km (62 miles) and lying along the south-central magnetic lineament in east-central Missouri, is also evident on the shaded-relief map. The geometry and location of the feature suggest that it reflects a pluton that developed in association with structures (possibly faults) related to the south-central magnetic lineament.

The complementary shaded-relief map compiled with an illumination direction of E 20 degrees S (Figure 7) enhances the expression of the Midcontinent magnetic-anomaly system, New York–Alabama lineament, Mississippi Valley graben anomaly, and northeast-trending features in Minnesota and elsewhere. An interesting, roughly circular feature is clearly evident straddling the Kansas–Missouri boundary. Some of the intense magnetic highs forming the feature's boundary are produced by 1.35-b.y. granitic plutons (Bickford et al., 1981).

Derivative analyses

The edges of anomalies become sharper or clearer if derivative operators are applied to the data. For example, the first vertical derivative of the magnetic field shown in Figure 8 enhances the signatures of the border faults along the horsts associated with the Midcontinent rift system. The presence of short-wavelength anomalies of high amplitude is evident north of the Great Lakes magnetic lineament, whereas south of the lineament anomalies are generally broader and less intense.

Maximum magnitudes of horizontal gravity gradients generally occur above lithologic contacts. Thus, lines drawn along ridges formed by enclosed gradient highs are useful in locating structural trends or contacts, a technique developed by Cordell and Grauch (this volume). This technique was applied to the magnetic field after transforming it to a pseudogravity field using Poisson's equation (Barnov, 1957). For this transformation, induced magnetization with an inclination of 65 degrees N and a declination of 0 degrees was assumed.

Lines representing interpreted lithologic or structural boundaries are shown on the map of the magnitude of the horizontal-pseudogravity-gradient field (Figure 9). These boundaries are also superimposed on the map showing the magnetic field draped onto Precambrian basement in Fig-

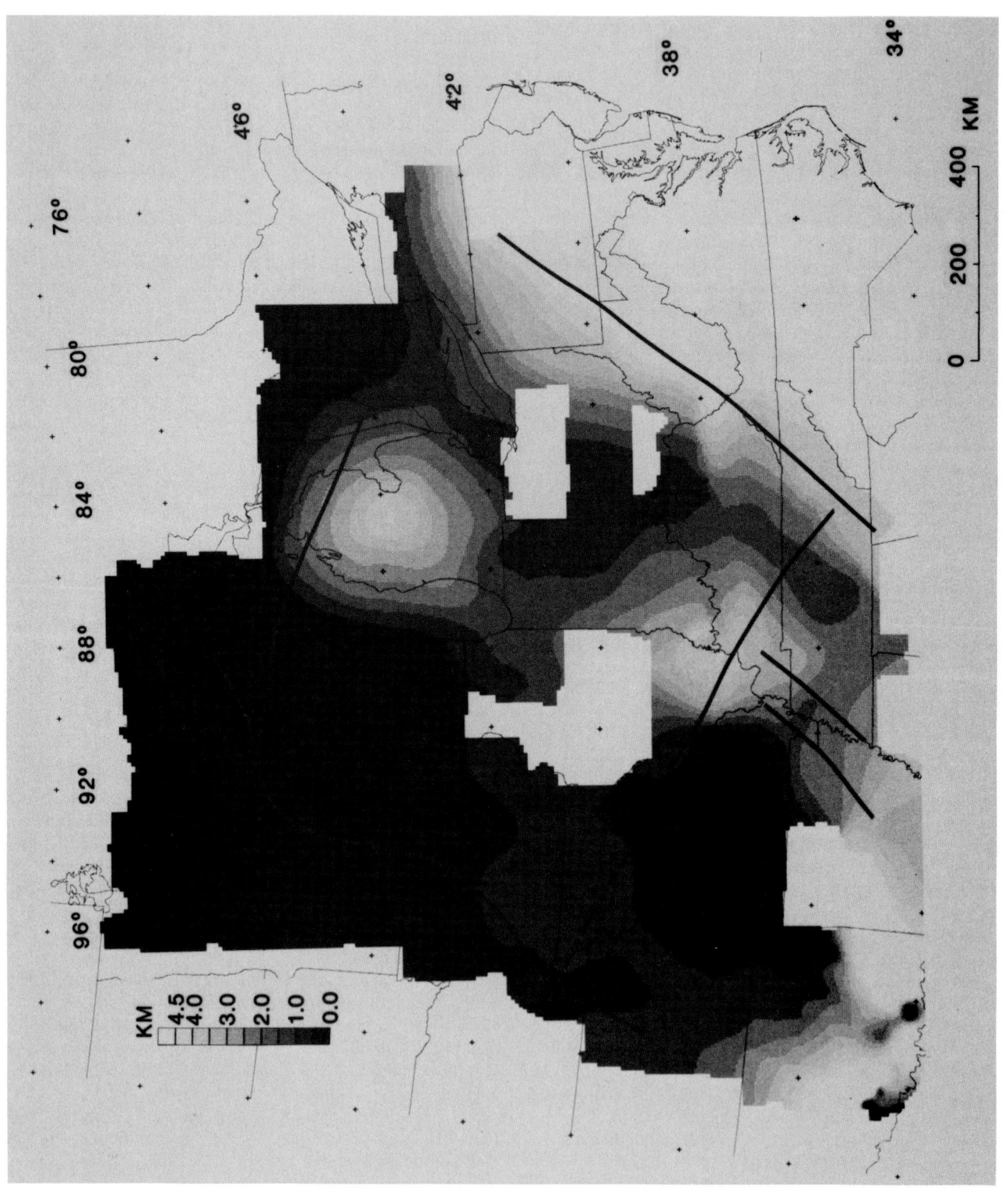

FIG. 4. Map showing depth to Precambrian basement (below ground surface). Gray shade interval is 0.5 km (0.31 miles). Heavy lines indicate major magnetic lineaments (dashed where less certain).

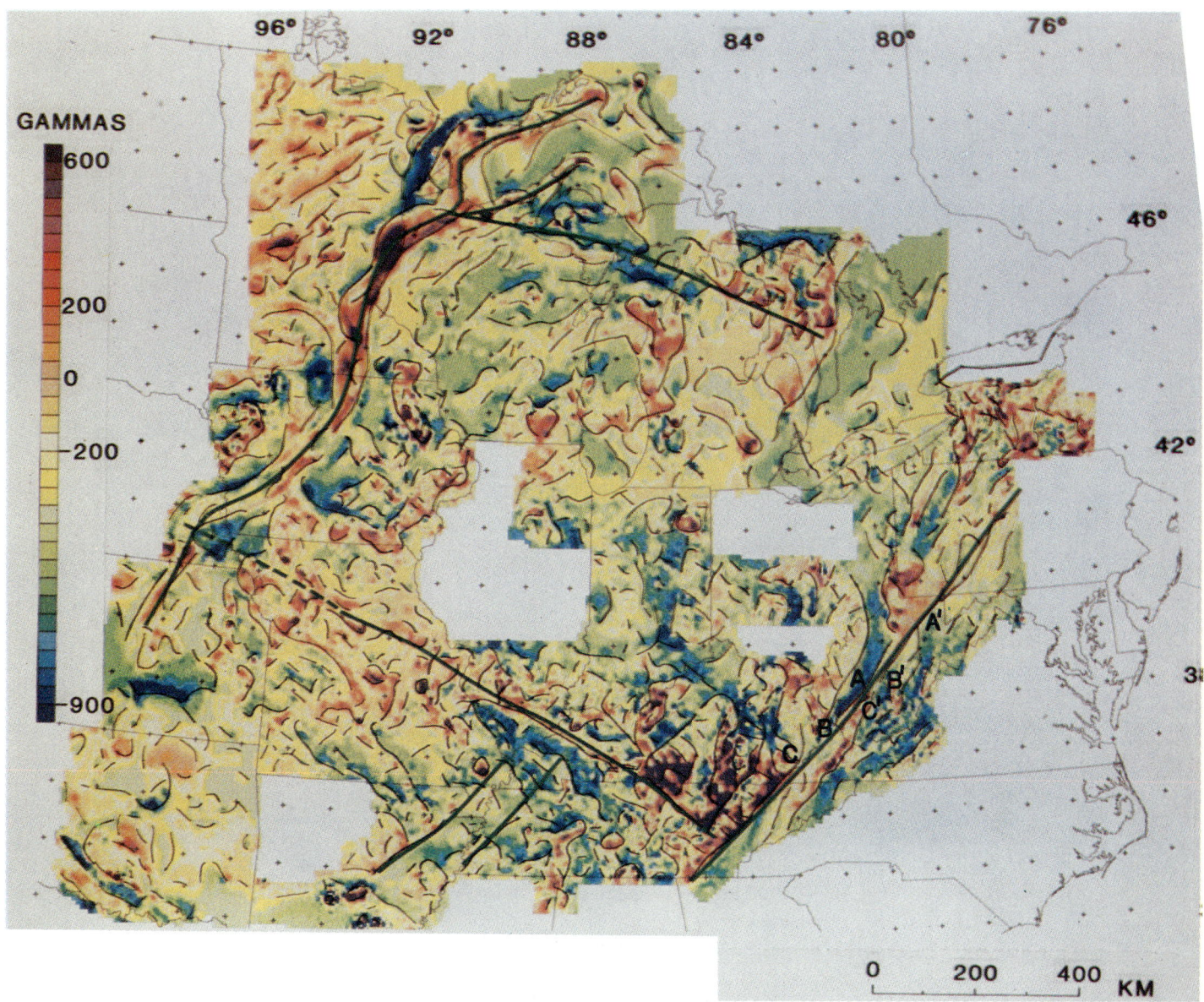

FIG. 5. Map showing the magnetic field draped directly onto the surface of Precambrian basement (shown in Figure 4). Color interval is 50 gammas. Heavy lines indicate major magnetic lineaments (dashed where location is less certain). Narrow lines represent interpreted lithologic and structural boundaries that were delineated by analyzing horizontal-gradient magnitudes of the pseudogravity field (Figure 9). On the east side of the map area along the New York–Alabama lineament, A′, B′, and C′ mark anomalies discussed in text that once may have been aligned with anomalies A, B, and C, respectively.

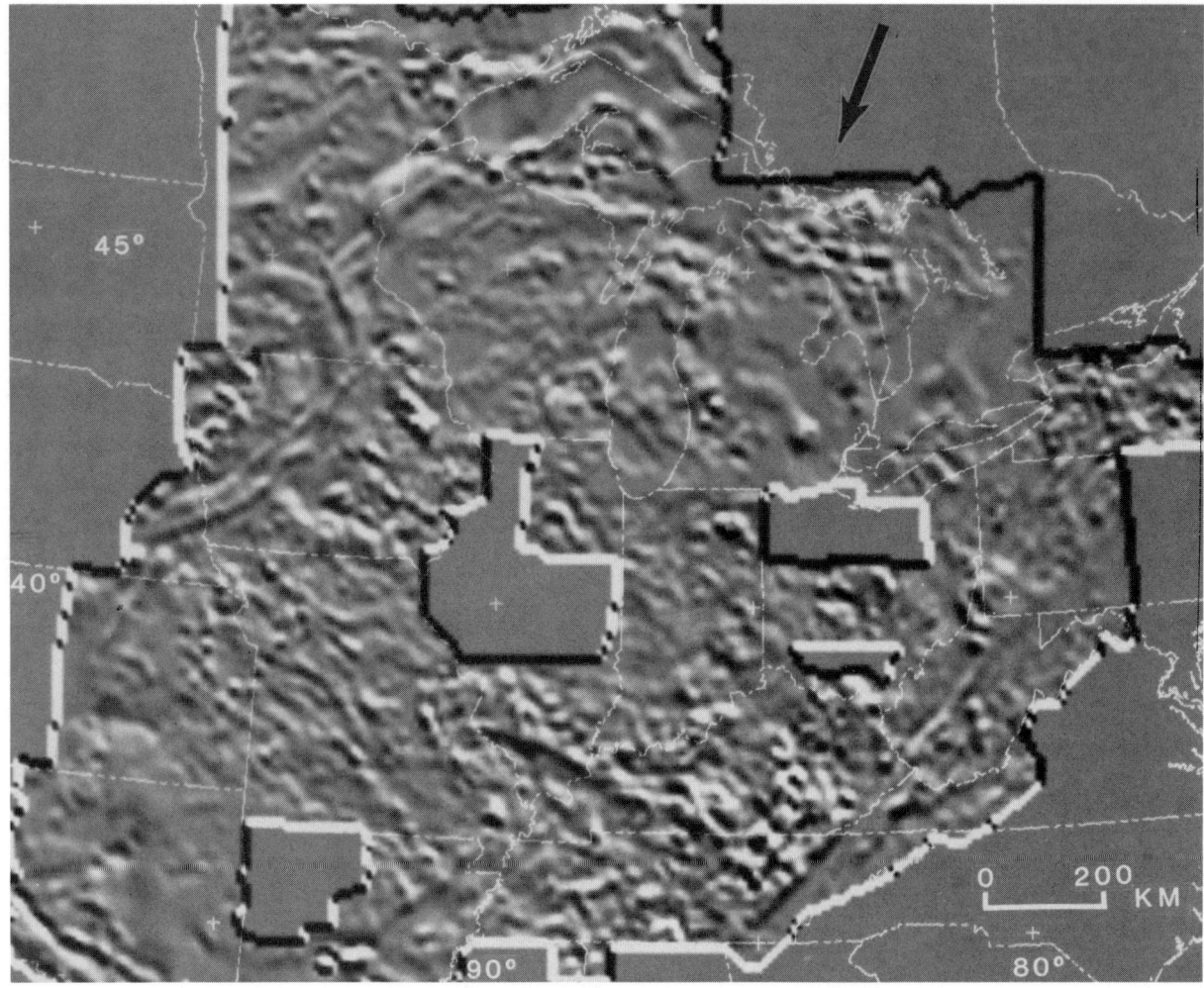

FIG. 6. Shaded-relief map of the magnetic field draped onto Precambrian basement (Figure 5). Sun elevation angle is 45 degrees and illumination azimuth (shown by arrow) is S 20 degrees W.

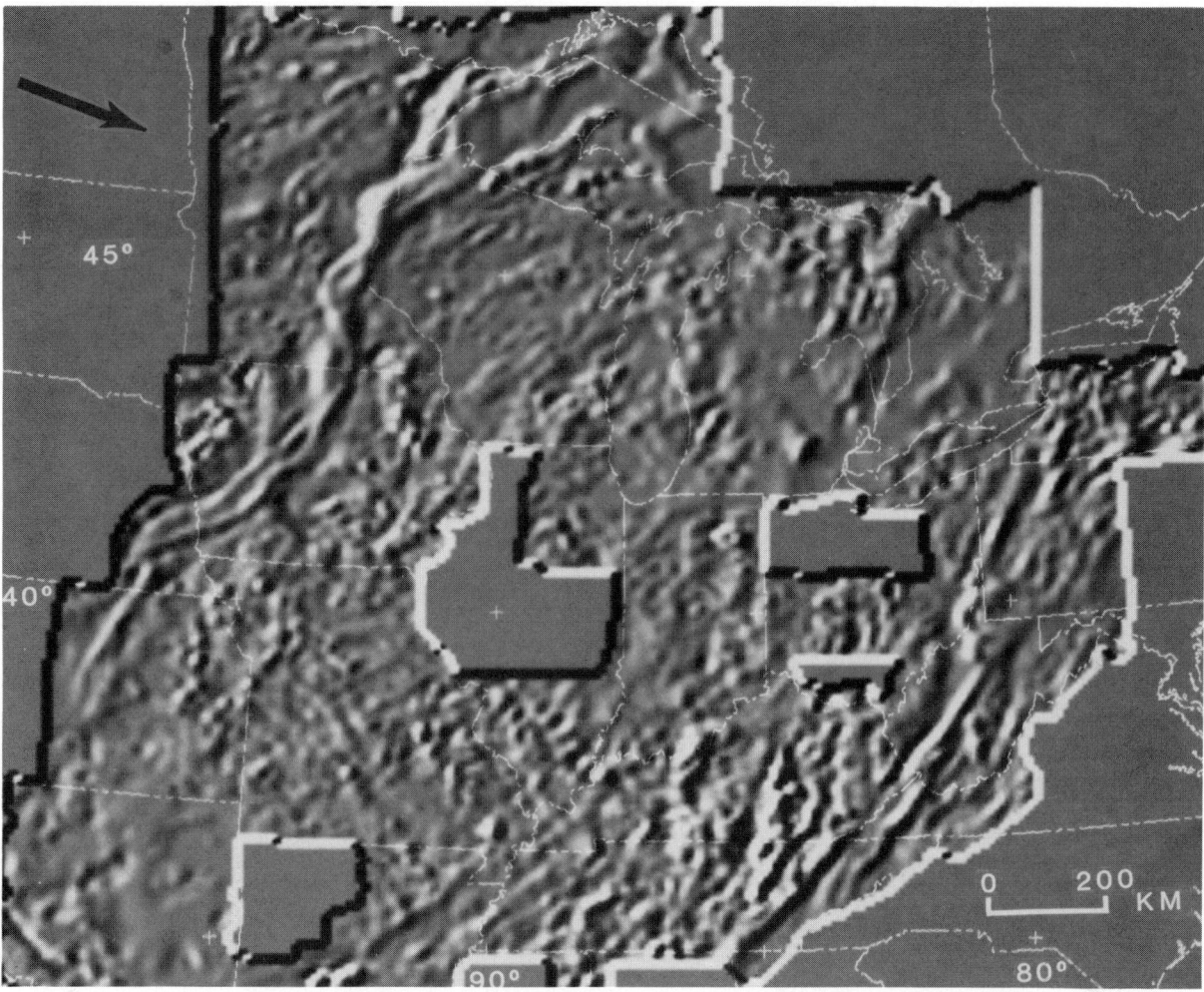

FIG. 7. Shaded-relief map of the magnetic field draped onto Precambrian basement (Figure 5). Sun elevation angle is 45 degrees, and illumination azimuth (shown by arrow) is E 20 degrees S.

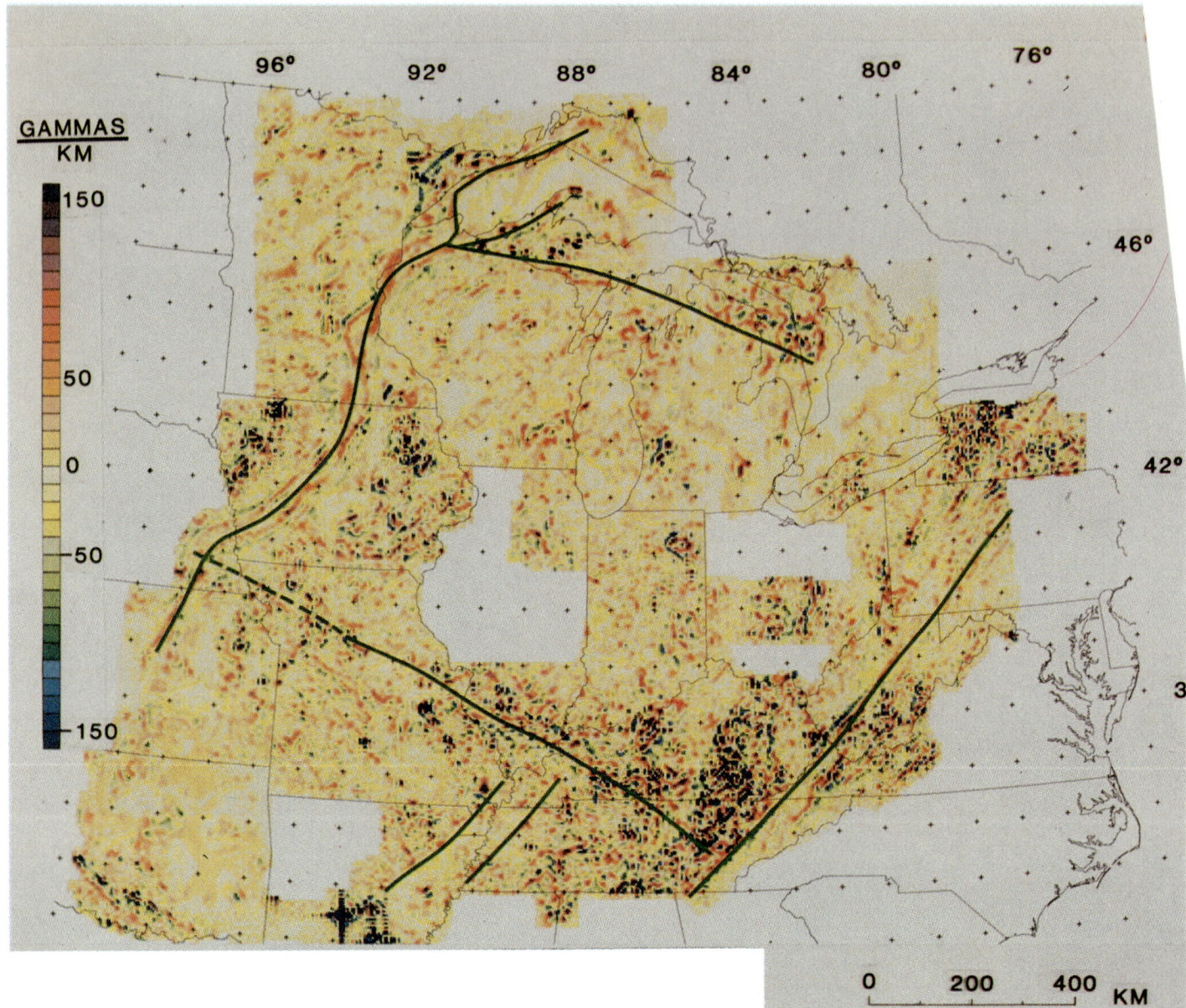

FIG. 8. Map showing the first vertical derivative of the magnetic field draped onto Precambrian basement (Figure 5). Color interval is 10 gammas/km. Heavy lines indicate major magnetic lineaments (dashed where less certain).

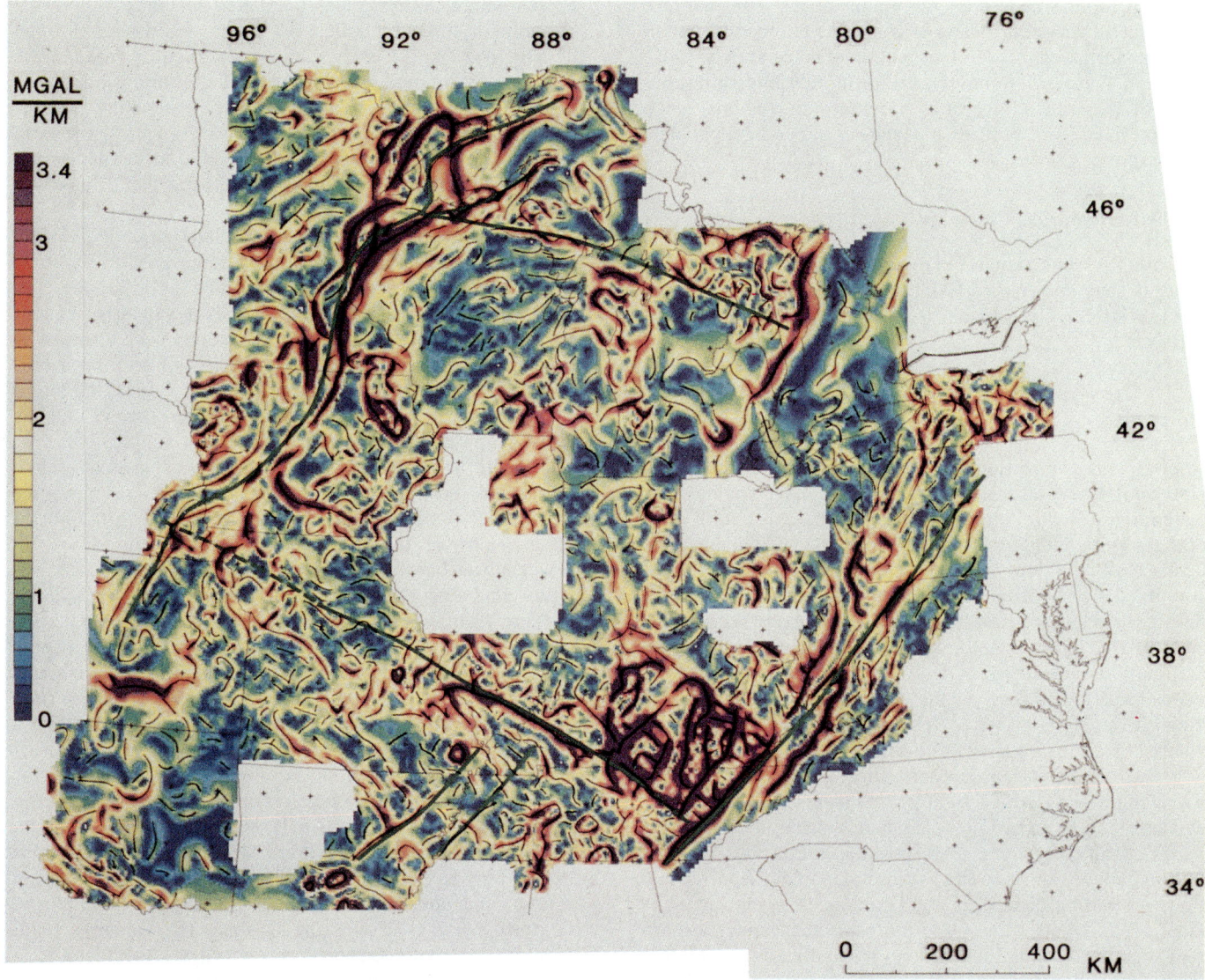

FIG. 9. Map showing horizontal-gradient magnitudes of the calculated pseudogravity field. Color interval is 0.1 mGal/km. Heavy lines indicate major magnetic lineaments (dashed where less certain). Narrow lines represent interpreted lithologic and structural boundaries.

ure 5. Although nearly all directions of the compass are represented, magnetic features trend predominantly northeast and northwest. Careful inspection of maps such as those shown in Figures 5–8 may be useful in determining boundaries of generalized tectonic provinces or regions, each characterized by structures of similar trend and geophysical characteristics.

Magnetization-density ratio

From Poisson's equation relating magnetic and gravity fields (Grant and West, 1965), the ratio of magnetization contrast to density contrast (Chandler et al., 1981) can be calculated:

$$\frac{\mathbf{I}}{\rho} = \frac{G\mathbf{M}}{\partial g/\partial z},$$

where $\mathbf{I}$ = magnetization contrast, ρ = density contrast, G = Universal Gravitational Constant, $\mathbf{M}$ = magnetic field reduced to the North Pole, and $\partial g/\partial z$ = first vertical derivative of the gravity field at the observational level of the magnetic field. This equation assumes that (1) a common isolated source of magnetization and density exists, (2) the ratio of density to magnetization for each causative body is constant, and (3) the direction of magnetization is uniform. These assumptions are stringent and are probably invalid within many localized regions in the Midcontinent. On the other hand, it was the purpose of this study to define large regions where the ratio remains fairly uniform. The magnetic and gravity fields were, therefore, filtered to remove short-wavelength (<64-km or 40-mile) anomalies related to near-surface, local sources. These localized sources result in large variations in ratio values over short horizontal distances and fail, in general, to obey the assumptions of Poisson's equation. As will be discussed, the resulting ratio map exhibits large areas where values are uniform, suggesting perhaps that Poisson's equation can be pragmatically applied in analyzing regional-scale anomalies.

Specific problems in analyzing regional-scale ratio anomalies arise and require discussion. For example, mantle and lower crustal sources at depths greater than the Curie point isotherm may be expressed in the gravity field but do not produce change in the magnetic field; thus the calculated ratio values in the area of these sources are meaningless. In addition, ambiguity arises because rock types with different densities and magnetizations may have the same ratio value. Nevertheless, the ratio-anomaly map of the Midcontinent exhibits regional patterns and correlations that seem worthwhile to point out.

For this study the magnetization vector was assumed to have a constant direction of 65 degrees N inclination and 0 degrees declination. The first vertical derivative of the gravity field used in the calculations is shown in Figure 10, and the resulting ratio map is shown in Figure 11. The large gaps between colors are regions where ratio values were determined to be unstable (that is, highly variable). In calculating the ratio, it was necessary to divide by the derivative of the gravity field. When the derivative approaches zero, ratio values became large and highly variable. A simple statistical scheme was used to define those regions where the ratio was unstable. A square window, 5 grid units or 40 km (25 miles) on a side, was passed through the data grid. Within each window, the median value of the ratio was determined and subtracted from the ratio value at the central point of the window. When the absolute value of this difference became large (>0.2 oersted/g/cm^3), the ratio value at the window's center was considered to vary significantly with respect to its neighboring ratio values and was removed from the data set.

Regions of large extent that are characterized by ratio anomalies of similar appearance are identifiable in Figure 11 and are shown as geophysical terranes 1–4. Because the boundaries of these terranes are not always clearly defined, the delineation of the interpreted geophysical terranes is somewhat arbitrary. Although terranes 1–4 may be subdivided into smaller areas of consistent ratio values, the intent in generating the ratio map was to define broad regions where geophysical terranes are somewhat similar.

Although uniformity of ratio values was used to define geophysical terranes 1–3, rock types and rock ages may vary within these terranes. For example, terrane 1 in the northwestern part of the map area coincides primarily with Keweenawan basalts associated with the Midcontinent rift system and underlying the Michigan basin (Hinze, 1963). Terrane 1, however, encompasses other basement rock types such as Lower Proterozoic igneous and metamorphic rocks and Middle Proterozoic sedimentary and igneous rocks in Wisconsin and Minnesota (Sims, 1976). These rock types apparently have mass and magnetization properties that produce ratio values similar to those of the Keweenawan basalts.

Geophysical terrane 2, which is characterized by low ratio values, lies within several small regions dispersed throughout the Midcontinent. In northern Missouri and northeastern Illinois, terrane 2 coincides roughly with the Mazatzal belt of metavolcanic and metasedimentary rocks (Van Schmus and Bickford, 1981). Coarse-grained granites (Kisvarsanyi, 1979) correlate well with terrane 2 in central and southern Missouri. Low ratio values near the intersection of the Mississippi Valley graben anomaly and the south-central magnetic lineament may reflect a major lithologic variation in basement.

High ratio values typify geophysical terrane 3. It is coincident with St. Francois batholithic rocks (primarily epizonal granite and rhyolitic ash-flow tuff) in southeastern Missouri and with basement rocks underlying the Appalachian basin along the eastern border of the map area. Mafic volcanics (Hinze et al., 1977) may be represented by high ratio values in eastern Tennessee and Kentucky and southern Ohio. Although lying within geophysical terrane 4, the New York–Alabama magnetic lineament is flanked by regions of high ratio values.

A significant part of the map is included in terrane 4, which is characterized by highly variable ratio values (that is, ratios that fail to define large regions of uniform values such as those in terranes 1–3). In the central and southern parts of the Midcontinent, terrane 4 may reflect the major belt of silicic volcanic rocks and related granitic plutons, which were described by Lidiak et al. (1966) as the buried basement of North America. Farther north, in central and northern Minnesota, terrane 4 is coincident with exposed Archean

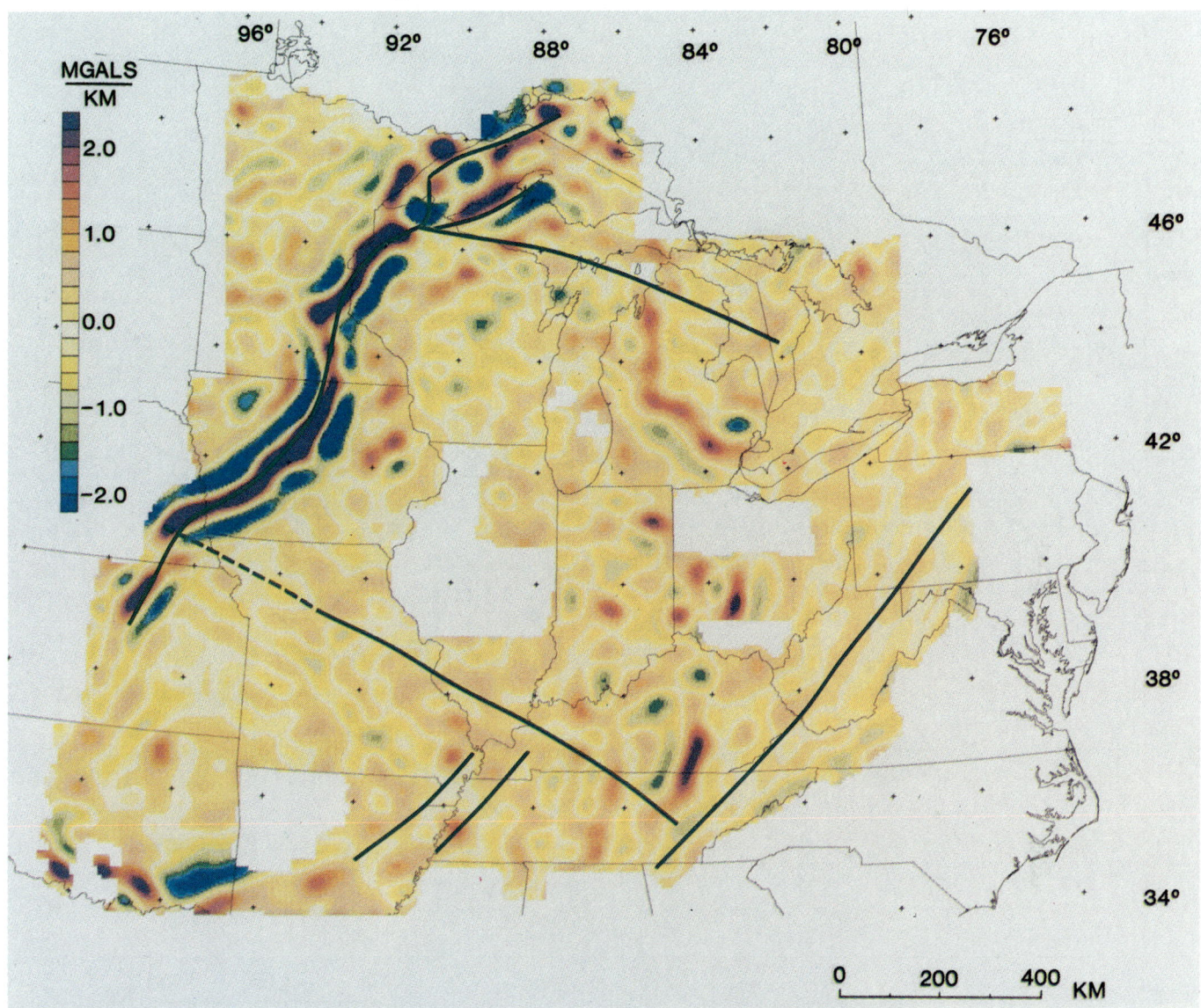

FIG. 10. Map showing first vertical derivative of the gravity field. The derivative gravity field has been upward continued to 1 000 ft (305 m) above ground level and filtered to remove wavelengths less than 64 km. Heavy lines indicate major magnetic lineaments (dashed where less certain).

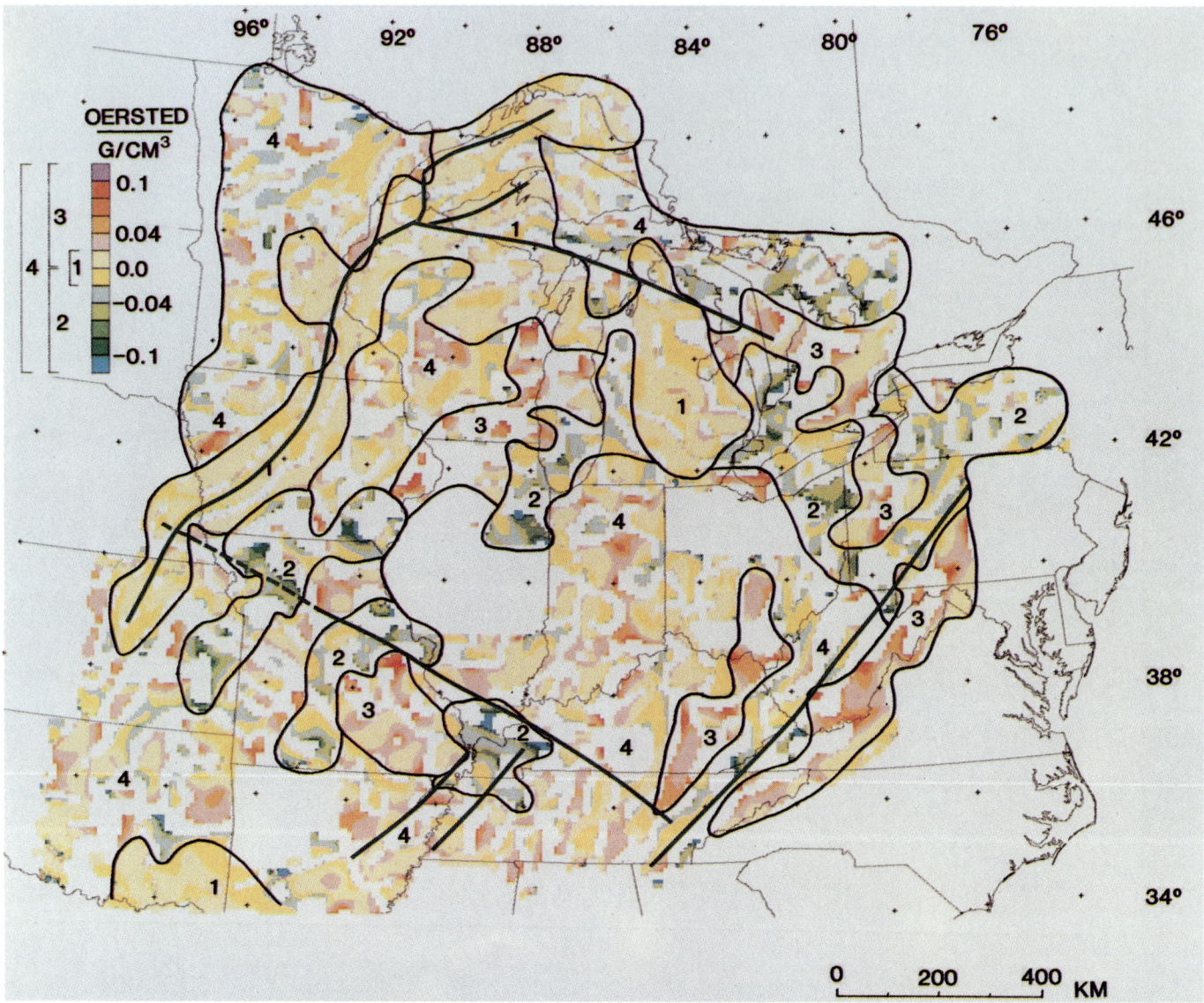

FIG. 11. Ratio (magnetization-contrast/density-contrast) map generated by using Poisson's equation. Color interval is 0.02 oersted/g/cm^3. Numbers 1–4 represent interpreted regions or geophysical terranes characterized by ratio anomalies of similar appearance. Range of ratio values for each geophysical terrane is indicated adjacent to the color scale. Heavy lines indicate major magnetic lineaments (dashed where less certain).

basement. An interesting correlation between rock types and ratio values occurs in this region. Archean granites correlate well with ratio lows, whereas Archean intermediate–mafic volcanic rocks appear to be associated with ratio highs. North of the Michigan basin, ratio anomalies are similar in appearance to those in northern Minnesota.

A close inspection of the ratio map and the geophysical terranes that it delineates may be helpful in understanding the extent of some major basement terranes within the Midcontinent region. For example, ratio anomalies within localized areas in Minnesota and Missouri suggest, in some instances, that geologic units can be extended to nearby regions where they are not mapped. On a regional scale, the map suggests that basement terranes trend, in general, roughly northeast and that the Keweenawan basalts underlying the Michigan basin form an axis of symmetry separating similar geophysical terranes.

SUMMARY

The magnetic-anomaly map of the Midcontinent region reveals the presence of numerous basement anomalies that may reflect major geologic features. The amplitudes and wavelengths of these anomalies vary considerably within the Midcontinent, suggesting that it has experienced a long and complex history that resulted in the formation of a variety of lithologies and structures. In general, magnetic features trend roughly northeast or northwest. The observed linearity of many magnetic features suggests that they formed from brittle failure of the crust as opposed to plastic flow. Magnetic anomalies correlate well with major Precambrian and Paleozoic tectonic features and aid in delineating the lateral extent of these features and associated structures.

The Midcontinent magnetic high and the New York–Alabama lineament are the most pronounced magnetic anomalies observed in the data. Two other megalineaments showing remarkable linearity are indicated on the map and geographically coincide with several structural and lithologic boundaries in the Midcontinent. Structures associated with these four megalineaments apparently influenced the evolution of some Phanerozoic tectonic features in the Midcontinent and consequently express major crustal discontinuities.

The availability of consistent digital data allowed the application of various analytical techniques that provided new information on lithologic and structural boundaries in the Midcontinent region. The selected analytical techniques that enhanced particular wavelengths or trends of anomalies were helpful in delineating features that were present on the unfiltered-anomaly map but not easily identifiable. Of particular interest are the magnetization–density ratio map (Figure 11) and the magnetic-lineament map (Figure 9) compiled by analyzing horizontal gradients of the pseudogravity field. These two maps contribute most in understanding the tectonic development of the Midcontinent. Pertinent research areas for future consideration are (1) calibration of the analytical techniques by carefully studying filtered magnetic data over known Precambrian geology in the Canadian shield and (2) correlation of lithologic and structural boundaries of the Midcontinent on a lineament map with the geophysical provinces delineated on a ratio map.

ACKNOWLEDGMENTS

I am especially grateful for the constructive criticisms of reviewers Lindrith Cordell, E. G. Lidiak, and Daniel H. Knepper, Jr. I thank Martin F. Kane, Don R. Mabey, and V.J.S. Grauch for their helpful suggestions. Don L. Sawatzky kindly furnished computer assistance in preparing the shaded-relief anomaly maps, and Robert P. Kucks provided valuable technical assistance.

REFERENCES

Ahbe, J. B., 1978, Southeastern Illinois magnetic anomaly map and its regional geologic interpretation: Purdue Univ., M.S. diss.

American Association of Petroleum Geologists and U.S. Geological Survey, 1962, Tectonic map of the United States: U.S. Geol. Surv., scale 1:2 500 000.

Barnov, V., 1957, A new method of interpretation of aeromagnetic maps—Pseudogravimetric anomalies: Geophysics, 22, 359–383.

Bayley, R. W., and Muehlberger, W. R., 1968, Basement rock map of the United States (exclusive of Alaska and Hawaii): U.S. Geol. Surv., scale 1:2 500 000, two sheets.

Bickford, M. E., Harrower, K. L., Hoppe, W. J., Nelson, B. K., Nussbaum, R. L., and Thomas, J. J., 1981, Rb–Sr and U–Pb and geochronology and distribution of rock types in the Precambrian basement of Missouri and Kansas: Geol. Soc. Am. Bull., pt. 1, 92, 323–341.

Braile, L. W., Keller, G. R., Hinze, W. J., and Lidiak, E. G., 1982, An ancient rift complex and its relation to contemporary seismicity in the New Madrid seismic zone: Tectonics, 1, 225–237.

Cain, J. C., Hendricks, S. J., Langel, R. A., and Hudson, W. V., 1967, A proposal model for the International Geomagnetic Reference Field—1965: J. Geomagn. Geoelect., 19, 335–355.

Chandler, V. W., Koski, J. S., Hinze, W. J., and Braile, L. W., 1981, Analysis of multisource gravity and magnetic anomaly data sets by moving-window application of Poisson's theorem: Geophysics, 46, 30–39.

Chase, C. G., and Gilmer, T. H., 1973, Precambrian plate tectonics—the midcontinent gravity high: Earth Planet. Sci. Lett., 21, 70–78.

Clegg, K. E., and Bradbury, J. C., 1956, Igneous intrusive rocks in Illinois and their economic significance: Rep. Invest. 197, Illinois Geol. Surv.

Coates, M. S., Haimson, B. C., Hinze, W. J., and Van Schmus, W. R., 1983, Introduction to the Illinois Deep Hole Project: J. Geophys. Res., 88, 7267–7275.

Erickson, R. L., and Blade, L. V., 1963, Geochemistry and petrology of the alkalic igneous complex at Magnet Cove, Arkansas: Prof. Pap. 425, U.S. Geol. Surv.

Gilliland, W. N., 1962, Possible continental continuation of the Mendocino fracture zone: Science, 137, 685–686.

Grant, F. S., and West, G. F., 1965, Interpretation theory in applied geophysics: McGraw-Hill.

Hayes, W. C., 1962, Configuration of the Precambrian surface showing major structural lineaments: Mo. Div. Geol. Surv. Water Resour., scale 1:1 000 000, 1 sheet.

Heyl, A. V., Jr., Brock, M. R., Jolly, J. L., and Wells, C. E., 1965, Regional structure of the southeast Missouri and Illinois–Kentucky mineral districts: Bull. 1202-B, U.S. Geol. Surv.

Hildenbrand, T. G., 1982, Model of the southeastern margin of the Mississippi Valley graben near Memphis, Tennessee, from interpretation of truck-magnetometer data: Geology, 10, 476–480.

Hildenbrand, T. G., Hendricks, J. D., and Kucks, R. P., 1981a, Aeromagnetic map of central and northeastern Arkansas: Open-File Rep. 81-758, U.S. Geol. Surv., scale 1:500 000.

Hildenbrand, T. G., Kane, M. F., and Hendricks, J. D., 1982, Magnetic basement in the upper Mississippi Embayment region—a preliminary report, in Pakiser, L., and McKeown, F. A., Eds., Investigations of the New Madrid, Missouri, earthquake region: Prof. Pap. 1236-E, U.S. Geol. Surv., 39–53.

Hildenbrand, T. G., Kane, M. F., and Stauder, W.[S.J.], 1977, Magnetic and gravity anomalies in the northern Mississippi Embayment and their spatial relation to seismicity: Misc. Field Stud. Map MF-914, U.S. Geol. Surv.

Hildenbrand, T. G., and Keller, G. R., 1983, Magnetic and gravity features of western Kentucky—their geologic significance: Open-File Rep. 83-164, U.S. Geol. Surv.

Hildenbrand, T. G., Kucks, R. P., and Johnson, R. W., Jr., 1981b,

Aeromagnetic map of east-central United States: Geophys. Invest. Map GP-948, U.S. Geol. Surv., scale 1:1 000 000.

Hildenbrand, T. G., Kucks, R. P., and Sweeney, R. E., 1983, Digital magnetic anomaly map of central United States—description of major features: Geophys. Invest. Map GP-955, U.S. Geol. Surv., scale 1:2 500 000.

Hinze, W. J., 1963, Regional gravity and magnetic anomaly maps of the southern peninsula of Michigan: Rep. Invest. 1, Mich. Geol. Surv. Div.

Hinze, W. J., Braile, L. W., Keller, G. R., and Lidiak, E. G., 1977, Tectonic overview of the central midcontinent: Rep. NUREG-0382; (RGA), U.S. Nucl. Regul. Comm.

Hoppe, W. J., Montgomery, C. W., and Van Schmus, W. R., 1983, Age and significance of Precambrian basement samples from northern Illinois and adjacent states: J. Geophys. Res., **88**, 7276–7286.

Jones, V. L., and Lyons, P. L., 1964, Vertical intensity magnetic map of Oklahoma: Map GM-6, Okla. Geol. Surv., scale 1:750 000.

Kane, M. F., Hildenbrand, T. G., and Hendricks, J. D., 1981, A model for the tectonic evolution of the Mississippi Embayment and its contemporary seismicity: Geology, **9**, 563–568.

Keller, G. R., Bland, A. E., and Greenberg, J. K., 1982, Evidence for a major late Precambrian tectonic event (rifting?) in the eastern midcontinent region, United States: Tectonics, **1**, 213–223.

Keller, G. R., Bryan, B. K., Bland, A. E., Greenberg, J. K., 1975, Possible Precambrian rifting in the southeast United States [abstr.]: Trans. Am. Geophys. Union, **56**, 602.

King, E. R., and Zietz, I., 1971, Aeromagnetic study of the midcontinent gravity high of central United States: Geol. Soc. Am. Bull., **82**, 2187–2207.

———1978, The New York–Alabama lineament—geophysical evidence for a major crustal break in the basement beneath the Appalachian Basin: Geology, **6**, 312–318.

Kisvarsanyi, E. B., 1974, Operation basement—buried Precambrian rocks of Missouri, their petrography and structure: AAPG Bull., **58**, 674–684.

———1979, Geologic map of the Precambrian of Missouri: Mo. Div. Geol. Land Surv., scale 1:1 000 000.

Koenig, J. B., 1956, The petrography of certain igneous dikes of Kentucky: ser. 9, Bull. 21, Ky. Geol. Surv.

Lidiak, E. G., Denison, R. E., Hinze, W. J., and Halpern, M., 1981, Precambrian rocks in the subsurface of Kentucky and Tennessee [abstr.]: Geol. Soc. Am. Abstr. Programs, **13**, 497.

Lidiak, E. G., Marvin, R. F., Thomas, H. H., and Bass, M. N., 1966, Geochronology of the midcontinent region, United States, Part 4—Eastern area: J. Geophys. Res., **71**, 5427–5438.

Lidiak, E. G., and Zietz, I., 1976, Interpretation of aeromagnetic anomalies between latitudes 37°N and 38°N in the eastern and central United States: Spec. Pap. 167, Geol. Soc. Am.

McCracken, M. H., 1971, Structural features of Missouri: Rep. Invest. 49, Mo. Geol. Surv. Water Resources.

Missouri Geological Survey and Water Resources, 1943, Magnetic map of Missouri showing anomalies of vertical intensity: scale 1 500 000.

Muehlberger, W. R., Denison, R. E., and Lidiak, E. G., 1967, Basement rocks in continental interior of United States: AAPG Bull., **51**, 2351–2380.

O'Hara, N. W., 1981, Geophysical and geological atlas of the Great Lakes region: Map Chart Ser. MC-41, Geol. Soc. Am., scale 1:2 500 000.

O'Hara, N. W., and Hinze, W. J., 1972, Basement geology of the Lake Michigan area from aeromagnetic studies: Geol. Soc. Am. Bull., **83**, 1771–1786.

———1980, Regional basement geology of Lake Huron: Geol. Soc. Am. Bull., pt. 1, **91**, 348–358.

Oray, E., Hinze, W. J., and O'Hara, N. W., 1973, Gravity and magnetic evidence for the eastern termination of the Lake Superior syncline: Geol. Soc. Am. Bull., **84**, 2763–2780.

Sims, P. K., 1976, Precambrian tectonics and mineral deposits, Lake Superior region: Econ. Geol., **71**, 1092–1118.

Sims, P. K., Card, K. D., Morey, G. B., and Peterman, Z. E., 1980, The Great Lakes tectonic zone—a major crustal structure in central North America: Geol. Soc. Am. Bull., pt. 1, **91**, 690–698.

Sims, P. K., and Peterman, Z. E., 1980, Geology and Rb–Sr age of lower Proterozoic granitic rocks, northern Wisconsin: Spec. Pap. 182, Geol. Soc. Am., 139–146.

———1981, Archean rocks in the southern part of the Canadian Shield—A review, in Glover, J. E., and Groves, D. E., Eds., Second International Archean Symposium, Perth, 1980: Spec. Publ. 7, Geol. Soc. Austr., 85–98.

Sims, P. K., and Zietz, I., 1967, Aeromagnetic and inferred Precambrian paleogeologic map of east-central Minnesota and part of Wisconsin: Geophys. Invest. Map GP-563, U.S. Geol. Surv.

Steiger, R. H., and Jager, E., 1978, Subcommission on Geochronology: Convention on use of decay constants in geochronology and cosmochronology, in Cohee, G. V., Glaessner, M. F., and Hedberg, H. G., Eds., Contributions to the geologic time scale: Stud. Geol. 6, AAPG, 67–71.

Thiel, Edward, 1956, Correlation of gravity anomalies with the Keweenawan geology of Wisconsin and Minnesota: Geol. Soc. Am. Bull., **67**, 1079–1100.

U.S. Geological Survey, 1973, Aeromagnetic map of southeastern Nebraska and parts of adjacent states: Open-File Rep., U.S. Geol. Surv., scale 1:250 000.

———1976, Aeromagnetic map of Iowa: Geophys. Invest. Map GP-910, U.S. Geol. Surv., scale 1:500 000.

Van Schmus, W. R., and Bickford, M. E., 1981, Proterozoic chronology and evolution of the midcontinent region, North America, in Kroner, A., Ed., Precambrian plate tectonics: Develop. Precambrian Geol. 4, Elsevier, 261–296.

Van Schmus, W. R., Card, K. D., and Harrower, K. L., 1975a, Geology and ages of buried Precambrian basement rocks, Manitoulin Island, Ontario: Can. J. Earth Sci., **12**, 1175–1189.

Van Schmus, W. R., Thurman, E. M., and Peterman, Z. E., 1975b, Geology and Rb–Sr chronology of Middle Precambrian rocks in eastern and central Wisconsin: Geol. Soc. Am. Bull., **86**, 1255–1265.

White, W. S., 1966, Tectonics of the Keweenawan basin, western Lake Superior region: Prof. Pap. 524-E, U.S. Geol. Surv.

Woollard, G. P., 1951, Annual report of the Special Committee on the Geophysical and Geological Study of Continents, 1950–1951: Trans. Am. Geophys. Union, **32**, 634–647.

———1958, Areas of tectonic activity in the United States as indicated by earthquake epicenters: Trans. Am. Geophys. Union, **39**, 1135–1150.

Yaghubpur, A., 1979, Preliminary geologic appraisal and economic aspects of the Precambrian basement of Iowa: Ph.D. diss., Univ. Iowa.

Yarger, H. L., Robertson, R. R., and Wentland, R. L., 1978, Aeromagnetic map of eastern Kansas: Open-File Rep., Kans. Geol. Surv., scale 1:500 000.

Zartman, R. E., 1977, Geochronology of some alkalic rock provinces in eastern and central United States: Earth Planet. Sci. Lett. Annu. Rev., **5**, 257–286.

Geologic interpretation of gravity and magnetic data for northern Michigan and Wisconsin

J. S. Klasner,* E. R. King,‡ and W. J. Jones‡

ABSTRACT

Three tectonic terranes, defined on geologic and geophysical evidence, are found in northern Michigan and Wisconsin. Each has its own characteristic gravity and magnetic signature. The northern terrane consists of basalt and sedimentary rocks of the Midcontinent rift system. Both gravity and magnetic maps have high-amplitude positive and negative anomalies in this area, reflecting the wide variation in physical rock properties. Both data sets have a strong northeast-trending fabric, which mimics the orientation of the major tectonic features of the rift.

The central terrane is underlain by rocks of the Penokean fold belt. The north half of this region consists largely of basins of Lower Proterozoic metasedimentary rocks, and the south half consists of volcanic rocks, granitoid plutons, and gneiss domes. The most prominent geophysical feature in the central terrane is a broad, long-wavelength gravity high along the fold belt. Shorter wavelength positive gravity and magnetic anomalies in the north part of the central terrane have a generally east-northeast-trending fabric that corresponds to troughs and basins in Archean rocks that are filled with iron-rich Lower Proterozoic sedimentary rocks. The south half of the central terrane is characterized by a gravity fabric showing numerous steep gradients relative to the north half, scattered orientations of elongate positive and negative anomalies, and a distinct magnetic signature. The south half is separated from the north half by sharp gradients in gravity and magnetic data and a geologically and geophysically mapped fault zone.

The southern terrane is underlain by the Wolf River batholith, which has a prominent north-northeast-oriented gravity low above it. Sharp, elliptically shaped magnetic highs and lows correlate with variations in rock types within the batholith. The strong north-northeast geophysical fabric in this region, which correlates with a north-northeast-trending zone of faulting on the west side of the batholith, distinguishes this terrane from that to the north. East–west–trending fabrics on either side of the gravity low likely reflect orientation of tectonic features associated with the Penokean fold belt.

Geologic features correlate well with anomalies on amplitude-filtered gravity maps. Early Proterozoic basins are within the area of the broad gravity high along the Penokean fold belt, but they are not large enough to account for the gravity high. Gneiss domes and granitic plutons, including the Wolf River batholith, produce gravity lows. Nodes of regional metamorphism are all within the regional gravity high but within zones of relatively lower gravity within the high.

A two-dimensional gravity model constructed across the broad gravity high suggests that it is caused by uplift of dense lower crustal rocks. But it may also be partly caused by the presence of relatively dense near-surface Archean gneiss beneath it.

Preliminary analysis of a combined gravity and magnetic profile across the Penokean fold belt suggests that geophysical data are compatible with a proposed rift, arc–continent–collision model for the origin of the Penokean fold belt.

INTRODUCTION

The study area (Figure 1) lies in northern Michigan and Wisconsin along the southern margin of the Canadian shield.

*Department of Geology, Western Illinois University, and U.S. Geological Survey, Macomb, IL 61455.
‡U.S. Geological Survey, National Center, Reston, VA 22092.

Much of it is underlain by Precambrian bedrock that ranges in age from roughly 1 000 m.y. or younger to more than 3 b.y. (see Morey et al., 1982), but younger Paleozoic bedrock covers the Precambrian rocks in the southern and eastern parts.

Parts of three major tectonic features of the north-central United States (Figure 1)—the Midcontinent rift system

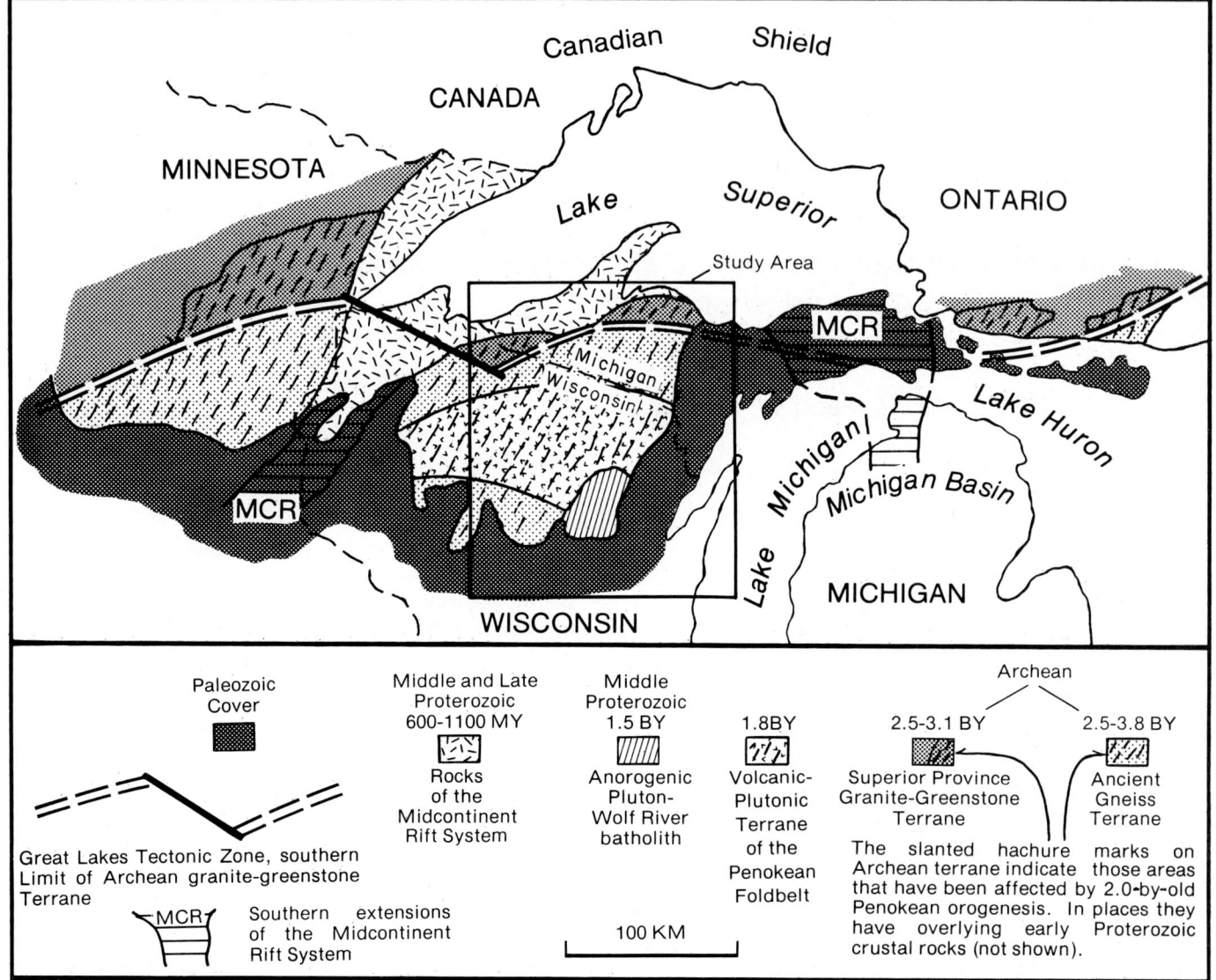

FIG. 1. Location and tectonic setting of the study area. All rocks older than 600 m.y. are part of the Canadian shield.

(MCR), the Penokean fold belt (PFB), and the Great Lakes tectonic zone (GLTZ), as well as a major pluton, the Wolf River batholith (WRB)—lie within the study area. Each of the tectonic features is separated from the other, at least in part, by location and by time of formation, so that they can be characterized as individual tectonic terranes. Thus, it seems logical that each terrane would have its own characteristic geophysical signature. In this paper we characterize the nature of the gravity and magnetic signature of each of the three terranes. Such a characterization should prove useful in extrapolating Precambrian geology into areas covered by Paleozoic rocks. We also examine the relationship between the gravity map and several individual geologic features such as Early Proterozoic basins, metamorphic aureoles, and granite plutons. In addition, we present a two-dimensional gravity model to explain the presence of the broad regional gravity high that lies along the Penokean fold belt. Finally, we briefly discuss the significance of the three terranes and associated geophysical signatures, especially as they apply to recently proposed plate-tectonic models for the origin of the Penokean fold belt.

TECTONIC SETTING AND GEOLOGY OF THE STUDY AREA

Each of the tectonic features that lies within the study area has a unique geologic history and a broad regional extent, as explained in the following paragraphs.

The MCR formed 1.1 b.y. ago (Van Schmus et al., 1982). In plan view it forms a concave-southward arch that extends for about 2 100 km from central Kansas to the Lake Superior region, and from there southward beneath the Michigan basin (a part of which is shown in Figure 1). Although not exposed along most of its length, its prominent gravity- and magnetic-anomaly signatures mark its position in the subsurface. Details of the geology and geophysics of the Lake

Superior basin portion of the MCR are given in a memoir on the Lake Superior basin edited by Wold and Hinze (1982).

The term Penokean fold belt (PFB) is used here in an informal sense to mean that east-trending belt of generally intensely folded sedimentary, metamorphic, and volcanic rocks and igneous intrusions that were formed during the Penokean orogeny of Goldich et al. (1961) and Cannon (1973). The term Penokean orogeny refers primarily to the western Lake Superior region, but Goldich et al. extended it to include a Lower Proterozoic (Huronian) terrane to the east in Ontario. It is used in that sense here. Tectonic activity associated with the Penokean orogeny peaked at about 1 800–1 900 m.y. ago in the western Lake Superior region, but it actually began well before 1 900 m.y. ago and spanned a long period of geologic time, including rhyolite-granite igneous activity in central Wisconsin at 1 760 m.y. and late-stage metamorphism at about 1 630 m.y. (Van Schmus, 1980).

Figure 1 shows that the PFB extends roughly 1 600 km from southern Ontario across northern Michigan and Wisconsin into Minnesota. A broad, regional gravity high lies along most of the length of the fold belt (Klasner and Bomke, 1977). Rock associated with the MCR cuts across the fold belt both east and west of the study area, and Paleozoic rocks of the Michigan basin cover the fold belt in northern Michigan and Wisconsin.

The GLTZ (Sims et al., 1980) lies along the PFB. It divides Archean crust beneath the fold belt into a granite–greenstone component on the north and a gneiss component on the south. Sims (1976) suggested that it controlled the tectonic evolution of the fold belt.

The WRB of north-central Wisconsin is one of several anorogenic plutons in the Great Lakes region. Studies by Van Schmus et al. (1975) indicated that it formed about 1.5 b.y. ago, and Silver et al. (1977) and Emslie (1978) suggested that it lies along a zone of anorogenic plutons that extends from the southwestern United States through the Great Lakes region to Canada. It is characterized by a prominent gravity low in central Wisconsin (Klasner et al., 1982).

Figure 2 shows the boundaries of the three different tectonic terranes in the study area (see also Klasner and Jones, 1983). The boundaries were drawn on the basis of variations in geology as well as variations in geophysical signature. They consist of (1) a northern terrane underlain by rocks mainly of the Midcontinent rift, (2) a central terrane underlain by rocks mainly of the PFB, and (3) a southern terrane underlain primarily by the Wolf River batholith and associated intrusive bodies. Each was formed under a different set of geologic conditions and at a different time in geologic history.

The Bedrock geology of each terrane and the physical properties that characterize the rocks in each terrane are shown in Figure 2 and in Tables 1 and 2, respectively.

Northern terrane

The northern terrane lies at the southern margin of the Lake Superior basin (Davidson, 1982) portion of the MCR. Inasmuch as the geology and geophysics of the basin have been discussed by Wold and Hinze (1982), they will not be covered in detail in this paper. It is important to point out, however, the existence of a prominent northeast-trending structural fabric in this area, as expressed by the orientation of the Keweenaw fault, basalt flows, and the Jacobsville trough, as shown in Figure 2.

Central terrane

The central terrane (Figure 2) consists of rocks of the Penokean fold belt. For discussion purposes and analysis of geophysical data, we have divided it into two parts. The north part is underlain mainly by basins and troughs of metasedimentary and metavolcanic rocks, which are unconformable above or in fault contact with Archean basement rocks. Archean basement is exposed in about one-third of the area. The south part consists of mainly poorly exposed mafic to felsic metavolcanic rocks, gneiss, and granitoid plutons. It is separated from the north part by a generally east-trending fault, the Niagara fault in the east (Figure 2).

Central terrane—north part.—Lower Proterozoic rocks in northern Michigan belong to the Marquette Range Supergroup (Cannon and Gair, 1970). They consist of thick sequences of slate, metamorphosed graywacke, quartzite, conglomerate, iron formation, mafic volcanic rocks, and dolomite. Primarily because of the presence of iron formation, iron-rich slates and graywackes, and mafic volcanic rocks, the basin fill has a high bulk density and high magnetic susceptibility relative to the underlying and adjacent Archean rocks (see Tables 1 and 2). In general, the Archean rocks are granitic to tonalitic (Sims, 1976). As mentioned above, the GLTZ divides the Archean upper crust into a granite–greenstone component to the north and a gneiss component to the south. Sims (1976) showed that the Great Lakes tectonic zone (GLTZ) also forms an important structural boundary along the Penokean fold belt (PFB). South of the zone, Lower Proterozoic supracrustal rocks lie within troughs and basins that have been infolded between uplifted fault blocks (Cannon, 1973) or between anatectically remobilized domes (Sims and Peterman, 1976) of Archean basement. Lower Proterozoic rocks are intensely folded (Cannon, 1973; Klasner, 1978) and regionally metamorphosed in nodal patterns (James, 1955) as shown in Figure 3. Block faulting of Archean basement also took place north of the GLTZ, but deformation of Lower Proterozoic rocks was not as intense in this region and no nodes of regional metamorphism are centered there, although some nodes overlap into the region.

The Early Proterozoic basins and troughs in this region form an east-trending belt that generally widens toward the east. Also, prominent features such as the Marquette trough and Iron River syncline are oriented east–west (Figure 2). Several other major structures, however, have trends that diverge from the east–west direction. These structures reflect the orientation of uplifted blocks of Archean basement (Cannon, 1973).

The depth of Lower Proterozoic troughs has been estimated at several places. Klasner and Cannon (1974) obtained depths of roughly 2.5 km for the Marquette trough and 1.6 km for the Republic trough from two-dimensional modeling of gravity data. Other unpublished depth estimates by Klas-

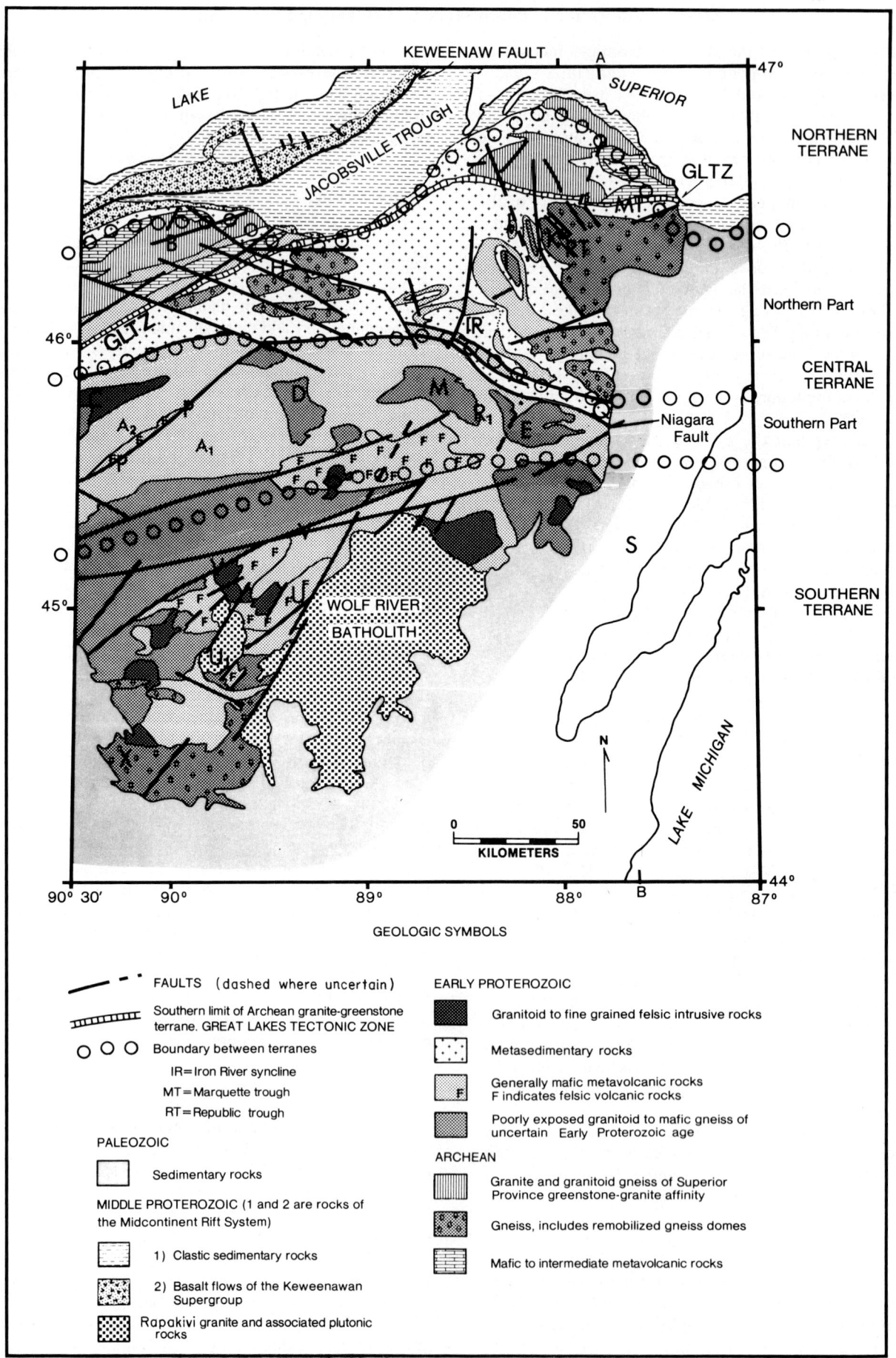

FIG. 2. Bedrock geology of the study area (modified from Morey et al., 1982). The area is divided into three tectonic terranes: northern, central, and southern, as discussed in the text. Bold letters explained in text. Profile A–B discussed in the text.

Table 1. Summary of geologic features and physical properties of rocks in the study area.

Terrane	Tectonic setting and major rock types	Comments on density and magnetic properties
Northern	*Midcontinent rift system*: 1.1 b.y. old. Mainly basalt and clastic sedimentary rocks such as sandstone, conglomerate, and shale.	*Basalt*: High density and strong magnetic susceptibility. Prominent remanent magnetization. *Sedimentary rocks*: Low density, low magnetic susceptibility.
Central	*Penokean fold belt*: 1.9 b.y. old. Divided into north and south parts. *North part*: Lower Proterozoic metasedimentary graywacke, slate, and iron formation with primarily gneissic Archean basement rock. *South part*: Predominantly mafic to felsic volcanic rock, areas of granitic to mafic gneiss, and granitoid plutons.	*Basins of metasedimentary rock*: Generally high bulk density because of iron formation, which has high magnetic susceptibility. *Archean basement*: Lower density and lower magnetic susceptibility than iron-rich Lower Proterozoic metasedimentary rocks. *Mafic volcanic rocks*: High density and magnetic susceptibility. Granite plutons and felsic volcanics have low density and susceptibility.
Southern	*Wolf River batholith*: 1.5 b.y. old. Mainly quartz monzonite pluton and associated alkalic plutons. Lower Proterozoic plutonic and volcanic rocks and Archean basement gneiss. Prominent fault zones of mylonitized rock.	*Plutonic rocks*: Generally low density, but appear to have a wide range of magnetic susceptibility with isolated zones of high susceptibility.

ner suggest that the Iron River syncline is about 4.8 km deep, as is the Early Proterozoic basin about 15 km northwest of the Iron River syncline. All depth estimates were made from two-dimensional modeling of gravity data.

Central terrane—south part.—In contrast to the sedimentary basins in the north half of the central terrane, Figure 2 shows that the south part is underlain by mostly mafic but also some areas of felsic volcanic rocks and gneiss of variable composition. Some of the granitoid plutons, or areas of uplifted granitic gneiss shown in Figure 2, were inferred from gravity and magnetic data. In fact, gravity and, especially, magnetic data played an important role in the

Table 2. Physical properties of stratigraphic units.

Age	Major rock types		Density (g/cm^3)	Magnetic susceptibility (cgs units)
Paleozoic	Sandstone, limestone, and shale		est. 2.5 (1)	nil
Middle Proterozoic	Keweenawan Supergroup	Sandstone, conglomerate, and shale	est. 2.5 (3)	nil
		Basalt (middle and lower)	2.92 (3)	$1.3–1.8 \times 10^{-3}$ (4) middle 3.0×10^{-3} (4) lower
	Quartz monzonite and syenitic plutons (Wolf River batholith)		2.67–2.80 (1)	$0–2.0 \times 10^{-3}$ (1)
Early Proterozoic	Granite and felsic volcanics		2.67 (1)	nil
	Metagraywacke and slate		2.57–2.84 (2)	nil
	Iron formation and iron-rich graywacke and slate		3.0–3.9	$0–385.0 \times 10^{-3}$ (5)
	Mafic metavolcanic rocks		avg. 2.92 (3)	$0.3–0.64 \times 10^{-4}$ (5) $0–0.75 \times 10^{-3}$ (6)
	Granitic gneiss, migmatite		2.5–2.8 (3)	nil
Archean	Granitic to mafic gneiss		2.5–3.33 (3)	$0.1–0.5 \times 10^{-4}$ (5) granitic $0–1.4 \times 10^{-3}$ (6) mafic

Sources: (1) Allingham and Bates (1961); (2) Klasner and Cannon (1974); (3) Klasner et al. (1979); (4) King (1975); (5) Case and Gair (1965); (6) Hall (1968).

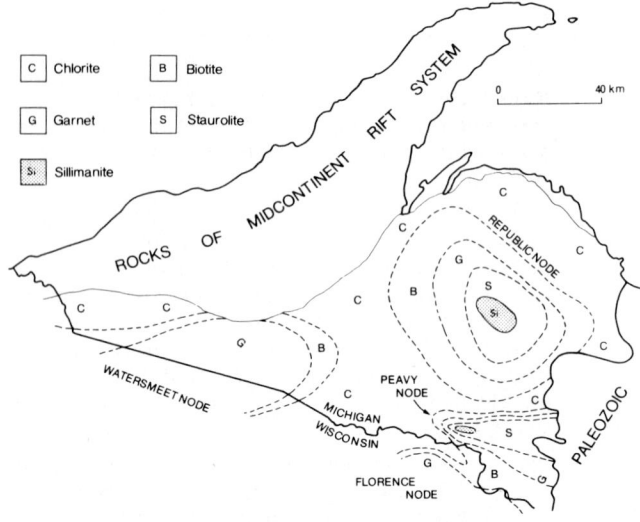

FIG. 3. Location of nodes of regional metamorphism in study area (from James, 1955).

preparation of the geologic map in northern Wisconsin because of the thick glacial cover in the region. In general, areas underlain by mafic volcanic rocks or mafic gneiss have large-amplitude positive gravity and magnetic signatures. Granite plutons and areas of granite gneiss have large-amplitude negative gravity and magnetic signatures.

Like the northern part of the PFB, the overall orientation of this zone is east–west, and most major faults in this part are oriented east–northeast. But the geologic fabric does not seem as well defined here as in the northern part. Several irregularly shaped areas of the foliated gneiss, with associated low-gravity anomalies (see anomalies D, E, and M in Figures 2 and 4), have more divergent orientations.

Southern terrane

This region is dominated by the elliptically shaped 1.5-b.y.-old Wolf River batholith and associated plutonic rocks with a north-northeast-trending long axis (Figure 2). The batholith is a composite feature consisting of several lithologic units of different ages (Van Schmus et al., 1975; Van Schmus, 1980). The dominant lithology is rapakivi-textured quartz monzonite, but it also includes granite, syenite, and isolated masses of anorthosite called the Tigerton anorthosite (Weiss, 1965; Van Schmus et al., 1975). The batholith intrudes Archean gneiss, Lower Proterozoic gneiss, and felsic and mafic volcanic rocks, as shown in Figure 2. Two smaller alkalic plutons, the Wausau and Stettin plutons (U_1 in Figure 2), lie about 10 km west of the major batholith.

A prominent north-northeast-trending zone of intensely faulted terrane, characterized by mylonitized rock, lies along the west edge of the Wolf River batholith (LaBerge, 1977). Remnants of Archean gneiss (Van Schmus and Anderson, 1977) and Lower Proterozoic volcanic rocks occur within this fault zone. Discontinuities in gravity and magnetic anomalies along the northeast projection of the fault suggest that it extends in that direction, along the trend of the proposed Trans Superior tectonic zone of Klasner et al. (1982).

GRAVITY AND MAGNETIC DATA

The Bouguer gravity-anomaly map (Figure 4) was compiled from 35 145 gravity stations. Specific parameters and sources of data are given in compilations by Ervin and Hammer (1974) for Wisconsin and Klasner et al. (1979) for the northern Michigan–Lake Superior region. Spacing of the gravity stations is irregular (Figure 5), ranging from less than 30 m to about 20 km. The Bouguer gravity-anomaly map (Figure 4) was hand contoured from the irregularly spaced data shown in Figure 5. Although it is slightly different than the *Gravity Anomaly Map of the United States* (Lyons and O'Hara, 1982, and this volume), which was machine contoured from computer-gridded data, the general configuration and amplitude of gravity anomalies are the same on both maps.

Gravity data were digitally filtered using a computer routine written by Hildenbrand (1979) to aid in differentiating the area into the three tectonic terranes and to emphasize the relationship between geology and the gravity signature. In this routine the data are first gridded, then transformed into the frequency domain using the fast Fourier transform, filtered in the frequency domain, and then transformed back into the space domain to produce the filtered maps. Two filtered maps were prepared: (1) a low-pass map that accepts wavelengths longer than 80 km and rejects shorter wavelengths (Figure 6a), and (2) a high-pass map that accepts wavelengths shorter than 80 km and rejects longer wavelengths (Figure 6b).

The magnetic map for the study area (Figure 7) is taken from the *Composite Magnetic Anomaly Map of the United States* (Zietz, 1982; Hinze and Zietz, this volume). This gray-tone shaded version is included in this paper for discussion purposes. Compilation parameters are discussed by Hinze and Zietz (this volume).

On the composite United States magnetic-anomaly map, as shown in Figure 7, there is an abrupt upward shift in level from north to south along the Michigan–Wisconsin border. This shift in magnetic level is suspect, because the U.S. magnetic map was prepared by tying together separate magnetic data sets from Wisconsin and Michigan. To check whether or not the shift in magnetic-anomaly level is real, we examined a magnetic profile obtained by the U.S. Naval Oceanographic Office's (NOO) Project MAGNET (see Sexton et al., 1982, for a report on these data). Although not corrected for diurnal variations, the NOO profile (Figure 8) provides continuous coverage from northern Michigan to south-central Wisconsin, along meridian 89° 40′ W. The position of the NOO profile is shown in Figure 7, and the corresponding magnetic profile from the U.S. magnetic map and Figure 7 are plotted along with the NOO profile in Figure 8. We find that at this meridian the profile from the U.S. magnetic map has no change in magnetic level at the border between Michigan and Wisconsin, although there is an abrupt north-to-south drop in magnetic level and change in character of the magnetic signature at the border on the NOO profile. This suggests that the tie between Michigan and Wisconsin magnetic data sets may be incorrect. How-

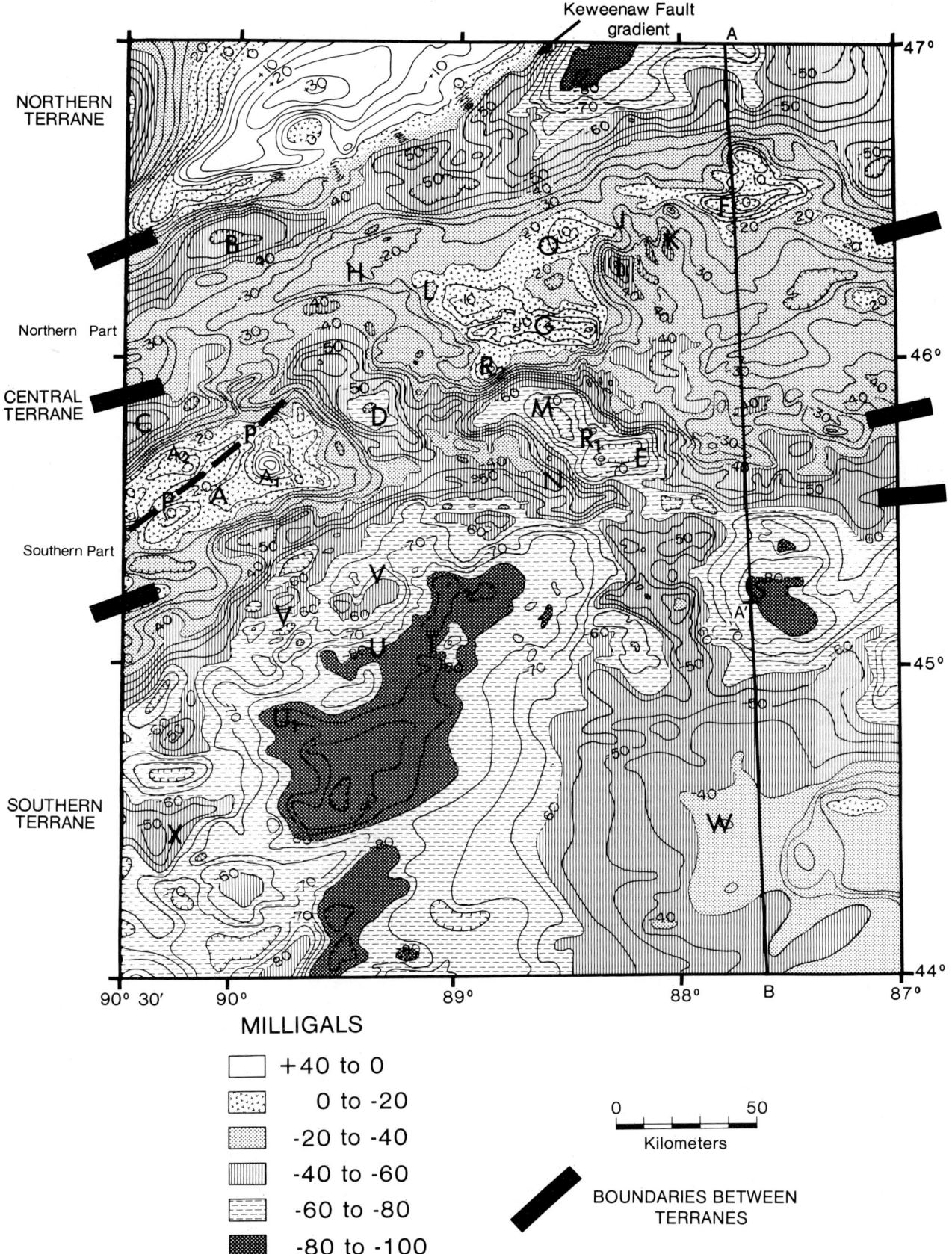

FIG. 4. Bouguer gravity-anomaly map of study area. Hand contoured, using the computer-contoured map as a guide. Countour interval, 5 mGal. Bold-lettered anomalies are discussed in text. Profile A–A' marks the location of the gravity model of Figure 10, and A–B marks the location of the profile shown in Figure 11.

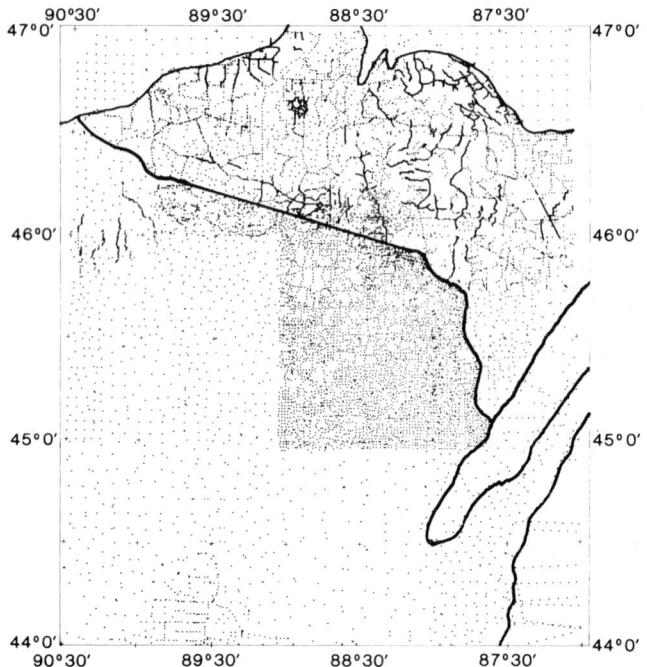

FIG. 5. Distribution of gravity stations used to prepare the Bouguer gravity-anomaly map of study area.

ever, there are no continuous magnetic profiles that lie in the Michigan–Wisconsin data sets in the eastern part of the study area, where the shift in level of the magnetic field is most pronounced. Therefore, it seems premature to conclude that the tie between Michigan and Wisconsin magnetic sets is wrong in its entirety on the basis of only one profile, but the tie should be checked. Other data, such as the high-pass gravity map (Figure 6b) and geologic mapping, indicate that there is indeed a structural and lithologic discontinuity along the position of the north-to-south shift in magnetic level (Figure 7), as discussed later in this paper.

CHARACTERIZATION OF GRAVITY AND MAGNETIC DATA BY TERRANE

General

Gravity maps reflect horizontal variations in crustal and possibly mantle densities. Many anomalies can be correlated directly with geologic features observed in Precambrian bedrock, herein called basement, but many others do not correspond to known basement geology and must be caused by density variations deep within the crust. Separation of the Bouguer anomalies into two wavelength components aids in determining which anomalies in the study area have near-surface sources and which may have deeper sources. As discussed below, the low-pass-filtered gravity map reflects density variations that may lie within the deep crust, or it may reflect broad-scale variation of near-surface basement rocks; whereas the short-wavelength map generally reflects near-surface density variations.

The magnetic map (Figure 7) provides an excellent picture of the Precambrian bedrock geology of the study area. The crystalline rocks of the Precambrian basement, which are largely concealed beneath a cover of glacial drift or Paleozoic sedimentary rocks, contain many strongly magnetic units. As mentioned above, correlation of the magnetic units with geologic data from available outcrops and with gravity data has been the basis of much of the mapping of bedrock geology in this area.

The major part of the magnetic patterns on the map (Figure 7) are produced by rocks in the upper part of the crystalline basement in contrast to the gravity data, which have a lower rate of attenuation and reflect deeper crustal structure. Therefore, the magnetic map shows the closest correlation with the high-pass-filtered gravity data (Figure 6b), which are related primarily to shallow density contrasts. The variable thicknesses of overlying glacial and sedimentary cover are virtually nonmagnetic and result in the slightly smoother character of the magnetic patterns in areas covered by Paleozoic rocks.

In all three terranes, magnetic anomalies are produced by a combination of the magnetization induced by the present field of the Earth and a remanent magnetization acquired when the rocks cooled below the temperature of the Curie point of the constituent magnetic minerals in the crystalline basement. In some cases this remanent magnetization is much larger than the induced magnetization and may have an orientation at a large angle to it or even in the opposite direction, depending on (1) the direction of the Earth's field at the time the magnetization was acquired, and (2) the tilting of the host rock in subsequent deformation.

Remanent magnetization for the Keweenawan basalts has been extensively studied, particularly by DuBois (1962), Books (1968, 1972), and Halls and Pesonen (1982) in the Lake Superior region. This magnetization has an average declination of 289 degrees and an average inclination of 35 degrees for the middle and upper Keweenawan flows and a reversed direction with an average declination of 47 degrees and an inclination of 71 degrees for most of the lower Keweenawan flows. This reversed magnetization was observed in the Baraga County dikes, an east-trending swarm of dikes that cut the older Precambrian terrane (DuBois, 1962) of northern Michigan and Wisconsin. These dikes produce characteristic narrow magnetic lows, many with amplitudes of 300 gammas or more, that can be traced for many miles along strike (Case and Gair, 1965; Meshref and Hinze, 1970). Most of these anomalies are too small to show on the regional magnetic map but are conspicuous on detailed aeromagnetic maps. They are observed as far south as the Wolf River batholith at 45° N and are probably present farther south but are difficult to detect under the thickening cover of Paleozoic sedimentary rock.

Many of the older Precambrian rocks also have strong remanent magnetizations, particularly some of the magnetic slates of the Marquette Range Supergroup, which produce narrow, linear, very high-amplitude anomalies that are far higher than could be produced solely by the induced magnetization. The magnetic properties of some of these rocks have been measured in parts of the northern peninsula of Michigan (Case and Gair, 1965).

The gravity and magnetic maps can be divided into three terranes, which correspond to the three tectonic terranes of Figure 2. They reflect the wide variation in physical proper-

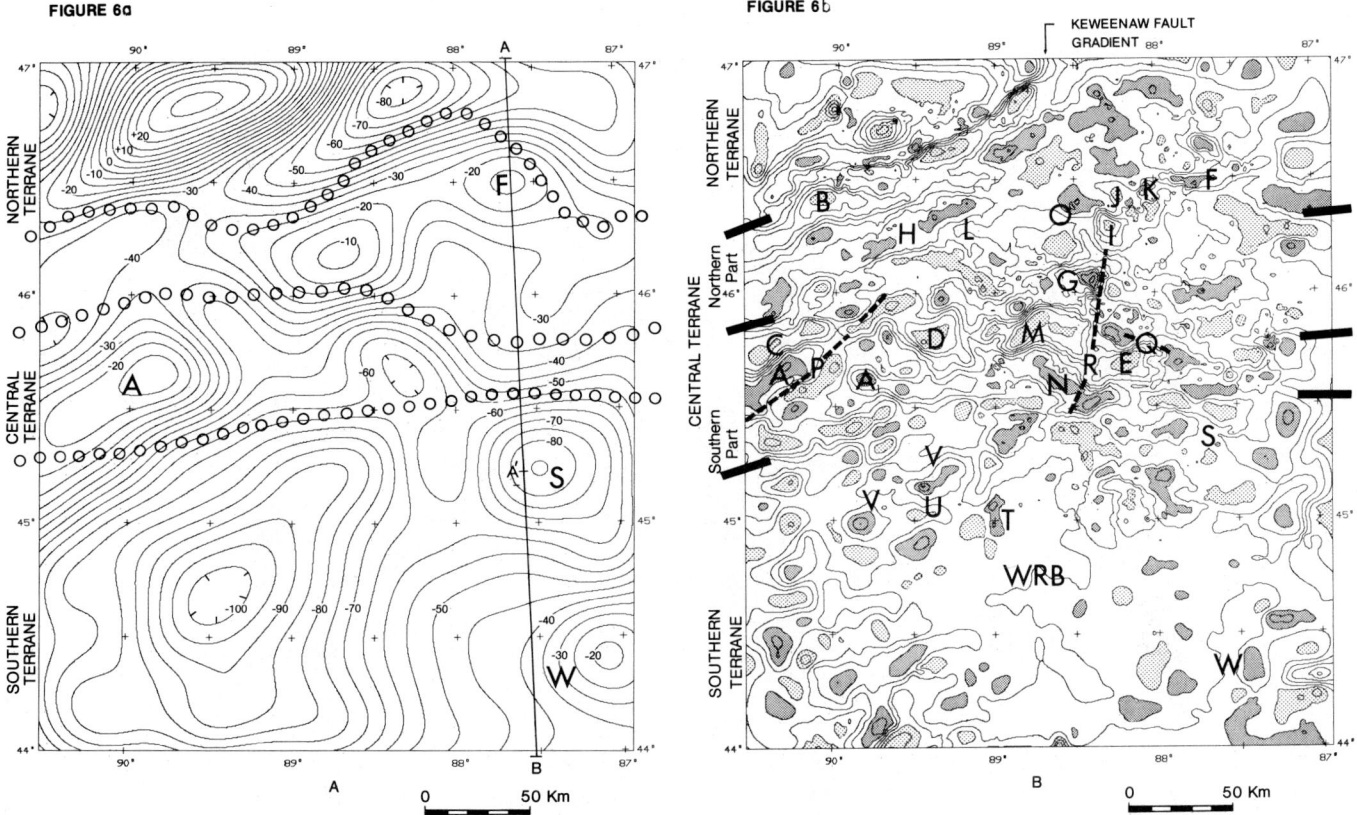

FIG. 6a. Low-pass filtered gravity map showing wavelengths longer than 80 km. Contour interval, 5 mGal. Profile A–A' marks the location of gravity model shown in Figure 10, and A–B, location of profile in Figure 11. Bold letters are explained in text.
 b. High-pass filtered gravity map showing wavelengths shorter than 80 km. Contour interval, 5 mGal. Contour values not shown in most places, but some gravity anomalies are highlighted by shading. Positive anomalies are darker shade, negative anomalies are lighter shade. Bold letters are explained in text. WRB refers to Wolf River batholith.

ties of the Precambrian crust in the study area. Table 2 provides some specific data on the physical properties of the rocks as they pertain to the magnetic and gravity data.

Northern terrane—gravity and magnetics

The most prominent features on both the gravity and magnetic maps in this terrane are the large-amplitude (both positive and negative) anomalies that trend in an east-northeast direction. Detailed interpretation of most of the geophysical features in this region have been covered by others. For example, Meshref and Hinze (1970) and King (in press) have interpreted magnetic data over parts of the area. Also, gravity interpretations were made for this area by Klasner and Jones (1979) and Miller (1966). A combined interpretation of both gravity and magnetic data was recently done by Hinze et al. (1982) for the Lake Superior basin. For this reason, the gravity and magnetic signatures of the northern terrane will not be discussed here other than to point out the prominent northeast-trending fabric in gravity and magnetic signatures that are related to an assemblage of MCR-associated geologic features that are distinctly different from the rest of the map. Most striking is the large variation in amplitude on both the gravity and magnetic maps on either side of the Keweenaw fault (Figures 4 and 7).

Central terrane—gravity

For discussion purposes, several geologic and geophysical features have been labeled in bold letters on the accompanying maps. The labels are the same for all of the maps, but not all of the maps have all of the labels. Figure 6a, for example, has features labeled A, F, S, W, whereas other maps have more labels.

The most prominent gravity feature of the central terrane is the broad, regional gravity-anomaly high (Figures 4 and 6a) that lies along the Penokean fold belt (PFB). On the low-pass map (Figure 6a), the broad high actually consists of three elliptical highs separated by two northwest-trending lows. As discussed above, the central terrane is divided into a north and south part. Mostly the broad gravity high encompasses the Early Proterozoic basins of the north part, but it passes into the south part (A in Figure 6a) near the west edge of the study area. This suggests that the broad gravity high is not caused by density variations in near-surface geology but rather by density variations in the deeper part of the crust.

In the east, the Great Lakes tectonic zone (GLTZ) lies just north of the peak of this regional gravity anomaly, but it diverges from it near the west edge of the map.

Figures 9a, b, and c illustrate the relationship between the

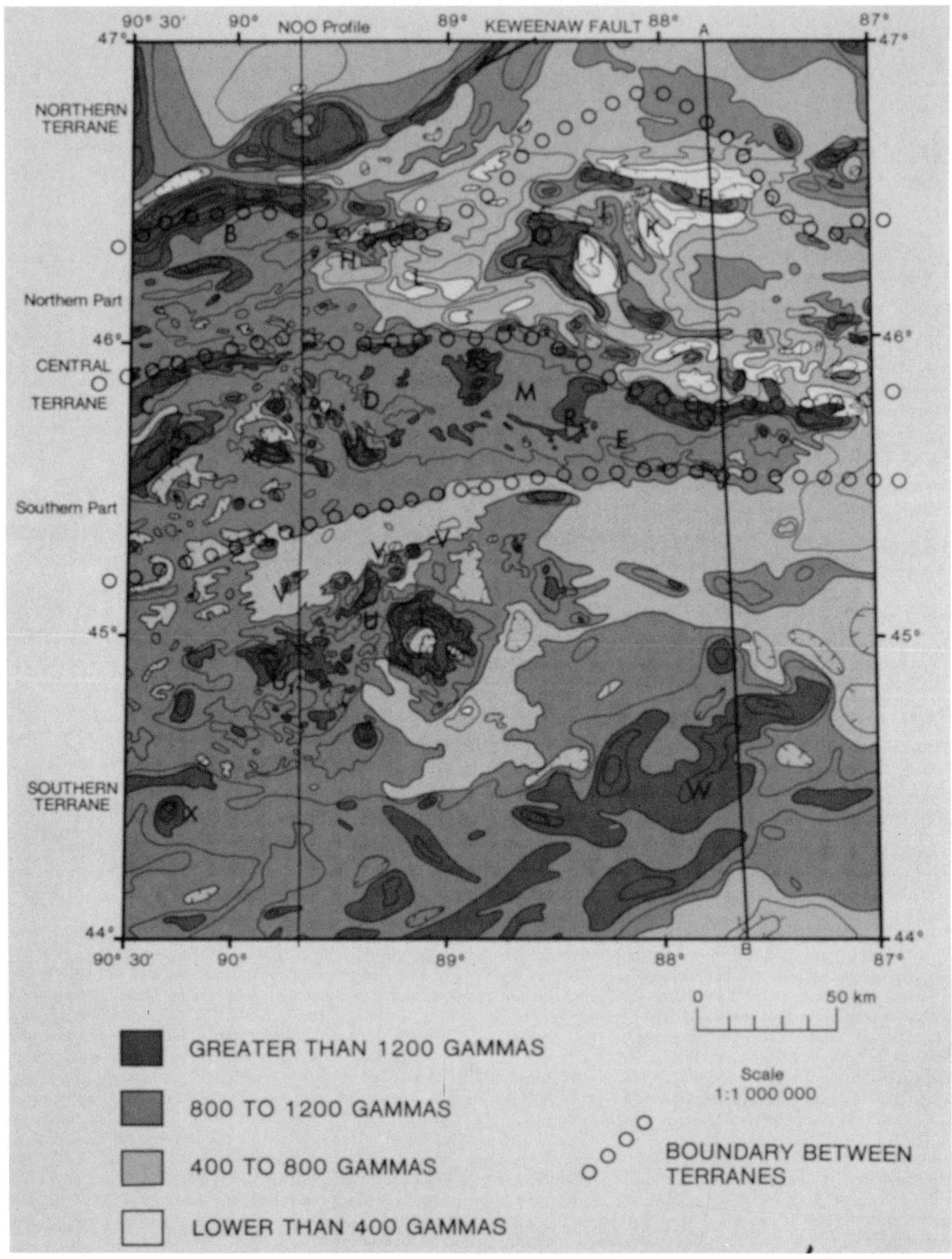

FIG. 7. Composite magnetic-anomaly map of the study area. Data for the map are from Zietz (1982). Contour interval, 200 gammas (nT). Values are in hundreds relative to an arbitary zero datum. Closed lows are indicated by hachures. Lettered anomalies are discussed in text. Location of project MAGNET NOO profile shown. Profile A–B discussed in text.

gravity anomalies and some major geologic features of the PFB. These figures are amplitude-filtered gravity maps, prepared by accepting only those gravity values that are greater than −40 mGal (Figures 9a and b) or less than −40 mGal (Figure 9c). The broad zone of gravity maxima that trend northeast across Figures 9a and 9b corresponds with the broad, regional gravity high mentioned in the previous paragraph. Figure 9a shows that, for the most part, Early Proterozoic basins of the PFB lie above the broad zone of high gravity, but the basins themselves cannot account for the (approximately 100-km-wide) regional gravity high beneath the fold belt, especially as seen on the low-pass map (Figure 6a). The basins likely extend farther east beneath the Paleozoic rocks (see Figure 2). That part of the regional gravity high at A in the west-central part of the area (Figures 9a and b) corresponds with terrane shown as mafic volcanic rock on the geologic map (Figure 2). There are relatively few bedrock exposures in this part of central Wisconsin, and the geologic map has been prepared mostly from magnetic and gravity data. As a result, the exact nature of bedrock in this area is not known.

Outside of the central terrane of the Penokean fold belt, the high gravity values in the northwestern corner of Figure 9a can be explained by subsurface bodies of dense Keweenawan basalt. Also, the cause of the high gravity values in the southeast corner of the map (W in Figures 9a and b) is not known.

James (1955) suggested that the Republic node of regional metamorphism (Figure 3) is caused by a subjacent pluton. If this is correct, then the gravity field may reflect it. Figure 9b shows the relationship between the gravity field and the locations of metamorphic nodes. All of the nodes are over the broad, long-wavelength gravity high. Within this zone of high gravity-anomaly values, however, the nodes lie in areas that have locally low gravity-anomaly values, and there are few areas lower than −35 mGal that do not have a metamorphic node in them. For example, both the Florence and Peavy nodes lie below the −35 mGal contour; the Republic node has a zone of low gravity, in part, beneath it; and most of the Watersmeet node lies outside the area of the −35 mGal gravity anomaly. This tends to confirm James's hypothesis that subjacent plutons exist beneath the nodes. Alternately, the nodes may be underlain by anatectically remobilized gneiss domes, as noted by Sims (1976) for the Watersmeet node. In either case, the rocks beneath the nodes appear to have low densities relative to the enclosing country rock.

Figure 9c illustrates that prominent gravity-anomaly lows in the central terrane are underlain by relatively low-density granitoid rocks. The low at B is associated with the Archean Puritan quartz-monzonite pluton of Schmidt (1976) and Sims et al. (1977). Scattered outcrops in the area of C consist of 1 760-m.y.-old biotite granite (P. K. Sims, personal communication, 1983). Rocks at D are also inferred to be granitic from gravity and magnetic data. The gravity low at E overlies granitoid rocks of the Dunbar gneiss dome of Sims et al. (1982). This gravity low was first mapped by Carlson (1972).

Other features outside the central terrane are also found in Figure 9c. To the north, the Jacobsville trough, a clastic basin of the northern terrane, has an associated gravity low.

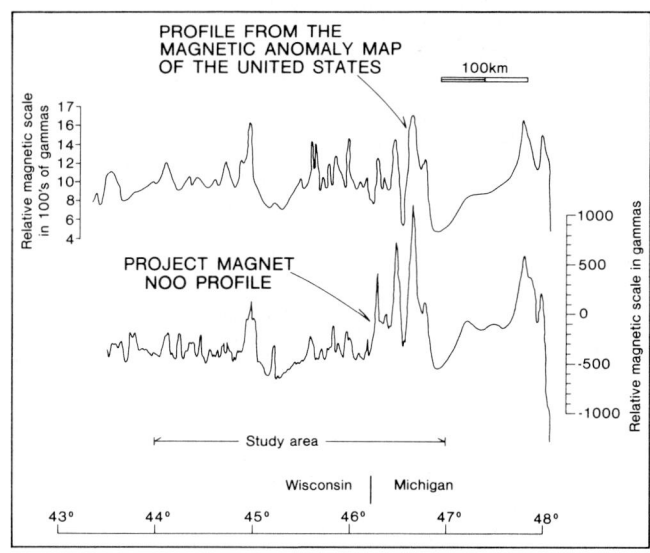

FIG. 8. Magnetic profiles along meridian 89°40′ W. Location of profiles shown in Figure 7. This figure compares magnetic data from the composite magnetic-anomaly map and a profile from the Naval Oceanographic Office (NOO) project MAGNET. See text for discussion.

The low in the northwest corner of the map also lies above a Keweenawan-age clastic basin. Further details concerning the origin of these low-gravity anomalies are discussed in Hinze et al. (1982). The broad area of low gravity to the south is discussed under the section on the southern terrane.

In the north part of the central terrane, the Bouguer gravity map (Figure 4) and high-pass map (Figure 6b) have a subdued, but nevertheless apparent, northeast-to-east-trending fabric parallel to the orientation of the PFB. It is best seen in the orientation of the long axes of gravity highs, such as F, O, and H (Figures 4 and 6b), that are caused by basins and troughs of dense Lower Proterozoic rocks. Other divergent trends, however, such as gravity lows I, J, and K, are caused by uplifted blocks of Archean gneiss. These correspond to first-order divergent basement structures pointed out by Cannon (1973, p. 258). The gravity low at L lies above a remobilized gneiss dome of Sims and Peterman (1976).

In general, the gravity signature over the north part of the central terrane has a strong east-trending fabric primarily expressed by a long-wavelength gravity high and several shorter wavelength highs that lie above basins of Lower Proterozoic rock. Divergent trends in short-wavelength anomalies correspond to some basement uplifts.

There is an abrupt change in character of gravity data at the boundary between the north and south portions of the central terrane. This change is most apparent on the high-pass gravity map (Figure 6b). The elongate, east-northeast-trending fabric of the north half changes abruptly to a less continuous fabric in the south half. This change in character corresponds to the apparent shift in level on the magnetic map (Figure 7). These changes in geophysical character, together with geologic data, provide independent evidence for the boundary between the north and south half of the central terrane.

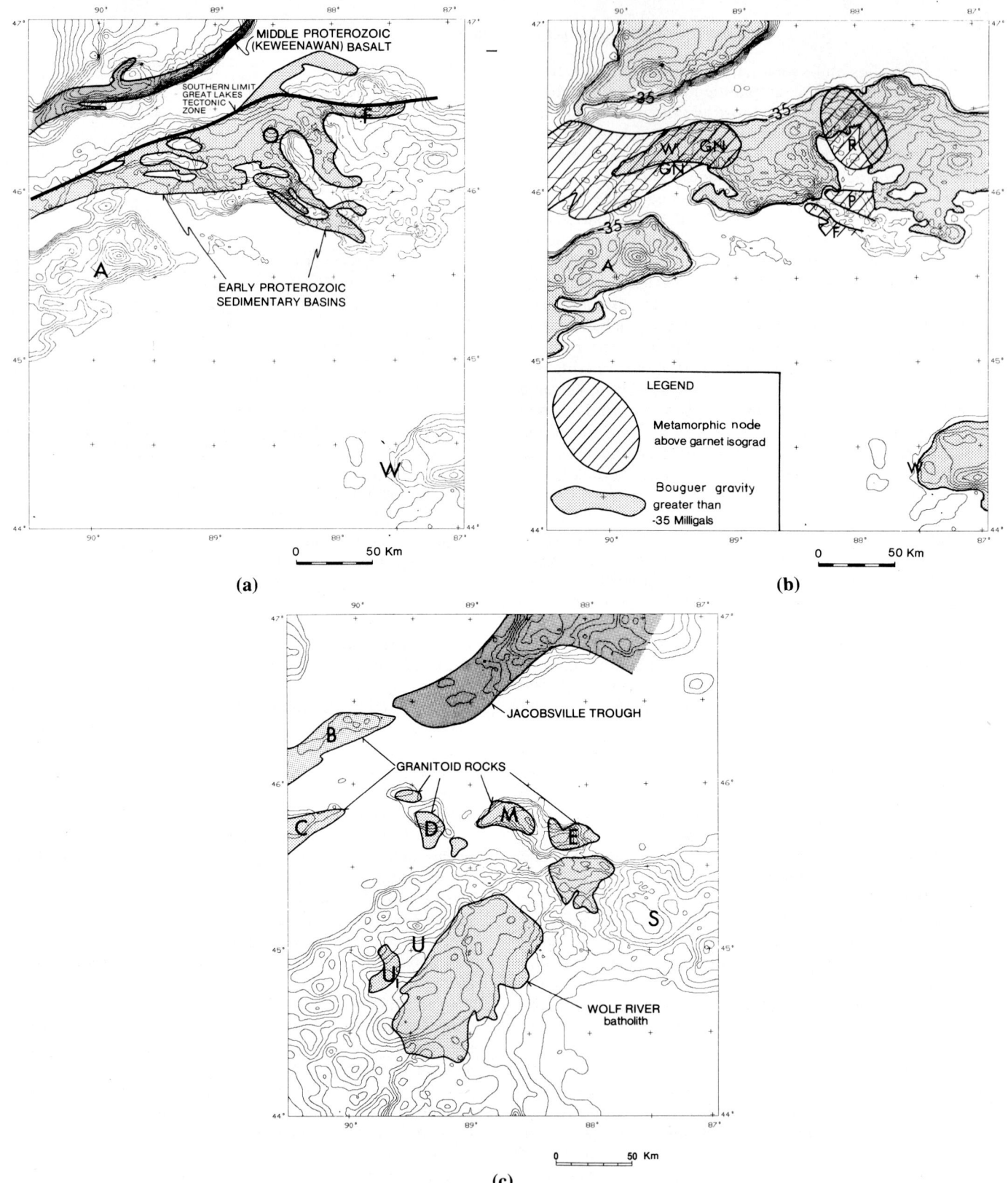

FIG. 9a. Amplitude-filtered gravity map showing gravity contours higher than −40 mGal. Contour interval, 5 mGal. Location of Early Proterozoic basins of metasedimentary rocks shown by light shading. Bold letters explained in text.

b. Amplitude-filtered gravity map showing gravity contours higher than −40 mGal. Contour interval, 5 mGal. Amplitudes higher than −35 mGal highlighted by light shading. Locations of regional metamorphic node above garnet isograd shown by slanted lines. R = Republic node; P = Peavy node; F = Florence node; W = Watersmeet node. GN marks general position of remobilized gneiss domes.

c. Amplitude-filtered gravity map showing gravity contours less than −45 mGal. Contour interval, 5 mGal. Jacobsville trough shown by dark shading. Locations of granitoid rock units shown by lighter shading.

The somewhat chaotic appearance of the gravity map in the south part of the central terrane is due to the fact that there are numerous steep gravity gradients and elongate anomalies that are diversely oriented in this region. For example, note the variation in trends of the long axes of gravity anomalies A, D, E, M, and N (Figures 4 and 6b). As mentioned above, anomaly A (A_1 and A_2 are local gravity highs within the broad gravity high shown by A) lies primarily above a large area of supposed mafic volcanic rocks (compare Figures 2 and 4). Prominent lows at D, E, and M lie above areas of granitoid rock. The origin of anomaly N is not known at this time. Geophysically mapped faults (Figure 2) in this area generally lie along gravity gradients, and in many cases positive linear magnetic anomalies lie above the gravity gradients. Most of the southern edge of this part of the central terrane lies along south-sloping gravity and magnetic gradients (see Figures 4 and 7).

Central terrane—magnetics

Although the tie in magnetic data sets between Michigan and Wisconsin needs to be checked, the character of the magnetic data for each data set reflects Precambrian basement geology. This is demonstrated by the fact that there are several good correlations in anomalies between the magnetic (Figure 7) and gravity maps, especially the high-pass gravity map (Figure 6b). A few of these correlations are discussed in the following paragraphs, but a complete discussion of magnetic and gravity features is beyond the scope of this paper.

The magnetic pattern of the north half of the central terrane has a dominant east-northeast grain that parallels the trend of the PFB. The excellent correlation of detailed structural patterns of folding and faulting with the magnetic data can best be seen at a larger scale and smaller contour interval than those of the magnetic map of the United States. In Michigan, the iron-rich Lower Proterozoic metasedimentary rocks overlie a basement of Archean granitic gneiss that has relatively little magnetic expression. Where this basement has been domed up and exposed it gives rise to many round to oval magnetic lows (I, J, K, and L in Figure 7) that lie in a broad east-northeast-trending zone. Many of the magnetic lows in this zone coincide with areas of low gravity such as anomalies I, J, and K (Figures 4 and 7), and many of these, such as I, J, K, and L, lie within regional metamorphic nodes (compare Figures 2 and 7). These magnetic lows and coincident gravity lows display a diverse pattern and delineate uplifted blocks or remobilized domes of Archean basement rocks. (As explained below, in the broad-scale sense, areas underlain by Archean gneiss tend to have regional gravity highs above them, but small uplifted blocks and domes of Archean gneiss that are surrounded by dense Lower Proterozoic sedimentary rocks have relatively small-amplitude gravity lows above them, even though they lie within the regional gravity highs.) Most positive magnetic anomalies lie above basins of Lower Proterozoic rocks (see O and F of Figures 7 and 9a).

Lidiak (1974) showed that there is a relationship between metamorphic grade and total magnetic intensity of iron formation and metasedimentary and metavolcanic rocks. He notes that total magnetic intensity of these rocks increases within the metamorphic nodes relative to areas outside the nodes. Thus, within the metamorphic nodes there are generally large-amplitude magnetic anomalies over Lower Proterozoic rocks that fringe regions of uplifted Archean basement rocks. These anomalies are most apparent on larger scale magnetic maps not shown in this report (see, for example, Case and Gair, 1965, and Meshref and Hinze, 1970).

The southern limit of the GLTZ is sharply defined in the eastern part of the magnetic map, where it follows the anomaly produced by the highly magnetic metasedimentary rocks of the Marquette trough (F in Figure 9). In the western part of the area the southern limit of the GLTZ follows major east-northeast-trending faults (see Figure 2), which were mapped from geologic and geophysical data (see Prinz, 1981; Sims and Peterman, 1980).

In the south part of the central terrane, especially in the east, the magnetic map tends to have a more regular east–west–trending fabric than the gravity map. Near the west edge of the map the magnetic pattern has a strong northeast grain, reflecting in part the presence of northeast-trending faults such as the Jump River fault (P in Figures 2 and 7) mapped by Sims et al. (1978). A prominent northeast-oriented magnetic high that occurs at A_2 near the west edge of the study area is bounded on the southeast edge by the Jump River fault (P in Figures 2 and 7). This area is mapped as predominantly metavolcanic terrane (Morey et al., 1982). The positive magnetic anomaly shows good correlation with a positive gravity anomaly, as shown on the high-pass-filtered map (A_2 in Figure 6b). These anomalies all lie within the broad regional high (A) on the Bouguer and low-pass gravity maps (Figures 4 and 6a).

South of the Jump River fault (P in Figure 7), the magnetic field has a generally lower amplitude than north of it. Although this area is also considered to consist of metavolcanic rocks, they may be thinner, or alternatively less magnetic, than those in the broad belt to the north, and the character of the basement rocks that lie beneath the postulated mafic volcanic terrane may change across the fault. In this area there are several small, circular magnetic highs, some of which coincide with gravity highs. These are likely caused by small plutons or plugs of mafic igneous rock. Other small, circular magnetic highs without a gravity analog are probably caused by small bodies of less dense but relatively magnetic plutonic rocks of intermediate composition. The prominent magnetic high at A_1 in Figure 7 coincides with a triangular-shaped area of high gravity (A_1 in Figure 4). It, too, probably represents an unexposed mafic plutonic body.

A zone of northwest-oriented linear magnetic highs lies between anomalies A and D. These highs lie along a steep gravity gradient (compare Figures 4 and 7) and may represent a magnetic fault anomaly or perhaps mineralization or mafic dikes along an intrusive contact or fault zone.

Toward the eastern part of the south-central terrane, several narrow magnetic anomalies are oriented in an east–west direction. For example, one at Q (Figure 7) lies along the north edge of the Niagara fault (Figure 2), an area where bedrock consists of mafic metavolcanic rocks, iron formation, and basalt flows (Bayley et al., 1966). There are two north-northeast-oriented highs at R_1 and R_2 in the eastern part of the south-central terrane. The anomaly at R_1 in Figure 7 is of particular interest. It lies between gneiss domes at E and M. Figure 6b shows that this magnetic anomaly corresponds with a prominent north-northeast-oriented linear gradient in the gravity data. This prominent gravity linear lies near the east edge of the Iron River trough and extends across the Niagara fault. It lies within the Trans Superior tectonic zone of Klasner et al. (1982). The linear, however, is not apparent on the magnetic map. The magnetic high at R_1 may indicate the presence of magnetite mineralization or mafic intrusion in a fault zone. The cause of the magnetic high at R_2 is not known but may be similar to that of R_1.

The southern edge of the central terrane is marked by an east–west–trending decrease from north to the south in the level of the magnetic field. This lies within the central Wisconsin magnetic survey of Zietz et al. (1977).

Southern terrane—gravity

The southern terrane is dominated by a prominent north-northeast-trending gravity low (Figures 4 and 6a) that is at least partly caused by the Wolf River batholith (WRB), a primarily quartz-monzonite and granite body with a density of 2.67–2.80 g/cm^3. The batholith intrudes mafic volcanic rocks and Archean gneiss of variable composition that have a higher bulk density than rocks of the batholith (see Table 2).

Numerous shorter wavelength gravity anomalies lie above intensely faulted terrane just west of the western edge of the batholith. The Wausau and Stettin plutons in this region have low gravity values over them (U_1 in Figure 4). Figure 9c, the amplitude-filtered gravity map with values less than −40 mGal, emphasizes that the Wolf River batholith lies in a broad area of low gravity. The region of low gravity extends to the east and southwest, well beyond the region of exposed batholithic rocks. This suggests that the subsurface extent of the batholith is much larger than its geologically mapped extent. Caution should be exercised, however, in inferring that all of the region of low gravity is underlain by 1.5-b.y.-old rocks that are genetically associated with the batholith. Geologic studies of central Wisconsin by LaBerge and Meyers (1984) and Van Schmus (1980) indicate that several plutonic events occurred in central Wisconsin. Smith (1978), for example, reports on 1 765-m.y.-old Fox River granites and rhyolite located roughly 40 km south of the south edge of the batholith. These rocks also lie within the broad gravity low, which extends well south of the 44°-latitude southern edge of the study area as shown on the *Gravity Anomaly Map of the United States* (Lyons and O'Hara, 1982, and this volume). Also, Van Schmus (1980) showed that the central Wisconsin region has undergone periods of granitoid plutonism at roughly 1 830 and 1 760 m.y. ago. Thus it may be concluded that the large gravity low in central Wisconsin is caused by composite low-density subjacent granitic bodies representing multiple intrusions of several different ages.

It is also significant that trends of long axes of gravity anomalies are oriented generally east–west on either side of the northeast-oriented gravity low. These east–west trends, including those near S in Figures 4, 6a, and 6b are interpreted to reflect orientation of tectonic features associated with the PFB.

Southern terrane—magnetics

Whereas the Bouguer gravity map has a pronounced north-northeast-trending low owing to the low-density rocks of the WRB and other granitic bodies, the magnetic map is dominated by two major positive magnetic areas: (1) a circular positive anomaly with a central low (T in Figure 7) and a zone of short-wavelength magnetic highs (near U in Figure 7), some of which are syenitic intrusions (U_1, for example); and (2) a broad, northeast-trending positive area to the southeast (W in Figure 7).

The circular positive anomaly at T with a central low lies above the area of mapped Tigerton anorthosite. Dutton and Bradley (1970) report that the contact zone between the anorthosite and granite that surrounds it has a high magnetite content. They interpret this as the likely cause of the magnetic anomaly. Also, Weiss (1965) reports that magnetite is common in the Tigerton anorthosite. This may also contribute to the observed positive magnetic anomaly.

A broad zone of short-wavelength magnetic highs and lows at U in Figure 7 lies west of the circular magnetic anomaly (T) outside the WRB. The largest of these anomalies (U_1) in both area and amplitude lies, in part, above the Wausau and Stettin plutons. Allingham and Bates (1961) report that the magnetic highs associated with these bodies occur above magnetite-rich material within the granitic rocks of the plutons. Table 2 gives the range of susceptibilities for rocks associated with these plutons. The numerous other small positive anomalies in this region occur in a northeast-trending zone that was mapped by LaBerge (1977) and LaBerge and Meyers (1984) as a region of intensely sheared and structurally complex crust, varying from granitic to ultramafic. The magnetic pattern in this region reflects the complex geology of the area. The zone is abruptly cut off by a northeast-trending zone of faulting at V in Figure 7, the magnetic-anomaly map.

A prominent belt of magnetic highs at W extends in a northeast direction across the southeast corner of the study area. It appears to have a double row of maxima along its northwest and southeast flanks. The smooth character of these anomalies is partly because the area is concealed under a thickening cover of Paleozoic sedimentary rocks and partly because magnetic flight lines in this region are more widely spaced and were recorded at a higher flight elevation (see Zietz, 1982). These positive magnetic anomalies roughly correspond to a broad area of high gravity in the southeastern part of the study area (W in Figure 4). However, the zone of magnetic highs extends to southeastern Wisconsin across mapped granite–rhyolite inliers of south-central Wisconsin (Van Schmus et al., 1975; Smith, 1978), and apparently beneath the southern margin of the prominent gravity low of central Wisconsin. Careful comparison of the national gravity and magnetic maps of the region, just south of the study area, shows that the northeast-trending belt of magnetic anomalies occurs where there are numerous gravity highs within the broad regional low. These gravity highs correspond with the magnetic highs. We think that this belt of magnetic highs may reflect the presence of scattered areas of Archean-gneiss terrane in the basement rocks of southern Wisconsin, analogous to Archean gneiss of northern Michigan. The rhyolite–granite of central Wisconsin would form a relatively thin veneer on top of or intrusions within the Archean gneiss. Sm–Nd studies of rhyolite and granite of central Wisconsin (Nelson and DePaolo, 1982), however, indicate that the rhyolite and granite are mantle derived with no involvement of Archean crust. This suggests that Archean crust is not present in central Wisconsin. But continuity of the southwest-trending magnetic high at W with mapped Archean rocks at X (Figures 4 and 7) strongly suggests that the magnetic anomaly is caused by subjacent Archean rocks in this region.

CAUSE OF THE LONG-WAVELENGTH GRAVITY ANOMALY BENEATH THE PENOKEAN FOLD BELT

Because the long-wavelength gravity anomaly beneath the PFB cannot be explained by obvious near-surface geologic features, there remain two possible reasons for its existence: (1) undetected changes in near-surface density of Archean basement rocks along with the presence of relatively dense rocks within overlying Early Proterozoic basins, and (2) lateral density changes in the deep crust.

In a recent study concerning the nature of the GLTZ in northern Michigan, Klasner and Sims (1984) suggested that the gravity high may be partly caused by an increase in density of Archean crust south of the tectonic zone (the GLTZ lies along the northern edge of the gravity high as mentioned above). This suggestion was made because Archean rocks north of the tectonic zone are granite–greenstone in contrast to a gneissic terrane south of the zone. In the Canadian shield north of Lake Superior, granite–greenstone belts tend to have a lower gravity-anomaly expression relative to gneiss belts (Hall, 1968). No systematic study of rock densities was made in the study area to test this idea, however.

Alternately, the gravity high was first explained as a manifestation of lower crust uplift in a gravity model constructed by Ocola and Meyer (1973). They based their hypothesis on seismic-refraction data as well as on a two-dimensional gravity model along a profile that crossed the gravity high at A (Figure 4) in the western part of this study area. Their model suggested that the broad gravity high is caused by uplift of dense (2.94 g/cm^3) and relatively high-seismic-velocity (6.4 km per s) lower crust to within about 5 km of the surface.

We constructed a two-dimensional gravity model (Figure 10) across the gravity high in the eastern part of the study area (see Figures 4 and 6a for location). The model tests the idea that the long-wavelength high is caused by density changes in the deep crust. A magnetic profile is included to show the relation between near-surface geology and magnetics.

The residual anomaly on the gravity profile was obtained by subtracting a regional profile from the observed gravity along the profile. Values for the regional profile were estimated from Woollard's chart (1962, p. VII-3) showing mean world Bouguer gravity values relative to elevation. A rough

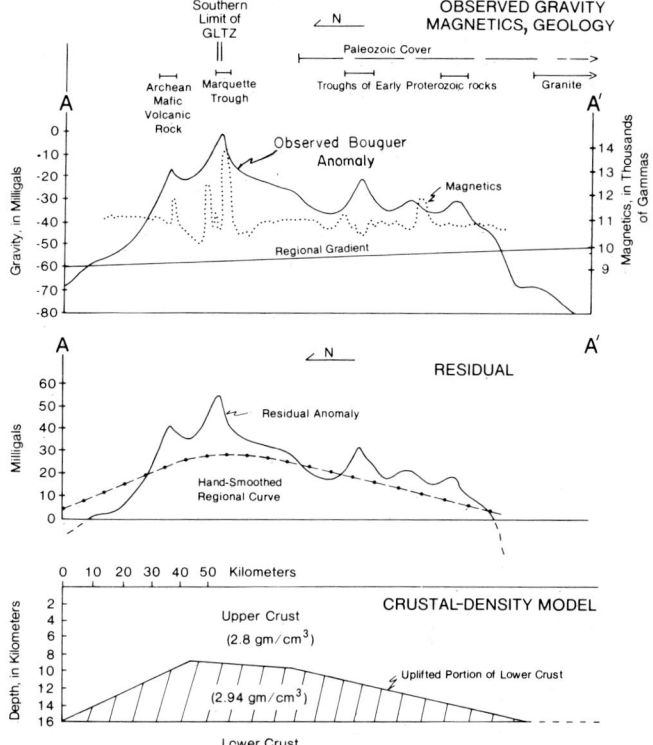

FIG. 10. Gravity and magnetic profile A–A' and two-dimensional model of the long-wavelength gravity high that lies along the Penokean fold belt. See Figures 4 and 6a for location.

approximation from Woollard's graph yields mean crustal gravity values of about -60 mGal on the north to -50 mGal on the south for the regional profile.

A hand-smoothed curve (dashed line in Figure 10) was drawn along the residual anomaly profile to approximate the shape of the long-wavelength anomaly. Variations in the observed residual from the hand-smoothed profile represent anomalies caused by density variations in the near surface. For example, the residual is higher than the smoothed profile at the locations of troughs of dense Lower Proterozoic rocks such as the Marquette trough.

One can compare these profiles to the low-pass and high-pass-filtered gravity maps. The low-pass map (Figure 6a) would be similar, at least in concept, to the hand-smoothed regional, and the high-pass map (Figure 6b) similar to the variations in the observed residual gravity from the hand-smoothed regional. However, the hand-smoothed regional does not match the low-pass map (Figure 6a), and a direct comparison is not possible. Gupta and Ramani (1980) showed that the graphic smoothing techniques as shown in Figure 10 are often more appropriate for geologic interpretation of gravity data than a wavelength-filtering technique as shown in Figure 6a.

The crustal-density model in Figure 10 assumes that the average density of the upper crust is 2.8 g/cm^3, a value within the density range for Archean crust (see Table 2). It extends to a depth of 16 km, where crustal density becomes 2.94 g/cm^3, similar to the model of Ocola and Meyer (1973). The

Table 3. Nature of tectonism in the study area.

Terrane	Tectonic event	Nature of tectonism
Northern	Formation of Midcontinent rift—1.1 b.y.	Rifting, associated basaltic magmatism, and clastic sedimentation in Lake Superior basin.
Central	Penokean orogeny—1.9 b.y.	Multiphase deformation, block faulting, and formation of basins and troughs in Archean crust; intrusive and extrusive igneous activity; metamorphism; anatectic remobilization of Archean and Proterozoic rocks.
Southern	Anorogenic plutonism and Penokean volcanism—1.5 b.y. to 1.75 b.y.	Intrusion of Wolf River batholith and associated alkalic plutonism. Rhyolite volcanism, associated granite intrusions, and sedimentation.

computed anomaly from the crustal-density model is represented by the dots along the hand-smoothed regional curve on the residual profile. The crustal-density model shows that the broad regional anomaly could be caused by uplift of the dense lower crust to within roughly 8 km of the surface beneath the highest part of the regional anomaly. This would be directly beneath the Marquette trough. It seems more reasonable, however, that the gravity high is caused by some combination of the two hypotheses mentioned above, that is, an increase in density of Archean crust beneath the area of high gravity and uplift of dense lower crustal rocks beneath the high.

TECTONIC IMPLICATIONS

In this paper we have outlined the major geophysical and geologic characteristics of three distinctly different tectonic terranes that lie within northern Michigan and Wisconsin (see Table 3). Hinze et al. (1982) showed that the large variation in amplitude and steep gradients in both gravity and magnetic data, such as found in the northern terrane, are compatible with structures formed by extensional tectonism of the continental crust during the Midcontinent rift event.

The large gravity low in central Wisconsin is, in part, caused by the relatively low-density rocks of the Wolf River batholith. As mentioned above, the batholith, which imparts the prominent northeast-trending fabric to the southern terrane, is one of several anorogenic plutons that extend in a belt from eastern Canada through the Great Lakes region and into the southwestern United States (Silver et al., 1977; Emslie, 1978). The unique nature of the batholith relative to surrounding areas makes it stand out geophysically from other features within the study area. We think that this unique signature, a prominent gravity low that does not fit the surrounding tectonic fabric, can be used as a model to detect other anorogenic plutons that may occur elsewhere in the Midcontinent, especially in areas of Paleozoic cover.

Although the Wolf River batholith pervades the southern terrane both in terms of geology and geophysical signature, it is important to reiterate that east–west–trending fabrics on either side of the batholith likely reflect tectonic features associated with the Penokean orogeny.

The central terrane is underlain primarily by the Penokean fold belt. As noted above, and as shown in Figures 4 and 11, this terrane consists of a broad positive gravity anomaly that has steep gradients sloping outward on either edge. The magnetic map (Figure 7) suggests that the fold belt can be divided into a north and south half. Table 4 shows that numerous plate-tectonic models have been proposed to explain the origin of the Penokean fold belt. It also shows that ideas concerning its origin are currently in a state of flux. All of the models, however, call upon some sort of plate-tectonic activity.

It has been previously recognized (Van Schmus, 1976) that Lower Proterozoic rocks of the PFB consist of primarily metasedimentary rocks in the north and metavolcanic–plutonic (magmatic) rocks in the south. The division occurs along the Niagara fault and its westward extension (Figure 2) and is reflected in gravity and especially magnetic data as shown in this paper. Larue and Sloss (1980) present a model, based on analysis of sedimentation in northern Michigan, in which initial rifting, suggested earlier by Cambray (1978), formed a passive continental margin in Michigan. This was followed by a "marginal event," which consisted of subduction of oceanic crust and concomitant volcanism to the south and ultimate suturing of Proterozoic island-arc and continental-rock assemblages along the boundary between sedimentary and magmatic parts of the PFB.

In our opinion the most compelling evidence for subduction in the region comes from geochemical trace-element studies of Schulz (1983), which show that the Early Proterozoic volcanic–plutonic belt in north-central Wisconsin (the south half of the central terrane of this paper) has geochemical affinities with modern island-arc volcanic and plutonic rocks. He also shows that these rocks are compositionally different from Lower Proterozoic basalts in northern Michigan (in the north half of the central terrane of this paper). Schulz et al. (1984) propose a rift-collision model based on these data.

It is important to determine whether or not the geophysical data are compatible with such a model. The following discussion shows that a combined gravity–magnetic profile across the central terrane (Figure 11) is not at variance with the proposed rift-suture model. This profile, with an accompanying geologic cross-section, lies mostly above rocks of the Penokean fold belt (see profile A–B in Figures 2, 4, and 7). Note that in the southern terrane the profile is outside the gravity low associated with the Wolf River batholith. The highest gravity anomalies on the profile lie above the north half of the central terrane where uplifted, relatively dense lower crust contributes to the regional gravity high as shown in Figure 10. Cambray (1978) suggests that this area represents a rifted continental margin and that block faulting and formation of troughs, such as the Marquette trough (Figure 2), formed in response to extension of the crust during rifting. We think that uplift of the lower crust may have accompanied or, in part, initiated the rift event. Large variations in amplitude of the magnetic data in

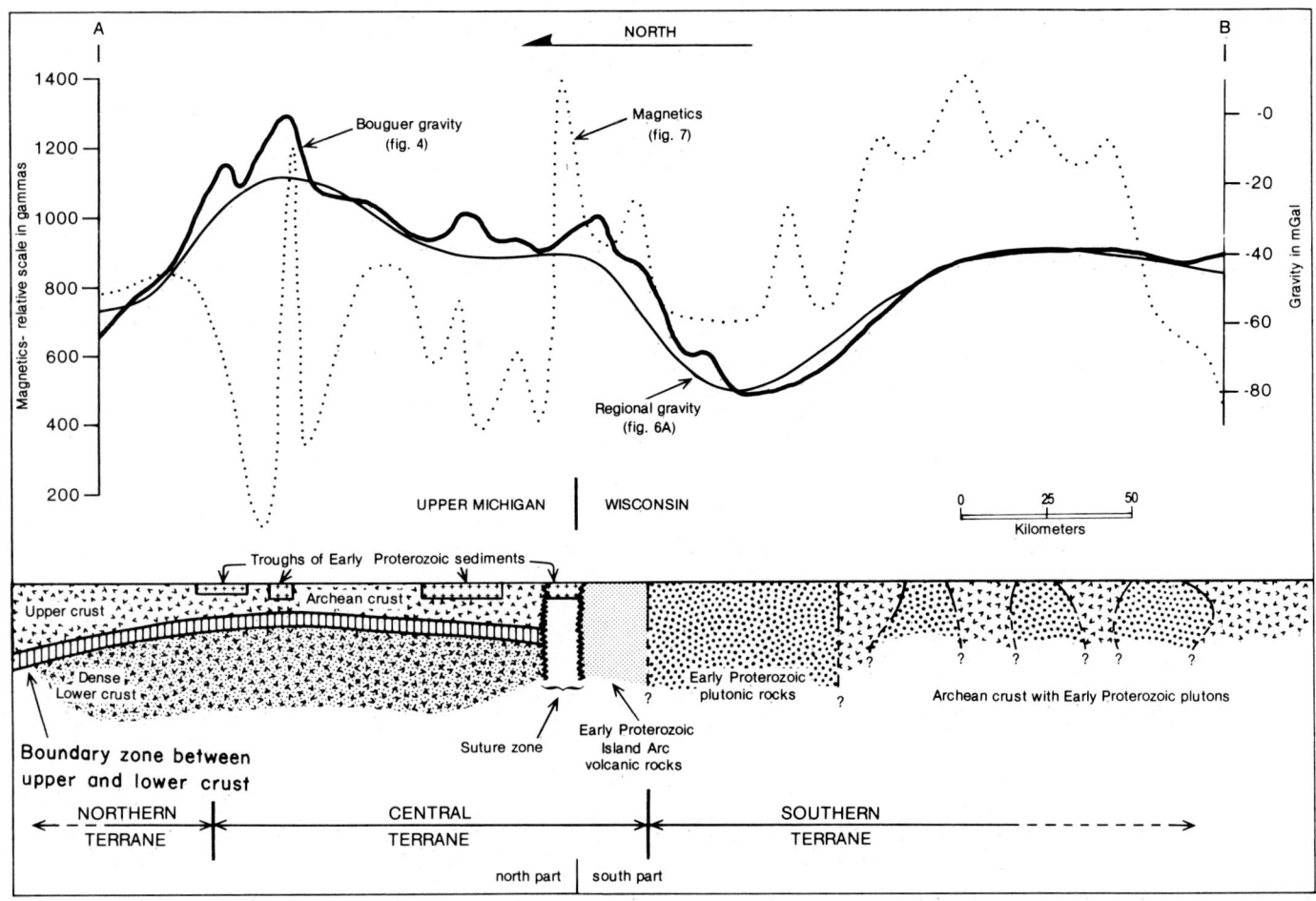

FIG. 11. Gravity and magnetic profile along line A–B of Figures 4, 6a, and 7, with diagrammatic crustal cross-section. Gravity and magnetic data are seen to be compatible with rift-suture tectonic model for origin of Penokean fold belt. See text for further discussion.

the north part of the central terrane reflect the juxtaposition of iron-rich Early Proterozoic metasedimentary basins and blocks of Archean gneiss.

The upper edge of a steep, south-dipping gravity gradient occurs at the boundary between the north and south halves of the central terrane and represents the proposed suture zone. A prominent low on the profile lies in the northern part of the southern terrane where Lower Proterozoic granitoid plutons occur (Figure 2). The magnetic-map pattern changes abruptly at the proposed suture and contains a unique signature that extends for 25 km to the southern boundary of the central terrane. This area consists of mafic volcanic rocks, which have been suggested by Schulz (1983) and Schulz et al. (1984) to be related to island-arc volcanism.

The gravity-anomaly values rise to about −40 mGal in the southern part of the profile. The cause of this rise in gravity is not known. However, it lies in the area of the northeast-trending magnetic high (W in Figure 7), near the 1 760-m.y.-old rhyolite–granite occurrences of Smith (1978). This area may represent a zone of subjacent Archean gneiss (beneath near-surface rhyolite–granitic rocks) that marks continental crust on the opposite side of the sutured zone. The Archean rocks at X in the southwest corner of the geologic map (Figure 4) are part of this zone but lie west of the profile.

The above discussion is preliminary and is meant to show that geophysical data of this study are not at variance with the proposed rift-suture models of Cambray (1978), Larue and Sloss (1980), and others, but they do not prove them. A similar model for the whole Early Proterozoic Circum-Superior belt was proposed by Baragar and Scoates (1981). Also, plate-tectonic–suture models have been proposed from analysis of gravity data for the boundary between the Churchill and Superior provinces of the Canadian shield (Gibb et al., 1983; Gibb, 1983; Gibb and Thomas, 1976) and for the Labrador trough (Keary, 1976). Further work is needed to prove such models, and many questions remain to be answered. In particular, does such a model agree with geologic and geophysical data found along the whole length of the Penokean fold belt? For example, the proposed rift-suture models mentioned above are mostly based on work done in the eastern half of the study area near profile A–B. The proposed suture lies along the juncture between sedimentary and volcanic–plutonic terranes of the Penokean fold belt. Yet, the long-wavelength gravity high crosses this

Table 4. Summary of plate-tectonic models regarding formation of the Penokean orogen (PFB).

Author(s) and year	Brief summary of model
Van Schmus (1976)	Division of PFB into sedimentary terrane on north and volcanic plutonic terrane on south. Formed by subduction of oceanic crust to north with island-arc assemblages and volcanic assemblages in Wisconsin and back-arc sedimentary basins in Michigan.
Cambray (1978)	Formation of troughs and basins filled with Early Proterozoic sediments by rifting processes. Subsequent compression caused by subduction and continental collision formed volcanic–plutonic terrane in northern Wisconsin.
Larue and Sloss (1980) Greenberg and Brown (1983)	Formation of sedimentary basins in upper Michigan by rifting and formation of a passive margin followed by subduction of oceanic crust to south and with ultimate collision of arc-related volcanics along Niagara fault zone.
Anderson and Black (1983)	Penokean orogenesis is characterized by two tectonic events: (1) tensional tectonics with formation of grabens and platform sedimentation, and formation of an ocean basin; and (2) ocean closure and suturing along the Niagara fault system.
Schulz et al. (1984)	Rifting, subduction, and collision model consisting of initial spreading, subsequent subduction and formation of volcanic arcs, and finally collision of arc volcanic rocks with Archean crust on the north and south.

juncture in the west. Does this mean that the gravity high at A may be underlain by Archean terrane, or is the positive gravity anomaly at this location caused in its entirety by dense, mafic volcanic rocks as suggested by the geologic map (Morey et al., 1982)? The gravity models of Ocola and Meyer (1973) suggest that the anomaly is caused by uplifted lower crust. Although beyond the scope of this paper, a series of crustal models should be prepared from gravity and magnetic data along the whole length of the PFB to test this and possibly other tectonic models.

SUMMARY

The northern Michigan–northern Wisconsin area can be divided into three tectonic terranes: (1) the northern terrane, which consists of basalt and clastic sedimentary rocks of the Midcontinent rift system (MCR); (2) the central terrane, which includes the Penokean fold belt (PFB), and is divided into a north part consisting of mostly Lower Proterozoic metasedimentary rocks and a south part underlain by mostly mafic to felsic volcanic rocks and gneiss of variable composition; and (3) the southern terrane, which consists mostly of the Wolf River batholith (WRB) and associated plutons and an intensely faulted region just west of the batholith.

Each terrane has a distinctive gravity and magnetic signature. Both gravity and magnetic data over the northern terrane have high-amplitude positive anomalies above Keweenawan basalt and low anomalies over the clastic basins. A steep south-sloping gradient lies along the Keweenawan fault on both maps. Both the gravity and magnetic data have a strong northeast-trending fabric that mimics the tectonic fabric of this part of the MCR.

The most prominent geophysical feature associated with the central terrane is an east-trending long-wavelength gravity maximum along the length of the PFB. It occurs largely in the north part of the central terrane but extends into the south part near the western edge of the study area. Shorter wavelength gravity anomalies and magnetic anomalies of the north part tend to be oriented east–west, parallel to the axis of the PFB. Some anomalies related to uplifted blocks of Archean basement, however, have divergent trends.

The south part of the central terrane is characterized by a somewhat chaotic gravity field that has numerous diversely oriented steep gravity gradients. The magnetic field appears to have a higher background amplitude than the north part of the central terrane. It also has many prominent positive anomalies that correlate with positive gravity anomalies, suggesting the presence of buried mafic bodies.

The southern terrane is dominated by a prominent gravity low that reflects, in part, the presence of the WRB. Its north-northeast orientation contrasts with the east-northeast orientation of the central terrane. Prominent short-wavelength high and low anomalies in the magnetic data reflect the distribution of rock types with contrasting magnetic properties within the WRB. East–west–oriented lineations in geophysical data on either side of the gravity low reflect the tectonic fabric of the PFB.

Comparison of amplitude-filtered gravity maps with mapped geologic features illustrates the relationship between surface geology and the gravity field. All basins of Lower Proterozoic rocks are within the area of the broad, regional gravity high associated with the PFB, but they do not account for the presence of the gravity high. The regional metamorphic nodes also are within the gravity high, but in areas of local, relatively low gravity-anomaly values. The relatively low gravity values associated with the nodes suggest that low-density subjacent plutons or gneiss domes lie beneath the nodes. All mapped granitic plutons and gneiss domes have associated local gravity lows. The gravity low above the WRB is much broader than the exposed part of the pluton, suggesting that the pluton may cover a larger area in the subsurface than its exposed extent. Geologic data indicate that the gravity low is also a composite feature caused by subjacent plutons that were intruded over a long period of time.

A two-dimensional gravity model was constructed to explain the source of the long-wavelength regional anomaly beneath the PFB. The model suggests that the anomaly may be caused by uplift of dense lower crustal rock to within about 8 km of the surface. The Great Lakes tectonic zone

(GLTZ) lies near the crest of the regional anomaly in the east but diverges from it near the west edge of the study area. A transition from granitic terrane north of the GLTZ to gneiss terrane on the south suggests that the regional anomaly may be partially caused by a density increase in near-surface rocks across the tectonic zone. The anomaly is probably caused by a combination of horizontal density variations in near-surface rocks and in the deeper crust.

Division of the study area into three distinct tectonic terranes with associated characteristic geophysical signatures indicates that each terrane has its own unique tectonic history. The geophysical fabric of the northern terrane reflects the nature of geologic features formed during the formation of the 1.1-b.y.-old Midcontinent rift. The strong north-northeast geophysical fabric of the southern terrane reflects the orientation of the 1.5-b.y.-old anorogenic Wolf River batholith. Preliminary analysis of a combined gravity and magnetic profile that extends across the whole study area, roughly orthogonal to the tectonic fabric, indicates that geophysical data are compatible with a proposed rift, followed by an arc–continent collision for the origin of the Penokean fold belt.

ACKNOWLEDGMENTS

This paper was reviewed by W. F. Cannon, K. J. Schulz, W. J. Hinze, and J. B. Bailey. Also, an anonymous reviewer made several suggestions that improved the paper considerably. Mrs. Betty Sargent kindly typed numerous versions of the paper, and cartographic work was done under the direction of Scott Miner. We are grateful to all for their help, but we assume full responsibility for any errors in interpretation or fact that might occur in this paper.

REFERENCES

Allingham, J. W., and Bates, R. G., 1961, Use of geophysical data to interpret geology in Precambrian rocks of central Wisconsin, in Geological Survey Research 1961: Prof. Pap. 424-D, U.S. Geol. Surv., 292–296.

Anderson, R. R., and Black, R. A., 1983, Early Proterozoic development of the southern Archean boundary of the Superior Province in the Lake Superior region [abstr.]: Geol. Soc. Am. Abstr. Programs, 15, 515.

Baragar, W.R.A., and Scoates, R.F.J., 1981, The Circum-Superior Belt: A Proterozoic plate margin?, in Kroner, A., Ed., Precambrian plate tectonics: Elsevier, 297–330.

Bayley, R. W., Dutton, C. E., and Lamey, C. A., 1966, Geology of the Menominee iron-bearing district, Dickinson County, Michigan, and Florence and Marinette Counties, Wisconsin: Prof. Pap. 513, U.S. Geol. Surv.

Books, K. G., 1968, Magnetization of the lowermost Keweenawan lava flows in the Lake Superior area: Prof. Pap. 600-D, U.S. Geol. Surv., D248–D254.

——— 1972, Paleomagnetism of some Lake Superior Keweenawan rocks: Prof. Pap. 760, U.S. Geol. Surv.

Cambray, F. W., 1978, Plate tectonics as a model for the environment of deposition and deformation of the early Proterozoic (Precambrian) of northern Michigan [abstr.]: Geol. Soc. Am. Abstr. Programs, 10, 376.

Cannon, W. F., 1973, The Penokean orogeny in northern Michigan, in Young, G. M., Ed., Huronian stratigraphy and sedimentation: Spec. Pap. 12, Geol. Assoc. Can., 253–271.

Cannon, W. F., and Gair, J. E., 1970, A revision of stratigraphic nomenclature of the middle Precambrian rocks in northern Michigan: Geol. Soc. Am. Bull., 81, 2843–2846.

Carlson, B. A., 1972, Bouguer gravity anomaly map of northeastern Wisconsin: Open-file rep., Wis. Geol. Nat. Hist. Surv.

Case, J. E., and Gair, J. E., 1965, Aeromagnetic map of parts of Marquette, Dickinson, Baraga, Alger, and Schoolcraft Counties, Michigan, and its geologic interpretation: Geophys. Invest. Map GP-467, U.S. Geol. Surv., scale 1:62 500.

Davidson, D. M., Jr., 1982, Geologic evidence relating to interpretation of the Lake Superior Basin structure, in Wold, R. J., and Hinze, W. J., Eds., Geology and tectonics of the Lake Superior Basin: Mem. 156, Geol. Soc. Am., 165–172.

DuBois, R. L., 1962, Paleomagnetism and correlation of Keweenawan rocks: Bull. 71, Geol. Surv. Can.

Dutton, C. E., and Bradley, R. E., 1970, Lithologic, geophysical, and mineral commodity maps of Precambrian rocks in Wisconsin: Misc. Geol. Invest. Map I-631, U.S. Geol. Surv., 6 sheets.

Emslie, R. F., 1978, Anorthosite massifs, rapikivi granites and late Proterozoic rifting of North America: Precambrian Res., 7, 61–98.

Ervin, C. P., and Hammer, S., 1974, Bouguer anomaly gravity map of Wisconsin: Wis. Geol. Nat. Hist. Surv., scale 1:500 000.

Gibb, R. A., 1983, Model for suturing of Superior and Churchill plates: An example of double indentation tectonics: Geology, 11, 413–417.

Gibb, R. A., and Thomas, M. D., 1976, Gravity signature of fossil plate boundaries in the Canadian Shield: Nature, 262, 199–200.

Gibb, R. A., Thomas, M. D., LaPointe, P. L., and Mukhopaday, M., 1983, Geophysics of proposed Proterozoic sutures in Canada: Precambrian Res., 19, 349–384.

Goldich, S. S., Nier, A. O., Baadsgaard, H., Hoffman, J. H., and Krueger, H. W., 1961, The Precambrian geology and geochronology of Minnesota: Bull. 41, Minn. Geol. Surv.

Greenberg, J. K., and Brown, B. A., 1983, Lower Proterozoic volcanic rocks and their setting in the southern Lake Superior district, in Maderis, G. L., Ed., Early Proterozoic geology of the Great Lakes region: Mem. 160, Geol. Soc. Am., 67–84.

Gupta, V. K., and Ramani, N., 1980, Some aspects of regional-residual separation of gravity anomalies in a Precambrian terrane: Geophysics, 45, 1412–1426.

Hall, D. H., 1968, Regional magnetic anomalies, magnetic units, and crustal structure in the Kenora District of Ontario: Can. J. Earth Sci., 5, 1277–1296.

Halls, H. C., and Pesonen, L. J., 1982, Paleomagnetism of Keweenawan rocks, in Wold, R. J., and Hinze, W. J., Eds., Geology and tectonics of the Lake Superior Basin: Mem. 156, Geol. Soc. Am., 173–201.

Hildenbrand, T. G., 1979, Preliminary documentation of Program FETFIL: U.S. Geol. Surv., unpubl. documentation.

Hinze, W. J., Wold, R. J., and O'Hara, N. W., 1982, Gravity and magnetic anomaly studies of Lake Superior, in Wold, R. J., and Hinze, W. J., Eds., Geology and tectonics of the Lake Superior Basin: Mem. 156, Geol. Soc. Am., 203–221.

James, H. L., 1955, Zones of regional metamorphism in the Precambrian of northern Michigan: Geol. Soc. Am. Bull., 66, 1455–1488.

Keary, P., 1976, A gravity survey of central Labrador Trough, northern Quebec: Can. J. Earth Sci., 14, 45–55.

King, E. R., 1975, A typical cross section based on magnetic data of lower and middle Keweenawan volcanic rocks, Ironwood area, Michigan: U.S. Geol. Surv. J. Res., 3, 543–546.

——— in press, Aeromagnetic map of Iron River 1° × 2° quadrangle, Michigan and Wisconsin: Misc. Geol. Invest. Map I-1360-F, U.S. Geol. Surv., scale 1:250 000.

Klasner, J. S., 1978, Penokean deformation and associated metamorphism in the western Marquette range, northern Michigan: Geol. Soc. Am. Bull., 89, 711–722.

Klasner, J. S., and Bomke, D., 1977, A regional gravity anomaly in northern Michigan and Wisconsin, possible extension around the Superior province, associated foldbelts [abstr.]: Eos, 59, 226.

Klasner, J. S., and Cannon, W. F., 1974, Geologic interpretation of gravity profiles in the western Marquette district, northern Michigan: Geol. Soc. Am. Bull., 85, 213–218.

Klasner, J. S., Cannon, W. F., and Van Schmus, W. R., 1982, The pre-Keweenawan tectonic history of the southern Canadian Shield, in Wold, R. J., and Hinze, W. J., Eds., Geology and tectonics of the Lake Superior Basin: Mem. 156, Geol. Soc. Am., 165–172.

Klasner, J. S., and Jones, W. J., 1979, Simple Bouguer gravity anomaly map and geologic interpretation, Iron River 1° × 2° quadrangle, northern Michigan and Wisconsin: Open-File Rep. OF 79-1564, U.S. Geol. Surv.

——— 1983, Geologic interpretation of gravity and magnetic data in northern Michigan and Wisconsin [abstr.]: Geophysics, 48, 451.

Klasner, J. S., and Sims, P. K., 1984, Geologic interpretation of gravity data, Marenisco–Watersmeet area, northern Michigan: Prof. Pap. 1292-B, U.S. Geol. Surv.

Klasner, J. S., Wold, R. J., Hinze, W. J., Bacon, L. D., O'Hara, N. W., and Berkson, J. M., 1979, Bouguer gravity anomaly map of the northern Michigan–Lake Superior region: Geophys. Invest. Map GP-930, U.S. Geol. Surv., scale 1:1 000 000.

LaBerge, G. L., 1977, Major structural lineaments in the Precambrian of central Wisconsin, in International Conference on the New Basement Tectonics proceedings: Publ. 5, Utah Geol. Assoc., 508–518.

LaBerge, G. L., and Meyers, P. E., 1984, Two Early Proterozoic successions in central Wisconsin and their tectonic significance: Geol. Soc. Am. Bull., 95, 246–253.

Larue, D. K., and Sloss, L. L., 1980, Early Proterozoic sedimentary basins of the Lake Superior region: Geol. Soc. Am. Bull., pt. 2, 91, 1836–1874.

Lidiak, E. G., 1974, Magnetic characteristics of some Precambrian basement rocks: J. Geophys. [Z. Geophys.], 40, 549–564.

Lyons, P. L., O'Hara, N. W., and others (compilers), 1982, Gravity anomaly map of the United States, exclusive of Alaska and Hawaii: SEG, scale 1:2 500 000, 2 sheets.

Meshref, W. M., and Hinze, W. J., 1970, Geologic interpretation of aeromagnetic data in western upper peninsula of Michigan: Rep. Invest. 12, Mich. Dep. Nat. Resources, Geol. Surv.

Miller, W. R., 1966, A gravity investigation of the Porcupine Mountains and adjacent area, Ontonagan and Gogebic counties, Michigan: M.S. thesis, Mich. State Univ.

Morey, G. B., Sims, P. K., Cannon, W. F., Mudrey, M. G., Jr., and Southwick, D. L., 1982, Geologic map of the Lake Superior region, Minnesota, Wisconsin, and northern Michigan: Minn. Geol. Surv., scale 1:1 000 000.

Nelson, B. K., and DePaolo, D. J., 1982, Crust formation age of the North American midcontinent [abstr.]: Geol. Soc. Am. Abstr. Programs, 14, 575.

Ocola, L. C., and Meyer, R. P., 1973, Central North American Rift System, 1, Structure of the axial zone from seismic and gravimetric data: J. Geophys. Res., 78, 5173–5194.

Prinz, W. C., 1981, Geologic map of the Gogebic range–Watersmeet area, Gogebic and Ontonagan Counties, Michigan: Misc. Geol. Invest. Map I-1365, U.S. Geol. Surv., scale 1:125 000.

Schmidt, R. G., 1976, Geology of the Precambrian W (lower Precambrian) rocks in western Gogebic County, Michigan: Bull. 1407, U.S. Geol. Surv.

Schulz, K. J., 1983, Geochemistry of volcanic rocks of northeastern Wisconsin [abstr.]: 29th Annu. Inst. on Lake Superior Geol., Houghton, Mich., 39–40.

Schulz, K. J., LaBerge, G. L., Sims, P. K., Peterman, Z. E., and Klasner, J. S., 1984, The volcanic plutonic terrane of northern Wisconsin, implications for early Proterozoic tectonism, Lake Superior region [abstr.]: Program Abstr., Geol. Assoc. Can., 9, 103.

Sexton, J. L., Hinze, W. J., von Frese, R.R.B., and Braile, L. W., 1982, Long wavelength aeromagnetic anomaly map of the conterminous United States: Geology, 10, 364–369.

Silver, L. T., Bickford, M. E., Van Schmus, W. R., Anderson, J. L., Anderson, T. H., and Medaris, L. G., Jr., 1977, The 1.4–1.5 b.y. transcontinental anorogenic perforation of North America [abstr.]: Geol. Soc. Am. Abstr. Programs, 9, 1176–1177.

Sims, P. K., 1976, Precambrian tectonics and mineral deposits, Lake Superior region: Econ. Geol., 71, 1092–1118.

Sims, P. K., Cannon, W. F., and Mudrey, M. G., Jr., 1978, Preliminary geologic map of part of northern Wisconsin: Open-File Rep. OF 78–318, U.S. Geol. Surv., scale 1:62 500.

Sims P. K., Card, K. D., Morey, G. B., and Peterman, Z. E., 1980, The Great Lakes tectonic zone—A major crustal structure in central North America: Geol. Soc. Am. Bull., 91, 690–698.

Sims, P. K., and Peterman, Z. E., 1976, Geology and Rb–Sr ages of reactivated Precambrian gneisses and granite in the Marenisco–Watersmeet area, northern Michigan: U.S. Geol. Surv. J. Res., 4, 405–414.

———1980, Geology and Rb–Sr age of lower Proterozoic granitic rocks, northern Wisconsin, in Morey, G. B., and Hanson, G. H., Eds., Selected studies of Archean gneisses and lower Proterozoic rocks, southern Canadian Shield: Spec. Pap. 182, Geol. Soc. Am., 139–146.

Sims, P. K., Peterman, Z. E., and Prinz, W. C., 1977, Geology and Rb–Sr age of Precambrian W Puritan quartz monzonite, northern Michigan: U.S. Geol. Surv. J. Res., 5, 185–192.

Sims, P. K., Peterman, Z. E., and Schulz, K. J., 1982, Dunbar gneiss dome and implications for the Proterozoic stratigraphy of northern Wisconsin [abstr.]: 28th Annu. Inst. on Lake Superior Geol., International Falls, Minn., 44.

Smith, E. I., 1978, Precambrian rhyolites and granites in south-central Wisconsin: Geol. Soc. Am. Bull., 89, 875–890.

Van Schmus, W. R., 1976, Early and middle Proterozoic history of the Great Lakes area, North America: Philos. Trans. R. Soc. London, 280A, 605–627.

———1980, Chronology of igneous rocks associated with the Penokean orogeny in Wisconsin, in Morey, G. B., and Hanson, G. H., Eds., Selected studies of Archean gneisses and lower Proterozoic rocks, southern Canadian Shield: Spec. Pap. 182, Geol. Soc. Am., 159–168.

Van Schmus, W. R., and Anderson, J. L., 1977, Gneiss and migmatite of Archean age in the Precambrian basement of central Wisconsin: Geology, 5, 45–48.

Van Schmus, W. R., Green, F. C., and Halls, H. C., 1982, Geochronology of Keweenawan rocks of the Lake Superior region: A summary, in Wold, R. J., and Hinze, W. J., Eds., Geology and tectonics of the Lake Superior Basin: Mem. 156, Geol. Soc. Am., 165–172.

Van Schmus, W. R., Medaris, L. G., and Banks, P. O., 1975, Geology and age of the Wolf River batholith, Wisconsin: Geol. Soc. Am. Bull., 86, 907–914.

Weiss, L. W., 1965, Origin of the Tigerton Anorthosite: Ph.D. thesis, Univ. Wis., Madison.

Wold, R. J., and Hinze, W. J., Eds., 1982, Geology and tectonics of the Lake Superior Basin: Mem. 156, Geol. Soc. Am.

Woollard, G. P., 1962, The relation of gravity anomalies to surface elevation, crustal structure, and geology: Final rep. contract AF 23 (601)-3455, Aeronaut. Chart Inf. Cent., U.S. Air Force.

Zietz, I. (compiler), 1982, Composite magnetic anomaly map of the United States, Part A: Conterminous United States: Geophys. Invest. Map GP-954-A, U.S. Geol. Surv., scale 1:2 500 000, 2 sheets.

Zietz, I., Karl, J. H., and Ostrom, M. E., 1977, Preliminary aeromagnetic map (in color) covering most of the exposed Precambrian terrane in Wisconsin: Open-File Rep. 77–598, U.S. Geol. Surv., scale 1:250 000.

Geologic significance of regional gravity and magnetic anomalies in the east-central Midcontinent

E. G. Lidiak,* W. J. Hinze,‡ G. R. Keller,§ J. E. Reed,‡ L. W. Braile,‡ and R. W. Johnson**

ABSTRACT

The compilation of regional Bouguer gravity-anomaly and magnetic-anomaly maps, the gridding of those data, and the development of numerous filtered maps, together with detailed petrographic analysis of basement-rock drill-hole samples, have provided significant insight into the Precambrian basement. The geophysical data have yielded important clues to the tectonic framework and the regional distribution of basement-rock types. In parts of the east-central Midcontinent the basement drill-hole data have been extremely useful in the evaluation of geophysical anomalies and in the interpretation of the Precambrian geology. However, in other parts of the region, correlation of the drill-hole data with the geophysical anomalies has been poor. This poor correlation has led to a consideration of factors that can produce ambiguities in correlating geophysical and geologic data.

This investigation has shown the value of an integrated geophysical and geologic approach to studying the tectonic framework of basement rocks in the east-central Midcontinent. As a result of this study, four principal basement zones are recognized: an anorogenic granite–rhyolite terrane, several basement rift zones underlain primarily by mafic volcanic rock, the southern continuation of the Grenville province, and the New Madrid rift complex.

INTRODUCTION

A critical aspect of the investigation of the Precambrian basement of the east-central Midcontinent is the utilization of recently compiled Bouguer gravity- and magnetic-anomaly maps, together with the development of a variety of filtered geophysical maps and petrographic analysis of drill-hole basement lithology to identify the main tectonic elements. This approach is useful not only in the identification of tectonic elements but also in the delineation of old zones of weakness in the crust, which help characterize potential earthquake hazards, especially when combined with information on the prevailing stress field and seismicity of the region. Previous work (Hinze et al., 1980; Braile et al., 1982a, b; Keller et al., 1983) has demonstrated the importance of an integrated geophysical and geologic approach to an understanding of the seismotectonics of the east-central Midcontinent. Gravity and magnetic data are particularly valuable in regions such as the east-central Midcontinent where drill holes to basement and basement outcrops are scarce.

In this paper we report on an integrated study of basement geology of the east-central Midcontinent covering the area approximately between 35°–39° N latitude and 82°–92° W longitude. Except for the core of the St. Francois Mountains in southeastern Missouri, the basement in the region is covered by younger, essentially flat-lying sediments and sedimentary rocks. However, recently compiled Bouguer gravity-anomaly and total-magnetic-intensity-anomaly maps are now available for this region and have been used to map the main lithologic and structural elements in the basement.

We also report on the relation between the *measured* physical parameters of buried basement-rock samples and the intensity of gravity and magnetic anomalies that occur in the immediate vicinity of drill holes to basement. It is commonly assumed that various physical parameters such as density and magnetic susceptibility identify specific rock types and can be used to characterize geophysical anomalies and as constraints to geophysical modeling. Thus, basalts and gabbros are relatively dense and magnetic and would be expected to coincide with gravity- and magnetic-anomaly highs; felsic igneous rocks such as granites and rhyolites are less dense and magnetic and should be associated with gravity- and magnetic-anomaly lows. However, there are many exceptions to such generalized correlations, and more detailed assessment of the problem is warranted. Consider-

---

*Department of Geology and Planetary Science, University of Pittsburgh, Pittsburgh, PA 15260.
‡Department of Geosciences, Purdue University, West Lafayette, IN 47907.
§Department of Geological Sciences, University of Texas at El Paso, El Paso, TX 79968.
**Division of Geology, Tennessee Department of Conservation, Knoxville, TN 37919.

ation of this problem has become increasingly more important with the recent availability of high-quality regional gravity- and magnetic-anomaly maps and improvements in their state-of-the-art processing, which make it possible to interpret with a good degree of accuracy the size, shape, and depth of an anomaly and its possible cause. For this reason we include an evaluation of factors that can lead to ambiguities in correlation between basement lithology obtained from drilled samples and geophysical anomalies.

GRAVITY- AND MAGNETIC-ANOMALY MAPS

The regional Bouguer gravity-anomaly map of Tennessee, Kentucky, and parts of adjoining states (Figure 1) was compiled by Keller et al. (1980). Gravity observations were made in selected areas to supplement existing data coverage to obtain stations along existing roads at roughly a 2-km interval. The resulting file of approximately 50 000 gravity measurements tied to the IGSN-71 gravity datum were reduced to simple Bouguer anomaly values using a sea-level datum, a reduction density of 2.67 g/cm^3, and the 1967 theoretical gravity formula (Morelli, 1976).

The total-magnetic-intensity-anomaly map of the region (Figure 2) was compiled by Johnson et al. (1980) from 28 individual aeromagnetic surveys. The data from these surveys were merged into a single digital set by visual comparison and manual adjustment of adjoining anomaly maps. The surveys were flown generally along flight paths spaced roughly 2 km apart at a mean elevation above the surface of approximately 300 m. The magnetic field derived from the Earth's core was removed from the observations by subtracting an appropriate, updated geomagnetic reference field. All data were adjusted upward by 1 000 gammas (nT) to minimize the occurrence of negative contour values.

The gravity and magnetic data sets were gridded on a registered 2-km orthogonal array. This grid was used for hand contouring the gravity-anomaly map, machine contouring the magnetic map, and filtering both data sets. The filtered magnetic maps are limited to an area west of longitude 84° and east of longitude 90°, as gridded magnetic data are not available for the westernmost and easternmost parts of Figure 2. Both data sets were filtered in a variety of ways in order to emphasize particular characteristics of the anomaly fields. The most useful examples of these filtered maps are shown in Figures 3–11.

The filtered maps are useful primarily for qualitative analysis, in identifying and extending subtle anomalies from either longer or shorter wavelength anomalies, and in modifying anomaly data to enhance the correlation of the gravity- and magnetic-anomaly fields. Filtering was performed in the wavenumber domain in which gridded spatial data were transformed by Fourier analysis to wavenumbers, multiplied by an appropriate filter, and transformed back to spatial data (Reed, 1980).

Figure 3 is a reduced-to-the-pole total-magnetic-intensity-anomaly map that is upward continued to 2 km. On this map the high-frequency anomalies are smoothed out, and the longer wavelengths are enhanced without destroying the physical reality of the map. The map shown as Figure 4 was prepared by passing gravity anomalies with wavelengths between roughly 8 to 100 km. On this map, local gravity anomalies within the upper crust are isolated, and the broad positive gravity anomaly over the Mississippi embayment and the regional negative anomaly associated with the Appalachian Mountains are removed. Removal of the longer wavelengths enhances the effects of density contrasts among the major rock types in the upper lithosphere. A third example, shown as Figure 5, is a Bouguer gravity-anomaly map in which all wavelengths from infinity to 80 km are band-passed. Figure 5 contrasts with Figure 4 in showing only the major regional anomalies without the effects of the shorter wavelengths. A companion long-wavelength magnetic-anomaly map is presented as Figure 6. Maps have been prepared that selectively reject wavelength components trending in a given direction. Strike-filtered gravity and magnetic maps that selectively remove either northwest or northeast azimuths are shown in Figures 7 through 10. On these maps there is little distortion of the trends that are not rejected; however, the strike-rejected trends are effectively eliminated. Maps have also been prepared that selectively pass wavelength components trending in a given direction. Figure 11 is a Bouguer gravity-anomaly map in which the northeast azimuths are passed. This filtering procedure produces large distortions in the anomalies but does enhance the anomalies that remain. Furthermore, not only are the northeast-trending anomalies amplified but more circular anomalies such as the one at latitude 36.8° N, longitude 89.8° W (Figure 1) also become elongate. The elongation of such anomalies can be useful in deducing true anomaly trends in an array of anomalies, but any such interpretation must be made with considerable caution.

The gravity-anomaly map (Figure 1) is dominated by a broad positive feature locally reaching absolute amplitudes greater than +25 mGal associated with the Mississippi embayment and negative values of less than −100 mGal related to the Appalachian Mountains. Upon removal of these long-wavelength components of the gravity field, four basic anomaly patterns emerge (Figure 4). The interpretation of the geologic significance of these patterns is assisted by lithologic information and isotopic age dates obtained from basement-rock samples retrieved from widely separated deep drill holes. Several basement geologic provinces are evident in the gravity and associated magnetic anomalies.

BASEMENT GEOLOGIC PROVINCES

Granite–rhyolite terrane

Several distinct basement geologic provinces are present in the east-central Midcontinent (Figure 12). The oldest of these is the 1 400–1 500-m.y.-old granite–rhyolite terrane of the Central province, which is extensively developed in the southern and central interior of the North American craton (Lidiak et al., 1966; Van Schmus and Bickford, 1981; Denison et al., 1984). Widely spaced wells to basement and sparse Precambrian outcrops, such as those of the St. Francois Mountains and Spavinaw, Oklahoma, indicate that the main rock types are epizonal and mesozonal granites and rhyolitic volcanic rocks. Diabase is also part of the terrane and is known to occur in the St. Francois Mountains and in a belt in central Tennessee. Some of the felsic volcanic rocks have apparent alkalic affinities that have been used to infer

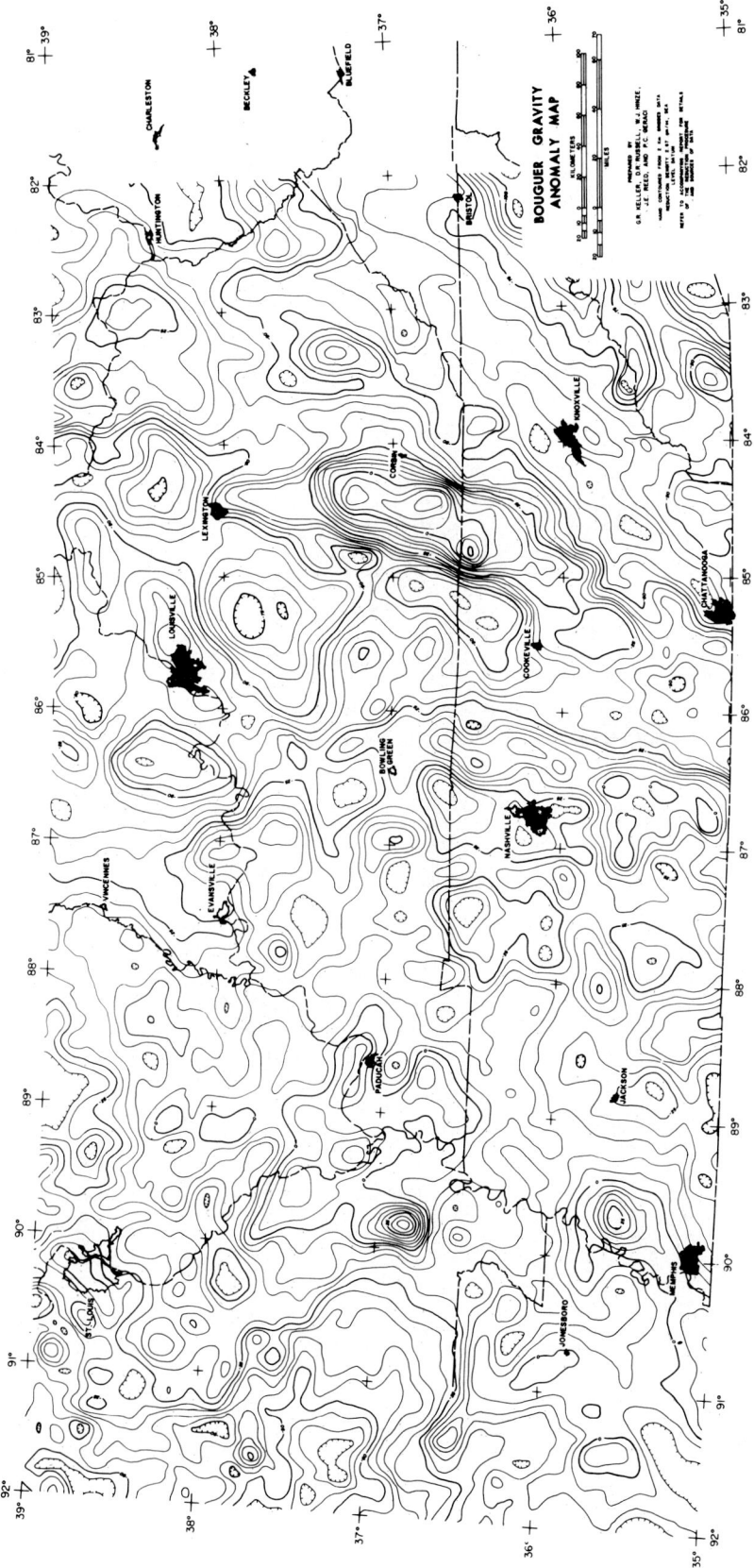

FIG. 1. Bouguer gravity-anomaly map of east-central Midcontinent of the United States. Contour interval, 5 mGal. After Keller et al. (1980).

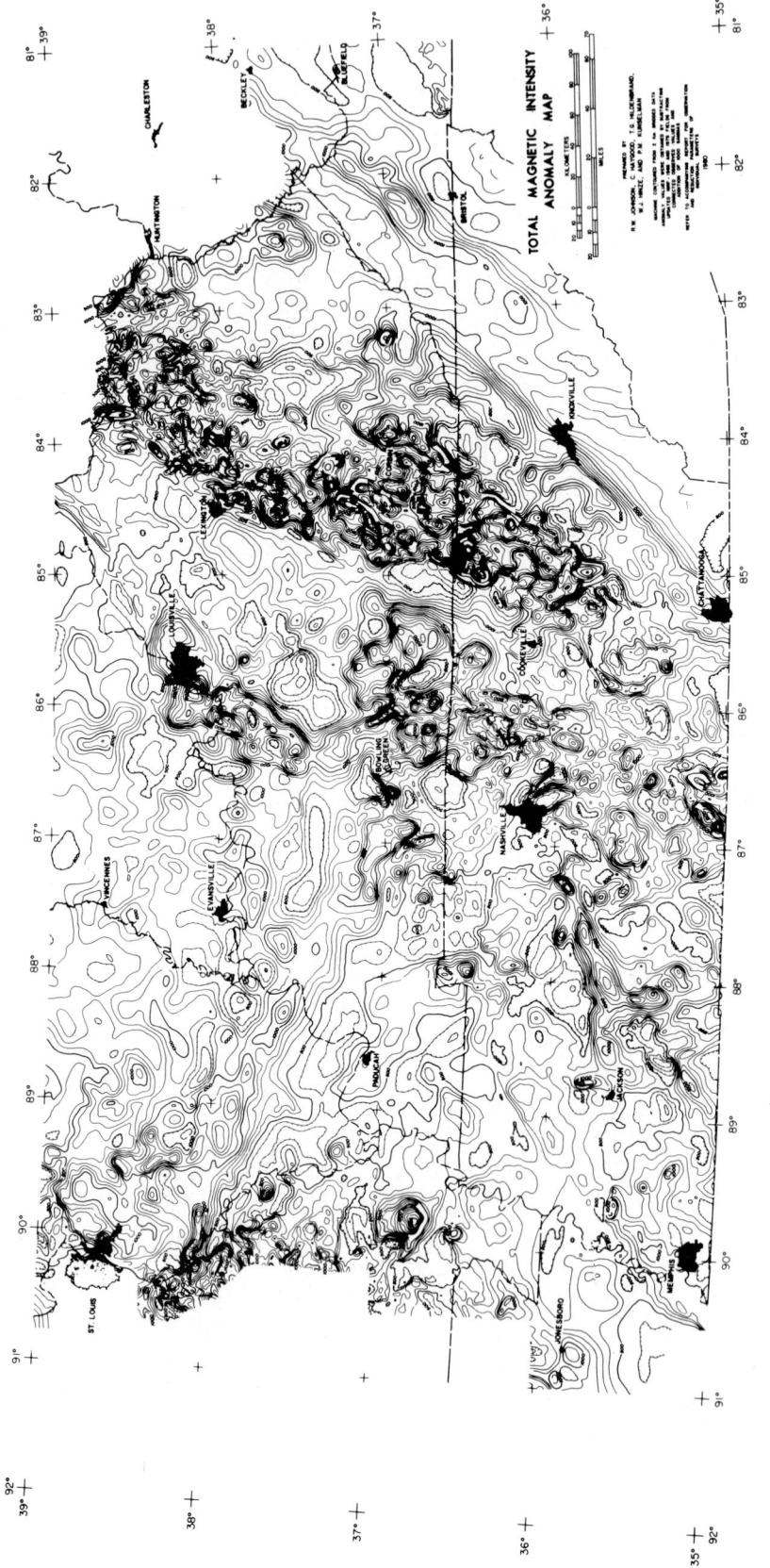

FIG. 2. Total-magnetic-intensity-anomaly map of east-central Midcontinent of the United States. Contour interval, 100 gammas (nT). After Johnson et al. (1980).

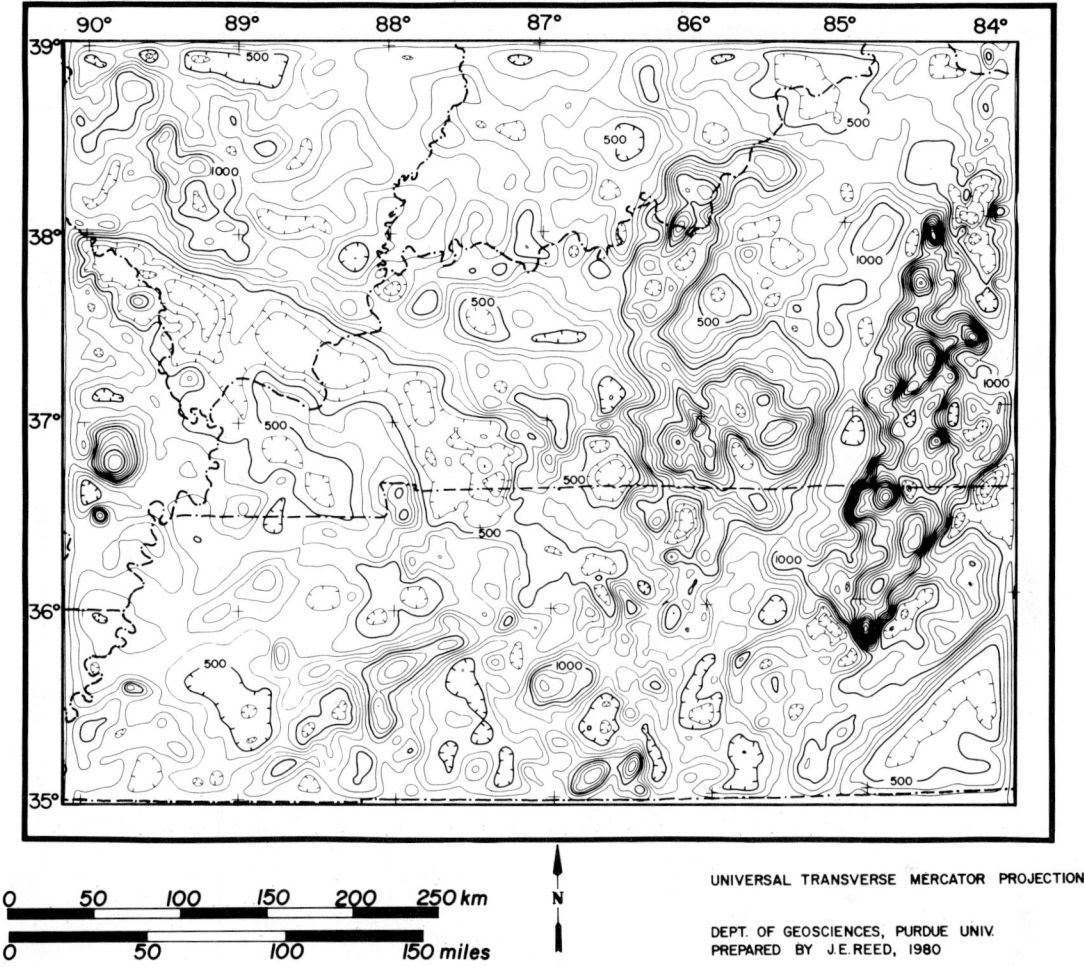

FIG. 3. Upward-continued to 2 km, reduced-to-the-pole, total-magnetic-intensity-anomaly map of east-central Midcontinent of the United States. Contour interval, 100 gammas (nT). After Reed (1980).

extrusion in a rift environment. However, such inferences should be made with extreme caution, as it has been demonstrated that the composition of similar rocks in Ohio has been modified by low-temperature alteration (Ceci and Lidiak, 1983; Faure and Barbis, 1983).

A west-northwest pattern of gravity and magnetic anomalies pervades the western part of the granite–rhyolite terrane shown in Figure 12. Generally this pattern is rather subtle, but a major anomaly having this trend strikes across the southern tip of Illinois into Missouri as well as into Kentucky and on into Tennessee. This pattern of anomalies may reflect petrologic variations in a more ancient basement that possibly underlies the felsic rocks of the granite–rhyolite terrane. The magnetic trends are on strike with northwest- and north-northwest-trending anomalies of the buried Churchill province of the western Dakotas and eastern Montana (Hinze and Zietz, this volume). Toward the east in Indiana, central Kentucky, and central Tennessee, both magnetic (Figure 2) and gravity anomalies (Figures 1, 4, 5) trend generally northeast. The northeast trend is especially enhanced by strike filtering. It can readily be seen from Figures 7 and 9, as well as from Figures 1–6, that northeast-trending

gravity and magnetic anomalies extend as far west as 87° W longitude. The causative source or sources of these anomalies is not immediately evident, and they may be caused by infrastructural variations in the crust. However, comparison of these anomalies with the basement-rock map (Figure 12) indicates that the anomalies between about 85° W and 87° W longitudes approximately parallel anomalies associated with the subsurface Grenville province (discussed subsequently). A possible interpretation thus is that the northeast trends may reflect the western extent of Grenvillian infrastructural orogenesis. Under this interpretation, the Grenville front occurs between 84° and 85° W longitude, and a foreland zone extends to about 87° W longitude (cf. Thomas, this volume). This foreland zone is envisaged as being part of the adjacent older granite–rhyolite terrane, which has a subtle and incipient overprint of Grenville deformation. The foreland formed in passive response of a craton to stresses developed during compressional orogenesis.

A second possible explanation of some of the northeast-trending anomalies is that they may be associated with rifting and the development of dense (mafic) crust. This aspect is discussed in the next section.

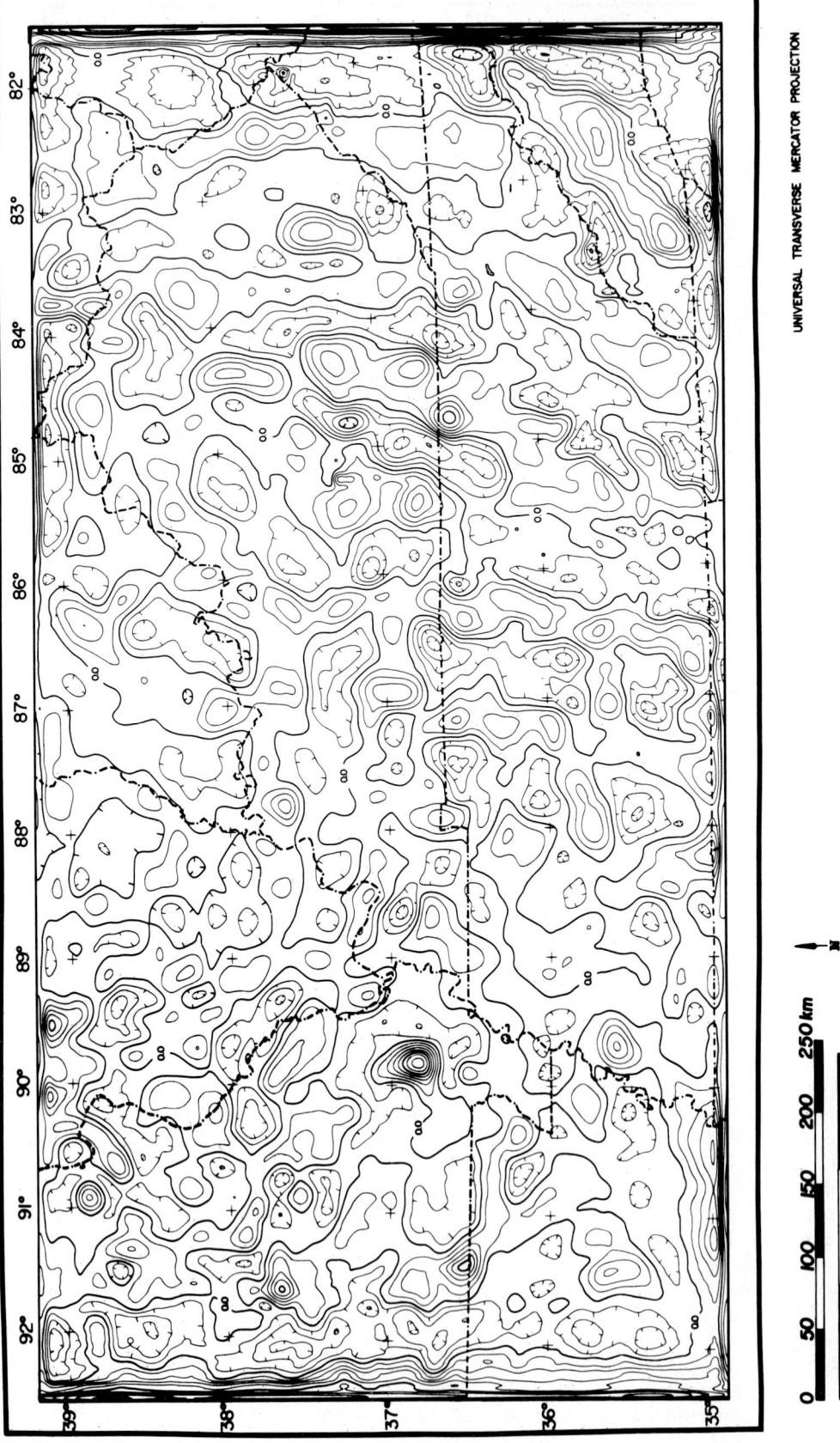

FIG. 4. Band-pass-filtered (8–100-km wavelengths) Bouguer gravity-anomaly map of east-central Midcontinent of the United States. Contour interval, 5 mGal. After Reed (1980).

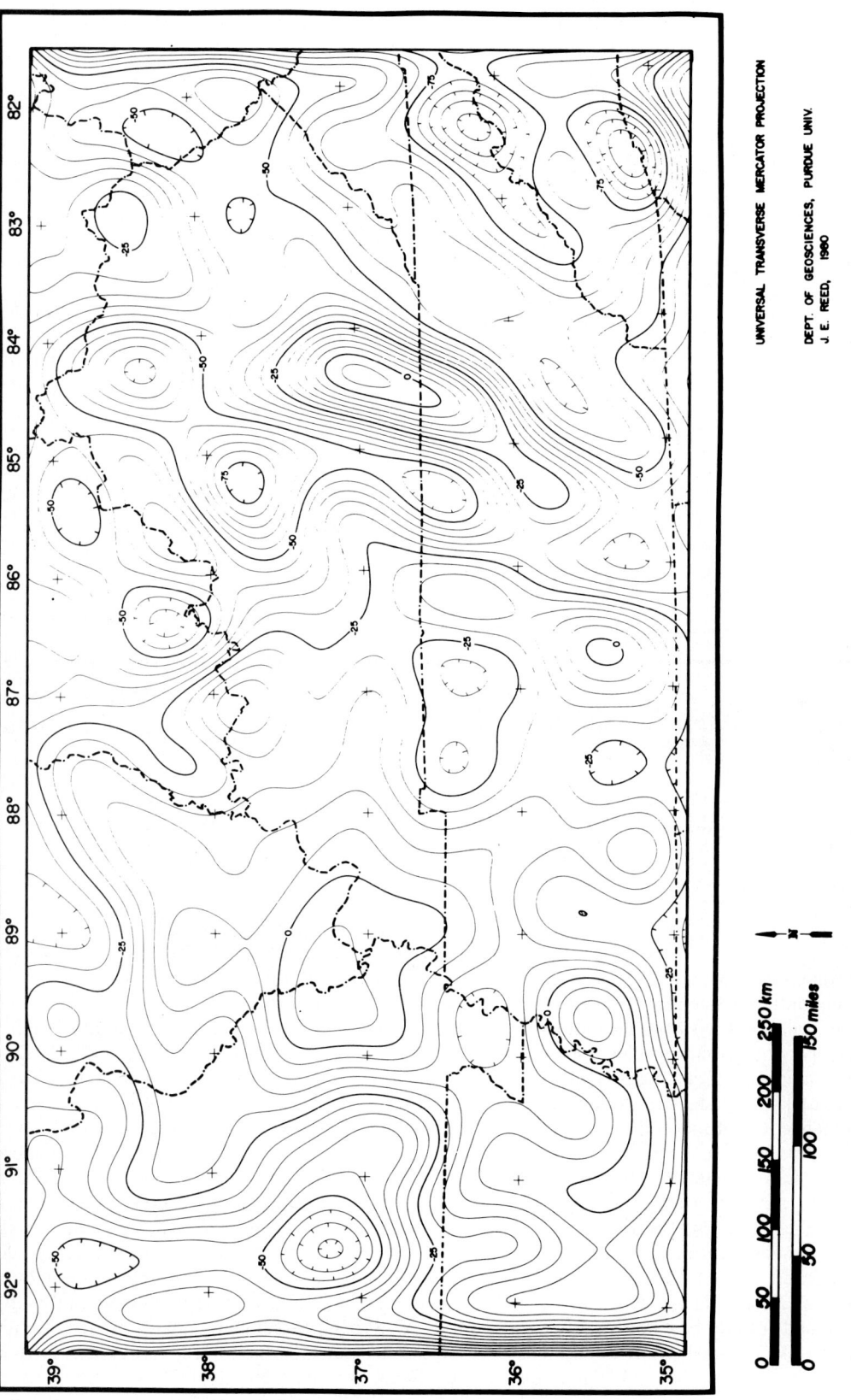

FIG. 5. Low-pass-filtered Bouguer gravity-anomaly map of east-central Midcontinent of the United States. Wavelengths from ∞ to 80 km are passed. Contour interval, 5 mGal. After Reed (1980).

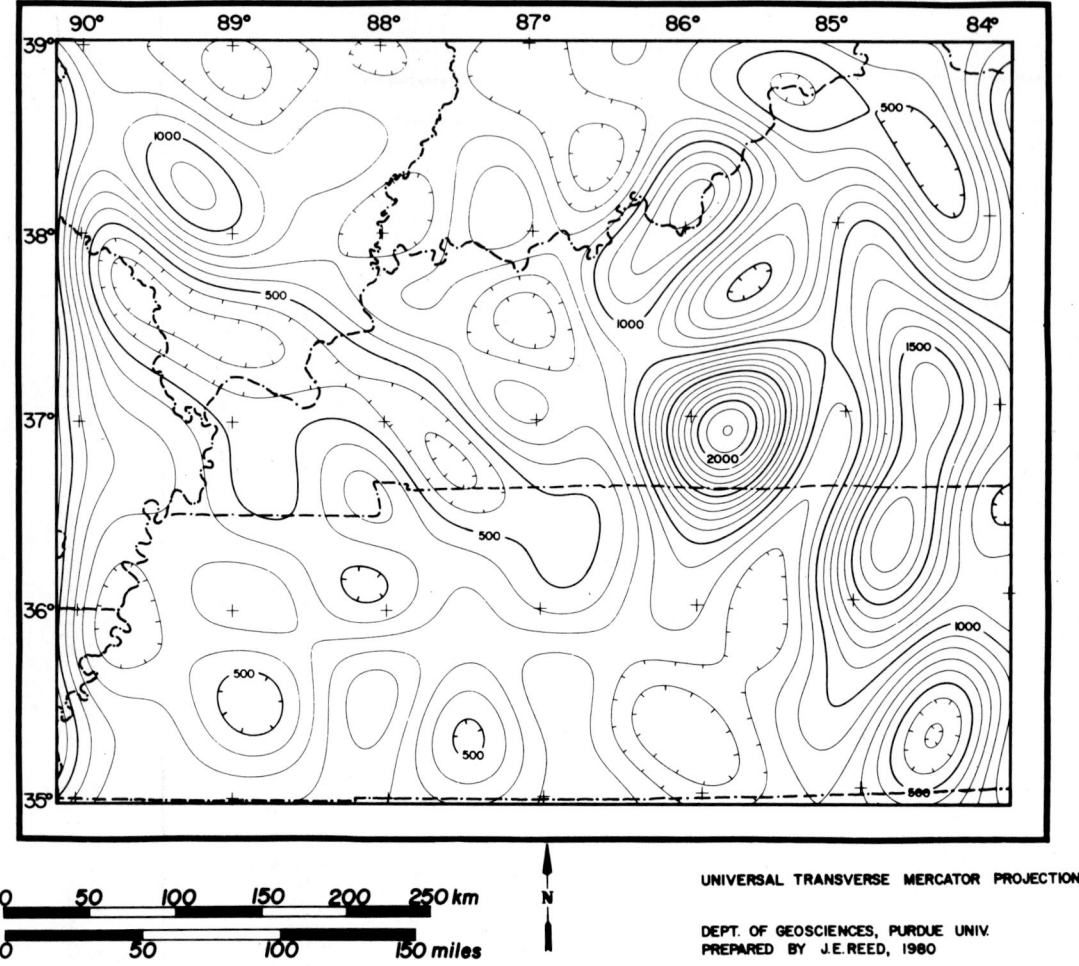

FIG. 6. Low-pass-filtered, reduced-to-the-pole, total-magnetic-intensity-anomaly map of east-central Midcontinent of the United States. Wavelengths from ∞ to 80 km are passed. Contour interval, 100 gammas (nT). After Reed (1980).

Mafic volcanic rocks

Basalts have been encountered in four widely spaced deep drill holes into the basement of the east-central Midcontinent (Figure 12). None of these drill holes has been located over a major geophysical anomaly. They occur instead along the flanks of major or minor Bouguer gravity highs. Only the basalt near Louisville, Kentucky (latitude 38.2° N, longitude 85.8° W), occurs over a magnetic high; the others lie along the flanks of highs or in proximity to complex magnetic anomalies. For this reason, it is difficult to infer the areal extent of the basalts or to correlate them with specific gravity and magnetic anomalies of regional extent.

Based primarily on linear gravity and magnetic anomalies, several Precambrian basaltic rift zones have been proposed to occur in the east-central Midcontinent (Keller et al., 1983). The most prominent and perhaps best known of these proposed rifts is the Kentucky anomaly (East continent gravity high). This anomaly, centered at about 38° N latitude and 84.4° W longitude (Figure 1), strikes north-northeast and has a Bouguer amplitude of more than 80 mGal (Keller et al., 1982). An intricate amoeboid pattern of magnetic highs and lows coincides in large part with the gravity high but has larger areal extent (Figure 2). The feature is thus both dense and magnetic and has been interpreted as a rift zone that is associated with mafic igneous rocks of inferred middle Keweenawan age, which extend to considerable depths in the crust (Lyons, 1970; Keller et al., 1975, 1982; Hawman, 1980; Mayhew et al., 1982). However, it needs to be pointed out that none of the deep wells to basement along this anomaly has bottomed in mafic igneous rock. A basalt (latitude 37.75° N, longitude 84.7° W) and a gabbro (latitude 35.8° N, longitude 84.95° W) occur along the flanks of the gravity high, but the other basement rocks consist of felsic volcanic rocks, chlorite schist, hornblende schist, and granitic gneiss (Figure 12). The basalt is interlayered with red arkosic sandstone, similar to the Keweenawan basalt–arkose association of the Lake Superior region, and is thus suggestive of a rift environment. Also of special note is the hornblende schist, which occurs along the northern nose of the main gravity high at 37.7° N latitude and 84.5° W longitude. This rock is a probable metabasalt, and its presence is a strong indication that the basaltic-rift-zone model does not fully describe the geologic relations, which are

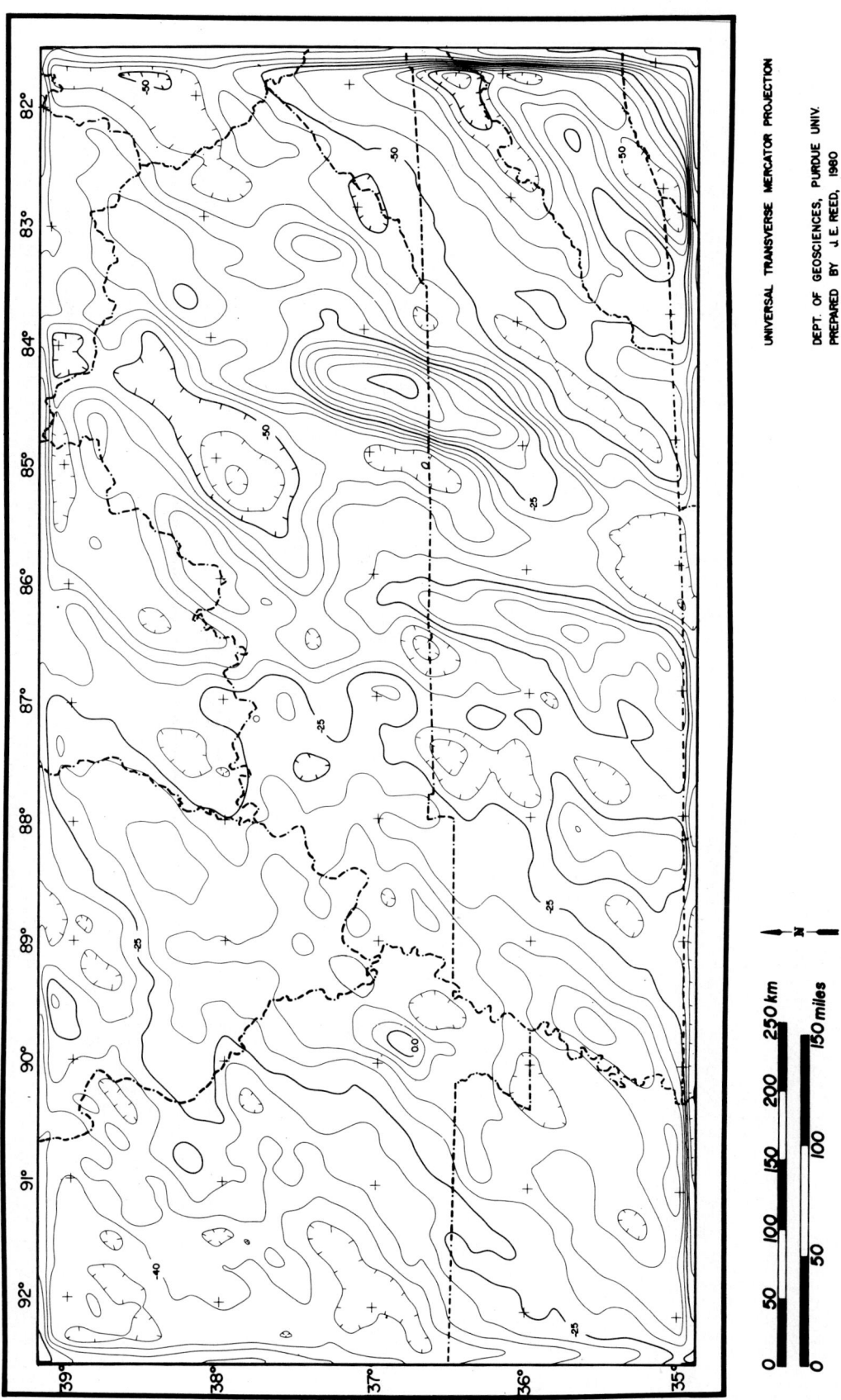

FIG. 7. Strike-filtered, upward-continued to 5 km, Bouguer gravity-anomaly map of east-central Midcontinent of the United States. Northwest (105°–165°) azimuths were rejected. Contour interval, 5 mGal. After Reed (1980).

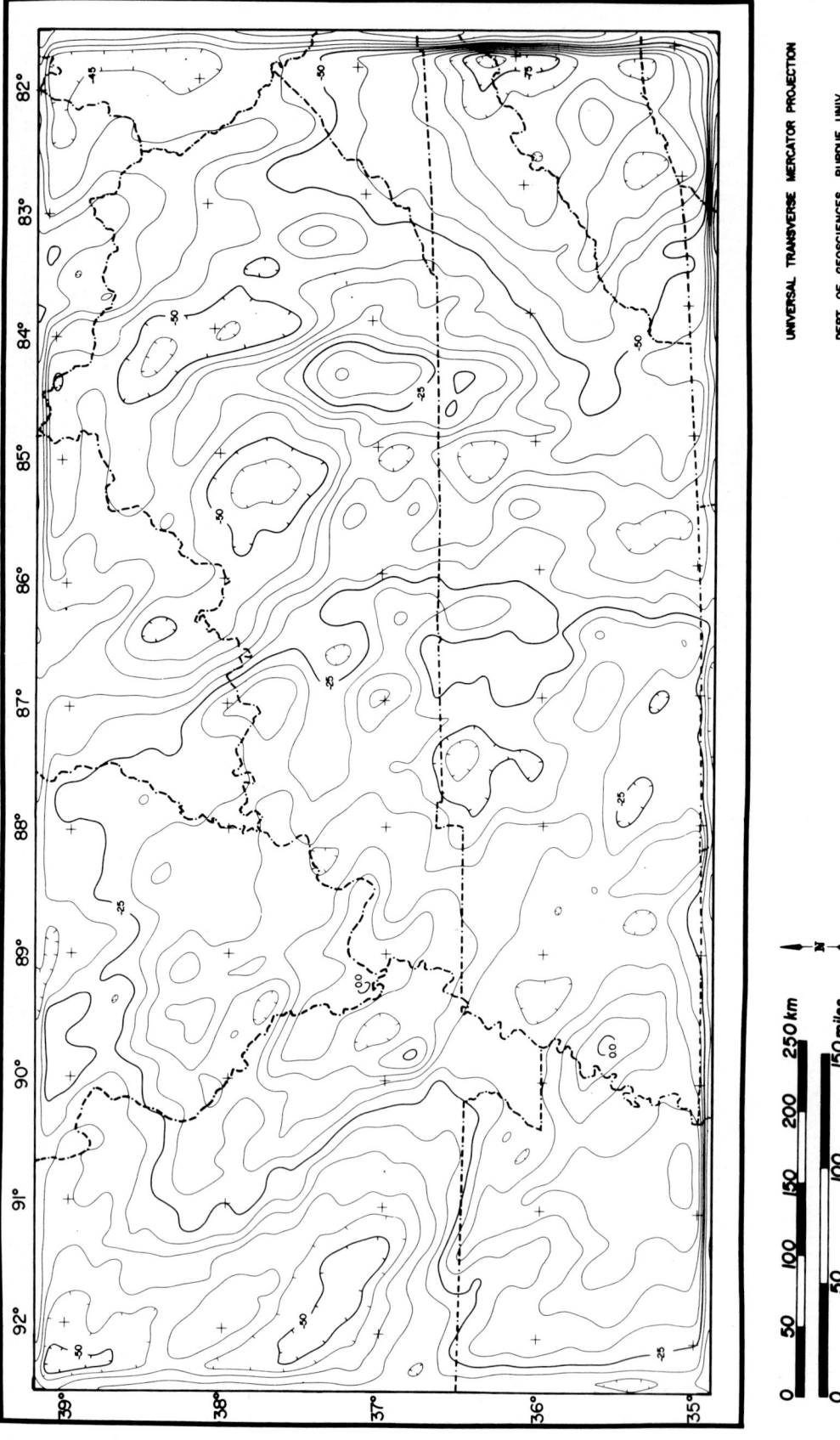

Fig. 8. Strike-filtered, upward-continued to 5 km, Bouguer gravity-anomaly map of east-central Midcontinent of the United States. Northeast (15°–75°) azimuths were rejected. Contour interval, 5 mGal. After Reed (1980).

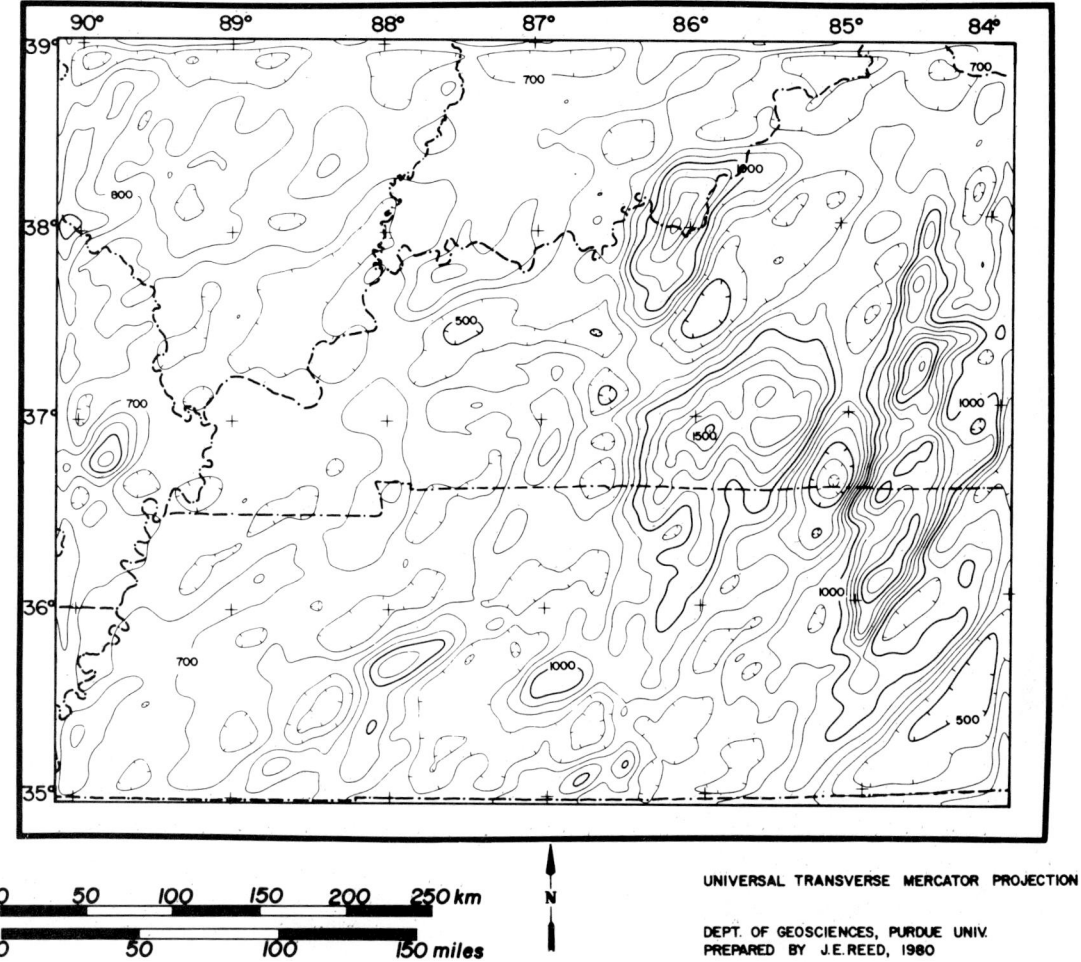

FIG. 9. Strike-filtered, upward-continued to 5 km, reduced-to-the-pole, total-magnetic-intensity-anomaly map of east-central Midcontinent of the United States. Northwest (105°–165°) azimuths were rejected. Contour interval, 100 gammas (nT). After Reed (1980).

more complex. Our current interpretation requires a two-stage model in which formation of the rift zone is followed by a major period of metamorphism and tectogenesis. We interpret this latter event as having occurred during the Grenville orogeny, and thus the basalts must predate this event. Lidiak and Zietz (1976) had previously mapped the Grenville front through this region between the basalt and hornblende-schist localities.

Another possible example of a rift zone is the linear anomaly east of Nashville, Tennessee (centered near latitude 36° N, longitude 86.4° W). This anomaly differs, however, in a number of aspects from the Kentucky anomaly. The Tennessee anomaly forms a relatively low-amplitude linear gravity high on both the Bouguer map (Figure 1) and the 8–100–km band-pass map (Figure 4). It also has an appreciable magnetic signature only in northern Tennessee, and this zone of high-amplitude complex magnetic anomalies is more extensively developed in southern Kentucky beyond the extent of the gravity high (Figure 2). It forms only a minor linear high on the long-wavelength magnetic map (Figure 6). A further characteristic of this region is that basalts do not appear to be present; the basement consists instead of an epizonal granitic complex intruded by diabase (Figure 12). Thus, the Tennessee anomaly does not appear to be a simple linear basaltic rift zone in the upper crust. However, several features are suggestive of a rift environment. The presence of riebeckite, commonly associated with rift volcanics, in the granitic complex (Keller et al., 1982) is supportive but not conclusive evidence of a rift setting. A second characteristic that is suggestive of a rift, and which this anomaly has in common with the Kentucky anomaly, is the presence of prominent linear gravity highs in both areas on the long-wavelength gravity map (Figure 5). On the basis of this evidence the Tennessee anomaly appears to be mainly a dense and deep crustal feature, possibly underlying an incipient rift zone. In this regard, it has some similarities to the upper Mississippi embayment, which von Frese et al. (1981) showed to be underlain by long-wavelength gravity highs and magnetic lows.

Basalt has been cored at three localities farther north: at Louisville, Kentucky (latitude 38.2° N, longitude 85.8° W); south of Cincinnati, Ohio, in Campbell County, Kentucky (latitude 38.9° N, longitude 84.5° W); and in Lawrence County, Indiana (latitude 38.8° N, longitude 86.4° W). All

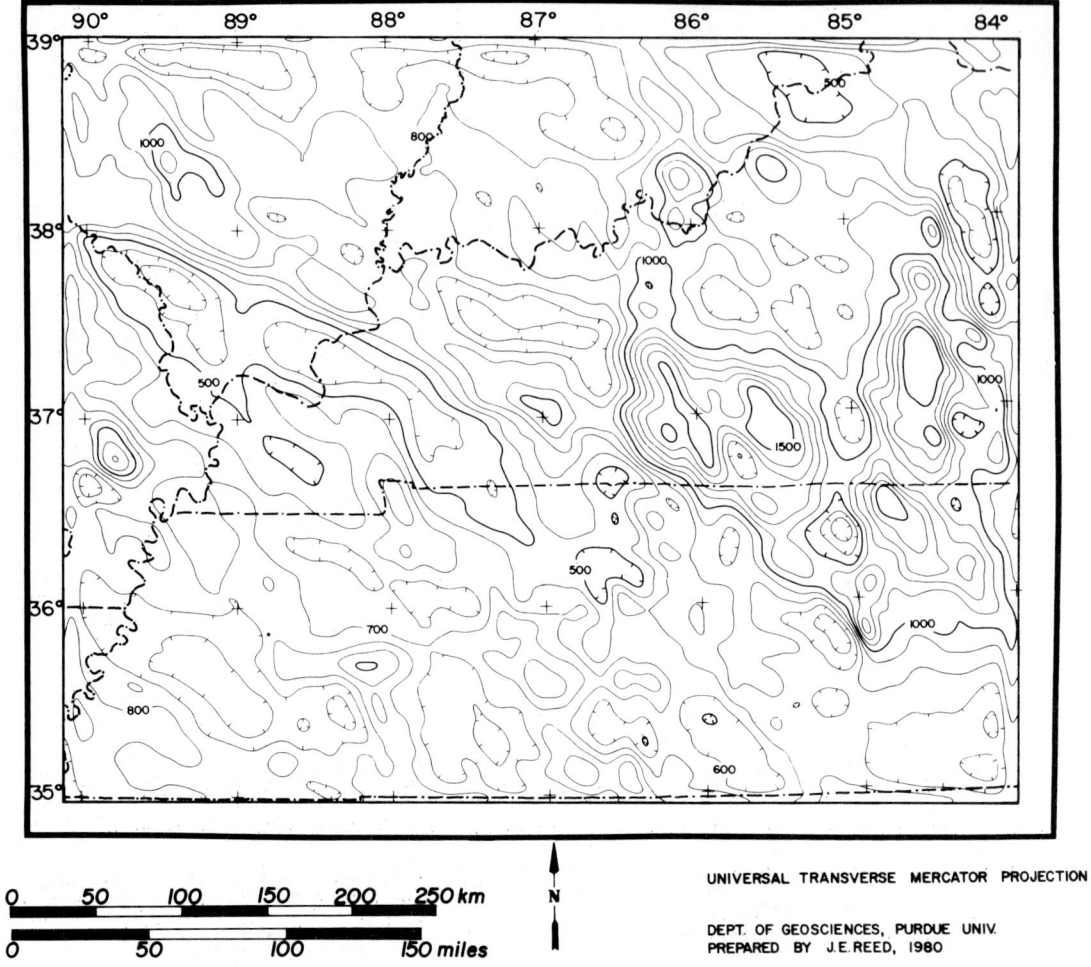

FIG. 10. Strike-filtered, upward-continued to 5 km, reduced-to-the-pole, total-magnetic-intensity-anomaly map of east-central Midcontinent of the United States. Northeast (15°–75°) azimuths were rejected. Contour interval, 100 gammas (nT). After Reed (1980).

three of the basalts have distinctive ophitic texture, which is strikingly similar to many Keweenawan basalts. The basalt at Louisville occurs in the saddle of a broad, rather indistinct northeast-trending gravity high, which has expression on both the short- (Figure 4) and long- (Figure 5) wavelength gravity maps. The magnetic anomalies form a complex series of both high-amplitude highs and lows (Figure 2) and longer wavelength highs (Figures 3 and 6) that occur in a broad zone that trends northeast through Louisville. The northeast linear trend is particularly evident in Figures 3 and 6. The trend is less distinct east of 85° W longitude and may terminate there. Farther northeast, the basalt in Campbell County, Kentucky, occurs along a narrow, northeast-trending magnetic high and a broad, east-northeast-trending gravity high. The lack of continuity between these geophysical anomalies and those in the vicinity of Louisville (cf. Figures 3, 4, 6, and 9) suggests that the two belts are not contiguous. They appear to be separated by a northwest-trending basement fault (Figure 12) that extends from Lexington to Bedford, Kentucky, near the Ohio River. This fault may account for the different geophysical signatures and trends in both regions. However, the linearity of the anomalies and the presence of basalts suggest that they may be related. One possible mode of origin is by rifting, but more detailed analysis of the region is necessary before a specific tectonic setting can be determined.

The basalt in Lawrence County, Indiana, is considered in the section on the New Madrid rift complex.

Grenville province

A third major basement-rock province is shown in Figure 12. This basement terrane, the subsurface extension of the Grenville province, occupies most of eastern Kentucky and eastern Tennessee and extends southward from the Canadian shield through southeastern Michigan and western Ohio. A proposed Grenville foreland zone is believed to be present and may extend as far west as about 87° W longitude. The foreland zone, which occurs within the older granite–rhyolite terrane, is characterized by subtle, apparently deep-seated northeast-trending gravity anomalies, which we attribute to the westernmost effects of Grenville orogenesis on the older craton. However, the Grenville front itself

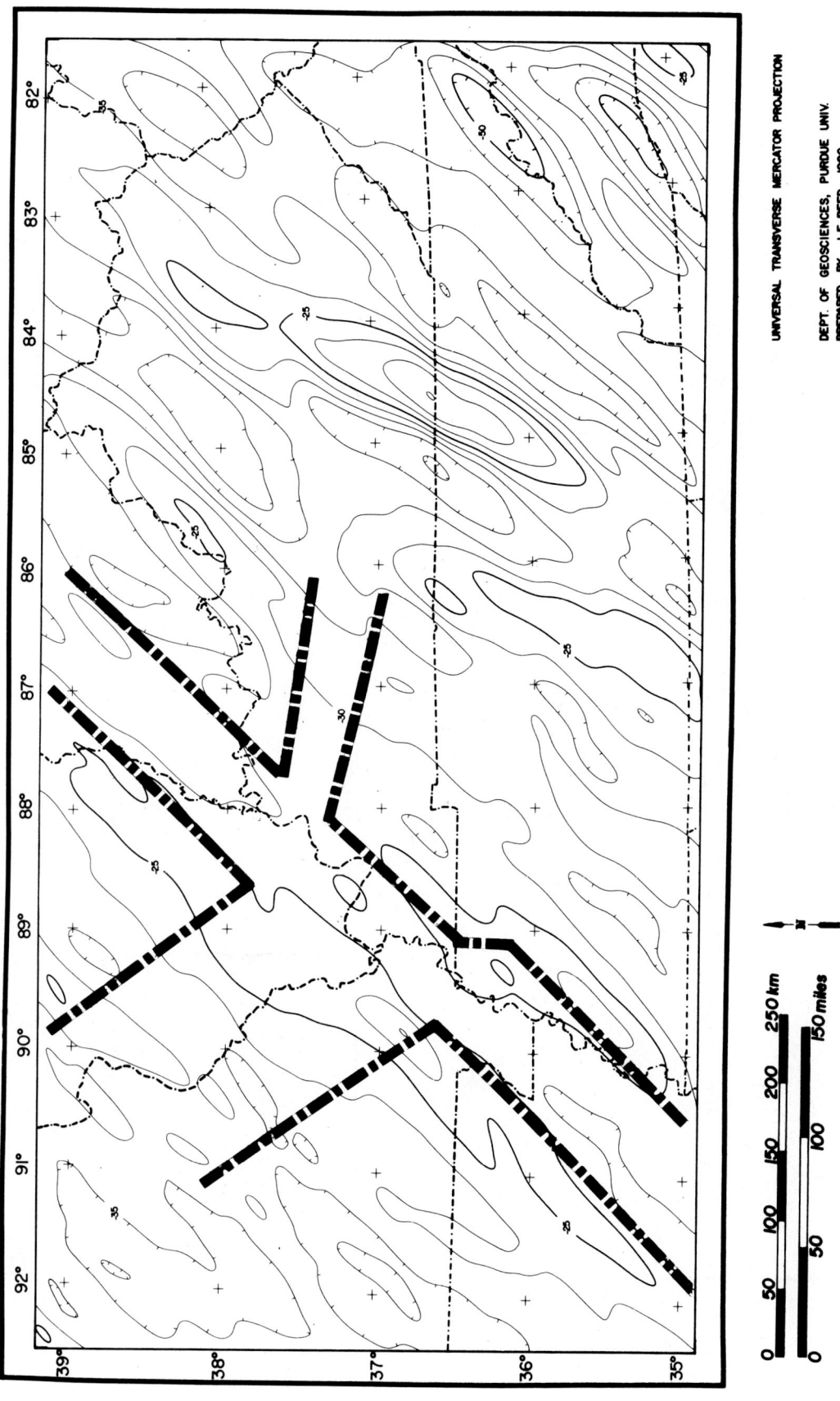

FIG. 11. Strike-filtered, upward-continued to 5 km, Bouguer gravity-anomaly map of east-central Midcontinent of the United States. Northeast (15°–75°) azimuths were passed. Contour interval, 5 mGal. After Reed (1980). Heavy dashed line delineates approximate boundary of New Madrid rift complex (see text).

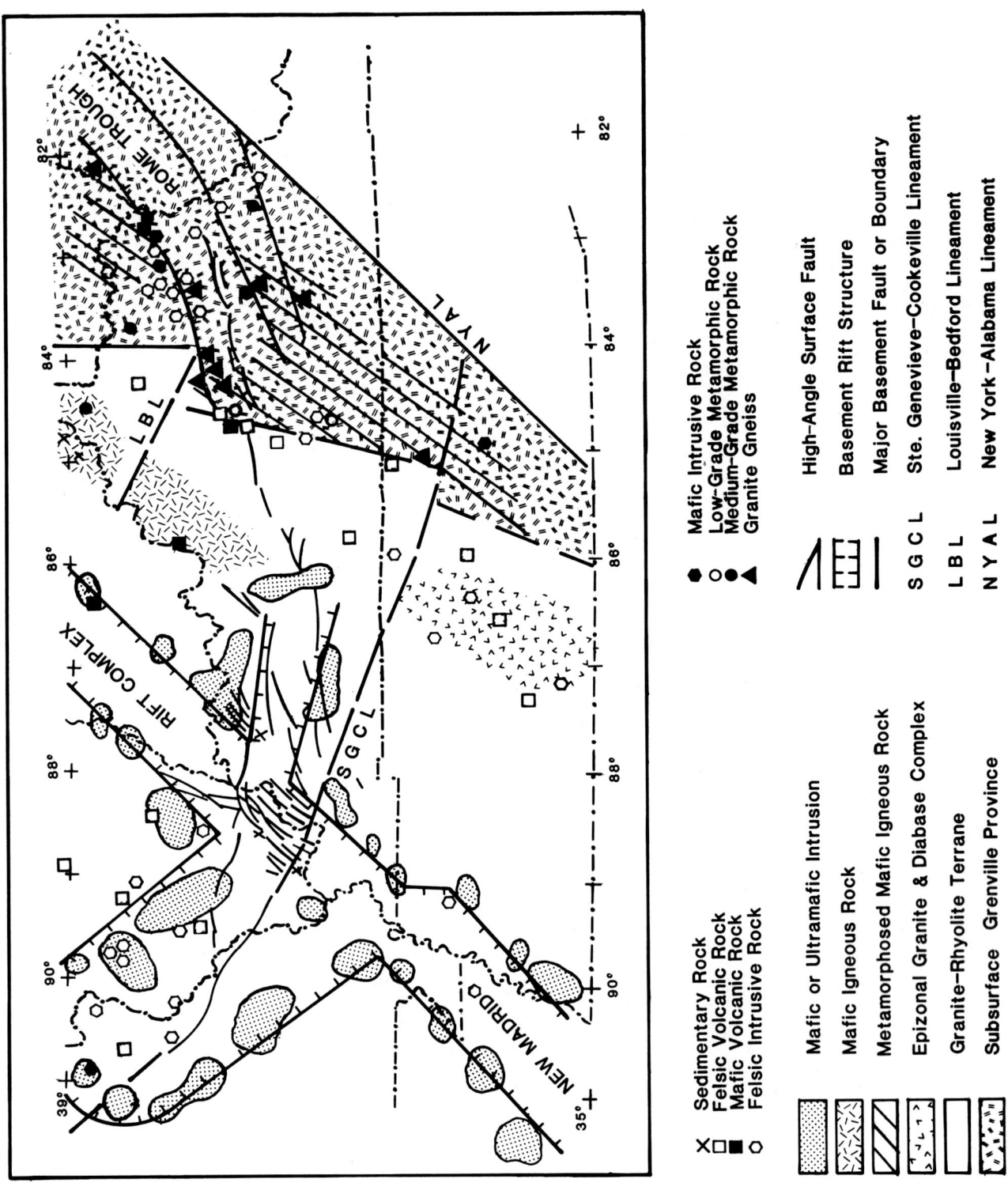

FIG. 12. Basement-rock map of east-central Midcontinent of the United States. Deep basement wells on map are indicated by symbols keyed to specific rock types in explanation.

probably represents a major lithologic, metamorphic, and tectonic boundary and lies between longitudes 84° W and 85° W along the western edge of the Kentucky anomaly. The amoeboid magnetic anomalies are more extensive than the linear gravity high and occupy a broad, north-northeast-trending belt between longitudes 84° W and 85° W at latitude 36° N and longitudes 83° W and 84° W at latitude 39° N. The trend and magnetic-anomaly pattern are characteristic of the Grenville basement to the north in Canada and in the Adirondack Mountains (Hinze and Zietz, this volume). The eastern limit of this anomaly pattern occurs along the New York–Alabama lineament.

The pronounced but less extensive Bouguer gravity high is the site of a proposed middle Keweenawan rift zone, which was discussed previously. Consistent with this interpretation is the presence along the gradients of the linear gravity anomaly of basalt interlayered with arkose and a gabbro. The inferred extent of mafic rocks is shown by a ruled pattern in Figure 12. However, these rocks occur east of the Grenville front and, under our interpretation, have been subjected to medium- and low-grade regional metamorphism. This metamorphism is most evident in northeastern and east-central Kentucky. It is less pronounced in southern Kentucky and Tennessee, but better basement well control is needed in these latter two areas to assess adequately an apparent change in metamorphic grade.

A variety of moderate- to deep-seated basement rocks, characteristic of the Grenville province, is present in eastern parts of Kentucky and Tennessee. The best well control is in northeastern Kentucky, and there are relatively few wells to basement in southeastern Kentucky and eastern Tennessee. The distribution of these rocks is shown in Figure 12. The Grenville basement consists of granitic gneiss, garnet hornblende gneiss, biotite hornblende gneiss, hornblende schist, biotite schist, chlorite schist, granite, diorite, gabbro, anorthosite, metarhyolite, and rhyolite. The rhyolite and metarhyolite occur near the Grenville front and are an indication that the front is not a metamorphic boundary throughout the entire region of Figure 12. Other unmetamorphosed rocks are the granites, diorites, and gabbro. The anorthosites show only low-grade metamorphic effects, as evidenced by the partial replacement of original pyroxenes by chlorite and actinolite and the preservation of original textures and unaltered primary plagioclase. The anorthosites form an apparent large body that occupies most of Boyd County in northeasternmost Kentucky. This diversity of rock types is a characteristic of the Grenville province. The presence of both metamorphosed and unmetamorphosed mafic and silicic rocks clearly demonstrates that this region has had a complex geologic history of magmatism and metamorphism.

The continuation of the Grenville province southward is somewhat speculative at present. South of about 35.5°–35.0° N latitude the strong north-northeast Grenville trends begin to lose definition as they converge with anomalies of the New York–Alabama lineament. The province probably continues into Alabama, but its precise location is not presently known.

New Madrid rift complex

A prominent subsurface failed-arm rift, the New Madrid rift complex (Braile et al., 1982a, b), extends into the craton of the southern Midcontinent and occupies most of the western part of Figure 12. The main rift structure is the Reelfoot rift (Ervin and McGinnis, 1975), which trends northeast from its intersection with the Ouachita front. The central graben was mapped by Hildenbrand et al. (1977) on the basis of two parallel trends of high-amplitude, circular gravity and magnetic anomalies, which are interpreted to be due to mafic intrusions. The left-lateral deflection in the rift at about 36.5° N latitude (Braile et al., 1982a) and the magnetic lows associated with the central rift are suggested in Figure 11. In southern Illinois, the rift splits into three distinct tectonic arms, which Braile et al. (1982b) refer to as the St. Louis arm, the southern Indiana arm, and the Rough Creek graben (Soderberg and Keller, 1981). Correlative circular gravity and magnetic anomalies delineate the margins of these arms, which also are interpreted as being due to mafic intrusions (Figure 12). The deep drill hole in Lawrence County, Indiana (latitude 38.8° N., longitude 86.4° W), provides support for this interpretation. This well occurs on the flank of the northeasternmost circular anomaly of the southern Indiana arm and penetrates ophitic basalt overlain by red arkosic sandstone. Sparse deep-well control in the vicinity of the complex indicates that pre–Upper Cambrian arkosic sedimentary rocks and basalts occur in the rift zone but are absent outside the rift (Lidiak et al., 1982). Recent drilling in the area interpreted as the main central graben encountered approximately 500 m of red arkosic sandstone lying directly on a basement of granitic gneiss (Denison, in press). This result also is in accord with a rift interpretation. The rift complex owes its origin to late Proterozoic–early Paleozoic plate-tectonic events. It is a reactivated structure that currently controls the location of earthquake epicenters in the New Madrid area and had localized intrusive and fault activity during Mesozoic and Cenozoic time (Braile et al., 1982b).

Rome trough

The Rome trough of eastern Kentucky and West Virginia is an apparent deep rift or graben structure that has had a complex history of concomitant sedimentation and movement along boundary faults and local fault blocks, which has produced great differences in sedimentary thicknesses, within the trough system (Woodward, 1961; McGuire and Howell, 1963; Webb, 1969, 1980; Silberman, 1972; Harris, 1975; Ammerman and Keller, 1979). The trough is bounded on the north by the Kentucky River fault zone and its eastward subsurface extension, and it is bounded on the south by several uplifts and basins (Ammerman and Keller, 1979). The boundary fault on the north has magnetic expression (Figures 2 and 12). The southern boundary fault is less distinct but does occur along several breaks in the magnetic anomalies. The central trough coincides with a northeast-trending gravity low, which can be seen on both the long-wavelength map (Figure 5) and the northwest-reject, strike-filtered map (Figure 7). The basement underlying the western part of the Rome trough has a distinct Grenville signature. This pattern becomes more subdued toward the northeast, reflecting the deepening of the trough as it enters West Virginia. Parts of the trough are associated with gravity and magnetic minima and contain as much as 6 km of sedimentary rock (Keller et al., 1983). Ammerman and Keller (1979) evaluate several possible modes of origin for the trough, and formation of a continental rift zone by extensional tectonism seems to be the most plausible of these.

Structural lineaments

One of the most prominent magnetic lineaments in eastern North America is the New York–Alabama lineament (King and Zietz, 1978). This lineament extends for more than 1 600 km and passes through eastern Tennessee and southeastern Kentucky (Figures 2, 3, 9, and 12). West of the lineament the anomalies have an overall Grenville signature of high-amplitude highs and lows. East of the lineament the magnetic anomalies are mainly negative and have broader wavelengths that parallel deep structures associated with the Appalachian Mountains. The steep magnetic gradient passing through Chattanooga and Knoxville coincides with the western edge of the Appalachian Bouguer gravity minimum (Figure 1) and with the western limit of the imbricate and foreland thrust provinces (Milici, 1980) of the Appalachian thrust belt. Comparison of Figures 4 and 5 shows that the lineament is better developed on the long-wavelength gravity map, indicating that it is associated with deep Appalachian structures rather than with shallower structural elements. The lineament is also clearly evident on both the northwest and northeast strike-filtered maps (Figures 7 and 8). King and Zietz (1978) propose that the lineament represents a major crustal break or suture that separates the southeastern edge of a stable crustal block from the Appalachian deformed belt.

The 38th-parallel lineament (Heyl, 1972; Lidiak and Zietz, 1976) is a major structural feature that coincides approximately with a series of generally west-trending surface faults that extend through Kentucky, southern Illinois, and eastern Missouri (Figure 12). Its relation to the New Madrid rift complex and other rift structures is discussed by Braile et al. (1982b). The lineament has both gravity and magnetic expression, indicating that it is a basement as well as a surface structure. In central Kentucky it occurs along breaks in the gravity and magnetic pattern of which highs predominate to the south and lows to the north. Toward the west the lineament merges with the strong northwest-trending anomalies that cross the New Madrid rift complex.

The prominent northwest-trending gravity and magnetic anomalies of western Kentucky, southern Illinois, and eastern Missouri (Lidiak and Zietz, 1976; Hildenbrand et al., 1977, 1982, 1983; Braile et al., 1982a, b) occur along a strong gradient that separates positive and negative anomalies of large magnitude and probably represents a major, steeply dipping crustal discontinuity. The gravity and magnetic anomalies on opposite sides of the gradient bear an inverse relation to one another. Figures 3, 6, 8, and 10 show clearly that linear magnetic highs predominate on the north side of the gradient and that linear gravity highs predominate on the south side. Drill-hole basement data are too widely spaced and insufficient to evaluate whether the feature is represented in the near-surface basement rocks; however, the linear gradient is so pronounced that it clearly reflects a major intracrustal structural break. This gradient may represent a boundary between early Proterozoic and later Proterozoic basement structures. The gradient crosses the New Madrid rift complex and appears to mark the northern boundary of major gravity-anomaly highs that are in part associated with the rift complex. The gravity highs may represent a broad contact zone along which mafic intrusions were pervasively emplaced.

The Ste. Genevieve–Cookeville lineament (Figure 12) is another extensive magnetic lineament that parallels the northwest-trending magnetic and gravity anomalies in eastern Missouri, southern Illinois, and western Kentucky. It coincides in part with the Ste. Genevieve fault and its subsurface extension (Lidiak and Zietz, 1976). It can be traced to the southeast north of Cookeville, Tennessee, along breaks in the gravity- and magnetic-anomaly patterns (Figures 1 and 2). Southeast of Cookeville it offsets anomalies associated with the Grenville front and appears to extend as far as the New York–Alabama lineament. The lineament is more evident on the 8–100–km–wavelength gravity map (Figure 4) than on the long-wavelength map (Figure 5). It also has expression along deflections and terminations of anomalies on the northeast strike-filtered map (Figure 11). Our interpretation of this lineament is that it represents a Precambrian basement fault zone, parts of which (Ste. Genevieve segment) were reactivated in Phanerozoic time.

A fourth structural lineament shown in Figure 12 is the Lexington–Bedford lineament. This feature subparallels the Ste. Genevieve–Cookeville lineament, but it is not as extensive. The lineament is also interpreted as being a Precambrian basement fault. Its major expression occurs north of Lexington, where the Grenville front is offset about 50 km. The lineament continues northwestward toward Bedford and the Ohio River along breaks and changes in trend of gravity- and magnetic-anomaly patterns. The lineament has expression on almost all of the geophysical maps (Figures 1–10). Particularly evident is the apparent right-lateral offset in the broad gravity (Figure 1) and magnetic (Figures 2, 3) lows that occur north and west of Lexington. This offset also can readily be seen on both the short- (Figure 4) and long- (Figure 5) wavelength gravity maps and on the long-wavelength magnetic map (Figure 6). Farther northwest the lineament appears to separate the prominent anomalies in the vicinity of Louisville from those toward the northeast in northernmost Kentucky (Figures 7–10).

Both the Lexington–Bedford lineaments and the previously described Ste. Genevieve–Cookeville lineament offset the Kentucky anomaly, which is associated with a proposed Keweenawan rift zone and the western edge of the Grenville front. They may thus be reactivated faults that originated as transforms which were part of that old rift zone.

RELATION BETWEEN DRILL-HOLE DATA AND GEOPHYSICAL ANOMALIES

This integrated study of the geology and geophysics of the main deep structures of the east-central Midcontinent provides an excellent opportunity to assess the degree to which basement rocks encountered in the deep drill holes correspond to the regional geophysical anomalies. In this section we discuss the relation between the physical properties of the basement rocks and specific geophysical anomalies.

Rock types and amplitudes of geophysical anomalies

Wells to basement were plotted on the aeromagnetic map (Figure 2) and on the Bouguer gravity map (Figure 4) from which wavelengths longer than 100 km were removed. Specific rock types are grouped into general rock categories to augment discussion. In these groupings, most of the felsic extrusive rocks are rhyolites, most felsic intrusive rocks are granites, most metamorphic rocks are mafic or pelitic schists, most mafic extrusive rocks are basalts, and most mafic intrusive rocks are gabbros or anorthosites. The well data are relatively sparse but are not atypical for many regions of the Midcontinent. The widely spaced basement data points are a clear indication that the use of geophysical data is critical in accurately interpreting the basement geology.

A comparison of Figures 2 and 4 with Figure 12 shows that some of the rock types are not associated with an expected geophysical signature. For example, not all of the basalts and gabbros coincide with gravity and magnetic highs. Similarly, some of the rhyolites occur in areas underlain by magnetic highs. It is also apparent that, in general, there is better agreement between geophysics and basement-rock type on the gravity map than on the magnetic map.

To gain a more quantitative insight into the degree to which correlations exist, the aeromagnetic- and Bouguer gravity-anomaly values that coincide with each well location in Figure 12 were determined. The intensities of the anomalies associated with each main type were then compiled. The results are summarized in Figures 13 and 14. In general, there is only moderate to poor correlation between rock type and both magnitude of total-intensity magnetic anomalies and Bouguer gravity anomalies. The magnetic results shown in Figure 13 indicate considerable overlap in total-intensity magnetic values for all the main rock types. Mafic intrusive rocks occur in areas of higher magnetic intensity, but, surprisingly, mafic extrusive rocks do not, possibly reflecting the fact that most mafic flows in the area are thin. Considering metamorphic rocks, which include both low and medium metamorphic grades and mafic and pelitic compositions, there is no relation to grade of metamorphism or composition. Also surprising is the fact that most felsic intrusive rocks occur in areas of relatively high magnetic intensity and that felsic extrusive rocks cover a wide range of magnetic-intensity response. The Bouguer gravity-anomaly data (100–8–km band-passed) listed in Figure 14 also show a wide variation for the main rock types. There is, however, some tendency for both extrusive and intrusive mafic igneous rocks to coincide with positive Bouguer anomalies and for felsic intrusive rocks and granitic gneisses, although variable, to be associated with lower Bouguer anomalies. Of the metamorphic rocks, two amphibolites reflect higher Bouguer anomalies than do lower grade rocks and more silicic compositions. The felsic extrusive and intrusive igneous rocks occur in areas of widely different Bouguer gravity anomalies, although many of the rocks do occur in areas of generally lower intensity anomalies than do the mafic rocks. The occurrence of about one-half of the felsic igneous rocks in areas of higher anomalies is a clear indication that denser basement rocks occur beneath some of the felsic igneous rocks.

Relation between magnetic and gravity anomalies

We have also plotted by rock type the total-intensity magnetic anomalies (Figure 2) versus the 100–8–km band-passed Bouguer gravity anomalies (Figure 4). The results are shown in Figure 15. As in the case with Figures 13 and 14, there is considerable overlap in geophysical anomalies among the rock types. Mafic igneous rocks occur in areas of consistently high gravity anomalies (0 to +11 mGal) but of wide variation in magnetic signature. Felsic volcanic rocks similarly occur in areas having widely different magnetic intensities. Gravity anomalies for most of these volcanics cluster at about 0 mGal. Felsic intrusive rocks show the greatest variation in gravity anomalies, most samples occurring between −10 mGal and +10 mGal.

Felsic intrusive rocks display a crude positive correlation between gravity and magnetic intensities, suggesting a relation between total magnetite content and the intensity of the anomalies. In contrast, felsic extrusive rocks show an apparent negative correlation between gravity and magnetic

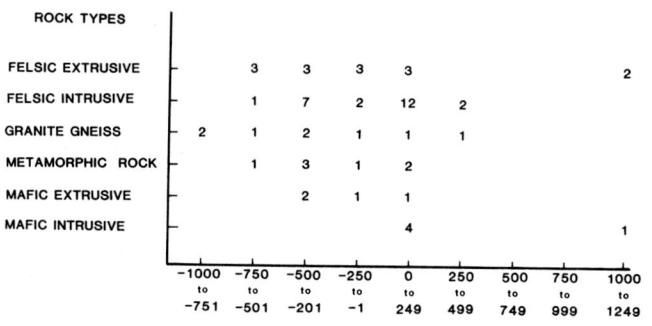

FIG. 13. Total-intensity magnetic anomalies (gammas) versus basement-rock types, east-central Midcontinent of the United States.

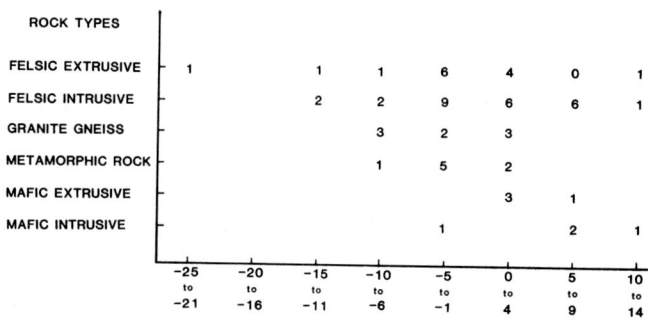

FIG. 14. 100–8–km band-passed Bouguer gravity anomalies (mGal) versus basement-rock types, east-central Midcontinent of the United States.

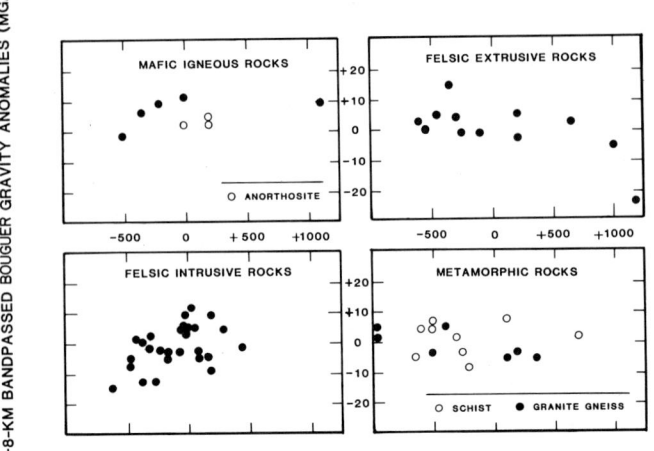

FIG. 15. Total-intensity magnetic anomalies (gammas) versus 100–8-km band-pass-filtered Bouguer gravity anomalies (mGal) associated with basement rocks encountered in drill holes, east-central Midcontinent of the United States.

anomalies. The cause of this inverse relation is not immediately evident, as a variety of factors can affect magnetic mineralogy (Haggerty, 1979). Among the possibilities are the generally higher Fe_3O_4 component in the magnetites of the less dense, more silicic extrusive igneous rocks and their enhanced NRM component relative to the total magnetic vector.

Metamorphic rocks also vary considerably, particularly in magnetic intensity. The mafic schists, both low and medium metamorphic grade, occur in areas of 0 to +6 mGal values and show no clear correlation with magnetic intensity. Most granitic gneisses of medium metamorphic grade occur in areas containing gravity anomalies between 0 and −10 mGal and show a negative correlation with the magnetic anomalies. This inverse relation may also be a compositional effect in which the magnetites of the more silicic rocks have a higher Fe_3O_4 component.

Relation between magnetic susceptibility and magnetic anomalies

Magnetic-susceptibility measurements were carried out on available basement-rock samples from the study area in order to evaluate the relation between measured susceptibility values and the regional aeromagnetic anomalies. Susceptibilities were measured on a Bison model 3101 Susceptibility Meter. Most of the samples studied were basement drill cuttings that were placed in a plastic cylinder having a diameter and length of 25.4 mm. Appropriate volume corrections were made. Cylindrical basement cores having a diameter and length of 25.4 mm were also used in the study. The results, plotted against total-intensity magnetic anomalies, are shown in Figure 16. Each rock type shows considerable variation within groups and overlapping values among groups, making characterization difficult. A major feature of Figure 16 is the excellent positive linear correlation of six of the eight metamorphic rocks. Such correlations are to be expected if the measured sample is representative of the basement-rock body in place and if the total-intensity magnetic value accurately reflects that body. Felsic igneous rocks show some tendency toward a broad, poorly defined positive correlation. Mafic igneous rocks may show a similar trend, but the data points are two few to be definitive. Felsic volcanic rocks show no direct relation between the two plotted parameters.

Causes of ambiguities in correlation

The preceding analysis shows that there is generally a moderate to poor correlation between drill-hole basement lithology, magnetic susceptibility, and geophysical anomalies. The reasons why the rocks do not display distinct magnetic and gravity signatures or magnetic susceptibility contrasts are varied. It must be kept in mind that rocks are not classified on the basis of susceptibility and density, and thus a 1:1 correlation should not be expected. This commonly results in contrasting rock types not having definitive geophysical signatures. There are other important factors as well. One of the possible causes of ambiguities in the interpretation of magnetic anomalies is the minimum associated with polarization effects caused by an inclined magnetic field (induced magnetization). This effect is only a minor problem on the maps presented in this paper. Comparison of Figures 2 and 3 demonstrates that there is no significant shift in anomalies on the reduced-to-the-pole map at this latitude. A second important factor is the NRM of Precambrian basement samples, where the total vector is not at a steep angle to the horizontal. In most Precambrian rocks the NRM effect is small (Hinze and Zietz, this volume), but the effect can be large and lead to errors. Both induced and NRM intensities need to be considered.

In this section, mainly we are not concerned with resolution of anomalous sources, latitude corrections, or inclined body problems but rather with more geologic aspects such as

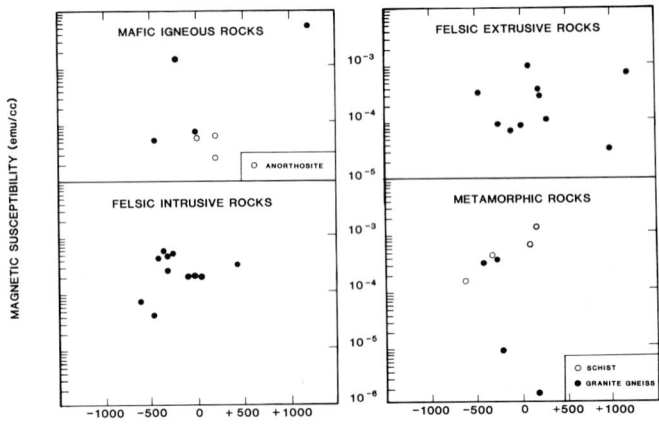

FIG. 16. Total-intensity magnetic anomalies (gammas) versus magnetic susceptibility (emu/cc) of basement rocks encountered in drill holes, east-central Midcontinent of the United States.

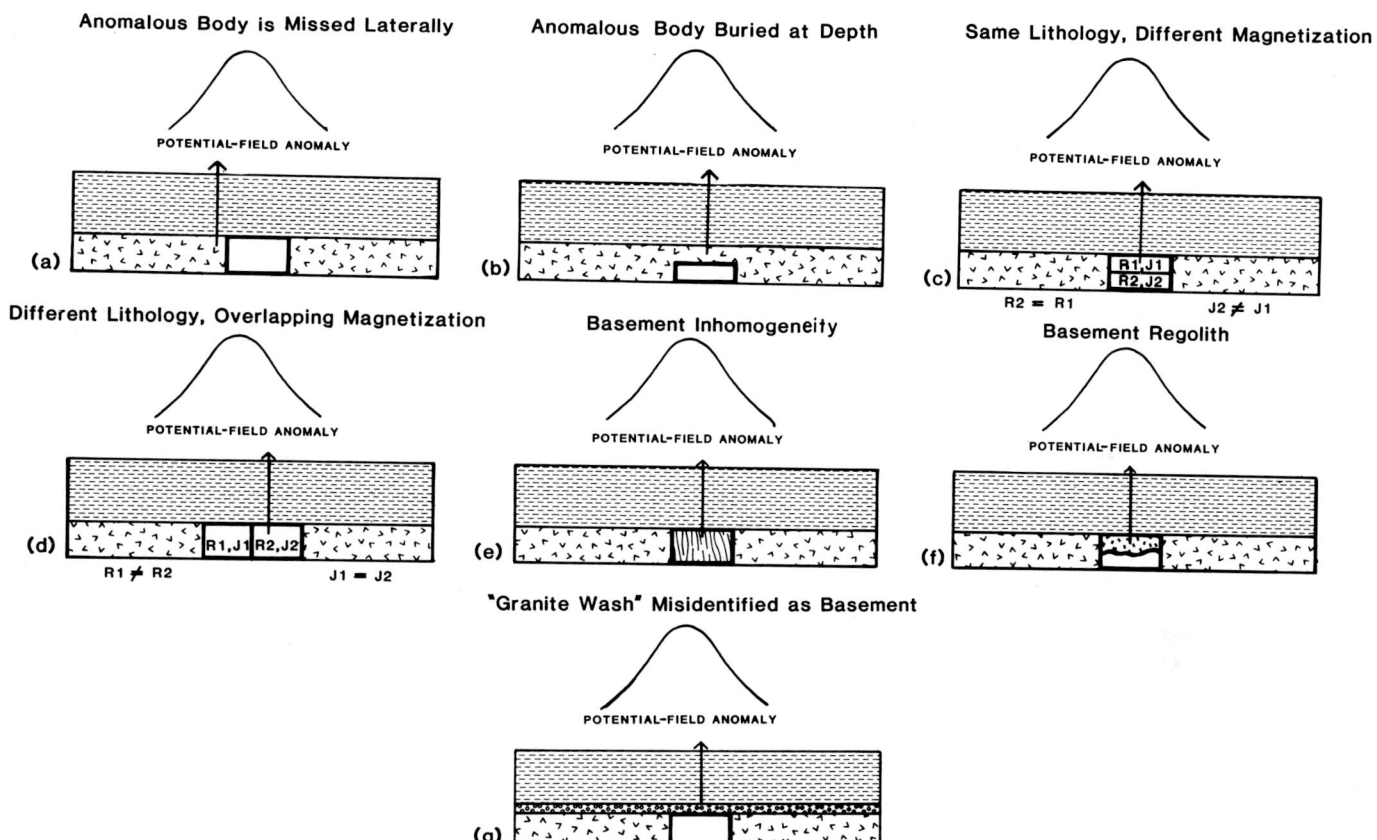

FIG. 17. Schematic illustration of ambiguities in correlation of geophysical anomalies with drill-hole basement lithology.

variations in basement lithology and sampling problems that lead to ambiguities in correlation. Examples of geologic factors that can result in incorrect interpretations are shown schematically in Figure 17.

Figure 17a depicts an anomalous body that produces a large regional potential-field anomaly. A problem in interpretation will result if a deep hole to basement is located within the domain of the anomaly but does not encounter the anomalous body itself. It penetrates instead the adjacent but different basement rock. If the geophysical anomaly has a highly characteristic shape, it is probable that modeling of the anomaly will result in a correct interpretation of the causative source. However, if the anomaly is more complex, the anomaly will be much more difficult to interpret and the nearby drill hole can lead to considerable ambiguity. This problem is compounded in areas where drill holes are widely spaced. We believe this is a particular problem in the east-central Midcontinent, where many of the anomalies are complex and sample control is sparse.

A second example (Figure 17b) involves an anomaly-causing body that is present at depth but does not reach the basement surface. A possible example is a mafic pluton emplaced into granitic country rock. A deep drill hole to basement over the anomaly penetrates the granite but is not deep enough to encounter the mafic intrusion. This is a major problem in interpretation because magnetic polarization and density variations occur throughout the thickness of the crust, whereas most deep wells penetrate less than 100 ft (30.5 m) of basement, commonly 10–20 ft (3–6 m) or less.

Figure 17c illustrates another type of problem in which two rock bodies—for example, two thick rhyolite flows—having closely similar silicate mineralogy and lithologic characteristics but strikingly different magnetic properties occur as adjacent bodies. One flow may be highly magnetic, the other essentially nonmagnetic. The problem can be further complicated if the two rocks are part of a layered sequence. Poor correlation will result if drilling encountered only the upper, nonmagnetic rhyolite and not the deeper equivalent that causes the magnetic-polarization variations. Another example of variable magnetic properties that is widespread in the Midcontinent involves magnetic and nonmagnetic granites. Granites can have widely different magnetic properties and intensities, and some are even associated with positive magnetic anomalies. It must always be kept in mind that the magnetic susceptibility of many rock types can vary by several orders of magnitude and that rock names are based on silicate mineralogy and not on magnetic mineralogy or geophysical properties.

A related problem is shown in Figure 17d. In this example, two juxtaposed rock bodies have different lithologies (gabbro and amphibolite) but closely similar magnetic-susceptibility contrasts, producing essentially identical magnetic-polarity variations. A single well drilled to only one of these bodies would result in an incorrect geologic interpretation that the anomaly is caused by one rock type only, when in fact both bodies contribute to the anomaly and both are widespread in the basement.

Figure 17e shows a potential-field anomaly developed in an area of inhomogeneous basement-rock lithology. Possible examples are migmatite terranes and steeply layered granitic gneiss and schist complexes. Anomalies in such terranes would probably be complex and occur on a regional scale. They could, however, be localized by the uplifting of a basement horst into more homogeneous terrane. If only one of the rock types in the basement complex is sampled by drilling, an incomplete assessment of the overall basement geology and geophysics results, and the anomaly will probably not be correctly evaluated.

Another cause of poor correlations between measured physical parameters and geophysical anomalies is the development of a thick residual basement regolith (Figure 17f). In the example shown in Figure 17f, the drill hole bottomed in the regolith zone and did not penetrate fresh rock. Fresh rock typical of the causative geophysical anomalies is, however, present at deeper levels, where the effects of chemical weathering are diminished. The recovered sample, although reflecting the correct lithology of the parent body, is not chemically or physically representative of it or of the geophysical anomaly it produces. Density or susceptibility measurements on the altered samples would not characterize the parent body and would result in generally poor correlation. Regolith zones at the basement surface of the east-central Midcontinent are not uncommon, but it is difficult at present to evaluate just how widespread these zones are. Almost certainly some of the poor correlations shown in Figure 16 can be attributed to this factor.

The final example shown in Figure 17 that can lead to poor or ambiguous correlation is the so-called granite-wash problem (Figure 17g). In some areas of the Midcontinent, arkosic sedimentary rock (granite wash) forms a thin layer that directly overlies the true basement. This arkose can be very similar to crystalline rock and thus can be misidentified as basement. This is particularly true if the only basement samples that are available are in the form of tiny cutting chips and if the arkosic minerals are only slightly altered. If drilling is terminated in this layer and the rock is not identified as being sedimentary, then a poor or erroneous correlation will result. Not only will the true basement not be encountered, but any anomaly-causing body in the basement cannot be correctly assessed.

There are thus a number of potential problems that can result in poor correlations. A consideration of the various factors discussed here is a clear indication that considerable caution needs to be exercised in comparing geophysical data with basement-rock samples obtained from widely separated drill holes.

SUMMARY AND CONCLUSIONS

Recently compiled Bouguer gravity- and magnetic-anomaly maps of the east-central Midcontinent (covering the area approximately between 35°–39° N latitude and 82°–92° W longitude) provided the opportunity to study the tectonic framework of the basement rocks that lie buried beneath generally gently dipping Phanerozoic sedimentary rocks. A variety of wavelength filters, including continuation, band-pass, derivative, and strike filters, are useful in isolating and identifying particular attributes of anomalies associated with the basement rocks. These maps in conjunction with basement geologic data obtained from widely distributed drill holes are used to define the principal basement zones. The oldest rocks are the 1 400–1 500-m.y.-old anorogenic granites and rhyolites of the Central province. These essentially unmetamorphosed felsic igneous rocks occur in the western part of the region. A generally subtle west-northwest pattern of anomalies pervades the Central province, probably owing to a more ancient basement that underlies the felsic rocks. These trends change in central Kentucky and Tennessee to northeast-striking anomalies that may represent a foreland zone of the Grenville province developed in the granite–rhyolite terrane. The Grenville front itself trends south through eastern Kentucky and Tennessee. The front marks the boundary between unmetamorphosed felsic igneous rocks to the west and generally medium-grade metamorphic rocks to the east. The northerly trends associated with the Grenville province terminate abruptly at the New York–Alabama lineament, where long, linear, northeast-striking anomalies that correlate with Appalachian structural trends are present.

Gravity and magnetic anomalies and sparse well control suggest that several Precambrian rift zones are also present in the east-central Midcontinent. The most prominent linear anomaly is the Kentucky anomaly (East continent gravity high), which appears to have formed as a basaltic rift zone. However, this feature occurs east of the Grenville front, and the rift structure is apparently complicated by overprinting of the Grenville orogeny. A second rift zone, the New Madrid rift complex, transects the western part of the region and is outlined by a series of parallel, correlative gravity and magnetic anomalies that mark the margins of the late Precambrian–Cambrian rift. Two other possible rifts occur east of Nashville and in the vicinity of Louisville.

A comparison of the drill-hole basement lithology with the gravity- and magnetic-anomaly maps indicated that many of the basement samples did not coincide with the expected gravity or magnetic signatures. A study was thus undertaken to determine why samples from the basement do not always reflect the magnetic and gravity anomalies and to evaluate the factors that lead to ambiguities in correlation. Wells to basement were plotted according to rock types on recently compiled aeromagnetic and long-wavelength-cut Bouguer gravity-anomaly maps of the east-central Midcontinent, and anomaly values coinciding with each well location were determined. In general, there is rather poor correlation between rock type and both magnitude of total-intensity magnetic anomalies and Bouguer gravity anomalies (100–8-km band-passed data). There is, however, some tendency for mafic igneous rocks to be associated with positive Bouguer anomalies and for felsic rocks to be associated with lower Bouguer anomalies. Plots by rock type of Bouguer anomaly against magnetic-anomaly values show that felsic intrusive rocks have a crude positive correlation and that felsic extrusive rocks and granitic gneisses have a negative correlation. The magnetic susceptibility of basement samples plotted against total magnetic intensity shows no clear distinction among the main rock types. Metamorphic rocks do display a positive correlation of these two parameters for most samples, as do, to a lesser extent, felsic intrusive rocks and possibly mafic igneous rocks. The causes of the generally poor correlations are varied and include such factors as drill holes not having encountered main causative anomalies both laterally and at depth, general basement inhomogeneity, inhomogeneity of physical properties, basement layering, sample alteration, and lack of definitive geophysical properties.

ACKNOWLEDGMENTS

We are pleased to acknowledge our colleagues of the New Madrid Seismotectonic Study Group, in particular Tom Buschbach, for helpful discussion regarding this work. The research was supported by U.S. Nuclear Regulatory Commission contract no. NRC-04-81-195-01.

REFERENCES

Ammerman, M. L., and Keller, G. R., 1979, Delineation of Rome trough in eastern Kentucky by gravity and deep drilling data: AAPG Bull., 63, 341–353.

Braile, L. W., Hinze, W. J., Keller, G. R., and Lidiak, E. G., 1982a, The northeastern extension of the New Madrid seismic zone: Prof. Pap. 1236-L, U.S. Geol. Surv., 175–184.

Braile, L. W., Keller, G. R., Hinze, W. J., and Lidiak, E. G., 1982b, An ancient rift complex and its relation to contemporary seismicity in the New Madrid seismic zone: Tectonics, 1, 225–237.

Ceci, V. M., and Lidiak, E. G., 1983, Chemical composition of Precambrian rocks from the subsurface of Ohio [abstr.]: Geol. Soc. Am. Abstr. Programs, 15, 216.

Denison, R. E., in press, Basement rocks in northern Arkansas, in McFarland, J. D., III, Ed., Contributions to the geology of Arkansas: Ark. Geol. Comm., 2.

Denison, R. E., Lidiak, E. G., Bickford, M. E., and Kisvarsanyi, E. B., 1984, Geology and geochronology of Precambrian rocks in the Central Interior region of the United States: Prof. Pap. 1241-C, U.S. Geol. Surv.

Ervin, C. P., and McGinnis, L. D., 1975, Reelfoot Rift: Reactivated precursor to the Mississippi Embayment: Geol. Soc. Am. Bull., 86, 1287–1295.

Faure, G., and Barbis, F. C., 1983, Detection of neoformed adularia by Rb–Sr age determinations of granitic rocks in Ohio, in Augustithis, S. S., Ed., Leaching and diffusion in rocks and their weathering products: Theophrastus Publications S. A. (Athens), 307–320.

Haggerty, S. E., 1979, The aeromagnetic mineralogy of igneous rocks: Can. J. Earth Sci., 16, 1281–1293.

Harris, L. D., 1975, Oil and gas data from the Lower Ordovician and Cambrian rocks of the Appalachian Basin: Misc. Geol. Invest. Map I-917D, U.S. Geol. Surv., scale 1:2 500 000.

Hawman, R. B., 1980, Crustal models for the Scranton and Kentucky gravity highs: regional bending of the crust in response to emplacement of failed rift structures: M.S. thesis, Penn. State Univ.

Heyl, A. V., 1972, The 38th parallel lineament and its relationship to ore deposits: Econ. Geol., 67, 879–894.

Hildenbrand, T. G., Kane, M. F., and Hendricks, J. D., 1982, Magnetic basement in the upper Mississippi embayment region—a preliminary report: Prof. Pap. 1236-E, U.S. Geol. Surv., 39–53.

Hildenbrand, T. G., Kane, M. F., and Stauder, W., 1977, Magnetic and gravity anomalies in the northern Mississippi embayment and their spatial relation to seismicity: Misc. Field Stud. Map MF-914, U.S. Geol. Surv.

Hildenbrand, T. G., Kucks, R. P., and Sweeney, R. E., 1983, Digital magnetic anomaly map of central United States—description of major features: Geophys. Invest. Map GP-955, U.S. Geol. Surv., scale 1:2 500 000.

Hinze, W. J., Braile, L. W., Keller, G. R., and Lidiak, E. G., 1980, Models for midcontinent tectonism, in Continental tectonics: Natl. Acad. Sci., 73–83.

Hinze, W. J., and Zietz, I., The composite magnetic-anomaly map of the conterminous United States, this volume.

Johnson, R. W., Jr., Haygood, C., Hildenbrand, T. G., Hinze, W. J., and Kunselman, P. M., 1980, Aeromagnetic map of the east-central midcontinent of the United States: Rep. NUREG/CR-1662, U.S. Nucl. Regul. Comm.

Keller, G. R., Bland, A. E., and Greenberg, J. K., 1982, Evidence for a major late Precambrian tectonic event (rifting?) in the eastern midcontinent region, United States: Tectonics, 1, 213–223.

Keller, G. R., Bryan, B. K., Bland, A. E., and Greenberg, J. K., 1975, Possible Precambrian rifting in the southeastern United States: Eos, 56, 602.

Keller, G. R., Lidiak, E. G., Hinze, W. J., and Braile, L. W., 1983, The role of rifting in the tectonic development of the midcontinent, U.S.A.: Tectonophysics, 94, 391–412.

Keller, G. R., Russell, D. R., Hinze, W. J., Reed, J. E., and Geraci, P. J., 1980, Bouguer gravity anomaly map of the east-central midcontinent of the United States: Rep. NUREG/CR-1663, U.S. Nucl. Regul. Comm.

King, E. R., and Zietz, I., 1978, The New York–Alabama lineament: Geophysical evidence for a major crustal break in the basement beneath the Appalachian basin: Geology, 6, 312–318.

Lidiak, E. G., Kersting, J. J., and Hinze, W. J., 1982, Basal sandstones in the subsurface of the central midcontinent, United States [abstr.]: Abstr. Programs, Geol. Soc. Am., 14, 547.

Lidiak, E. G., Marvin, R. R., Thomas, H. H., and Bass, M. N., 1966, Geochronology of the midcontinent region, United States, Part 4, Eastern area: J. Geophys. Res., 71, 5427–5438.

Lidiak, E. G., and Zietz, I., 1976, Interpretation of aeromagnetic anomalies between latitudes 37° N and 38° N in the eastern and central United States: Spec. Pap. 167, Geol. Soc. Am.

Lyons, P. L., 1970, Continental and oceanic geophysics, in Johnson, H., and Smith, B. L., Eds., The megatectonics of continents and oceans: Rutgers Univ. Press, 147–166.

Mayhew, M. A., Thomas, H. H., and Wasilewski, P. J., 1982, Satellite and surface geophysical expression of anomalous crustal structure in Kentucky and Tennessee: Earth Planet. Sci. Lett., 58, 395–405.

McGuire, W. H., and Howell, P., 1963, Oil and gas possibilities of the Cambrian and Lower Ordovician in Kentucky: Spindletop Res. Cent. (Lexington, Ky.).

Milici, R. D., 1980, Relationship of regional structure to oil and gas producing areas in the Appalachian Basin: Misc. Geol. Invest. Map I-917F, U.S. Geol. Surv., scale 1:2 500 000.

Morelli, C., 1976, Modern standards for gravity surveys: Geophysics, 41, 1051.

Reed, J. E., 1980, Enhancement/isolation wavenumber filtering of potential field data: M.S. thesis, Purdue Univ.

Silberman, J. D., 1972, Cambro–Ordovician structural and stratigraphic relationships of a portion of the Rome trough: Spec. Publ. 21, Ky. Geol. Surv., ser. 10, 35–45.

Soderberg, R. K., and Keller, G. R., 1981, Geophysical evidence for deep basin in western Kentucky: AAPG Bull., 65, 226–234.

Thomas, M. D., Gravity studies of the Grenville province: Significance for Precambrian plate collision and the origin of anorthosite, this volume.

Van Schmus, W. R., and Bickford, M. E., 1981, Proterozoic chronology and evolution of midcontinent region, North America, in Kroner, A., Ed., Precambrian plate tectonics: Elsevier, 261–296.

von Frese, R.R.B., Hinze, W. J., Braile, L. W., and Luca, A. J., 1981, Spherical earth gravity and magnetic anomaly modelling by Gauss–Legendre quadrature integration: J. Geophys., 49, 234–242.

Webb, E. J., 1969, Geologic history of the Cambrian System in the Appalachian basin: Spec. Publ. 18, Ky. Geol. Surv., ser. 10, 7–15.

——— 1980, Cambrian sedimentation and structural evolution of the Rome trough in Kentucky: Ph.D. diss., Univ. Cincinnati.

Woodward, H. P., 1961, Preliminary subsurface study of southeastern Appalachian interior plateau: AAPG Bull., 45, 1634–1655.

Studies of gravity anomalies in Georgia and adjacent areas of the southeastern United States

Leland Timothy Long* and Anton M. Dainty*

ABSTRACT

Summaries of simple Bouguer gravity surveys in Georgia and their interpretations by the School of Geophysical Sciences, Georgia Institute of Technology, are presented. These surveys, primarily at a spacing of about 1 km, cover parts of the folded Appalachians, the Inner Piedmont, the Charlotte and Carolina Slate belts, and the Coastal Plain. In the folded Appalachians, positive anomalies of up to 5 mGal occur over outcrops of the carbonate members of the sedimentary sequence, indicating that these members are denser than the rest of the sequence. Few Bouguer anomalies are found over the fault contact between the folded Appalachians and the Blue Ridge, and these seem to correlate with topography. There is a general lack of anomalies owing to near-surface sources in the Inner Piedmont, although a large regional anomaly there is caused by lower crustal sources. In contrast, large Bouguer anomalies of 20-mGal amplitude are associated with surface exposures of mafic rocks (positive anomalies) and granites (negative anomalies) in the Charlotte and Carolina Slate belts adjacent to the Inner Piedmont. The fault contact between the Inner Piedmont to the northwest and the Charlotte or Carolina Slate belt to the southeast is the site of a steep gradient in the Bouguer gravity, owing to denser, more mafic rocks in the upper 5 km of the crust on the southeast side. In the Coastal Plain, many Bouguer gravity anomalies and total-magnetic-field anomalies are caused by Triassic rift structures. Another feature of the Bouguer gravity field in the Coastal Plain is a correlation of gravity and topography, suggesting that basement structures control topography in some regions.

INTRODUCTION

Gravity coverage in the southeastern United States consists of regional coverage (about 10 km spacing) that extends

*School of Geophysical Sciences, Georgia Institute of Technology, Atlanta, GA 30332.

over all of the region, together with a number of more detailed studies of particular regions, usually at about 1-km spacing. In general, these gravity readings have been reduced using simple Bouguer corrections only, although in some proprietary studies terrain corrections have also been made.

The regional gravity readings taken reside primarily in the files of the U.S. Defense Mapping Agency and for Georgia were incorporated in the Bouguer map of Georgia (Long et al., 1972). Large-scale regional compilations are to be found in the Society of Exploration Geophysicists' *Bouguer Gravity Map of North America* and the Appalachian orogen map of Haworth et al. (1980). There are similar regional compilations of the aeromagnetic data.

Detailed studies have been undertaken by, among others, Georgia Institute of Technology, Georgia Southwestern College, University of Georgia, University of South Carolina, University of North Carolina, Tennessee Valley Authority, and various oil companies. Many of these studies are proprietary and not generally available. In this paper we shall concentrate on detailed studies conducted in Georgia by the School of Geophysical Sciences, Georgia Institute of Technology (Figure 1). These studies encompass most of the publicly available work in this state.

We have examined the gravity anomalies in four general regions in the southeastern United States (see Figure 1). From northwest to southeast, they are (1) the folded Appalachians, (2) the Inner Piedmont, (3) the Charlotte and Carolina Slate belts, and (4) the Coastal Plain. The characteristics of the gravity anomalies are distinctive for each region and require that acquisition and analysis techniques be adapted to the study of each region. The folded Appalachian Mountains are bounded on the southeast by the Cartersville or Great Smoky fault system. The Inner Piedmont extends from the Brevard fault zone southeastward to the Towaliga and/or Middleton–Lowndesville faults. The Charlotte and Carolina Slate belt equivalents in Georgia extend from the Towaliga or Middleton–Lowndesville fault to the Coastal Plain.

Regional Bouguer anomalies in Georgia are controlled by two trends, the Paleozoic Appalachian trend and a Mesozoic trend associated with the rifting that formed the present-day Atlantic Ocean. In the folded Appalachians and the Blue

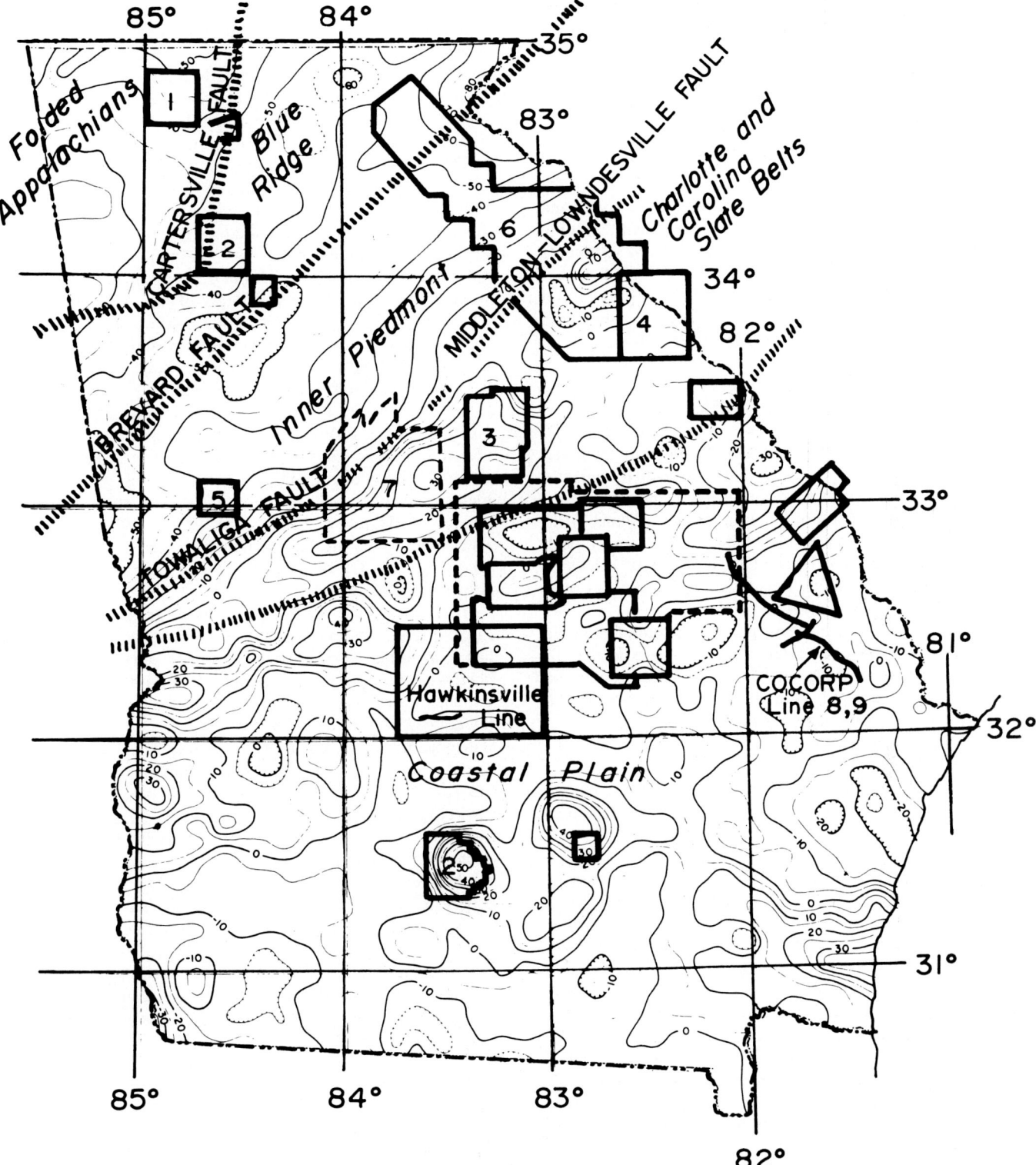

FIG. 1. Locations of gravity surveys obtained throughout Georgia. Dashed outlines indicate areas of data spaced at about 2.5 km. Solid lines indicate areas with data spaced at 0.5 to 1.0 km. References to areas are: 1—Guinn and Long (1978); 2—Long (1974); 3—O'Nour (1982); 4—Long et al. (1976); 5—Rothe and Long (1975); 6—Obaoye (1979); 7—Carpenter and Prather (1971). All other areas are unpublished internal reports on file at Georgia Tech or are proprietary surveys. Gravity base map from Long et al. (1972).

Ridge, regional Bouguer anomalies are negative, −40 to −90 mGal, and are appropriate for isostatic compensation of the mountainous topography. In the Inner Piedmont a large regional gradient is the transition from the negative values to the northwest to regional values near zero in the Charlotte and Carolina Slate belts to the southeast. The area of negative values in the Inner Piedmont indicates overcompensation.

The Charlotte and Carolina Slate belts exhibit 15- to 20-mGal positive anomalies associated with mafic rocks and 15- to 20-mGal negative anomalies associated with granitic rocks. Over the Coastal Plain, the unconsolidated sediments of Cretaceous and Tertiary age affect the gravity-field expression only in a minor way. The most striking gravity and magnetic features of the Coastal Plain are associated with relict rifts and volcanics derived from the Triassic–Jurassic opening of the Atlantic.

FOLDED APPALACHIANS

The folded Appalachians consist of an unmetamorphosed Paleozoic sedimentary sequence that has been folded and overthrust to the northwest. Gravity data near Maryville, Tennessee, and near Dalton, Georgia (area 1, Figure 1), in the folded Appalachians are spaced at about 1.0 km. In both areas the regional field produces Bouguer gravity values of between −40 and −60 mGal. Seismic data (Kean and Long, 1980) indicate a crustal thickness of 45 km.

Near Maryville, Tennessee (Figure 2), residual positive anomalies of 3 to 5 mGal correlate with surface exposures of Cambrian–Ordovician limestones and dolomites. Negative anomalies are associated with surface exposures of the Cambrian clastic sedimentary sequence. The shape of the anomaly indicates a southeastward dip of the (more dense) limestones and dolomites.

In the Dalton area map (Figure 2), near-surface occurrences of the Cambrian–Ordovician Knox Dolomite (shaded) correlate with a residual positive gravity anomaly of 2 to 4 mGal. The correlation is best in the east to northeast part of the Dalton area, where local negative anomalies occur over a syncline in which the younger, less dense rocks are exposed. The trend of increasingly negative regional Bouguer anomaly field toward the north-central part of the area has been interpreted as being partially caused by a thickening of the sedimentary rocks above the crystalline basement. Evidence from a refraction line (Guinn and Long, 1978) indicates deepening of the basin toward the north, with a total depth of 2.0 km to basement. Approximately 5 km east of the Dalton area map, Blue Ridge rocks are thrust over the sedimentary rocks. Near Carters Dam, Georgia, the thrust plane of the Cartersville fault is nearly horizontal, and the corresponding gravity expression of −2 mGal can be explained by the topography (Figure 3).

GRAVITY ANOMALIES OF THE PIEDMONT

The Piedmont province in Georgia is underlain by metamorphosed rocks that extend from the Blue Ridge to the Coastal Plain. The province can be divided into northeast–southwest–trending belts based on metamorphic grade and rock types (King, 1959). From northwest to southeast (see Figure 4), these belts are the Inner Piedmont, the Kings Mountain belt, the Charlotte belt, the Carolina Slate belt, the Kiokee belt, and the Belair belt. We shall concentrate on the Inner Piedmont and the combined area of the Charlotte and Carolina Slate belts, which were the site of detailed gravity surveys by Obaoye (1979), Frazier (1982), Long et al. (1976), O'Nour (1982), and Rothe and Long (1975).

The Inner Piedmont is underlain by rocks that may originally have been mixed sedimentary and volcanic rocks and that have been metamorphosed to sillimanite grade. The Bouguer gravity field is dominated by a distinctive gradient, here called the Piedmont gravity gradient, in which Bouguer anomaly values increase from about −50 mGal to the northwest in the Blue Ridge to values of about 0 mGal in the Charlotte and Carolina Slate belts to the southeast (Figure 4). This gradient is relatively smooth and extends over a distance of about 50 km in this area, indicating a source at lower crustal depths. One interpretation of the gradient is that it marks the edge of late Precambrian–early Paleozoic North America following Iapetan rifting (Long, 1979; Frazier, 1982; Cook and Oliver, 1981; Hatcher and Zietz, 1980) and that this ancient margin (possibly with its shelf sediments) has been subsequently buried by thrusting to the northwest (Hatcher, 1972; Cook et al., 1979, 1981). The Piedmont gravity gradient extends the length of the Appalachian orogen. Other interpretations have been proposed (Best et al., 1973).

Apart from the Piedmont gravity gradient, the Inner Piedmont is notable for a lack of distinctive gravity anomalies exceeding 2 mGal associated with surface rocks. The northwest boundary of the Inner Piedmont, a thrust fault known as the Brevard zone, has little gravity expression (Obaoye, 1979), although there is a more marked change in the magnetic field (Zietz et al., 1980). The lack of gravity signature suggests that the crust is similar on either side of the Brevard zone. This zone may not be a major crustal boundary, as suggested by Rankin (1975), among others. Similarly, the Towaliga fault in the Inner Piedmont has little expression in the Bouguer gravity field, although its trace can be seen in magnetic data. Frazier (1982) investigated the Elberton Granite, a post-metamorphic Paleozoic pluton near the boundary between the Inner Piedmont and the Charlotte and Carolina Slate belts, and found a maximum Bouguer gravity anomaly of 0.3 mGal associated with this body (Figure 4). This small anomaly contrasts markedly with the large anomalies of −10 to −20 mGal seen over post-metamorphic granites in the Charlotte and Carolina Slate belts and demonstrates both a different mode of emplacement (the Elberton Granite is a thin body, the other plutons are stocks) and a different origin for the parent magma (the Elberton Granite magma was formed by anatexis of nearby rocks, leading to a small density contrast; magma of the other plutons was formed at depth).

One body where significant gravity and magnetic anomalies can be measured, however, is the Meriwether diabase dike, running north–south in west-central Georgia counter to the regional trend. Rothe and Long (1975) (Figure 5) observed a Bouguer gravity anomaly of 1 to 2 mGal and a total-field magnetic anomaly of 1 000 gammas at ground level associated with this dike. They noted that an aeromagnetic survey with a flight-line spacing of 1 mile (1.6 km) at an

Gravity Anomalies in Georgia

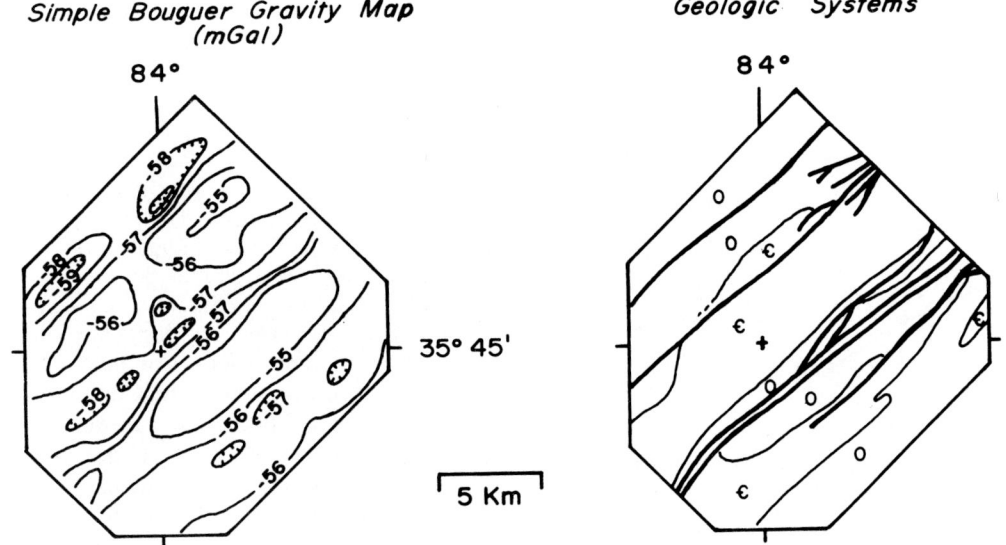

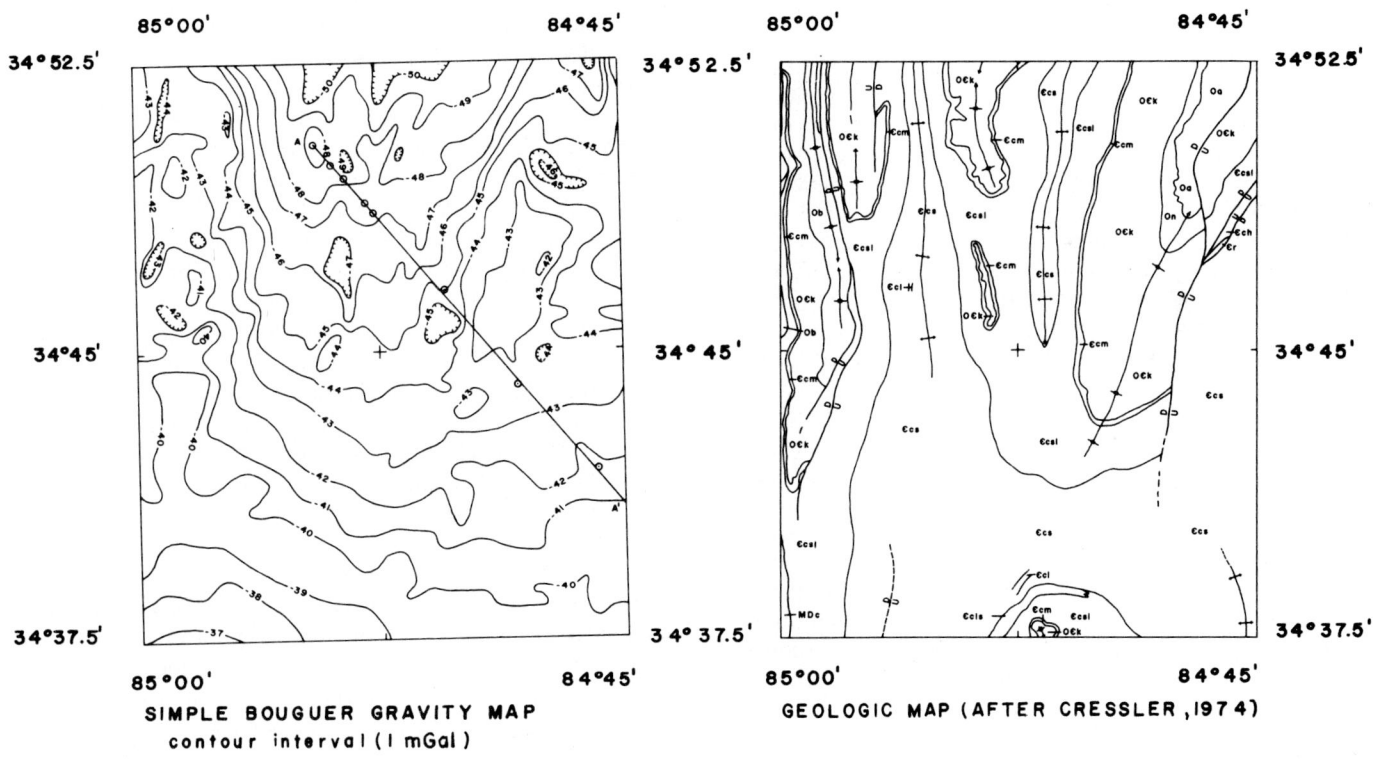

FIG. 2. Comparison of gravity anomalies and near-surface geology in the Maryville, Tennessee, and Dalton, Georgia, areas. C = Cambrian remnant; O = Ordovician.

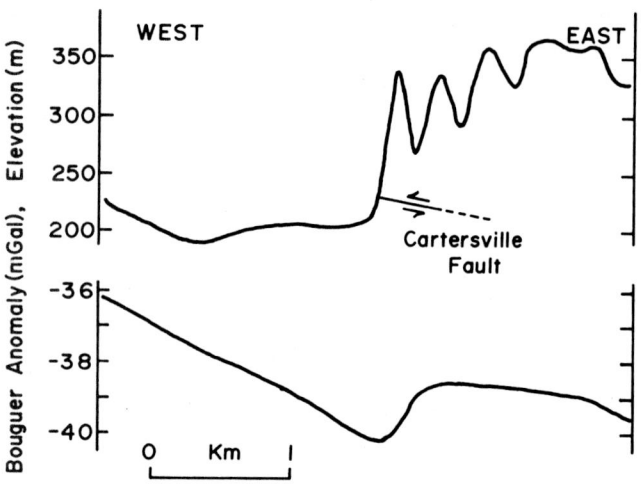

Fig. 3. Comparison of elevation and gravity data across the Cartersville fault at Carters Dam, Georgia.

altitude of 500 ft (152 m) above mean terrain showed evidence of anomalies associated with the dike but that these sharp anomalies were not correlated between flight lines because of the wide spacing.

Although the regional negative Bouguer anomalies imply that the Inner Piedmont contains a major buried structure, little evidence of surface structure is seen in the details of the Bouguer gravity field, with the exception of the (post-metamorphic) Triassic dike just discussed. The magnetic field also does not show many anomalies, although the Triassic dike, the Brevard fault, and the Towaliga fault show magnetic expression. The two faults, however, do not have an associated Bouguer gravity anomaly of greater than 2 mGal. We interpret the apparent homogeneity of the surface rocks as being due to the intense metamorphism suffered by the Inner Piedmont.

The gravity and magnetic anomalies in the Charlotte and Carolina Slate belts southeast of the Inner Piedmont are more closely correlated with surface structure. Large anomalies in both Bouguer gravity and total magnetic field are very often associated with rock units observed at the surface. Rocks of these belts, like those of the Inner Piedmont, have been subjected to metamorphism but to a lesser grade. The rocks of the Charlotte belt have a pronounced plutonic aspect, as shown by the presence of both felsic rocks such as granite and granite gneiss and mafic rocks such as metagabbro. Carolina Slate belt rocks are only lightly metamorphosed and consist mainly of a sequence of volcanic and associated sedimentary rocks of late Precambrian–early Paleozoic age. There is some controversy concerning the relationship between the Charlotte belt and the Carolina Slate belt. We follow the interpretation of Hatcher (1972) in which the Carolina Slate belt lies over the Charlotte belt in normal synclinal stratigraphic succession and treat these two belts as a single geologic province.

One reason for regarding the Charlotte and Carolina Slate belts as a single province can be seen in the detailed Bouguer gravity map of Figure 4. A sharp local gradient is evident, running from northeast to southwest across the middle of the map (this gradient should not be confused with the Piedmont gravity gradient, a much broader gradient to the northwest; the −4 mGal contour is in the middle of the sharp local gradient). This gradient coincides with the Middleton–Lowndesville fault zone (Rozen, 1973), which is the boundary between the Inner Piedmont and the Charlotte and Carolina Slate belts. This fault-contact boundary brings the Charlotte belt in contact with the Inner Piedmont in some places and the Carolina Slate belt in contact with the Inner Piedmont in other places. The gravity gradient, however, is present in both places, indicating that the Charlotte and Carolina Slate belts should be treated together and contrasted to the Inner Piedmont as a single unit.

We interpret the gradient across the Middleton–Lowndesville fault zone as indicating that denser, more mafic rocks are present in the Charlotte and Carolina Slate belts as compared to the Inner Piedmont. Since the gradient is very sharp, these denser rocks must be close to the surface. Model calculations indicate that the dense rocks extend from the surface to a depth of 5 km (Frazier, 1982). The inclusion of denser, more mafic rocks in the Charlotte and Carolina Slate belts is supported by evidence from the aeromagnetic data of Zietz et al. (1980), which show more pronounced anomalies (and thus greater magnetization) in the Charlotte and Carolina Slate belts. We interpret these anomalies as an indication of more mafic material.

Another striking feature of the Charlotte and Carolina Slate belts seen in Figure 4 is the presence of numerous pronounced Bouguer gravity anomalies resulting from near-surface sources, a characteristic that is in marked contrast with the homogeneous Inner Piedmont. Also, the aeromagnetic map of Zietz et al. (1980) has more and larger anomalies in the Charlotte and Carolina Slate belts compared to the Inner Piedmont. Residual positive Bouguer gravity anomalies (up to 25 mGal) are associated with intrusive and extrusive upper Precambrian–lower Paleozoic mafic units, while negative Bouguer anomalies (to −20 mGal) are associated with upper Paleozoic post-metamorphic granites. Many of the anomalies may be associated with units exposed at the surface, but the negative anomaly of −16 mGal centered at 33°48′ N, 82°53′ W (Figure 4) may represent a buried, unexposed granite pluton. Frazier (1982) used three-dimensional modeling to examine this anomaly and the +24 mGal and −20 mGal anomalies to the northeast. He concluded that the +24 mGal anomaly is due to mafic units in the Carolina Slate belt in the form of a body elongated parallel to the Middleton–Lowndesville fault zone and of a maximum depth extent of 5 km, if a density contrast of +0.2 g/cm^3 is assumed. The −20 mGal and −16 mGal anomalies are interpreted as granite stocks; using a density contrast of −0.1 g/cm^3, the −20 mGal body (Danburg Granite; Fullager and Butler, 1979) extends to a depth of 15 km and the −16 mGal body to a depth of 8 km.

There do not appear to be large-scale regional Bouguer gravity anomalies resulting from deep-seated sources in the Charlotte and Carolina Slate belts; Frazier (1982) assumed a flat regional field of 0.0 mGal Bouguer for the modeling discussed above. The crustal thickness is a normal 33 km (Kean and Long, 1980).

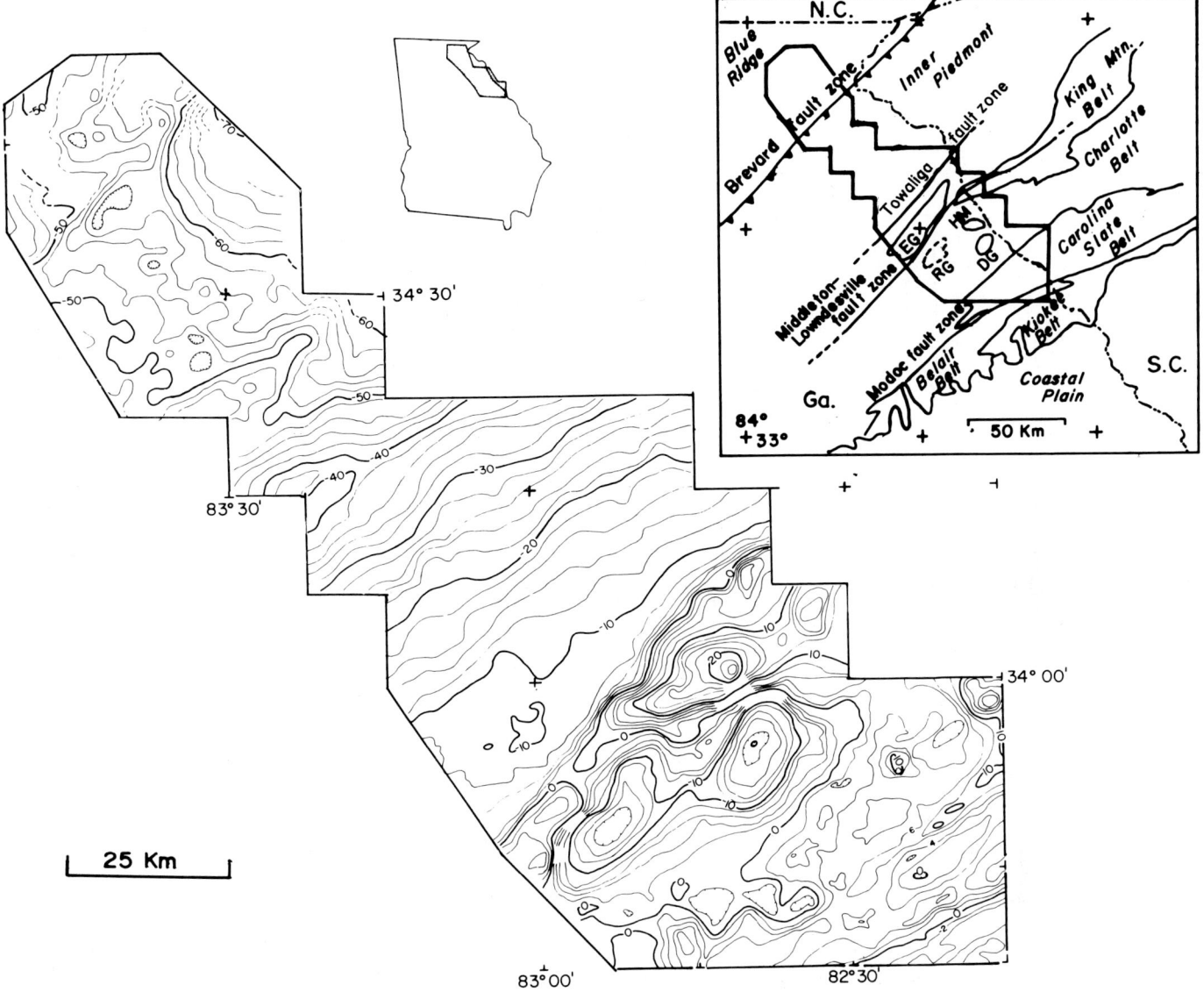

FIG. 4. Simple Bouguer gravity-anomaly map of northeastern Georgia, with sketch map of general surface geology. Contour interval, 2 mGal. Geologic contacts and faults modified from Frazier (1982) and King and Beikman (1974). HM = Heardmont Metagabbro; DG = Danburg Granite; RG = proposed Rayle Granite; EG = Elberton Granite.

COASTAL PLAIN

The Fall Line marks the boundary between the Cretaceous and Tertiary Coastal Plain sedimentary rocks and the Piedmont crystalline rocks. The Piedmont rocks continue beneath the Coastal Plain sedimentary rocks, but Triassic rift basins and associated faults have been discovered in east-central Georgia through the interpretation of magnetic and gravity data. Confirmation of the Triassic age of these rifts has come from well cores. In Figure 6 we show an interpretation of a gravity profile across one of these rift basins and compare it at the same scale to the interpretations of COCORP seismic-reflection data (Brown et al., 1980) across the Rio Grande graben. It is apparent that our interpretation of this rift basin in south Georgia is consistent with that of a continental rift system. The interpreted rift is present just south of the Piedmont–Coastal Plain boundary. A gravity anomaly that is characteristic of granite intrusives of the central Piedmont is present on the edge of the interpreted rift. A system of rifts may be inferred to underlie this region by combining the magnetic and gravity anomalies (Figure 7). The southeast edge of the Riddleville basin, the northernmost basin shown in Figure 7, may represent a horst, as modeled in Figure 6. Southeast of the horst the gravity and magnetic anomalies are smoother and may indicate a zone of more extensive rifting. In the quiet zone the rift structures are not as apparent in the magnetic data as in the gravity data and may indicate a zone of more uniform crustal structure or greater thicknesses of sedimentation. This zone is the northern edge of the landward extension of the east coast magnetic anomaly entering south Georgia.

Three large positive gravity and magnetic anomalies that

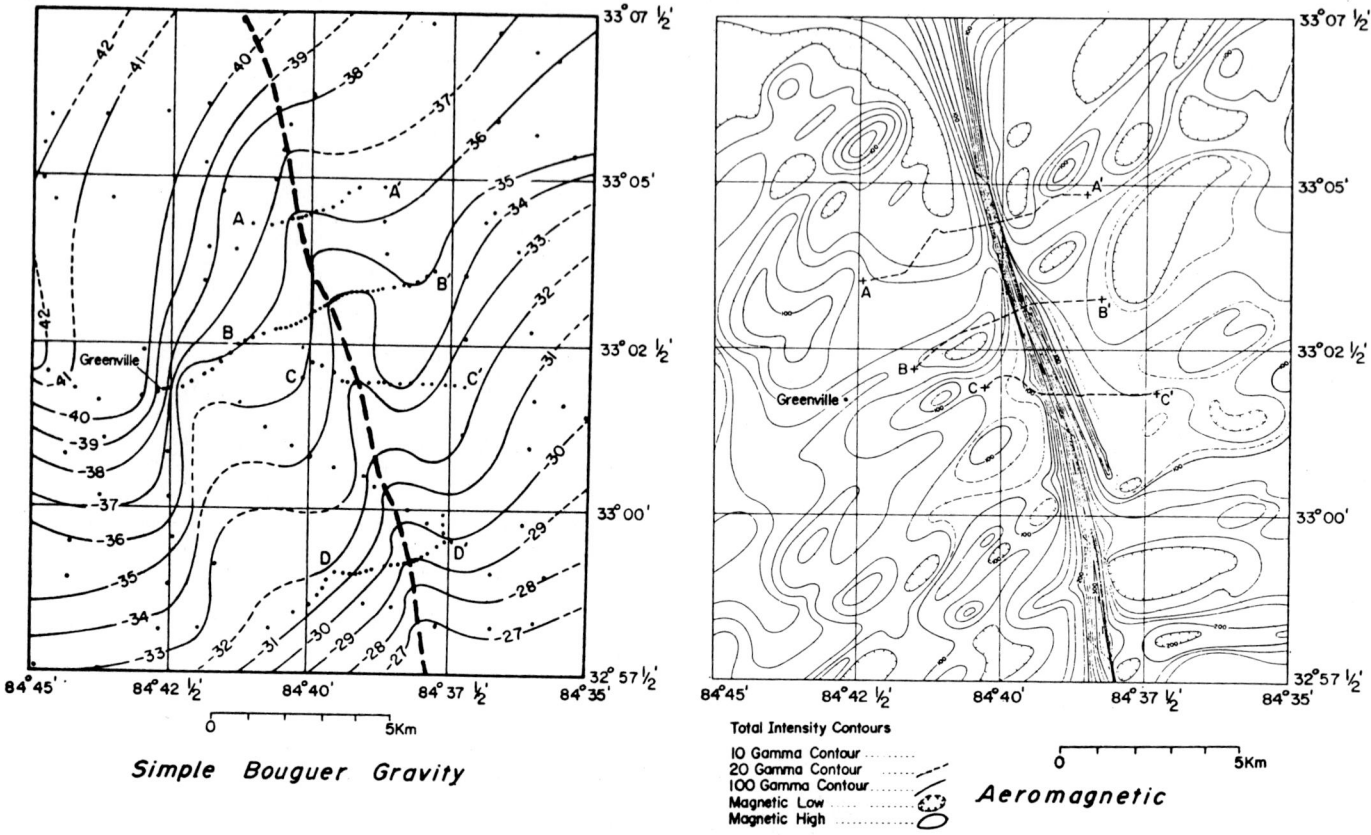

FIG. 5. Simple Bouguer gravity map of Meriwether dike area (contour interval, 1 mGal) and aeromagnetic map (total field) of the same area, recontoured, assuming the Meriwether dike to be a continuous magnetic feature. (Anomalies are with reference to 55 000-gamma datum.)

dominate south-central Georgia (Figure 8) have been interpreted as a Triassic mafic volcanic complex (Herbert, 1980). To model these, we used spheres of constant susceptibility and density contrast. This is equivalent to placing magnetic dipoles and point masses at the centers of the spheres. The sizes and locations of the spheres were adjusted until a fit was obtained within a 10-percent tolerance. Both the gravity and magnetic data require, if one uses susceptibilities of the order of 0.001 and density contrasts of 0.3 g/cm^3 or less, structures that extend down close to the Moho at an estimated depth of 30 km. The dipoles used to model the shallower structures show that the positive susceptibility material extends to the base of the Coastal Plain sediments in most areas. We have interpreted the shallow negative anomalies that exceed the expected dipole effect on the north edge of the intrusives as further evidence for a basin that is coincident with the zone of increased sedimentation or more extensive rifting noted above. The gravity and magnetic models for the westernmost anomaly show a caldera-like low at the top with a diameter of about 10 km and lead us to interpret this feature as a remnant of a volcano.

One of the more intriguing problems of the gravity anomalies in the Coastal Plain is the association of anomalies with minor changes in elevation and the apparent correlation in some areas of basement structures with topography. We present in Figures 9 and 10 two examples of elevation versus gravity. In most areas we can observe the expected correlation of minor short-wavelength anomalies with elevation. These are minimized by an expected density contrast of 2.0 to 2.5 g/cm^3. The first profile is near Hawkinsville, Georgia (Figure 9). The Bouguer reduction density needed to eliminate the influence of irregularities in topography on the Bouguer anomalies (see anomalies at 50–55 km, Figure 9) is the expected 2.0 g/cm^3. However, the gravity along this profile is also associated with changes in elevation that occur at the crossing of the Ocmulgee River and its 15-km-wide flood plain. The correlation with elevation in the center of the flood plain is minimized at a Bouguer reduction density near zero, making the reduction equivalent to a free-air anomaly and implying a shallow condition of isostatic equilibrium. In other areas of the Coastal Plain of Georgia and South Carolina (Long, 1974), the apparent correlation of elevation with gravity implies a range of -6 to 10 g/cm^3 as the Bouguer reduction density required to minimize anomalies correlated with topography.

In the second example, along COCORP Georgia line 8 (Figure 10) we again see a correlation of the gravity anomalies with elevations. The north end of the line shows a 25-m elevation drop into the area of the eastern extension of the Riddleville basin. Moving to the southeast along the profile, the elevation data indicate four terraces. The northernmost terrace corresponds to the horst interpreted to exist just

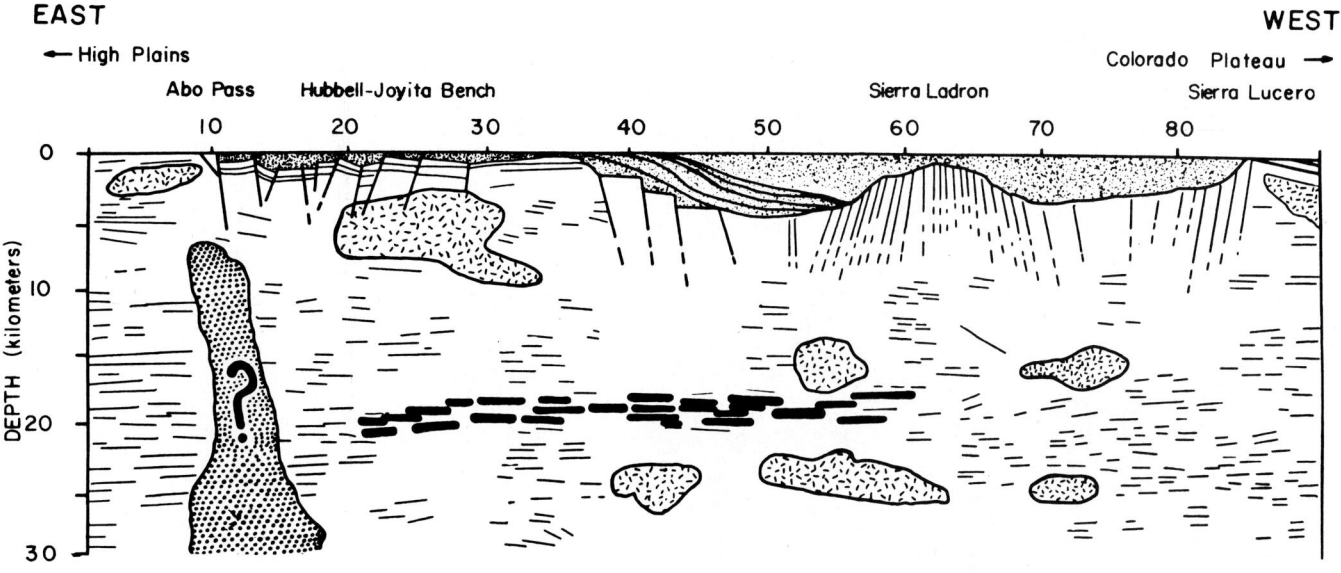
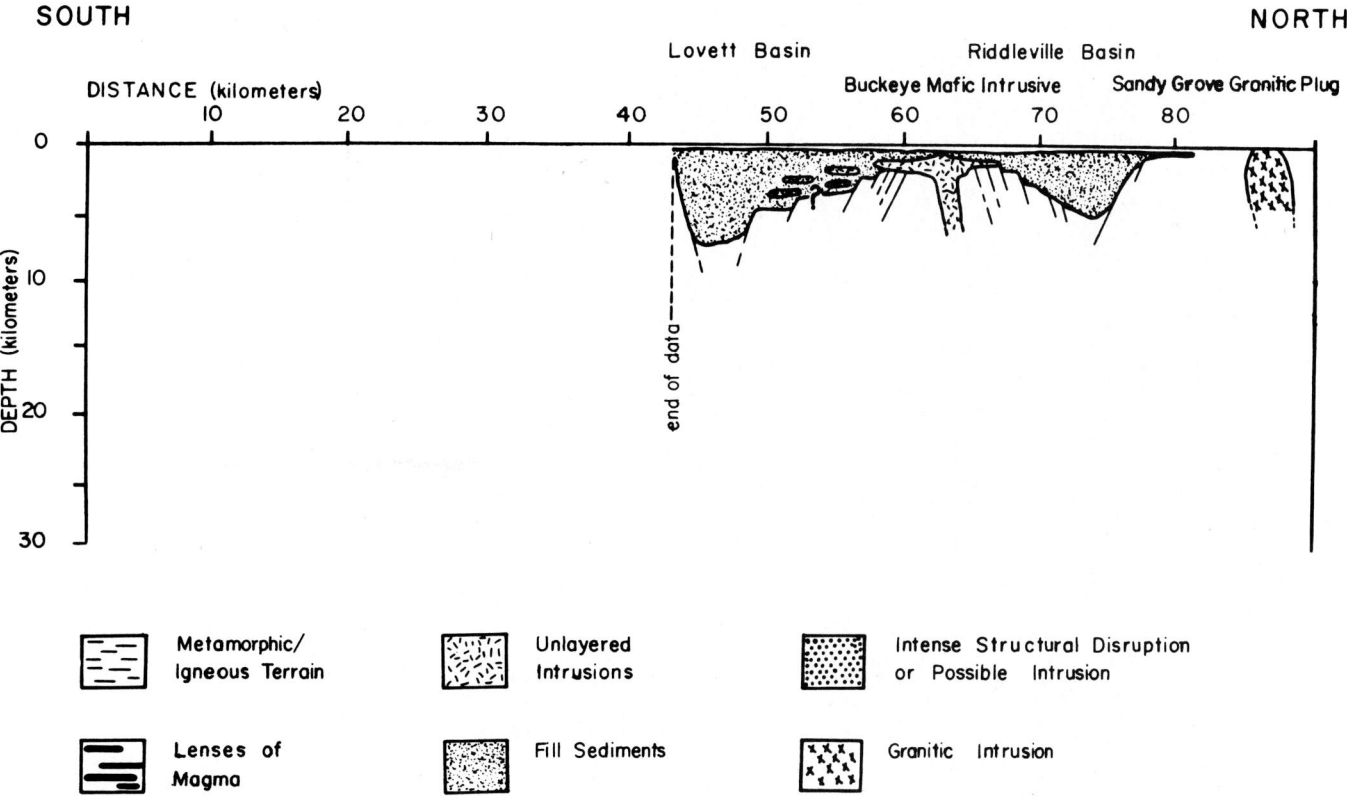

FIG. 6. Comparison of the Rio Grande rift COCORP profile (Brown et al., 1980) with structure of the Lovett and Riddleville basins interpreted from gravity data.

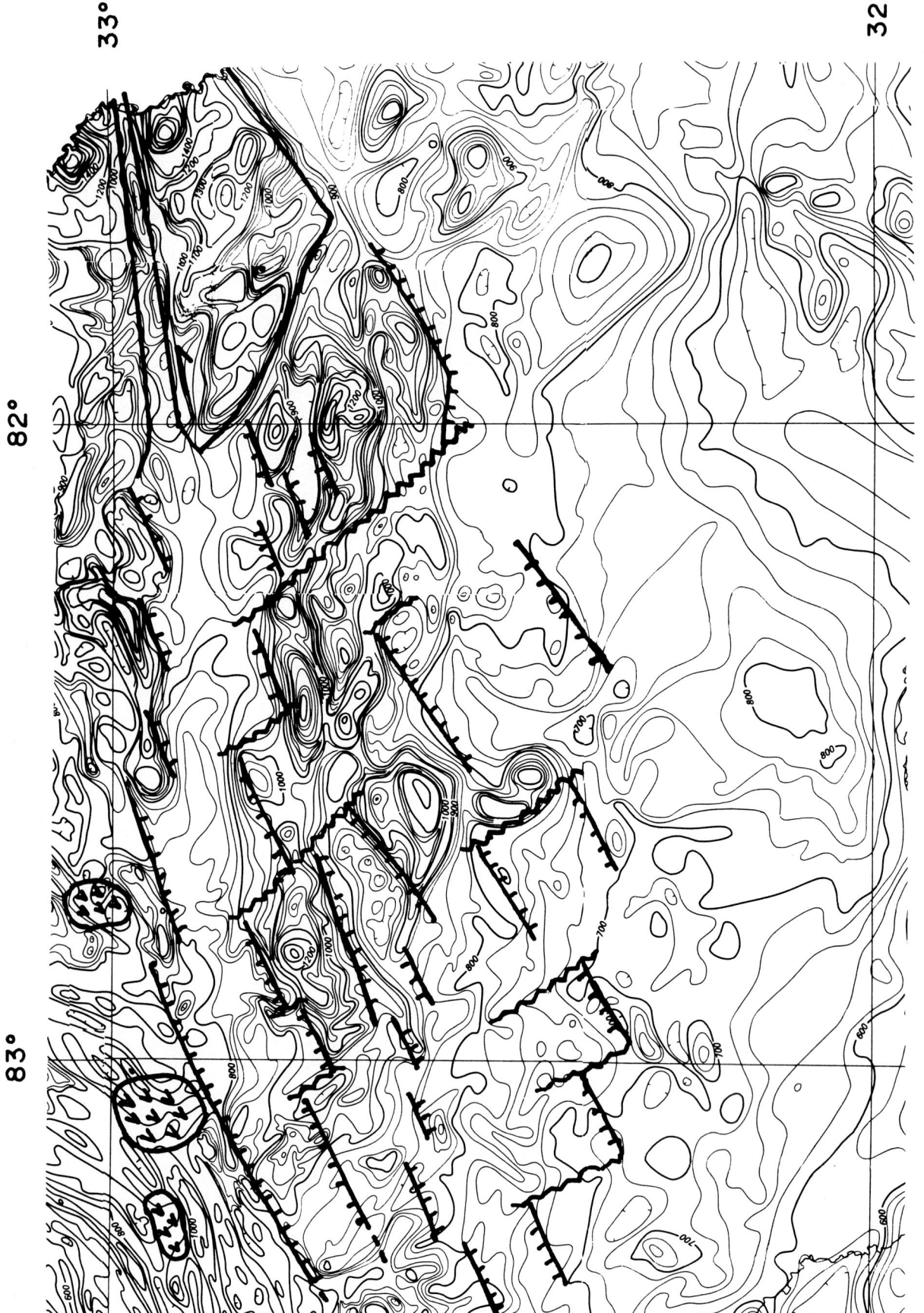

FIG. 7. Rift basins and crustal blocks under the south Georgia Coastal Plain interpreted from gravity and magnetic data. The rift basins are superimposed onto the magnetic map of Georgia (Zietz et al., 1980).

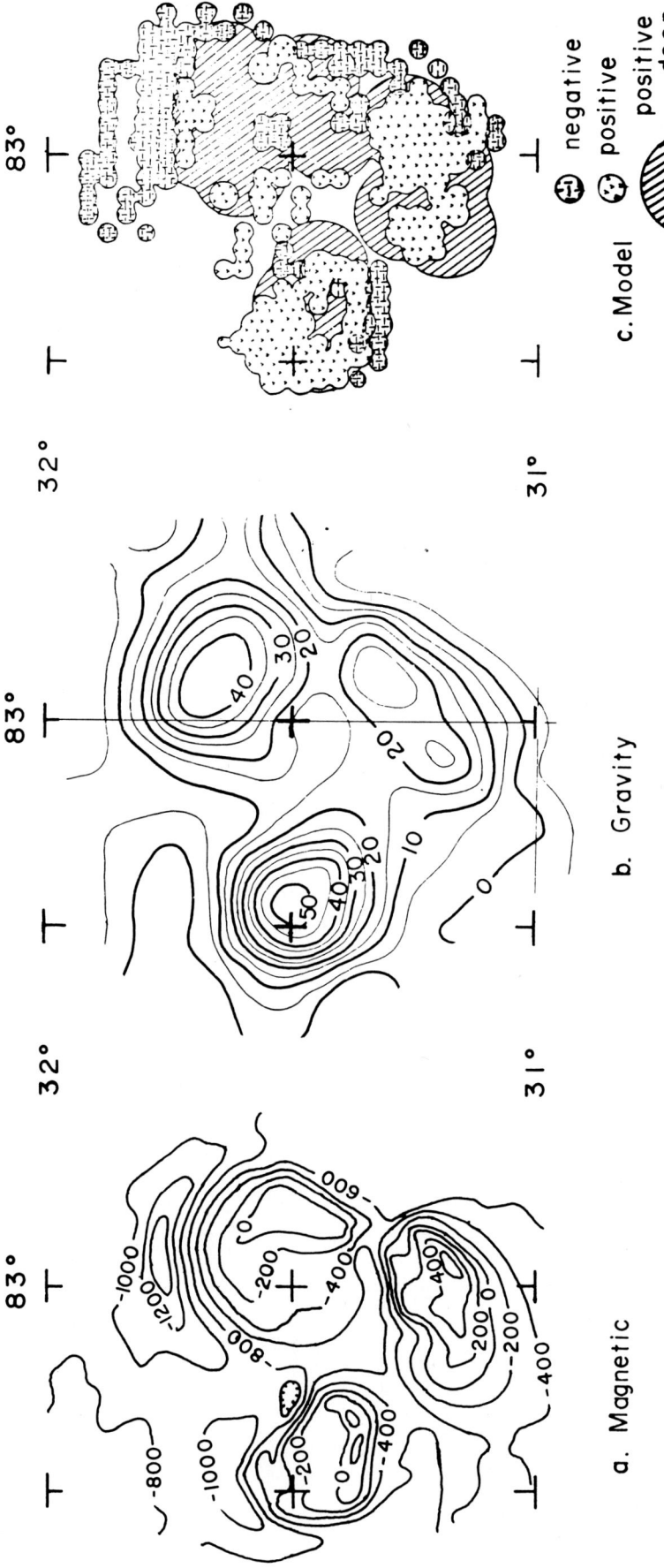

Fig. 8. a. Aeromagnetic map of three large anomalies in south-central Georgia interpreted as a Triassic mafic volcanic complex. Contour interval, 200 gammas. b. Gravity-anomaly mpa of same area. Contour interval, 5 mGal. c. Areal distribution of dipoles (spheres) used to model magnetics and gravity) for the three anomalies. Both figures are adapted from Herbert (1980). See text for discussion of model.

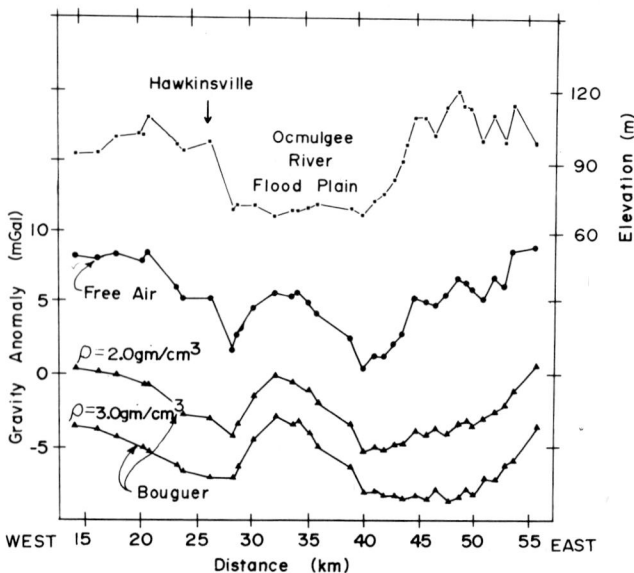

FIG. 9. Detailed west-to-east gravity profile near Hawkinsville, Georgia, showing both the anomalous correlation of elevations with Bouguer anomalies and the effect of a correct Bouguer reduction density of 2.0 g/cm^3 near the 50th kilometer of the line. See Figure 1 for location.

south of the Riddleville basin. This first terrace is an area of higher elevation. In the second terrace, from 30 to 45 km, the elevation and gravity anomalies are flat. From 45 to 50 km, the elevation and gravity anomalies decrease to a third terrace. In the fourth terrace the elevations are smooth, but the gravity data indicate larger deep-seated structures. At 45 km, the correlation with elevation implies a Bouguer reduction density of 5 g/cm^3. In the Coastal Plain, the changes in elevation in general correspond to changes in the character or magnitude of the gravity anomaly.

The complete explanation for this unusual association of elevations and gravity anomalies is not yet resolved. One explanation could be an association with continued minor crustal uplift. Another could be anomalies from structures generated by faulting during or soon after deposition of the Coastal Plain sediments.

ACKNOWLEDGMENTS

The data set of about 20 000 gravity values obtained by the Georgia Institute of Technology represents projects with diverse objectives and the efforts of most of our geophysics graduate students. Many students were supported in projects with objectives directed toward oil or gas exploration. More often, however, the objectives of the projects were directed toward earthquake studies and the search for structures related to seismicity. Some objectives were simply student exercises or data for theses. The authors express appreciation to the many students who have contributed to these studies through their hard work in the field, collecting data, and through their independent efforts in course work and their studies at the Georgia Institute of Technology.

REFERENCES

Best, D. M., Geddes, G. H., and Watkins, J. S., 1973, Gravity investigation of the depth of source of the Piedmont gravity gradient in Davidson County, North Carolina: Geol. Soc. Am. Bull., **84**, 1213–1216.

Brown, L. D., Chapin, C. E., Sanford, A. R., Kaufman, S., and Oliver, J., 1980, Deep structure of the Rio Grande rift from seismic reflection profiling: J. Geophys. Res., **85**, 4773–4800.

Carpenter, R. H., and Prather, P., 1971, A gravity survey of the south-central Georgia Piedmont: Inf. Circ. 42, Ga. Geol. Surv.

Cook, F. A., Albaugh, D. S., Kaufman, L. D., Oliver, J. E., and Hatcher, R. D., Jr., 1979, Thin-skinned tectonics in the crystalline southern Appalachians: COCORP seismic reflection profiling of the Blue Ridge and Piedmont: Geology, **7**, 563–567.

Cook, F. A., Brown, L. D., Kaufman, S., Oliver, J. E., and Petersen, T. A., 1981, COCORP seismic profiling of the southern Appalachian orogen beneath the Coastal Plain of Georgia: Geol. Soc. Am. Bull., **92**, 738–748.

Cook, F. A., and Oliver, J. E., 1981, The late Precambrian–early Paleozoic continental edge in the Appalachian orogen: Am. J. Sci., **281**, 993–1008.

Frazier, J. E., 1982, Analysis of the Bouguer gravity anomalies in the region surrounding the Elberton and Danburg granites in east-central Georgia: M.S. thesis, Ga. Inst. Tech.

Fullager, P. D., and Butler, J. R., 1979, 325 to 265 m.y.-old granitic plutons in the Piedmont of the southeastern Appalachians: Am. J. Sci., **279**, 161–185.

Guinn, S. A., and Long, L. T., 1978, A gravity survey of the Dalton, Georgia, area, in Short contributions to the geology of Georgia: Bull. 93, Ga. Geol. Surv., 78–81.

Hatcher, R. D., Jr., 1972, Developmental model for the southern Appalachians: Geol. Soc. Am. Bull., **83**, 2735–2760.

Hatcher, R. D., Jr., and Zietz, I., 1980, Tectonic implications of regional aeromagnetic and gravity data from the southern Appalachians, in Wones, D. R., Ed., The Caledonides in the U.S.A.: Mem. 2, Va. Polytech. Inst. State Univ., Dep. Geol. Sci., 235–244.

Haworth, R. T., Daniels, D. L., Williams, H., and Zietz, I., 1980, Bouguer gravity anomaly map of the Appalachian orogen: Memorial Univ. Newfoundland, scale 1:1 000 000, 2 sheets.

Herbert, J. C., 1980, Modeling of crustal structures in southwest Georgia from magnetic data: M.S. thesis, Ga. Inst. Tech.

Kean, A. E., and Long, L. T., 1980, A seismic refraction line along the southern Piedmont and crustal thicknesses in the southeastern United States: Earthquake Notes, **51**, 4, 3–14.

King, P. B., 1959, The evolution of North America: Princeton Univ. Press, 43–53.

King, P. B., and Beikman, H. M., 1974, Geologic Map of the United States: U.S. Geol. Surv., scale 1:2 500 000.

Long, L. T., 1974, Bouguer gravity anomalies of Georgia, in

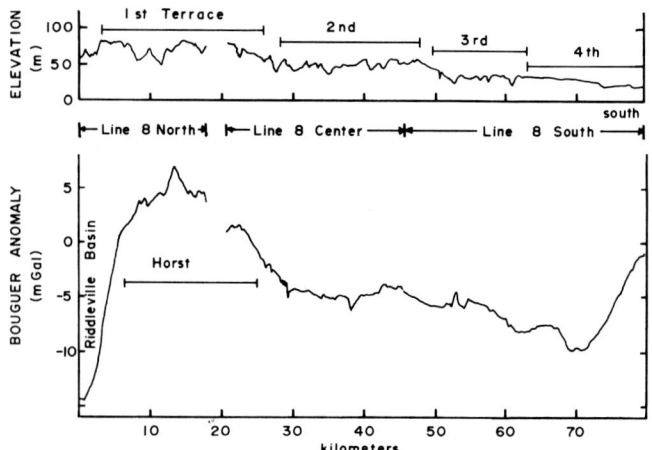

FIG. 10. Correlation of topography and Bouguer anomaly along COCORP line 8 (8 north, 8 center, and 8 south). See Figure 1 for location.

Symposium on the Petroleum Geology of the Georgia Coastal Plain: Bull. 87, Ga. Geol. Surv., 141–166.

———1979, The Carolina Slate belt—Evidence of a continental rift zone: Geology, 7, 180–184.

Long, L. T., Bridges, S. R., and Dorman, L. M., 1972, Simple Bouguer gravity map of Georgia: Ga. Geol. Surv.

Long, L. T., Denman, H. E., Hsiao, H. E., and Marion, G. E., 1976, Gravity and seismic studies in the Clark Hill reservoir area, in Chowns, T. M., Ed., Stratigraphy, structure, and seismicity in slate belt rocks along the Savannah River: Guideb. 16, Ga. Geol. Surv., 33–41.

Obaoye, M. O., 1979, Interpretation of detailed gravity traverses across northeastern Georgia: M.S. thesis, Ga. Inst. Tech.

O'Nour, I. M., 1982, Gravity anomalies in central Georgia: M.S. thesis, Ga. Inst. Tech.

Rankin, D. W., 1975, The continental margin of eastern North America in the southern Appalachians: The opening and closing of the proto-Atlantic Ocean: Am. J. Sci., **273-A**, 1–40.

Rothe, G. H., and Long, L. T., 1975, Geophysical investigation of a diabase dike swarm in west-central Georgia: Southeast. Geol., **17**, 67–79.

Rozen, R. W., 1973, The geology of the Elberton East quadrangle: M.S. thesis, Univ. Ga.

Zietz, I., Riggle, F. E., and Gilbert, F. P., 1980, Aeromagnetic map of Georgia: U.S. Geol. Surv.

Some effects of regional metamorphism and geologic structure on magnetic anomalies over the Carolina Slate belt near Roxboro, North Carolina

E. S. Robinson,* P. V. Poland,* L. Glover III,* and J. A. Speer*

ABSTRACT

Aeromagnetic anomalies over the Carolina Slate belt near Roxboro, North Carolina, are influenced by geologic structure and regional metamorphism. Anomaly amplitudes diminish with the transition from greenschist- to amphibolite-facies rocks along a northwest-trending metamorphic gradient. The northeast-trending Virgilina synclinorium is the principal structure in the area. A greenschist-facies metasedimentary rock unit is the principal magnetic-anomaly source in the northeast part of the synclinorium. Magnetization of this unit appears to diminish in a southwesterly direction along the structure, apparently because of the alteration of magnetite to hematite by oxidation.

Field-intensity profiles crossing linear anomalies were compared with profiles computed for two-dimensional models to determine the subsurface configuration of the Virgilina synclinorium. These comparisons indicate that the greenschist-grade metasedimentary rocks extend to depths of at least 2.5 km along the axis of the synclinorium. This implies that the amphibolite–greenschist isograd surface, which is exposed approximately 8 km northwest of the synclinorium axis, cannot dip toward the southeast at less than 20 degrees.

GENERAL

The subsurface contrasts in rock magnetism that cause magnetic anomalies can be produced in different ways. These contrasts can result from the distribution of chemically different lithologic units. Equally important in some regions are the effects of metamorphism on the magnetic mineralogy of chemically similar suites of rock. Such effects

*Department of Geological Sciences, Virginia Polytechnic Institute and State University, Blacksburg, VA 24061.

are evident in rocks of the Carolina Slate belt near Roxboro, North Carolina.

Glover and Sinha (1973) described the geology of this area (Figure 1). Stratified rocks include a tuffaceous epiclastic metasandstone unit interbedded with felsic and mafic metavolcanic units. Regional metamorphism produced greenschist-facies rocks in the southeast that are transitional into amphibolite-facies rocks of similar chemistry in the northwest. Prior to metamorphism, these stratified rocks were folded and thrust faulted. The prominent regional structure produced at that time is the northeast-trending Virgilina synclinorium. Pre-Mesozoic granitic and gabbroic plutons plus diabase dikes of Mesozoic age intrude the area.

MAGNETIC ANOMALIES

Magnetic-field intensity (Figure 1) was recorded at an elevation of 120 m above the land surface along east–west flight lines spaced at 0.6-km intervals (U.S. Geological Survey, 1971). Local variations of field intensity exceed 500 gammas in the southeast and diminish to less than 100 gammas in the northwest. This indicates that magnetic minerals stable in the greenschist facies were consumed by amphibolite-grade metamorphic reactions.

Field-intensity variations exceed 400 gammas across the northeast-trending linear anomalies over the tuffaceous epiclastic metasandstone exposed in the northeastern part of the Virgilina synclinorium. Anomalies are much smaller over the bordering metavolcanic rocks. This pattern is inverted over the southwestern part of the synclinorium, where metavolcanic rocks rather than the tuffaceous epiclastic metasandstone are the principal magnetic-anomaly sources.

Magnetic anomalies are observed over plutonic-rock exposures. Field intensity increases by more than 2 500 gammas over a poorly exposed gabbro stock near the southeastern border. Narrow linear anomalies over north- to northwest-trending diabase dikes occur in the northern part of the area. Other large anomalies are situated close to the border of the Roxboro pluton, which is the granodiorite exposed in the southwestern part of the area.

ROCK MAGNETISM

Magnetic susceptibility, remanent magnetism, and opaque-mineral petrography of 45 specimens indicate ways that regional metamorphism may have influenced magnetic-anomaly patterns. All of the principal rock units (Figure 1) were sampled, but most of the specimens were moderately to severely weathered. For this reason, measurements of the strength of rock magnetism are of questionable value except for confirming which lithologic units are the most likely sources of magnetic anomalies.

In that part of the Virgilina synclinorium extending northeast of latitude 36°30′ N, volume fractions of magnetite as high as 5 percent were found in the eight specimens of tuffaceous epiclastic metasandstone. Euhedral to subhedral magnetite grains (Figure 2a) occur with anhedral hematite grains having ragged and irregular borders (Figure 2b). Magnetic susceptibility and remanent magnetism were detected in all but one of these specimens. The susceptibility values, ranging between 270×10^{-6} (cgs) and $1\,370 \times 10^{-6}$ (cgs), would produce induced magnetism stronger than the remanent magnetism in these specimens. The directions of remanent magnetism varied widely and without obvious pattern, suggesting effects of chemical remanent magnetization (CRM) possibly related to greenschist-grade metamorphism.

Seven specimens of tuffaceous epiclastic metasandstone from locations south of latitude 36°30′ N were analyzed for rock magnetism. No magnetite was found in these specimens. Instead, hematite is the dominant opaque mineral, occurring both in anhedral grains and in sparsely distributed euhedral pseudomorphs after magnetite (Figure 2c). Magnetic susceptibility was detected in only one of these specimens. These observations indicate that alteration of magnetite to hematite may partly explain why the magnetic anomalies over the tuffaceous epiclastic metasandstone in the northeastern part of the Virgilina synclinorium do not extend farther southwest. Reasons for such alteration are not clear.

Three specimens of mafic metavolcanic rock were obtained from localities near the center of profile A–A′ (Figure 1) close to the axis of the synclinorium. Their magnetic susceptibilities of 0, 90×10^{-6}, and 170×10^{-6} (cgs) are considerably lower than values measured from specimens of the bordering tuffaceous epiclastic metasandstone in this part of the synclinorium. In these mafic metavolcanic rocks, grains of hemanoilmenite occur as rims and oriented lamellae with epidote pseudomorphs after an originally octahedral mineral. These intergrowths are believed to have been produced from original igneous titanomagnetite, which exsolved into magnetite and ilmenite by oxidation during the early stage of regional metamorphism. Later, the magnetite was partially consumed in the formation of hydrous iron silicates, such as the epidote, by metamorphic reactions of the type:

$$2\underset{\text{magnetite}}{Fe_3O_4} + 12\underset{\text{anorthite}}{CaAl_2Si_3O_8} + 12\underset{\text{quartz}}{SiO_2} + 6\underset{\text{water}}{H_2O} + \underset{\text{oxygen}}{O_2} \rightarrow$$

$$\rightarrow 3\underset{\text{epidote}}{Ca_2FeAl_3Si_3O_{12}(OH)}. \quad (1)$$

The exsolved ilmenite remained largely unaffected. This reaction was identified near the core of the synclinorium.

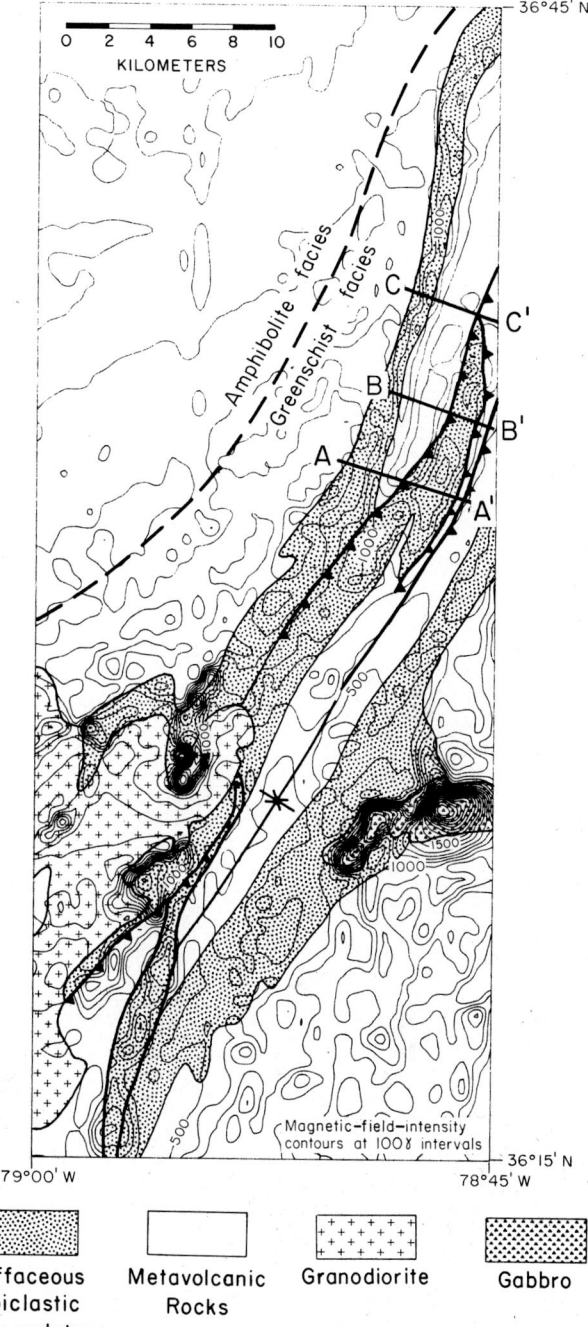

FIG. 1. Generalized geology (modified from Glover and Sinha, 1973) and aeromagnetic field-intensity variations near Roxboro, North Carolina (modified from U.S. Geological Survey, 1971).

Similar rock textures and reactions were described by Cann (1969), Zen (1974), and Kuniyoshi and Liou (1976). At higher metamorphic grade the ilmenite may be consumed in the formation of titanium iron silicates by reactions of the type:

$$\underset{\text{ilmenite}}{FeTiO_3} + \underset{\text{calcite}}{CaCO_3} + \underset{\text{quartz}}{SiO_2} \rightarrow$$

$$\rightarrow \underset{\text{titanite}}{CaTiOSiO_4} + CO_2 + FeO. \quad (2)$$

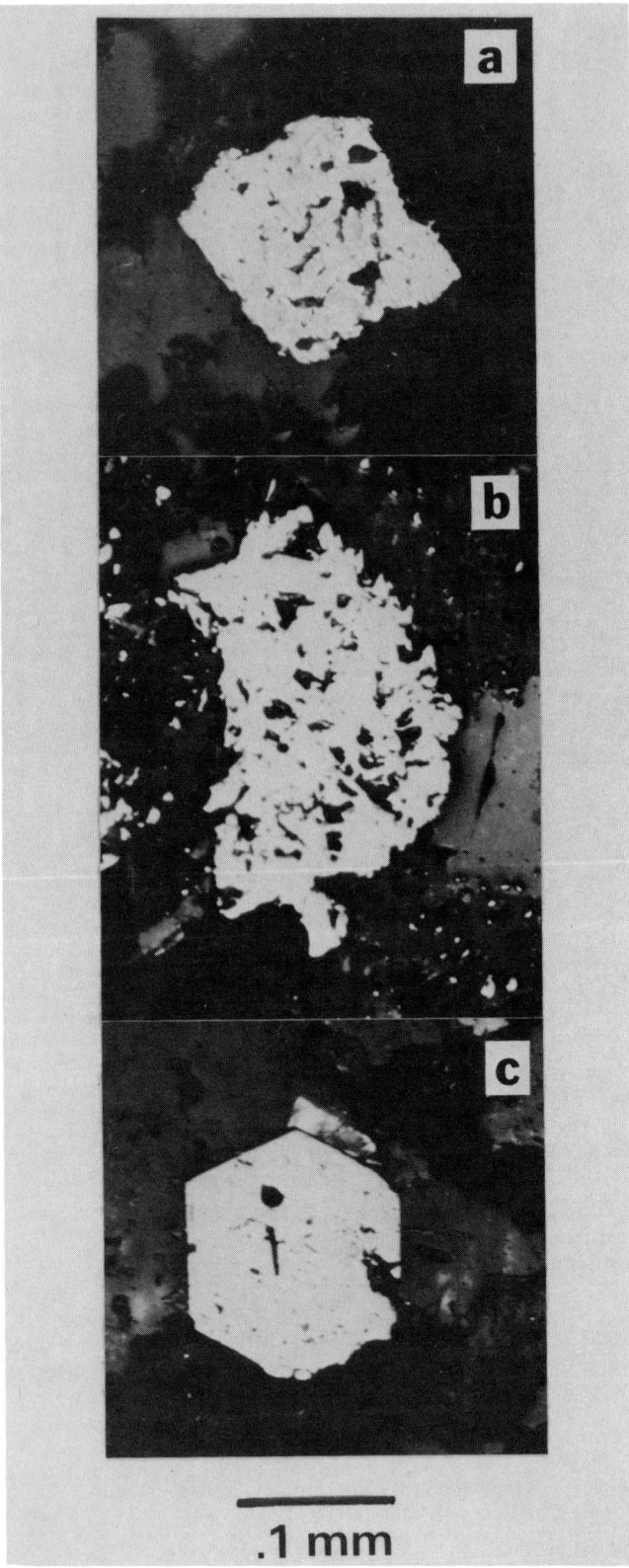

FIG. 2. Grains of (a) euhedral magnetite, (b) anhedral hematite, and (c) hematite occurring as a pseudomorph after magnetite in tuffaceous epiclastic metasandstone.

This reaction was also recognized in exposed mafic metavolcanic rocks. The bordering tuffaceous epiclastic rocks probably lacked sufficient calcite or oxygen fugacity for these reactions to occur. Instead, the magnetite recrystallized into the euhedral and subhedral grains observed in this rock unit.

In the greenschist environment the partial pressures of H_2O and O_2 influence the direction of the reaction expressed in equation (1). The possibility exists that local variations of these partial pressures in the southeastern part of the area may have alternately contributed to the production or consumption of magnetite. Magnetic susceptibilities of 15 specimens of greenschist-grade metavolcanic rock from this area ranged from 0 to $4\ 000 \times 10^{-6}$ (cgs) and displayed no obvious spatial pattern of variation. The numerous magnetic anomalies over this area of otherwise undifferentiated metavolcanic rocks may result from contrasts in rock magnetism produced by metamorphism of chemically similar lithologies as well as by the distribution of chemically different units. However, in the amphibolite environment that existed in the northwest part of the area, the reaction expressed in equation (1) would go to completion toward the right, consuming the magnetite.

Elsewhere in the study area the Roxboro pluton was found to contain isolated unaltered magnetite grains and other smaller grains partially altered to hematite. The total-volume proportion of magnetite in this metagranitic rock is approximately 5 percent. It intrudes host rocks that contain small hematite grains and local pyrite.

Rocks exposed in the southeastern part of the area near 36°25′ N–78°48′ W indicate a gabbro stock that possesses strong rock magnetism. Examination of one weathered specimen revealed clumps and stringers of unaltered magnetite grains with subhedral and anhedral shapes. This gabbro specimen was so severely weathered that a reliable estimate of the volume proportion of magnetite could not be made, but it must be large to explain the intense anomaly over this stock.

SYNCLINORIUM STRUCTURE

The strongly magnetized tuffaceous epiclastic metasandstone is a prominent stratigraphic unit in the Virgilina synclinorium. The shape of the synclinorium should be revealed by the subsurface configuration of this unit, which can be estimated from analysis of the configurations of the aeromagnetic anomaly. This was done by comparing anomalies along three profiles crossing the synclinorium with field-intensity variations over two-dimensional models computed by the method of Talwani and Heirtzler (1964). Because the synclinorium is a linear structure, two-dimensional models are suitable for representing its shape along profiles transverse to its axis.

The two-dimensional models and the associated magnetic-field-intensity profiles are shown in Figure 3. These models are the product of 40 iterations that tested the sensitivity of the magnetic field to depth, shape, thickness, and magnetism of the two-dimensional units. In each model, unit A represents that part of the tuffaceous epiclastic metasandstone which possesses significant rock magnetism. The magnetic-anomaly shape precludes a thicker magnetized unit. The

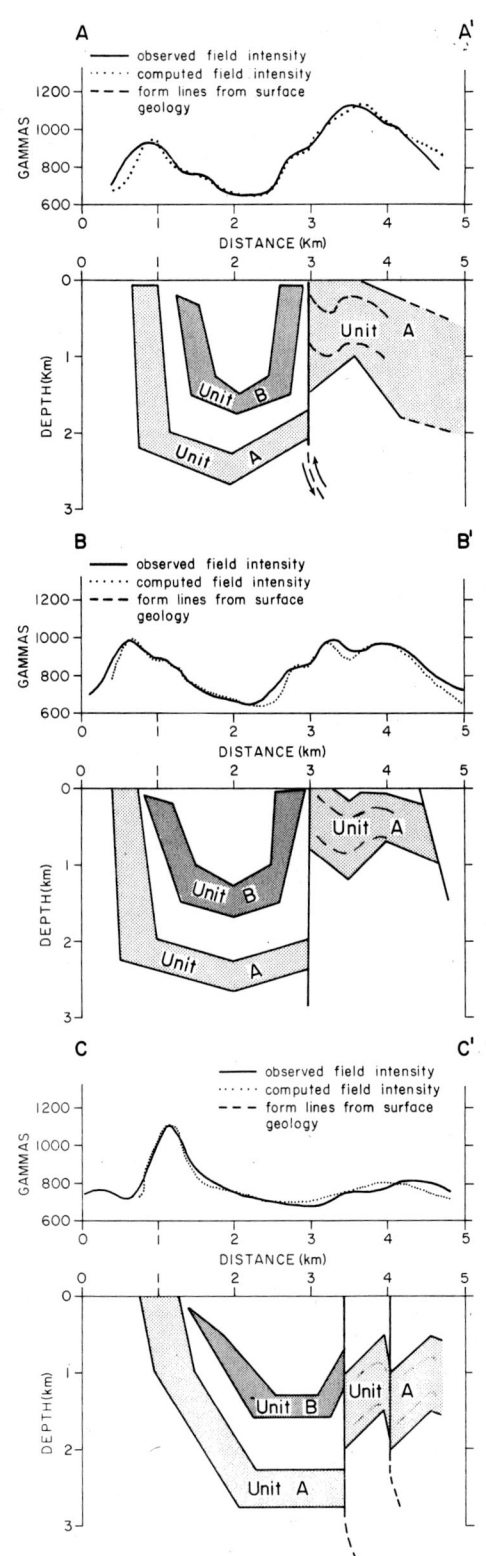

FIG. 3. Comparison of observed magnetic-field-intensity variation along three profiles and field-intensity variation computed for two-dimensional models of magnetized rock units. Profile positions are shown in Figure 1.

depth to which unit A extends is the minimum required to reproduce the anomaly shape. Numerous configurations of unit A alone were tested without success before the introduction of unit B, which proved necessary to reproduce the measured field-intensity variation with a model consistent with other constraints determined from geologic mapping. Unit B in each model indicated that a magnetized zone must exist within the metavolcanic rocks that make up the core of the synclinorium. There is no evidence of this magnetized zone in rocks exposed at the surface. A magnetic moment per unit volume of 110 gammas in the direction of the present main field for both unit A and unit B is required to reproduce the measured amplitude of field-intensity variation. In the absence of remanent magnetism, this would imply a susceptibility of $2\,060 \times 10^{-6}$ (cgs).

The distribution of magnetized rock units in the two-dimensional models indicates that greenschist-facies metamorphic rocks extend to a depth of at least 2.5 km along the axis of the synclinorium. Magnetization would not exist in rocks metamorphosed to amphibolite grade. Therefore, the amphibolite–greenschist isograd surface indicated by rocks exposed approximately 8 km north of the synclinorium axis cannot dip toward the southeast at less than 20 degrees.

CONCLUSIONS

Amplitudes of local magnetic anomalies diminish dramatically across the boundary separating greenschist-grade rocks of the Carolina Slate belt from chemically similar amphibolite-grade rocks near Roxboro, North Carolina. Apparently the constituents of magnetic minerals stable in the greenschist environment were incorporated into iron silicate minerals by amphibolite-grade metamorphic reactions. Even within the Carolina Slate belt, local magnetic anomalies probably result from contrasts in rock magnetism produced by metamorphic reactions in chemically similar lithologies as well as from contrasts between rock units of differing chemical composition. Comparison of magnetic anomalies measured over the Virgilina synclinorium and field-intensity variations over two-dimensional models indicates that greenschist-grade rocks exposed along the axis of the structure in the South Boston quadrangle extend to depths of at least 2.5 km.

ACKNOWLEDGMENTS

The U.S. Geological Survey sponsored the aeromagnetic survey and provided partial support for data reduction and interpretation. Kenneth Books of the U.S. Geological Survey made the remanent-magnetism measurements. Computer costs were provided by Virginia Polytechnic Institute and State University.

REFERENCES

Cann, J. R., 1969, Spilites from the Carlsberg Ridge, Indian Ocean: J. Petrol., **10**, 1–19.

Glover, L., and Sinha, A. K., 1973, The Virgilina deformation, a late Precambrian to early Cambrian (?) orogenic event in the central piedmont of Virginia and North Carolina: Am. J. Sci., Cooper v., **273-A**, 234–251.

Kuniyoshi, S., and Liou, J. G., 1976, Contact metamorphism of Karmutsen Volcanics, Vancouver Island, British Columbia: J. Petrol., **17**, 73–79.

Talwani, M., and Heirtzler, J. R., 1964, Computation of magnetic anomalies caused by two dimensional structures of arbitrary shape: Computers in the mineral industries, part 1: Stanford Univ. Publ., Geol. Sci., **9**, 1, 464–480.

U.S. Geological Survey, 1971, Geophysical Investigations Maps GP747 (South Boston) and GP749 (Roxboro).

Zen, E-An, 1974, Prehnite and pumpellyite bearing mineral assemblages, west side of Appalachian metamorphic belt, Pennsylvania to Newfoundland: J. Petrol., **15**, 197–242.

Structure of the U.S. Atlantic continental margin from derivative and filtered maps of the magnetic field

John C. Behrendt* and Muriel S. Grim‡

ABSTRACT

The availability of a high-sensitivity digital aeromagnetic survey over the U.S. Atlantic continental margin and new graphical color-display techniques enabled us to compile residual maps of total magnetic intensity, and of its second-vertical-derivative and wavelength-filtered maps, which show structural elements that are not obvious in the original aeromagnetic maps.

A regional tilt in the original total-intensity map was removed by fitting a quadratic surface to the data. The resultant residual map reveals broad positive anomalies over the Baltimore Canyon trough, Carolina trough, and Blake Plateau basin. A regional negative anomaly overlies the Georges Bank basin and Long Island platform.

The visual effect of the second-vertical-derivative map is to emphasize the gradients, thereby sharpening and resolving anomalies of small areal extent; the high frequencies are enhanced. The low-frequency anomalies define the major basins that are presently being explored for petroleum resources, such as the Georges Bank basin, Baltimore Canyon trough, Carolina trough, and Blake Plateau basin, as well as a number of smaller basins. The second-vertical-derivative map indicates complex structure associated with the East Coast Magnetic Anomaly (ECMA). For example, two positive lineaments between 36° and 37°30′ N suggest that although a simple edge effect associated with the oceanic crust may account for a large part of the total anomaly of 200–600 nT, more complex structures are also present. We compiled a map showing tectonic elements of the continental margin from the second-vertical-derivative map.

High-pass filters of 60-, 40-, and 20-km wavelength illustrate successively shallower sources of anomalies and provide a useful first approximation of depth to magnetic basement.

INTRODUCTION

In 1975–76 a high-sensitivity aeromagnetic survey was made of the U.S. Atlantic margin (Figure 1), roughly from mid-Florida to Canada and from the coast to 4-km water depth. A magnetic-anomaly map (Behrendt and Klitgord, 1979) published at a scale of 1:1 000 000 and having a 2-nT contour interval was previously interpreted and discussed by Klitgord and Behrendt (1979). A number of petroleum-exploration holes have been drilled, but as of this date there are no producing fields. Data availability, the 14–15-km relief on magnetic basement (Klitgord and Behrendt, 1979), and the general understanding of the tectonics of the transition from continental to oceanic crust made the U.S. Atlantic margin an excellent area in which to examine the usefulness of various digital filtering techniques using the aeromagnetic data.

In this report we describe new analyses of the 1975–76 aeromagnetic data that allow us to define in more detail sedimentary basins and structural lineations associated with the regional tectonics. Using computer programs (Hildenbrand, 1983; Hildenbrand et al., 1982) and graphical color-display techniques (R. H. Godson, written commun., 1981), we recompiled the total-magnetic-intensity map, with the international geomagnetic reference field and a quadratic surface removed (Figure 2), and produced a second-vertical-derivative magnetic map (Figures 3–6) and various wavelength-filtered maps (Figures 7–9). An interpretation of these data is shown in Figure 1, which also indicates the locations of geologic features referred to in the text. The digital magnetic data, having a relative precision (but not an absolute accuracy) of about ±0.2 nT (Behrendt and Klitgord, 1980), were compiled on a north-oriented 4-km grid

---

*U.S. Geological Survey, Box 25046, MS 903, Denver Federal Center, Denver, CO 80225.
‡Formerly U.S. Geological Survey, Box 25046, MS 964, Denver Federal Center, Denver, CO 80225; presently Naval Ocean Research and Development Activity, NSTL Station, MS 39529.

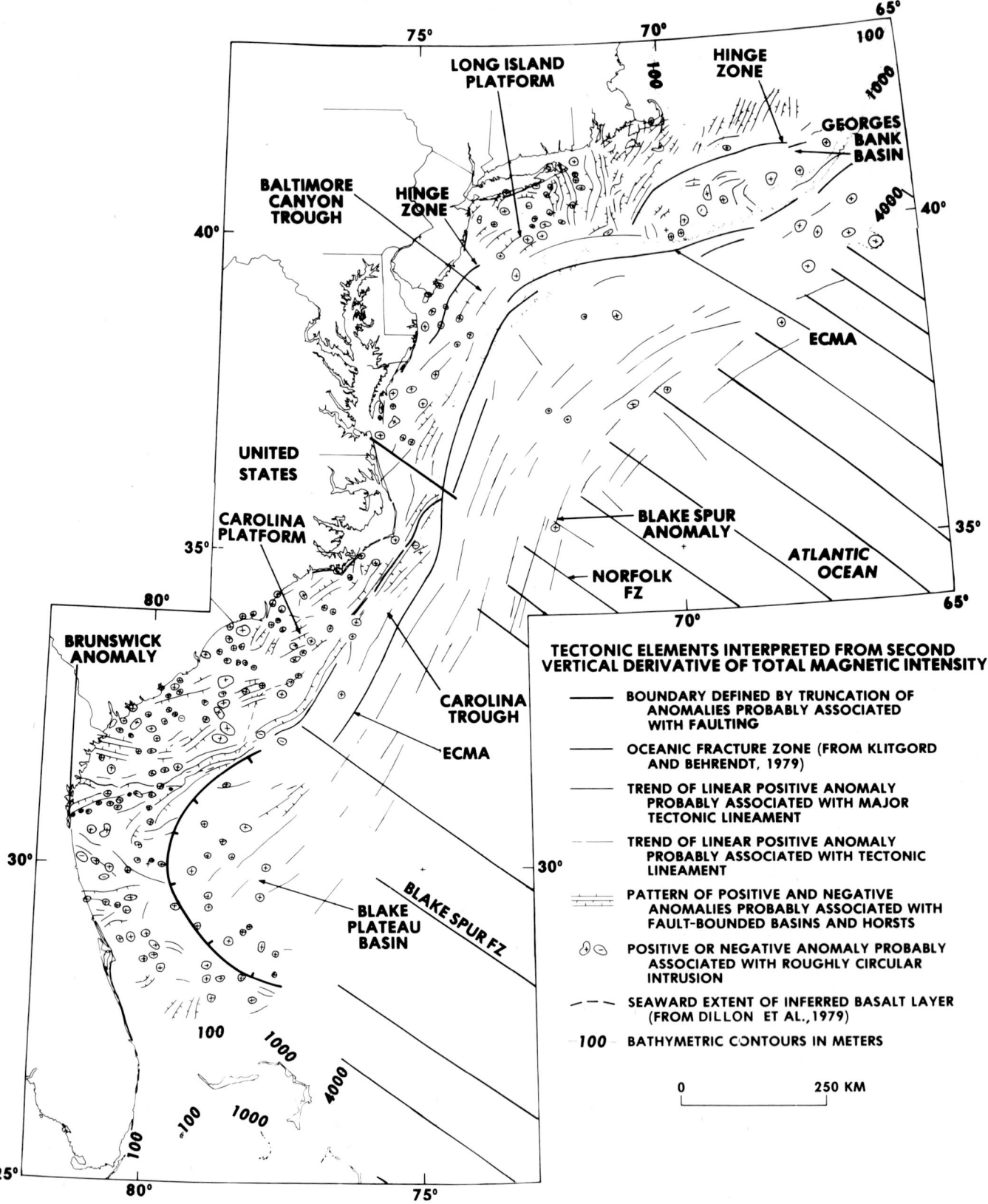

FIG. 1. Map showing tectonic elements of the U.S. Atlantic continental margin, interpreted from the second-vertical-derivative map of Figure 3, compared with the total-intensity map of Figure 2 at large scales. Although many of the features shown were known from previous interpretations, all shown here were compiled solely from this data set except the oceanic fracture zones and the boundary of the basalt layer.

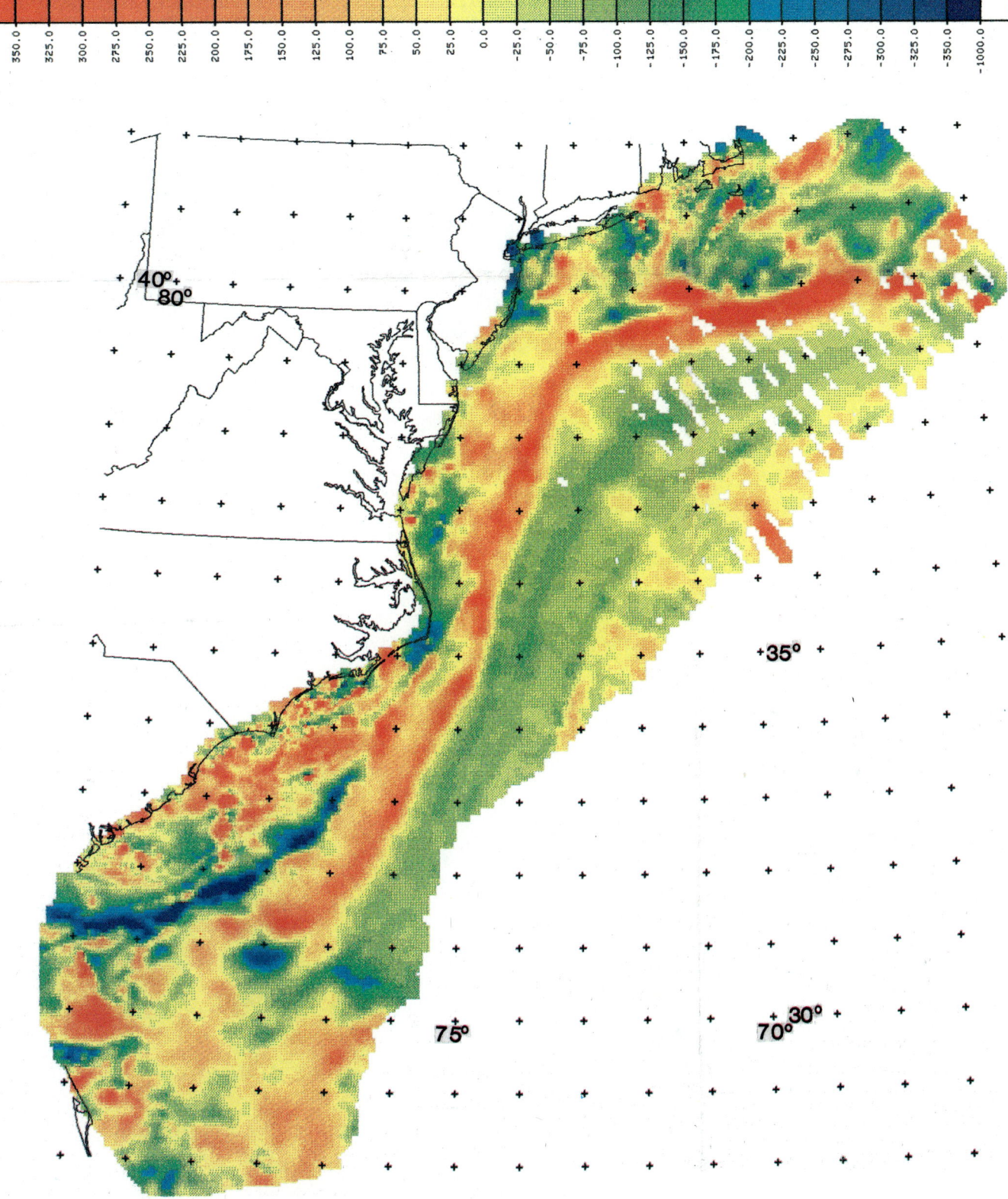

FIG. 2. Total-magnetic-intensity map compiled from the high-sensitivity survey over the area of Figure 1. The IGRF and a second-degree quadratic surface have been removed. Contour interval is 25 nT; the color bar at the top indicates interval values. Grid ticks in this figure and in Figures 3–9 are at a 1° latitude and longitude interval. The white bands in this and in Figures 3, 4, 7, 8, and 9 seaward of the ECMA in the northeast are the result of data gaps owing to wide flight-line spacing.

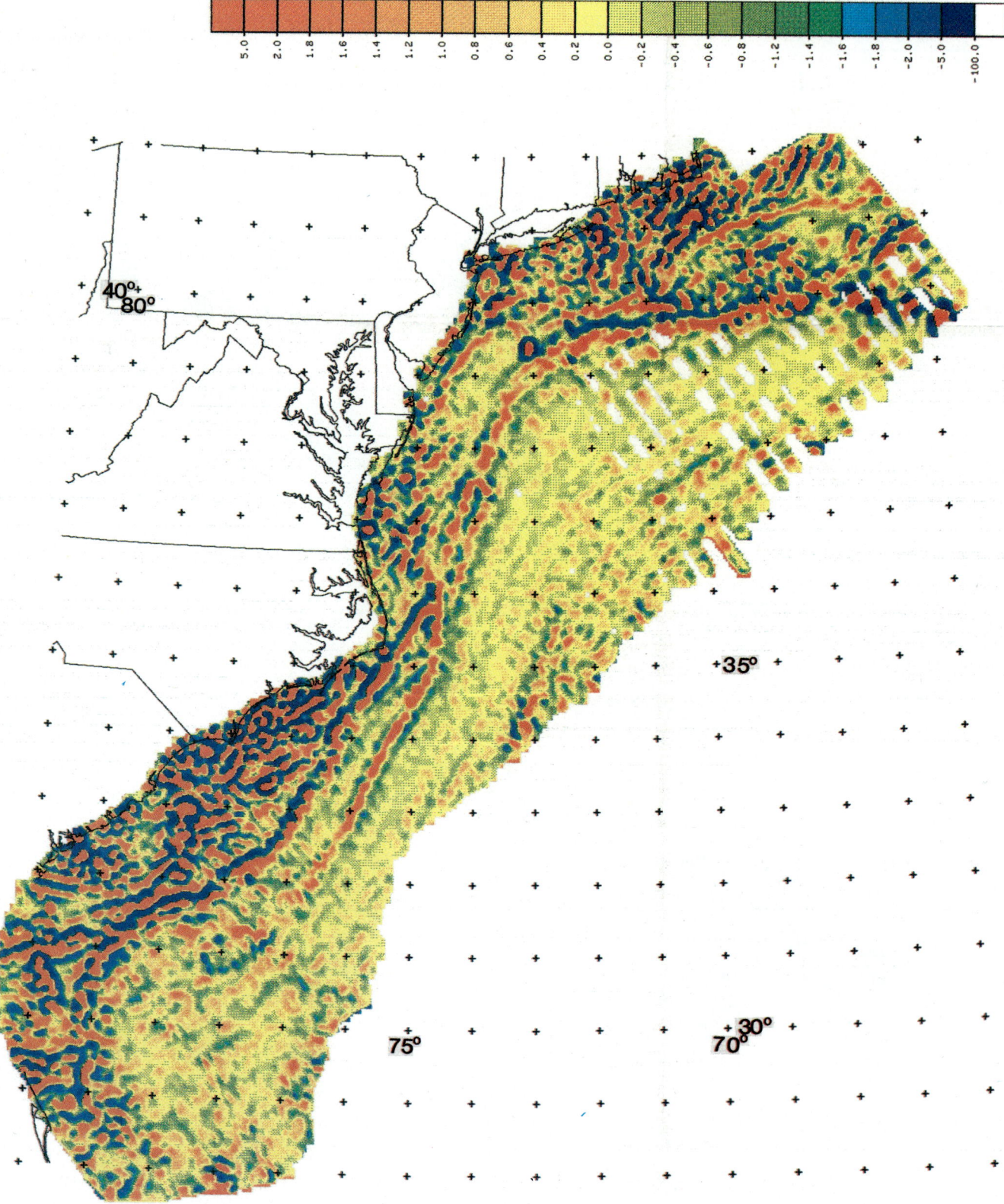

FIG. 3. Map showing second vertical derivative of total magnetic intensity over the area of Figure 1. The low-curvature areas (shades of yellow) overlie the thickest sections of sedimentary rock. The major basins currently being explored for oil and gas (Georges Bank basin, Baltimore Canyon trough, Carolina trough, and Blake Plateau basin) are apparent. Contour interval is 0.2 nT/km^2; the color bar at the top indicates interval values.

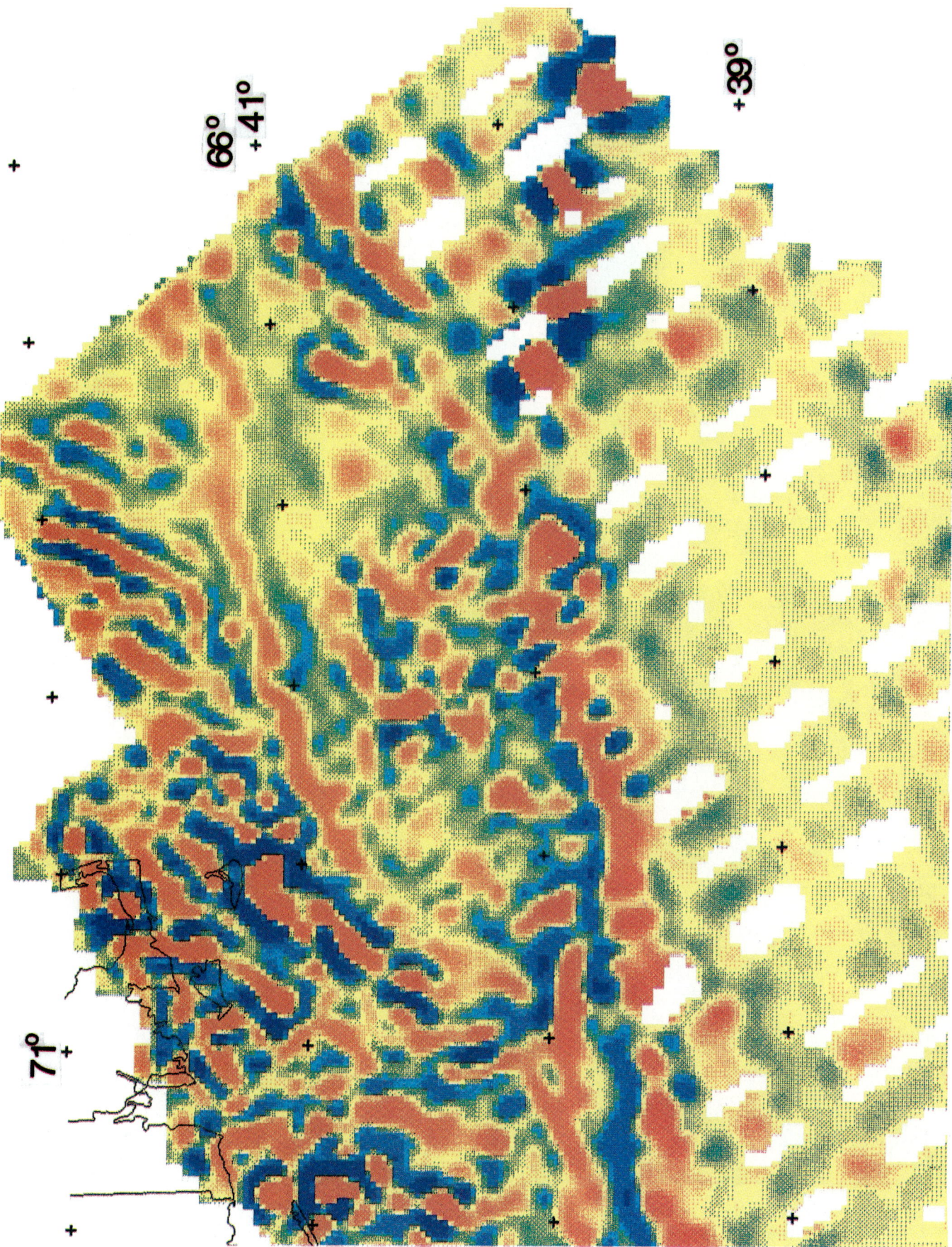

FIG. 4. Map showing part of second vertical derivative of total-magnetic-intensity map of Figure 3 at a larger scale; contour interval is same as for Figure 3. The area shown covers part of the Georges Bank basin and the Long Island platform. Compare with the tectonic interpretation of Figure 1.

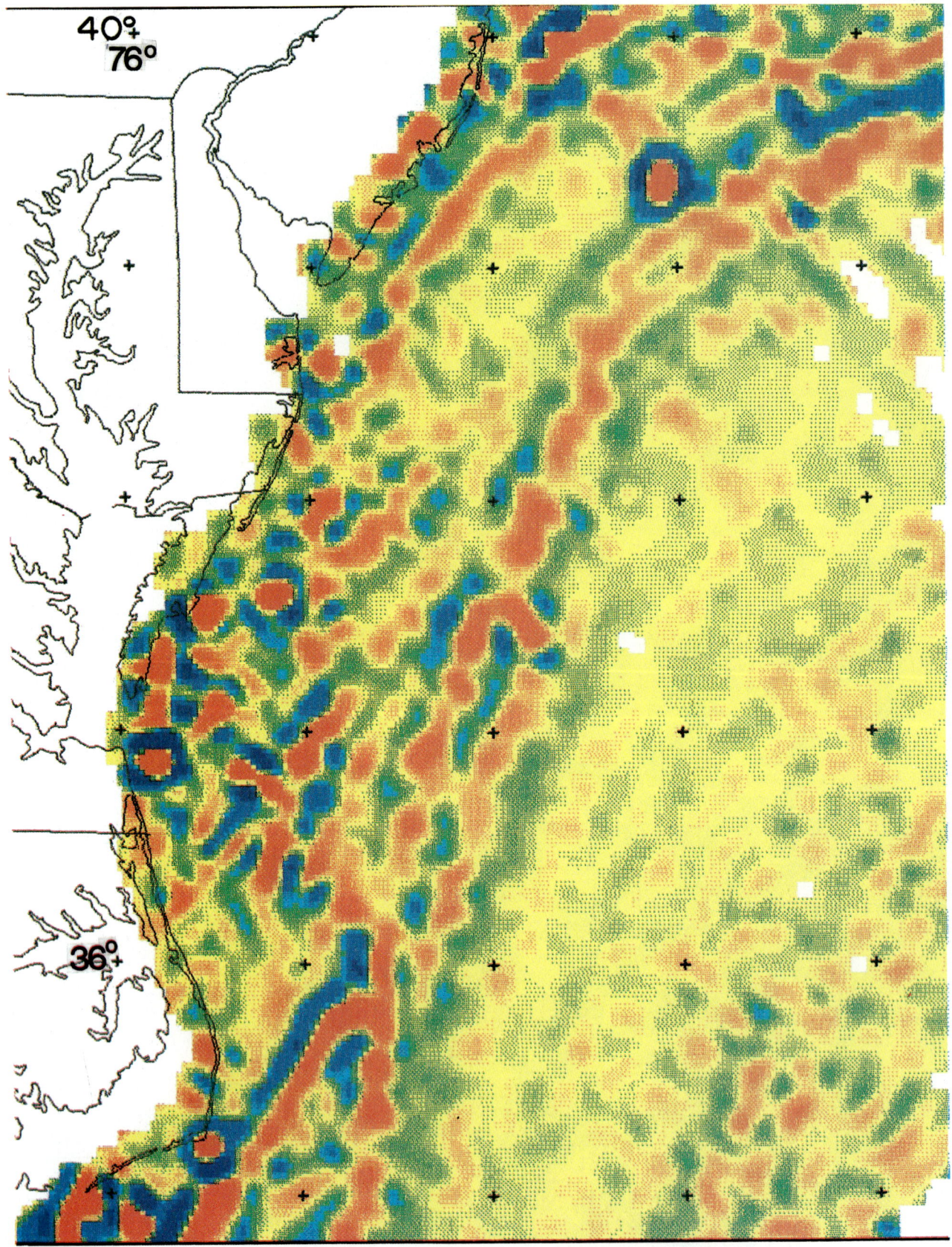

FIG. 5. Map showing part of second vertical derivative of total-magnetic-intensity map of Figure 3 at a larger scale; contour interval is same as for Figure 3. The area shown covers part of the Baltimore Canyon trough. Compare with the tectonic interpretation of Figure 1.

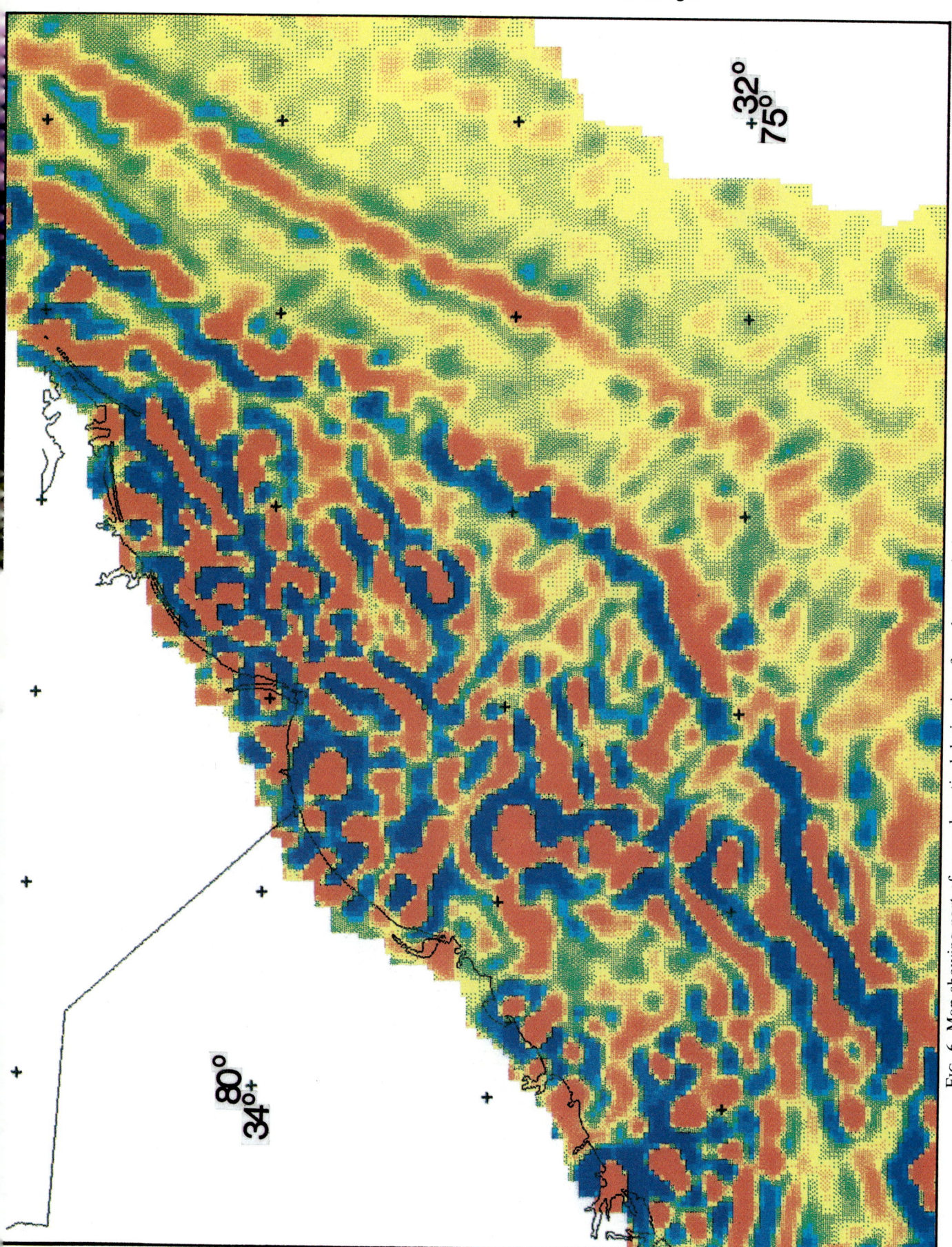

FIG. 6. Map showing part of second vertical derivative of total-magnetic-intensity map of Figure 3 at a larger scale; contour interval is same as for Figure 3. The area shown covers part of the Carolina platform and Carolina trough. Compare with the tectonic interpretation of Figure 1.

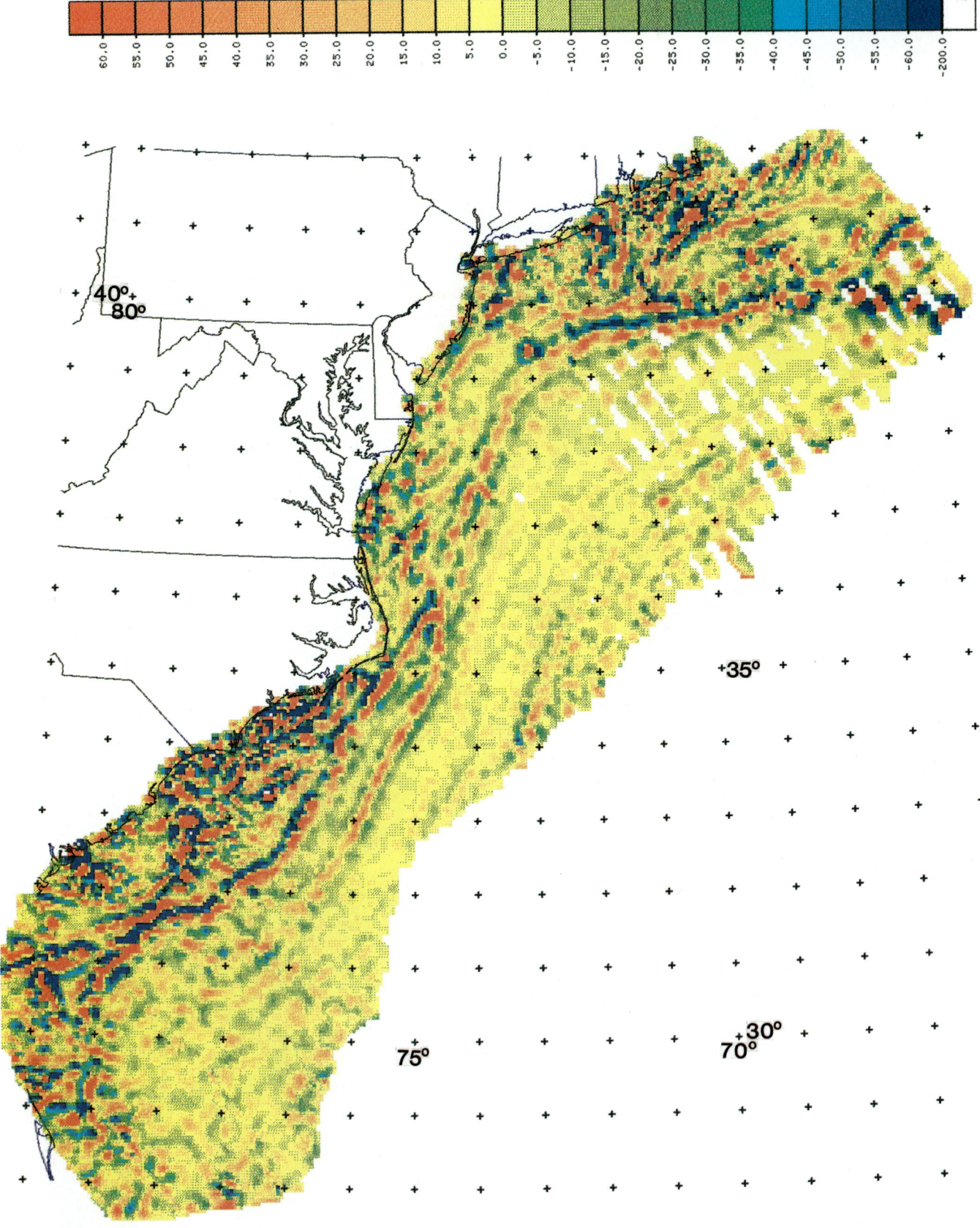

FIG. 7. Residual total-magnetic-intensity map over area of the survey shown in Figure 1. Contour interval is 5 nT; color bar at top indicates interval values. Anomalies showing wavelengths greater than 60 km are removed; generally, anomalies having sources deeper than 15 km have been removed.

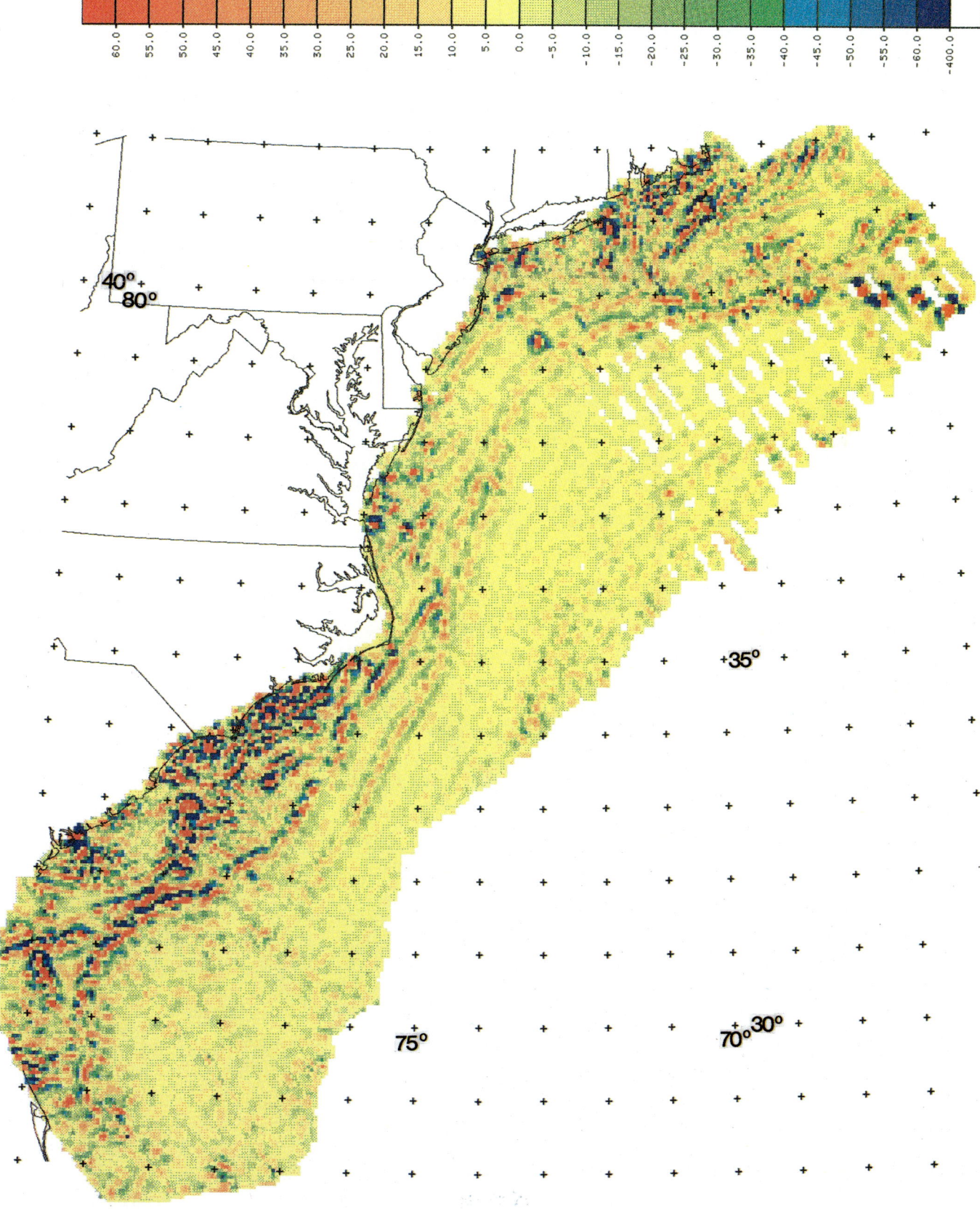

FIG. 8. Residual total-magnetic-intensity map over area shown in Figure 1. Contour interval is 5 nT; color bar at top indicates interval values. Anomalies having wavelengths greater than 40 km are removed; generally, anomalies having sources deeper than 10-km depth are removed.

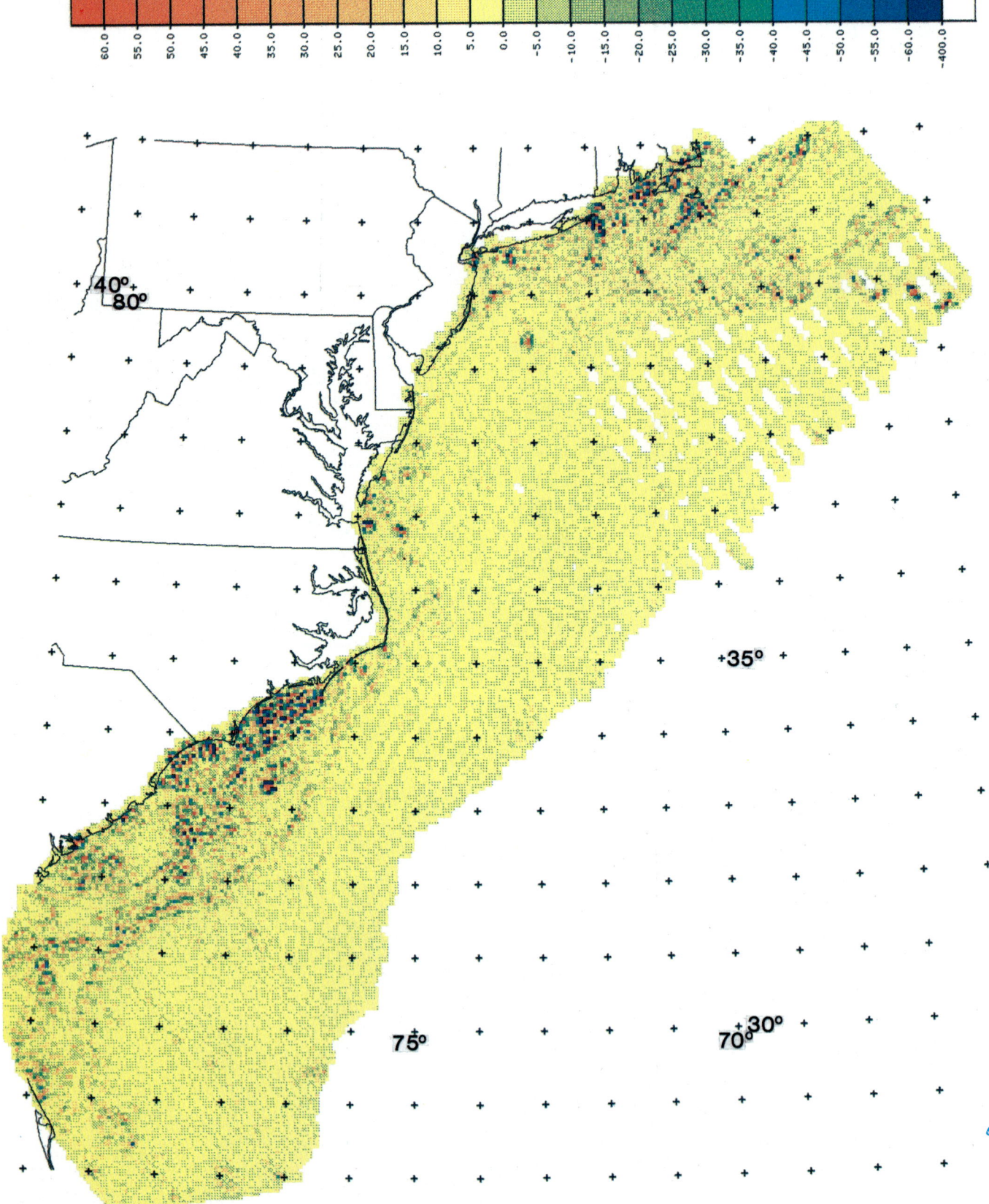

FIG. 9. Residual total-magnetic-intensity map over area shown in Figure 1. Contour interval is 5 nT; color bar at top indicates interval values. Anomalies having wavelengths greater than 20 km are removed; generally, anomalies having sources deeper than 5 km are removed. Only anomalies from platform areas and the ~3.5-km-deep intrusion in the Baltimore Canyon trough remain.

using an iterative minimum-curvature program (Webring, 1981). The 4-km grid interval is less than the depth to magnetic basement over about 80 percent of the area so that aliasing is largely eliminated. The grid is generally compatible with the flight-line spacing, although wider line spacing over the north end of the surveyed area caused gaps in magnetic-data contouring (shown in white on the maps in this report). We used the gridded data to make machine-contoured magnetic-anomaly maps at various scales from 1:10 000 000 to 1:250 000 and with 100-, 50-, 25-, and 5-nT contour intervals. The tectonic interpretation (Figure 1), the total-magnetic-intensity map (Figure 2), and the second-vertical-derivative map (Figure 3) are also published at a scale of 1:2 500 000 (Behrendt and Grim, 1983).

TOTAL-MAGNETIC-INTENSITY MAP

The previously published total-magnetic-intensity map (Behrendt and Klitgord, 1979) (2-nT contour interval, 1:1 000 000 scale) presented the data with the 1965 International Geomagnetic Reference Field (IGRF) updated to 1975 (IAGA, 1969) removed. In this area the IGRF is substantially tilted so that the southern part of the survey is more negative than the northern. The total-magnetic-intensity map, prepared by removal of a second-degree least-squares quadratic surface after correcting for the IGRF (Figure 2), shows the regional magnetic anomalies more clearly than the previous compilation. We used the 25-nT color contour interval shown on the map (Figure 2) to emphasize the lower amplitude regional anomalies; the short-wavelength and high-amplitude anomalies are better shown at the 1:1 000 000 scale.

The total-magnetic-intensity map (Figure 2) reflects the major tectonic features indicated in Figure 1. The East Coast Magnetic Anomaly (ECMA) (Taylor et al., 1968) marks the edge of oceanic crust (Klitgord and Behrendt, 1979). South of the bend in the ECMA (about 39° N), the Baltimore Canyon trough, Carolina trough, and Blake Plateau basin (Figure 1) are marked by positive anomalies, whereas the Georges Bank basin and the adjacent Long Island platform north of the bend are marked by a regional negative anomaly. We do not have a definitive explanation for these features. A possible explanation is that the three southern basins are underlain by a greater amount of magnetic rocks that originated at the time of rifting and initial opening of the Atlantic. Small positive anomalies are present over the Georges Bank basin and may be caused by post-rift intrusions, but the cause of the generally low field level is unknown. The Brunswick anomaly, largely positive over land (Taylor et al., 1968), has a strongly negative component offshore, in contrast to the positive anomaly over the adjacent Carolina trough. Other features of the total-intensity map are discussed by Klitgord and Behrendt (1979).

SECOND-VERTICAL-DERIVATIVE MAGNETIC MAP

The second-vertical-derivative magnetic map (Figure 3) reveals a great amount of detail not obvious on the total-magnetic-intensity map (Figure 2). The results presented here were compiled using the procedure of Hildenbrand (1983). According to Laplace's equation for potential-field data,

$$\frac{d^2f}{dx^2} + \frac{d^2f}{dy^2} + \frac{d^2f}{dz^2} = 0,$$

where f = total magnetic intensity and x, y, and z are the distances along the two horizontal and the vertical axes, respectively. Upon rearranging, the second vertical derivative

$$\frac{d^2f}{dz^2} = -\left(\frac{d^2f}{dx^2} + \frac{d^2f}{dy^2}\right).$$

In other words, the second vertical derivative is the negative sum of the curvature of the field in the x and y direction. The sign change means that positive peaks in the second derivative will occur over positive peaks in the total intensity.

Many of the tectonic elements that are readily apparent in Figure 3 and indicated in Figure 1 have been previously reported on the basis of earlier interpretations of other forms of the aeromagnetic data (Klitgord and Behrendt, 1979), and from seismic-reflection profiles (Schlee et al., 1976; Grow et al., 1979; Dillon et al., 1979) and by a number of earlier researchers. The low-frequency anomalies in Figure 3 define the major basins that are currently being explored for petroleum resources, such as the Georges Bank basin, Baltimore Canyon trough, Carolina trough, and Blake Plateau basin, as well as a number of smaller basins. Examples from the second-vertical-derivative data of Figure 3 are shown at larger scale in Figures 4–6. The Jurassic–Tertiary sedimentary rocks in these basins are as thick as 14–15 km, as determined by multichannel seismic-reflection profiles and depths interpreted from the aeromagnetic data (Klitgord and Behrendt, 1979; Schlee et al., 1976; Grow et al., 1979; Dillon et al., 1979).

The prominent curvilinear feature extending along the upper continental slope from the northern end of the survey to about 31°30′ N is the high-frequency component of the ECMA. While the large-amplitude, long-wavelength anomaly is generally considered to be caused by the magnetization contrast between oceanic crust to the east and transitional continental crust to the west (Klitgord and Behrendt, 1979), there is a difference of opinion as to whether sources associated with the short-wavelength components of the ECMA lie at a 7–10–km depth (Klitgord and Behrendt, 1979), possibly within the sedimentary section, or whether the sources are all at a basement depth of 14–15 km (Grow, 1980). The second-vertical-derivative magnetic map does not answer this question quantitatively. It does, however, indicate complex structure associated with the ECMA; for example, see Figure 5. Two positive components between 36° and 37°30′ N suggest that although a simple edge effect associated with the oceanic crust may account for a large part of the total anomaly of 200–600 nT (Figure 2), more complex structures such as intrusions within (i.e., younger than) the sedimentary rocks or a ridge on the underlying basement (Klitgord and Behrendt, 1979) (i.e., older than the sedimentary rocks) probably are responsible for the high-frequency components of the ECMA.

There are additional prominent features of the second-vertical-derivative map. The areas underlain by thin sections of sedimentary rock (Klitgord and Behrendt, 1979), such as the Long Island platform (Figures 3, 4) and the Carolina platform (Figures 3, 6), are characterized by short-wavelength, predominantly northeast-trending anomalies, probably associated with Triassic rifting and syntectonic igneous activity just prior to the opening of the Atlantic. In the northern part of the Georges Bank area (Figure 4, at about 41° N) the northeast-trending structures appear to be separated from the rest of the Georges Bank basin by a roughly east-trending linear anomaly nearly 300 km long, which we interpret as marking a hinge zone. The thickest section in the basin and the area of greatest subsidence is seaward of the hinge zone. A similar anomaly marks the northwest edge of the Baltimore Canyon trough (Figures 3, 5) and the west edge of the Carolina trough (Figures 3, 6). The Brunswick anomaly (Taylor et al., 1968), which parallels the ECMA, extends from 35° N to the coast at about 31° N (Figures 3, 6). A pronounced arcuate pattern outlines the western edge of the Blake Plateau basin (Figure 3). The curvature of this feature suggests a meteorite-impact origin, but there is little evidence to support this idea, with the possible exception of the abrupt truncation of the ECMA at the south end (Figures 2, 3).

The magnetic quiet zone east of the ECMA is characterized by low-amplitude northeast-trending anomalies (Figure 3) that parallel the ECMA. These anomalies are the type that would be expected from topographic relief on a uniformly magnetized oceanic crust (Klitgord and Grow, 1980). Northeast of the Blake Spur anomaly, a feature presumably caused by a failed spreading center (Klitgord and Behrendt, 1979), higher amplitudes are apparent over the quiet zone (Figure 3).

The many northwest-trending oceanic fracture zones known to exist (Klitgord and Behrendt, 1979) in the area do not, in general, have obvious associated anomalies on either the total-magnetic-intensity, the second-vertical-derivative, or the wavelength-filtered magnetic maps. This may be because of the similarity between the direction of the flight lines and trends of the fracture zones. Marine seismic, gravity, and magnetic profiles oriented oblique to fracture zones, where seismic-reflection and magnetic data are coincident, show associated magnetic anomalies (Klitgord and Grow, 1980). Northwest-trending features correlating with flight-line direction may in some cases (e.g., southeast Blake Plateau) be due to a certain amount of low-amplitude noise caused by navigation error.

One particular structure trending northwest from the ECMA at 36° N is, however, reflected in the data (Figures 3, 5). This feature, if projected to the southeast, coincides with the less obvious Norfolk fracture zone (Figure 1) on the ocean side of the ECMA. The interpreted structure may represent a fault or preexisting zone of weakness in the continent, which localized the oceanic fracture zone at the time of the Atlantic opening. A fault is recognized northwest of this structure on land (Cederstrom, 1945), which extends at least as far west as the Fall Line and has been correlated with the central Virginia seismic zone (Sykes, 1978).

We interpret many roughly circular (mostly positive) anomalies as having been caused by mafic intrusions. Although difficult to see in the small scales of Figures 3–6, we have been able to identify them with confidence (Figure 1) on the larger scale plots of the second-derivative maps and in combination with the total-intensity data.

WAVELENGTH-FILTERED MAPS

A series of high-pass filters was applied to the total-intensity data. Three maps were generated that retained wavelengths less than 60 km (Figure 7), less than 40 km (Figure 8), and less than 20 km (Figure 9). Because these maps are meant to be used as a visual aid to interpretation rather than as a quantitative presentation, choice of filter taper was made largely on map appearance. The similarity between these residual maps and the second-vertical-derivative map (compare Figures 3 and 7) is expected because both are dependent on the curvature of the anomalies (Dobrin, 1976, p. 447). Therefore, the features described above in Figures 3–6 are apparent in Figure 7 and to lesser extents with the successively higher frequencies passed in Figures 8 and 9.

Considering the half-width approximation of depth to the source of a magnetic anomaly (Dobrin, 1976, p. 543), we infer that we are looking at sources progressively shallower than 15-, 10-, and 5-km depth for the less than 60-, 40-, and 20-km high-pass-filtered maps, respectively, of Figures 7–9. In Figure 7, for example, we see that most of the 200–600 nT amplitude of the ECMA (probably caused mostly by the bulk magnetization contrast between oceanic and continental crust) has been removed. When Figure 7 is compared with the total-intensity map of Figure 2, only the short-wavelength component (probably caused by sources shallower than 15 km) is seen. Note that the contour interval in Figures 7–9 is 5 nT, compared with 25 nT in Figure 2. The broad positive anomalies over the Baltimore Canyon trough, Carolina trough, and Blake Plateau basin are removed (compare Figures 2 and 7), but the short-wavelength features, such as the hinge zones bounding the Georges Bank basin and the Baltimore Canyon trough on the continent side, are retained. On the 40-km high-pass map of Figure 8, we see that the effect of many of the deeper sources has been removed. The residual effect from the ECMA is only a few tens of nanoteslas, whereas the deeper sources below the basins and the oceanic crust of the magnetic quiet zone east of the ECMA are almost totally removed. Some effect of the Blake Spur anomaly remains. All of the shallow-source anomalies over the platforms toward the continent from the hinge zones remain, as does the prominent circular anomaly caused by an intrusion (Behrendt and Klitgord, 1980) in the Baltimore Canyon trough.

Figure 9, the 20-km high-pass map, shows only effects from the shallowest sources. The ECMA is entirely removed, but the circular anomaly in the Baltimore Canyon trough remains. The inferred circular intrusion causing this anomaly has a calculated depth of 3.5 km (Behrendt and Klitgord, 1980), which is supported by seismic-reflection data. The effects of sources at shallow depths beneath the platforms remain, but the anomalies caused by the hinge zones bounding the Georges Bank basin and the Baltimore Canyon trough have been removed.

STRUCTURAL-ELEMENTS MAP

Although many of the features shown in Figure 3 have been recognized previously and are described elsewhere (e.g., Klitgord and Behrendt, 1979; Schlee et al., 1976; Grow et al., 1979; Dillon et al., 1979), the interpretation shown in Figure 1 is based entirely on the second-vertical-derivative map except for the oceanic fracture zones outside the area of the magnetic survey, and the extent of a basalt layer discussed below. This presentation indicates the inferred structural trends without regard to depths to sources or amplitudes of anomalies. Depths to magnetic basement are shown in Klitgord and Behrendt (1979). Only structural lineations associated with positive anomalies are indicated in Figure 1 (compare to Figures 2, 3). Where there is other evidence that suggests an association of positive and negative anomalies with horsts and grabens or some analogous structures, such as in the northernmost part of the survey, these structural elements are shown. Long, linear structures such as the ECMA are indicated by heavier lines, but many of these linear features (compare Figures 3–6) also have associated circular anomalies, suggesting circular intrusions along the trends. These circular intrusions may be tectonically significant, but it is not possible to show them clearly at the scale of Figure 1. The heaviest lines in Figure 1 show structural elements that are not associated with specific anomalies but rather with truncation of anomalies, probably resulting from faulting. Examples of these lineaments include the arcuate western boundary of the Blake Plateau basin and the presumed fault having the same trend as the Norfolk fracture zone beneath the continental shelf between 37° and 36° N.

The basins indicated in Figure 1 are better observed as tones of yellow on the second-vertical-derivative map (Figure 3); the low-curvature areas (yellow on the color scale) are generally indicative of thick nonmagnetic sedimentary rock. Hinge zones well delineated in Figures 3–6 are indicated in the Georges Bank basin and Baltimore Canyon trough in Figure 1.

The dashed line in Figure 1 just landward of the Blake Plateau basin indicates the seaward extent of a Jurassic basalt layer that has been recognized as a strong seismic reflector at the base of the Cretaceous sedimentary rocks and overlies the post-rift unconformity (Dillon et al., 1979). It is not possible to observe seismic reflections that originate beneath this layer (Behrendt et al., 1983). The aeromagnetic data and the second-vertical-derivative map, in particular, define older basement structure, because the basalt layer is a magnetically thin sheet and thus in effect magnetically transparent.

In compiling Figure 1 we have avoided northwest lineaments unless corroborated by other geological and geophysical evidence because of the northwest direction of the flight lines. Undoubtedly, we have excluded some real northwest-trending structural elements like those associated with the fracture zones. The fracture zones shown seaward of the aeromagnetic survey are taken from Klitgord and Behrendt (1979).

We used large-scale maps in the compilation of Figure 1, so the small circular features indicated can be accepted with confidence. There is a much greater occurrence of circular second-derivative anomalies in the continental terrane than over oceanic crust. The intrusion near 39°15′ N, 73° W, in the vicinity of the Baltimore Canyon trough is known to be Early Cretaceous in age (Schlee et al., 1976) from seismic-reflection interpretations. The New England sea mounts trending east from about 40° N, 68°30′ W, are also of Early Cretaceous age (Vogt and Tucholke, 1979). The Blake Plateau area, which probably is underlain by continental transitional crust (Klitgord and Behrendt, 1979), contains numerous circular anomalies caused by intrusions as compared with fewer in the oceanic crust to the north, but east of the ECMA.

SUMMARY

In this paper we have demonstrated the usefulness of digital filtering of a high-sensitivity aeromagnetic survey over a large region. The basins and numerous other structures are well delineated by the filter computations and color-plotting techniques used. Had the aeromagnetic results been available in this form prior to carrying out the regional seismic-reflection surveys, planning of seismic profiles would have been more efficient.

All of the structural elements shown in Figure 1 (except the fracture zones and the boundary of the basalt layer) were interpreted from the second-vertical-derivative maps at large scales. Each of the features so delineated was also recognized upon reinspection of the original contoured total-intensity data. However, many of these features were not originally recognized (e.g., the two parallel components of the ECMA between 36° and 37°30′ N and the hinge zones of the Georges Bank basin and Baltimore Canyon trough indicated in Figures 4 and 5). These results demonstrate the usefulness of the second-vertical-derivative calculation in greatly speeding up the tectonic interpretation.

The wavelength-filtered maps, while similar in appearance to the second-vertical-derivative results, allow a qualitative estimation of depth to magnetic basement on a regional basis. Therefore, they provide immediate, useful information that could be obtained in detail only by much more laborious processes along individual profiles such as those employed for this data set by Klitgord and Behrendt (1979).

If a high-sensitivity survey similar to that discussed here were available in a frontier area of the world prior to other geophysical surveys, rapid tectonic interpretation would greatly accelerate the understanding of the region and the planning of additional, more expensive studies.

ACKNOWLEDGMENTS

We thank K. D. Klitgord, R. W. Simpson, and T. G. Hildenbrand for many helpful discussions. R. H. Godson and M. W. Webring provided invaluable assistance in using their computer programs for the data processing.

REFERENCES

Behrendt, J. C., and Grim, M. S., 1983, Maps showing second vertical derivative, total magnetic intensity and tectonic elements of the U.S. Atlantic continental margin: Geophys. Invest. Map GP-956, U.S. Geol. Surv., scale 1:2 500 000.

Behrendt, J. C., Hamilton, R. M., Ackermann, H. D., Henry, V. J., and Bayer, K. C., 1983, Marine multichannel seismic-reflection evidence for Cenozoic faulting and deep crustal structure near Charleston, South Carolina, in Gohn, G. S., Ed., Studies related to the Charleston, South Carolina, earthquake of 1886—tectonics and seismicity: Prof. Pap. 1313, U.S. Geol. Surv., J1–J29.

Behrendt, J. C., and Klitgord, K. D., 1979, High resolution aeromagnetic anomaly map of the U.S. Atlantic continental margin: Geophys. Invest. Map GP-931, U.S. Geol. Surv., scale 1:1 000 000.

——1980, High sensitivity aeromagnetic survey of the U.S. Atlantic continental margin: Geophysics, **45**, 1813–1846.

Cederstrom, D. J., 1945, Structural geology of southeastern Virginia: AAPG Bull. **29**, 71–95.

Dillon, W. P., Paull, C. K., Buffler, R. T., and Fail, P., 1979, Structure and development of the Southeast Georgia Embayment and Northern Blake Plateaus; preliminary analyses, in Watkins, J. S., Montadert, L., and Dickerson, P. W., Eds., Geological and geophysical investigations of continental margins: Mem. 29, AAPG, 27–41.

Dobrin, M. B., 1976, Introduction to geophysical prospecting: McGraw–Hill.

Grow, J. A., 1980, Deep structure and evolution of the Baltimore Canyon Trough in the vicinity of the COST no. B-3 Well, in Scholle, P. A., Ed., Geological studies of the COST No. B-3 Well, United States Mid. Atlantic continental slope area: Circ. 833, U.S. Geol. Surv., 117–125.

Grow, J. A., Mattick, R. E., and Schlee, J. S., 1979, Multichannel seismic depth sections and interval velocities over outer continental shelf and upper continental slope between Cape Hatteras and Cape Cod, in Watkins, J. S., Montadert, L., and Dickerson, P. W., Eds., Geological and geophysical investigations of continental margins: Mem. 29, AAPG, 65–83.

Hildenbrand, T. G., 1983, Preliminary documentation of program "FFTFIL": Open-File Rep. 82–237, U.S. Geol. Surv.

Hildenbrand, T. G., Kane, M. F., and Hendricks, J. D., 1982, Magnetic basement in the upper Mississippi Embayment region—a preliminary report, in McKeown, F. A., and Pakiser, L. C., Eds., Investigations of the New Madrid, Missouri, earthquake region: Prof. Pap. 1236, U.S. Geol. Surv., 39–53.

IAGA, 1969, Commission Two, Working Group 4, Analysis of the geomagnetic field, International Geomagnetic Reference Field, 1965: J. Geophys. Res., **74**, 4402–4408.

Klitgord, K. D., and Behrendt, J. C., 1979, Basin structure of the U.S. Atlantic margin, in Watkins, J. S., Montadert, L., and Dickerson, P. W., Eds., Geological and geophysical investigations of continental margins: Mem. 29, AAPG, 85–112.

Klitgord, K. D., and Grow, J. A., 1980, Jurassic seismic stratigraphy and basement structure of western Atlantic magnetic quiet zone: AAPG Bull., **64**, 1658–1680.

Schlee, J. S., Behrendt, J. C., Grow, J. A., Robb, J. M., Mattick, R. E., Taylor, P. T., and Lawson, B. J., 1976, Regional geologic framework off northeastern United States: AAPG Bull., **60**, 926–951.

Sykes, L. R., 1978, Intra-plate seismicity, reactivation of pre-existing zones of weakness, alkaline magmatism, and other tectonism post-dating continental separation: Rev. Geophys. Space Phys., **16**, 621–687.

Taylor, P. T., Zietz, Isidore, and Dennis, L. S., 1968, Geologic implications of aeromagnetic data for the eastern continental margin of the United States: Geophysics, **33**, 755–780.

Vogt, P. R., and Tucholke, B. E., 1979, The New England sea mounts—testing origins, in Tucholke, B. E., and Vogt, P. R., Deep Sea Drilling Project, Leg 43 of the cruises of the drilling vessel Glomar Challenger; Istanbul, Turkey to Norfolk, Virginia, June–August, 1975: Initial Rep. 43, Deep Sea Drill. Proj., 847–856.

Webring, M. W., 1981, MINC, a gridding program based on minimum curvature: Open-File Rep. 81–1223, U.S. Geol. Surv.

The change in the magnetic-anomaly pattern at the ocean–continent boundary

J. R. Heirtzler*

ABSTRACT

The magnetic-anomaly pattern shows characteristic changes across the ocean-continent boundary. At the Northeast Atlantic passive margin, linear oceanic anomalies disappear at the edge of the continent as that edge is defined by seismic means. In the western North Atlantic off Florida there is an anomalous strip of ocean crust, but a change in character of the magnetic anomalies can still be seen. At the Aleutian active margin, oceanic anomalies are destroyed along a line that lies about 100 km landward of the trench axis. At the Japan active margin a similar situation exists. It is not clear why oceanic anomalies are so drastically quenched at this point near the trenches, but this condition may be related to the fact that the oceanic plate is carrying the magnetic layer through the Curie isotherm or to the seismic activity.

INTRODUCTION

Active margins and passive margins are the main types of margins that separate continental from oceanic crust. Continental crust differs from oceanic crust in its thickness, age, average density, and composition. Accordingly, one might expect to find, and does find, the character of magnetic anomalies differing on opposite sides of the ocean–continent boundary. This change is not the same as that encountered over seamounts, fracture zones, and most other structural features in the oceans or as that encountered in going from one continental feature to another.

Although the difference in the continental and oceanic crust has been recognized for many years, it is only within the last few decades that attempts have been made to locate the exact boundary. This search took on renewed vigor when global-tectonics theory indicated that there should be a geometrical fit of the continents bordering the passive mar-

*Woods Hole Oceanographic Institution, Woods Hole, MA 02543. Contribution no. 5375 of Woods Hole Oceanographic Institution.

gins. Bullard et al. (1965) made a fit of the continents bordering the Atlantic by matching the 500-fathom contour lines on the opposite sides. Since that time Sproll and Dietz (1969), Smith and Briden (1977), Barron et al. (1978), and other authors showed that passive margins can be matched in a general way using bathymetric contours. Such matches, and the ocean–continent transition lines that result from them, will not be precise, since they are determined by the surficial sediment configuration of the sea floor and not by the deeper structure that truly distinguishes continental from oceanic crust.

Precise continental fits of active margins are not attempted, since these margins are continually eroded and deformed by the subduction process.

The oceanic–continental boundary is usually thought to be best defined by seismic means. The depth to Moho (8 km/s horizon) is at about 30 km under the continental crust and of the order of 10 km below the oceanic basaltic basement. Technical limitations, however, prevent the definition of deep continental and oceanic crust with great spatial resolution. Seismologists resort to other characteristics of the shallow seismic structure (e.g., the loss of the oceanic-basement reflector near the continent) to define the boundary better. In fact, oceanic crust underlies some continental crust, as at subduction zones, or continental crust may be intermixed with oceanic crust, as may be the case at passive margins if the continental crust is stretched during the initiation of continental breakup and sea-floor spreading (McKenzie, 1978; Karner and Watts, 1982). Thus it may be a mistake to think of the ocean–continent transition as being a sharp line on a map. In this article I shall rely on the marine seismologist's location of the passive-margin boundary.

For active margins I shall use the results of marine seismic profiling and the location of earthquake epicenters to define the geometry of the downgoing oceanic crust. Everything above the downgoing slab except the accretionary wedge close to the trench is assumed to be continental.

There are relatively few margins where marine seismic and magnetic measurements have been made with sufficient density to undertake a sensible comparison of the two data sets. Two passive margins will be discussed in the Atlantic, and two active margins in the Pacific.

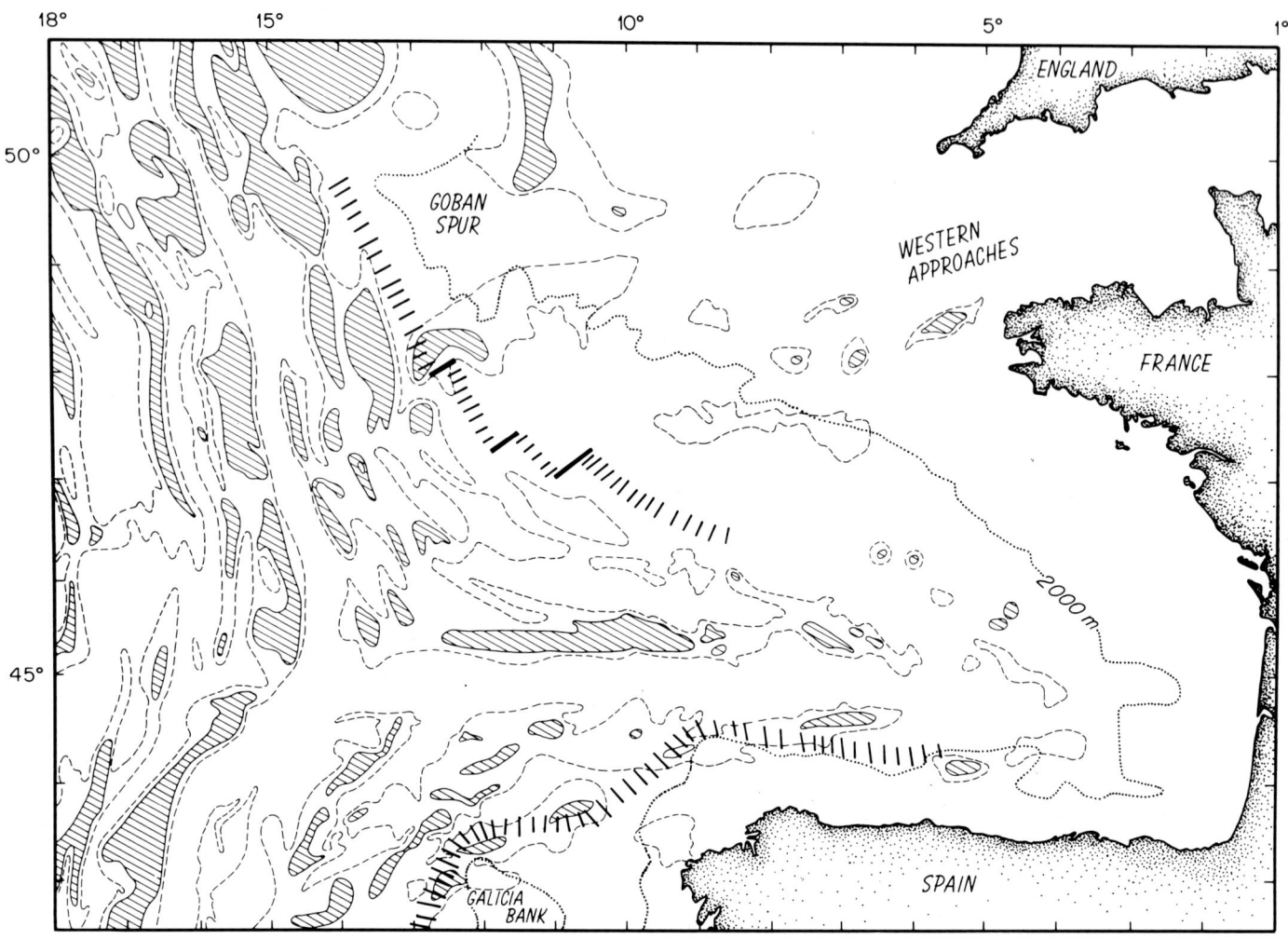

FIG. 1. Magnetic anomalies and the ocean–continent boundary in the Northeast Atlantic. Anomalies with amplitudes of 50 nT or more are shaded, and the −50 nT contour line is shown as a dashed line (after Guennoc et al., 1978). The broken hachured line identifies ocean–continent boundary as identified by seismic studies (Montadert et al., 1979). The 2 000-m isobath is shown.

NORTHEAST ATLANTIC PASSIVE MARGIN

Seismic studies by Montadert et al. (1979) for the Goban Spur and Galicia Bank, and several holes drilled by the Glomar Challenger in this area (Montadert and Roberts, 1979), indicate an ocean–continent transition to be about where the hachured line is drawn in Figure 1. This transition is in water depths of 2 km or greater and does not closely follow the 2-km or any other isobath. The westernmost continental crust appears to be only about 11 km thick, thinned from its usual thickness of 30 km, and it is subsided. This thinned crust is broken into tilted blocks that have been rotated by listric faults which increase in depth to seaward and form half-grabens along their upper surfaces.

A magnetic-anomaly map of this area was made by Guennoc et al. (1978) (Figure 1). The lineated character of the oceanic anomalies is shown by the shaded areas, which have amplitudes of 50 nT or more, and by the dashed line, which is the −50 nT contour line. Except for minor lineations in the Western Approaches and north of the Goban Spur, there are no anomaly lineations on the continental crust. The termination of the magnetic lineation defines the transition to within an uncertainty of about 100 km. The east–west-striking oceanic anomalies in the Bay of Biscay, between France and Spain, are believed to have been produced when the Iberian Peninsula rotated in a counterclockwise direction away from France.

WESTERN NORTH ATLANTIC PASSIVE MARGIN

Much geophysical work has been undertaken off the east coast of the United States in recent years. Figure 2 shows part of that area. Magnetic anomalies greater than 50 nT in amplitude are shaded, and the −50 nT contour line is shown as a dashed line. The hachured line is where the ocean–continent transition is generally believed to be, under the East Coast Magnetic Anomaly (ECMA), although some investigators feel that this transition is more likely at the Blake Spur Anomaly (BSA).

A recent review of all marine seismic data in this area that are available to the public (Bryan and Heirtzler, in press; Ewing and Rabinowitz, in press) shows that oceanic base-

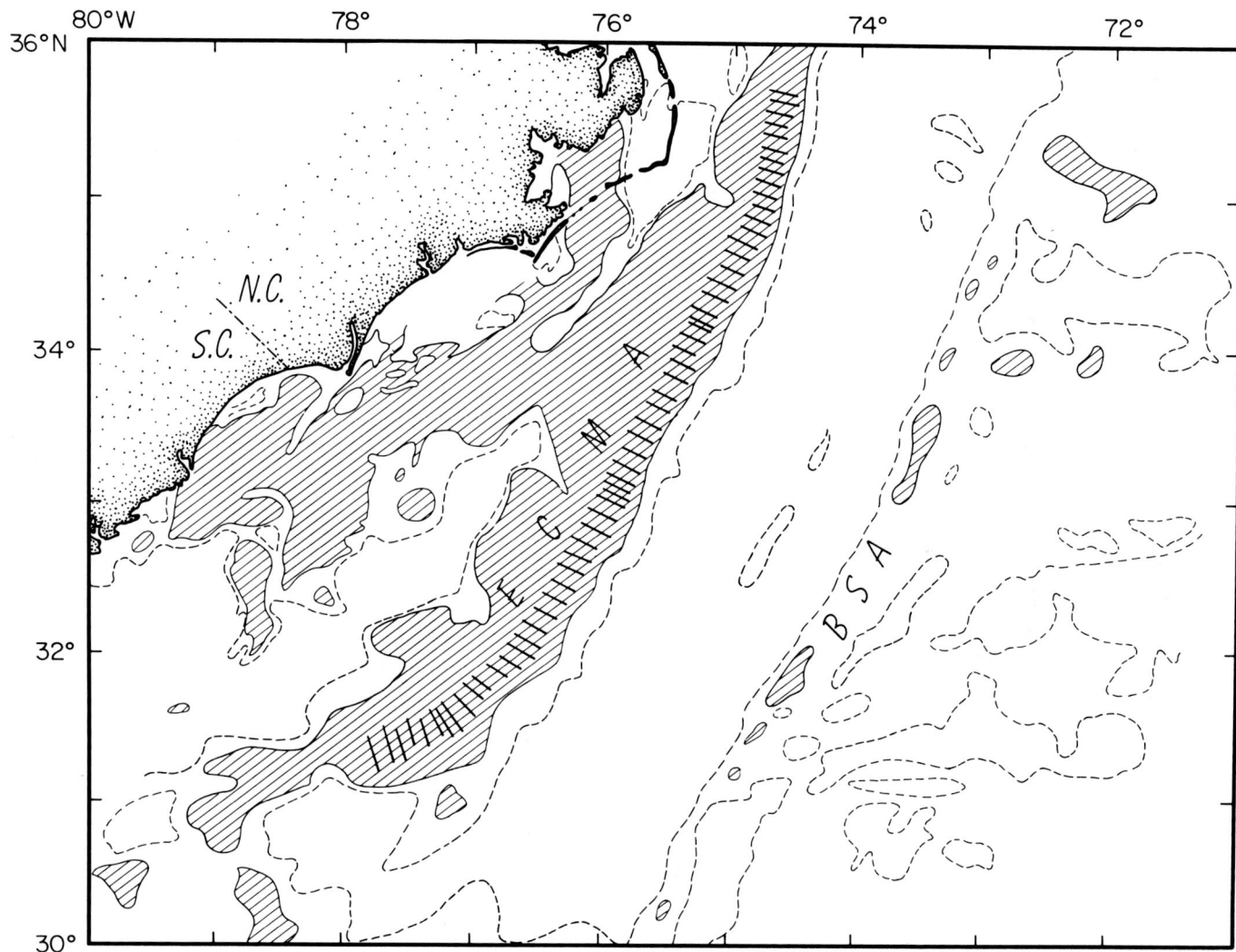

FIG. 2. Magnetic anomalies and the ocean–continent boundary in the western North Atlantic Ocean. Magnetic anomalies with amplitudes of 50 nT or more are shaded, and the −50 nT contour line is shown as a dashed line. (After Bryan and Heirtzler, in press; Ewing and Rabinowitz, in press.) ECMA and BSA identify the East Coast Magnetic Anomaly and the Blake Spur Anomaly, respectively.

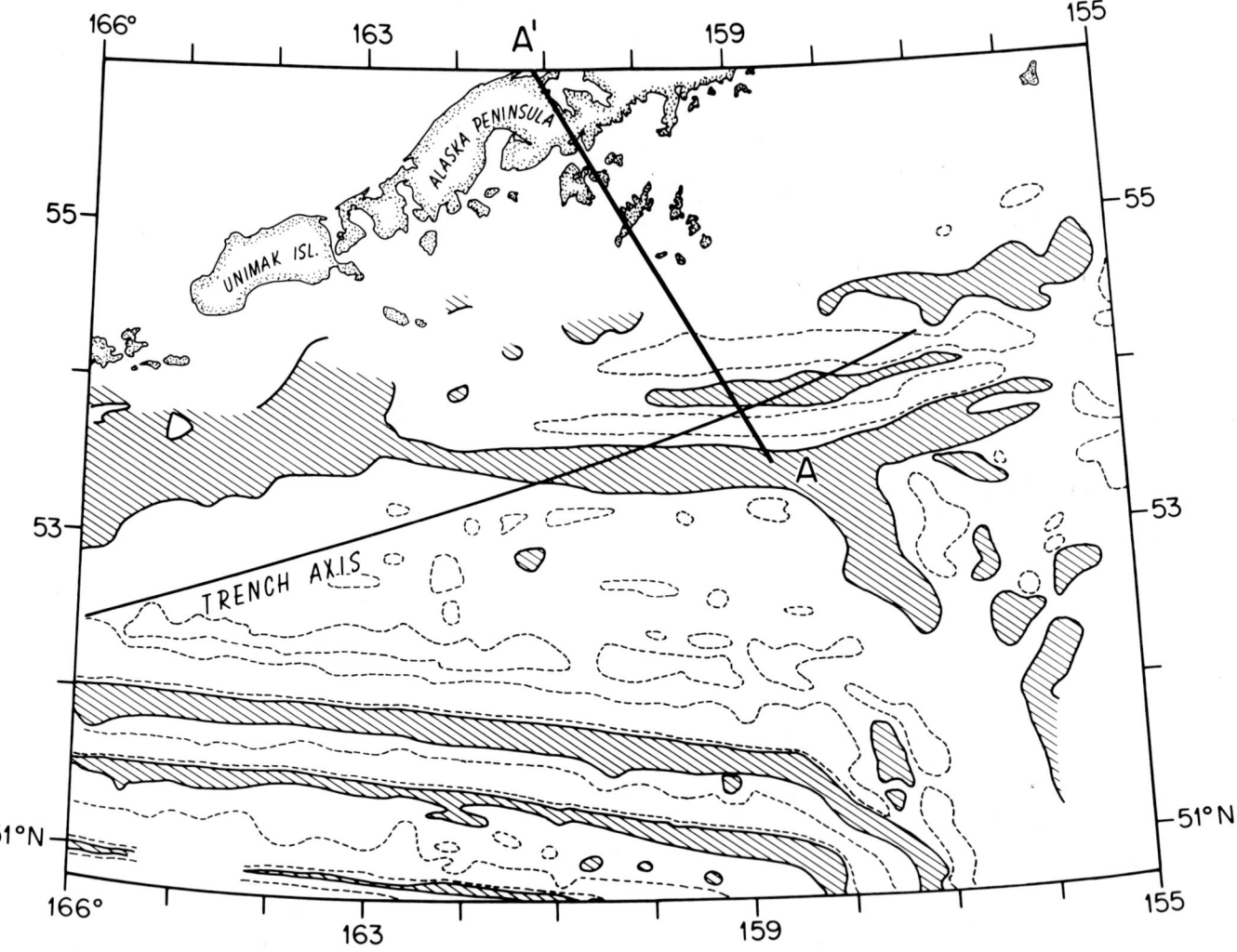

FIG. 3. Magnetic anomalies and the Aleutian trench axis off the eastern Aleutian Arc (after NOAA Sea Maps NOS 15248-14, 1973, and 20242-12M, 1974). Earthquake epicenters projected along section A–A' are shown in Figure 4. Shaded areas are greater than +100 nT; dashed lines are −100 nT contours.

ment can be followed to approximately the ECMA, where it disappears at a major Jurassic carbonate bank. Between the BSA and the ECMA the basement has some structural characteristics that are different from the more usual appearance of oceanic basement to the east of the BSA. The chief feature of this region is a median rise or ridge that some have called an extinct spreading center. There is little if any magnetic expression of this ridge.

A basement flexure or hinge also lies beneath the BSA, dropping the basement to the west. The Blake Spur fracture zone is believed to offset the oceanic and continental basement at the bottom of this figure.

Sea-floor-spreading magnetic anomalies are not strongly evident in the seaward portion of this area, because the basement there is of Jurassic age, created when there were no geomagnetic field reversals (Larson and Hilde, 1975). On the continental side of the ocean–continent transition the anomalies are large, having an overall long wavelength component that follows the general trend of the Appalachians. Thus the transition is a line of demarcation for the types of magnetic anomalies.

ALEUTIAN ACTIVE MARGIN

The sharp extinction of oceanic magnetic lineations at the subduction zone of the Aleutian active margin was first illustrated by Hayes and Heirtzler (1968) from isolated ships' magnetic profiles. Since that time, good magnetic-anomaly maps have been published for the Gulf of Alaska and the eastern Aleutian Islands (NOAA Sea Map series; Taylor and O'Neill, 1974; Schwab and Bruns, 1979; Godson, 1982). Magnetic anomalies shown in Figure 3 were taken from those sources.

Numerous studies also have been made of the seismicity of the area. Seismicity occurs at the top of the Moho (8 km/s velocity) and defines the Benioff zone. The magnetic-source layer (layer 2) occurs above Moho and just below the sea-floor sediments (velocity 3.5–6.5 km/s). A cross-section along line A–A' is shown in Figure 4. These epicenters define the top of the downgoing slab for depths below about 50 km (about 180 km on horizontal scale). The magnetic anomalies disappear at about 260 km on the horizontal scale, where the epicenters are shallow and scattered. It is possible that earthquake activity may destroy the magnetization by fracturing the magnetic material above the epicenters and rotating the fragments. If a relationship exists between the location of epicenters and the disappearance of magnetic anomalies, that relationship is not a definitive one. That possible relationship does not hold farther to the west, where ocean-bottom seismographs have located small epicenters immediately *south* of the trench axis (Frolich et al., 1980; Frolich et al., 1982); and yet magnetic lineations there are not destroyed.

The Curie point of known earth minerals (575°C for magnetite) is at a depth of about 25 km (Stacey, 1969; McElhinny, 1973), where the lineations disappear north of the trench axis. Thus it may be temperature rather than seismic activity that destroys the magnetization. Uyeda (1978) illustrated that, in general, heat flow from behind the trench is about twice the average from the ocean floor in front of the trench. However, quite near the trench the heat

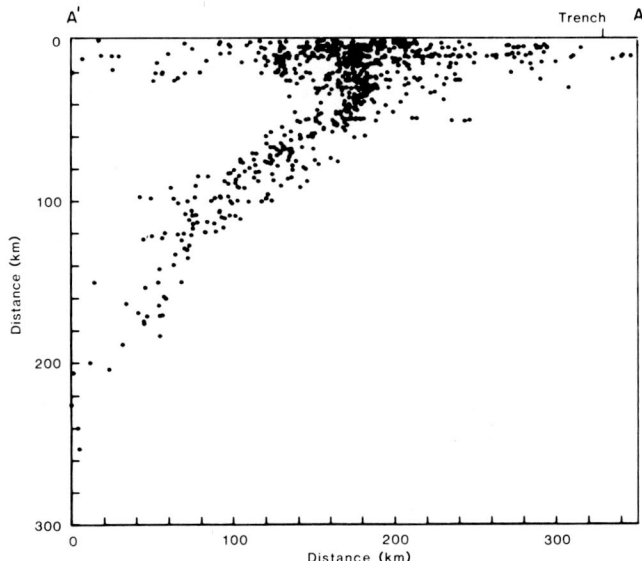

FIG. 4. Earthquake epicenters projected on the plane A–A'. These epicenters were detected between the longitudes 157° and 164° W during 1979. (After Beavan et al., 1983.)

flow may be suppressed to even below-normal ocean values because the oceanic plate is cold as it starts descending (Minear and Toksöz, 1970; Oxburgh and Turcotte, 1970). The temperature of the plate, in contrast to the heat flow through the sediments, will continually increase as the plate subducts.

JAPAN ACTIVE MARGIN

Another area where good magnetic, seismic, and other geophysical data have been gathered is near the Japan trench off Honshu (Figure 5). Magnetic lineations extend well landward of the trench axis and show a change in character north of the latitude where the Japan trench meets the Kuril trench and a change in strike of the trench axis occurs. The lineations extend 100 to 120 km west of the Japan trench axis.

The configuration of the seismic horizons near the trench axis has been defined by a sonobuoy-refraction program, and the configuration of the Moho defined by seismicity (Murauchi and Ludwig, 1980). Figure 6 shows the epicenters. At about 142°45′, where the magnetic lineations disappear, the depth to Moho is about 40 km. Some epicenters occur near the trench axis—where there are magnetic lineations. These observations suggest that there is little, if any, relationship between where the lineations disappear and the location of seismic activity.

Heat-flow measurements were made in deep-sea drill holes at sites 438A and 439 (Langseth and Burch, 1980). These two nearby holes were drilled at about 40°38′ N, 143°14′ and 19′ E, 75 km west of the juncture of the Kuril and Japan trenches and within 25 km of the line where the linear anomalies are extinguished (Von Huene et al., 1982). Extrapolating the temperature found there to the 575°C isotherm gives a depth of 16 to 18 km for the Curie isotherm. This compares to a depth of 10 km for the dipping oceanic

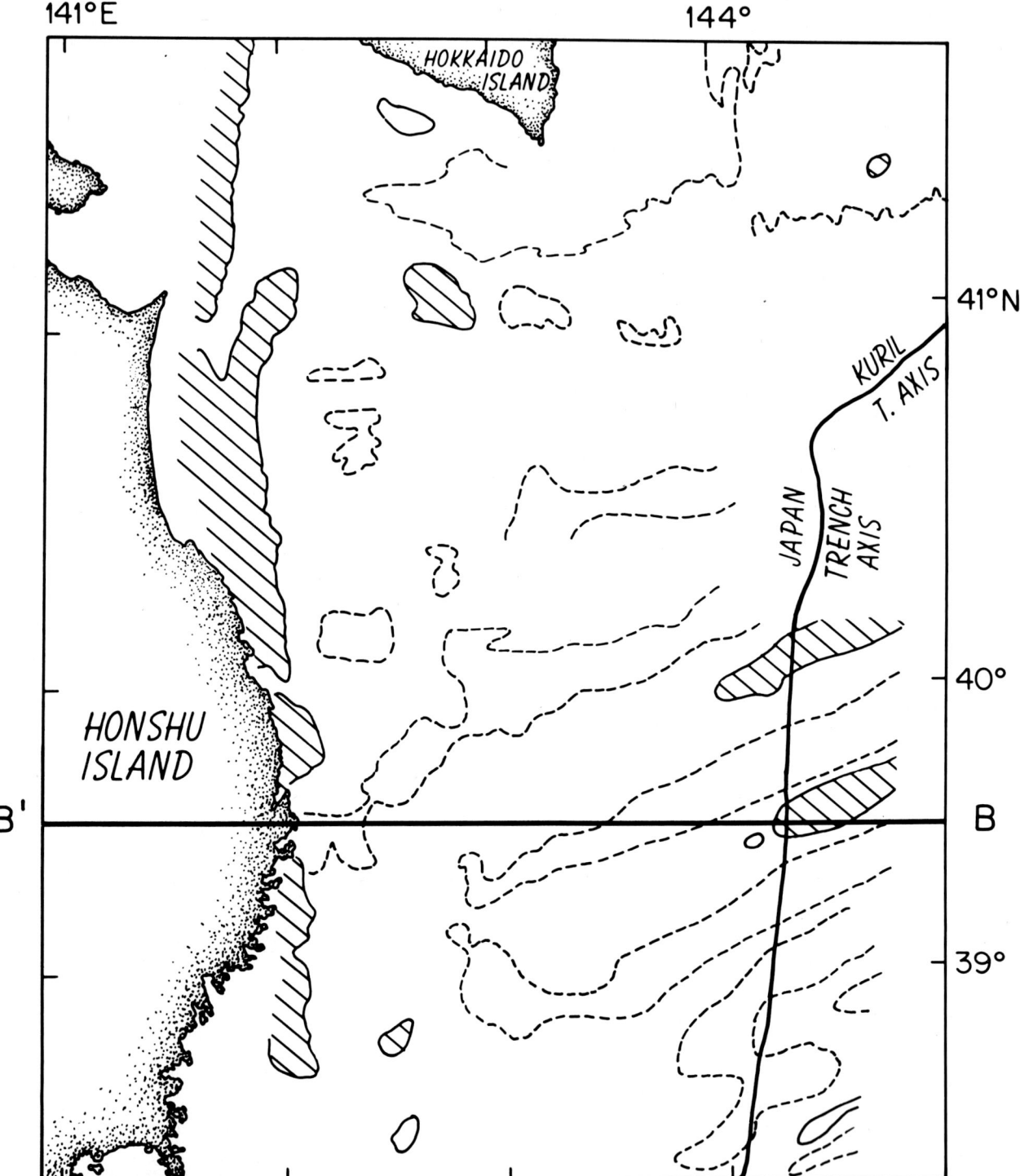

FIG. 5. Magnetic anomalies and the Japan trench axis off Japan (after Oshima et al., 1975). Earthquake epicenters projected along section B–B' are shown in Figure 6. Shaded areas are greater than +100 nT; dashed lines are −100 nT contours.

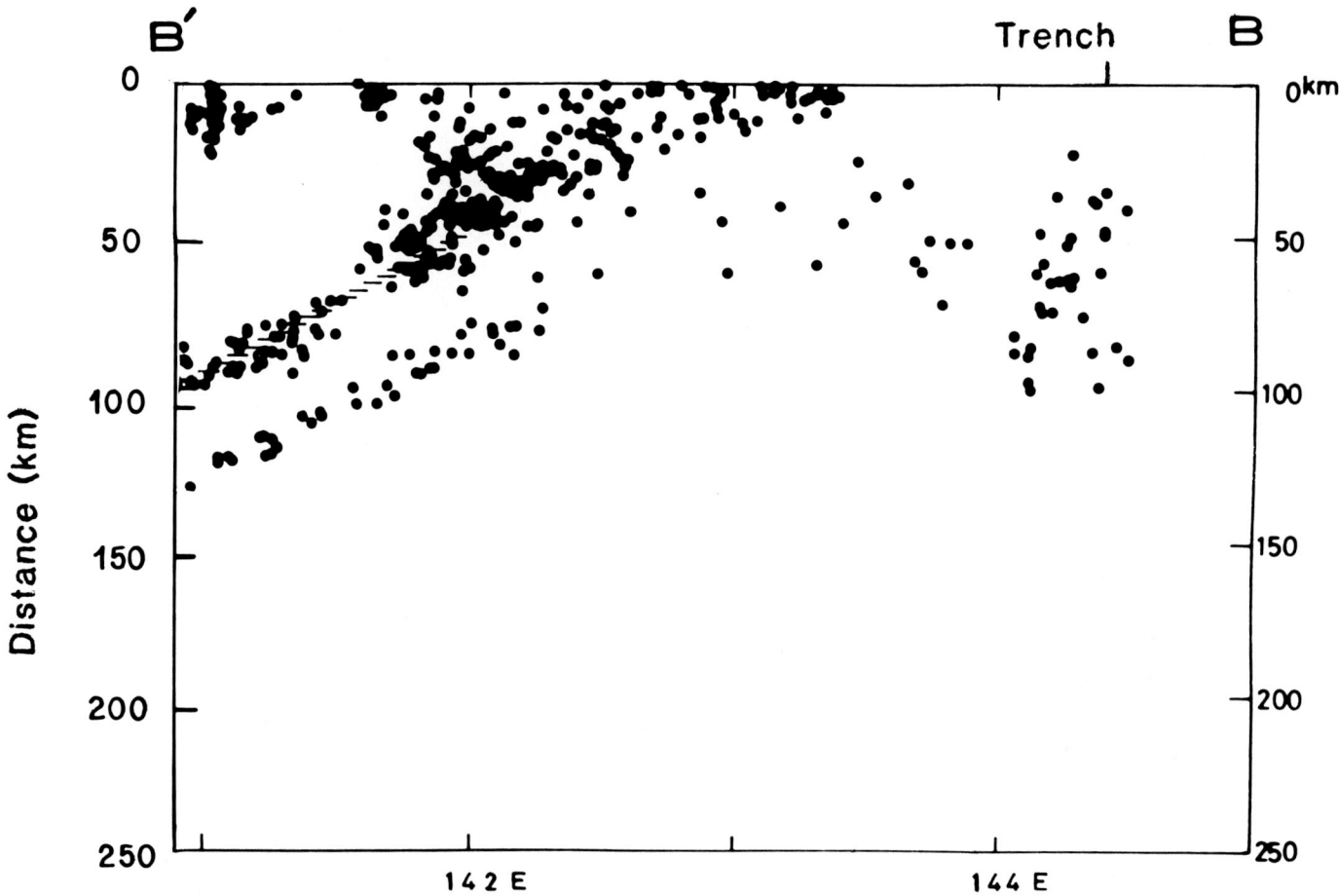

Fig. 6. Earthquake epicenters projected on the plane B–B′. These epicenters occurred between 39° and 40° N during April–December 1975. Hachured line shows the top of the subducting slab. (After Hasegawa et al., 1978.)

basement (as shown by seismic-reflection records) and to about 40 km for the Moho, as just mentioned.

CONCLUSIONS

For passive margins around the Atlantic Ocean, and probably for other passive margins as well, the character of the magnetic-anomaly pattern changes across the ocean–continent boundary. The nature of this change may be different from one part of the margin to the next, but the change is usually characterized by a disappearance of oceanic magnetic lineations.

The active margins show that the oceanic-anomaly pattern is maintained for about 100 km landward of the trench axis. Why it is destroyed is not resolved by this study, but the magnetic layer approaching the Curie isotherm appears to be significant. Further heat-flow measurements near the point where the oceanic anomalies disappear may firmly establish this relationship.

REFERENCES

Barron, E. J., Harrison, C. G. A., and Hay, W. W., 1978, A revised reconstruction of the southern continents: Trans. Am. Geophys. Union, 59, 436–449.

Beavan, J., Hauksson, E., McNutt, S. R., Bilham, R., and Jacobs, K. H., 1983, Tilt and seismicity changes in the Shumagin Seismic Gap: Science, 222, 322–325.

Bryan, G., and Heirtzler, J. R., in press, Ocean Margin Drilling Program, regional data synthesis, 28°–36°N, 70°–82°W: Mar. Sci. Int.

Bullard, E. C., Everett, J. E., and Smith, A. G., 1965, The fit of the continents around the Atlantic, in Blackett, P.M.S., Bullard, E., and Runcorn, S. K., Eds., A symposium on continental drift: Philos. Trans. R. Soc. London, A256, 41–51.

Ewing, J. I., and Rabinowitz, P., in press, Ocean Margin Drilling Program, regional data synthesis, 34°–41°N, 68°–78°W: Mar. Sci. Int.

Frolich, C., Caldwell, J. G., Malahoff, A., Latham, G. V., and Lawton, J., 1980, Ocean bottom seismograph measurements in the central Aleutians: Nature, 286, 144–145.

Frolich, C., Billington, S., Engdahl, E. R., and Malahoff, A., 1982, Detection and location of earthquakes in the central Aleutian subduction zone using island and ocean bottom seismograph stations: J. Geophys. Res., 87, 6853–6864.

Godson, R. H., 1982, Composite magnetic anomaly map of the United States, Part B—Alaska and Hawaii: Open-File Rep. 82-970, U.S. Geol. Surv.

Guennoc, P., Jonquet, H., and Sibuet, J–C., 1978, Carte magnetique de l'Atlantique nord-est: CNEXO.

Hasegawa, Akira, Umino, Norihito, and Takagi, Akio, 1978, Double-planed deep seismic zone and upper-mantle structure in the northeastern Japan Arc: Geophys. J. R. Astron. Soc., 54, 281–296.

Hayes, D. E., and Heirtzler, J. R., 1968, Magnetic anomalies and their relation to the Aleutian Island Arc: J. Geophys. Res., 73, 4637–4646.

Karner, G. D., and Watts, A. B., 1982, On isostasy at Atlantic-type continental margins: J. Geophys. Res., **87**, 2923–2948.

Langseth, M., and Burch, T., 1980, Geothermal observations on the Japan Trench transect: Initial Rep. Deep Sea Drill. Proj., **56, 57**, pt. 2, U.S. Govt. Print. Off., 1207–1210.

Larson, R. L., and Hilde, T.W.C., 1975, A revised time scale of magnetic reversals for the Early Cretaceous and Late Jurassic: J. Geophys. Res., **80**, 2586–2594.

McElhinny, M. W., 1973, Paleomagnetism and plate tectonics: Cambridge Univ. Press.

McKenzie, D. P., 1978, Some remarks on the development of sedimentary basins: Earth Planet. Sci. Lett., **40**, 25–32.

Minear, J. W., and Toksöz, M. N., 1970, Thermal regime of a downgoing slab and new global tectonics: J. Geophys. Res., **75**, 1397–1419.

Montadert, L., de Chapal, O., Roberts, D., Guennoc, P., and Sibuet, J., 1979, Northeast Atlantic passive continental margins: rifting and subsidence processes, *in* Talwani, M., Hay, W., and Ryan, W.B.F., Eds., Deep sea drilling in the Atlantic Ocean: continental margins and paleoenvironment: Am. Geophys. Union.

Montadert, L., and Roberts, D. G., 1979, Initial Reports of the Deep Sea Drilling Project: **48**, U.S. Govt. Print. Off.

Murauchi, S., and Ludwig, W. J., 1980, Crustal structure of the Japan Trench: the effect of subduction of ocean crust: Initial Rep. Deep Sea Drill. Proj., **56, 57**, pt. 1, U.S. Govt. Print. Off., 463–469.

Oshima, S., Kondo, T., Tsukamoto, T., and Onodera, K., 1975, Magnetic anomalies at sea around the northern part of Japan: Rep. Hydrograph. Res. 10, Hydrograph. Off., Japan, 39–44.

Oxburgh, E. R., and Turcotte, D. L., 1970, Thermal structure of island arcs: Geol. Soc. Am. Bull., **81**, 1665–1688.

Schwab, W. C., and Bruns, T. R., 1979, Preliminary residual magnetic map of the northern Gulf of Alaska: Misc. Field Stud. Map MF-1054, U.S. Geol. Surv.

Smith, A. G., and Briden, J. C., 1977, Mesozoic and Cenozoic paleocontinental maps: Cambridge Univ. Press.

Sproll, W. P., and Dietz, R. S., 1969, Morphological continental drift fit of Australia and Antarctica: Nature, **222**, 345–348.

Stacey, F. D., 1969, Physics of the Earth: John Wiley.

Taylor, P. T., and O'Neill, N. J., 1974, Results of an aeromagnetic survey in the Gulf of Alaska: J. Geophys. Res., **79**, 719–723.

Uyeda, S., 1978, The new view of the Earth: W. H. Freeman.

Von Huene, R., Langseth, M., Nasu, N., and Okada, H., 1982, A summary of Cenozoic tectonic history along the IPOD Japan Trench Transect: Geol. Soc. Am. Bull., **93**, 829–846.

An isostatic residual gravity map of California—A residual map for interpretation of anomalies from intracrustal sources

Robert C. Jachens* and Andrew Griscom*

ABSTRACT

An isostatic residual gravity map of California effectively separates gravity anomalies caused by intracrustal and near-surface density inhomogeneities from the large Bouguer gravity anomalies that result from isostatic compensation of the topography. This regional–residual separation reveals some anomalies that are not easily recognized on the Bouguer gravity map and converts others that are difficult to interpret quantitatively on the basis of the Bouguer gravity data into anomalies that can readily be analyzed. Major residual anomalies in the first group include (1) a gravity anomaly caused by the Gorda plate subducted beneath northern California, and (2) a pattern of linear gravity highs along the western margins and gravity lows in the eastern parts of both the Sierra Nevada and Peninsular Ranges batholiths. Two examples of prominent anomalies in the second group are (1) gravity highs defining a major detached thrust sheet within the western Klamath Mountains, and (2) a gravity low caused by a combination of low-density sedimentary rocks in the Ventura basin and the associated mantle upwarp accompanying isostatic compensation of the basin fill.

INTRODUCTION

Regional Bouguer gravity maps with good coverage, many of which are discussed in this volume, are now available for large sections of the United States. These maps contain a wealth of information about crustal geology. However, any attempt to interpret them in terms of geology must include consideration of the problems posed by the gravitational effects of isostasy, namely, that high topography tends to be supported by deep-seated mass deficiencies which produce regional gravity lows. This is especially true in areas of extreme topographic relief such as the western United States, where gravity anomalies arising from isostatic compensation of topography tend to dominate or mask the smaller anomalies that reflect mid- to upper-crustal geology. The dominant nature of topography-related Bouguer gravity anomalies is well illustrated in the transcontinental gravity, topographic, and geologic profiles presented by Woollard (1966).

The Bouguer gravity map of California (Oliver et al., 1980) displays considerable gravity relief, with a range of more than 300 mGal (Figure 1). Much of this gravity relief, especially at longer wavelengths, strongly correlates with averaged topography (Figure 2) in a manner that is compatible with the principle of isostasy (Oliver, 1980). In California as elsewhere many gravity anomalies caused by density inhomogeneities in the crust and upper mantle are superposed on larger anomalies related to topography and isostasy. As a means of isolating the smaller anomalies, an isostatic residual gravity map (Figure 3) was prepared by computing an approximation to the regional gravity field produced by isostatic compensation of topography and removing it from the Bouguer gravity map (Jachens and Roberts, 1981; Roberts et al., 1981).

This approach to regional–residual separation has proved useful in California for several reasons. First, it is based on a physical process, isostasy, that is known from empirical evidence to obtain to a high degree throughout most of the world (Heiskanen and Vening Meinesz, 1958). Second, the method is stable, as demonstrated by tests which show that large changes in the isostatic model parameters cause only small, long-wavelength changes in the isostatic residual gravity map. Third, the part of the gravity field that is removed while generating the residual gravity map has an associated physical model in which the density contrasts and geometries are completely specified; this feature is particularly useful during quantitative interpretation of anomalies from the residual gravity map because it may be necessary to examine, and sometimes alter, the physical model on which the regional removal was based. Fourth, the residual gravity map provides the basis for comparing anomalies that occur in widely differing topographic regimes.

In the sections that follow, we discuss how the map in Figure 3 was constructed, describe the results of tests

*U.S. Geological Survey, 345 Middlefield Road, Menlo Park, CA 94025.

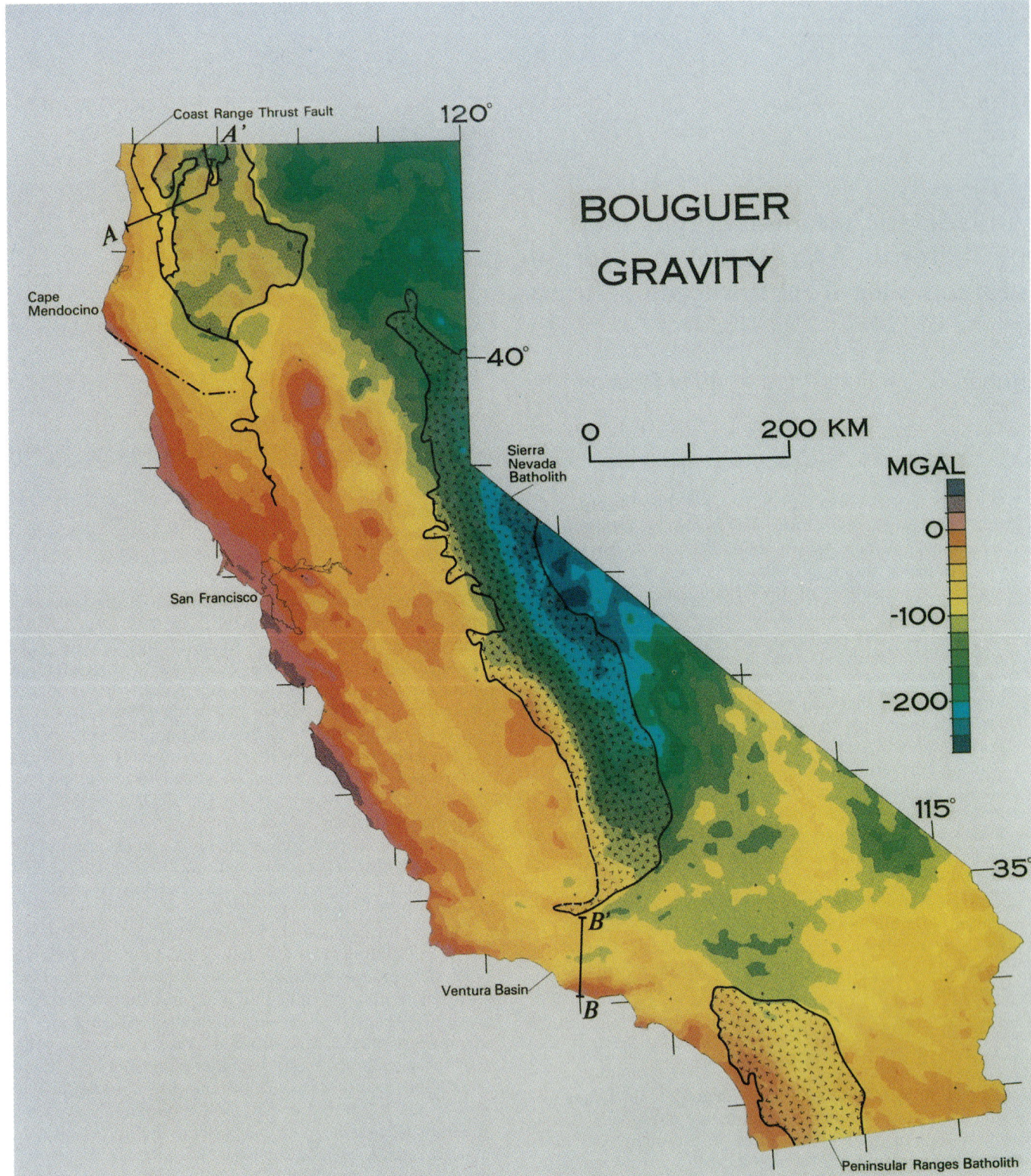

Fig. 1. Bouguer gravity map of California after Oliver et al. (1980). Stipple pattern in northwestern California indicates areas where the western Paleozoic and Triassic belt and the older belts of the Klamath Mountains province are exposed. "V" pattern indicates areas of exposed granitic rocks of the Sierra Nevada and Peninsular Ranges batholiths. Dashed part of boundary of Sierra Nevada batholith indicates westernmost exposure of granitic rocks in area where these rocks extend westward beneath Cenozoic sedimentary deposits. Thrust faults are shown with teeth on the upper plates. Dash-dot line indicates inferred location of south edge of subducted Gorda plate.

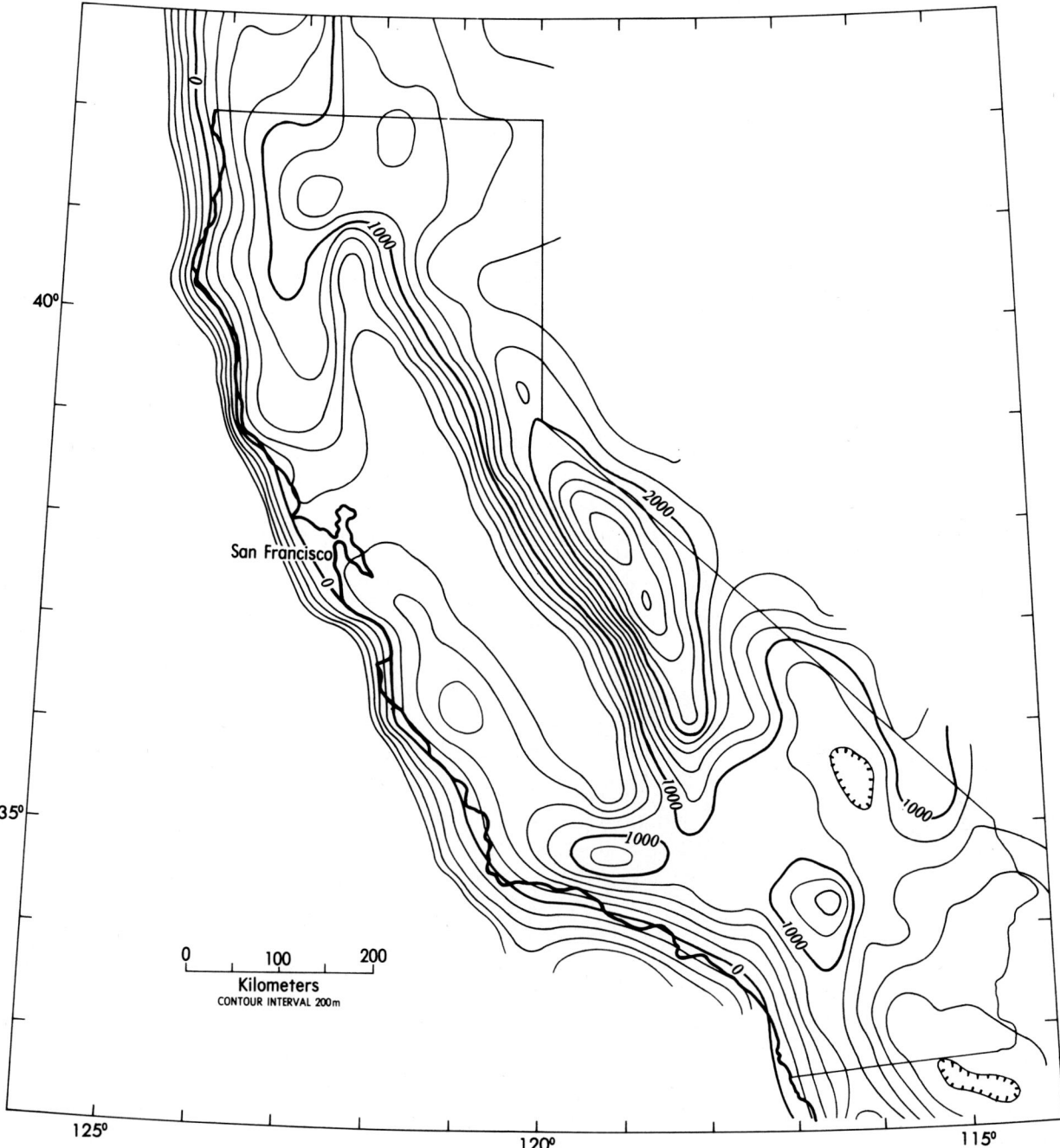

FIG. 2. Average elevation map of California and surrounding regions from Jachens and Griscom (1983). The map was generated by contouring values of average elevation of 30- × 30-minute compartments. Digital elevation data from Robbins et al. (1973).

designed to assess the stability of the map in terms of its sensitivity to imprecise knowledge of the parameters in the isostatic model, and present four examples of qualitative and quantitative interpretations based on this map that illustrate the utility of this method of regional–residual separation.

MAP CONSTRUCTION

The isostatic residual gravity map (Figure 3) was prepared from the Bouguer gravity map of California by removing the gravity attraction of a model crust of variable thickness, assuming complete local isostatic compensation of topography according to the Airy–Heiskanen system (Heiskanen and Vening Meinesz, 1958). In this system the Earth's crust rests on a denser substratum (the model mantle), and lateral variations in the height of the topography are accompanied by lateral variations in crustal thickness (Figure 4) such that hydrostatic equilibrium prevails at some depth beneath each column of crust. Given the distribution of topography, the values of three model parameters (density of the topography,

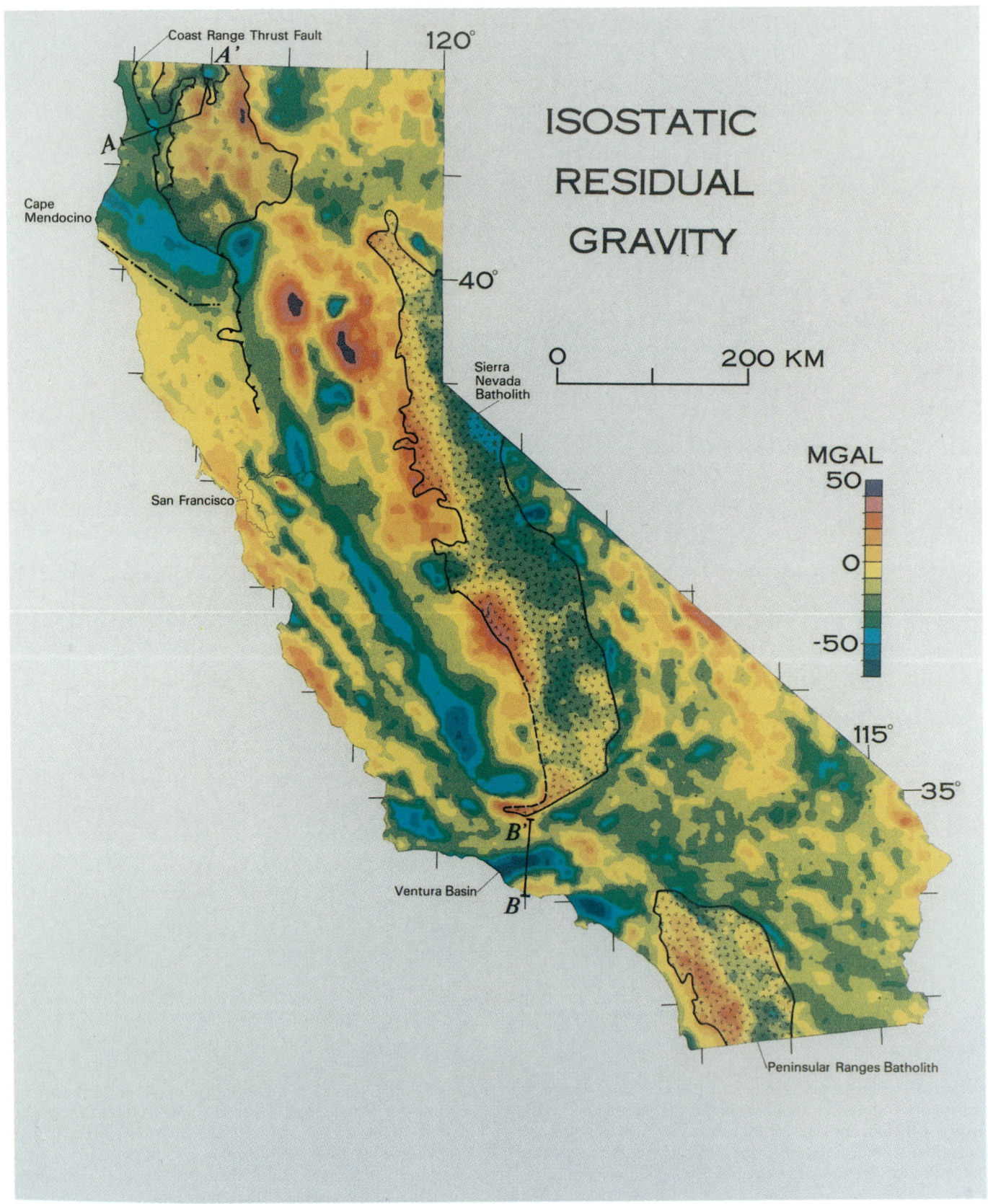

FIG. 3. Isostatic residual gravity map of California after Roberts et al. (1981). Model parameters are (1) crustal thickness at sea level, 25 km; (2) density of topography, 2.67 g/cm^3; and (3) density contrast across base of crust, 0.4 g/cm^3.

ρ; crustal thickness at sea level, T; density contrast across the base of the crust, $\Delta\rho$) completely determine the three-dimensional configuration of the model crust. The actual mode of compensation in California is undoubtedly more complex than this simple model implies. Fortunately for the purposes of calculating isostatic regional gravity fields, however, the attractions calculated from markedly different isostatic models tend to be similar in value (see Heiskanen and Vening Meinesz, 1958, Chapter 7, for examples of isostatic anomalies based on various isostatic models). In addition, McNutt (1980) showed that whatever mode of isostatic compensation of topography exists in the western United States, its general gravitational expression is difficult to distinguish from that of local Airy isostasy.

In preparing the residual gravity map shown in Figure 3, the effects of worldwide topography and its compensation were taken into account. The combined effects of topography and isostasy at each gravity station owing to all regions lying more than 166.7 km from the station were interpolated from the maps of Karki et al. (1961). For regions within 166.7 km of each station, digital topographic models had already been used during the Bouguer reduction to calculate terrain corrections (Oliver, 1980). The same topographic data were used to calculate the effect of isostatic compensation at each station (Jachens and Roberts, 1981).

In calculating isostatic effects from this model, the standard sea-level crust yields no contribution, but departures from the standard crustal thickness contribute, both positively and negatively, to the calculated effect. The model is computationally equivalent to a crust with three layers, all with laterally uniform densities. From top to bottom, these are (1) a layer of density, ρ, that includes all material between the highest and lowest points of the topography; (2) a layer of arbitrary density and uniform thickness whose upper and lower surfaces pass through the lowest point of the topography and the highest point of the mantle; and (3) a lower crustal layer whose density is specified only in terms of its contrast, $\Delta\rho$, with the underlying mantle material.

Numerical values of the model parameters used to construct the residual gravity map (Figure 3) are (1) topographic density, ρ, of 2.67 g/cm^3, to maintain compatibility with the density used in the Bouguer reduction; (2) sea-level crustal thickness, T, of 25 km, based on the results of seismic-refraction surveys near San Francisco (Healy, 1963); and (3) a density contrast, $\Delta\rho$, across the base of the crust of 0.4 g/cm^3. The value of the assumed density contrast is close to the 0.385 g/cm^3 proposed by Woollard (1966) on the basis of extensive seismic-refraction measurements and yields depths to the base of the model crust that agree with seismically inferred depths to the Mohorovičić discontinuity beneath the Sierra Nevada (Eaton, 1966; Pakiser and Brune, 1980) and offshore California (Shor et al., 1968). Topographic data were taken from the digital elevation model of California (Robbins et al., 1973) and represent elevations averaged over 3- × 3-minute compartments. Karki et al. (1961) used 1- × 1-degree elevation data and model parameters ρ = 2.67 g/cm^3, T = 30 km, and $\Delta\rho$ = 0.6 g/cm^3. Although the crustal thickness at sea level and the density contrast across the base of the crust are larger in the crustal model of Karki et al. (1961) than in our model, results given

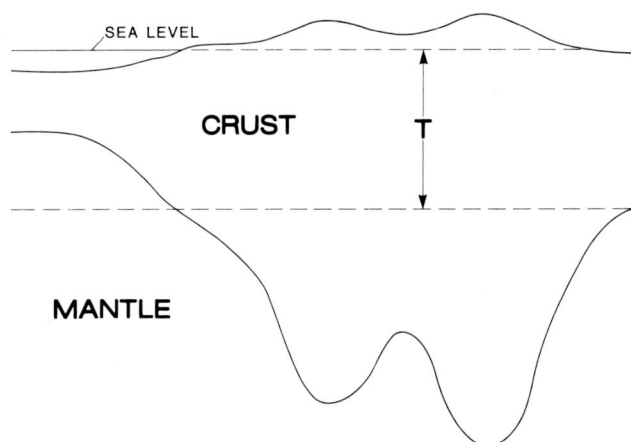

FIG. 4. Schematic diagram of the local Airy–Heiskanen isostatic-compensation model used to compute regional gravity field resulting from variations in crustal thickness.

in the next section suggest that the two models yield similar results.

The model parameters used to prepare the isostatic residual gravity map probably typify general conditions in California, but deviations from these values undoubtedly exist both because our selections are based on imprecise measurements and because the values of all three parameters probably vary from place to place. With such uncertainties in mind, we conducted tests along two east–west profiles, at 37° N and 41° N, to determine how sensitive the computed isostatic effect, and therefore the residual gravity map, are to changes in the model parameters. The tests consisted of determining the changes in the computed isostatic effect that resulted when the model parameters were varied, one at a time, through ranges that probably encompass the actual values of each parameter. The values tested were: ρ = 2.50, 2.67, 2.85 g/cm^3; T = 20, 25, 30 km; and $\Delta\rho$ = 0.2, 0.4, 0.6 g/cm^3. The results of these tests, shown in Figure 5, indicate that all sets of parameters yield similar results. For the two least reliable parameters, changes with respect to the standard model parameters of ±0.2 g/cm^3 in $\Delta\rho$ and of ±5 km in T (percentage changes of ±50 and ±20 percent, respectively) resulted in average differences in the computed isostatic effect of only ±5 percent or less and maximum differences of about ±10 percent over the crest of the Sierra Nevada (about longitude 118.6° W on lower profile of Figure 5). As a further test, we compared the computed isostatic effects along the profiles based on the parameters of Karki et al. (1961) with those based on our standard-model parameters. The average magnitude of the difference between the two calculations was 2.7 mGal, and the largest discrepancy was 4.2 mGal. The small changes in the computed isostatic regional field along the profiles that result from large changes in the isostatic model parameters suggest that the residual gravity map is very stable with respect to possible uncertainties in these parameters.

We have not computed isostatic regional gravity fields in California based on models of compensation that differ from the Airy–Heiskanen model. However, some sense of the possible gravity differences between various models can be

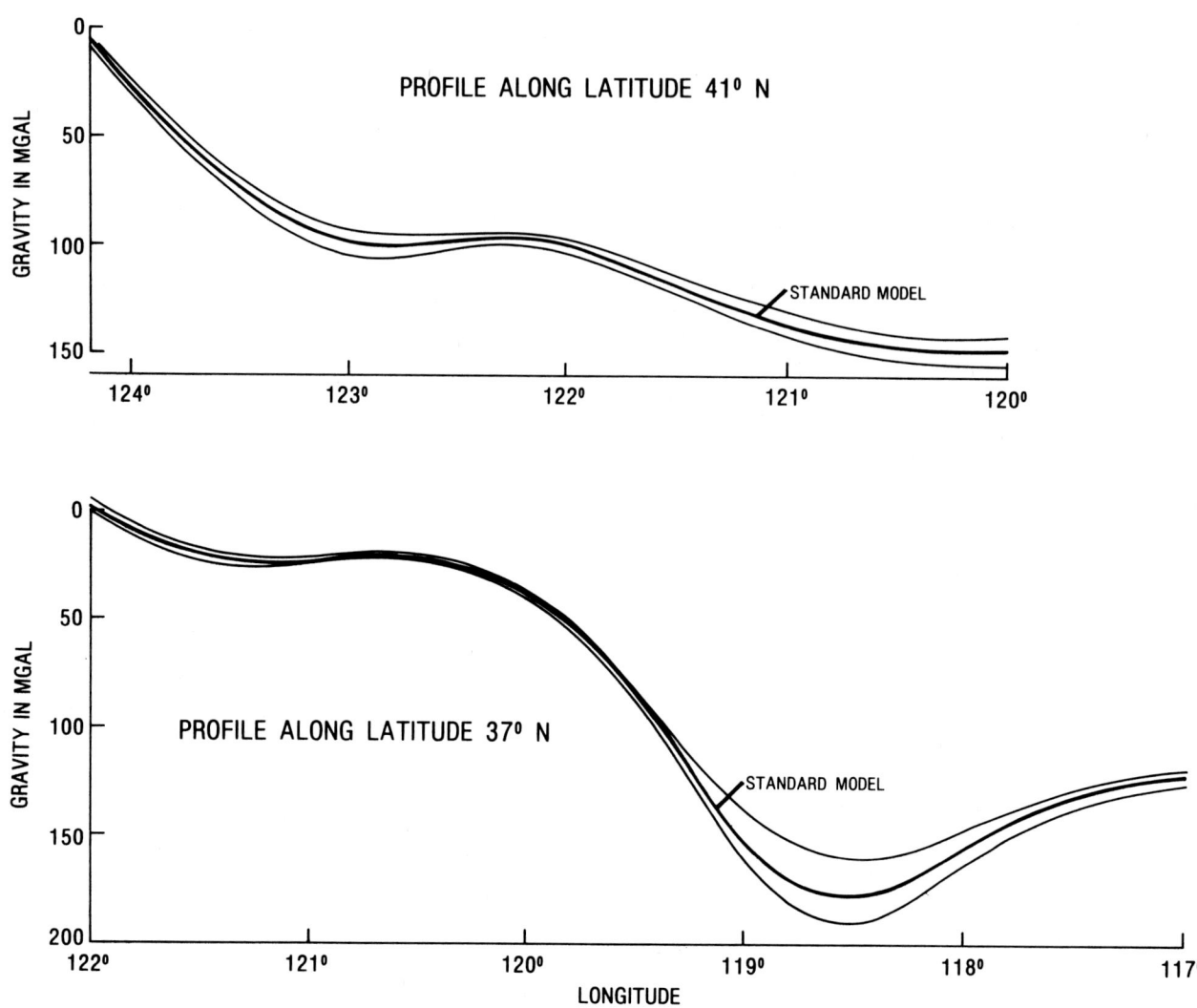

FIG. 5. Computed regional gravity fields along two east–west profiles based on Airy–Heiskanen model. The standard curves (thick lines) result from models with $\rho = 2.67$ g/cm^3, $T = 25$ km, and $\Delta\rho = 0.4$ g/cm^3. All regional fields resulting from other model parameters discussed in the text lie between the two curves (thin lines) that bound each standard curve.

obtained by examining the results presented by Oliver (1973) of computed regional fields along three profiles that cross the southern Sierra Nevada. He computed regional-field values based on Airy–Heiskanen models with $T = 20$, 30, and 60 km and on a Pratt–Hayford model with a depth of compensation of 113.7 km. Interpolating from his tables to obtain the regional gravity values for an Airy–Heiskanen model with $T = 25$ km, and comparing these with the Pratt–Hayford model values, show that the computed regional fields based on these two substantially different models differ on average by only about 10 percent. Also, the differences tend to be of long wavelength so that, although base levels may change, the shapes of short-wavelength residual anomalies produced by upper-crustal sources tend to be little changed by the use of different isostatic models. This comparison suggests that the isostatic residual gravity map shown in Figure 3 also is stable with respect to possible departures from the assumed mode of compensation.

RESIDUAL GRAVITY FIELD OF CALIFORNIA

The isostatic residual gravity field over most of California contains numerous anomalies (Figure 3), a fact which indicates that the combined Bouguer and isostatic models used to reduce the gravity data are too simple to account for all the variations in the observed gravity. It does not indicate, however, that the remaining gravity anomalies necessarily reflect areas that are out of isostatic balance (see Woollard, 1966, for an extensive discussion of the causes of isostatic anomalies). In fact, most of the anomalies probably result from density distributions in the mid to upper crust that are more complicated than the layers with laterally uniform densities implicit in the Bouguer and isostatic models. We emphasize: crustal-density inhomogeneities give rise to isostatic residual gravity anomalies, even if they are in perfect isostatic balance.

A simple example of a shallowly buried body that has a

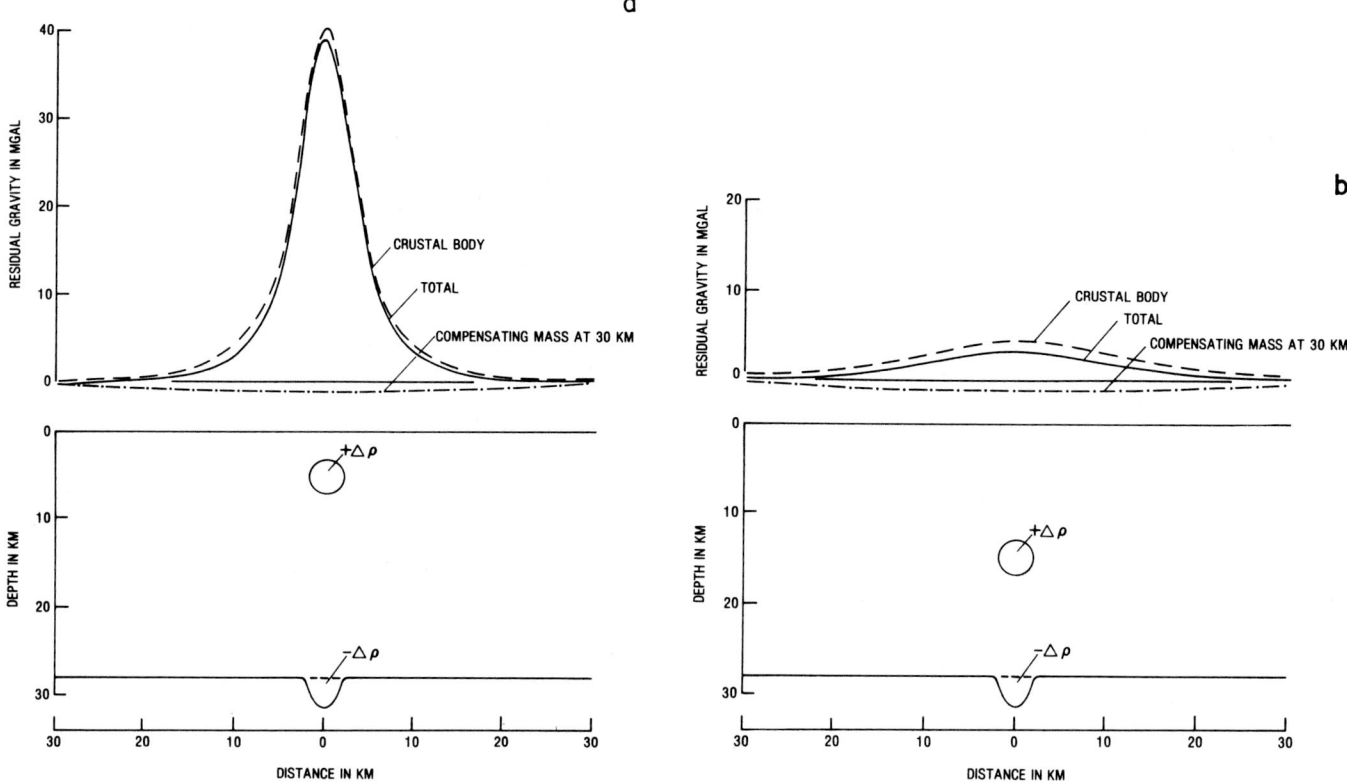

FIG. 6. Theoretical isostatic residual gravity anomalies at surface over perfectly compensated spherical density inhomogeneities. a. Center of sphere at depth of 5 km. b. Center of sphere at depth of 15 km. Center of compensating mass at depth of 30 km for both models. Although it is not obvious from the figures, it should be noted that the total "volume" under the crustal-body anomaly is equal and opposite to the "volume" under the compensating body anomaly. The combined anomaly has a moat of negative values around the central high and has a "volume" that integrates to zero.

higher density than its surroundings and is compensated by a downwarp of the crust into the higher density mantle is shown in Figure 6a. Because there is no topography, neither the Bouguer reduction nor the isostatic correction for topography affects the gravity anomalies from the intracrustal source and its compensating mass. The residual gravity field over such an anomalous mass would be that which results from combining the two anomalies. The large positive anomaly results because the surface of observation is closer to the crustal body than to its compensating mass. More deeply buried bodies would have total anomalies of smaller amplitude (Figure 6b), but only in the limiting case of an intracrustal body lying near the base of the crust would the gravity effects of the body and its compensating mass nearly cancel.

Residual gravity anomalies also would be observed over perfectly compensated density inhomogeneities with large lateral extent (Figure 7). Although it can be seen from Figure 7 that in general the gravity anomaly caused by a density distribution that compensates a broad anomalous region is more effective in reducing the amplitude of the total anomaly than one that compensates a narrow region, a residual anomaly remains in all cases shown. For example, the total anomaly over a compensated thin, 10-km-radius disc exposed at the surface is about 95 percent of that over an equivalent uncompensated disc, whereas the total anomaly over a similar 100-km-radius disc is only about 30 percent of that over its uncompensated equivalent. The important point

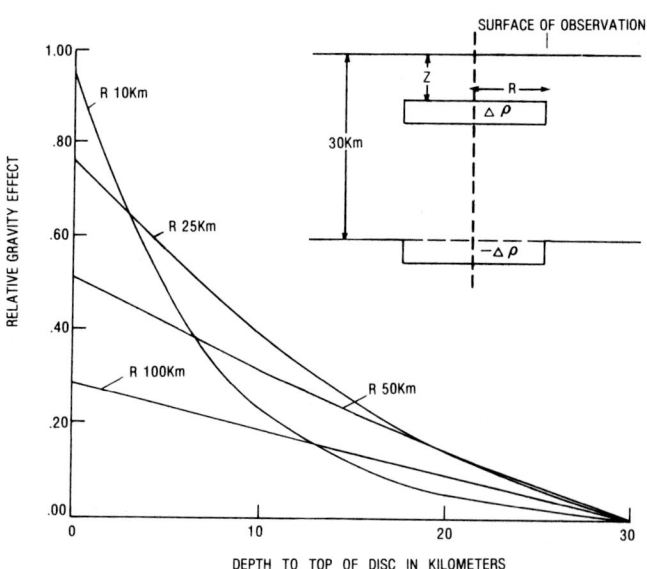

FIG. 7. Theoretical isostatic residual gravity anomalies of right-circular discs and their compensation for various radii (R) as a function of depth of burial (Z). Gravity for each different geometry was calculated at the surface on the axis of the discs and normalized by the gravity anomaly of an equivalent uncompensated disc with its top at the surface.

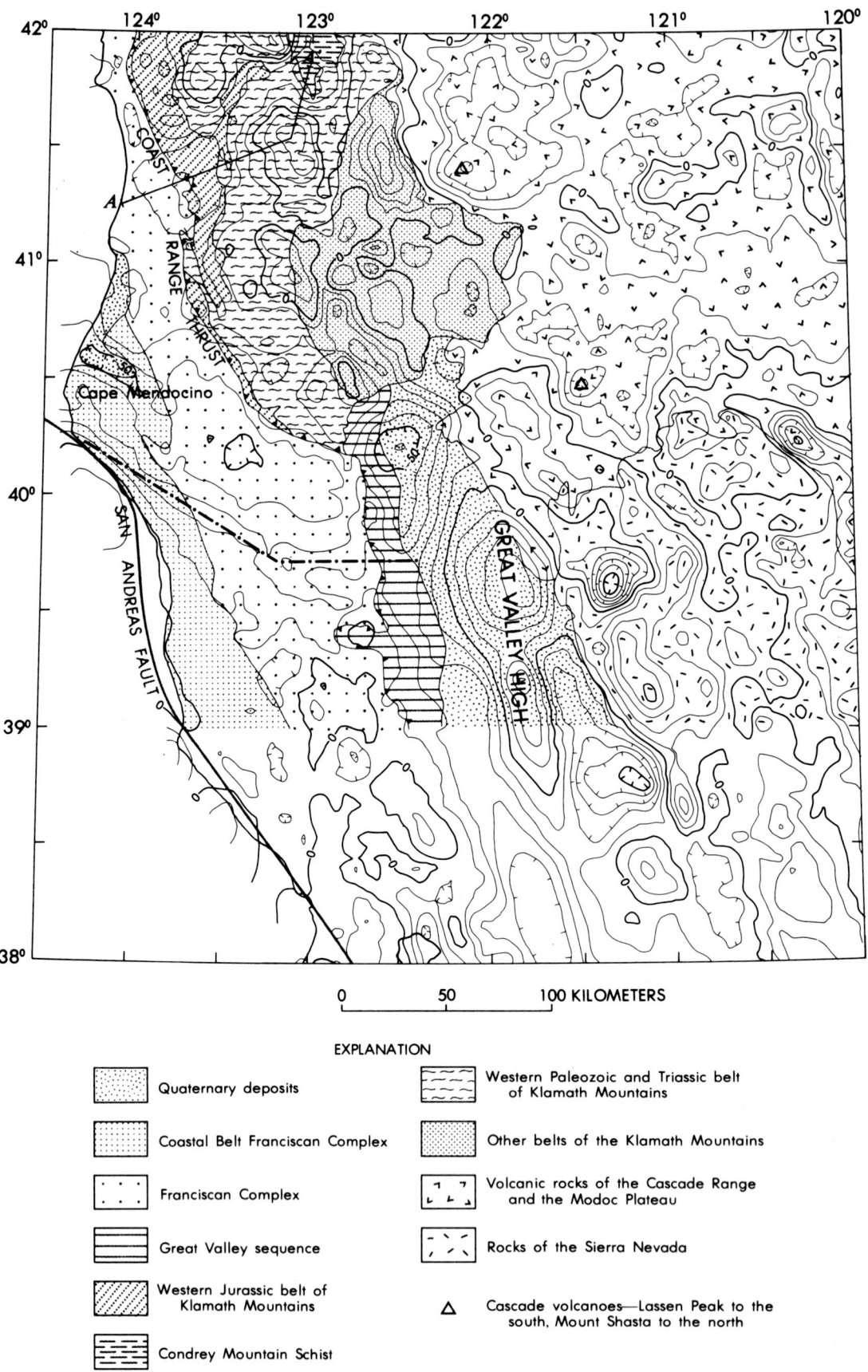

FIG. 8a. Isostatic residual gravity map of northern California. Generalized geology from Jennings et al. (1977), Irwin (1981), and compilation by C. G. Barnes (written commun., 1982). Dash-dot line indicates inferred location of south edge of Gorda plate. Data were contoured by computer with grid spacing of 5 km. Contour interval, 10 mGal; hachures indicate direction of lower gravity.

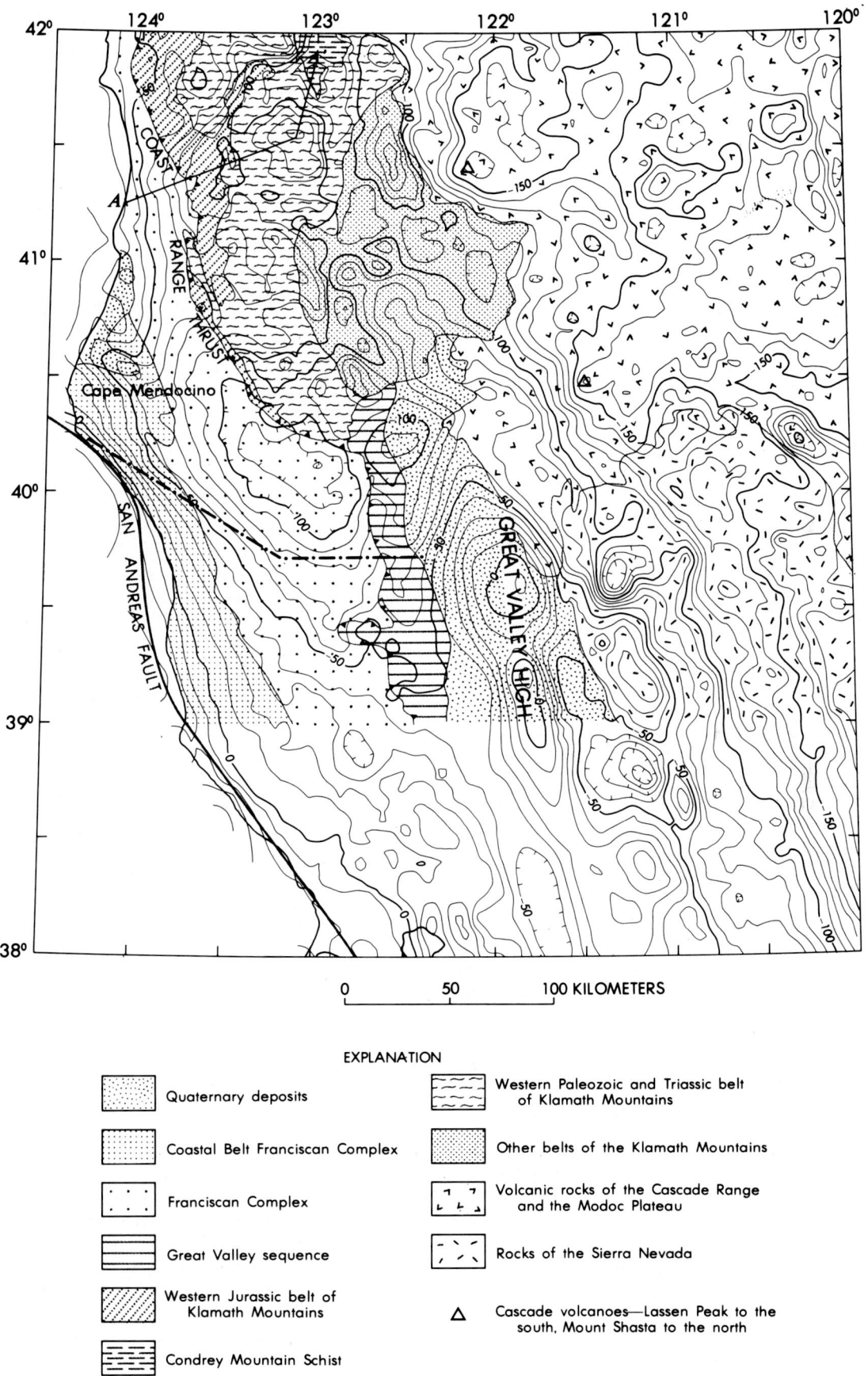

FIG. 8b. Bouguer gravity map of northern California and generalized geology. Data were contoured by computer with grid spacing of 5 km. Contour interval, 10 mGal; hachures indicate direction of lower gravity.

is, however, that even perfectly compensated anomalous regions having lateral dimensions as large as 200 km or more can cause significant anomalies in the isostatic residual gravity field.

The simple examples discussed above illustrate that compensated lateral density variations within the crust cause isostatic residual gravity anomalies. This does not, of course, rule out the possibility that some anomalies in Figure 3 could reflect regions that are out of isostatic balance or regions that are compensated according to an isostatic mechanism different from the Airy–Heiskanen model.

A cursory examination of the isostatic residual gravity map (Figure 3) reveals many major gravity anomalies that are correlated both with major geologic features and with the densities of rocks exposed at the surface. Most large gravity lows lie over low-density plutons or over structural depressions filled with low-density sedimentary or volcanic rocks. Similarly, numerous gravity highs occur over exposures of high-density metamorphic or plutonic rocks. Some of these anomalies are clearly evident on the Bouguer gravity map, but others are confused or masked by the effects of compensation of the topography.

In the following sections, we discuss four areas in California for which we have performed interpretations based on the isostatic residual gravity map. In order to illustrate the utility of this map for geologic interpretations, we compare it in each area with the corresponding Bouguer gravity map and point out how this method of regional–residual separation helped to identify the important features of the gravity field or facilitated their interpretation. The first two examples represent cases in which the important gravity features are difficult to recognize on the Bouguer gravity map, and the second two examples illustrate cases in which the major anomalies are clearly evident on the Bouguer map but are difficult to interpret.

GORDA PLATE ANOMALY

In northern California (Figures 3 and 8a) a pronounced residual gravity low extends northward along the coast from the vicinity of Cape Mendocino to the Oregon border (latitude 42° N) and eastward at least as far as the Coast Range thrust (Figure 8a). This low is terminated on the south by a southeastward-trending zone (centered about the dash-dot line on Figures 3 and 8a) across which gravity values increase toward the south. Gravity values north of this zone and west of the Coast Range thrust are typically 25–50 mGal lower than values to the south, even though similar rocks with comparable densities are exposed at the surface in both areas. Along the southern boundary of this gravity low, the transition from high to low gravity values is narrowest and steepest near the coast and becomes progressively wider and more gentle toward the east. The locus of points of steepest gradient (dash-dot line, Figure 8a) trends S 60 degrees E from the coast for approximately 120 km, then nearly due east to the Coast Range thrust.

We believe that this gravity low reflects the presence at depth of the subducted Gorda plate (Jachens and Griscom, 1983), a plate that lies offshore north of Cape Mendocino. The south edge of the plate is inferred from magnetic data and the distribution of seismicity to pass beneath the coast at the point where the dash-dot line and the coastline intersect (Figure 8a). In this interpretation, the buried south edge of the Gorda plate lies beneath the dash-dot line, and the eastward dip of the plate is reflected by the form of the gravity anomaly above its south edge.

The critical features of the gravity field that either led to or enhanced this interpretation are (1) the location of the gravity low adjacent to the oceanic part of the Gorda plate, (2) the spatial coincidence at the coastline of the south edge of the Gorda plate and the transition from high to low gravity values, (3) the parallelism between the trend of the transition zone and the present direction of motion of the Gorda plate relative to the Pacific plate (S 60 degrees E), and (4) the progression from steep to gentle gravity gradients eastward along the transition zone. Although all of these features are clearly depicted on the isostatic residual gravity map (Figure 8a), even the existence of the gravity low was not recognized on the Bouguer gravity map (Figure 8b) because it occurs in an area that contains large lateral gravity changes related to isostasy. In addition, the gravity gradient that defines the south boundary of the gravity low, a feature whose characteristics are crucial to the interpretation, is masked on the Bouguer gravity map because there it is superposed obliquely across a higher gradient that reflects compensation of the change in topography associated with the transition from oceanic to continental crust.

SIERRA NEVADA AND PENINSULAR RANGES BATHOLITHS

The isostatic residual gravity field over granitic rocks of the Sierra Nevada and Peninsular Ranges batholiths (Figure 3) shows a pattern of linear highs near the west margins of the batholiths and broad lows over the eastern parts (Griscom and Oliver, 1980); the gravity difference between the western and eastern parts typically is 40–60 mGal. Local highs along the crest of the western anomaly reach values of +10 to +40 mGal, whereas residual gravity levels over the eastern parts are less than −30 mGal in many places and consistently are less than −15 mGal. Most of the transition from high to low values occurs across a zone about 20 km wide that extends for nearly the entire length of both batholiths, a total distance of more than 750 km.

Although this major regional gravity feature has not been studied in detail, these anomaly patterns imply similar distributions of density units in both batholiths, an interpretation supported by density measurements and gravity modeling. Densities of surface rocks from the Sierra Nevada (Oliver et al., 1961; Coatney, 1965; Oliver, 1977) and the Peninsular Ranges (Baird et al., 1979) show trends of higher (+0.05 to +0.15 g/cm^3) values toward the west. Furthermore, many local gravity highs along the west margin of the Sierra Nevada batholith are associated with small mafic plutons with relatively high densities. Modeling of gravity data along profiles that cross the Sierra Nevada (Oliver, 1977; Griscom and Oliver, 1980) and the Peninsular Ranges near the Mexican border (Kovach and others, 1962) indicates that at these locations the surface-density contrasts extend to depths of about 10–15 km.

The residual gravity field over the batholiths also is correlated with strontium and oxygen isotope data. In the

Sierra Nevada, the gravity highs lie west of the 0.706 contour of $^{87}Sr/^{86}Sr$ (Kistler and Peterman, 1978), whereas in the Peninsular Ranges they lie west of the 0.705 contour and the $\delta^{18}O$ "step" at about +8.5 (Taylor and Silver, 1978). These isotopic values are all lower to the west. The relatively simple distribution of rock densities within the Sierra Nevada batholith contrasts sharply with the complex distribution of ages of Mesozoic plutons (Kistler, 1974), suggesting that the composition (and density) of the batholithic rocks is partly determined by the composition of the crust through which they passed, oceanic in the western parts of the batholith and continental in the eastern.

The spatially consistent form and regional nature of the gravity-anomaly patterns over the two batholiths are obvious on the isostatic residual gravity map (Figure 3) but are difficult to recognize on the Bouguer gravity map (Figure 1), even with hindsight. The characteristic anomaly pattern is evident in the Bouguer gravity data from the Peninsular Ranges batholith, because here the regional topographic relief is moderate and consequently the variation of the isostatic regional field is small. In the Sierra Nevada, however, the anomaly is superposed on more than 175 mGal of gravity relief, most of which is related to isostatic compensation of topography. Removing a regional field based on isostasy facilitated the comparison of gravity data in the widely diverse topographic regimes of the two batholiths and permitted correlation of the residual gravity with density and isotopic data, comparisons and correlations that are not possible from the Bouguer gravity data alone.

KLAMATH MOUNTAINS

The western part of the Klamath Mountains province in northern California is composed of two eastward-dipping imbricate thrust slices or belts, separated by a regional thrust fault (Irwin, 1981). The structurally highest unit, the western Paleozoic and Triassic belt, overlies the western Jurassic belt, which, in turn, overlies the Franciscan complex along the eastward-dipping Coast Range thrust fault (Figure 8). The Condrey Mountain Schist is exposed in a window in the western Paleozoic and Triassic belt (Figure 8), and although it occupies the same structural position as the western Jurassic belt, its relationship to this belt is not clear (Klein, 1977; Coleman et al., 1983).

The deep structure of the western Klamath Mountains, especially the geometry of the fault that separates the western Paleozoic and Triassic belt from the western Jurassic belt, is well suited for study by gravity techniques, because characteristic densities are associated with rocks of the major geologic units. The western Paleozoic and Triassic belt is composed of rocks having densities that typically are greater than 2.75 g/cm^3, whereas rocks of the units below it (Franciscan complex, western Jurassic belt, and Condrey Mountain Schist) typically are less than 2.65 g/cm^3 (Irwin, 1961; Jachens et al., 1985). As expected, the residual gravity values (Figures 3 and 8a) correlate with the density distributions—low values (< −25 mGal) over areas where low-density rocks are exposed, and closed gravity highs over exposures of the western Paleozoic and Triassic belt. Straightforward modeling of the residual gravity data indicates that the western Paleozoic and Triassic belt is a thin

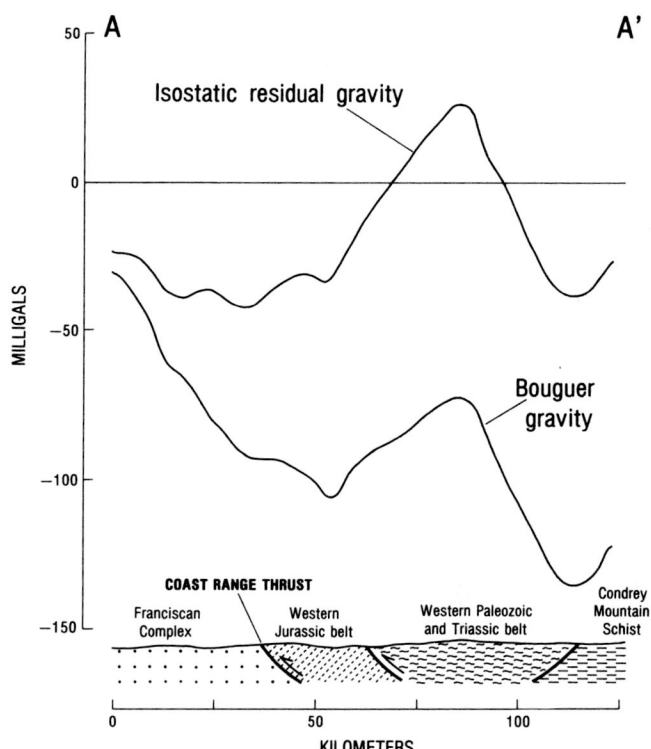

FIG. 9. Comparison of Bouguer and isostatic residual gravity along profile A–A'. See Figures 3 and 8 for location of profile. Geology is schematic.

plate (generally less than 6 km thick) bounded at its base by a flat-lying fault and underlain by the western Jurassic belt and the Condrey Mountain Schist, which are connected at depth (Jachens and Elder, 1983; Jachens et al., 1985).

The key element that led to this interpretation of the residual gravity data was the recognition that the residual gravity-field values over exposures of the low-density rocks of the Franciscan complex, the western Jurassic belt, and the Condrey Mountain Schist are uniformly low and of roughly similar magnitude, whereas those over the western Paleozoic and Triassic belt are generally much higher. Although local gravity highs over the western Paleozoic and Triassic belt are evident on the Bouguer gravity map (Figures 1 and 8b), on a regional scale the relationship between the gravity field over this belt and the field over the structurally lower units is severely distorted by large, lateral gravity variations associated with isostasy. Isostatic effects cause Bouguer gravity values over some parts of the Franciscan complex to be more than 100 mGal higher than those over the Condrey Mountain Schist and 25–50 mGal higher than local highs over the western Paleozoic and Triassic belt (Figure 9).

VENTURA BASIN

The residual gravity field over the Ventura basin shows a narrow, 70-mGal low (Figure 3) that reflects a 15-km-wide, 200-km-long basin (including its extension offshore) filled by about 15 km of layered sedimentary rocks. Residual gravity

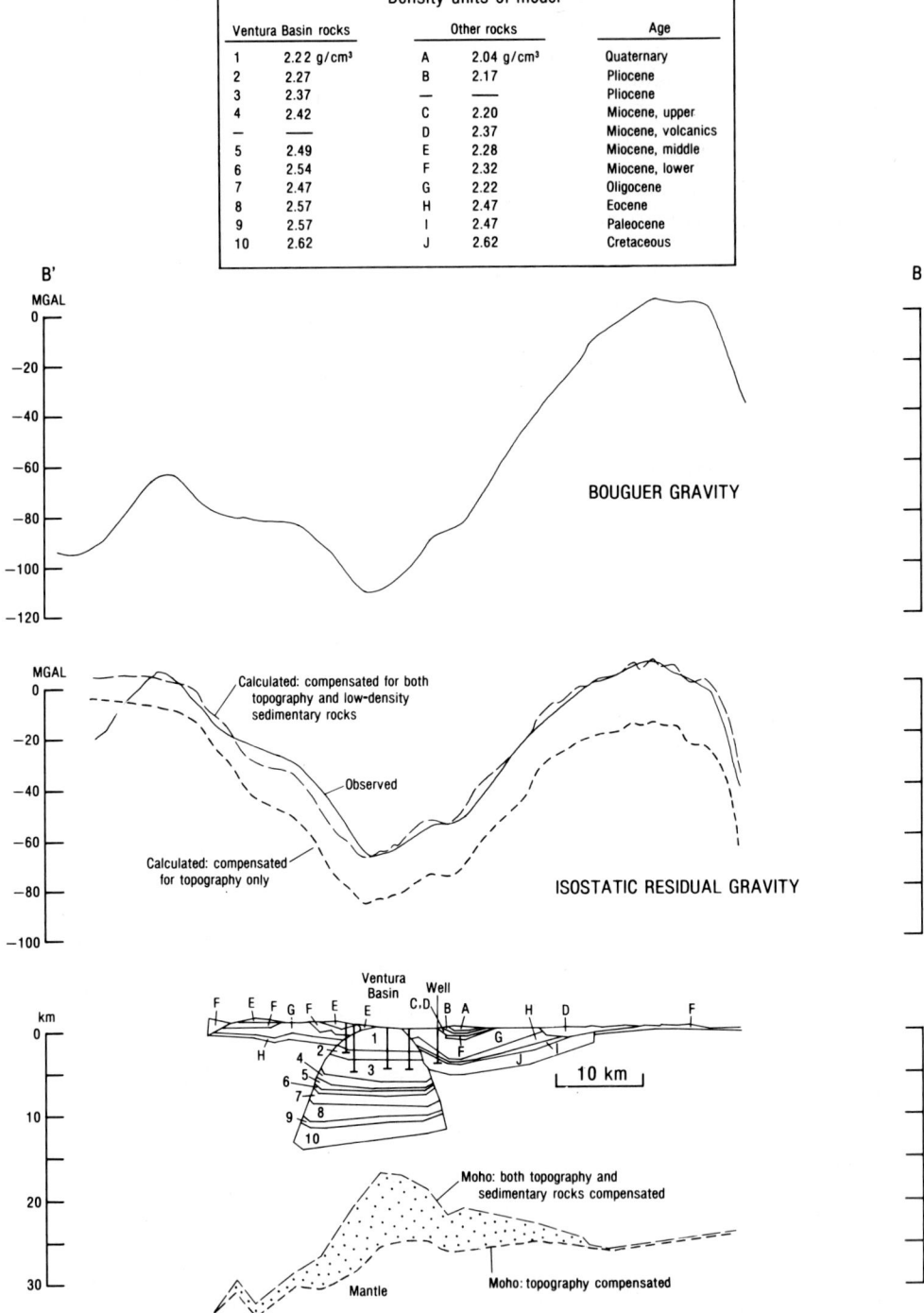

FIG. 10. Observed and calculated gravity along profile B–B' across the Ventura basin (see Figure 3 for location of profile). Calculated curves are given for two configurations of the Mohorovičić discontinuity (Moho), one assuming that only the topography is compensated and the other assuming that both the topography and the basin are compensated. Density of the crust surrounding the basin model and above the bottom of the basin is 2.67 g/cm³. Density contrast across the Moho is 0.4 g/cm³. Density of the layer between the bottom of the basin and the top of the mantle is unspecified but is laterally homogeneous. Density model used for calculations extends offshore south of B–B' and includes an approximation of the density structure in the offshore Santa Monica basin plus isostatic compensation of the basin deposits. Letters and numbers shown on model identify layers of constant density.

values over granitic and volcanic rocks exposed north and south of the basin, respectively, are near 0 mGal. In cross-section the gravity anomaly is nearly symmetrical about the center of the basin (Figure 10).

Griscom and Grafft (1982) constructed a gravity model of the basin along profile B–B' (Figure 3) using the geologic cross-section of Nagle and Parker (1971) as an initial model. They assigned densities to the various geologic units based on well control, inferences from detailed gravity profiles, and general density–depth functions for thick sedimentary sections. The gravity calculated from this model agreed with the residual gravity in shape and amplitude, but the datum levels of the observed and calculated gravity disagreed by about 20 mGal (Figure 10). Subsequent modification of the Ventura basin crustal model to include isostatic compensation, not only of the topography (which had been accomplished during the regional–residual separation) but also compensation of the low-density basin fill by a 10-km protrusion of the mantle directly below the basin, accounted for the datum discrepancy (Figure 10). Because profile B–B' crosses the basin near its eastern end and because the mantle protrusion is deep, a two-dimensional model is not an adequate representation of the mantle configuration in this area. Griscom and Grafft found that a three-dimensional representation of the geometry of the Mohorovičić discontinuity beneath the basin was required in order to satisfy the gravity data, because the two-dimensional model caused the calculated gravity to exceed the observed gravity by about 10 mGal. The relatively high-density Cretaceous and Paleocene rocks in the lower part of the Ventura basin model contribute only about 4 mGal to the total anomaly so that their presence is not proved by the gravity model. The small distance (about 4 km) between the bottom of the basin and the calculated position of the Mohorovičić discontinuity implies that (1) the basin is floored by oceanic crust, and (2) as much as 4 km of assumed section in the lower part of the basin may in fact not be present.

Two aspects of the Ventura basin gravity interpretation illustrate the utility of the isostatic reduction as a means for regional–residual separation. First, the process of generating the residual map from the Bouguer gravity map removed approximately 75 mGal of north–south gravity relief across the basin and converted an asymmetric basin anomaly into one that is roughly symmetrical and amenable to quantitative interpretation (Figure 10). Other techniques could have been used to isolate the narrow gravity anomaly over the basin. However, because the regional field relief across this feature is of comparable magnitude to the basin anomaly, it is doubtful that the 20-mGal datum discrepancy between the observed and calculated gravity along profile B–B' would have been identified or recognized as significant on the basis of interpretations from residual gravity maps generated by empirical means. Second, once the datum discrepancy was recognized, the fact that the regional field originally removed from the Bouguer gravity data was based on a physical model with completely specified densities and geometries permitted straightforward modification of the complete crustal model in order to account for the observed gravity. This modification could have been accomplished by modeling directly from the Bouguer gravity data, but such an interpretation would have required a fully three-dimensional model covering a large region, a formidable modeling task.

SUMMARY AND CONCLUSIONS

In the four examples discussed, the regional–residual separation based on isostasy revealed important gravity features that are not apparent on the Bouguer gravity map and successfully converted other anomalies that are difficult to interpret into ones that are readily amenable to quantitative analysis. Based on these four examples and our experience with qualitative and quantitative interpretations from other areas of California, we feel that the process of converting the Bouguer gravity map into the isostatic residual gravity map represents an effective additional step in a gravity-reduction process whose goal is to enhance our ability to understand gravity data in terms of geologic features.

The method used to generate the residual gravity map in Figure 3 is analogous to the standard Bouguer reduction used routinely in gravity-data processing. In both the Bouguer and the isostatic reductions, simple layered models with laterally uniform densities are used to approximate and remove first-order features of the observed gravity field. In most situations, neither the Bouguer model nor the combined Bouguer and isostatic models are sufficient to account completely for the observed gravity variations. The remaining anomalies reflect departures from the simple physical models and can be interpreted in terms of lateral density variations and/or departures from the assumed mode or degree of isostatic compensation. A distinct advantage of the isostatic reduction over other methods of regional–residual separation, such as wavelength filtering, graphic methods, and methods based on smoothed topography, is that the regional field removed by the isostatic reduction results from a physical model in which the density contrasts and geometries are completely prescribed. This model can be changed at any time as part of the interpretive procedure.

Our experience with interpretations based on the isostatic residual gravity map of California suggests that regional–residual separation based on isostasy should be effective in other areas that contain considerable topographic relief. Isostatic anomaly maps already are available for some large areas, for example, the United States (Woollard, 1966; Oliver et al., 1982; Simpson et al., 1983b; Saltus, 1984), Switzerland (Klingele and Kissling, 1982; Kissling, 1982), and New Zealand (New Zealand Department of Scientific and Industrial Research, 1977). For other areas where Bouguer gravity data are available, an efficient method now exists for generating isostatic gravity maps from large data sets (Simpson et al., 1983a).

REFERENCES

Baird, A. K., Baird, K. W., and Weldy, E. E., 1979, Batholithic rocks of the northern Peninsular and Transverse Ranges, Southern California: chemical composition and variation, *in* Abbot, P. L., and Todd, V. R., Eds., Mesozoic crystalline rocks: Peninsular Ranges batholith and pegmatites, Point Sal ophiolite: San Diego State Univ., Dep. Geol. Sci., 111–133.

Coatney, R. L., 1965, Modal analysis of the granite rocks of the northern Sierra Nevada between Yosemite and Lake Tahoe, California: Ph.D thesis, Penn. State Univ.

Coleman, R. G., Helper, M. D., and Donato, M. M., 1983, Geologic map of the Condrey Mountain roadless area, Siskiyou County, California: Misc. Field Stud. Map MF-1540-A, U.S. Geol. Surv., scale 1:50 000.

Eaton, J. P., 1966, Crustal structure in northern and central California from seismic evidence, in Bailey, E. H., Ed., Geology of northern California: Bull. 190, Cal. Div. Mines Geol., 419–426.

Griscom, A., and Grafft, K. S., 1982, A model of the Ventura basin, California, from gravity data [abstr.]: Geol. Soc. Am. Abstr. Programs, **14**, 168.

Griscom, A., and Oliver, H. W., 1980, Isostatic gravity highs along the west side of the Sierra Nevada and Peninsular Ranges batholiths, California: Eos, **61**, 1126.

Healy, J. H., 1963, Crustal structure along the coast of California from seismic refraction measurements: J. Geophys. Res., **68**, 5777–5787.

Heiskanen, W. A., and Vening Meinesz, F. A., 1958, The earth and its gravity field: McGraw–Hill.

Irwin, W. P., 1961, Specific gravity of sandstones in the Franciscan and related Upper Mesozoic formations of California: Prof. Pap. 424-B, U.S. Geol. Surv., B189–B191.

——— 1981, Tectonic accretion of the Klamath Mountains, in Ernst, G. W., Ed., The geotectonic development of California: Prentice–Hall, 29–49.

Jachens, R. C., Barnes, C. G., and Donato, M. M., 1985, Map showing geophysical interpretation of the Marble Mountain Wilderness, Siskiyou County, California: Misc. Field Stud. Map MF-1452-C, U.S. Geol. Surv., scale 1:62 500.

Jachens, R. C., and Elder, W. P., 1983, Aeromagnetic map and interpretation of geophysical data from the Condrey Mountain Roadless Area, Siskiyou County, California: Misc. Field Stud. Map MF-1540-B, U.S. Geol. Surv., scale 1:50 000.

Jachens, R. C., and Griscom, A., 1983, Three-dimensional geometry of the Gorda plate beneath northern California: J. Geophys. Res., **88**, 9375–9392.

Jachens, R. C., and Roberts, C. W., 1981, Documentation of a FORTRAN program, 'isocomp', for computing isostatic residual gravity: Open-File Rep. 81-574, U.S. Geol. Surv.

Jennings, W. W., Strand, R. G., and Rogers, T. H., 1977, Geologic map of California: Cal. Div. Mines Geol., scale 1:750 000.

Karki, P., Kivioja, L., and Heiskanen, W. A., 1961, Topographic-isostatic reduction maps for the world for Hayford Zones 18-1, Airy-Heiskanen system, T=30 km: Internat. Assoc. Geod., Isostatic Inst., **35**.

Kissling, E., 1982, Aufbau der Kruste und des oberen Mantels in der Schweiz: Geod.-Geophys. Arb. Schweiz, **35**, 37–126.

Kistler, R. W., 1974, Phanerozoic batholiths in western North America: A summary of some recent work on variations in time, space, chemistry, and isotopic composition: Annu. Rev. Earth Planet. Sci., **2**, 403–418.

Kistler, R. W., and Peterman, Z. E., 1978, Reconstruction of crustal blocks of California on the basis of initial strontium isotopic compositions of Mesozoic plutons: Prof. Pap. 1071, U.S. Geol. Surv.

Klein, C. W., 1977, Thrust plates of the north-central Klamath Mountains near Happy Camp, California: Spec. Rep. 129, Cal. Div. Mines Geol., 23–26.

Klingele, E., and Kissling, E., 1982, Zum Konzept der isostatischen Modelle in Gebirgen am Beispiel der Schweizer Alpen: Geod.-Geophys. Arb. Schweiz, **35**, 3–36.

Kovach, R. L., Allen, C. R., and Press, F., 1962, Geophysical investigations in the Colorado delta region: J. Geophys. Res., **67**, 2845–2871.

McNutt, M. K., 1980, Implications of regional gravity for state of stress in the earth's crust and upper mantle: J. Geophys Res., **85**, 6377–6396.

Nagle, H. E., and Parker, E. S., 1971, Future oil and gas potential of onshore Ventura basin, California: Mem. 15, AAPG, 254–297.

New Zealand Department of Scientific and Industrial Research, 1977, Gravity map of New Zealand, North Island: Scale 1:1 000 000, 4 sheets.

Oliver, H. W., 1973, Principal facts, plots, and reduction programs for 1753 gravity stations in the southern Sierra Nevada and vicinity, California: Natl. Tech. Inf. Serv., U.S. Dep. Commerce, NTIS-PB 231185.

——— 1977, Gravity and magnetic investigations of the Sierra Nevada batholith: Geol. Soc. Am. Bull., **88**, 445–461.

——— 1980, General introduction, in Oliver, H. W., Ed., Interpretation of the gravity map of California and its continental margin: Bull. 205, Cal. Div. Mines Geol., 1–8.

Oliver, H. W., Pakiser, L. C., and Kane, M. F., 1961, Gravity anomalies in the central Sierra Nevada, California: J. Geophys. Res., **66**, 4265–4271.

Oliver, H. W., Chapman, R. H., Biehler, S., Robbins, S. L., Hanna, W. F., Griscom, A., Beyer, L. A., and Silver, E. A., 1980, Gravity map of California and its continental margin: Cal. Div. Mines Geol., scale 1:750 000, 2 sheets.

Oliver, H. W., Saltus, R. W., Mabey, D. R., and Hildenbrand, T. G., 1982, Comparison of Bouguer anomalies and isostatic residual gravity maps of the southwestern Cordillera [abstr.]: Tech. Program Abstr. Biograph., 52nd Annu. Int. SEG Meet., Oct. 17–21, Dallas, Tex., 306–308.

Pakiser, L. C., and Brune, J. N., 1980, Seismic models of the root of the Sierra Nevada: Science, **210**, 1088–1094.

Robbins, S. L., Oliver, H. W., and Plouff, D., 1973, Magnetic tape containing average elevations of topography in California and adjacent regions for areas of 1 × 1 minute and 3 × 3 minutes in size: Natl. Tech. Inf. Serv., U.S. Dep. Commerce, NTIS-PB 219794.

Roberts, C. W., Jachens, R. C., and Oliver, H. W., 1981, Preliminary isostatic residual gravity map of California: Open-File Rep. 81-573, U.S. Geol. Surv., scale 1:750 000, 5 sheets.

Saltus, R. W., 1984, A description of colored gravity and terrain maps of the southwestern Cordillera: Open-File Rep. 84-95, U.S. Geol. Surv.

Shor, G. G., Jr., Dehlinger, P., Kirk, H. K., and French, W. S., 1968, Seismic refraction studies off Oregon and northern California: J. Geophys. Res., **73**, 2175–2194.

Simpson, R. W., Jachens, R. C., and Blakely, R. J., 1983a, AIRYROOT: A FORTRAN program for calculating the gravitational attraction of an Airy isostatic root out to 166.7 km: Open-File Rep. 83-883, U.S. Geol. Surv.

Simpson, R. W., Saltus, R. W., Jachens, R. C., and Godson, R. H., 1983b, A description of colored isostatic gravity maps and a topographic map of the conterminous United States available as 35 mm slides: Open-File Rep. 83-884, U.S. Geol. Surv.

Taylor, H. P., and Silver, L. T., 1978, Oxygen isotope relationships in plutonic igneous rocks of the Peninsular Ranges batholith, southern and Baja California: Open-File Rep. 78-701, U.S. Geol. Surv., 423–426.

Woollard, G. P., 1966, Regional isostatic relations in the United States, in Steinhart, J. S., and Smith, T. J., Eds., The earth beneath the continents: Geophys. Mono. 10, Am. Geophys. Union, 557–594.

Analysis of gravity data in volcanic terrain and gravity anomalies and subvolcanic intrusions in the Cascade Range, U.S.A., and at other selected volcanoes

David L. Williams* and Carol Finn*

ABSTRACT

Gravity data were investigated to reveal the presence of subvolcanic intrusions. With few exceptions, these intrusions produce a detectable gravity anomaly. In the past, these gravity anomalies have often been overlooked or misinterpreted because the data reduction procedure was inadequate. A pragmatic method for reducing and interpreting reconnaissance gravity data from volcanoes as well as gravity models of a variety of volcanoes is developed.

Large calderas (diameters greater than 15 km) have relatively low-density intrusions beneath them. All other large volcanic systems that would include small calderas (diameters less than 15 km) have relatively high-density intrusions beneath them. The density contrasts that produce the observed anomalies occur between the intrusion, whose density is usually greater than 2.6 g/cm^3, and the country rock. Commonly, the shallow country rock is an older volcanic layer with a density less than 2.5 g/cm^3. The result of the contrast is a positive anomaly over the intrusion. For larger calderas, the surrounding volcanic layer is usually thin and overlies dense metamorphic and plutonic country rocks. In this case, we find the intrusion commonly less dense than country rock. The result is a negative anomaly.

In modeling volcanoes of the Cascade Range, gravity data and geologic considerations required a bottom on the intrusion. This may be an actual bottom or the depth at which the density contrast between the intrusion and the country rock disappears. The tops of the intrusions are usually shallow and are significantly wider than overlying craters or calderas. Calderas are associated with wider intrusions. Some intrusions are single cooling units, but more commonly they are an accumulation of the unerupted portions of individual magmatic injections. These injections could occur periodically throughout the life of the volcano, and would generally be accompanied by eruption. Comparing the volume of the intrusion and the volume of the volcanic edifice indicates that only a small part of a magma injection erupts, although some of the apparent intrusive material may be reworked older volcanics. Exceptions to the general discussion presented tend to be related to the nature of the country rock.

INTRODUCTION

This report deals with the subject of subvolcanic intrusions as revealed by gravity data. Data and models for several volcanoes of the Cascade Range are presented and compared to a wide variety of volcanoes. Volcanoes included range in age from young, active ones to old, eroded volcanoes and in size from relatively small ones like Mount St. Helens (< 20 km diameter) to giants such as Mauna Loa (> 200 km diameter) and the Yellowstone caldera (60 × 40 km).

A portion of this report is devoted to a discussion of how to reduce and interpret reconnaissance gravity data from volcanic terrain. We did not develop, nor are we the first to use, the methods described. Investigators such as Woollard (1951) and Nettleton (1939) introduced the fundamentals

*U.S. Geological Survey, Box 25046, MS 964, Denver Federal Center, Denver, CO 80225.

which we utilize. No single reference for this information is available in the literature of gravity studies in volcanic terrain; numerous examples occur in which inappropriate methods of data reduction and interpretation have been utilized.

Ours is a pragmatic approach with a goal of gleaning as much as is reasonably possible from the data. We try to avoid getting bogged down in the problems related to ambiguity and superposition. The balance of the report deals with some of the interesting observations that result from comparing the consistently handled data from a variety of volcanoes. Work is currently under way to improve the data coverage of several volcanoes. Also, we are employing a number of other geophysical techniques to constrain our models in a better manner.

GENERAL COMMENTS ON REDUCTION AND INTERPRETATION OF GRAVITY DATA IN VOLCANIC TERRAIN

General procedure

Procedures for reduction and interpretation of gravity data are briefly discussed in numerous texts (e.g., Grant and West, 1965), but these discussions are inadequate for a study of subvolcanic intrusions.

Several corrections are applied to gravity data to produce a Bouguer gravity-anomaly map. It is important to understand that these corrections are designed to account for things we know about. Gravity anomalies remaining after the application of these corrections are those not adequately accounted for by the corrections or those related to horizontal subsurface-mass variations. The corrected data are not reduced to a datum level. The Bouguer gravity data are data draped onto the observation surface.

In this study it is usually prudent to remove a regional field. The removal of the regional anomaly tends to be a subjective process, but several techniques will yield a useful residual anomaly map. Regardless of the method utilized, it is important to make the best effort possible to ensure that the regional field removed contains all of the regional field and none of the anomaly sought.

Problematical corrections and anomaly separations

The Bouguer correction.—The Bouguer reduction density is usually chosen by one of three methods. The first is to use what could be called the traditional or standard density. Most large regional maps have traditionally been reduced using a value of 2.67 g/cm^3; the greatest consistency and comparability with older adjacent or regional surveys is achieved if that value is used.

The second method, originally suggested by Nettleton (1939; also see Vajk, 1956), determines a Bouguer reduction density that minimizes the correlation between the complete Bouguer gravity anomaly and topography. Although the latter method is widely used, it will fail in either one of two situations. First, it will fail if there is correlation between topography and geologic structures beneath the topography. An example would be a dense intrusion directly beneath a less dense volcanic cone. This failure can easily go undetected but becomes evident if the wavelength of the anomalies relating to topography is significantly different from those produced by the underlying bodies, or if the method produces a clearly unrealistic Bouguer reduction density. A second reason the method fails is more pragmatic than theoretical. If a rugged topographic feature has an average density substantially different than the selected Bouguer reduction density, the gravity anomaly associated with it will be of high amplitude and will be highly variable on the Bouguer anomaly map. It can easily obscure a much smaller but significant anomaly produced by underlying mass. Generally the chance of this failure occurring is reduced as the area to which the method is applied is also reduced, simply because the larger the area, the less likely it is that a single density will be appropriate for all the terrain.

The third method of choosing the Bouguer reduction density is to measure the density of representative rock samples. It is difficult to obtain a suite of rock samples that is truly representative. Furthermore, in volcanic terrain the bulk density of a formation is primarily a function of porosity. Porosity resulting from vesicles will be reflected in rock-sample measurements, but voids resulting from fractures and flow contacts can easily be overlooked. As a result, this method has never been dependable for determining the Bouguer reduction density in volcanic terrains.

We find it most useful to make maps at a variety of Bouguer reduction densities. These allow the study of various features of the gravity field at different Bouguer reduction densities, thereby reducing the chance of overlooking important anomalies or exaggerating the importance of others.

Terrain corrections.—The correction density used in terrain corrections is usually the Bouguer reduction density. An inappropriate density can create false anomalies or obscure important anomalies. As with the Bouguer correction, the quality of the terrain correction tends to improve as the area from which the Bouguer reduction density is calculated is reduced.

Anomaly separation.—Anomalies can be separated approximately on the basis of their apparent widths or characteristics. We have found in volcanic terrain that these qualities are difficult to observe on contour maps, even colored ones. It is far more productive to work with profile data. To illustrate the relationship between anomaly sources and topography, gravity-anomaly profiles are plotted above topographic profiles in several of our figures. We found it useful to display the gravity profiles with a variety of Bouguer reduction densities.

Once these profiles are constructed, they must be carefully studied. We have found that in volcanic areas superimposed anomalies are the rule. After the individual anomalies have been identified, those of interest should be extracted or separated. We have tried to accomplish this using numerous computer techniques involving filtering, surface fitting, and the like. None has been highly successful, probably because of the subjective nature of separation. In small areas with well-defined anomalies, it is generally better to do the separation by inspection, that is, to use our geologic knowledge of the area to form a subjective judgment and force this judgment on the separation process. We seek a residual anomaly that is due wholly to some specific and distinct body of interest. It is prudent to inspect the gravity field remaining after the residual is extracted, as errors in the separation process will usually be evident.

Modeling

The interpretation of gravity data involves the development of a model, which can either be entirely descriptive or incorporate any amount of quantitative computation. It should include all other relevant and reliable data that might help constrain the model. Strictly speaking, the resulting model can only describe a geologically possible density distribution. As such, it can confirm and refine other interpretations, contradict them, or reveal new anomalous masses.

Unfortunately, many gravity interpretations involving volcanic terrains are too descriptive and include little or no quantitative confirmation. They are based on education, experience, and intuition, which in our opinion are insufficient to produce a reliable model. We have found that our most revealing discoveries are a result of our being forced to resolve the differences between the gravity anomaly calculated from our best descriptive model and the observed gravity anomaly.

GRAVITY ANOMALIES ASSOCIATED WITH SUBVOLCANIC INTRUSIONS

In the previous section, we outlined a procedure by which we could produce a residual gravity anomaly associated with an identified source. In this section, we will discuss the residual gravity anomalies associated with subvolcanic density contrasts. Several major sources of gravity anomalies seem to exist on volcanoes, including the subvolcanic intrusion we seek, the low-density edifice of the volcano, low-density fill associated with calderas or other depressions, and regional trends. Our strategy is to choose a Bouguer reduction density as close to the bulk density of the volcanic edifice as possible. We then take the resulting Bouguer gravity map and remove a regional gravity field. The sources of the regional anomaly field include the high-amplitude broad low associated with the "root" of the mountain range. This root provides the means for isostatic compensation of the excess mass of the range. Also included are the large anomalies associated with continental margins and other broad anomalies whose sources are not well known but are probably related to density variation in the crust or upper mantle. Finally, if necessary, we model low-density caldera fill and remove its gravity effect. The remaining gravity field contains anomalies associated with subvolcanic intrusions and other unknown sources. Because we have removed most of the broad anomalies, these other sources usually cause localized anomalies. The localized anomalies are due to small, shallow density contrasts within the volcanic edifice or immediately below it. If they are small, they can be removed by filtering or by some other smoothing technique. If they are large, a further anomaly-separating step may be necessary. Of course, these small bodies may be extremities of the intrusion, in which case we may not want to remove their effect from the residual anomaly.

To gain a full understanding of gravity anomalies associated with subvolcanic stocks or intrusions, we must return to first principles. Lateral changes in the complete Bouguer gravity-anomaly maps reflect *lateral* density changes. The density contrast in our particular problem is the difference in density between the intrusion and the country rock. In the simplest case, the intrusion and the country rock each have a single but different density. If the intrusion is more dense than the country rock, we will observe a gravity high above the intrusion. If the density contrast is negative, we will observe a gravity low. This simple case can be complicated in three ways. First, if there is little or no density contrast, there is little or no associated gravity anomaly. Second, if the densities of one or both bodies are variable, we can expect the density contrast to vary. Finally, if the contrasting body of rock is small or distant, it may not be detected. There may be different *lateral* density contrasts at different *vertical* positions, and these contrasts may range from highly positive to negligible to highly negative.

For example, imagine a single-density intrusion extending from the top of a volcano in a volcanic mountain range to well into the asthenosphere. This intrusion will probably be denser than the volcanic rocks of the volcano's edifice. It may have a density similar to other upper crustal rocks, and it may be lighter than the other rocks of the lower crust and mantle. As such, it could produce a narrow, high-amplitude positive anomaly superimposed on a broad, low-amplitude low anomaly. Assuming the system is in a region that is in a state of near isostatic equilibrium, this latter anomaly may be indistinguishable from or become lost in the anomaly, owing to the "root" of the mountain range.

In this report we are dealing only with narrow anomalies resulting from density contrasts in the upper crust. We do not know the density of the intrusive material, although we can often make estimates that we feel are accurate to ± 0.1 g/cm^3. We do have some idea of the density of the country rock. In the Cascades the young, low-density volcanoes lie on top of a layer of older volcanic rocks. This "volcanic layer" may be several kilometers thick. It typically has a density of about 2.45 g/cm^3 (Pitts, 1979). Beneath the volcanic layer the density increases sharply to a value similar to that of the subvolcanic intrusions. The gravity anomalies evident in Figure 1 are principally due to dense intrusive material lying within the less dense volcanic layer and to a varying extent within the volcanic edifice.

We reduced (see Cordell et al., 1982) and separated the anomalies from gravity data from volcanoes of the Cascade Range. The gravity data are from gravity-data files of the U.S. Geological Survey, Oregon State University (Couch, written commun.), and Puget Sound University (Danes, written commun.). Samples of profiles of topography and gravity at different Bouguer reduction densities are illustrated in Figure 1. A set of profiles is produced for each volcano, and from these a Bouguer reduction density is chosen. The location of each profile should be selected carefully. Each profile should show the broad features of the topographic and gravity fields and avoid obvious locally anomalous areas. Figure 2 is a collection of idealized sketches in which we illustrate the gravity anomaly of the volcanic edifice and the subvolcanic intrusions at various reduction densities for the volcanic edifice. It is clear from this figure that the choice of reduction density has a dramatic

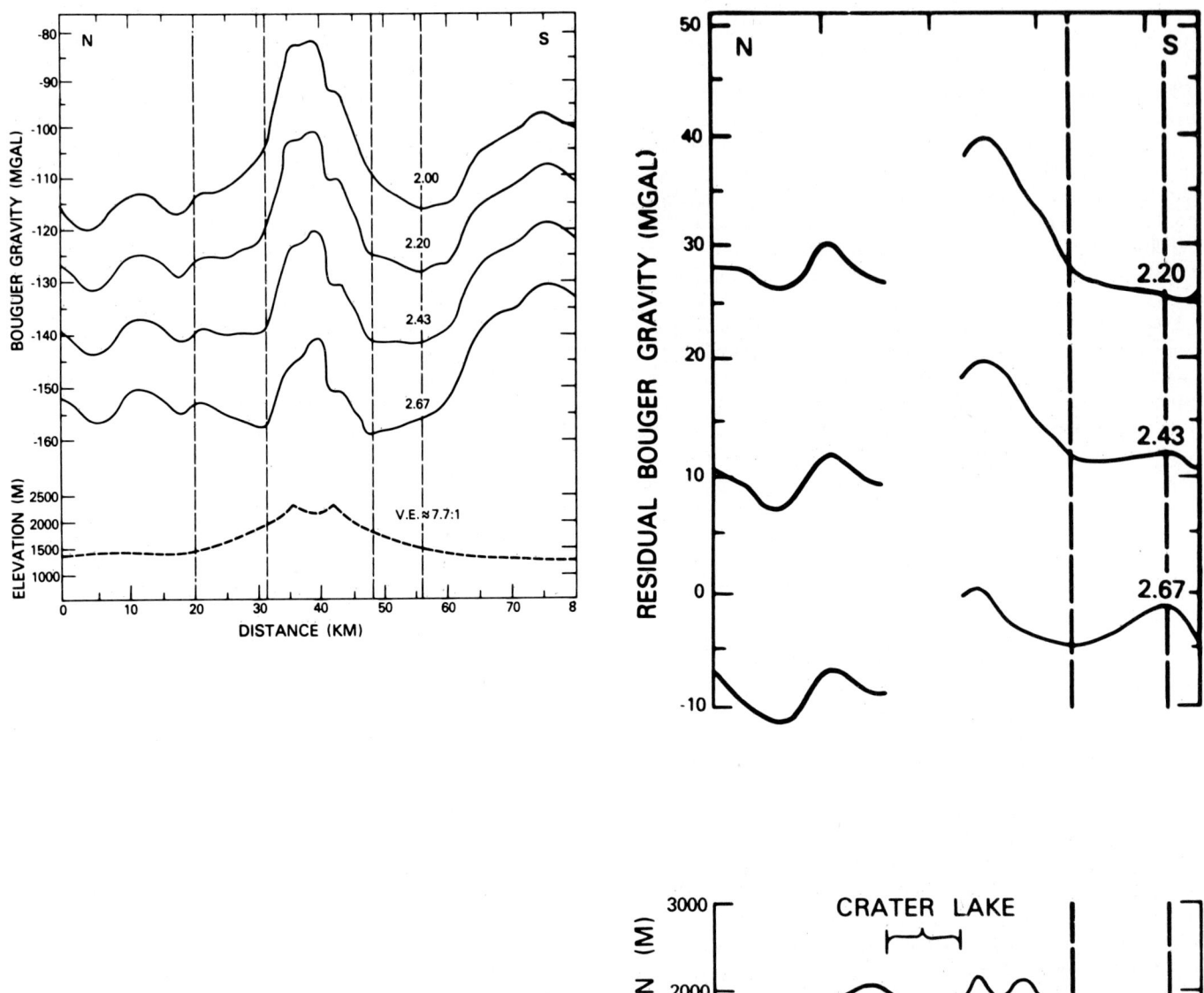

FIG. 1. Complete Bouguer gravity and topographic profiles for five volcanoes of the Cascade Range. The profiles cross the approximate center of the cones and are smoothed. a, Medicine Lake (above, left); b, Crater Lake (above, right; no gravity data are available from the lake); c, Newberry; d, Mount Hood (Couch and Gemperle, 1979); e, Mount Shasta; f, Mount St. Helens (prior to May 18, 1980; no gravity data were available on the south slope of the mountain). Our sources of gravity data for these figures are compilations of various published and unpublished data furnished by R. Chapman, California Division of Mines and Geology; R. Couch, Oregon State University; and F. Danes, University of Puget Sound. Pairs of dashed lines indicate the parts of the anomalies that we feel are the most definitive in choosing the correct Bouguer reduction density.

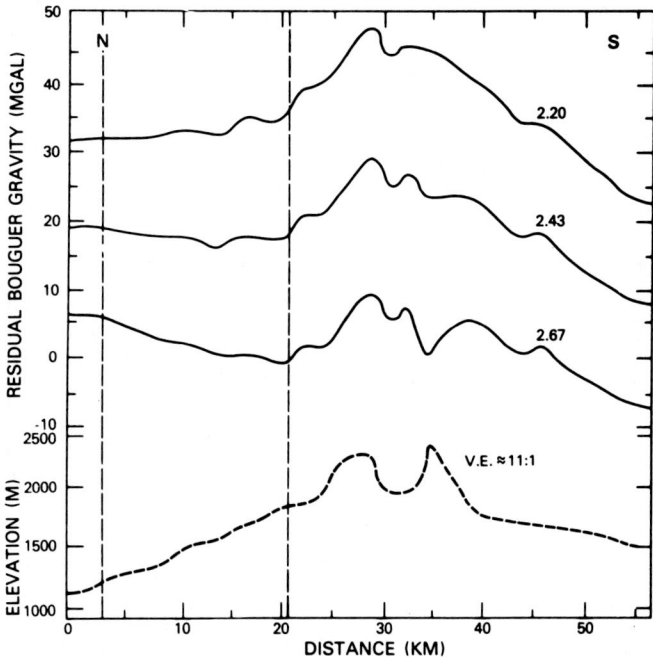

FIG. 1c. Newberry.

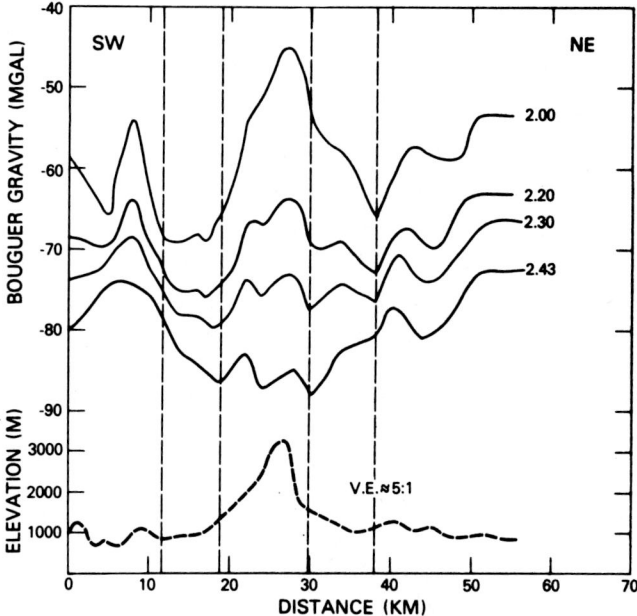

FIG. 1d. Mount Hood.

effect on the superimposed anomalies. A gravity survey of a volcano presumably contains these two anomalies. Our goal is to separate these anomalies, and we utilize Figure 2 as a guide. In Figure 2 we start at a high reduction density and, step by step, approach the assumed correct reduction density of 2.2 g/cm³. At the correct reduction density the tails or flanks of the anomaly slope smoothly outward. This is the main diagnostic feature. At the 2.4 g/cm³ reduction density, the tails are flat or sloping inward. At 2.0 g/cm³, the effect of the edifice again is evident as a positive correlation between gravity and topography.

On any given volcano we inspect several profiles, but it is not reasonable to include them all here. In Figure 1 we have provided pairs of dashed lines to help the reader discern the part of the anomaly used to determine the representative reduction density. This procedure is based on the study of many profiles from several volcanoes. On any given volcano, the results can be less than convincing. We have deduced a bulk density for the volcanic rock of the edifice of 2.2 ± 0.1 g/cm³. The results seem consistent from volcano to volcano. If for any reason the density cannot be ascertained at a specific volcano, it is best to choose the Bouguer reduction density by analogy with similar volcanoes. The density of 2.2 g/cm³ is representative of all the Cascade volcanoes presented in this report. Braman (1981) derived a best reduction density value of 2.43 g/cm³ for the northern Oregon Cascades. Much of his study was of older volcanic rocks, and his methods ignored any dependence on geology. Pitts (1979) estimated a reduction density of 2.67 g/cm³ for the Newberry (Figure 1b) volcano. We believe Pitts failed to recognize the presence of a dense subvolcanic intrusion and erroneously assumed that the gravity high at Newberry was due to a dense edifice. Griscom and Roberts (1983) obtained a density of 2.25 g/cm³ for the edifice of Newberry from density measurements from cores.

After the correct Bouguer reduction density is applied to the observed data and the other corrections are made, the anomaly-separation procedure is complete. In most cases the resulting anomaly is symmetrical and has a fairly regular shape.

These residual anomalies were modeled using, where appropriate, a three-dimensional computer program (Cordell and Henderson, 1968) or a two-and-a-half-dimensional program (Cady and Sweeney, 1980). Figure 3 shows the results of the modeling of the volcanoes illustrated in Figure 1.

DISCUSSION AND INTERPRETATION OF MODELS OF SUBVOLCANIC INTRUSIONS

The six subvolcanic intrusions modeled show many common characteristics and a few distinct differences (Figure 3). The models of the Medicine Lake (Figure 3a) and Newberry (Figure 3c) volcanoes are similar. They are both shield volcanoes with shallow calderas. The intrusions lie about 1.5 km beneath the caldera floors. The intrusions have broad, elliptical tops roughly 5 × 9 km in dimension. They have generally sloping sides, and lower parts of the intrusions extend laterally at least 25 km, which puts them well beyond the 10- × 6-km caldera rim. The thicknesses are greater than the thickness of the volcanic layer that they intrude, and they penetrate the base of the volcanic edifice. The modeled density contrast between the intrusions and the volcanic layer is about 0.4 g/cm³. We estimate the intrusions to be mafic with a density of 2.8–2.9 g/cm³, and the older volcanic layer has a density of 2.4 to 2.6 g/cm³. The volumes of the models at the Medicine Lake and Newberry intrusions are 420 and 413 km, respectively, compared with 340 and 300

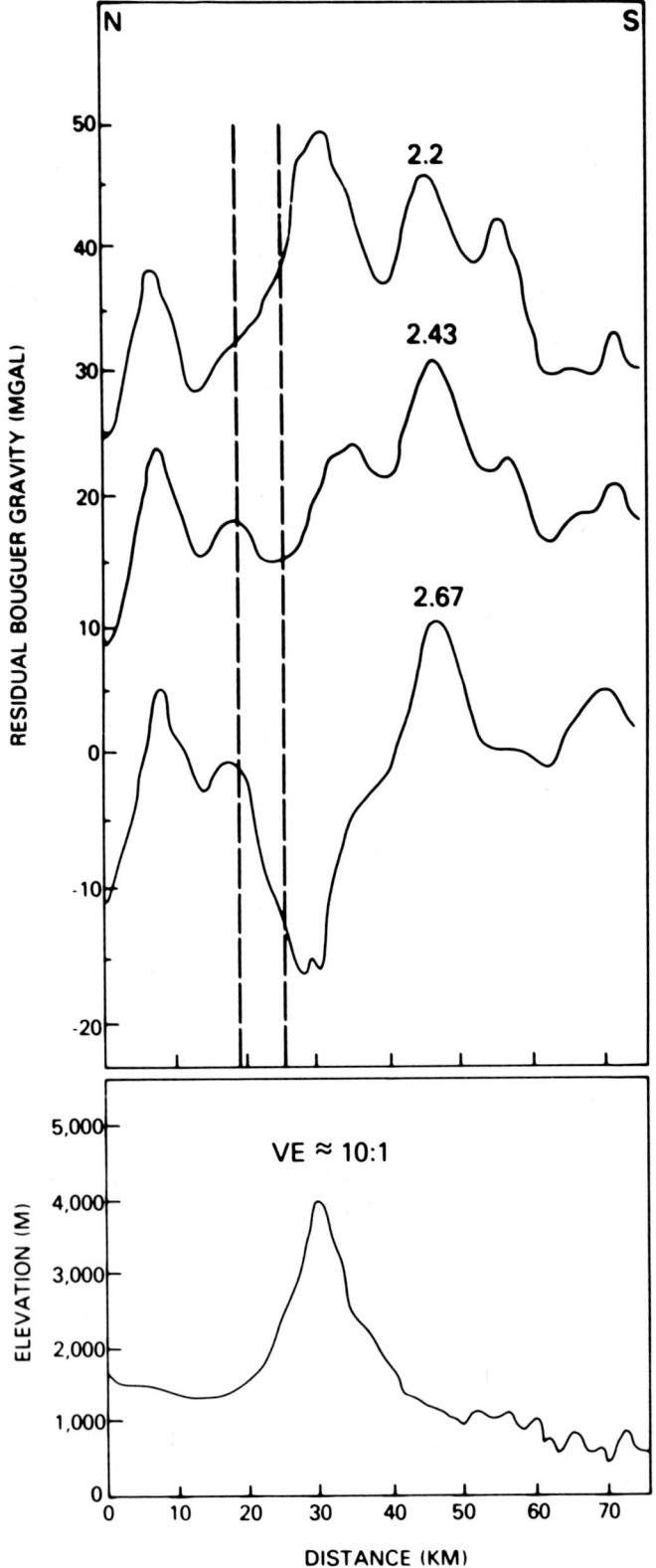

Fig. 1e. Mount Shasta.

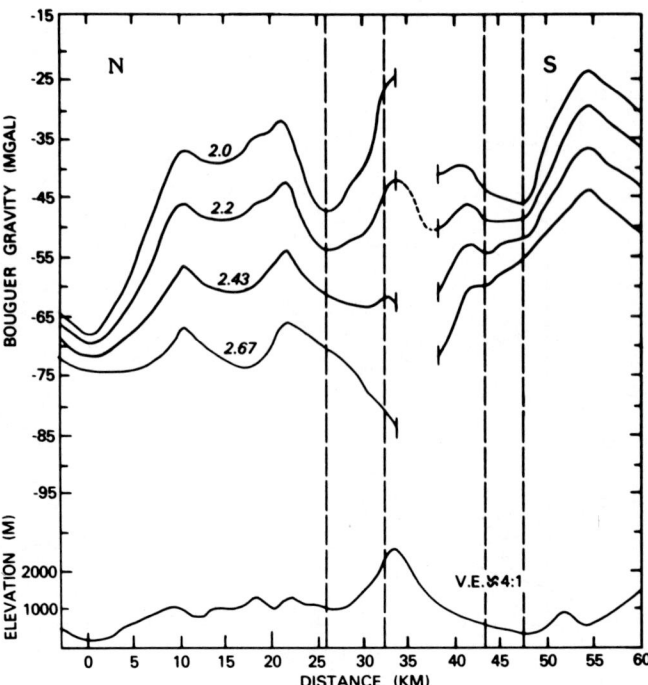

Fig. 1f. Mount St. Helens.

km³ for the volumes of the volcanic edifices. The Newberry volcano model shows evidence of a low-density caldera fill, which is less than the 2.2 g/cm³ used for the bulk density of the edifice. The dashed lines in Figure 3c are intended to approximate the intrusion, with the effect of the caldera fill removed.

Crater Lake, Oregon (Figure 3b), has an intrusion beneath it similar to Medicine Lake and Newberry volcanoes. It has a broad top (10 km wide) and gently sloping sides, and its base is roughly 26 km wide. Here, the caldera fill has a dramatic effect on the gravity field. The modeled intrusion is a crude approximation, because we had to separate the large gravity anomaly associated with the caldera fill without the benefit of gravity data on the lake itself.

The modeled intrusions beneath stratovolcanoes of Mount Hood (Figure 3d), Mount Shasta (Figure 3e), and Mount St. Helens (Figure 3f) have a distinctly different shape from those found beneath the shield volcanoes. Mount Hood has an intrusion whose upper boundary is 3 km wide within the cone at 3 km above sea level. It has steep sides, and the intrusion expands to a width of about 9 km at the base of the layer of older volcanic rocks 3 km below sea level. The modeled density contrast between the intrusions and the layers of older volcanic rocks at Mount Hood and Mount St. Helens is about 0.2 g/cm³ or about half that observed at the Mount Shasta, Medicine Lake, Newberry, and Crater Lake volcanoes. This implies that older volcanic rocks beneath

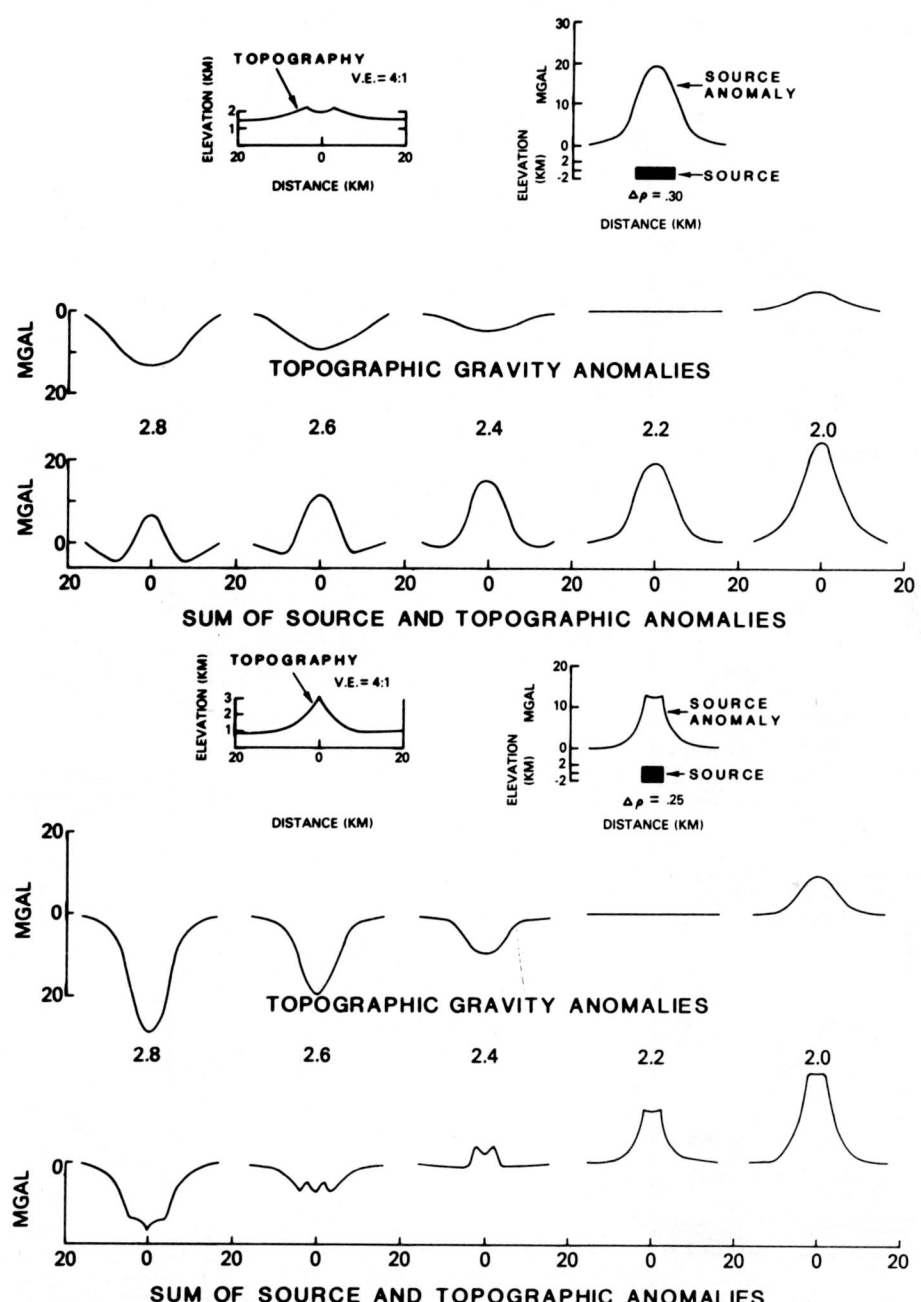

FIG. 2. Idealized sketches illustrating the selection of the correct Bouguer reduction density. The upper sketch is a shield volcano similar to the Medicine Lake volcano, and the lower sketch is a stratocone volcano similar to Mount Hood. For each there is a topographic profile and a hypothetical source body with its calculated gravity anomaly, as would be observed on the topographic surface. Below these are calculated profiles of the gravity anomaly associated with the edifice for each of four Bouguer reduction densities (g/cm^3), again as would be observed on the topographic surface. Below those are four calculated gravity anomalies representing the sum of the gravity anomaly associated with the edifice and that of the source body. In both cases, 2.2 g/cm^3 is the correct Bouguer reduction density. The practical use of the sketches is to illustrate the characteristics of the summed anomalies useful in selecting the correct Bouguer reduction density.

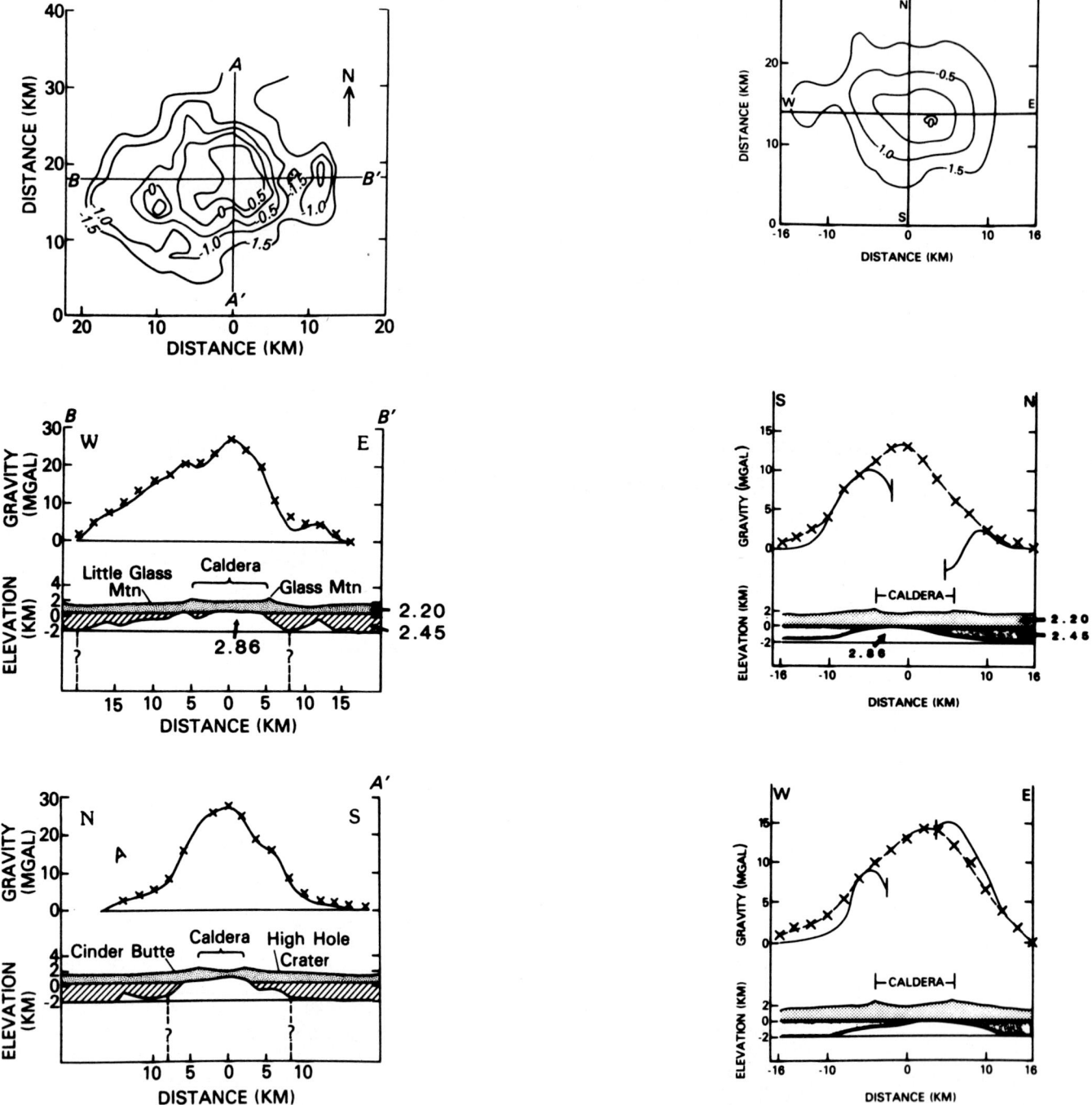

FIG. 3. Three-dimensional source models calculated for the anomalies of Figure 1: a, Medicine Lake (above, left); b, Crater Lake (above, right); c, Newberry; d, Mount Hood; e, Mount Shasta; f, Mount St. Helens. The data coverage at Crater Lake and Mount St. Helens (prior to May 18, 1980) was not sufficient to define the anomaly adequately. Anomaly shape is assumed in the dashed parts of the gravity profiles. All gravity profiles have had a regional gravity field removed. The modeling technique assumes density contrasts between the various bodies (i.e., edifice, older volcanic layer, and intrusions). We can convert these estimates to estimates of the actual density of the bodies. This was possible in five of the models. The exception is Mount Shasta, where the intrusion is most evident in the edifice and the thick pile of young volcanic rocks beneath the edifice. The density contrast appears to remain fairly constant, which implies that both the density of the intrusion and the density of the young volcanic rocks increase with increasing depth. This conclusion is supported by recent seismic data (Zucca et al., 1981) that show a similar increase in velocity in the young volcanic rocks beneath the flanks of Mount Shasta. At Mount St. Helens the data coverage is poor, and much of the model is constructed by analogy with the other five modeled volcanoes.

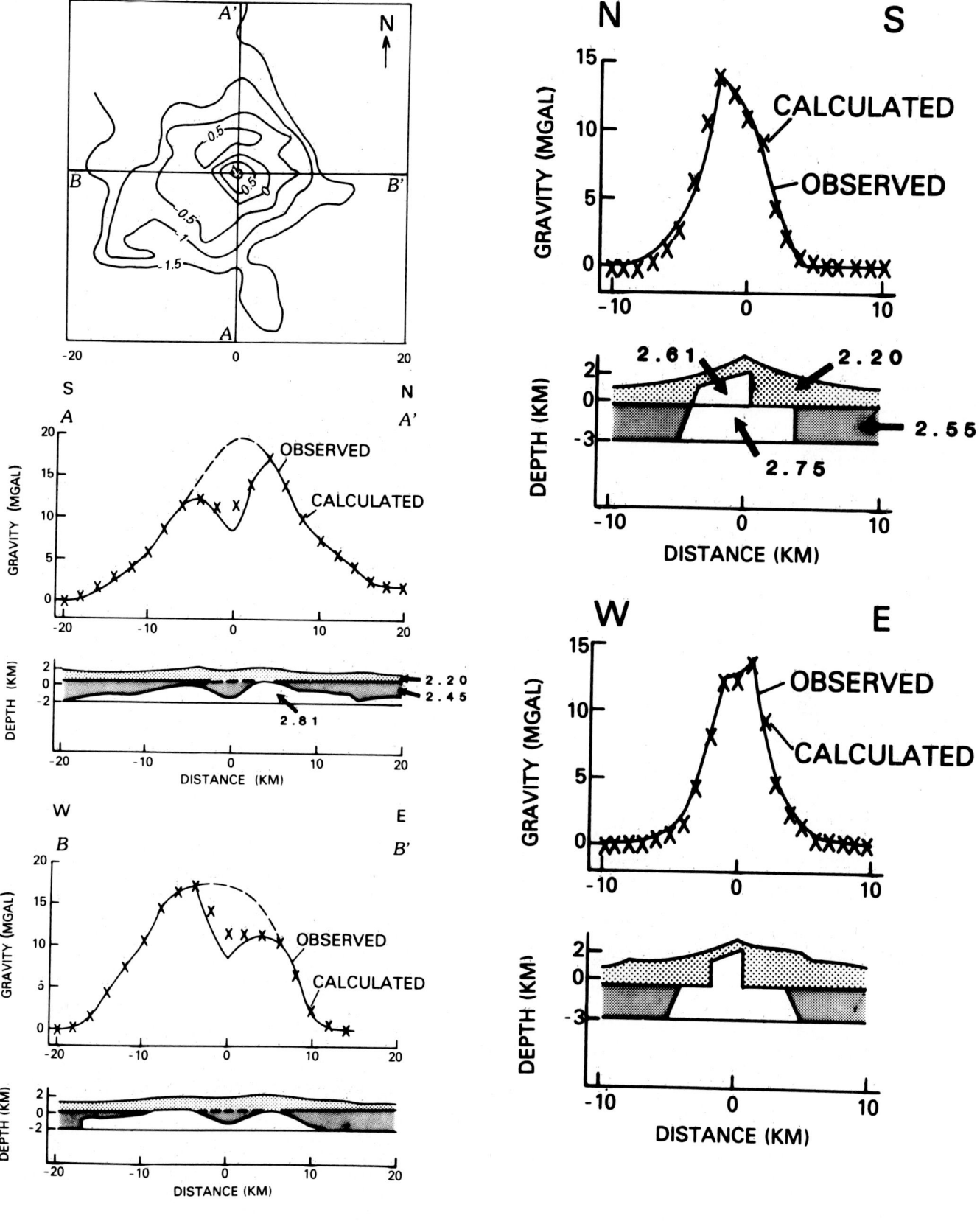

Fig. 3c. Newberry.

Fig. 3d. Mount Hood.

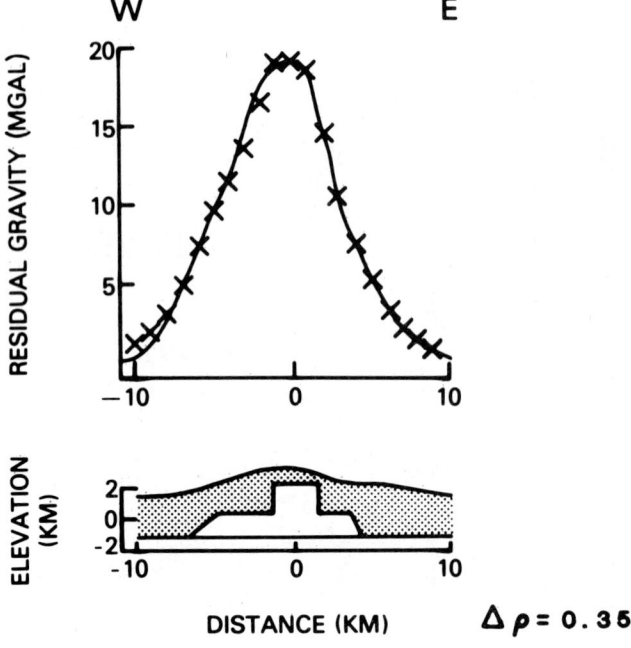

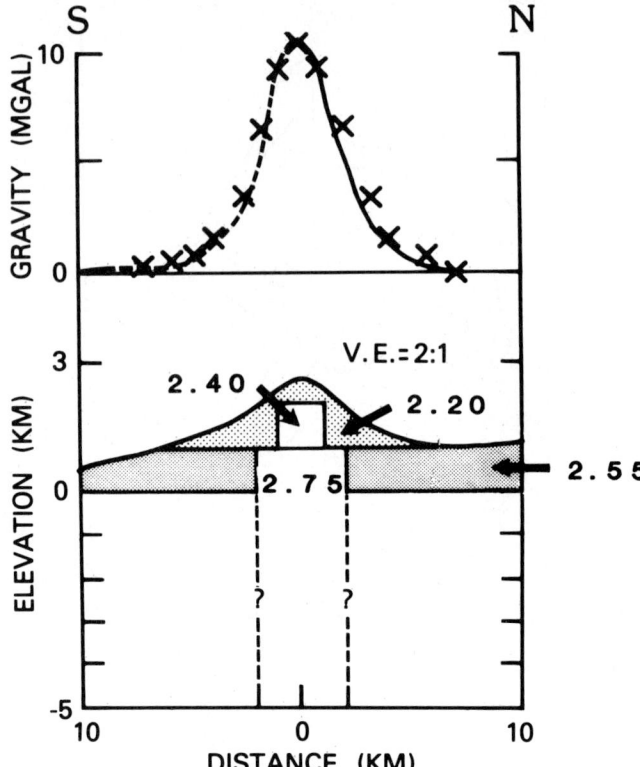

Fig. 3f. Mount St. Helens.

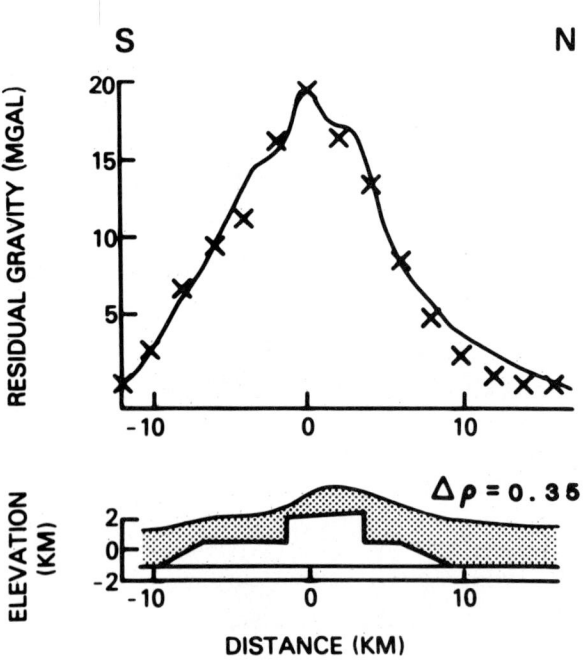

Fig. 3e. Mount Shasta.

those stratocones are somewhat more dense. If the intrusions are dioritic and have a density of 2.7–2.8 g/cm^3, then the older volcanic rocks have a density of 2.5 to 2.6 g/cm^3. Recent drilling on the flanks of Mount Hood has revealed the presence of highly indurated basalts and andesites beneath the edifice. This indicates that the older volcanic rocks adjacent to the Mount Shasta, Medicine Lake, Newberry, and Crater Lake volcanoes are probably less indurated. A recent gravity study of Mount Adams (Rowley, 1982), a stratovolcano in Washington, has revealed that it, too, is underlain by a dense intrusion.

Some gravity data are available over most of the exposed or suspected plutons in the Cascade Range. Most of these plutons appear to be of dioritic composition, for example, the Laurel Hill and Still Creek plutons southwest of Mount Hood. We have not attempted to model these, but they all show a local gravity high, implying that they are also intruded into a layer of less dense rock.

At all the volcanoes modeled except Mount St. Helens, some data on the location and thickness of the "volcanic layer" were available, and we constrained the intrusions to be within this layer. By "volcanic layer" in the Cascade Range, we mean a layer of volcanic material lying beneath the base of the volcanic edifice being investigated. These volcanic rocks are older, are commonly of Tertiary age, are more electrically conductive (Stanley, in press), have a higher seismic velocity, about 5 km/s (Zucca, 1981; Kohler et al., 1982), and have a higher density, about 2.4 to 2.6 g/cm^3 (Pitts, 1979), than the material contained in the edifice. Although they are more dense than the volcanic edifice, they are significantly less dense than the underlying basement rock. These basement rocks are presumably older intrusions of various diorite, gabbro, and in some cases more silicic rocks or metamorphic rocks. It is probably not reasonable to assume that the young subvolcanic intrusions would be 0.2 to 0.4 g/cm^3 denser than these older intrusions. Therefore, we are forced to place our subvolcanic intrusions in the volcanic layer or volcanic edifice, where such a density contrast is geologically reasonable.

The volumes of the subvolcanic intrusion are usually substantially larger than the volumes of the volcanic edifice. Even these volume estimates are a minimum, because the intrusion may extend deeper than the level at which we observe the density contrast, yielding a volume larger than the one we modeled.

Other volcanoes

We conducted a literature search for reports that contained the results of gravity surveys on volcanoes. Our objective was to see if any pattern could be found, and our goal was to investigate subvolcanic stocks. Because this was not necessarily the goal of the original investigators, the data often were not reduced in a way in which to enhance the signature of such stocks. We were often forced to glean evidence of subvolcanic stocks without benefit of tabulated data and in some cases without a knowledge of the density that was used in the Bouguer reduction. In spite of these difficulties, a clear pattern is evident.

Mafic volcanoes

Large mafic volcanoes, especially basaltic oceanic islands, are the type of volcano with the most extensive gravity coverage. Woollard (1951) reported the results of a reconnaissance gravity survey of the Island of Oahu. This was not the first gravity survey of an oceanic island or even of the Island of Oahu, but it set a standard. His method of interpretation as well as the results have been repeated for numerous oceanic islands. He found the average bulk density of the island above sea level to be 2.3 g/cm^3 (exclusive of the intrusives throat). When that reduction density is used, the individual large volcanic centers are Bouguer gravity highs, which are caused by the density contrast between the surficial rocks (vesicular basalts or corals) and high-density rocks beneath the major volcanic vents, whose density is closer to 2.8 or 3.0 g/cm^3. Below sea level, the bulk of the volcanic edifice on the Island of Hawaii is composed of the high-density rocks (Zucca et al., 1982; Kinoshita, 1965). Similar results were obtained for the Hawaiian Swell (Strange et al., 1965) and for the Hawaiian Islands of Maui (Kinoshita and Okamura, 1965), Molokai (Moore and Krivoy, 1965), Lanai (Krivoy and Lane, 1965), Kahoolawe (Furamoto, 1965), Kauai (Krivoy et al., 1965), and Niihau (Krivoy, 1965). Similar results were found also for the Leeward Islands of the Hawaiian Ridge and Johnston Island (Kroenke and Woollard, 1965); American Samoa and the Society Islands (Machesky, 1965); Jasper seamount (Harrison and Brisbin (1959); the Galapagos Islands (Case et al., 1973; Watts and Cochran, 1974); the islands of Oosima and Miyake (Yokoyama, 1969); the Cook Islands (Robertson, 1967a, 1967b, 1970); Ambrym Island (McCall et al., 1970); numerous Polynesian islands and atolls (H. G. Barsczus, pers. comm., 1981) including Maiao, Tetioroa, Huakine, Tahaa, Bora-Bora, Maupiti, and Scilly, all in the Society Islands; Ripa, Raivave, Tubuai, Rurutu, and Rimatara in the Austral or Tubuai Islands; Mataiva, Maniki, Rangiroa, and Makatea in the Tuamotus Islands; and Hiva-Oa, Ua-Pou, and Nuku-Hiva in the Marquesas and Gambier Islands.

Most of the oceanic-island gravity surveys are conducted at or above sea level and suffer because they miss the large part of the volcanic edifice that lies under water. However, there are cases where marine gravity data are available (e.g., Woollard, 1951; Zucca et al., 1982). Many of the island portions of these volcanoes are composed of low-density volcanic material with shallow intrusions occurring beneath the major vent areas. Below sea level this characterization changes dramatically. Most of the volcanic edifice, from the base of the low-density surficial volcanics to the Moho (which is warped down beneath the volcano), is composed of material with a density closer to 2.8 to 3.1 g/cm^3. Presumably these rocks were once volcanics similar to the surficial rocks but have been repeatedly intruded and altered and so compacted that they now have a density nearer that of intrusive rock.

Ramberg (1976) showed several examples of gravity surveys of mafic and ultramafic necks in the Oslo graben. Here, the low-density volcanic edifices, if they ever existed, have been completely eroded. Nevertheless, a gravity high is found. Some of these necks intrude gneisses and some intrude sedimentary rocks, but all the necks are more dense than the country rocks.

Composite cones

The next group of volcanoes we discuss are the basaltic or andesitic composite cones. Those with adequate gravity coverage include Mount Fuji (Huzi) (Yokoyama and Tajima, 1960), Mount Asama (Yokoyama, 1971), Mount Etna (Medi and Morelli, 1952), volcanoes of the Avachinsky group (Steinberg and Zubin, 1965), Tahia volcano (Sissons, 1981), and the other composite cones of the Cascade Range described earlier. It is particularly difficult to remove the effects of the volcanic edifices of composite cones. The edifices are large and of low density, and typically the main cone is built atop earlier ones. In addition, the gravity anomaly associated with a subvolcanic stock has a dimension close to that produced by the cone.

A very interesting study of rock densities and Tahia volcano, New Zealand, was reported by Sissons (1981). He compared gravity data obtained from a water-diversion tunnel bored through the volcano with data taken immediately above the tunnel on the surface of the volcano. With these data, he computed the bulk density of the volcanic cone and found values similar to those we find for the Cascade volcanoes. In addition to the similarity in bulk densities, Sisson's model also reveals the presence of a large, dense subvolcanic body. He did not mention this feature, since his was a study of the density of the cone, but its presence was clearly required in order to fit his data.

Small calderas

Many of the mafic volcanoes mentioned in the previous paragraphs have summit calderas. The calderas may have diameters up to 10 km, but they all are small in comparison with the volcanic edifice on which they occur. These calderas seem to have little or no influence on the gravity field. A report by MacDonald (1965) contains an informative discussion on Hawaiian calderas. Where the caldera is large enough to be the dominant feature of the edifice, the gravity picture is different. We have chosen a 15-km caldera diameter as the dividing line between small and large calderas. Examples of small calderas include Batur caldera in Bali, Indonesia (10 km diameter) (Yokoyama and Suparto, 1970); Kuttara caldera, Japan (2.4 km diameter) (Yokoyama, 1967); Crater Lake, Oregon (9 km diameter); Santorin caldera, Greece (or Thera, 10 km diameter) (Yokoyama and Bonasia, 1978); Krakatau caldera, Indonesia (8 km diameter) (Yokoyama and Hadikusumo, 1969); Toya caldera, Japan (11 km diameter) (Yokoyama, 1963a); Hakone caldera, Japan (11 km diameter) (Yokoyama, 1963a); Medicine Lake caldera, California (8 km diameter) (Finn and Williams, 1983a); Newberry caldera, Oregon (9 km diameter) (Griscom and Roberts, 1983); and Menengai (10 km diameter), Longonot (12 km diameter), and Suswa (11 km diameter) calderas, Kenya (Searle, 1970).

Crater Lake, Batur, Kuttara, Medicine Lake, Newberry, Menengai, Longonot, and Suswa all display evidence of a dense subvolcanic stock. The evidence of the remaining calderas is not as clear. At Santorin and Krakatau, the data coverage is limited to island rims of the caldera, and both calderas are mantled by a deposit of pumice of unknown but substantial thickness.

Large calderas

Large calderas include volcanoes with calderas having a diameter larger than 15 km. Most of the large calderas for which we have adequate gravity data have had one or more large, silicic ash-flow eruptions. They include Aso caldera, Japan (16–23 km diameter) (Yokoyama, 1963b), Kuttyaro caldera, Japan (22 km diameter) (Yokoyama, 1958); Aira caldera, Japan (15 km diameter) (Yokoyama, 1961); Sikotu caldera, Japan (15 km diameter) (Yokoyama and Aota, 1965); Akan, Japan (13–24 km diameter) (Yokoyama, 1958); Valles caldera (20 km diameter) (Cordell, 1976b); Long Valley caldera, California (15–30 km diameter) (Kane et al., 1976); Yellowstone caldera, Wyoming (45–70 km diameter) (Eaton et al., 1975); and Timber Mountain caldera, Nevada (40 km diameter) (Kane et al., 1981). Each of these calderas produces a large Bouguer gravity low.

One large mafic caldera, Ngorongoro caldera, Tanzania (18 km diameter) (Searle, 1971), produces a gravity low.

The gravity lows associated with these young, large calderas are difficult to interpret because the calderas commonly have a substantial thickness of low-density fill. Based on these data alone, an argument can be made that the entire gravity low can be explained by the low-density fill. However, if we look at older, more eroded calderas in which most or all of the fill has been removed, we see a different picture.

Gravity data from granitic intrusions interpreted to be the roots of old silicic calderas in the Oslo graben (Ramberg, 1976); in Great Britain, Ireland, New Brunswick, and New England (Bott et al., 1958; Bott, 1953; Bott, 1956); at the Questa caldera, New Mexico (Cordell, 1976a); at the Bursum and Gila Cliff Dwellings calderas, New Mexico (Ratté et al., 1979); and at the Creede caldera, Colorado, all reflect Bouguer gravity lows. The source of the low is generally the density contrast between relatively light rocks of the granitic pluton and denser (usually metamorphic) country rocks. In a few of these calderas, some low-density intracaldera tuff and other fill remain and increase the magnitude of the gravity low.

We found an example of a suspected large silicic intrusion with an accompanying large Bouguer gravity low that has not produced a caldera. In the Geysers region of California, Isherwood (1976) interpreted extensive gravity data as indicating the presence of a large (20 km diameter) buried intrusion or magma chamber whose top lies within 10 km of the surface. In contrast to all other eruptive centers, large calderas are associated with gravity lows (Figure 4).

Exceptions

In the previous paragraphs we have presented the results of gravity surveys at numerous volcanoes out of which an interesting pattern has emerged. The pattern is related to the type and size of the volcano, but as we discussed earlier, the gravity anomaly associated with a subvolcanic stock is as much dependent on the density of the country rocks as it is on the density of the intrusion. Therefore, exceptions to this observed pattern may occur.

We have yet to see a large volcanic system that convincingly lacks a gravity signature, although such systems may exist. One possible example is the Coso volcanic field in California. Plouff and Isherwood (1980) carefully interpreted gravity data from this region and concluded that they could see no indication of a suspected silicic intrusion. The regional gravity anomaly is particularly complicated here and cannot be removed with much confidence. Also, the suspected intrusion would underlie the Coso Mountains, which are primarily silicic intrusive rocks for which it is difficult to pick an accurate and representative Bouguer reduction density. As a result, a small gravity anomaly may exist and be undetected, but it is safe to say that no large anomaly is associated with the suspected intrusion. The intrusion may exist, but the density contrast that would be expected from a young silicic magma intruding an older silicic intrusion is small. Therefore, there is no reason to expect a large gravity anomaly at the Coso volcanic field.

Another possible exception is a large caldera (30 × 40 km) in which a silicic intrusion has apparently risen into the volcanic pile of the Snake River Plain in Idaho (L. McBroome, pers. comm., 1982). The intrusion has been interpreted as the stock of a large, old silicic caldera for which 2.6 to 2.7 g/cm^3 would be an appropriate density. As the volcanic rocks should have a bulk density of 2.2–2.4 g/cm^3, we would expect a gravity high associated with this particular silicic intrusion, and this is in fact what is observed.

We conducted a gravity survey of Diamond Peak in the Oregon Cascades (Finn and Williams, 1983b). Diamond Peak is smaller than any of the volcanoes illustrated in Figures 1 and 4. Owing to the rugged topography and the reconnaissance nature of the survey, our data are probably only accurate to about ±2 mGal. At this level of resolution, there is no observable subvolcanic intrusion. This leads us to believe that the smaller volcanic systems may not produce an anomaly large enough to be detected by anything short of a highly detailed survey.

SOME REMAINING PROBLEMS

We have detected the presence of an anomalous mass beneath almost every volcano studied. Where the volcanic edifice has been partly removed by erosion, an intrusion is revealed. Are these intrusions single cooling units, or are they a result of the solidification of the unerupted portions of magma injected during the numerous eruptive cycles in the life of a volcano? Examples of both are found at intrusions exposed by erosion. Gravity data alone will not answer this question for the young active volcanoes. However, it is difficult to believe that the 420 km^3 intrusion modeled beneath the Medicine Lake volcano was ever a single magma chamber. Similarly, the intrusion beneath Mount Hood is much shallower than is suggested by petrologic studies (e.g., White, 1980). These two examples suggest that the underlying bodies are at least in part the solidification of intrusions from earlier stages in the evolution of these volcanoes.

We also do not know what percentage of the anomalous mass is intrusive rock directly associated with the existing volcano. Magma could intrude into the older volcanic layer as a myriad of dikes and sills. This could result in these older volcanic rocks being heated and severely altered or even melted. The density of the resulting formation could easily approach that of intrusive rocks. Seismic velocities and electrical resistivities might also resemble those of intrusive rocks.

Another problem raised by the gravity data is that the modeled intrusions for composite volcanoes all have distinct bottoms. Three alternative explanations are possible. The first is that this is built into our method of selecting the Bouguer reduction density. If we choose a lower reduction density, the modeled bottom becomes deeper. Nevertheless, in several cases we found it difficult to fit a significantly deeper bottom, even with a Bouguer reduction density of 1.8 g/cm^3. Also, it is not reasonable to assume that the large density contrasts persist to great depth. The second explanation is that this is the actual bottom of the intrusion and that only a small feeder pipe or dike extends below the bottom. The third is that the intrusion does not drastically neck down below this bottom, but that instead the country-rock density increases to close to that of the intrusive, making the density contrast nearly vanish. We favor this latter model, but all three present a room problem. Tectonic activity along intersections of faults could provide the kind of radial extension implied by the sizes and shapes of the intrusions. Uplift or uplift and erosion seem unlikely to have produced the room. Erosion is minimal at the young volcanoes, and little uplift can be assumed. At some volcanoes, like Mount Hood, subsidence can be documented (Williams et al., 1982). Making room by magmatic stoping or the plastic flow of material down around the magma calls for the stoped blocks or flowing material to sink by gravity. In the volcanic layer the country rock is lighter than the intrusion and will not sink unless metamorphically altered to a point at which

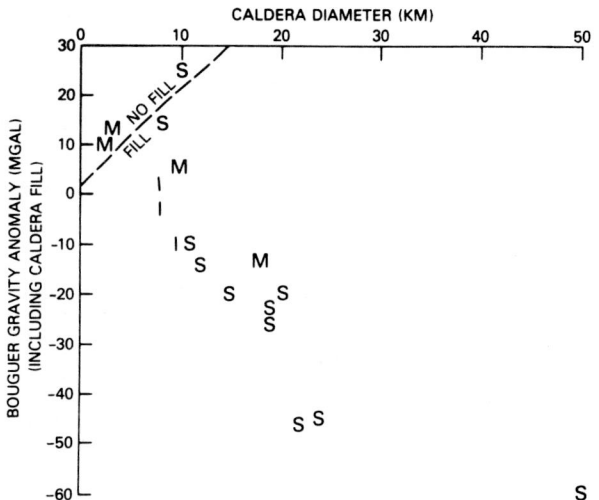

FIG. 4. Diameter versus Bouguer gravity anomaly for 18 calderas. Caldera diameters are generally measured from the caldera rim, and axes are averaged in the case of elliptically shaped calderas. M = mafic; I = intermediate; S = silicic. A caldera is considered silicic if it has a significant component of silicic eruptive material, even though most of the eruptive material may be more mafic. Only calderas with adequate gravity coverage are plotted. In the upper left corner above the dashed line, three calderas are shown that have little or no caldera fill. The remaining calderas have low-density caldera fill that has some significant influence on the gravity anomaly. Usually the larger the caldera, the larger the influence of the caldera fill. The break between calderas that produce negative and positive anomalies would occur at a diameter of about 15 km if a crude correction is made for the effect of caldera fill. This conclusion is different than that reached by some previous investigators (e.g., Williams and McBirney, 1979).

its density is sufficiently large. This requires temperatures in the neighborhood of 300°C.

A final problem relates to isostatic compensation. If the volcanic chain is associated with a topographic swell, it should have a low-density "root" that not only will compensate for the mass excess of the elevated topography but also must compensate for the mass excess associated with these large, dense subvolcanic bodies. This isostatic compensation probably occurs in various degrees from very shallow depths to well into the mantle and can be expected to be markedly different at different localities.

REFERENCES

Bott, M.H.P., 1953, Negative gravity anomalies over acid "intrusions" and their relation to the structure of the Earth's crust: Geol. Mag., **90**, 257–267.

——— 1956, A geophysical study of the granite problem: Q. J. Geol. Soc. London, **112**, 45–67.

Bott, M.H.P., Day, A. A., and Masson-Smith, D., 1958, The geological interpretation of gravity and magnetic surveys in Devon and Cornwall: Philos. Trans. R. Soc. London, ser. A, **251**, 161–191.

Braman, D., 1981, Interpretation of gravity anomalies observed in the Cascade Mountain Province of northern Oregon: M.S. thesis, Oreg. State Univ.

Cady, J. W., and Sweeney, R. E., 1980, Program ZHDPOT for 2½ dimensional gravity and magnetic modeling: SEG suppl. to Cady, J. W., Calculation of gravity and magnetic anomalies of finite-length right rectangular prisms: Geophysics, **45**, 1507–1512 (and comput. tape).

Case, J. E., Ryland, S. L., Simkin, T., and Howard, K. A., 1973, Gravity anomalies in the Galapagos Islands area: Science, **181**, 1040.

Cordell, L., 1976a, Preliminary complete Bouguer gravity anomaly map of the Taos basin section of Rio Grande graben, New Mexico: Open-File Rep. 76-358, U.S. Geol. Surv.

——— 1976b, Aeromagnetic and gravity studies of the Rio Grande graben in New Mexico between Belen and Pilar, in Tectonic and mineral resources of southwestern North America: Spec. Pap. 6, N.Mex. Geol. Soc., 62–70.

Cordell, L., and Henderson, R., 1968, Iterative three-dimensional solution of gravity data using a digital computer: Geophysics, **33**, 596–601.

Cordell, L., Keller, G. R., and Hildenbrand, T. G., 1982, Bouguer gravity map of the Rio Grande rift: Geophys. Invest. Map GP-949, U.S. Geol. Surv.

Couch, R., and Gemperle, M., 1979, Gravity anomalies in the area of Mt. Hood, Oregon, in Riccio, J., Ed., Geothermal resource assessment of Mount Hood: Oreg. Dep. Geol. Mineral Ind., RLO-1040, 137–189.

Eaton, G. P., Christiansen, R. L., Iyer, H. M., Pitt, A. M., Mabey, D. R., Blank, H. R., Zietz, I., and Gettings, M. E., 1975, Magma beneath Yellowstone National Park: Science, **188**, 787–796.

Finn, C., and Williams, D. L., 1983a, Gravity evidence for a shallow intrusion under Medicine Lake volcano: Geology, **10**, 503–507.

———1983b, Principal facts for fifty-six gravity stations near Diamond Peak, Oregon: Open-File Rep. 83-177, U.S. Geol. Surv.

Furamoto, G., 1965, A gravity survey of the Island of Kahoolawe, Hawaii: Pac. Sci., **19**, 349–350.

Grant, F., and West, G., 1965, Interpretation theory in applied geophysics: McGraw-Hill.

Griscom, A., and Roberts, C., 1983, Gravity and magnetic interpretation of Newberry Volcano, Oregon: Open-File Rep. 83-657, U.S. Geol. Surv.

Harrison, J. C., and Brisbin, W. C., 1959, Gravity anomalies of the west coast of North America. 1: Seamount Jasper: Geol. Soc. Am. Bull., **70**, 929–934.

Isherwood, W., 1976, Gravity and magnetic studies of the Geysers–Clear Lake geothermal region, California, USA: Proc. Sec., U.N. Symp., Dev. Use Geotherm. Resources, U.S. Gov. Print. Off., 1065–1073.

Kane, M. F., Mabey, D. R., and Brace, R., 1976, A gravity and magnetic investigation of the Long Valley caldera, Mono County, California: J. Geophys. Res., **81**, 754–762.

Kane, M., Webring, M., and Bhattacharyya, B., 1981, A preliminary analysis of gravity and aeromagnetic surveys of the Timber Mountain area, Southern Nevada: Open-File Rep. 81-189, U.S. Geol. Surv.

Kinoshita, W., 1965, A gravity survey of the Island of Hawaii: Pac. Sci., **19**, 339–340.

Kinoshita, W., and Okamura, R., 1965, A gravity survey of the Island of Maui, Hawaii: Pac. Sci., **19**, 341–342.

Kohler, W., Healy, J., and Wegener, S., 1982, Upper crustal structure of the Mount Hood, Oregon, region as revealed by time-term analyses: J. Geophys. Res., **87**, 339–355.

Krivoy, H., 1965, A gravity survey of the Island of Niihau, Hawaii: Pac. Sci., **19**, 359–360.

Krivoy, H., Baker, M., and Moe, E., 1965, A reconnaissance gravity survey of the Island of Kauai, Hawaii: Pac. Sci., **19**, 350–358.

Krivoy, H., and Lane, M., 1965, A preliminary gravity survey of the Island of Lanai, Hawaii: Pac. Sci., **19**, 346–348.

Kroenke, L., and Woollard, G., 1965, Gravity investigations on the Leeward Islands of the Hawaiian Ridge and Johnston Island: Pac. Sci., **19**, 361–366.

MacDonald, G., 1965, Hawaiian calderas: Pac. Sci., **19**, 320–334.

Macheskey, L., 1965, Gravity relations in American Samoa and the Society Islands: Pac. Sci., **19**, 367–373.

McCall, G., LeMaitre, R., Malahoff, A., Robinson, G., and Stephenson, P., 1970, The geology and geophysics of the Ambrym caldera, New Hebrides: Bull. Volcanol., 681–696.

Medi, E., and Morelli, C., 1952, Rilievo gravimetrics della Sicilia [in Italian, with English summary]: Ann. Geofis., **5**, 209–245.

Moore, J., and Krivoy, H., 1965, A reconnaissance gravity survey of the Island of Molokai, Hawaii: Pac. Sci., **19**, 343–345.

Nettleton, L., 1939, Determination of density for reduction of gravimeter observations: Geophysics, **4**, 176–183.

Pitts, S., 1979, Interpretation of gravity measurements made in the Cascade Mountains and adjoining Basin and Range Province in central Oregon: M.S. thesis, Oreg. State Univ.

Plouff, D., and Isherwood, W., 1980, Aeromagnetic and gravity surveys in the Coso Range, California: J. Geophys. Res., **85**, 2491–2501.

Ramberg, I., 1976, Gravity interpretation of the Oslo Graben and associated igneous rocks: Bull. 325, Nor. Geol. Unders., 1–194.

Ratté, J., Gaskill, D., Eaton, G., Petterson, D., Stotelmeyer, R., and Meeves, H., 1979, Mineral resources of the Gila Primitive Area and Gila Wilderness, New Mexico: Bull. 1451, U.S. Geol. Surv.

Robertson, E. I., 1967a, Gravity effects of volcanic islands: N.Z. J. Geol. Geophys., **10**, 1466–1483.

——— 1967b, Gravity survey in the Cook Islands: N.Z. J. Geol. Geophys., **10**, 1484–1497.

——— 1970, Additional gravity surveys in the Cook Islands: N.Z. J. Geol. Geophys., **13**, 184–198.

Rowley, S. H., 1982, A geophysical investigation of Mt. Adams, Washington: Eos, **63**, 174.

Searle, R., 1970, Evidence from gravity anomalies for thinning of the lithosphere beneath the rift valley in Kenya: Geophys. J. R. Astron. Soc., **21**, 13–31.

——— 1971, A gravity survey of Ngorongoro caldera, Tanzania: Bull. Volcanol., **34**, 1–8.

Sissons, B. A., 1981, Densities determined from surface and subsurface gravity measurements: Geophysics, **46**, 1568–1571.

Stanley, W., in press, Magnetotelluric survey of the Cascade volcanoes region, Pacific Northwest: SEG.

Steinberg, G., and Zubin, M., 1965, Geological structure of the Avachinsky group of volcanoes according to the geophysical data: Bull. Volcanol., ser. 2, **28**, 1–8.

Strange, W., Woollard, G. P., and Rose, J., 1965, An analysis of the gravity field over the Hawaiian Islands in terms of crustal structures: Pac. Sci., **19**, 381–389.

Vajk, R., 1956, Bouguer corrections with varying surface density: Geophysics, **21**, 1004–1020.

Watts, A. B., and Cochran, J. R., 1974, Gravity anomalies in the Galapagos Islands area: Science, **184**, 808–809.

White, C., 1980, Geology and geochemistry of Mt. Hood volcano: Spec. Pap. 8, Oreg. Dep. Geol. Mineral Ind.

Williams, D. L., Hull, D. A., Ackermann, H. D., and Beeson, M. H., 1982, The Mt. Hood region: Volcanic history, structure and geothermal energy potential: J. Geophys. Res., **87**, 2767–2781.

Williams, H., and McBirney, A., 1979, Volcanology: Freeman, Cooper & Co.

Woollard, G., 1951, A gravity reconnaissance of the Island of Oahu: Trans. Am. Geophys. Union, **32**, 358–368.

Yokoyama, I., 1958, Gravity survey on Kuttyaro caldera lake: J. Phys. Earth, **6**, 75–79.

——— 1961, Gravity survey on the Aira caldera, Kyusyu, Japan: Nature, **191**, 966–967.

——— 1963a, Structure of caldera and gravity anomaly: Bull. Volcanol., **26**, 67–72.

——— 1963b, Gravity survey on the Aso caldera: Geophys. Pap. Dedicated to Prof. K. Sassa, 687–692.

——— 1967, Gravity anomaly on Kuttara caldera lake in Hokkaido: Geophys. Bull., Hokkaido Univ., **17**, 23–31. [In Japanese with English abstract.]

——— 1969, The subsurface structure of Oosima Volcano, Izu: J. Phys. Earth, **17**, 55–68.

——— 1971, Gravimetric, magnetic and electrical methods, in The surveillance and prediction of volcanic activity: UNESCO, 75–101.

Yokoyama, I., and Aota, M., 1965, Geophysical studies on Sikotu Caldera, Hokkaido, Japan: J. Fac. Sci., Hokkaido Univ., **2**, 103–122.

Yokoyama, I., and Bonasia, V., 1978, Gravity anomalies on the Thera Island: Thera and the Aegean World J., 144–150.

Yokoyama, I., and Hadikusumo, D., 1969, A gravity survey on the Krakatau Islands, Indonesia: Bull. Earthquake Res. Inst., **47**, 991–1001.

Yokoyama, I., and Suparto, S., 1970, A gravity study on and around Batur Caldera, Bali: Bull. Earthquake Res. Inst., **48**, 317–329.

Yokoyama, I., and Tajima, H., 1960, A gravity survey on Volcano Huzi, Japan, by means of a Worden gravimeter: Geofis. Pura Appl., **45**, 1–12.

Zucca, J. J., Catchings, R. D., Fries, G. S., and Mooney, W. D., 1981, Preliminary report of a seismic refraction study in the Mt. Shasta region of Northern California [abstr.]: Eos, **62**, 1089.

Zucca, J. J., Hill, D. P., and Kovach, R. L., 1982, Crustal structure of Maura Lava volcano, Hawaii, from seismic refraction and gravity data: Seis. Soc. Am. Bull., **72**, 1535–1550.

Interpretation of Precambrian geology in Minnesota using low-altitude, high-resolution aeromagnetic data

Val W. Chandler*

ABSTRACT

High-resolution aeromagnetic data recently acquired over northeastern and east-central Minnesota are useful in interpreting the Archean and Proterozoic bedrock beneath a widespread mantle of Pleistocene glacial materials. The new aeromagnetic data, gathered at a 400-m line spacing and a 150-m ground clearance, allow a much improved structural interpretation of many Archean and Proterozoic features, including the Archean migmatitic rocks of the Vermilion Granitic Complex, the Vermilion fault zone, the Keweenawan (Middle Proterozoic) lavas on the St. Croix horst, and the Keweenawan Duluth Complex. The data have delineated two poorly exposed dike swarms, one of Early Proterozoic age striking northwest in north-central Minnesota, and another of Keweenawan affinity striking northeast in east-central Minnesota. Several prominent anomaly lineaments occur over the region and may indicate previously unknown faults. Extremely weak but spatially coherent maxima define curvilinear and dendritic patterns over the subdued anomaly signature of the Lower Proterozoic Animikie basin and reflect near-surface sources, perhaps within the overlying Cretaceous or Pleistocene (glacial) materials.

Detailed interpretation using frequency-domain filtering was conducted on data grids from northern (Hibbing area), east-central (Carlton County area), and northeastern (Gabbro Lake area) Minnesota, and a variety of magnetic interpretational problems were addressed. High-pass filtering in the Hibbing area revealed internal structures within a large Archean batholith and also defined several Lower Proterozoic dikes. Strike-sensitive filtering over the Lower Proterozoic Biwabik Iron Formation in the Hibbing area was useful in separating stratigraphy-related anomaly components along strike from components associated with crosscutting structures containing natural ore zones. Calculation of the second vertical derivative over the Carlton County aeromagnetic grid was useful in tracing and determining the magnetic polarity of Keweenawan dikes. The second-derivative data also enhanced anomalies that may be related to large-scale folds in Lower Proterozoic rocks, and possibly in some cases to sources in the overlying younger materials. Reduction to the pole of the Gabbro Lake aeromagnetic data was effective in eliminating the effect of a strong remanent component, which is characteristic of many Keweenawan igneous rocks, and significantly improved the correspondence of anomaly patterns to known geology. The results observed over the Gabbro Lake area demonstrate the potential of this method for investigating the Keweenawan terranes of the Lake Superior region.

Although much work remains to be done, the preliminary results of this study demonstrate the great utility and geologic resolution of the data and should provide some general guidelines for future studies using filtering–enhancement techniques.

INTRODUCTION

Minnesota is underlain by a complex and unique assemblage of Precambrian rocks that are of considerable scientific

*Minnesota Geological Survey, University of Minnesota, 2642 University Avenue, St. Paul, MN.

and economic importance. The northern two-thirds of the state is underlain by 2 700-m.y.-old rocks of the Superior province and is characterized by belts of granitic rocks and greenschist-facies metavolcanic and metasedimentary rocks (units Ag, Amv, and Amg, respectively, in Figure 1). A markedly different terrane underlies the southern part of the state, which is characterized by upper amphibolite- to gran-

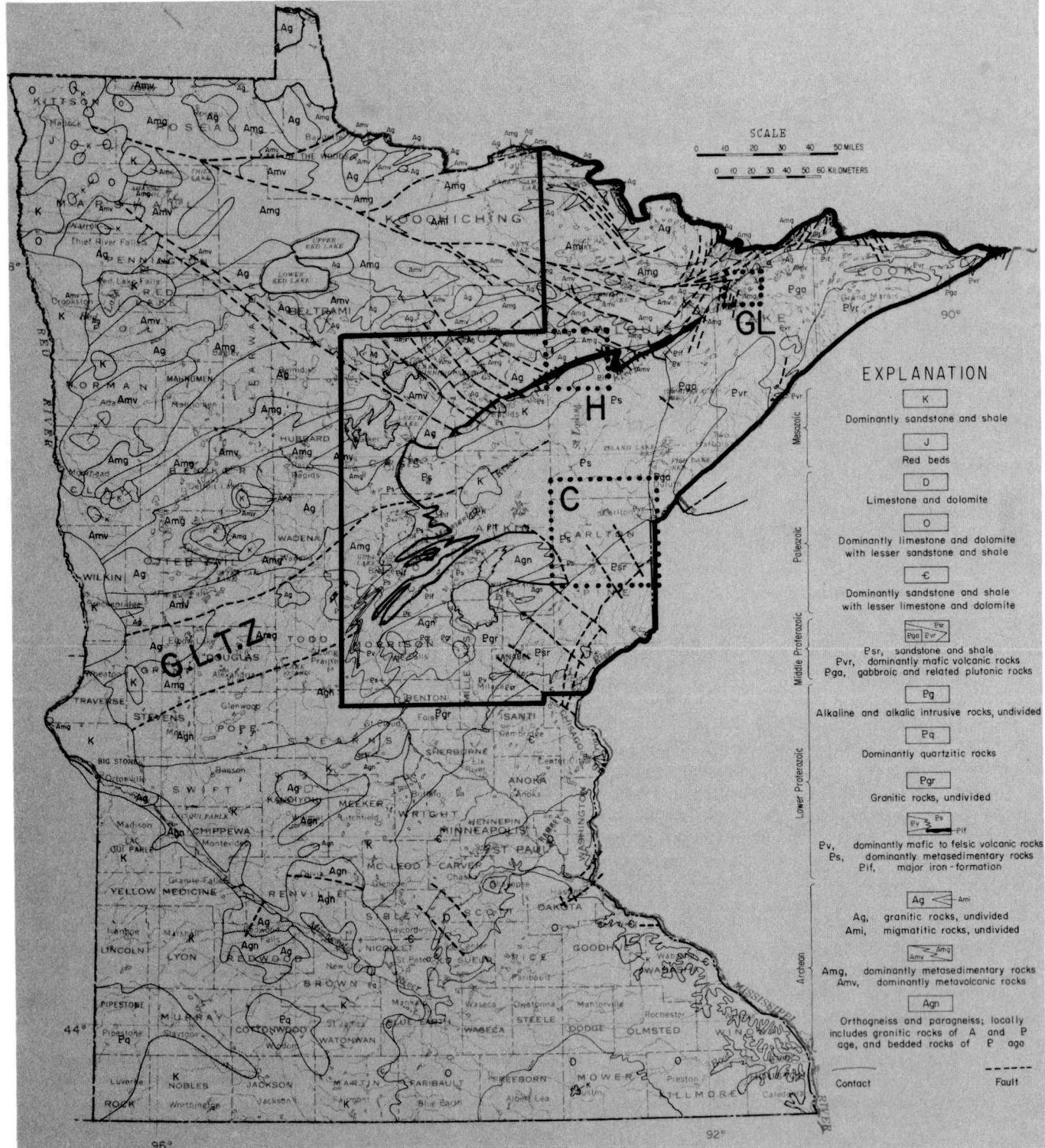

FIG. 1. Generalized bedrock geologic map of Minnesota (modified from Morey, 1976). Northeastern area covered by detailed surveying outlined by heavy line. Dotted lines delineate Hibbing (H), Carlton County (C), and Gabbro Lake (GL) study areas. Great Lakes tectonic zone is indicated by bold initials.

ulite-facies gneisses (unit Agn in Figure 1) that have components older than 3 500 m.y. The Great Lakes tectonic zone, a major crustal discontinuity described by Sims et al. (1980), forms the boundary between the differing Archean terranes (Figure 1). In Early Proterozoic time the zone was the locus of extensive deposition and subsequent orogeny, as evidenced by the variably deformed metavolcanic, metasedimentary, and plutonic rocks (units Pv, Ps, and Pgr) of that age in east-central Minnesota. Keweenawan (Middle Proterozoic) rocks associated with the Midcontinent rift system underlie northeastern and part of east-central Minnesota; they consist of predominantly mafic lavas, mafic intrusions, and terrestrial clastic rocks (units Pvr, Pga, and Psr in Figure 1). In total, the Precambrian rocks of Minnesota allow a unique opportunity for study of the early crustal evolution of the Midcontinent. They also contain deposits of iron, copper, and nickel and have further potential for undiscovered mineral deposits.

What we can directly learn about these Precambrian rocks, however, is severely restricted by a widespread mantle of Pleistocene glacial materials (Figure 2). Outcrop control adequate for geologic mapping is restricted to the extreme northeastern part of the state and a few small, widely scattered areas elsewhere. Furthermore, drill holes to bedrock with reliable lithologic data are unevenly distributed and are absent over many large areas. Consequently, geophysical data have commonly supplemented geologic studies in Minnesota. Because the Precambrian rocks of the area vary greatly as to density and magnetization, gravity and magnetic methods have been especially useful. Indeed, the geologic map in Figure 1 was compiled in large part from a qualitative analysis of regional-scale gravity and aeromagnetic data.

Accelerated geophysical activity in Minnesota occurred immediately following World War II, primarily in response to the advent of airborne geophysical methods. The most significant aeromagnetic surveying to be carried out in the public sphere was the statewide mapping program of the U.S. Geological Survey (summarized by Zietz and Kirby, 1970), which was flown for the most part at a 1-mile line spacing and a terrain clearance of 1 000 ft (305 m). Although this surveying was originally designed to locate major iron formations, the resulting data showed a significant correspondence to general bedrock type and have been of tremendous value in regional-scale interpretation (Sims, 1972a, b). These data, however, are of limited use in detailed geologic mapping, and a digital base is not available for the computer-based processing and interpretive schemes that have been developed over the last 20 years.

In 1979 the Minnesota Geological Survey (MGS), funded by the state Legislative Commission on Minnesota Resources (LCMR), began a long-range program to acquire low-altitude, high-resolution aeromagnetic coverage over the state and to make the resulting analog and digital data available to the public. The objectives of this program are (1) to further the MGS mission in detailed mapping of Precambrian bedrock, (2) to provide data to assist in mineral evaluation and regulation of state-owned lands, (3) to provide academic institutions with high-quality geophysical data for studies on Precambrian geology, and (4) to encourage mineral exploration in the state by providing private companies with detailed geophysical data at low cost.

The new aeromagnetic surveying has specifications similar to those used in mineral exploration but differs greatly in areal scope. Flight lines were flown north–south at a 400-m spacing and a mean terrain clearance of 150 m. Tie-lines were flown at either 2-km or 4-km intervals. GeoMetrics model G803 and G813 proton precession magnetometers with a stinger mount were used with doppler-based sampling set at either 50 or 75 m. Note that the term "high resolution" used in this paper refers to tight data control (close flight lines at low altitude) and not to the use of optically pumped magnetometers (often referred to as high-resolution magnetometers). Magnetometer sensitivities varied from 0.1 to 0.50 gamma (1 gamma = 1 nT), depending on the magnetometer used and the sampling interval selected. Flying was restricted to days when the temporal variations caused less than a 3-gamma departure over a 5-minute chord.

The data were reduced by standard procedures and were delivered in several digital and analog forms. Regional field removal for all data used the updated 1975 American World Charts geomagnetic field (Peddie and Fabiano, 1976), which was slightly favored over the 1975 International Geomagnetic Reference Field Model (IGRF) based on a discussion by Cain (1979). The 1980 IGRF was not available at the start of the program. Digital data consisted of tapes of all recovered flight-path data, including all appropriate locational and temporal information, and tapes of gridded data used in contouring. Analog data included 1:24 000-scale magnetic-intensity contour maps (10-gamma minimum interval) with flight-path recovery, and 1:250 000-scale summary contour maps (minimum interval, 50 gammas). Color-coded anomaly-intensity maps at a 1:250 000 scale have recently been added to the list of available products (Chandler, 1983a, b, c, d).

More than 174 000 line-km of data have been gathered over large tracts of northeastern and east-central Minnesota (Figure 1), and plans are continuing toward statewide coverage. When completed, the survey will be one of the largest aeromagnetic databases of its kind in the world and should be useful for geologic studies in Minnesota for many decades.

This paper examines the new aeromagnetic data in two sections. The first section gives a cursory examination of the data acquired to date and discusses the geologic implications of some of the more prominent anomalies. The second section examines in detail the aeromagnetic data over three small test areas—the Hibbing, Carlton County, and Gabbro Lake areas (Figure 1)—and demonstrates how a wide variety of interpretational problems can be approached by computer-based filtering and graphics.

GENERAL DISCUSSION OF ANOMALY DATA

The great wealth of geologic information contained within the new aeromagnetic data is evident from Figure 3, which shows all data acquired to date as a shaded relief map. Comments will be restricted to some prominent anomalous features and will emphasize those that delineate geologic features that previously were unknown or poorly defined.

Extending eastward from bracket I in Figure 3 are belts of

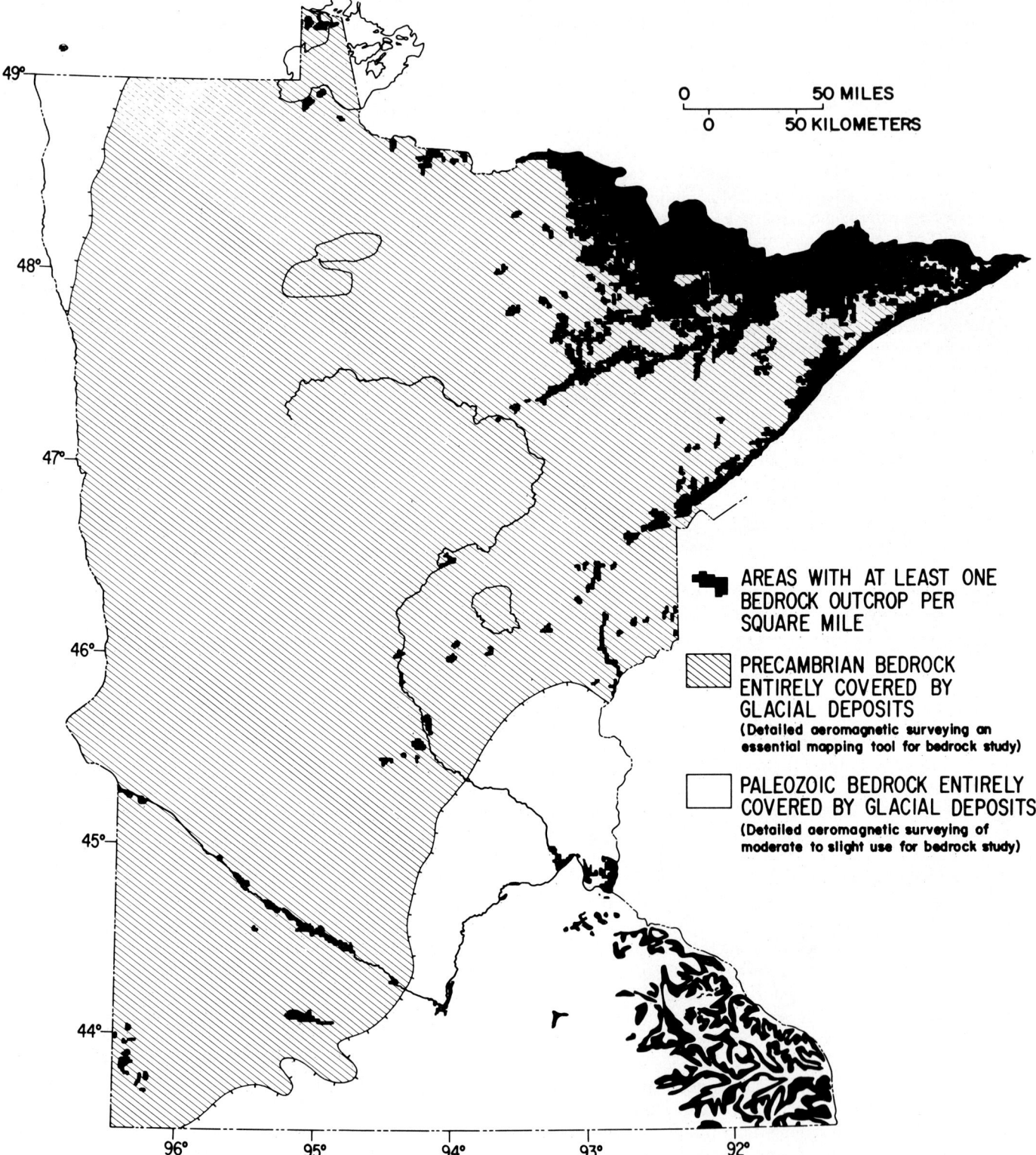

FIG. 2. Generalized bedrock outcrop map of Minnesota (modified from Morey, 1981).

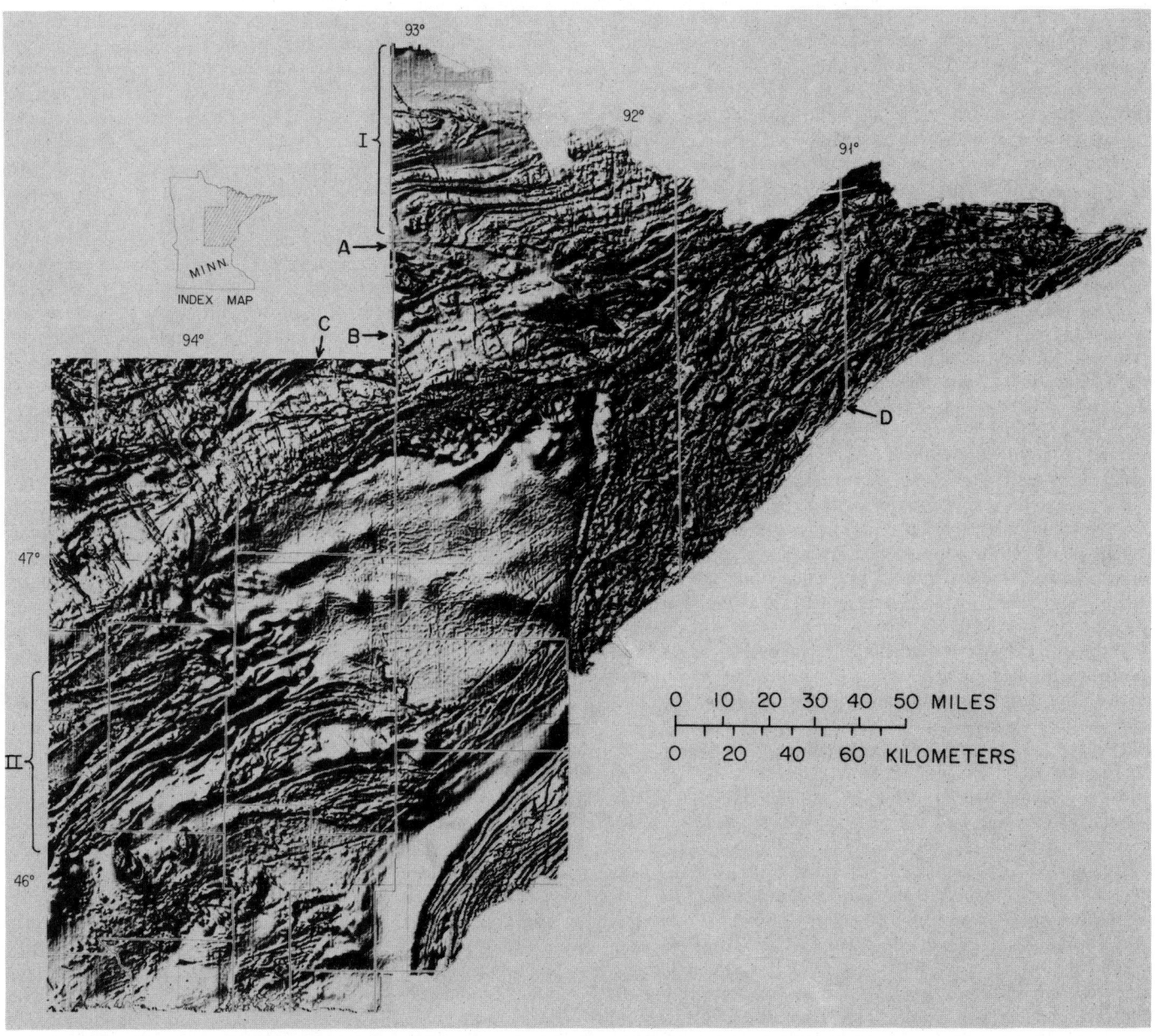

FIG. 3. Computer-generated shaded-relief aeromagnetic map of northeastern and east-central Minnesota. The hypothetical illumination is from the northwest at an inclination of 45 degrees. White lines represent county boundaries.

intermediate-intensity (100–300 gamma) magnetic maxima that define broad east–west fold patterns within Archean rocks of the Vermilion Granitic Complex (units Ami and Ag in Figure 1). This broad folding was already known from reconnaissance mapping (Southwick and Ojakangas, 1979) and structural studies (Bauer, 1981) in the western and central parts of the complex, where the rocks are stratiform migmatites. Eastward, the layering in the migmatite becomes weaker and more nebulitic until, as seen in the field, an area of hundreds of square miles appears to be essentially massive granite; but the stratiform magnetic signature typical of migmatitic rocks extends eastward across the entire terrane. The eastward continuation of the anomaly banding across this area of massive granite (the Lac La Croix Granite of Southwick and Ojakangas, 1979; unit Ag near the international boundary at 92° W in Figure 1) implies that this pluton retains within it the signature of a layered structure, probably owing to small variations in magnetite content, while the silicate-mineral assemblage has been virtually homogenized. North–south anomaly lineaments occur over the Lac La Croix Granite, and minor offsets along the east–west anomaly banding suggest faulting. This discussion demonstrates how the new aeromagnetic data can provide new insights and pose new questions, even within reasonably well-exposed terranes.

The structures of the Vermilion Granitic Complex are truncated to the south by the Vermilion fault zone (Sims et al., 1970), which appears in Figure 3 as a series of sharp anomaly breaks extending east-southeast from arrow A. South of this fault zone are belts of Archean metavolcanic, metasedimentary, and granitic rocks (units Amv, Amg, and Ag, respectively, in Figure 1). The granitic rocks are generally associated with a moderate intensity (100–500 gamma) anomaly expression, whereas the bulk of the metasedimentary and metavolcanic belts are nonmagnetic. Localized iron formations within the metavolcanic rocks, however, can give rise to narrow, curvilinear maxima with extreme amplitudes (>5 000 gammas). A magnetic lineament extending for almost 70 km eastward from B in Figure 3 appears to be associated with a belt of small dioritic and lamprophyric plutons (Sims et al., 1970), and the change in magnetic-anomaly character across this lineament implies possible faulting. No fault had been mapped previously (Figure 1; Sims et al., 1970), but outcrop control along this feature is generally poor, and the area should be reinvestigated. Another possible fault is indicated by the anomaly lineament extending southwest from arrow C (Figure 3). This lineament trends into a mapped fault to the northeast and trends southwestward toward the north boundary of the Great Lakes tectonic zone (Sims et al., 1980; Southwick and Chandler, 1983). Moreover, linear, northeast-striking anomalies between arrows A and B in Figure 3 indicate that this structure may also extend as far northeast as the Vermilion fault zone.

A major dike swarm that cuts across the Archean rocks of northern Minnesota appears in Figure 3 as a prominent series of northwest-striking, linear maxima that have generally moderate amplitudes (100–500 gammas). Prior to the surveying, this swarm was poorly known from widely scattered outcrops in the north-central and northeastern parts of the state (Sims and Mudrey, 1972). The dikes are Early Proterozoic in age and are generally 10 to 70 m in width (Southwick and Day, 1983).

In Figure 3 the magnetic expression of the Great Lakes tectonic zone, which extends northeastward from bracket II, is obscured and dominated by a cover of Lower Proterozoic metasedimentary rocks in the Animikie basin, the large area of unit Ps in Figure 1. The high-amplitude (>1 000-gamma) linear anomalies that extend northeastward from the south half of bracket II in Figure 3 correspond to folds in the iron-formation-bearing Lower Proterozoic rocks of the Cuyuna iron range. The Biwabik Iron Formation crops out near the base of the Animikie sequence along the northern rim of the basin and, where dominated by oxide-rich facies, corresponds to a belt of high-amplitude (>1 000-gamma) anomalies. The anomaly expression of the Biwabik Iron Formation is discussed further in the section on the Hibbing area. Most of the basin, however, is characterized by a subdued anomaly signature, reflecting a thick fill of nonmagnetic slate and graywacke. Magnetic-depth estimates over the central part of the basin (Chandler, 1982) indicate that this nonmagnetic section may be more than 3 km thick. Weak (10–20 gamma) variations over the central parts of the Animikie basin form narrow, east-trending dendritic patterns that apparently arise from near-surface sources, although the patterns generally show little correspondence to present topography or drainage. All glacial advances in the Pleistocene passed over nearby, oxide-rich rocks, either to the north (Biwabik Iron Formation) or to the northeast (Duluth Complex), and any channel deposits within the till therefore may contain heavy-mineral concentrates of magnetic iron oxides. It is also possible that the weak anomalies reflect Cretaceous channel deposits in a westward-flowing drainage system developed on a Cretaceous peneplain, because near-shore-marine rocks of that age are known over the northern and western Animikie basin (Austin, 1972). Clearly, the problem of these weak anomalies should be investigated further.

The poorly exposed terrane south of the Great Lakes tectonic zone and the Animikie basin is associated with highly varied magnetic anomalies and may prove to be one of the more fruitful areas for future aeromagnetic interpretation. The terrane consists of Archean gneissic rocks that have infolds of Lower Proterozoic supracrustal rocks, all of which have been intruded by mafic to felsic plutons of Early Proterozoic age (Morey et al., 1981). The extremely complicated east-trending maxima patterns (200–1 000–gamma amplitude) that lie immediately west and south of 46°30′ N and 93° W in Figure 3 correspond to Lower Proterozoic metavolcanic and metasedimentary rocks that are complexly interfolded with Archean gneissic rocks. Broad, irregular maxima (300–1 000–gamma amplitude) are associated with Lower Proterozoic plutons of granodioritic composition, whereas the numerous small, circular maxima (100–300 gamma amplitude) appear, from limited outcrops and drilling data, to be related to dioritic to gabbroic stocks of similar age (Morey et al., 1981; D. L. Southwick, pers. comm.). Relatively subdued magnetic signatures occur over widely scattered exposures of Lower Proterozoic metasedimentary and granitic rocks and over migmatized rocks of undetermined age (Morey et al., 1981).

Keweenawan (Middle Proterozoic) rocks associated with the Midcontinent rift system produce a complex and high-amplitude (>1 000-gamma) anomaly expression in the extreme southeastern and northeastern parts of Figure 3. The belt of narrow, northeast-trending maxima in the southeast corner of Figure 3 is caused by faulted and tilted basalt flows along an uplifted horst, fault bounded on the northwest against a wedge of nonmagnetic clastic rocks. The flow-related anomalies reveal many structural details that cannot be derived from the sparse outcrops, and further anomaly studies should prove fruitful here. Northeast-trending linear anomalies (50–200 gamma amplitude) are caused by diabase dikes in older rocks away from the rift. The anomaly expression of these dikes is discussed further in the section on the Carlton County area. The extremely complicated and high-amplitude (500–1 000–gamma) anomaly expression over the northeastern prong of Figure 3 reflects the complex geometry and intrusive history of anorthositic, gabbroic, and troctolitic rocks in the Duluth Complex (unit Pga in Figure 1) and its roof of associated volcanic rocks. Attempts to unravel this highly complex anomaly signature are being made by the author and co-workers and will be reported in later papers. The complicating effect of remanent magnetization in Keweenawan rocks is discussed briefly below in the section on the Gabbro Lake area. A prominent anomaly lineament extends across the complex west-northwest from arrow D in Figure 3, arising from an unknown source. Anomaly truncations and/or offsets along its trace indicate that the lineament may reflect a major fault.

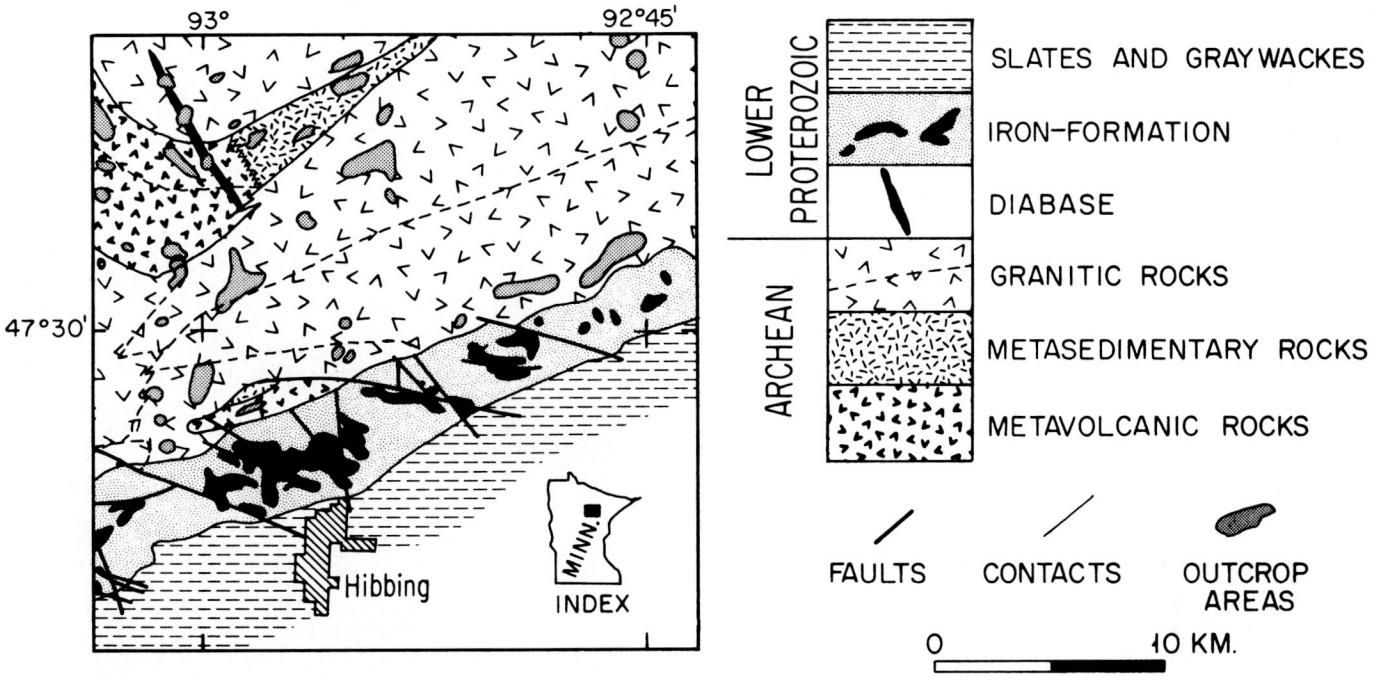

FIG. 4. Generalized bedrock geologic map of the Hibbing area (after Sims et al., 1970; Morey, 1972). Width of diabase dike has been exaggerated. Irregular black areas within the iron formation represent major natural ore deposits. Subunits within the granitic and metavolcanic rocks, as determined from composition, texture, and field relations (Sims and Viswanathan, 1972), are bounded by dashed lines.

DETAILED INVESTIGATIONS OF THE DATA

The Hibbing, Carlton County, and Gabbro Lake areas (Figure 1) were selected for detailed study because together they include a diverse sampling of rocks and illustrate a variety of interpretational problems common in northern Minnesota. The Hibbing and Carlton County areas have poor outcrop control and serve as good examples of how detailed interpretation of the aeromagnetic data can enhance geologic mapping. In contrast, the Gabbro Lake area, within the northwestern Duluth Complex, has extensive outcrop control and has already been mapped in considerable detail. The problem here, however, is the lack of close correspondence of the anomalies to the geology, apparently caused by a strong remanent magnetization oblique to the present geomagnetic field.

Frequency-domain filtering of the aeromagnetic data was investigated for all three test areas. The programs used were first described by Bowman et al. (1978) and later revised by Reed (1980) at Purdue University. The programs include wavelength filtering, strike-sensitive filtering, continuation, derivative calculation, and reduction of magnetic data to the pole. The reduction-to-the-pole program was modified to include source polarization oblique to the inducing field. A lag of about one-quarter of the square root of the total number of grid points yielded a suitable filter response and was used in all filtering runs. The results were presented on a DIANZA cathode-ray tube that has display options of color, gray tone, and shaded relief. This mode of presentation is faster, less expensive, and generally more revealing than standard contour-map presentation. This report will use gray tone and/or shaded relief, depending on which format presents the more effective display. Because of computer core limitations during processing, the grid sizes were limited to 128 by 128 data points. To provide a more continuous image display, the data grids were interpolated to a denser sampling using a cubic convolution with moving-window smoothing.

Hibbing area

Archean rocks that underlie the northern two-thirds of the Hibbing study area (Figure 4) include granitic rocks of the Giants Range batholith (Sims and Viswanathan, 1972) and lesser amounts of greenschist-facies metavolcanic and metasedimentary rocks. On the basis of composition, texture, and field relations, the granitic and metavolcanic units in the study area have been further divided into six and two subunits, respectively (Sims et al., 1970). The boundaries between subunits are shown by light dashed lines in Figure 4. The northwest-striking diabase dike in the northwestern part of Figure 4 is part of the previously mentioned Lower Proterozoic swarm that crosses north-central Minnesota; this particular dike is 40–70 m in width (Viswanathan and Ojakangas, 1974a, b). Field relations elsewhere (Southwick and Day, 1983) indicate that the Lower Proterozoic dike swarm is pre-Animikie in age. In the southern part of the study area the Archean rocks are unconformably overlain by the shallow-dipping rocks of the Lower Proterozoic Animikie Group. A thin basal quartzite is overlain by the Biwabik Iron Formation, which in this area is highly magnetic. The iron formation is overlain by nonmagnetic argillite, slate, and graywacke of the Virginia Formation (Figure 4). Unlike the Archean rocks, the Animikie Group is little metamorphosed or deformed except for minor folding and block faulting on reactivated Archean faults.

The aeromagnetic data, shown in gray tone (Figure 5a) and shaded relief (Figure 5b), delineate structural details not evident from the poor outcrop control, although interference among various anomaly components locally hampers interpretation. Over the Archean rocks, curvilinear, east- to northeast-trending anomalies reflect compositional variations within the granitic rocks and flanking belts of metamorphosed volcanic and sedimentary rocks. Localized maxima within the metavolcanic rocks may reflect stringers of iron formation. The diabase dike in the northwestern part of the area is associated with a subtle linear maximum, and similar north- to northwest-trending anomalies over much of the Archean terrane may have similar sources.

In light of the mild deformation and simple magnetic stratigraphy, consisting commonly of two magnetic members (Bath, 1962), the anomaly expression of the Biwabik Iron Formation is surprisingly complicated. Much of the complexity results from crosscutting faults and folds, which, in spite of their minor displacement, played a major role in localizing highly oxidized and essentially nonmagnetic natural ores (White, 1954; Figure 4). Additional complexity may result from Archean sources—most notably iron formation in Archean metavolcanic rocks—beneath the Animikie Group. The small area of Archean metavolcanic rocks just north of the Proterozoic rocks in Figure 4 lies along an east-striking belt of locally intense magnetic maxima that cuts across the expression of the Biwabik Iron Formation (Figures 5a, b). This belt of magnetic maxima in turn lies along the nose of a broad gravity maximum that extends across much of the Animikie basin (Figure 6) and resembles the gravity-anomaly expression of Archean metavolcanic belts described elsewhere in Minnesota (Sims, 1972a). Thus, the magnetic-anomaly pattern observed in the Hibbing area represents a jumble of sources that vary in age as well as in size, depth, and strike direction.

To help unravel the magnetic expression in the Hibbing area, high-pass and strike-sensitive filtering were applied. Figure 7 shows a shaded-relief presentation of the magnetic data following 2.5-km high-pass filtering, which attenuated the interference from broad and deep anomaly sources. The detailed anomaly expressions of the Biwabik Iron Formation and, to a lesser degree, of the Archean rocks, have been clarified. The east- to northeast-trending-anomaly banding over the granitic rocks corresponds closely with foliations observed in the field (Viswanathan and Ojakangas, 1974a, b). The expression of the two magnetic units of the Biwabik Iron Formation has been enhanced over the eastern part of the area, and the effect of several crosscutting structures has also been significantly enhanced.

The magnetic-anomaly expression of the Biwabik Iron Formation was further investigated by strike-sensitive filtering, which can isolate the anomaly components related to along-strike stratigraphy from those related to the crosscutting structures. Shown in Figure 8 are the 2.5-km high-passed data, following a N 65° E strike pass, using a ±30° wedge. The high-pass filter was included in the analysis to

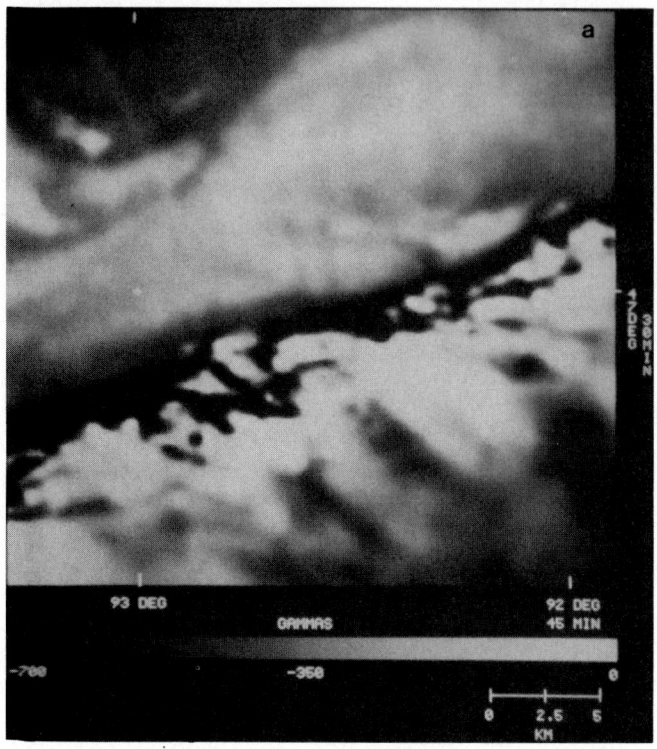

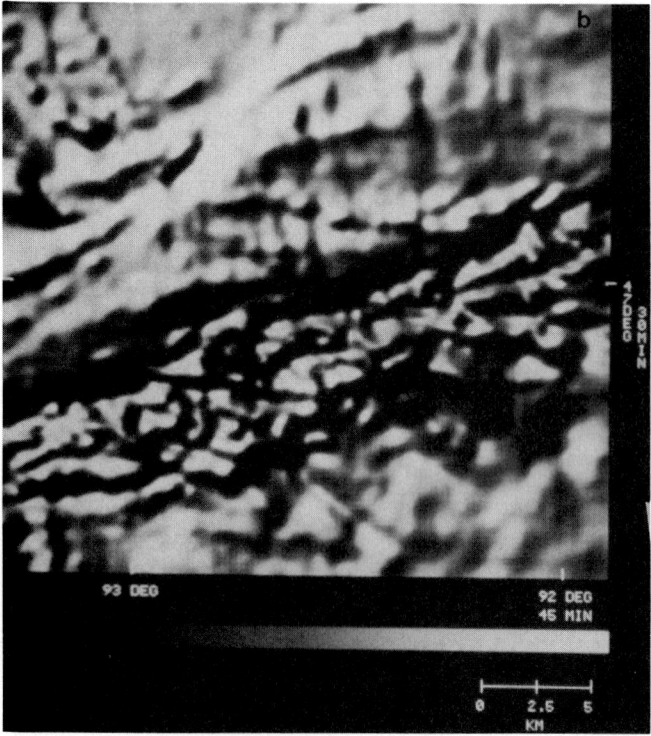

Fig. 5. Total-magnetic-intensity anomaly data over the Hibbing area. a = gray-tone image; b = shaded-relief image with illumination from the northwest at 45-degree inclination. Data based on a 110 by 110 grid using a spacing of 244 m.

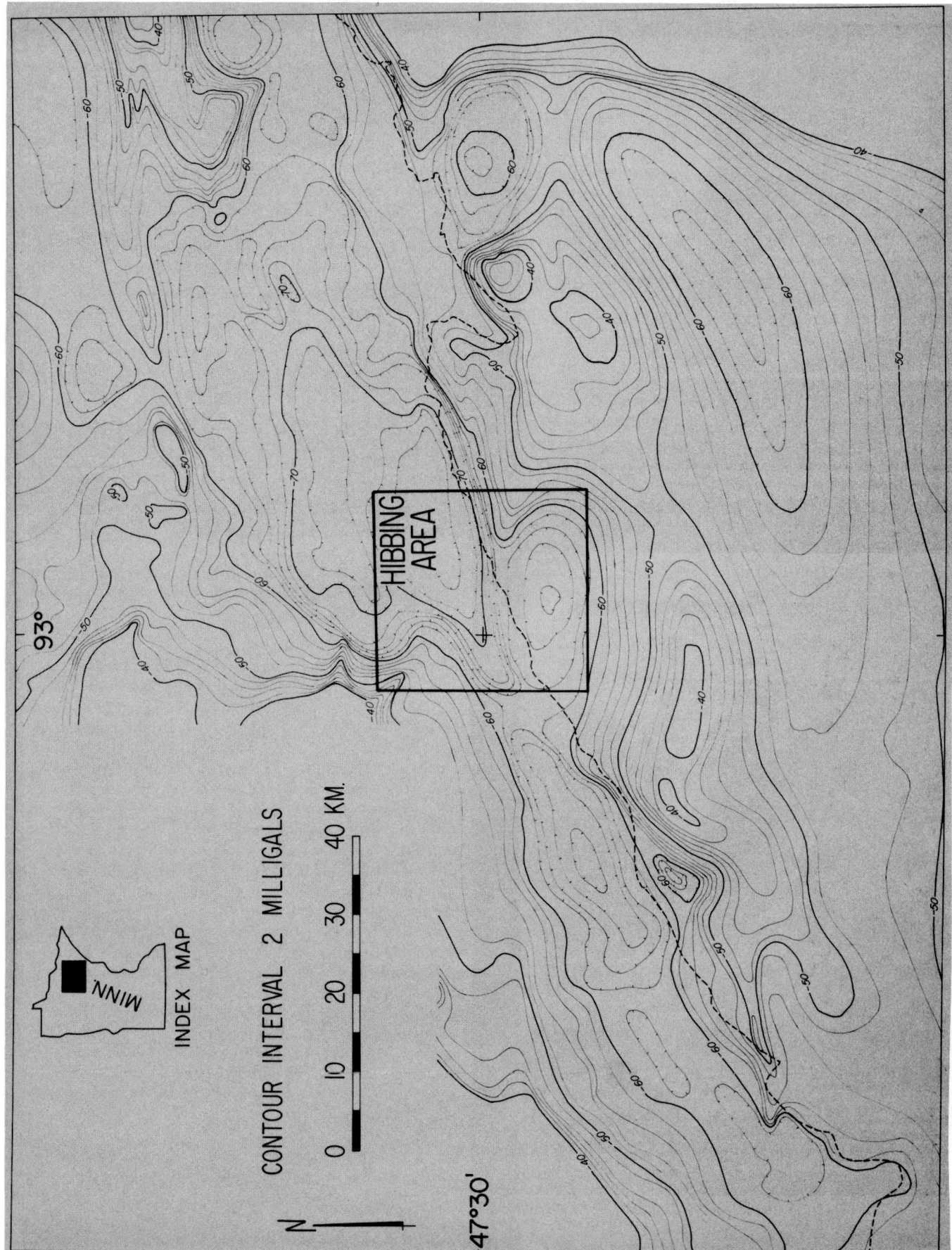

FIG. 6. Bouguer gravity-anomaly map of the northeastern Animikie basin (after Ikola, 1968). Northern edge of Animikie rocks is shown as a dashed line.

attenuate interference from longer wavelength sources beneath the Animikie rocks. In Figure 8 the two magnetic horizons in the Biwabik Iron Formation can be traced confidently across offsets to the central part of the area. The anomaly expression of the iron formation is more diffuse in the vicinity of Hibbing (Figure 4) and probably reflects nearly complete destruction of magnetite in unusually large "oxidized" ore zones of this area.

The effect of structures that cut across the Biwabik Iron Formation was investigated by a strike-reject version of the filter used above. Because these features may have some relationship to deeper structures in the Archean basement, the original anomaly data were used instead of the high-pass-filtered data. Most of the resulting anomalies over the iron formation (Figure 9) show either a west-northwest or northwest strike. Several of these crosscutting anomalies extend to or align with anomalies over the Archean rocks. The extensive "oxidized" ore zones near Hibbing are characterized by pronounced crosscutting minima about 5 km south and southeast of 47°30′ N and 93° W.

Shown in Figure 10 is a reinterpretation of the geology in the Hibbing area, using all of the filtered aeromagnetic data. Magnetic maxima imply that granitic rocks invade the belt of Archean metavolcanic rocks in the northwestern corner of the area. Localized stringers of iron formation have been interpreted within some of the Archean metavolcanic rocks and, in the central part of the area, have been interpreted to extend beneath the Lower Proterozoic Biwabik Iron Forma-

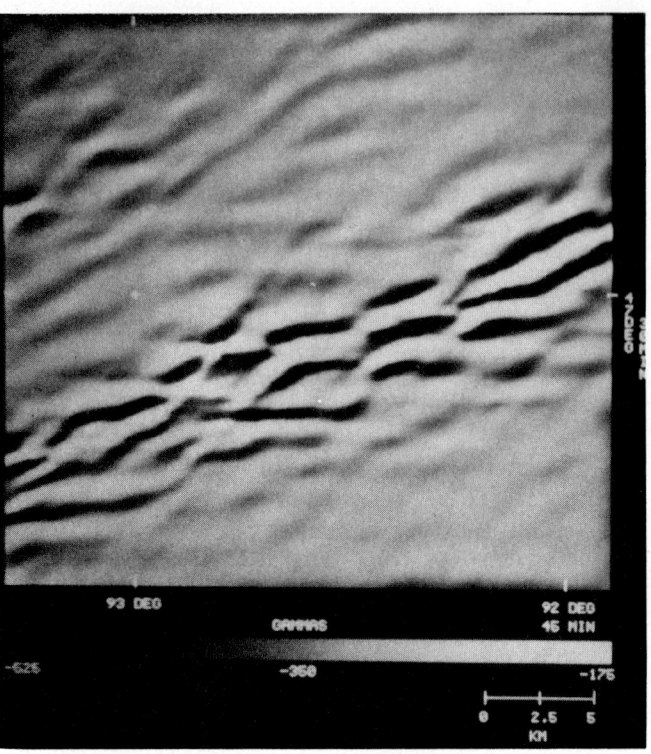

FIG. 8. The N 65° E strike-pass-filtered magnetic-anomaly data over the Hibbing area, gray-tone image.

tion. The structural fabric of the granitic rocks as defined by magnetic-anomaly patterns shows little correspondence to the individual granitic subunits outlined in Figure 4. In fact, any reinterpretation of the contacts of these internal units was difficult within the framework of the anomaly fabric. Perhaps the units defined by Sims et al. (1970) in this area do not actually represent large, spatially coherent plutons but instead represent narrow, intermixed bands within a somewhat migmatized body. Numerous north- to northwest-striking diabase dikes have been interpreted to cut discontinuously across all Archean rock units.

The mapped surface positions of the magnetic horizons within the Biwabik Iron Formation were based primarily on the strike-passed data (Figure 8) and, because of shallow southward dip of the iron formation, were placed along the steep northern flanks of the associated maxima. The positions of crosscutting, nonmagnetic zones in the Biwabik Iron Formation correspond generally to the natural ore zones (Figure 4), although significant exceptions occur in detail. Several northwest-trending structures have been traced from the Biwabik Iron Formation northward into the Archean terrane.

Carlton County area

Bedrock in the area roughly centered over Carlton County is poorly exposed (Figure 11) but is known to consist mostly of the Lower Proterozoic Thomson Formation, a sequence of slate, graywacke, and minor metavolcanic rocks, which is an Animikie Group unit correlated with the Virginia Forma-

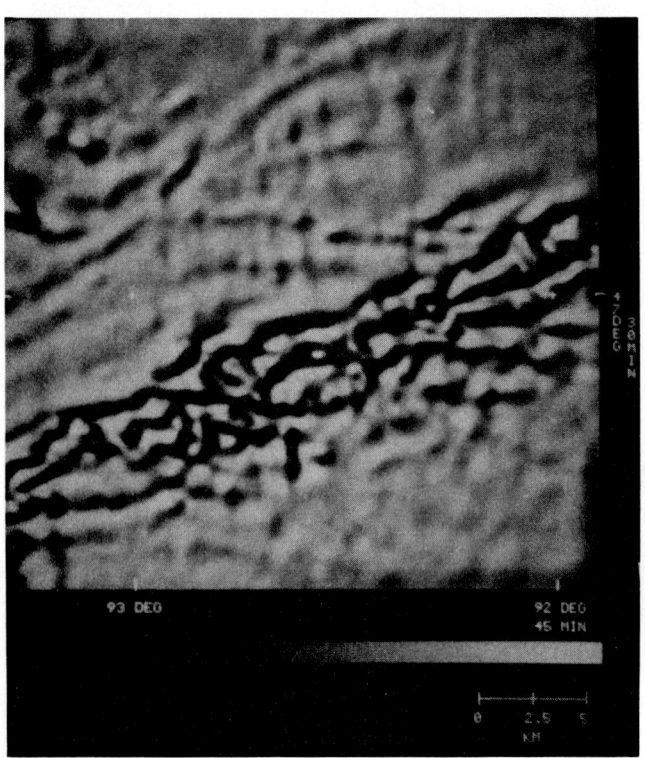

FIG. 7. The 2.5-km high-pass-filtered magnetic-anomaly data over the Hibbing area; shaded-relief image; illumination from northwest at 45-degree inclination. Maximum anomaly amplitudes are about 500 gammas.

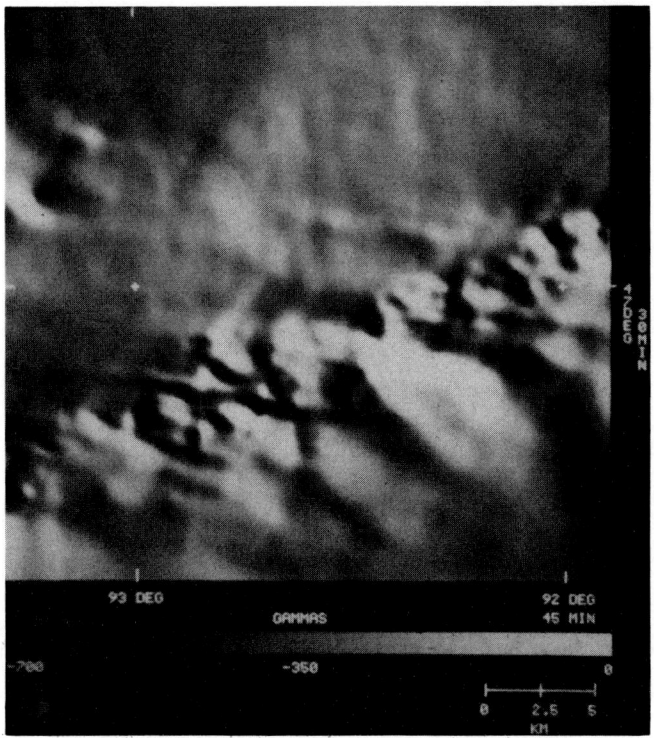

FIG. 9. The N 65° E strike-reject-filtered magnetic-anomaly data over the Hibbing area, gray-tone image.

tion in northern Minnesota (Morey, 1978). The slate and graywacke in the Carlton County area have been metamorphosed to at least greenschist facies and pervasively deformed about east-trending axes (Keighin et al., 1972). The southwestern part of the area includes pre-Animikie metasedimentary rock assigned to the Lower Proterozoic Mille Lacs Group (Morey et al., 1981). The southeastern corner of the area includes part of the Keweenawan Midcontinent rift system and includes part of the uplifted St. Croix horst of basalt and flanking wedge of clastic rocks. Numerous diabasic dikes believed to be related to the rifting are exposed west of the clastic rocks, but outcrop control is not adequate to define their regional distribution and extent.

The magnetic-anomaly data in Figures 12a and 12b show a subdued expression over much of the Thomson Formation, consistent with its low magnetic susceptibility (Mooney and Bleifuss, 1953). The broad anomaly across the north-central part of the area reflects a source either deep within or beneath the Thomson Formation, and graphical depth estimates from contoured data of the anomaly yield a depth to the top of the source of about 3 km. A few anomaly sources, however, do exist locally within the Thomson Formation, as evidenced by the irregular, east- to northeast-striking maxima of moderate amplitude (100 to 500 gammas) in the west-central and central parts of the area. Small amounts of magnetite were reported by Connolly (1981) for some phyllitic and metavolcanic units in the Thomson Formation. In addition, carbonaceous zones in the Thomson Formation locally contain pyrrhotite (G. B. Morey, pers. comm.).

Several of these belts define tightly curved arcs in Figure 12 and are inferred here to reflect folding.

The Keweenawan dikes give positive- and negative-anomaly signatures that imply normal and reversed polarities. Reversed and normal polarities are typical of Keweenawan rocks elsewhere (Halls and Pesonen, 1982), with the reversed polarity generally assumed to be the older. A large-amplitude (>4 000-gamma) minimum lies along a reversed dike in the northeastern corner of the area and reflects a small, reversely polarized gabbroic stock (Morey et al., 1981).

The second-vertical-derivative calculation most effectively enhanced the Carlton County data (Figure 13). A reinterpretation of the geology using the second-derivative data is shown in Figure 14. The distribution of the dikes and their individual polarities were clarified by the filtering. The surface positions of the normal dikes in Figure 14 were interpreted to lie along the steep western flank of linear maxima, because this is consistent with a polarization along the average normal direction of 290-degree declination and 40-degree inclination for the Keweenawan field (Halls and Pesonen, 1982). The second-vertical-derivative data indicate that reversed dikes tend to occur farther northwest, farther from the rift axis, than do normally magnetized dikes. The dikes are curvilinear locally and strike from N 20° E to about N 50° E, although a few reversed dikes appear to strike northwest. Detailed outcrop mapping of dikes east of 92°30′ (Wright et al., 1970) indicates that many of the dikes interpreted in Figure 14 as single large dikes may actually be tight, braided swarms of smaller dikes. The strongly reversed gabbroic body in the northeastern corner of the area is interpreted to lie at the intersection of two reversed dikes. The Keweenawan dikes and other magnetic rocks are buried to the southeast by an increasingly thick wedge of nonmagnetic sedimentary rocks. The edge of these sedimentary rocks is characterized by a slight loss in sharpness of many of the dike-related second-derivative anomalies (Figure 13), and this pattern has been used to revise the edge of the sedimentary rocks in Figure 14. The dike-related anomalies continue to decrease in sharpness toward the southeast, and finally, about 5 km closer to the rift, they abruptly attenuate. This latter change may indicate a fault-related escarpment beneath the clastic sequence.

The second-vertical-derivative data also enhance several features of the Lower Proterozoic rocks. The irregular belts of maxima over the west-central and central parts of the area display tightly convoluted patterns that probably reflect intricate folding. A major fold has been interpreted to extend across the central part of the area (Figure 14). Because older rocks are believed to lie westward along its axis, this large fold is interpreted as an anticline with a northeastward plunge.

Weak linear and curvilinear maxima over the northwestern and north-central parts of Figure 14 are enhanced by the second-derivative filter, and some of the better defined patterns are presented in Figure 14. As mentioned in the general discussion above, these subtle anomalies occur over much of the Animikie basin, and the nature of their sources is as yet unknown. Similar anomalies could cover many parts of the map area, but they may be visible only where anomalies from stronger sources do not obscure them. When

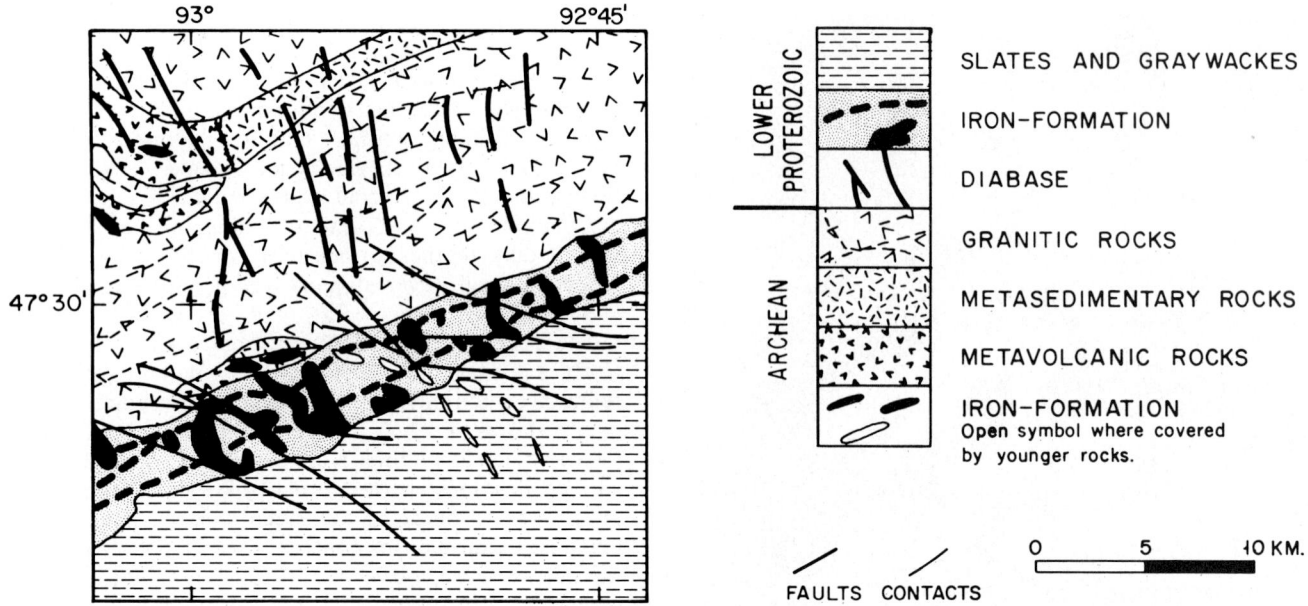

FIG. 10. Magnetic-interpretation map of the Hibbing area. Light dashed lines over granitic rocks follow axes of localized magnetic maxima. Heavy dashed lines are magnetic horizons in the Biwabik Iron Formation, whereas irregular black areas within the iron formation are parts of "oxidized" ore zones where there is nearly total conversion of magnetite to nonmagnetic minerals.

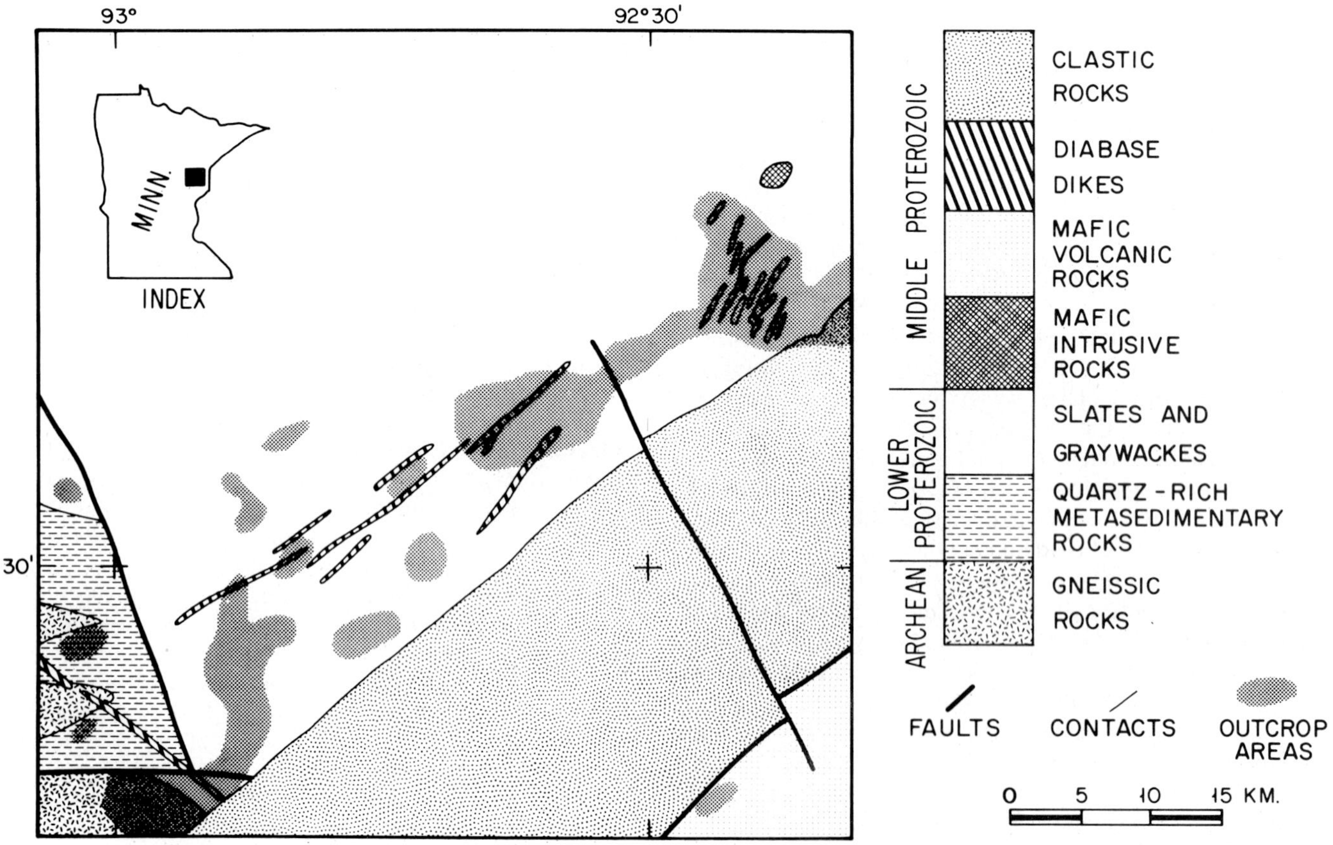

FIG. 11. Generalized bedrock geologic map of the Carlton County area (after Morey et al., 1981). Widths of the diabase dikes have been exaggerated.

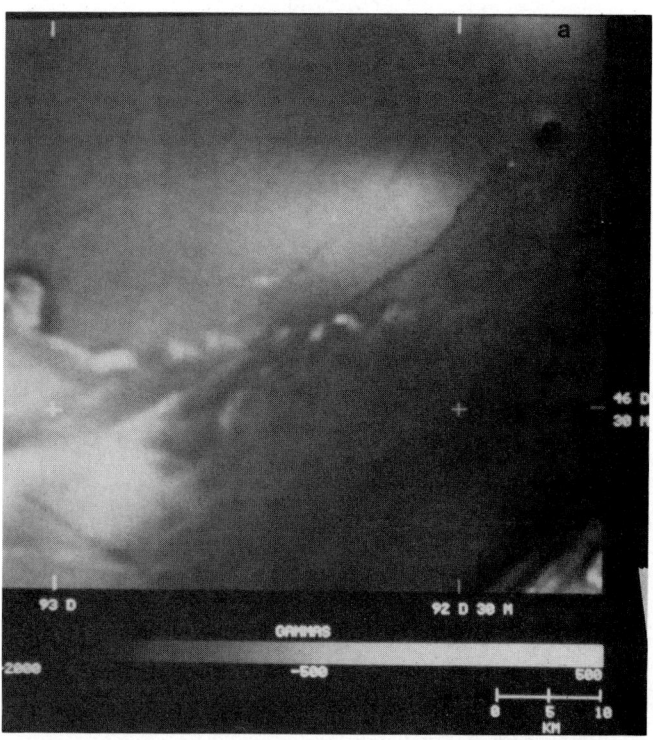

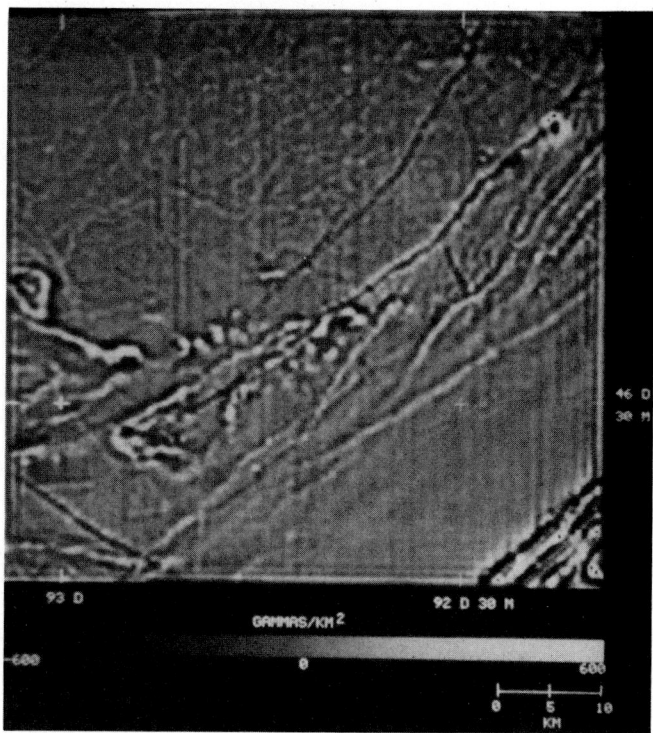

FIG. 13. Second vertical derivative of the magnetic-anomaly data over the Carlton County area; gray-tone image.

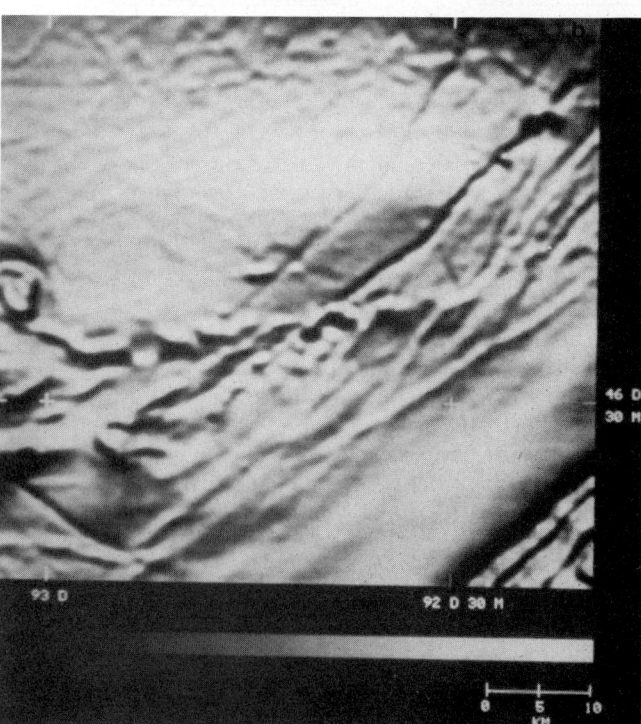

FIG. 12. Total-magnetic-intensity anomaly data over the Carlton County area. a = gray-tone image; b = shaded-relief image with illumination from north at 45-degree inclination. Data based on a 128 by 128 grid using a spacing of 457 m.

compared with the Quaternary geology (Hobbs and Goebel, 1982), a few of the weak anomalies in the northwestern corner of Figure 14 appear to correlate in part with some esker deposits. The vast bulk of these lineaments, however, shows little relationship to the Quaternary geology as mapped. It is conceivable that the lineaments represent magnetite-bearing channel deposits in pre-glacial or pre–Late Cretaceous drainage systems.

Gabbro Lake area

The Gabbro Lake area (Figure 1) lies within the Duluth Complex. Unlike much of the Duluth Complex, however, the Gabbro Lake area has generally good outcrop control and has been mapped (Figure 15) in detail (Weiblen, 1965; Green et al., 1966). The area includes the basal contact zone of the complex, which dips moderately to the southeast and rests primarily on Archean granitic rocks. However, small hornfelsic inclusions of slate and iron formation in the gabbroic rocks imply that rocks of the Animikie basin are involved in the contact zone. This area of the Duluth Complex includes rocks from both the anorthositic suite and the slightly younger troctolitic suite of Weiblen and Morey (1980). A large, elliptically shaped pluton with a troctolite ring surrounding a gabbro core, named the Bald Eagle intrusion (Weiblen, 1965), occupies the east-central and southeastern parts of the area.

The magnetic-anomaly expression of the Gabbro Lake area (Figure 16) corresponds poorly with known geology.

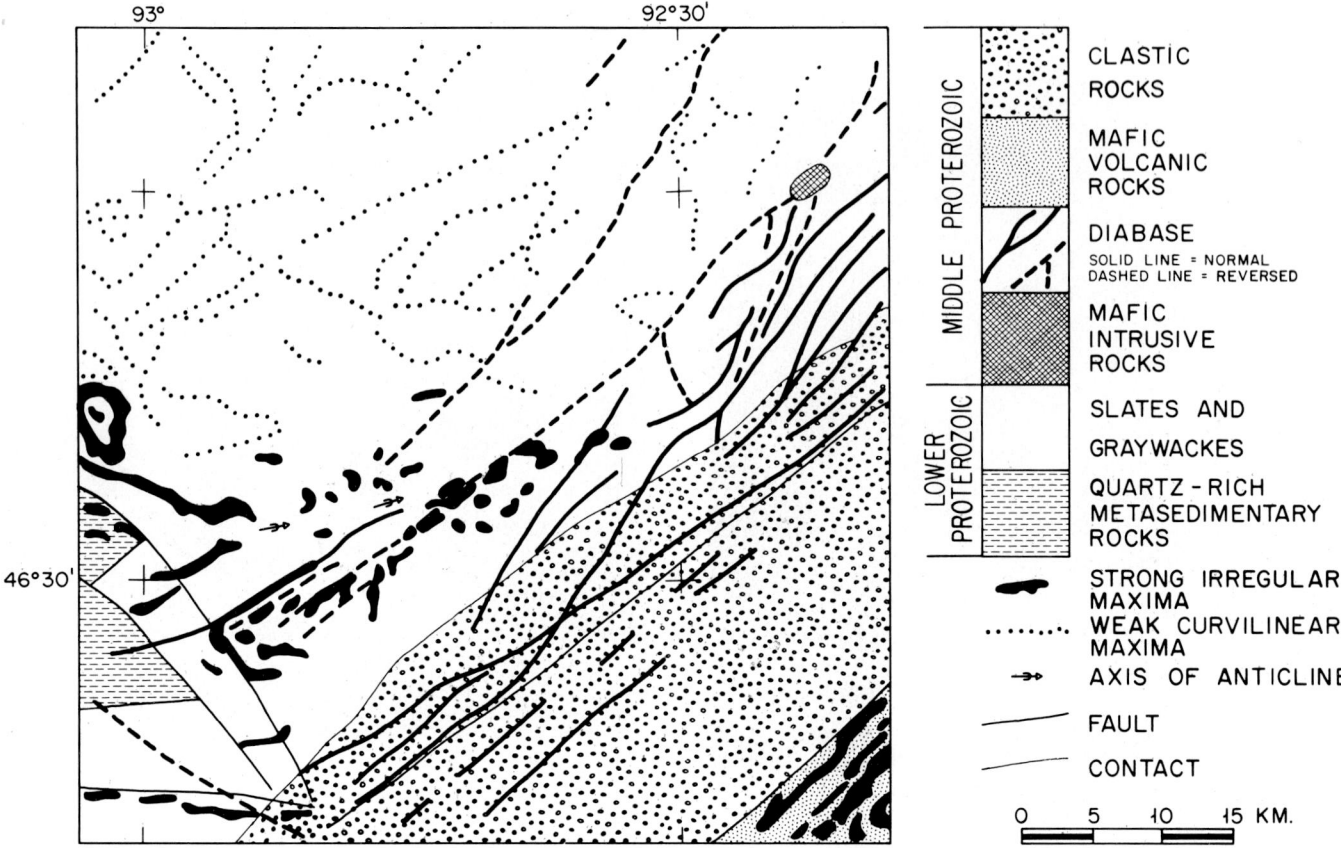

FIG. 14. Magnetic-interpretation map of the Carlton County area. Middle Proterozoic clastic rocks are presumed to bury the interpreted diabase dikes.

The basal contact zone generally corresponds to a southeastward decrease in anomaly values but has little diagnostic signature itself. Maxima do not occur directly over ironformation hornfels near the contact zone but appear to be shifted eastward. The contact between anorthositic and troctolitic rocks generally lacks an anomaly signature. The anomaly signatures over the Bald Eagle intrusion appear to be shifted toward the east. In short, the results here demonstrate the serious pitfalls of using total-magnetic-intensity anomaly data as a mapping tool in the less exposed parts of the Duluth Complex.

Structural and mineralogic attributes of the Duluth Complex, including shallow-dipping layers and a widely varying distribution of oxides within individual members, can certainly account for some of the perplexing magnetic-anomaly signatures. However, a far more important factor could be the contribution from remanent magnetization. Previous paleomagnetic studies in the Duluth Complex (Jahren, 1965; Beck, 1970; Beck and Lindsley, 1969) report reverse and normal remanent magnetizations that have declination/inclination values of about 110/−70 degrees and 290/40 degrees, respectively, and are commonly many times stronger than the induced components. Preliminary measurements on several samples from the Bald Eagle intrusion revealed normal remanent magnetizations that were commonly more than 5 times stronger than the corresponding induced magnetizations. Thus, it seems reasonable to assume that the total (induced plus remanent) polarization lies very close to the Keweenawan normal direction. The highest remanent values were associated with the gabbroic core of the intrusion, although the observed magnetic susceptibilities of both the gabbro and outer troctolite were generally low (1×10^{-3} emu/cm^3 = 1A/m) and nearly equal.

The aeromagnetic data were reduced to the pole, assuming geomagnetic-field declination/inclination values of 3° W/75° N and normal polarization declination/inclination values of 290°/40° N. The results in Figure 17 show a much improved correspondence between anomalies and known geology. The anomaly expression of the basal contact zone, although still locally obscure, commonly lies just southeast of an enhanced belt of minima. The correspondence of maxima to hornfelsic rocks along the basal contact zone is improved, and a broad maximum has shifted to correspond better with the anorthositic rocks in the central part of the area. A significant improvement also has occurred over the Bald Eagle intrusion, where a ring minimum corresponds with the troctolitic rocks and an elongate maximum occurs over the gabbro core.

To eliminate long-wavelength interference and emphasize local structure, the data reduced to the pole were high-pass filtered at 2.0 km (Figure 18). This shows the basal contact to be actually characterized by a narrow, curvilinear belt of maxima over much of its length. Curvilinear anomaly patterns within the anorthositic rocks may reflect flow or

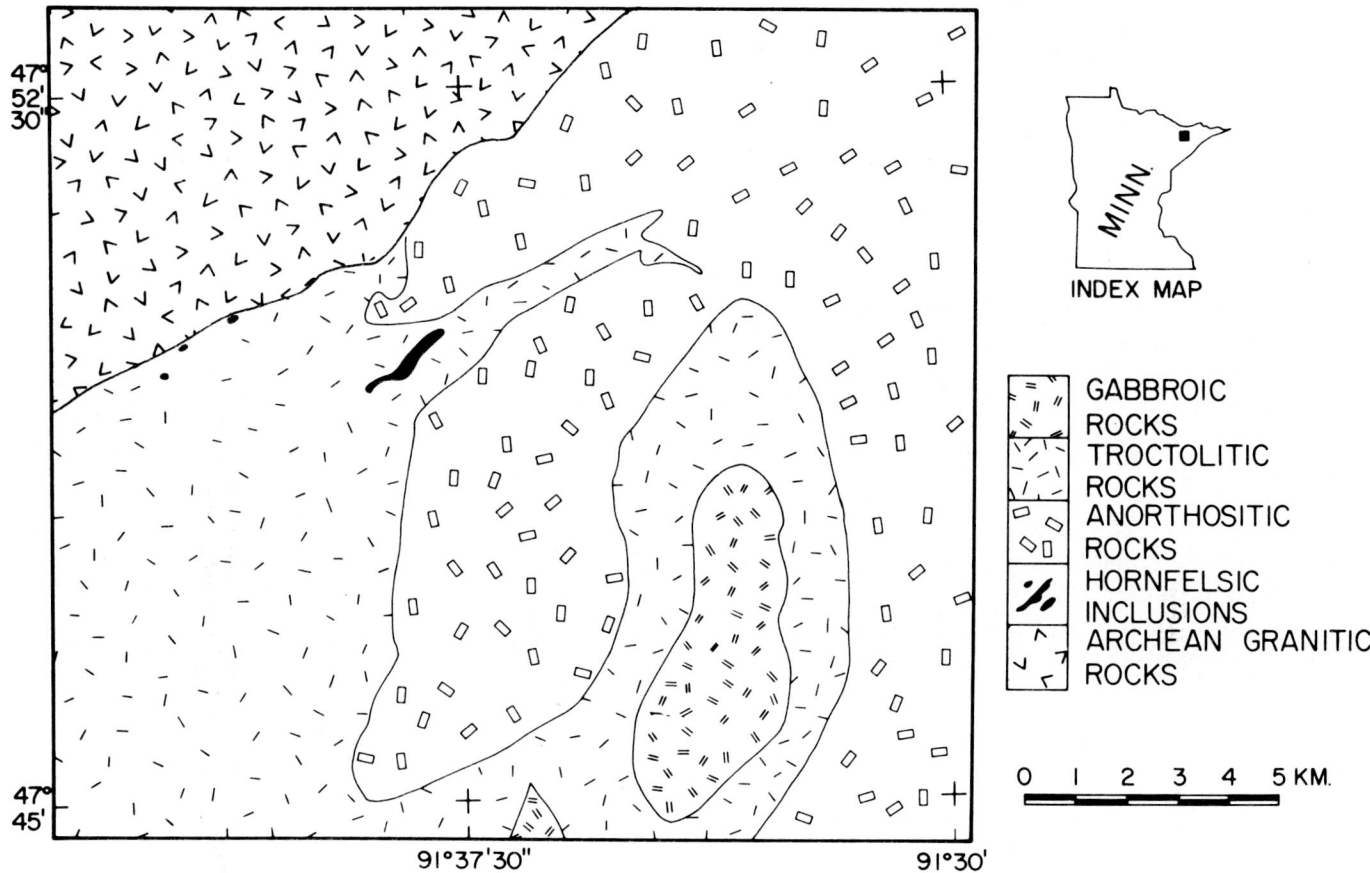

FIG. 15. Bedrock geologic map of the Gabbro Lake area (after Weiblen, 1965; Green et al., 1966).

cumulate structures. The Bald Eagle intrusion is characterized by short-wavelength maxima along or near most of its contacts. The southern part of the gabbro core is associated with a ring maximum along its contacts, and the northern part, with two circular maxima. This change in expression coincides with a northwest-trending lineament that extends out of the complex into Archean rocks and may thereby reflect a fault. Flow-foliation patterns south of the lineament in the gabbro core imply a shallow basin configuration, whereas nearly vertical foliation is reported to the north (Weiblen, 1965). Perhaps the northern half of this gabbro has been uplifted along the proposed fault and therefore has the form and structure of a feeder zone.

SUMMARY AND CONCLUSIONS

High-resolution aeromagnetic surveying recently begun in Minnesota is a powerful tool for investigating the poorly exposed Precambrian geology of the area. To date, more than 174 000 line-km of data have been acquired over northeastern and east-central Minnesota, and the results show geologic detail not attainable from outcrop or previously existing geophysical data. The data are available to the public in several digital and analog forms and should have significant bearing on geologic investigations in the region for decades to come.

The data acquired to date reveal many major geologic features that were either unknown or poorly defined prior to the surveying. For example, enhanced features of the Archean greenstone–granite terrane include (1) broad east–west folding and other internal structures within the Vermilion Granitic Complex, (2) the trace of the Vermilion fault and related structures, and (3) the configuration of a northwest-striking, Lower Proterozoic dike swarm. Anomaly lineaments over the greenstone–granite terrane may reflect previously unknown faults including (1) one that extends northeastward from about 46°45′–94°15′ to the Vermilion fault, (2) one that extends eastward from about 47°45′–93°37′30″, and (3) a tight series of north–south lineaments over the Lac La Croix Granite near the International Boundary. Except for iron formations along its rim, the Lower Proterozoic Animikie basin is associated with a subdued anomaly expression, reflecting a thick upper section of nonmagnetic slate and graywacke. Extremely weak but spatially consistent curvilinear maxima superimposed over this magnetic quiet zone arise from unknown sources but may be channel deposits of Cretaceous or Pleistocene age, which may be enriched in magnetic oxides owing to their likely provenance. Over the Keweenawan rocks the new data reveal (1) a major dike swarm extending southeastward from the Duluth Complex along the St. Croix horst, (2) intricate structures within lavas in the St. Croix horst, and (3) an extremely complex structure in the Duluth Complex. A prominent lineament extending west-northwest from

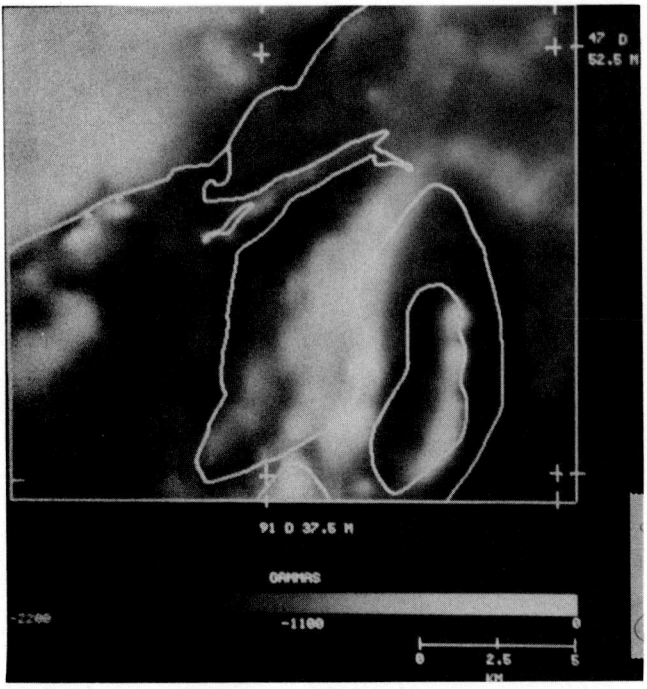

FIG. 16. Total-magnetic-intensity anomaly data over the Gabbro Lake area; gray-tone image. Geologic contacts from Figure 14 are shown as white lines. Data based on an 84 (east-west) by 77 (north-south) grid using a spacing of 213 m.

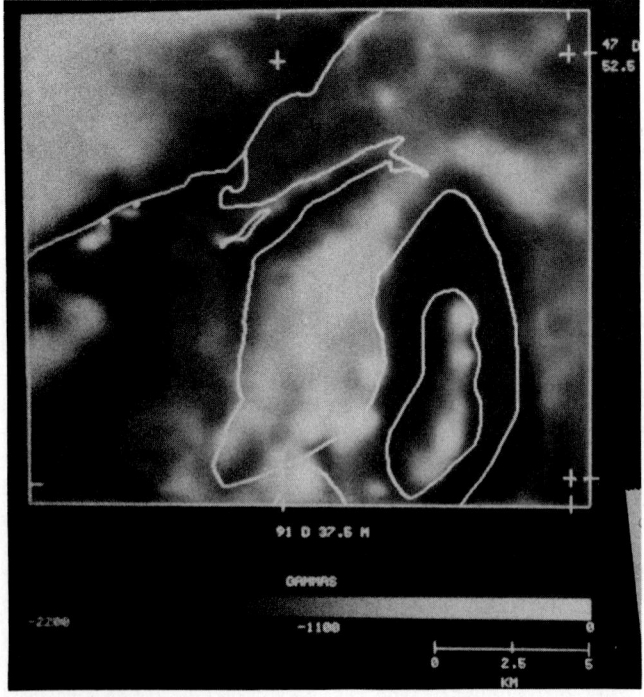

FIG. 17. Total-magnetic-intensity data over the Gabbro Lake area reduced to the pole, assuming a total polarization along a declination of 290 degrees and inclination of 40 degrees; gray-tone image. Geologic contacts from Figure 14 are shown as white lines.

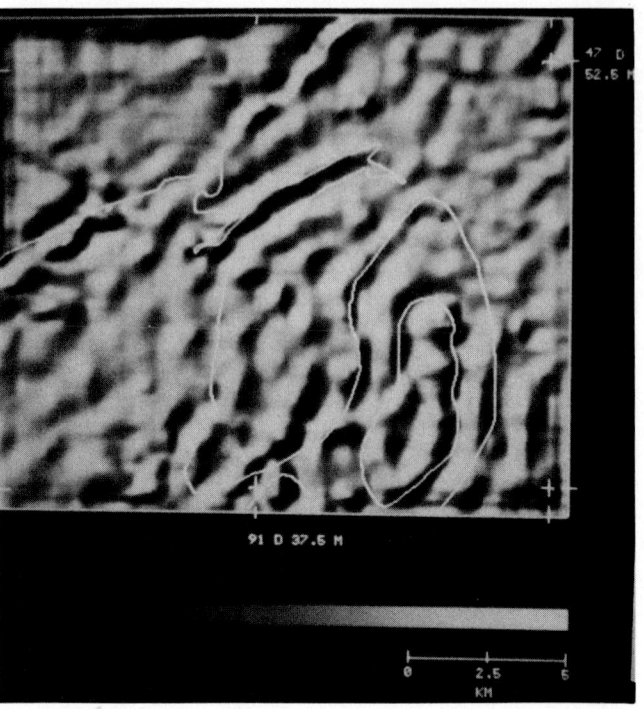

FIG. 18. The 2.0-km high-pass-filtered magnetic-anomaly data, reduced to the pole, assuming a total polarization along a declination of 290 degrees and inclination of 40 degrees; shaded-relief image. Maximum anomaly amplitudes are about 500 gammas. Geologic contacts from Figure 14 are shown as white lines.

47°30′ N–91° W cuts across the entire complex and may reflect a major fault.

Frequency-domain filtering significantly enhances detailed interpretation of the data. Data from the Hibbing, Carlton County, and Gabbro Lake areas were selected for analysis because the three areas together include a diverse suite of rocks with interpretational problems typical of northern Minnesota. High-pass filtering and calculations of the second vertical derivative delineate internal structures within large rock units and enhance crosscutting faults and dikes. In the Carlton County area the second-derivative filtering improves the definition of Keweenawan dike swarms and also enhances some structures in the Lower Proterozoic Thomson Formation. Strike-sensitive filtering in the Hibbing area isolates stratigraphy-related anomalies along the strike of the Biwabik Iron Formation from the interfering effects of crosscutting structures and presumed iron formation in the underlying Archean basement. Reduction-to-the-pole processing in the Gabbro Lake area demonstrates the usefulness of this approach in dealing with Keweenawan rocks whose magnetic-anomaly expression is complicated by a strong remanent magnetization oblique to the Earth's present field. The results of the detailed study not only demonstrate the use of computer processing to optimize magnetic interpretation but also provide examples of how computer graphics, which can present the results in optimum fashion rapidly and inexpensively, can become an integral part of an interpretational program.

ACKNOWLEDGMENTS

This aeromagnetic-surveying/interpretation program is funded by the Legislative Commission on Minnesota Resources. Acquisition and compilation of data were done by GeoMetrics, Inc., of Sunnyvale, California. G. B. Morey, D. L. Southwick, and M. S. Walton of the Minnesota Geological Survey provided constructive criticism. All computer graphics were conducted at the Minnesota Land Management Information Center under supervision of Earl Nordstrand and Steve Anderson.

REFERENCES

Austin, G. S., 1972, Cretaceous rocks, in Sims, P. K., and Morey, G. B., Eds., Geology of Minnesota: A centennial volume: Minn. Geol. Surv., 509–514.
Bath, G. D., 1962, Magnetic anomalies and magnetizations of the Biwabik Iron Formation, Mesabi area, Minnesota: Geophysics, **27**, 627–650.
Bauer, R. L., 1981, The petrology and structural geology of the western Lake Vermilion area, northeastern Minnesota: Ph.D. diss., Univ. Minn., Minneapolis.
Beck, M. E., Jr., 1970, Paleomagnetism of Keweenawan intrusive rocks, Minnesota: J. Geophys. Res., **75**, 4985–4996.
Beck, M. E., Jr., and Lindsley, N. C., 1969, Paleomagnetism of the Beaver Bay complex, Minnesota: J. Geophys. Res., **74**, 2002–2013.
Bowman, P. L., Braile, L. W., Chandler, V. W., Hinze, W. J., Luca, A. J., and Von Frese, R.R.B., 1978, Magnetic and gravity anomaly correlation and its application to satellite data: Computer codes: Final rep. NASA Contract NAS522816.
Cain, J., 1979, Main field and secular variation: Rev. Geophys. Space Phys., **17**, 273–277.
Chandler, V. W., 1982, Interpretation of high-resolution aeromagnetic data in northeastern Minnesota [abstr.]: 50th Annu. SEG Meet., Dallas, Tex., 288–291.
———1983a, Aeromagnetic map of Minnesota, Cook and Lake Counties: Aeromagn. Map A-1, Minn. Geol. Surv., scale 1:250 000, 2 sheets.
———1983b, Aeromagnetic map of Minnesota, St. Louis County: Aeromagn. Map A-2, Minn. Geol. Surv., scale 1:250 000, 2 sheets.
———1983c, Aeromagnetic map of Minnesota, Carlton and Pine Counties: Aeromagn. Map A-3, Minn. Geol. Surv., scale 1:250 000, 2 sheets.
———1983d, Aeromagnetic map of Minnesota, east-central region: Aeromagn. Map A-4, Minn. Geol. Surv., scale 1:250 000, 2 sheets.
Connolly, M. R., 1981, The geology of the Middle Precambrian Thomson Formation in southern Carlton County, east-central Minnesota: M.S. thesis, Univ. Minn., Duluth.
Green, J. C., Phinney, W. C., and Weiblen, P. W., 1966, Geologic map of Gabbro Lake quadrangle, Lake County, Minnesota: Misc. Map M-2, Minn. Geol. Surv., scale 1:31 680.
Halls, H. C., and Pesonen, L. J., 1982, Paleomagnetism of Keweenawan rocks, in Wold, R. J., and Hinze, W. J., Eds., Geology and tectonics of the Lake Superior basin: Memoir 156, Geol. Soc. Am., 173–201.
Hobbs, H. C., and Goebel, J. E., 1982, Geologic map of Minnesota, Quaternary geology: Map S-1, Minn. Geol. Surv., scale 1:500 000.
Ikola, R. J., 1968, Simple Bouguer gravity map of Minnesota, Hibbing sheet: Misc. Map M-3, Minn. Geol. Surv., scale 1:250 000.
Jahren, C. E., 1965, Magnetization of Keweenawan rocks near Duluth, Minnesota: Geophysics, **30**, 858-874.
Keighin, C. W., Morey, G. B., and Goldich, S. S., 1972, East-central Minnesota, in Sims, P. K., and Morey, G. B., Eds., Geology of Minnesota: A centennial volume: Minn. Geol. Surv., 240–255.
Mooney, H. M., and Bleifuss, R. L., 1953, Analysis of field results, pt. 2 of Magnetic susceptibility measurements in Minnesota: Geophysics, **18**, 383–393.
Morey, G. B., 1972, Mesabi range, in Sims, P. K., and Morey, G. B., Eds., Geology of Minnesota: A centennial volume: Minn. Geol. Surv., 204–217.
———1976, Geologic map of Minnesota, bedrock geology: Misc. Map M-24, Minn. Geol. Surv., scale 1:3 168 000.
———1978, Lower and Middle Precambrian stratigraphic nomenclature in east-central Minnesota: Rep. Invest. 21, Minn. Geol. Surv.
———1981, Geologic map of Minnesota, bedrock outcrops: Map S-10, Minn. Geol. Surv., scale 1:3 168 000.
Morey, G. B., Olsen, B. M., and Southwick, D. L., 1981, Geologic map of Minnesota, east-central Minnesota, bedrock geology: Minn. Geol. Surv., scale 1:250 000.
Peddie, N. W., and Fabiano, E. B., 1976, A model of the geomagnetic field for 1975: J. Geophys. Res., **81**, 2539–2547.
Reed, J. E., 1980, Enhancement/isolation wave number filtering of potential field data: M.S. thesis, Purdue Univ.
Sims, P. K., 1972a, Regional gravity field, in Sims, P. K., and Morey, G. B., Eds., Geology of Minnesota: A centennial volume: Minn. Geol. Surv., 581–584.
———1972b, Magnetic data and regional magnetic patterns, in Sims, P. K., and Morey, G. B., Eds., Geology of Minnesota: A centennial volume: Minn. Geol. Surv., 585–592.
Sims, P. K., Card, K. D., Morey, G. B., and Peterman, Z. E., 1980, The Great Lakes tectonic zone—a major crustal structure in central North America: Geol. Soc. Am. Bull., **91**, 690–698.
Sims, P. K., Morey, G. B., Ojakangas, R. W., and Viswanathan, S., 1970, Geological map of Minnesota, Hibbing sheet: Minn. Geol. Surv., scale 1:250 000.
Sims, P. K., and Mudrey, M. G., Jr., 1972, Diabase dikes in northern Minnesota, in Sims, P. K., and Morey, G. B., Eds., Geology of Minnesota: A centennial volume: Minn. Geol. Surv., 256–259.
Sims, P. K., and Viswanathan, S., 1972, Giants Range batholith, in Sims, P. K., and Morey, G. B., Eds., Geology of Minnesota: A centennial volume: Minn. Geol. Surv., 120–139.
Southwick, D. L., and Chandler, V. W., 1983, Subsurface investigations of the Great Lakes tectonic zone, west-central Minnesota [abstr.]: Geol. Soc. Am. Abstr. Programs, **15**, 692.
Southwick, D. L., and Day, W. C., 1983, Geology and petrology of Proterozoic mafic dikes, north-central Minnesota and western Ontario: Can. J. Earth Sci., **20**, 622–638.
Southwick, D. L., and Ojakangas, R. W., 1979, Geologic map of Minnesota, International Falls sheet, bedrock geology: Minn. Geol. Surv., scale 1:250 000.
Viswanathan, S., and Ojakangas, R. W., 1974a, Reconnaissance geologic map of Stingy Lake quadrangle, Itasca and St. Louis Counties, Minnesota: Misc. Map M-20, Minn. Geol. Surv., scale 1:24 000.
———1974b, Reconnaissance geologic map of Dewey Lake quadrangle, St. Louis County, Minnesota: Misc. Map M-22, Minn. Geol. Surv., scale 1:24 000.
Weiblen, P. W., 1965, A funnel-shaped, gabbro–troctolite intrusion in the Duluth Complex, Lake County, Minnesota: Ph.D. diss., Univ. Minn., Minneapolis.
Weiblen, P. W., and Morey, G. B., 1980, A summary of the stratigraphy, petrology and structure of the Duluth Complex: Am. J. Sci., **280-A**, pt. 1, 88–133.
White, D. A., 1954, The stratigraphy and structure of the Mesabi range, Minnesota: Bull. 38, Minn. Geol. Surv.
Wright, H. E., Mattson, L. A., and Thomas, J. A., 1970, Geology of the Cloquet quadrangle, Carlton County, Minnesota: Geol. Map GM-3, Minn. Geol. Surv.
Zietz, I., and Kirby, J. R., 1970, Aeromagnetic map of Minnesota: Geophys. Invest. Map GP-725, U.S. Geol. Surv., scale 1:1 000 000.

Mineral-exploration aspects of gravity and aeromagnetic surveys in the Sudbury–Cobalt area, Ontario

V. K. Gupta* and F. S. Grant‡

ABSTRACT

A detailed gravity survey comprising approximately 11 800 gravity stations has been carried out in a 33 000-km^2 region of the southern Canadian shield between the Sudbury and Kirkland Lake mining camps in east-central Ontario. The purpose of the survey was to shed some light on the economic mineral potential of the less well-explored parts of this very important mining district of Canada.

This article describes an interpretation of gravity and aeromagnetic data in the Sudbury–Cobalt area, the aim of which has been to derive information of exploration interest for base and precious metals. In addition to using standard procedures such as vertical gradient mapping to increase the resolution of the Bouguer gravity data, a new analytical method called the "apparent-density map" has been introduced as a lithologic-mapping tool. The apparent-density algorithm is used to define rock-unit boundaries and to assign average densities to these units for the purpose of preparing a basement-lithology map. The results suggest that the mineral-rich Archean Abitibi greenstone belt, consisting mainly of mafic rocks, extends beneath the Lower Proterozoic sedimentary cover much farther than had previously been realized. The volumes, depth extent, and major stratigraphic subdivisions of Archean greenstones, together with a lithologic map of the Archean basement, all derived from the gravity and magnetic data, are of material benefit in outlining prospective regions for mineral exploration. Three-dimensional gravity models of the Round Lake Batholith and the adjacent steeply dipping mafic and ultramafic metavolcanics suggest a favorable environment for gold exploration. The gravity and magnetic maps have also been helpful in locating possible diabase feeder zones and in determining their volumes and depths—information that is important for outlining potential silver-cobalt prospects.

It is concluded that gravity and aeromagnetic surveys provide an informative, cost-effective approach to regional exploration for mineral-prospective zones in the Sudbury–Cobalt area, and that these surveys should be followed up by direct-search exploration strategies.

INTRODUCTION

This article is concerned with an interpretation of a regional gravity survey carried out by the Ontario Geological Survey (OGS) between 1977 and 1980 in the Sudbury–Cobalt region. The project area lies within the southwestern portion of the Abitibi greenstone belt, which extends through east-central Ontario and west-central Quebec (Figure 1). This survey was the second of its type to be carried out by the OGS; the first covered the Red Lake–Uchi Lake greenstone belt in northwestern Ontario (Gupta and Wadge, 1980). Both the Red Lake–Uchi Lake and the southwestern Abitibi greenstone belts are perceived to be regions having high potential for still-undiscovered economic mineralization, but prospecting and mapping in both regions are restricted by thick glacial overburden. The previous success of the gravity method in defining the structure of the Sudbury Igneous Complex (Popelar, 1972) gave some encouragement to the idea that gravity surveys might be used effectively to outline areas that offer the prospect of being favorable zones for more costly follow-up exploration strategies. In the Red Lake–Uchi Lake greenstone belt the gravity method proved also to be an extremely cost-effective reconnaissance technique for mapping hidden volcanic stratigraphy (Gupta et al.,

---

*Ontario Geological Survey, 711-77 Grenville Street, Toronto, Ont. M5S 1B3, Canada.
‡Paterson, Grant and Watson, Ltd., 111 Richmond Street W., Toronto, Ont. M5H 2G4, Canada.

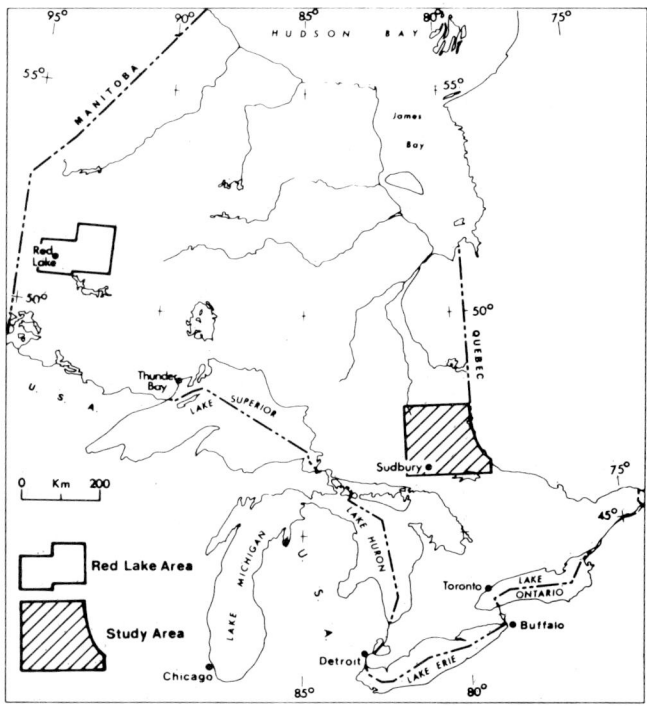

FIG. 1. Map of Ontario and surrounding region.

1982; Gupta and Ramani, 1982); and based upon this result, it was decided to enlarge the program to include the geologically complex, mineral-rich Sudbury–Cobalt area.

The Sudbury–Cobalt gravity survey has a high station density maintained over a large area; also, a large number of rock-density measurements were made during the course of the field operations. From past experience, it has proved necessary to collect a statistically representative number of rock densities for each major lithologic unit to support the interpretations. Altogether, 10 825 gravity stations were surveyed within an area that measures approximately 33 000 km^2. The average station density in the areas of highest mineral potential (mostly metavolcanic terranes) is approximately one station per 1.5 km^2. About 7 200 stations were established by the Ontario Geological Survey; the remaining 3 625 were supplied from earlier surveys by the Earth Physics Branch, Canadian Department of Energy, Mines and Resources. OGS observers made more than 3 400 rock-density measurements during the course of the field work. This survey is unique in our experience in terms of the amount of detail that is provided throughout such a large area in the Canadian shield. A station-location map is shown in Figure 2.

The aim of the interpretation is to extend the mapping of the bedrock in sparsely mapped and sediment-covered areas, to the extent that gravity methods will allow, and to identify areas having higher than normal potential for economic mineralization.

GENERAL AND ECONOMIC GEOLOGY

The survey area includes part of the Superior, Southern, and Grenville provinces of the Canadian shield in northeastern Ontario. The Superior province is characterized by metavolcanic–metasedimentary and intrusive rocks of Archean age; the Southern province, by metasedimentary and minor metavolcanic rocks of the Huronian Supergroup of Early Proterozoic age, by diabase intrusions, and by the rocks of the Sudbury Igneous Complex (Figure 3). The Lower and Middle Proterozoic rocks of the Grenville province consist of metasedimentary–metavolcanic and metaigneous gneissic rocks. Structurally the area is dominated by the Grenville front tectonic zone, which separates the relatively little deformed, low-grade metamorphic terrane of the Superior and Southern provinces from the strongly deformed, high-grade metamorphic terrane of the Grenville province.

The Sudbury and Temagami–Cobalt mining camps lie within the project area, and immediately to the north lie the Timmins and Kirkland Lake gold and base-metal mining districts.

Much exploration interest is currently focused on the Lower Proterozoic metasedimentary rocks of the Huronian Supergroup that lie within the "Cobalt embayment," where clastic sediments were deposited within a broad depression in the Archean basement. The geological environment is similar in a number of respects to the Witwatersrand Basin in South Africa, where the world's largest gold deposits are located. The Cobalt embayment therefore is thought to be an area with significant gold potential.

There is also thought to be some potential for finding additional silver–cobalt (Ag–Co) vein deposits of the type that made the town of Cobalt one of the world's most important silver-mining camps during the early years of this century. Good possibilities still exist for finding new gold and new base-metal sulfide deposits in the Archean greenstones, which are partly covered by Lower Proterozoic metasediments.

As a background to the conceptual thinking that guided the interpretation of the gravity survey, we shall briefly describe ore-deposit types that are found within the study area and current hypotheses relating to their genesis. The first commodity, and the one that currently commands the highest interest, is gold, which is currently mined in the greenstone belts of the Superior province. There are two general classes of deposit, the lode type and the paleoplacer type. The two types are usually proximal to each other, suggesting that they may be related. Each type has its own environmental parameters.

The lode-type deposits of the Timmins and Kirkland Lake districts apparently have five regional geological associations that are common to all important concentrations (Pyke, 1981): (1) proximity to major faulting, (2) proximity to ultramafic volcanics (komatiites), (3) proximity to (usually felsic or alkaline) plutons, (4) carbonatization of the host rock, and (5) presence of nearby sedimentary rocks. Without going deeply into Pyke's theory of the genesis of the ores, the ultramafic rocks are thought to be the source of the gold, which is released during the carbonatization process (Pyke, 1976). Major faulting is necessary to provide a zone of fracturing or dilatancy of suitably large dimensions to concentrate the gold by hydrothermal activity. The sedimentary rocks are a source of CO reductant, which determines favorable precipitation sites. The thermal energy that is

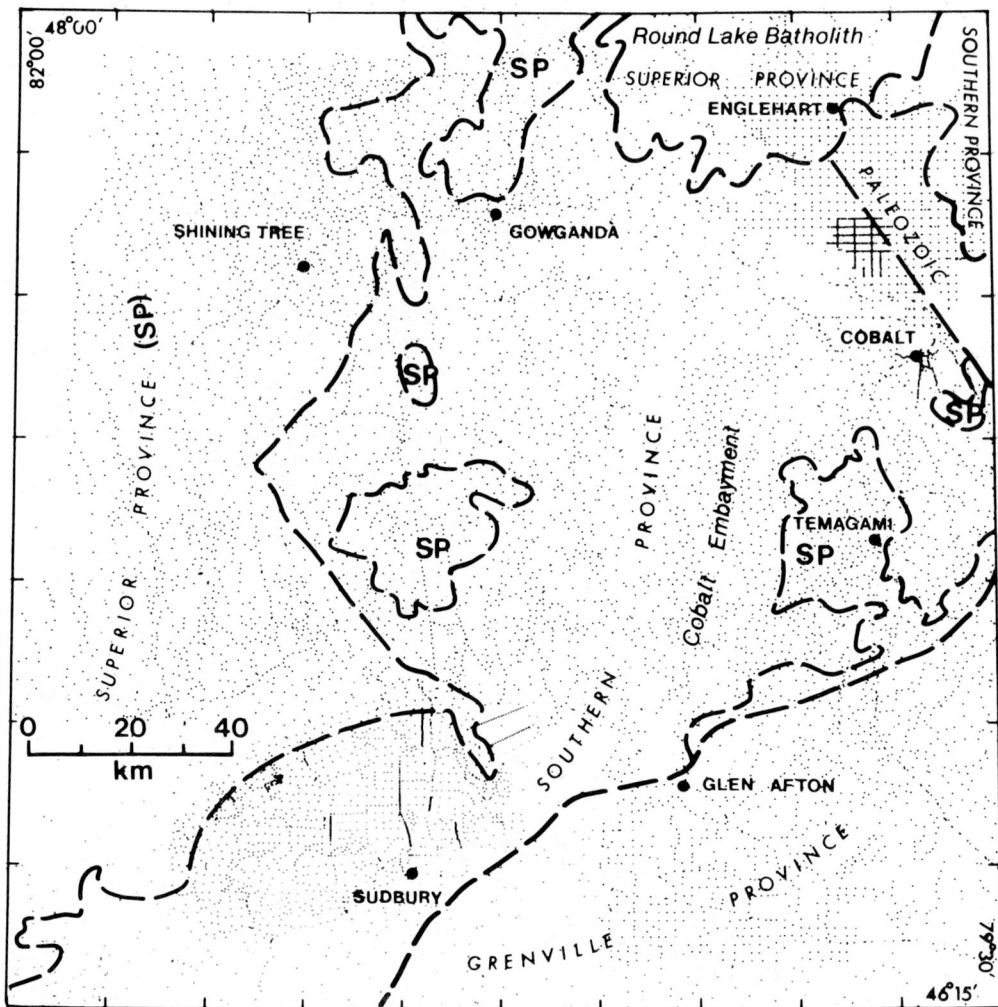

FIG. 2. Gravity-station locations. Boundaries of major geological provinces are also shown.

needed to drive the fluids is provided by nearby intrusions of quartz-feldspar porphyry or syenite.

Witwatersrand-type gold deposits might conceivably exist within the Cobalt embayment. In South Africa these paleoplacer deposits occur in large paleodeltas and drainage systems (Pretorius, 1976) proximal to Archean source rocks. Similar paleoenvironments (ancient high-energy drainage systems traversing Archean greenstone belts) may exist in the Cobalt embayment and may very well host economic gold deposits.

With regard to the Ag–Co deposits of the Cobalt region, evidence points to mafic, Archean volcanic flows as the original provenance, redistribution of the metals having occurred during the Huronian period of sedimentation and subsequent reconcentration by Nipissing diabase intrusive activity during Early Proterozoic time (Patterson, 1979). Accordingly, favorable regional indices are (1) proximity to mafic volcanics, (2) proximity to a fault/fracture system, (3) proximity to a major diabase "feeder," and (4) proximity to sedimentary rocks. The environment described above is similar in many ways to the type that we are seeking for gold deposits.

Few economic deposits of uranium or of base metals (other than the Sudbury nickel ores and the Agnew Lake uranium mine) are at present known in the survey area; however, the Elliot Lake–Blind River conglomerate-hosted uranium deposits lie not very far to the southwest. To the north, the Abitibi greenstone belt is a veritable storehouse of volcanogenic deposits of base and precious metals.

GRAVITY SURVEY

Statement of objectives

With the regional characteristics of ore environments discussed above in mind, the following objectives were established for the gravity program:

1. To interpret the lithology of the Archean basement rocks beneath the Huronian (Lower Proterozoic) metasediments, and in particular to determine the extent of the greenstone belts.
2. To map the thickness of the Huronian metasediments

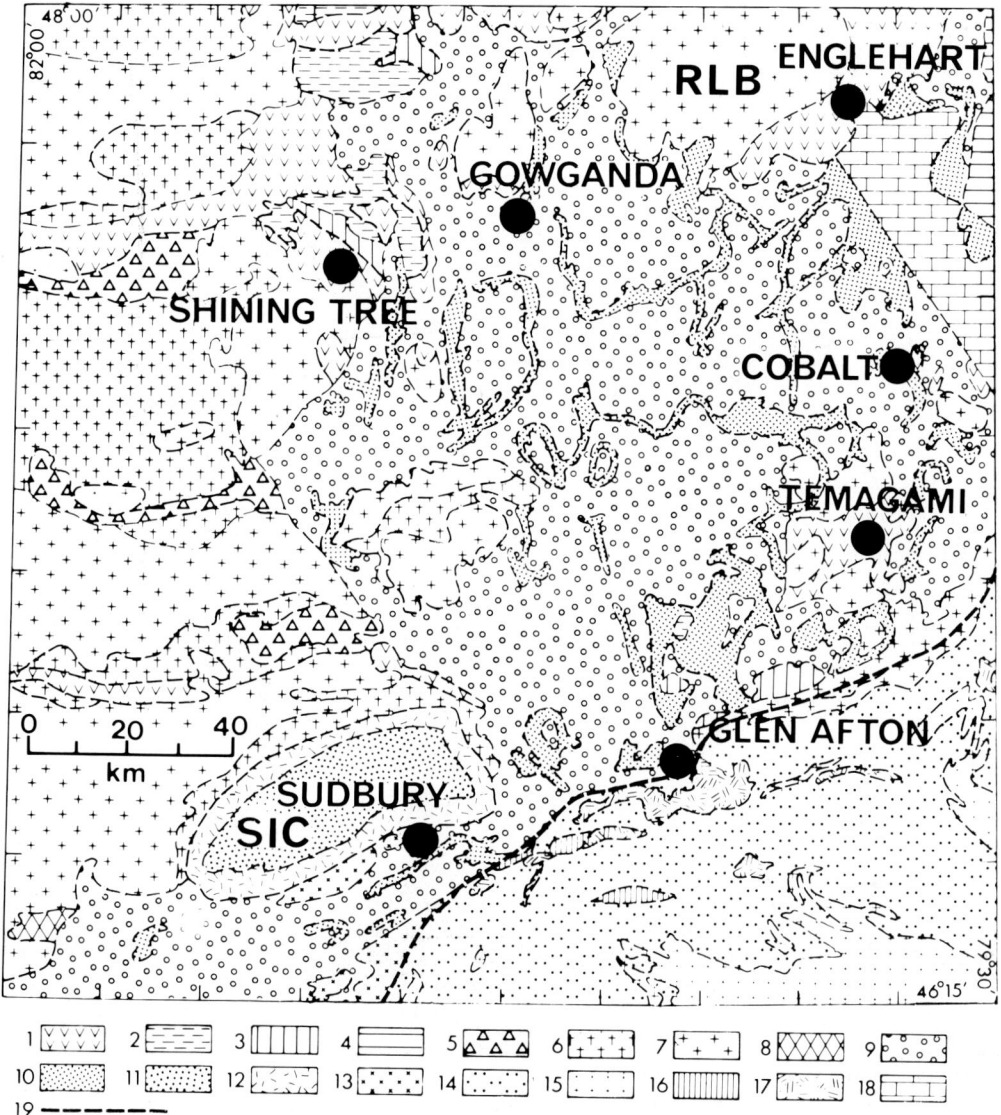

Fig. 3. General geology: 1 = mafic–intermediate metavolcanics; 2 = felsic–intermediate metavolcanics; 3 = metasediments; 4 = paragneiss and associated migmatitic rocks; 5 = mafic gneiss and amphibolite; 6 = orthogneiss and migmatitic rocks; 7 = felsic plutonic rocks; 8 = mafic intrusives; 9 = Huronian Supergroup; 10 = Nipissing diabase; 11 = Whitewater Group (breccia, mudstone, wacke); 12 = Sudbury Igneous Complex; 13 = felsic intrusives; 14 = metasediments; 15 = felsic intrusives; 16 = mafic–ultramafic intrusives; 17 = anorthosite-suite intrusives; 18 = Paleozoic sediments; 19 = Grenville front boundary. SIC = Sudbury Intrusive Complex; RLB = Round Lake Batholith.

and, if possible, to develop from this a generalized topographic map of the Archean basement surface.
3. To determine the volumes and, if possible, the internal stratigraphic boundaries of the major greenstone belts.
4. To determine the volumes and depths of significant diabase–gabbro bodies.
5. To determine the forms of major intrusive bodies such as the Round Lake Batholith and its relation to surrounding rock units.
6. To relate the gravity data to major fault zones and, if possible, to find out which are associated with rock-alteration effects and which appear to have significant depth extent.
7. To determine the nature of the Grenville–Superior province boundary and to develop a crustal model that will explain the gravity data in a regional manner.

The remainder of this paper is a discussion of the degree to which the gravity program succeeded or failed in its stated objectives, and the implications of the interpretations to mineral exploration.

Bouguer gravity map

The map of Bouguer gravity (Figure 4) reflects the geologic complexity of the Sudbury–Cobalt area. It shows a number of positive features, including the Sudbury Igneous Complex (SIC) (Gupta et al., 1984), the Glen Afton anortho-

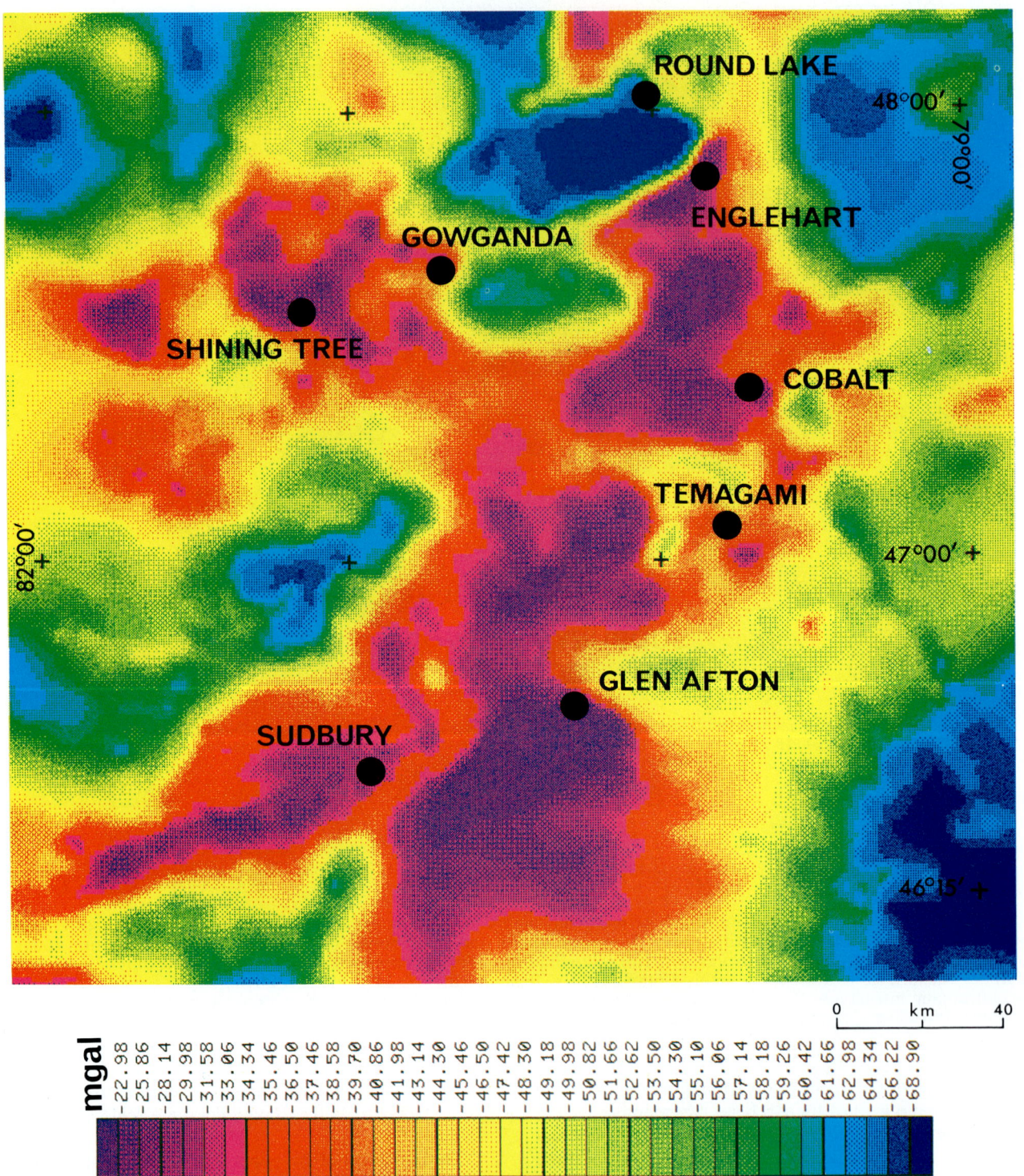

Fig. 4. Bouguer gravity map.

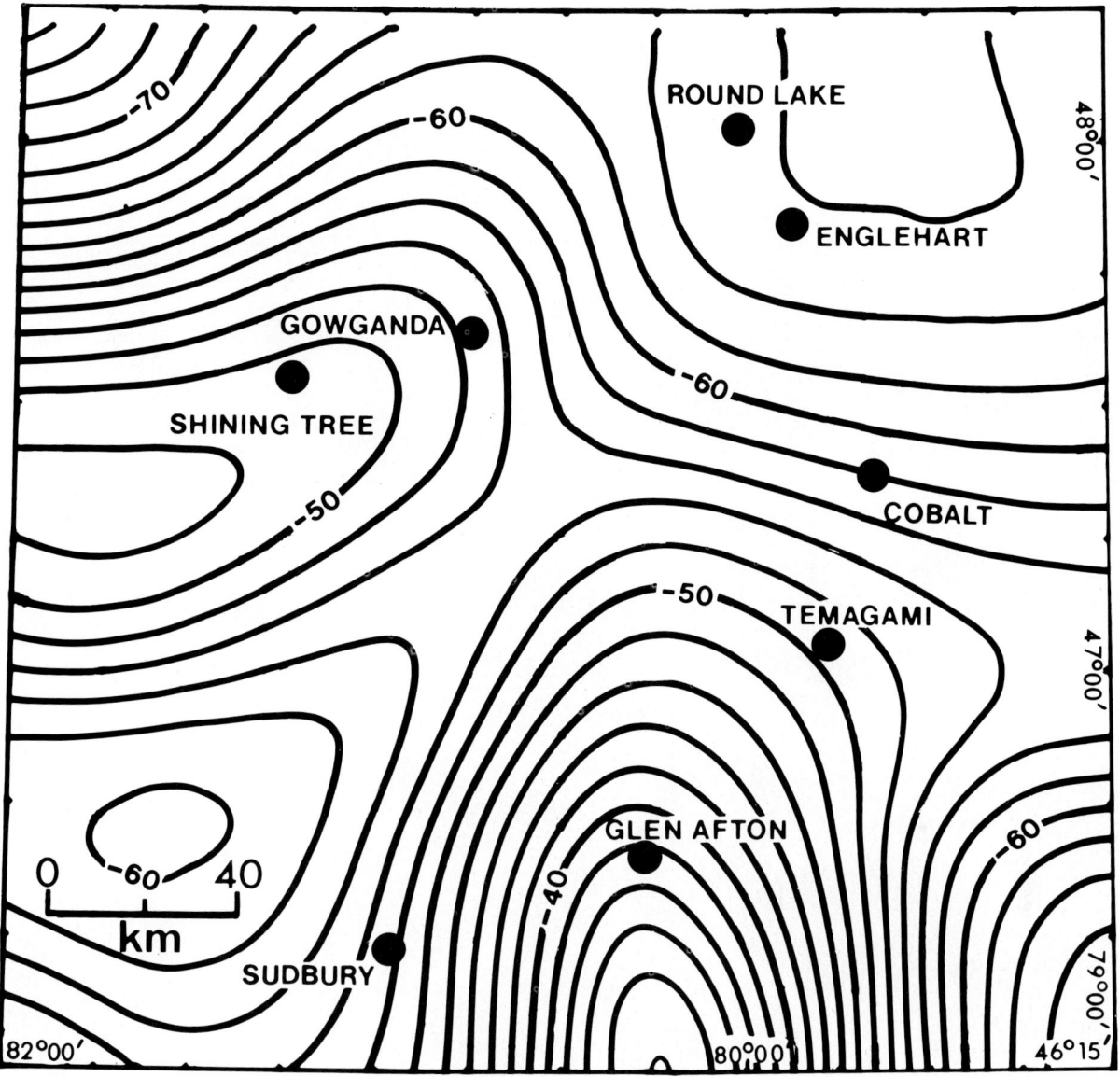

FIG. 5. Regional component of the Bouguer gravity map. Contour interval is 2 mGal.

site-gabbro complex, and the Shining Tree greenstone belt. Negative anomalies occur over known felsic intrusions such as the Round Lake Batholith (RLB), located immediately west of Englehart, and also over a number of other, unnamed plutonic bodies. The contours shown in Figure 4 are well controlled by the close station spacing of this survey, and they undoubtedly give an accurate portrayal of the Bouguer gravity field. The density that was used in making the Bouguer reductions was 2.67 g/cm^3. Topographic relief throughout the entire area is moderate to gentle, and the maximum error from irregular topography would not exceed 0.1 mGal.

Regional–residual analysis

Although it is not conspicuous in Figure 4, a regional-anomaly component in the Bouguer gravity field distorts and sometimes masks the relationship that exists between the shapes of the anomalies and the near-surface geology. The regional anomaly (Figure 5) is determined by fitting smoothly varying base levels to local anomalies along north–south and east–west profiles, forming a 9-km × 9-km intersecting grid, using known geology as a guide in choosing the background levels. The residual Bouguer gravity field, which is obtained by subtracting the regional anomaly from the Bouguer gravity map, is shown in Figure 6. Although it may not appear to be very different from Figure 4, the residual gravity map is more closely related to the surface geology than Bouguer gravity, in spite of the fact that extensive overlapping of the gravity effects of neighboring rock units still exists. The regional anomaly shown in Figure 5, on the other hand, is at least an order of magnitude broader and smoother than the residual gravity effects. Two-dimensional computer modeling of the positive regional anomaly in the south-central part

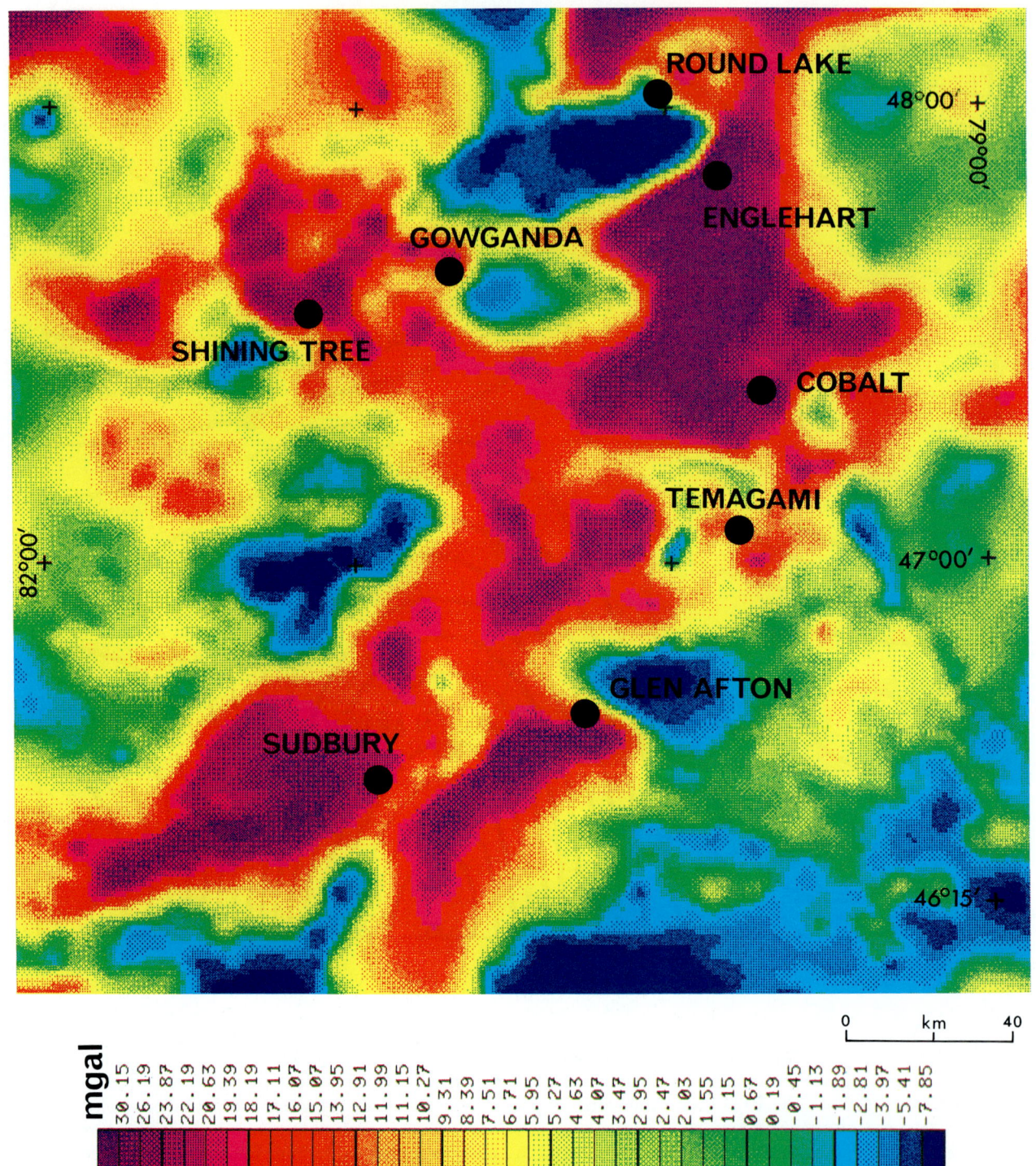

FIG. 6. Residual component of the Bouguer gravity map.

of Figure 5 indicates that this feature is caused by rocks having an average density of at least 3.1 g/cm^3, which extend well down into the lower part of the crust and possibly even into the upper mantle. We do not propose to speculate upon the origin of these dense rocks but would point out two obvious facts: (1) the only relationship that can be seen between the regional gravity field and the geology is the presence of the Glen Afton anorthosite-gabbros on the flank of the large anomaly, and (2) the Grenville front tectonic zone is in no obvious way indicated by the regional gravity contours. This latter observation suggests that the average density of the Grenville and Superior crust must be very nearly the same, a conclusion that is supported by the density measurements that were taken on rock specimens in the field.

Vertical gravity-gradient map

The residual-gravity map is an improvement on the Bouguer gravity map from the standpoint of defining geological boundaries; but as we have said, the details are obscured by the overlapping of gravity effects. In order to improve the resolution, we have prepared a map of the vertical gradient (Figure 7) of the Bouguer gravity field. This was done by applying a vertical-gradient operator in the frequency domain. The logic here is simply that the rate of change of gravity with elevation is much more sensitive to changes in rock densities occurring near the ground surface than to changes occurring at depth; thus lithologic boundaries should be revealed in the gravity-gradient map with greater precision than in the Bouguer gravity map itself. The improvement in resolution is evident in Figure 7. It is particularly noticeable with regard to density zoning within the Shining Tree greenstone belt, the Sudbury Igneous Complex, and the Round Lake Batholith with its southern satellite body. What we want to point out in particular on this contour map, apart from the sharper definition of geologic boundaries, are the weak but nonetheless real gravity-gradient lows (A and B in Figure 7), which follow a northwest–southeast trend across the eastern half of the map area. These trends coincide with known fault systems. Neither can be seen (or they can be perceived only with considerable imagination) in the Bouguer gravity map (Figure 4) or the residual-gravity map (Figure 6), so that some form of mathematical enhancement such as the vertical-gradient calculation is necessary to make these features visible. The fact that the faults are outlined by vertical gravity-gradient lows indicates that a reduction has taken place in the densities of the rocks surrounding the faults; and an obvious (although not necessarily correct) suggestion would be that the rocks have in some way been altered, possibly by silicification or carbonatization, so as to make them less dense. An increase in SiO$_2$ content of 20 percent, for example, would reduce the average density by approximately 0.05 g/cm^3. Whatever the reason for the gravity-gradient lows, there can be little doubt about their reality, or about the fact that they coincide with known major fault systems. The fact that they cross the Grenville front tectonic zone suggests that the processes that gave rise to the reduction in rock density are probably post-Grenville in age.

Density measurements and the apparent-density map

The vertical gravity gradient has been helpful in drawing the boundaries between different rock-density units. The problem now is to decide what these boundaries represent in terms of the regional geology. To make this interpretation it is most desirable to know the average densities of the rocks within the different units. Rock densities are largely determined by chemical composition and by major mineralogy. These attributes also play important, but not exclusive, roles in determining the names by which rocks are identified in the field. Lithologic nomenclature also depends on factors that are unrelated to rock density, such as texture and mode of emplacement. Accordingly, there is not a simple, one-to-one relationship between lithology and bulk density. Several different types of rocks may have similar densities, and, by the same token, a given rock type (e.g., a gabbro or a wacke) may have a constant chemical composition but a varying mineralogy, depending on metamorphic grade, and hence a varying density. However, we observe that within a given region, trends in mineralogy and major chemistry do tend to conform to lithologic boundaries, and therefore apparent rock density should be useful as a first-order mapping device in areas where the rocks themselves are hidden from view.

To complete the lithologic mapping of the Archean basement, therefore, we have prepared a map of apparent density of the bedrock (Figure 8) from the residual Bouguer gravity map. In order to do this, we have assumed that the bedrock density varies laterally but not vertically to a depth of 6 km. We chose 6 km to be the average depth extent of the gross aspects of the surface geology, because this seemed to be close to the upper limit on the depth extent of the major units calculated from the Bouguer gravity, using mean densities taken from the rock-sample measurements. In addition, 6 km appears to be about the maximum depth extent that is attained by supracrustal rocks of the greenstone belts within the Superior province, according to various lines of evidence (Gupta et al., 1982). The survey area may then be thought of as consisting of an array of vertical-sided right-rectangular prisms measuring 1.5 km × 1.5 km horizontally and extending to a depth of 6 km, each containing material that has a constant but unknown density. The densities that are required to make the combined gravity effect of all of the prisms reproduce the residual Bouguer gravity field may be calculated, and the contour map of these values we term an "apparent-density map." The principle is similar to that which is used to calculate apparent-magnetic-susceptibility maps from airborne-magnetometer data (Grant, 1973), except that, in the case of gravity data, theory requires that the prisms must be depth-limited (see Appendix).

Figure 8 shows the apparent-density contours for the project area. The contour interval used in this drawing is 0.05 g/cm^3. The lowest values are approximately 2.60 g/cm^3, and they occur within the central and eastern part of the Round Lake Batholith. The highest values occur along an arc bordering the Round Lake structure on the southeast, and these reach almost 3.0 g/cm^3. Elsewhere, the apparent-density values lie between these two limits. From the vertical gravity gradient and the apparent-density maps, we are able to draw rock-unit boundaries and assign average

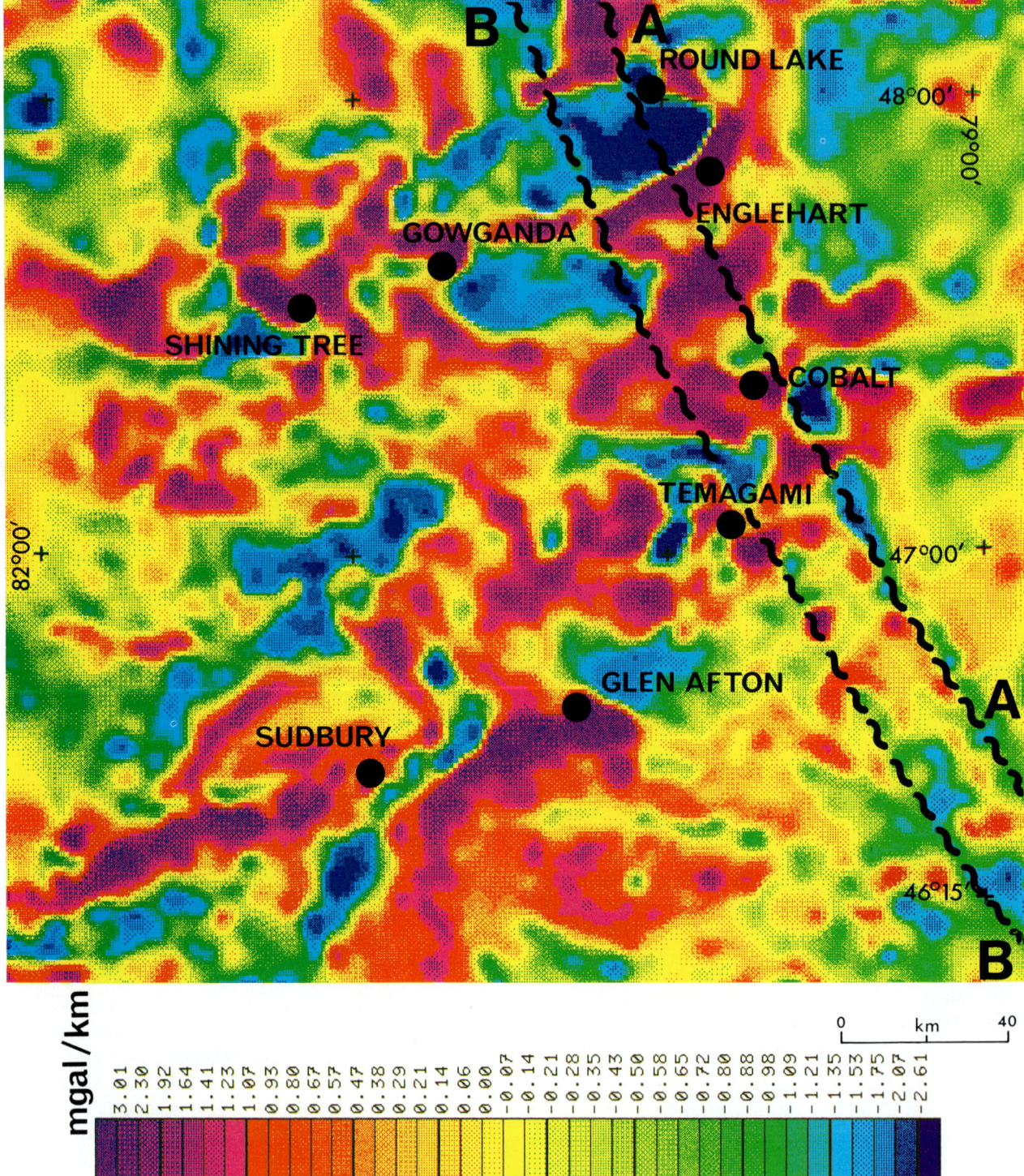

FIG. 7. Vertical gravity-gradient map. A and B are the two faults interpreted from the gradient map.

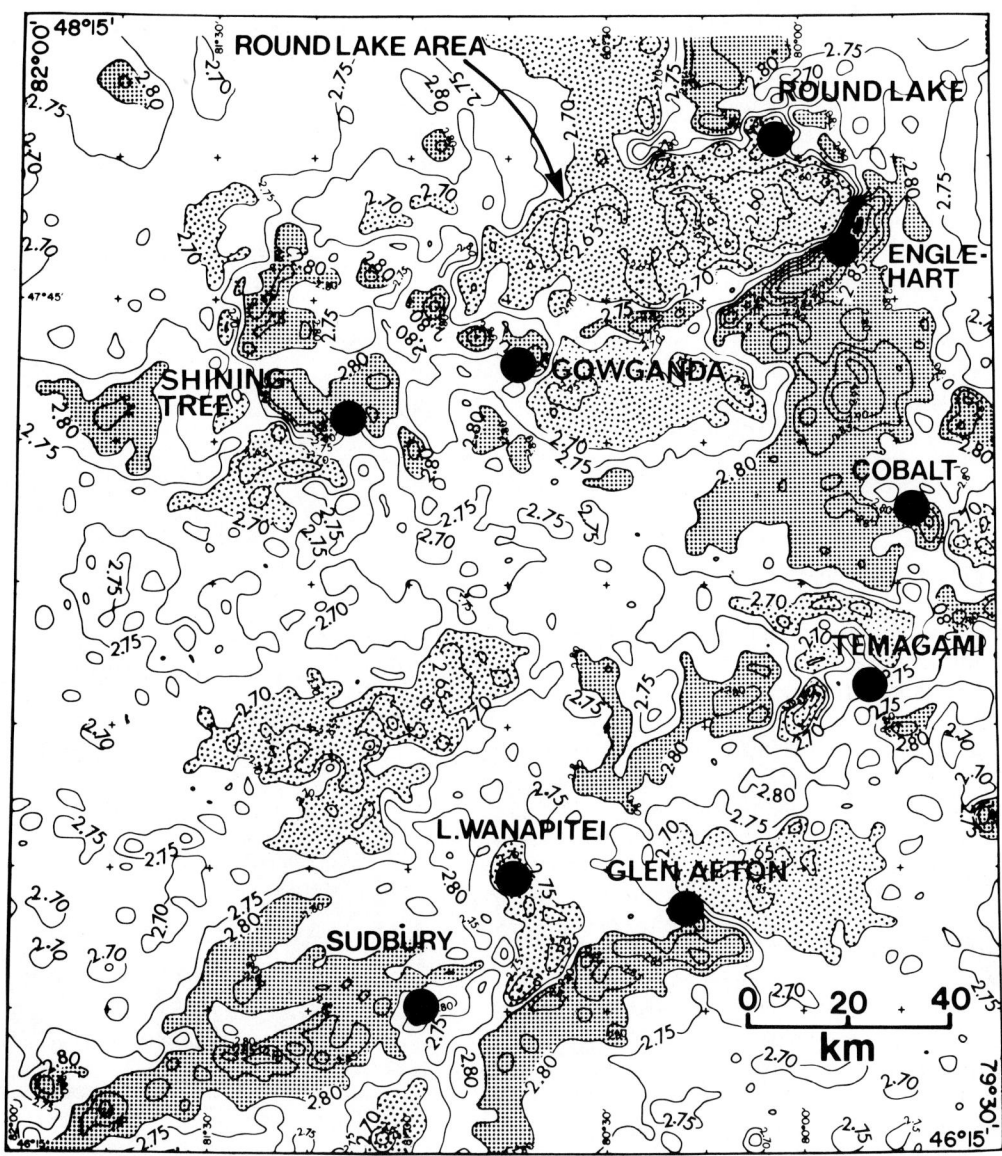

FIG. 8. Apparent-density map. Contour interval is 0.05 g/cm³.

densities to these units. The next step is to translate the density units into a map of the interpreted lithology.

This final, interpretive step involves the use of density measurements that were made in the field. A statistical tabulation of these densities is given in Table 1. However, for the purpose at hand, the densities of the major rock types have been organized into frequency plots, which are shown in Figure 9, and which convey most of the information that we need. A direct comparison of measured with apparent densities, using the rather general rock-classification system shown in Table 1, indicates a surprising degree of correlation.

Interpreted map of basement geology

Table 1 has been used as a basis for interpreting the density units throughout the project area in compositional-lithologic terms. The interpretation map of the Archean basement complex (with Lower Proterozoic sedimentary rocks removed) and the Grenville gneissic complex are shown in Figure 10. The following are some general comments relating to this map.

Outside the Cobalt embayment, where controls in the form of surface exposures exist, the gravity-derived lithologic contacts are close to their mapped locations, and the interpretation has added information in areas of poor rock exposure. As particular examples (see Figure 10), we draw attention to the subdivision of the Shining Tree greenstone belt into regions of thick and of thin to very thin accumulations of felsic to intermediate metavolcanics; to the indication of a central, lighter core within the Round Lake Batholith (also suggested by Gibb and Van Boeckel, 1970); and to the existence, beneath thick overburden, of anomalously dense units, probably ultramafic rocks, that flank the Round

Lake Batholith on its eastern and southeastern margins. We also note a distinct zonation of the Sudbury Igneous Complex and the existence of a number of felsic plutons scattered throughout the project area, some of which had not been mapped previously.

The Cobalt embayment region is sharply divided into an eastern sector of comparatively high apparent density (2.80–2.85 g/cm^3) and a western sector of lower apparent density (2.70–2.75 g/cm^3) along a line that runs north-northeast from Sudbury to Englehart (Figures 8, 10). West of this line the rocks have an apparent density that is only slightly above the unweighted average of density measurements taken on rock samples from the Huron Supergroup (2.70 ± 0.06 g/cm^3) and is very nearly the same as the average density (2.71 ± 0.02 g/cm^3) of the intrusive and gneissic rocks of felsic to intermediate composition that border the Huronian metasediments to the west. From this we infer that mafic units are not present in significant proportions within this sector, at least within the upper 6 km of crust; and accordingly, the basement rocks beneath the Huronian sedimentary rocks are probably compositionally similar to the granodioritic and trondhjemitic rocks that lie adjacent to the Huronian sedimentary rocks along their western margin. The minor increases in apparent density (to about 2.75 g/cm^3) that occur within restricted areas are probably due, at least in part, to local intrusions of Nipissing diabase–gabbro, since these apparent-density increases commonly coincide with magnetic anomalies.

East of the Sudbury–Englehart line, the apparent-density values (Figure 8) lie generally within the range 2.80–2.85 g/cm^3, with local "anomalies" as high as 2.95 g/cm^3 and as low as 2.65 g/cm^3. Analysis of magnetic anomalies (see below) and a small number of drilling records from this region indicate that the Huronian sedimentary formations have a total thickness on the order of 2 km. If that estimate is even roughly correct, it implies that about two-thirds of the 6-km column involved in the apparent-density calculation is made up of rocks in which average density is approximately 2.85 g/cm^3. The unweighted average density of rock samples from Superior metavolcanic units of all compositions is about 2.81 ± 0.13 g/cm^3; however, if the samples are weighted by their areal distribution, the average value becomes somewhat higher than this, owing to the higher proportion of mafic members. We believe that the average density for all greenstone-belt rock types would be close to the unweighted average for intermediate to mafic metavolcanic rocks, i.e., about 2.85 g/cm^3 (Gupta et al., 1982). In short, there is a clear implication from the apparent-density contours (Figure 8) that the basement rocks beneath the eastern sector of the Cobalt embayment are generally greenstone-belt-type metavolcanics of mafic to intermediate composition. There are two additional reasons why we believe that this part of the Cobalt embayment is underlain by metavolcanic rocks: (1) the zone of high apparent density makes a smooth join between the outcropping Temagami greenstones and the Kirkland Lake–Larder Lake greenstones to the north, and (2) aeromagnetic data show clear evidence of buried iron formations along the interior margin of this great, southwesterly trending arc. Accordingly, the Kirkland Lake–Larder Lake and the Temagami greenstone belts appear to be part of a major metavolcanic–

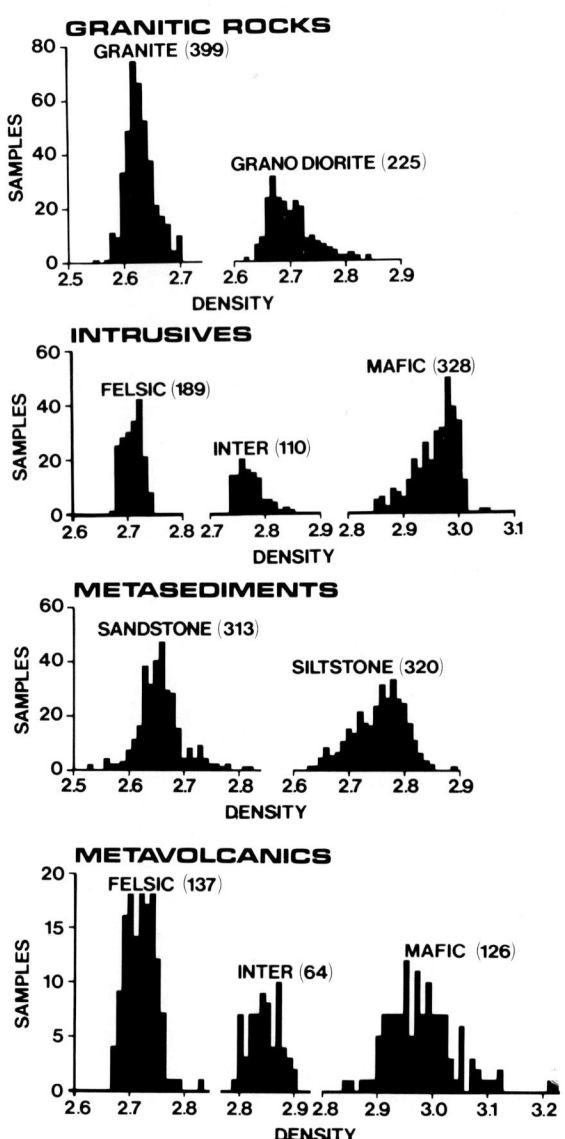

FIG. 9. Density histograms. Number of samples given in parentheses.

metasedimentary complex extending beneath the Huronian sedimentary rocks, which continues southwest of Lake Temagami to Lake Wanapitei (Figure 10). Apparent-density "anomalies" indicate a number of felsic and mafic or ultramafic units along the length of this arc.

The above are the most important results of interpretation in terms of the genral geology. What the gravity data do *not* show are changes in the thickness of the Lower Proterozoic sedimentary-rock units. The average density of the Huronian sedimentary rocks is nearly the same as that of the granodioritic rocks with which they are in contact to the west, and by which they are presumably largely underlain in the western part of the Cobalt embayment. Thus, no gravity effect is produced by this contact. If one wants to map the thickness of the Lower Proterozoic sedimentary rocks, he must look to methods other than gravity.

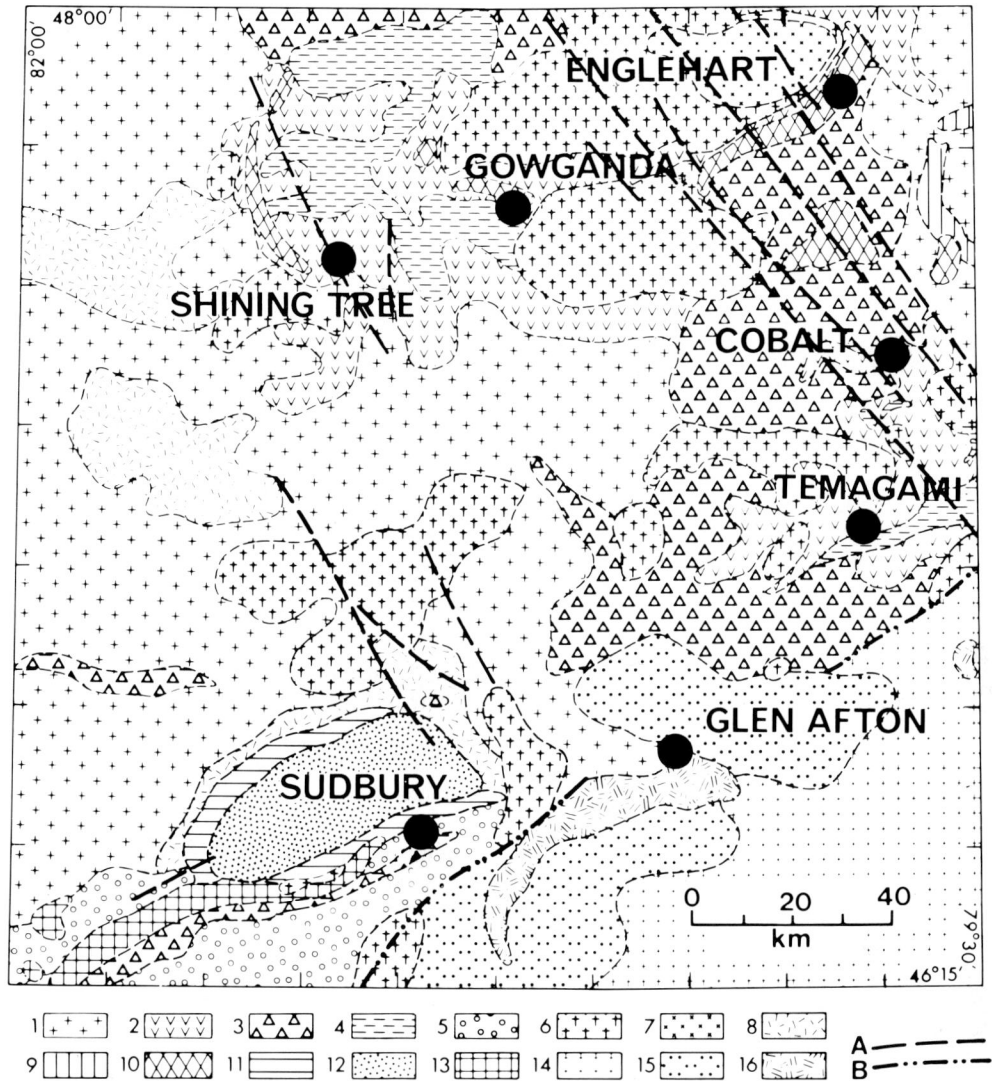

FIG. 10. Interpretation map of basement (Archean) lithology. Symbols: A = fault; B = Grenville front. For geologic explanation, see Table 1.

AEROMAGNETIC SURVEY

Since the gravity method fails to enable us to calculate the thickness of the sedimentary rocks of the Huronian Supergroup, and since this remains an important objective of our economic study, it is logical that we should turn to aeromagnetics. A total-field aeromagnetic survey of the region was flown in 1960 by the Geological Survey of Canada, using a fluxgate magnetometer. The mean line spacing was 0.8 km, and the nominal terrain clearance was 300 m. The total-field aeromagnetic contours within the Cobalt embayment region are shown in Figure 11. The magnetic field contains numerous anomalies, many of which are caused by diabase intrusions that occur both within the Huronian sedimentary rocks and within the Archean basement. The plan was to calculate depths for as many intra-basement magnetic bodies as possible, with a view to constructing from these depths a basement-elevation map; but unfortunately, it was impossible to tell from the character-

istics of the magnetic anomalies themselves whether they originate from within the basement or the overlying sedimentary rocks. The only exceptions were anomalies from Archean iron formations, which are easily recognizable by their amplitudes. Of the more than 200 anomalies analyzed by computer, using least-squares inversion methods with 2½-D prismatic models, only 15 or 20 gave estimates of basement depth that were considered reliable; and these estimates turned out to be in the range 0.8–2.0 km, agreeing reasonably well with the sparse drilling information that exists. A large proportion of the anomalies turned out to be due to bodies (probably diabase intrusions) within the Huronian sedimentary rocks, and the remainder (except for the 15 or 20 referred to) could not be assigned unambiguously either to the basement or to the sedimentary section. The few basement-depth estimates considered reliable did not provide a sufficient density of information to attempt a contour presentation.

In an effort to extract additional information from the

Table 1. Interpretation explanation and comparison of densities.

Map unit (Figures 10, 13)	Apparent density (g/cm³)	Measured density ± s (g/cm³)	Rock type
Superior province			
1	2.70	2.71 ± 0.02	Felsic to intermediate intrusives, gneiss
2	2.86	2.84 ± 0.03	Intermediate to mafic metavolcanics
3	2.75–2.85	—	Unsubdivided metavolcanics
4	2.70	2.72 ± 0.03	Felsic to intermediate metavolcanics
5	2.70–2.75	2.70 ± 0.07	Metasediments with minor diabase
6	2.67	2.67 ± 0.04	Felsic stocks/batholiths (granite–granodiorite)
6a	2.77	—	Hybrid rocks
7	2.60–2.65	2.63 ± 0.03	Felsic stocks (granite–syenite)
8	2.80	2.82 ± 0.04	Intermediate to mafic intrusives, gneiss
9	2.92	2.97 ± 0.06	Diabase or gabbro (Nipissing Formation)
10	2.97	2.99 ± 0.06	Ultramafic intrusives or metavolcanics
Southern province			
11	2.90	—	Intermediate to mafic intrusives (Sudbury Igneous Complex)
12	2.75–2.80	—	Sudbury basin rocks (Whitewater Group)
13	2.95	2.95–3.02	Mafic to ultramafic intrusives
Grenville province			
14	2.70	2.70 ± 0.04	Felsic to intermediate intrusives, gneiss
15	2.65–2.70	2.70 ± 0.07	Metasediments, paragneiss
16	2.90	2.86 ± 0.12	Anorthosite and anorthositic gabbro

aeromagnetic survey, we prepared a regional magnetic-contour map of the Cobalt embayment area by re-contouring the data at an interval of 100 nT and smoothing out local disturbances (Figure 12). This map shows magnetic effects which we believe are caused by magnetic minerals (mostly iron formations, probably with some altered ultramafic rocks) within the basement complex, and it gives a remarkable confirmation of the gravity-derived hypothesis of a basement divided into an eastern metavolcanic–metasedimentary sector and a western granitoid sector. Since all known iron formations and ultramafic bodies within the project area are associated with Archean greenstone belts, their presence is taken as confirmation of the volcano-sedimentary nature of the basement complex beneath the eastern half of the Cobalt embayment. The high apparent densities (2.90–2.95 g/cm³) of some of the rocks within this region, combined with their low magnetic response, lead us to theorize that the rocks are made up largely of komatiitic or magnesium-rich tholeiitic flows, resembling in this respect the volcanic units of the Kirkland Lake area.

DETAILED STUDIES: ROUND LAKE AREA

One region which is thought to have high potential for gold mineralization is the eastern part of the Round Lake Batholith and the adjacent area underlain by Archean metavolcanics. This area contains at least four of the five regional geologic indices favorable for gold concentrations, according to Pyke (1976): major faulting, komatiitic flows, nearby sedimentary rocks, and felsic intrusions. Only the fifth characteristic element, carbonate alteration, is not detectable by geophysics. However, there is some evidence, mentioned previously, for believing that the rocks adjacent to major faults that traverse this region have been altered by silicification or carbonatization or by some other process that has reduced their density. Unfortunately, the favorable area for gold is covered by heavy overburden, so that prospecting will be difficult. Geophysics, therefore, has to play an especially active role in exploring for gold in this region.

Exploration strategy requires adding a third dimension to the geological-interpretation map. We need to know not only where the different apparent-density units intersect the bedrock surface but also their geometry and depth extent. To add the third dimension, we subdivided the area into 1.5-km × 1.5-km vertical-sided prisms whose densities are assumed from the apparent-density map but whose depth extent is unknown. A map of the Round Lake area showing the prism array in plan is shown in Figure 13. This array, consisting of approximately 4 000 prisms, was computer analyzed and the thicknesses of the prisms were calculated by repeated adjustments until a sufficiently good match was obtained between the residual Bouguer gravity values and the calculated gravity effect of the prisms at the 4 000-odd grid points. A "sufficiently good match" was deemed to have been reached when the calculated and residual-gravity values agreed to within ± 2 mGal at 98 percent or more of the grid points. The calculated depths after six forward cycles of computation are shown as depth contours in Figure 14. The larger discrepancies were subsequently eliminated by altering the apparent-density unit boundaries as indicated by the broken lines in Figure 13.

There are several features of interest in Figure 14. We shall mention only two:

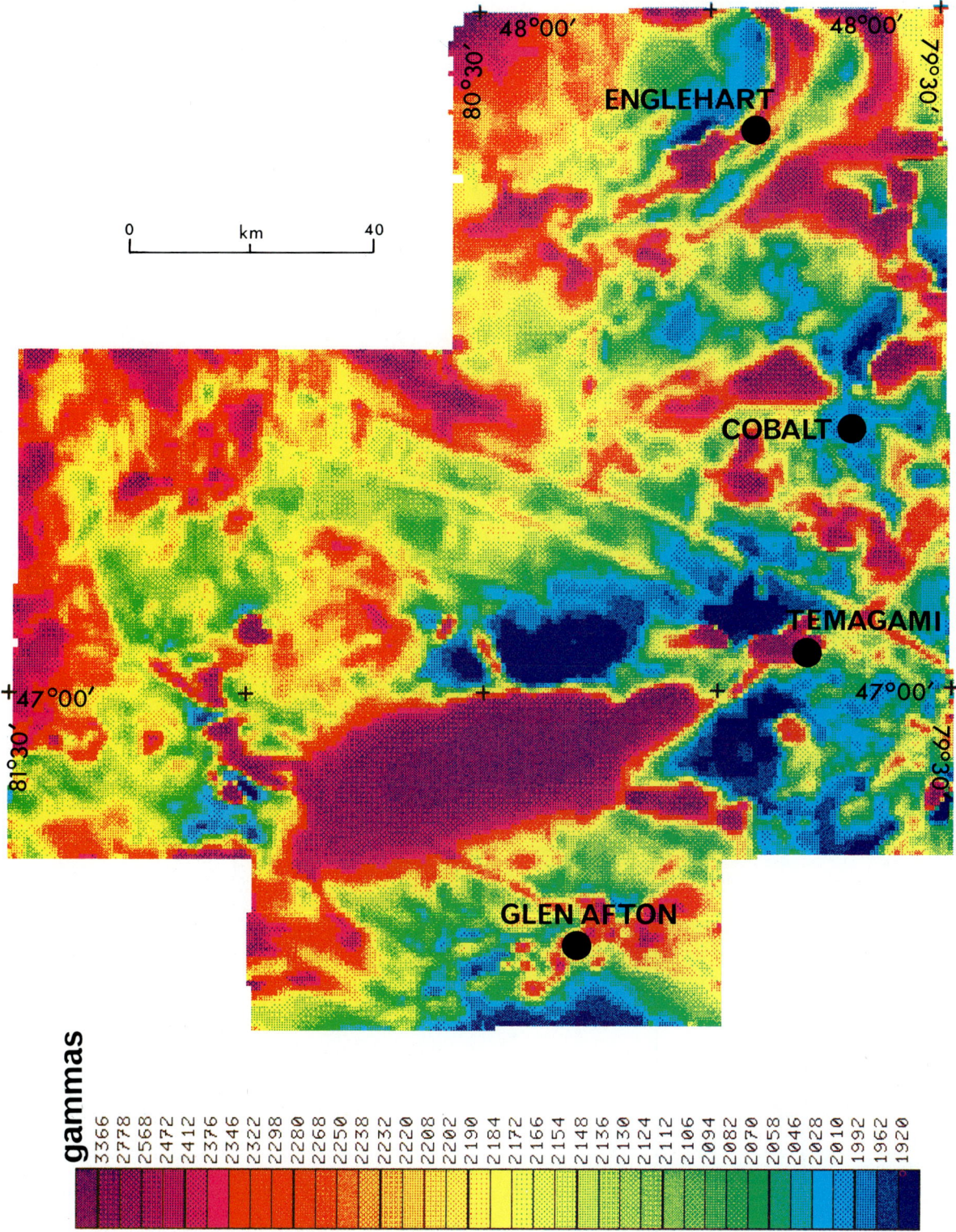

FIG. 11. Total-field aeromagnetic contours of Cobalt embayment area.

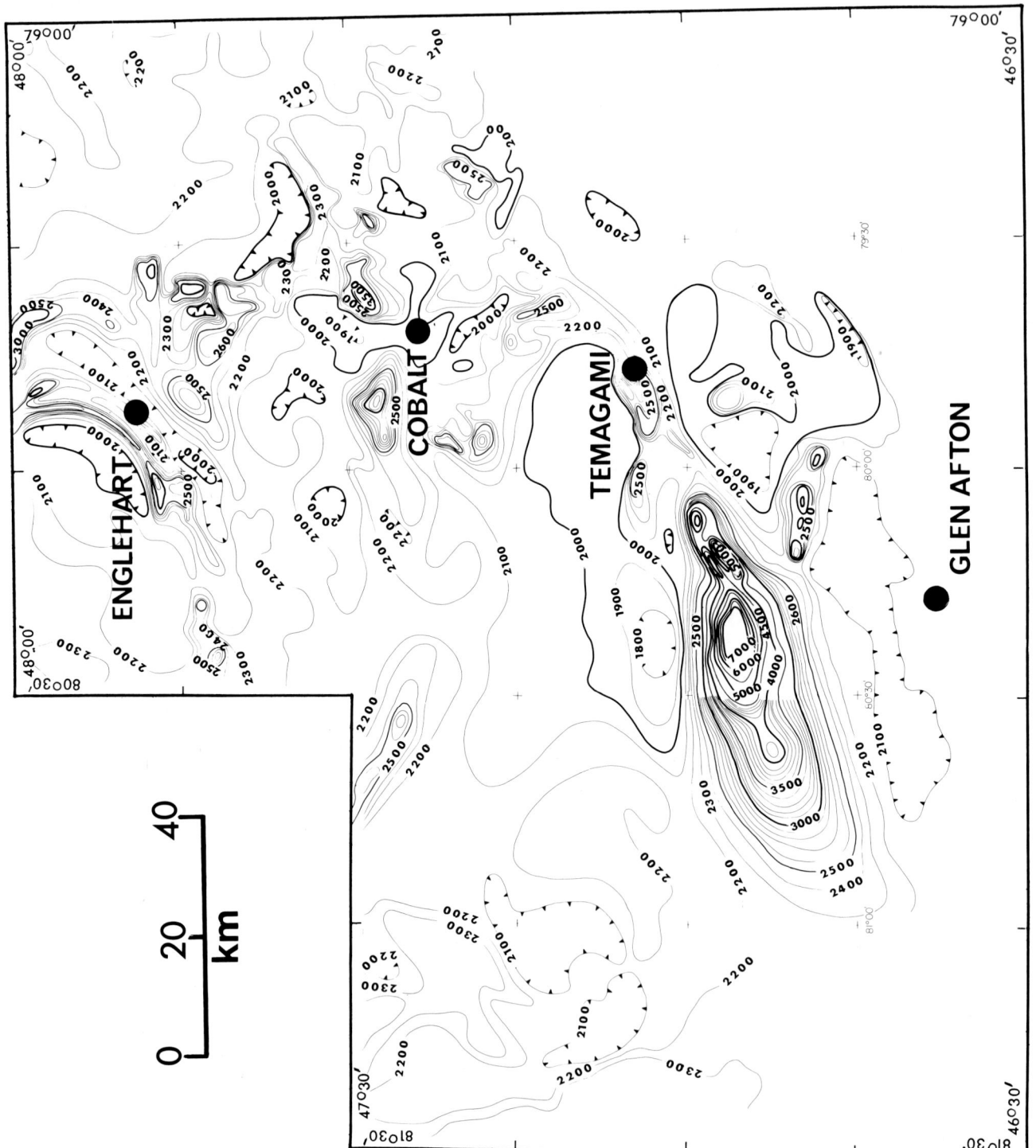

FIG. 12. Regional aeromagnetic contours of Cobalt embayment area. Contour interval is 100 nT.

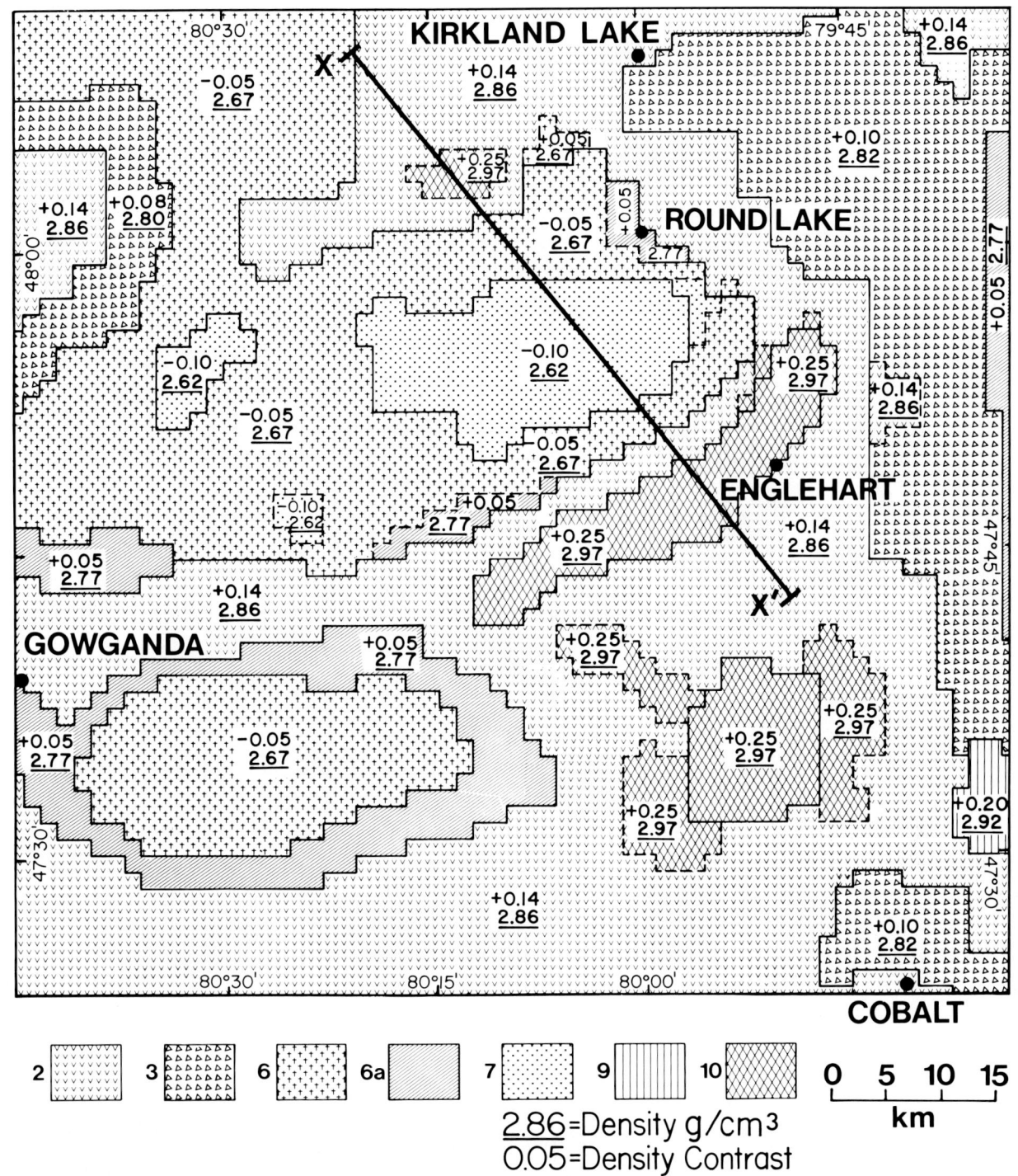

FIG. 13. Digital density boundaries of Round Lake area outlined from apparent-density map for the purpose of 3-D interpretation (location of profile X–X′ is shown). Broken density boundaries indicate adjustments to 3-D model required during computer modeling. For geologic explanation, see Table 1.

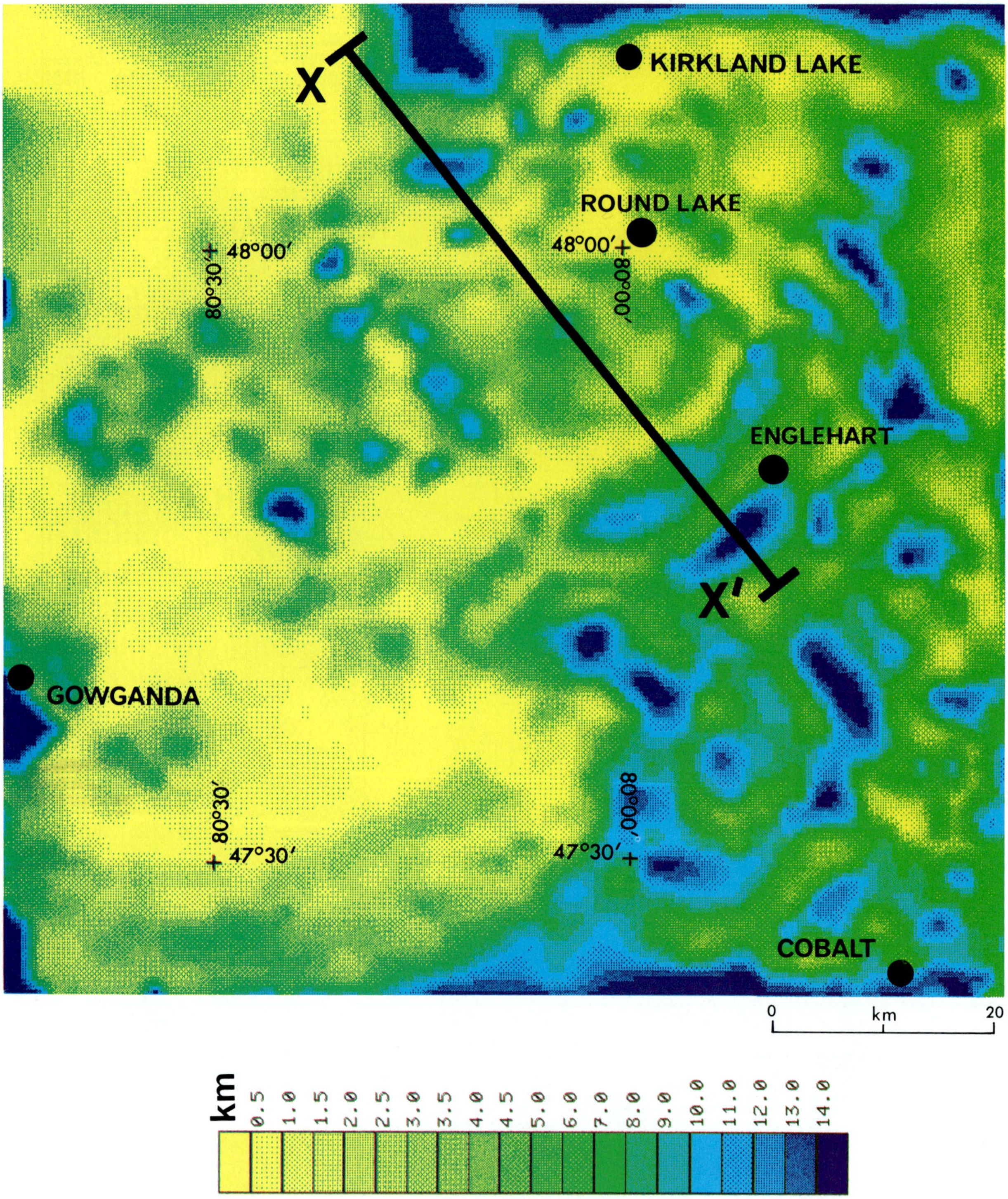

FIG. 14. Calculated depth contours of Round Lake area. Location of profile X–X' is shown.

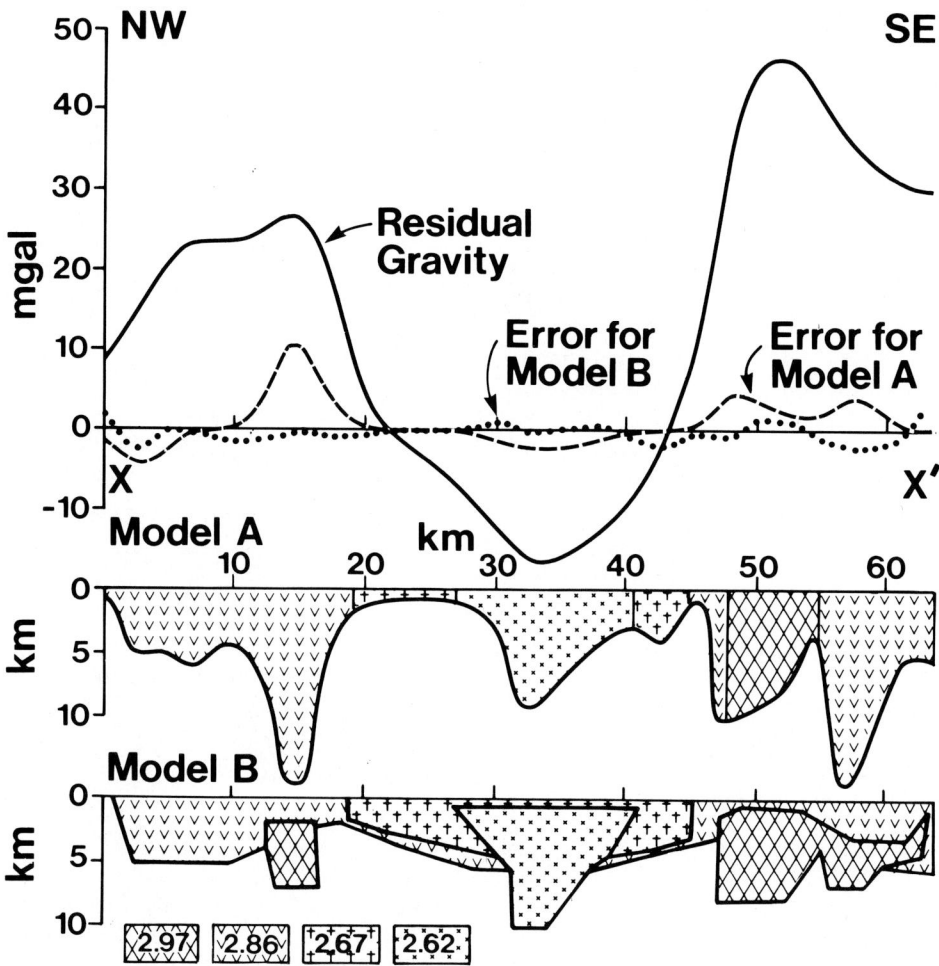

FIG. 15. Profile X–X'. Model A shows in section the results of 3-D model (from Figure 14) computed, using digital density boundaries (from Figure 13), Round Lake area. Model B is the geophysical model obtained from 2½-D interactive inversion. For profile location X–X', see Figures 13 and 14. For geologic explanation, see Table 1.

1. The Round Lake Batholith is apparently zoned. The gravity data cannot be satisfied without introducing a "core" of lighter rock (having a density that is approximately 2.62 g/cm^3) within the eastern half, and two smaller "cores" having a similar density in the western half. These core units probably have a chemical composition that is roughly similar to that of syenite or quartz monzonite and are thought to consist principally of trondhjemites.
2. Archean metavolcanic units along an arc parallel to the southeastern margin of the Round Lake Batholith have an apparent density of approximately 3.0 g/cm^3 and extend to a depth of more than 12 km. The rocks are strongly magnetic where they are in contact with the surrounding rock units, but not in the middle. They are believed to be partially serpentinized ultramafic rocks, probably steeply dipping to overturned komatiites, based on geologic mapping north of the study area (Jensen, 1981). These rocks are also present on the north and east sides of the batholith, where they are largely concealed by thick overburden.

A cross-section drawn from Figure 14 and a geophysical interpretation along this cross-section, using 2½-D interactive-inversion software, is shown in Figure 15. The important features to note are the funnel shape of the Round Lake Batholith and the steeply dipping ultramafic masses that surround it. These are features that are determined fairly unambiguously from analysis of the gravity data, and they suggest a favorable environment for gold exploration, according to Pyke's (1981) regional geologic indices. Without three-dimensional analysis, the gravity interpretation would not have uncovered the evidence needed to apply Pyke's systematics.

DETAILED STUDIES: SHINING TREE AREA

Among the known greenstone belts within the project area, the Shining Tree belt is the largest and probably the most complex. It contains numerous occurrences of base- and precious-metal mineralization, but no currently producing mines. We would like to know more about the Shining Tree complex, particularly with regard to the volumes of the metavolcanic units. To this end, we have calculated thickness-contour (isopach) maps of the felsic-intermediate and intermediate-mafic volcanic members, which are shown in Figure 16. The algorithm that we used was once again to divide the region into a 1.5-km × 1.5-km grid of vertical-sided prisms, whose densities are in this case made equal to the difference between the mean sample densities of felsic to intermediate intrusive rocks (2.71 g/cm^3) and either intermediate to mafic metavolcanics (2.84 g/cm^3) or felsic to intermediate metavolcanics (2.72 g/cm^3), and to calculate the depths of the prisms by matching their gravity effect to the residual Bouguer gravity values at the grid points. The

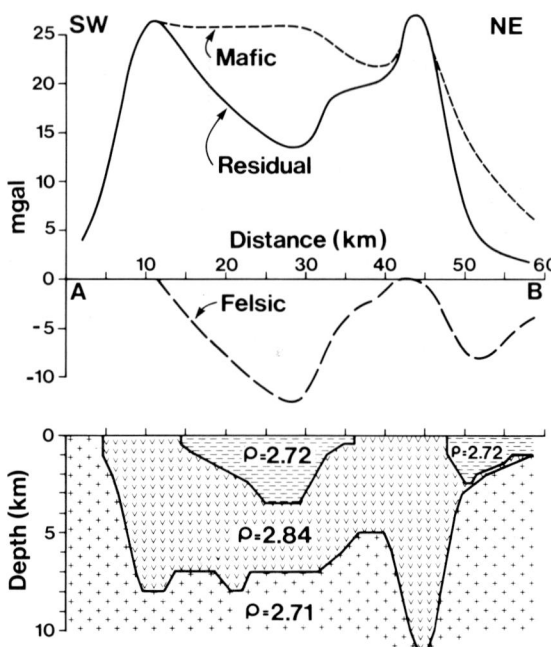

FIG. 17. Profile A–B, showing the residual Bouguer anomaly profile, separated into a positive component from mafic rocks (density = 2.84 g/cm^3) and a negative component from felsic rocks (density = 2.72 g/cm^3), and the 3-D-model section drawn from Figure 16.

differences in method between this and the Round Lake study are twofold: (1) in this study the prisms have a uniform density, and (2) the residual Bouguer gravity field must first be separated into positive and negative components. Intermediate to mafic metavolcanics (density difference = +0.13 g/cm^3) are fitted to the positive residual-gravity field on the assumption that they displace intrusive and gneissic rocks, and felsic to intermediate metavolcanics (density difference = −0.12 g/cm^3) are fitted to the negative residual-gravity field on the assumption that they displace intermediate to mafic metavolcanic rocks. A cross-section of the composite interpretation is shown for easier viewing in Figure 17.

The Shining Tree metavolcanic complex, as interpreted from the gravity survey, shows many of the features typical of Superior province greenstone belts: It contains local accumulations of felsic to intermediate metavolcanics up to 3 km in depth extent, and the maximum depth extent of the metavolcanics appears to be about 6 km, as shown in Figure 16. The apparent "trough" close to the western margin, where the depth extent exceeds 8 km (Figure 16a), is more likely, in our view, to be an indication of local density increase rather than a greater depth extent for the metavolcanics. As an alternate interpretation, buried mafic to ultramafic bodies (density ~2.95 g/cm^3) of limited depth extent can easily be modeled under the two positive peaks of the residual anomaly in Figure 17. The same is also true of the two small stocks of mafic–ultramafic material that have been interpreted in the north-central part of the Shining Tree area.

By presenting the volcanic stratigraphy of the Shining Tree belt in three-dimensional form, the gravity interpretation makes a useful contribution to the development of an

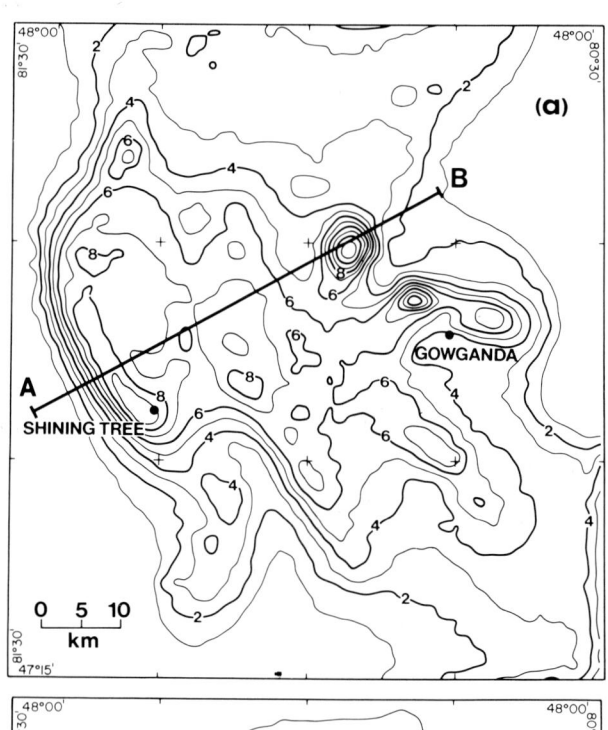

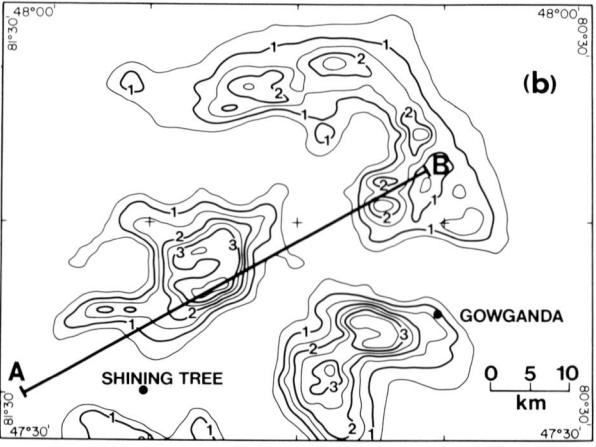

FIG. 16. Computed 3-D-model depths for Shining Tree greenstone belt. a = depth contours (in km) for mafic rocks; b = depth contours (in km) for felsic rocks. Location of profile A–B is shown.

exploration strategy for this area, particularly with respect to prospecting for massive, volcanogenic base-metal deposits.

DETAILED STUDIES: OTHER AREAS

In addition to the Round Lake Batholith and Shining Tree greenstone belt, computer-modeling studies have been carried out on four discrete, positive gravity anomalies in other areas. These anomalies lie within a 40-km radius of the town of Cobalt, and they are believed to be due to massive intrusions of diabase–gabbro, which might have served as feeders for the extensive sill-like intrusions of Nipissing diabase that characterize the geology of the eastern half of the Cobalt embayment. The anomalies were modeled successfully by using strike-limited bodies that are oblong in plan, up to 4 km wide and extending to depths of up to 5 km, and which come close to the present erosional surface. The density difference that was assumed for these bodies was +0.23 g/cm^3 in all cases, being the difference between the densities of the Nipissing diabase (2.95 g/cm^3) and the country rock (2.72 g/cm^3). We mention these results only in order to confirm that this particular objective of the gravity survey was in fact met. Of the four anomalies studied, one and possibly two are thought to represent areas of significant potential for Ag–Co vein-type deposits because they meet the criteria defined by Patterson (1979).

DISCUSSION

The following are what we consider to be the significant achievements and the failures of gravity-based geological interpretation in this area.

1. The study has indicated that Archean greenstone-belt rocks extend beneath the Huronian Supergroup, much farther southward than previously recognized. Thus, the Abitibi subprovince appears to extend southward at least as far as Lake Wanapitei in the Sudbury area.
2. It has indicated that the Archean greenstones beneath the Huronian cover sequence are mostly mafic in composition.
3. It has enabled us to determine the depth extent and the major stratigraphic subdivisions of the Shining Tree portion of the Abitibi belt and other greenstone belts.
4. It has enabled us to determine the three-dimensional shape and depth extent of the Round Lake Batholith and the depth extent of the surrounding mafic and ultramafic metavolcanic formations, with useful implications for gold exploration.
5. It has enabled us to locate and determine the volumes of possible diabase feeder zones, which have some significance in outlining potential Ag–Co prospects.
6. It has been directly helpful in locating favorable areas for gold exploration in the Sudbury and Englehart regions.
7. Gravity data did not provide information that could be used for calculating the elevation on the Archean basement surface beneath the Huronian cover sequence in the Cobalt embayment. This was, however, accomplished by analysis of aeromagnetic data, but not in the desired detail.
8. Gravity data did not allow us to model the nature of the Grenville front tectonic zone. The gravity data suggest that the Archean basement rocks of the Superior province have densities similar to those of the gneissic rocks of the Grenville province.

On balance, we believe that gravity surveying has been a successful and cost-effective method for outlining mineral-prospective areas within the Sudbury–Cobalt region, and that the interpretive methodology described in this paper could well serve as a model for mineral reconnaissance studies in other areas of the Canadian shield.

REFERENCES

Gibb, R. A., and Van Boeckel, J., 1970, Three-dimensional gravity interpretations of the Round Lake batholith, northeastern Ontario: Can. J. Earth Sci., **7**, 156–163.

Grant, F. S., 1973, Magnetic susceptibility mapping: The first year's experience [abstr.]: Presented at 43rd Annu. Int. SEG Meet., Mexico City.

Gupta, V. K., Grant, F. S., and Card, K. D., 1984, Gravity and magnetic characteristics of the Sudbury structure, in Pye, E. G., Naldrett, A. J., and Giblin, P. E., Eds., The geology and ore deposits of the Sudbury structure: Spec. Vol. 1, Ont. Geol. Surv.

Gupta, V. K., and Ramani, N., 1982, Optimum second vertical derivative in geologic mapping and mineral exploration: Geophysics, **47**, 1706–1715.

Gupta, V. K., Thurston, P. C., and Dusanowskyj, T. H., 1982, Constraints upon models of greenstone belt evolution by gravity modelling: Birch–Uchi greenstone belt, northern Ontario: Precambrian Res., **16**, 233–255.

Gupta, V. K., and Wadge, D. R., 1980, Gravity study of the Birch–Uchi and Red Lakes area, District of Kenora (Patricia portion), Ontario: Open-File Rep. 5278, Ont. Geol. Surv.

Jensen, L. S., 1981, A petrogenic model for the Archean Abitibi belt in the Kirkland Lake area, Ontario: Ph.D. thesis, Univ. Saskatchewan, Saskatoon.

Patterson, G. C., 1979, Metallogenic relationships of base-metal occurrences in the Cobalt area, in Milne, V. G., et al., Eds., Summary of field work, 1979: Misc. Pap. 90, Ont. Geol. Surv., 222–229.

Popelar, J., 1972, Gravity interpretation of the Sudbury area: Spec. Pap. 10, Geol. Assoc. Can., 103–115.

Pretorius, D. A., 1976, The nature of the Witwatersrand gold–uranium deposits, in Wolf, K. H., Handbook of stratabound and stratiform ore deposits: Elsevier, **7**, 29–88.

Pyke, D. R., 1976, On the relationship of gold mineralization and ultramafic volcanic rocks in the Timmins area, northeastern Ontario: Bull. 69, Can. Inst. Min. and Metall., 79–87.

——— 1981, Relationship of gold mineralization to stratigraphy and structure in Timmins and surrounding area, in Pye, E. G., and Roberts, R. G., Eds., Genesis of Archean, volcanic-hosted gold deposits: Misc. Pap. 97, Ont. Geol. Surv., 1–16.

APPENDIX

A simple algorithm for apparent-density mapping

Assumptions:

1. Gravity data are available on a (x, y) grid.
2. Gravity data are in residual Bouguer form. That is, regional anomalies have been removed. It may be assumed that the residual anomalies are caused by masses that lie within the uppermost few (say, ~ 10) km of the Earth's crust.
3. Ground surface is horizontal.

Model:

Bedrock density ρ varies with location (x, y), but not with depth (z), down to a depth, h, below which significant contributions to the residual Bouguer gravity field do not arise. This implies that the geology is not horizontally layered. *This bedrock model is appropriate for steeply dipping geology.*

Theory:

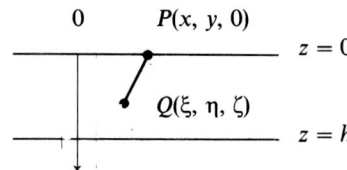

The gravity anomaly at $P(x, y, 0)$ that is caused by the excess or deficiency in the mass of the bedrock between the ground surface $(z = 0)$ and depth, h, is:

$$[\Delta g(x, y)]_{z=0}$$

$$= -G \left\{ \frac{\partial}{\partial z} \int_{-\infty}^{\infty} \int_{-\infty}^{\infty} \int_0^h \frac{\Delta\rho(\xi, \eta)\, d\xi\, d\eta\, d\zeta}{\sqrt{(x-\xi)^2 + (y-\eta)^2 + (z-\zeta)^2}} \right\}_{z=0}$$

$$= G \int_{-\infty}^{\infty} \int_{-\infty}^{\infty} \frac{\Delta\rho(\xi, \eta)\, \partial\xi\, \partial\eta}{\sqrt{(x-\xi)^2 + (y-\eta)^2}}$$

$$- G \int_{-\infty}^{\infty} \int_{-\infty}^{\infty} \frac{\Delta\rho(\xi, \eta)\, d\xi\, d\eta}{\sqrt{(x-\xi)^2 + (y-\eta)^2 + h^2}}, \quad \text{(A-1)}$$

where $\Delta\rho(x, y) = \rho(x, y) - \rho_0$, ρ_0 is the local background density of the crust. Take the two-dimensional F-transform of equation (A-1):

$$\Delta g(u, v) \equiv \int_{-\infty}^{\infty} \int_{-\infty}^{\infty} [\Delta g(x, y)]_{z=0}\, e^{i(ux+vy)}\, dx\, dy$$

$$= 2\pi G \Delta\rho(u, v) \left[\frac{1}{\omega} - \frac{e^{-h\omega}}{\omega} \right], \quad \omega = \sqrt{u^2 + v^2};$$

i.e.,
$$\Delta\rho(u, v) = \frac{1}{2\pi G} \times \frac{\omega}{1 - e^{-\omega h}} \times \Delta g(u, v)$$

Thus, apparent density

$$\rho(x, y) = \rho_0 + \frac{1}{2\pi G} F^{-1} \left\{ \frac{\omega}{1 - e^{-\omega h}} \times \Delta g(u, v) \right\}.$$

If the ground surface is not flat, the calculations must be performed in the spatial domain by forward iteration, using an array of square-ended, vertical-sided prisms of varying heights and of uniform depth extent.

Geological interpretation of a high-resolution aeromagnetic survey in the Amos–Barraute area of Quebec

Sun Yunsheng,* D. W. Strangway,‡ and W.E.S. Urquhart‡

ABSTRACT

A high-resolution aeromagnetic survey of a part of the Abitibi Archean greenstone belt of the Amos–Barraute area in Quebec was studied to attain a better understanding of the geology of the area using various processing methods. The apparent-susceptibility map offers higher resolution than the total-intensity map, and many rock types in the area can be mapped using contrasts in apparent susceptibility. The second-vertical-derivative map gives some useful information about the structure as well as the stratigraphy of the rock units. An interpreted geologic map was developed using these maps.

INTRODUCTION

The Canadian shield has had a complicated history of intrusion, extrusion, sedimentation, and metamorphism. One of the major features of the shield is the Superior province, which contains numerous greenstone belts. The Abitibi greenstone is one of the most extensively studied and explored of these belts.

The study area shown in Figure 1 extends from latitude 48°15′ N to 48°45′ N and from longitude 77°30′ W to 78°15′ W. This was one of the first regions in the northern part of Quebec to be prospected for exploitable minerals. Prospecting activity in the area has been extensive since the beginning of the century, first for gold and then for base metals and nickel (Brett et al., 1976). The southern part of the area, LaMotte–Lacorne townships, is of special interest owing to commercial quantities of lithium, molybdenum, bismuth, and nickel (Tremblay, 1950). To assist in prospecting, the Quebec Department of Natural Resources carried out geologic mapping in which geophysical data were used to outline poorly exposed ultramafic rocks.

In 1948 the Geological Survey of Canada (GSC) carried out an airborne magnetic survey in the area. Mining companies also performed a variety of ground magnetic surveys in the area. The area was flown again with the GSC's high-sensitivity magnetometer system during the 1971 field season. The survey was flown at an elevation of 1 000 ft (305 m) with a flight-line spacing of 1 000 ft (Sawatzky and Hood, 1972). Maps with a contour interval of 5 nT were digitally processed and published in 1974.

Since the late 1960's, computer processing of aeromagnetic data has been extensively developed. Particular attention has been given to the application of Fourier-transform methods for analysis of data in the wavenumber domain. Upward and downward continuation, first and second derivatives, and reduction to the pole are all well-known techniques. In order to evaluate the use of the resulting high-resolution aeromagnetic data for detailed geologic mapping, a project was initiated by the GSC in Godfrey Township in 1972. In that work, downward continuation; first, second, and third vertical derivatives; and bandpass filter were computed and tested (Kornik et al., 1975). Recent work, using apparent-susceptibility mapping, has shown that this technique is particularly useful for geologic mapping (Letros, 1980; Urquhart and West, 1979; Letros et al., 1983; Grant, 1974).

The principal goal of the present work is to provide a geological interpretation of the Amos–Barraute area, Quebec, by interpreting the high-resolution aeromagnetic data with modern data-processing techniques and using measurements on samples for control.

GEOLOGY

The Amos–Barraute area covers nine townships and parts of 16 other townships and is underlain by early Precambrian (Archean) volcanic rocks that are part of the Abitibi greenstone belt (Figure 1). The belt is composed of volcanic rocks ranging in composition from ultramafic to rhyolitic and of lesser amounts of sedimentary rocks.

The oldest rocks of the area are volcanic rocks, which were originally mapped as the Kinojevis Group in the southern part of the area, and in the northern part as Keewatin-type volcanics (Table 1); they are now considered

---

*Department of Applied Geophysics, Changchun College of Geology, China.
‡Department of Geology, University of Toronto, Toronto, Ont. M5S 1A1, Canada.

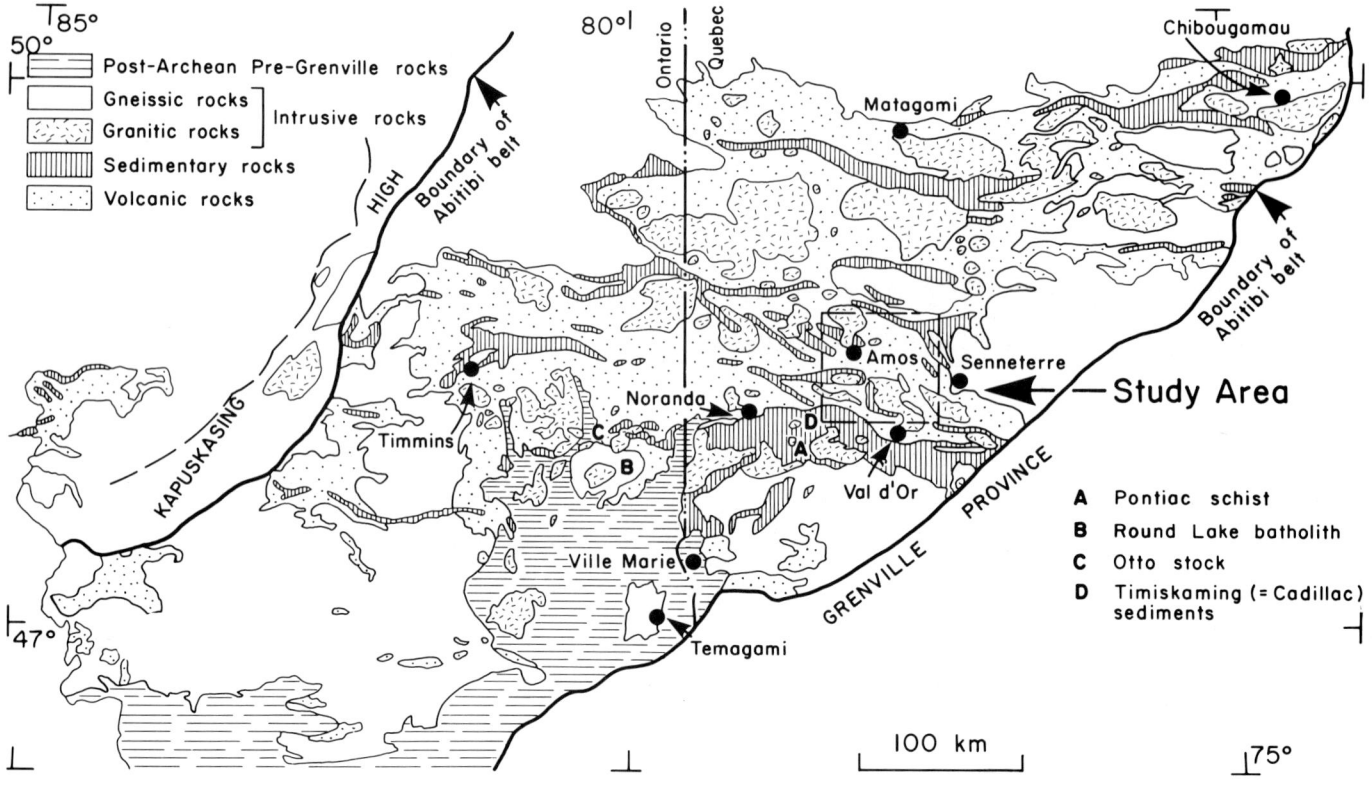

FIG. 1. Map showing part of the Abitibi greenstone belt (by Jolly, 1978).

to be equivalent (Weber and Latulippe, 1964). In the southwest quarter, another region of volcanic rocks is called the Malartic Group, although it is possible, because of folding and faulting, that the Kinojevis volcanic rocks are a repetition of the Malartic Group of the area (Tremblay, 1950). Overlying these volcanic rocks are the sedimentary strata of the Keewagama Group. Both the volcanic and sedimentary rocks are intruded by plutons ranging in composition from ultramafic to felsic. Their relative ages are not known everywhere, but some of them are nearly contemporaneous with the lavas and some are clearly younger. The final period of intrusion in the late Precambrian is marked by diabase and olivine gabbro; all other intrusives are probably Archean (Weber and Latulippe, 1964). Figure 2 is a geologic map of the area. The following description is taken from Tremblay (1950), Weber and Latulippe (1964), Brett (1960), Brown (1958), Jones (1964), McDougall (1965), Sharpe (1961), and Brett et al. (1976).

The volcanic rocks have been divided into an upper and a lower series. The lower series consists mainly of basaltic and andesitic flows interbanded with minor amounts of rhyolite, trachyte, and pyroclastic rocks. The rocks of the upper series are mainly of intermediate composition. The sedimentary rocks appear to overlie the Kinojevis and Malartic Groups of volcanic rocks conformably, and the contact zone suggests an interbedded sequence of lava flows, tuffs, agglomerate, and sedimentary rocks. The sedimentary rocks in the northern part are made up of slates, cherts, graywackes, argillites, quartzites, and carbonate rocks. There are also many beds of fine-grained ferrodolomites. In the southern part the sedimentary rocks have been converted to quartz-biotite schist. Numerous types of concordant and discordant intrusives ranging from ultramafic to felsic occur throughout the volcanic and sedimentary rocks.

In LaMotte township, a series of ultramafic rocks is intruded and intercalated with lavas and sedimentary rocks. They occur as sills in the volcanic rocks and form one of the largest concentrations of ultramafic rocks in western Quebec. Recent work on these rocks has shown that they may, at least in part, be ultramafic lavas (Brett et al., 1976). The

Table 1. Geologic units in the Amos–Barraute area.

Pleistocene	Clay, sand, gravel
Unconformity	
Upper Precambrian	Diabase dikes
Lower Precambrian	Intrusive rocks: granite, granodiorite, gabbro, peridotite, pyroxenite
	Keewagama Group: slates, graywacke, biotite schist
	Kinojevis Group (equivalent to Keewatin Group) Upper volcanic series: intermediate and siliceous lavas interbanded with narrow layers and lenses of mafic lava
	Lower volcanic series: mafic lavas (basalt and andesite) with minor amounts of rhyolite

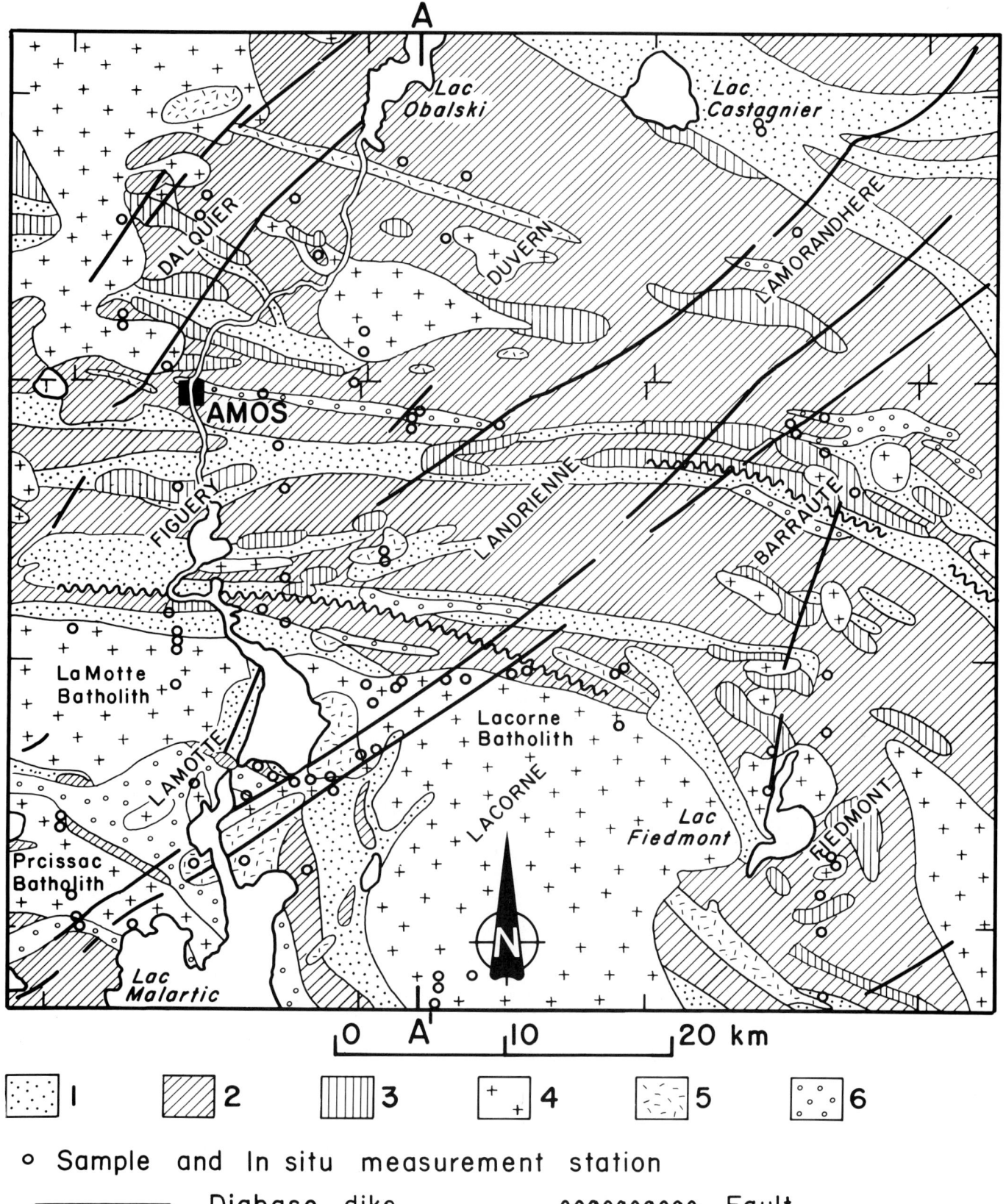

FIG. 2. Geologic map of the study area, showing sample locations.

Preissac–Lamotte–Lacorne batholiths form a band of intrusive rocks about 30 miles (48 km) long, in an east–west direction, by 14 miles (22 km) wide in the southwestern quarter.

Large granitic bodies are present in Dalquier and Duverny townships. Intrusive bodies of lamprophyre, aplite, pegmatite, quartz porphyry, and quartz gabbro are found within the granitic stocks. Close to the northern contact of the Lacorne and LaMotte batholiths, a belt of lens-like peridotite sills was traced within the volcanic rocks. They are now composed mainly of talc or serpentine, chlorite, and tremolite. In the northern half of the map area, there are three gabbro–peridotite–pyroxenite complexes. One lies along the northern border of Figuery and Landrienne townships; the others are in Barraute and Lamorandiere townships, respectively. The main rock type is a medium- to coarse-grained gabbro associated with dike-like bodies of peridotite and pyroxenite that represent multiple injections within the gabbro. In addition, other intrusive bodies of gabbro and altered gabbro are present in Dalquier, Duverny, and Lamorandiere townships. In the remaining part, many intrusive rocks occur as sills, plugs, and irregular masses. All the intrusive rocks above are believed to be of Archean age.

Late Precambrian diabase dikes are numerous in the area. They can be grouped into two distinct sets, striking N 50° E to N 60° E and north to N 20° W. All the dikes are in sharp contact with adjacent rocks, are continuous for long distances, and are relatively straight, although some variations in their strike produce locally sinuous forms.

The metamorphism is dominantly greenschist facies (Jolly, 1977, 1978), although locally it has only reached prehnite-pumpellyite facies. Contacts associated with intrusive bodies may have higher grade amphibolite facies.

APPARENT-SUSCEPTIBILITY MAPPING

Studies have shown (Letros, 1980; Urquhart and West, 1979; Letros et al., 1983) that a combination of linear filters known as apparent-susceptibility mapping (SUSC) lends itself particularly well to geological interpretation. The apparent-susceptibility map gives a better picture of geologic boundaries than the total-field map does, because a regional field has been removed and the data are downward continued. In addition, the map values are in apparent-susceptibility units, which are related to measurable rock properties. These are highly useful for geologic mapping. Because of these significant advantages, a review of the calculation and application of the apparent-susceptibility method follows.

Several basic conditions are assumed (Urquhart, 1976) in the calculation of the apparent-susceptibility map.

1. The measured magnetic field is caused by an assemblage of bodies of rectangular cross-section one grid cell in dimension, with a body centered on each grid point.
2. Magnetization is by induction only.
3. The bodies are vertically sided.
4. The bodies extend to infinite depth.

Deviations from these assumptions do not prevent one from

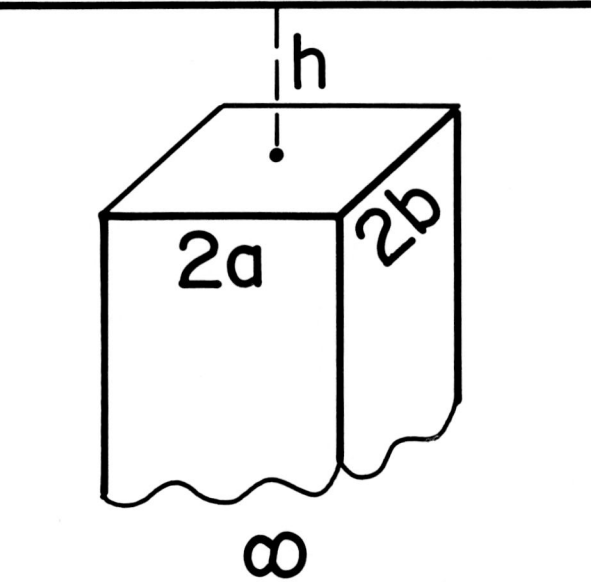

FIG. 3. Prism model used to develop apparent-susceptibility map. Depths below observing plane, 2a, 2b, are dimensions of the individual prisms.

deriving apparent-susceptibility maps, but characteristic features that require careful interpretation are produced.

The magnetic field of a prism model (Figure 3) is defined in the frequency domain as:

$$T(u, v) = 2KT_0 M(u, v) H(u, v) S(u, v) D(x, y; u, v), \quad (1)$$

where

$M(u, v) = [-\ell L u^2 - mMv^2 + nN(u^2 + v^2)$
$\quad - (Lm + M\ell)uv + i(Ln + N\ell)u(u^2 + v^2)$
$\quad + i(Mn + Nm)v(u^2 + v^2)](u^2 + v^2)^{-1}$ – magnetization factor;

$H(u, v) = \exp[-h(u^2 + v^2)^{1/2}]$ – depth factor;

$S(u, v) = 4 \sin(ua) \sin(vb)$ – size factor;

$D(x, y; u, v) = \exp[-i(ux + vy)]$ – displacement factor;

K – susceptibility;

T_0 – intensity of the Earth's field;

ℓ, m, n – direction cosines of the Earth's magnetic field;

L, M, N – direction cosines of the total polarization vector;

a, b – half-width sides of the prisms;

u, v – wavenumbers in the x, y direction.

If only induced magnetization is present, then $L = \ell$, $M = m$, $N = n$, and the magnetization factor can be rewritten as:

$$M(u, v) = \left[-\frac{\ell^2 u^2 - m^2 v^2}{(u^2 + v^2)} + n^2 - \frac{2\ell muv}{(u^2 + v^2)} + i\frac{2\ell nu + 2mnv}{(u^2 + v^2)^{1/2}} \right].$$

Now, it is assumed that the surface geology is equivalent to a number of independent prisms with data-interval dimensions, extending infinitely deep and centered under their respective data points. Then, the aeromagnetic map is con-

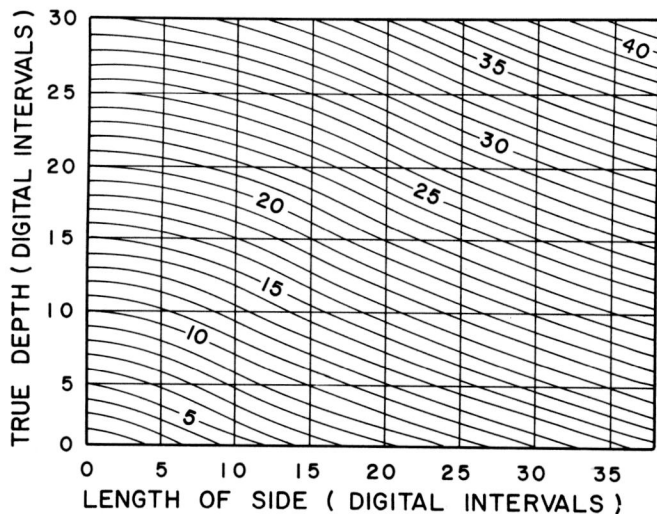

FIG. 4. Nomogram to correct the effect of finite horizontal size for bodies of varying length. Contour lines are of apparent depth.

sidered to be a result of the superposition of the anomalies owing to each small prism; i.e.,

$$T(u, v) = 2\pi T_0 \sum_{j=1}^{n} K_j M_j(u, v) H_j(u, v) S_j$$
$$\times (u, v) D_j(x, y; u, V), \quad (2)$$

where j is the index of each small prism, and n is the number of prisms. Because all prisms are assumed the same, M, H, S will be independent of the index, and

$$T(u, v) = 2\pi T_0 M(u, v) H(u, v) S(u, v) \sum_{j=1}^{n} K_j D_j(x, y; u, v). \quad (3)$$

Here, the complex $T(u,v)$ is given as an array of discrete values corresponding to the fast Fourier transform (FFT) of the aeromagnetic data. If $T(u,v)$ is divided by M, H, and S, then

$$T_n(u, v) = 2\pi T_0 \sum_{j=1}^{n} K_j D_j(x, y; u, v). \quad (4)$$

Taking the inverse FFT gives

$$T_n(x, y) = 2\pi T_0 \sum_{j=1}^{n} K_j d_j(u, v; x, y), \quad (5)$$

where $d_j(u,v;x,y)$ corresponds to the inverse FFT of $D_j(x,y;u,v)$. Since reduction to the pole, removal of the shape of the bodies and downward continuation to the surface have been done, and each data point is so close to the center of the surface of its corresponding prism that there is no significant contribution from other prisms, equation (5) becomes

$$T_n(x, y) = 2\pi T_0 K(x, y). \quad (6)$$

Dividing by $2\pi T_0$, we obtain the apparent susceptibility of each point, i.e., the apparent-susceptibility map.

EFFECT OF SIZE ON DEPTH ESTIMATES

In the apparent-susceptibility method, data are downward continued to the top of prisms. The distance from the flying height to the top of the prisms has to be estimated first. For the purpose of surface geological interpretation, however, the flight elevation is used. If the rocks in an area do not crop out, the depth is often determined by using a radial logarithmic energy-density spectrum (LnEDS). Bhattacharyya (1966) gave the expression of EDS of a vertical prism of infinite extent and suggested that depth estimates to magnetic sources could be determined by using the decay rate of the radial LnEDS. Spector (1968) and Spector and Grant (1970) calculated the average depth of an ensemble of small prisms using the same method. Magnetic bodies are usually of finite size, and the slope of the spectral curve is, in part, controlled by the size of the causative sources (Spector and Parker, 1979); the larger the source, the longer the wavelength spectral components, and, therefore, the greater the slope of the spectral curve. For this reason, Spector (1968) gave an approach to correcting for the size effect. The curves shown in Figure 4 have been calculated in a similar manner and can be used to correct the estimated depth, based on the size of the body.

The radial LnEDS for the Amos–Barraute area is given in Figure 5. From this plot the depth of the shallower sources was estimated to be about 11 grid cells, i.e., about 700 m. This is larger than the terrain clearance of 300 m. Six small, sharp anomalies on the aeromagnetic map were selected for analysis. With a depth of 11 grid cells (700 m) and a rough approximation of 16 grid cells (1 000 m) for the size of these bodies, the depth is 5 grid cells or 350 m. This value corresponds to the flying height, and we can conclude that the size estimate was reasonable.

EFFECT OF PRISM DIP

The model used to derive the apparent-susceptibility map assumes an ensemble of vertically sided prisms. Deviations from this assumption lead to incorrect positioning of the body and to an erroneous estimate of the susceptibility.

For a vertical sheet-like body extending to infinity, the total-field anomaly, ΔT, when reduced to the pole, is the same as the vertical-field anomaly, ΔZ_1, if the magnetization is vertical. Let ΔH_1 be the horizontal component owing to vertical magnetization, and α the angle between the magnetization direction and the sheet face. The vertical component (ΔZ_1) owing to a model with arbitrary magnetization is

$$\Delta Z_1 = \Delta Z_1 \cos \alpha + \Delta H_1 \sin \alpha. \quad (7)$$

If the dip is not vertical, reduction to the pole gives the value ΔZ rather than ΔZ_1. If α is small, then

$$\Delta Z \approx \Delta Z_1 \cos \alpha, \quad (8)$$

which implies that the magnitude of the calculated apparent susceptibility will be reduced in proportion to the cosine deviation from the vertical.

In the studied area, most contacts and the dikes dip steeply, typically about 70–80 degrees, so that the magnitude will be reduced by up to 6 percent and the displacement of

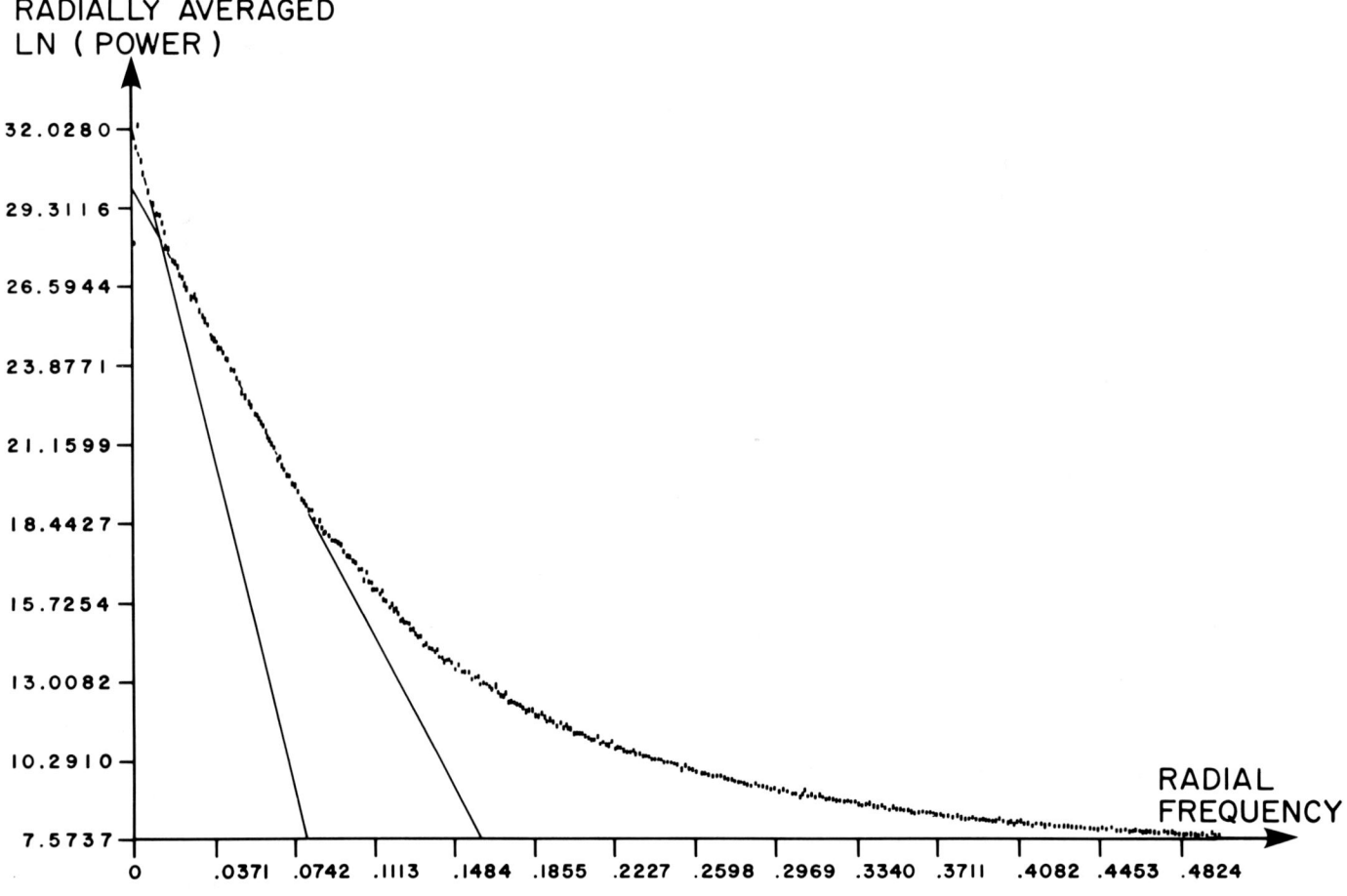

Fig. 5. Radial average logarithmic energy density spectrum of the Amos–Barraute area.

the peak of the anomaly from the source will typically be less than 150 ft (46 m).

The aeromagnetic data collected on north–south flight lines were interpolated into a square grid, with about 280-ft (85-m) spacing. The flight-line spacing was 1 000 ft (305 m), and the interval of measured points was about 50 ft (15 m) along each line. Therefore, the data density is much higher along the flight line than perpendicular to the line. The differing densities mean that the interpolated grid contains east–west–trending noise, which becomes obvious when simple enhancement techniques are applied. Figure 6a shows the total-field map owing to a northwest-striking diabase dike. Figure 6b shows the calculated apparent-susceptibility map using a simple low-pass filter. The apparent-susceptibility map shows a series of east–west–striking anomalies distributed along the dike. The anomalies are caused by the enhancement of the interpolation error. Therefore, a strike-sensitive filter that suppresses higher frequencies in the east–west direction, more than in the north–south direction, was designed so that the enhancement of the interpolation error could be suppressed. Figure 6c is the resulting calculated apparent-susceptibility map using a strike-sensitive filter. It is evident that using a strike-sensitive filter improves the results considerably.

GEOLOGICAL INTERPRETATION OF THE DATA

In the study, five filtered maps (first and second vertical derivative, apparent susceptibility, regional field, and apparent susceptibility of the residual field) were produced in addition to the total-field map. The total-field map, the apparent-susceptibility map, and the second-vertical-derivative map were color contoured as an aid to interpretation.

Seventy-five samples of various rock types were collected (Figure 2), and their magnetic susceptibilities (Ks) were measured as a control on the interpretation. The susceptibilities were measured on a standard AC bridge, and the remanence on a cryogenic magnetometer. The susceptibilities are shown in Figure 7, and the susceptibilities and remanences tabulated in Table 2.

MAJOR MAGNETIC-ANOMALY FEATURES

Intense anomalies seen on the total-field map (Figure 8) correspond to mafic or ultramafic intrusives, such as peridotite, serpentinite, and pyroxenite, especially in the southwestern corner, where the ultramafic rocks cause aeromagnetic anomalies stronger than 1 500 nT. Near the northern border of Figuery and Landrienne townships there is a long are due to peridotite distributed along the northern edge of the Lacorne–LaMotte Batholiths.

There are a number of individual diabase dikes in the area, but only a few of them are clearly seen, because most are obscured to some degree by superposition. One distinctive linear anomaly and one weaker anomaly cross the whole map in a northeasterly direction. These anomalies are caused by upper Precambrian Abitibi diabase dikes. At the northwestern corner, another poorly defined northeast-trending linear anomaly may be due to a dike. It will be shown later that the apparent-susceptibility map indicates the dikes that are shown on geologic maps as well as a few dike-related anomalies that have not been previously mapped.

All the magnetic minima and flat patterns are associated with granitic, volcanic, and sedimentary rocks, although generally it is not possible to distinguish between them on the basis of the total-field map alone; a possible exception is the Lacorne Batholith.

FILTERED MAPS

Five separate filters, including strike sensitive, reduction to the pole, downward continuation, size correction, and band-pass, were employed for the calculation of the apparent-susceptibility map (Figure 9). In addition to the greater resolution, the data are presented in a form more appropriate for direct interpretation. The highest observed apparent-susceptibility value (more than 120×10^{-3} SI), occurs over the mafic and ultramafic units underlying the area. The contours of the SUSC map can be used to outline the contacts much more precisely than those of the total-field map. The SUSC amplitude of the ultramafic rocks in La-Motte, Barraute, Amos, and Landrienne townships is

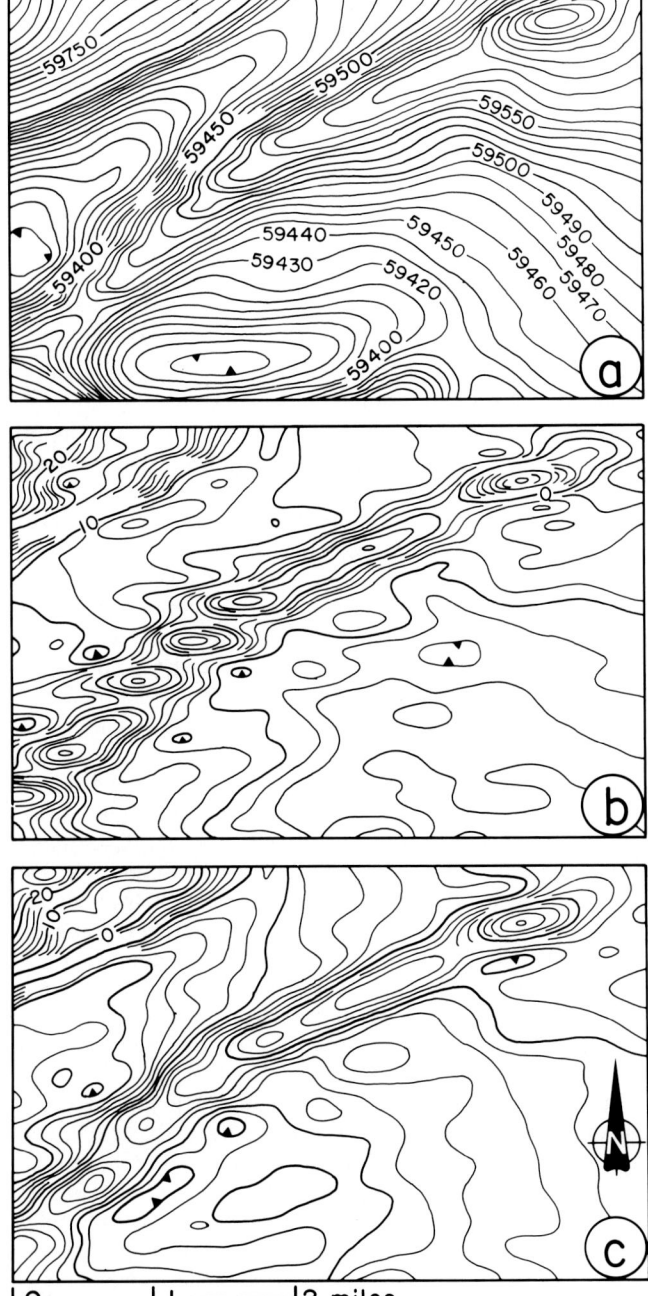

FIG. 6. Diabase-dike anomaly: a. Total-field anomaly. b. Apparent-susceptibility anomaly without strike-sensitive filter. c. Apparent-susceptibility anomaly with strike-sensitive filter.

(about 11 miles or 18 km), linear anomaly that coincides with peridotite. At the northeastern corner of Barraute township, another peridotite body is the cause of a strong anomaly. In Barraute and Lamorandiere townships, and at the northern edge of the map, similar anomalies may also be caused by ultramafic rocks. A series of anomalies that are less intense than those described previously occur along the northern border of Lacorne and LaMotte townships. These anomalies

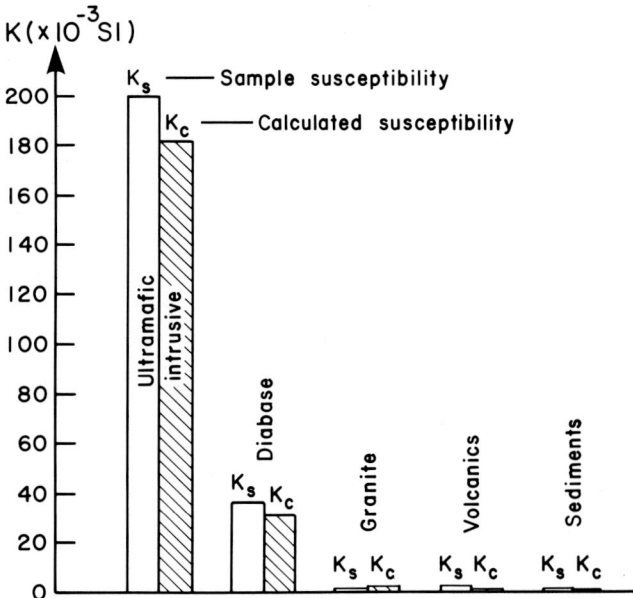

FIG. 7. Magnetization of various rock types: histograms of susceptibilities (Ks) and equivalent susceptibilities calculated from remanent magnetization.

Table 2. Magnetic properties of various rock types.

Rock type	No. of samples	K_s ($\times 10^{-3}$ SI)	J_r ($\times 10^{-3}$ SI)	K_c ($\times 10^{-3}$ SI)
Mafic intrusive	7	200	856	160–190
Diabase	12	33	198	13–38
Granite	14	2	<20	2.5–5
Volcanic	20	3	<5	<1
Sedimentary	3	<0.6	<3	<0.5

K_s—sample susceptibility.
K_c—calculated susceptibility.
J_r—remanent magnetization.

160–190 $\times 10^{-3}$ (SI). This is very close to the sample susceptibility (K) values, which average 200 $\times 10^{-3}$ (SI). The correspondence implies that the induced magnetization is primarily responsible for the strongest aeromagnetic anomalies. It is quite possible that the strong apparent-susceptibility anomalies in the middle of the eastern boundary of Barraute township and at the northern border of the map area also indicate the existence of mafic or ultramafic rocks, although the geologic maps show the presence of volcanic rocks. Susceptibility values of about 15–40 $\times 10^{-3}$ (SI) are seen over the diabase dikes. These values are also consistent with the measured sample values of 33 $\times 10^{-3}$ (SI). Several northeast-trending linear anomalies can be recognized. They are probably associated with upper Precambrian dikes, not all of which are shown on the geologic maps. For instance, northeast of Amos, a N 45° E linear anomaly can be seen. The two long, linear northeast-striking anomalies that are prominent on the total-magnetic-intensity map are outlined much more precisely. Between these two prominent dikes a third parallel one can be recognized on the apparent-susceptibility map.

The volcanic rocks and granites are both characterized by low apparent-susceptibility values, but there are some differences among them. In the east and southeast corner, mainly mafic to intermediate volcanic rocks are apparent, and their magnetization is variable. Along an outcrop in the middle of Fiedmont township, four in-situ measurements 50 m apart were made, giving sample values of 1.9, 17, 1.6 and 1.6 ($\times 10^{-3}$ SI), while about 3 miles (4.5 km) to the south an in-situ susceptibility value of 41 $\times 10^{-3}$ (SI) was measured on a volcanic outcrop. This variability makes the contour pattern irregular, with an amplitude of about 12 $\times 10^{-3}$ (SI). These features are believed to be largely associated with stratigraphic variations in volcanic rocks. Granitic rocks have low or even slightly negative values of apparent susceptibility as well as a smoother appearance. Apparent-susceptibility values found over the Lacorne Batholith are about 5–8 $\times 10^{-3}$ (SI), whereas those over the LaMotte Batholith are somewhat higher. The susceptibilities measured on samples from the batholiths, shown in Table 3, are consistent with the apparent-susceptibility values determined from the map. The Lacorne Batholith is described as hornblende granodiorite and hornblende monzonite, whereas the rocks in the LaMotte Batholith are described as biotite granodiorite and quartz monzonite (Brett et al., 1976). The wall rocks adjacent to the granitic intrusives display a metamorphic aureole. These alteration haloes cause local anomalies around the contact zones, such as is seen around the intrusive in the northwest corner of the map sheet. It should be mentioned that south of Amos the apparent-susceptibility contours are slightly negative and smooth. This pattern is similar to that of granitic intrusives in both value and appearance. It is possible that there is a granitic or dioritic intrusive body at depth, although volcanic rocks have been mapped on the geologic map.

In places, extremely negative apparent susceptibilities are seen. In the southwestern corner, the lowest total-field intensity appears associated with the highest anomalies in the area. A peridotite zone striking east–west from Amos is highly magnetic. The large contrast in total field creates an oscillation in the apparent-susceptibility map (Bambrick et al., 1982).

In addition to its use in outlining rock types, the apparent-susceptibility map shows structural features. Figure 10 is a local example of an anomaly, illustrating the difference in apparent resolution between the apparent-susceptibility map and the total-field map. The displacement of the magnetic anomaly seen on the apparent-susceptibility map suggests the presence of a fault, whereas this offset is not seen on the total-field map. It is inferred from offsets in the apparent-susceptibility map (Figure 9) that four faults cut the Amos east–west–trending peridotite at the northwestern corner of Landrienne township.

The second-vertical-derivative map is shown in Figure 11. On this map, all the northeast-trending diabase dikes are easily recognized. Slight differences between batholiths and volcanic rocks can be detected. In the areas of volcanic and sedimentary rocks, the second vertical derivative shows strong lineation, probably associated with stratigraphic variations in the volcanic–sedimentary sequence. For instance, in the northeastern part of the map the patterns appear to change gradually from a N 40°–50° W trend to an east–west trend. This appears to follow the strike of the volcanic rocks, suggesting stratigraphic sequences in the volcanics. Thus, the study of this map gives some useful information on the volcanic rocks. In the batholith area, no similar trends can be seen. In the northeastern corner, the map shows a band of sedimentary rocks, but the pattern is similar to that seen in the southeastern part of the map, where volcanics are mapped. It is therefore considered that the band mapped as sedimentary rocks is probably a sequence of interbedded volcanic and sedimentary rocks.

Based on the geophysically processed high-resolution aeromagnetic data, the sample magnetic-property measurements, and the existing geologic data, an interpreted geologic map was developed (Figure 12). This map gives more

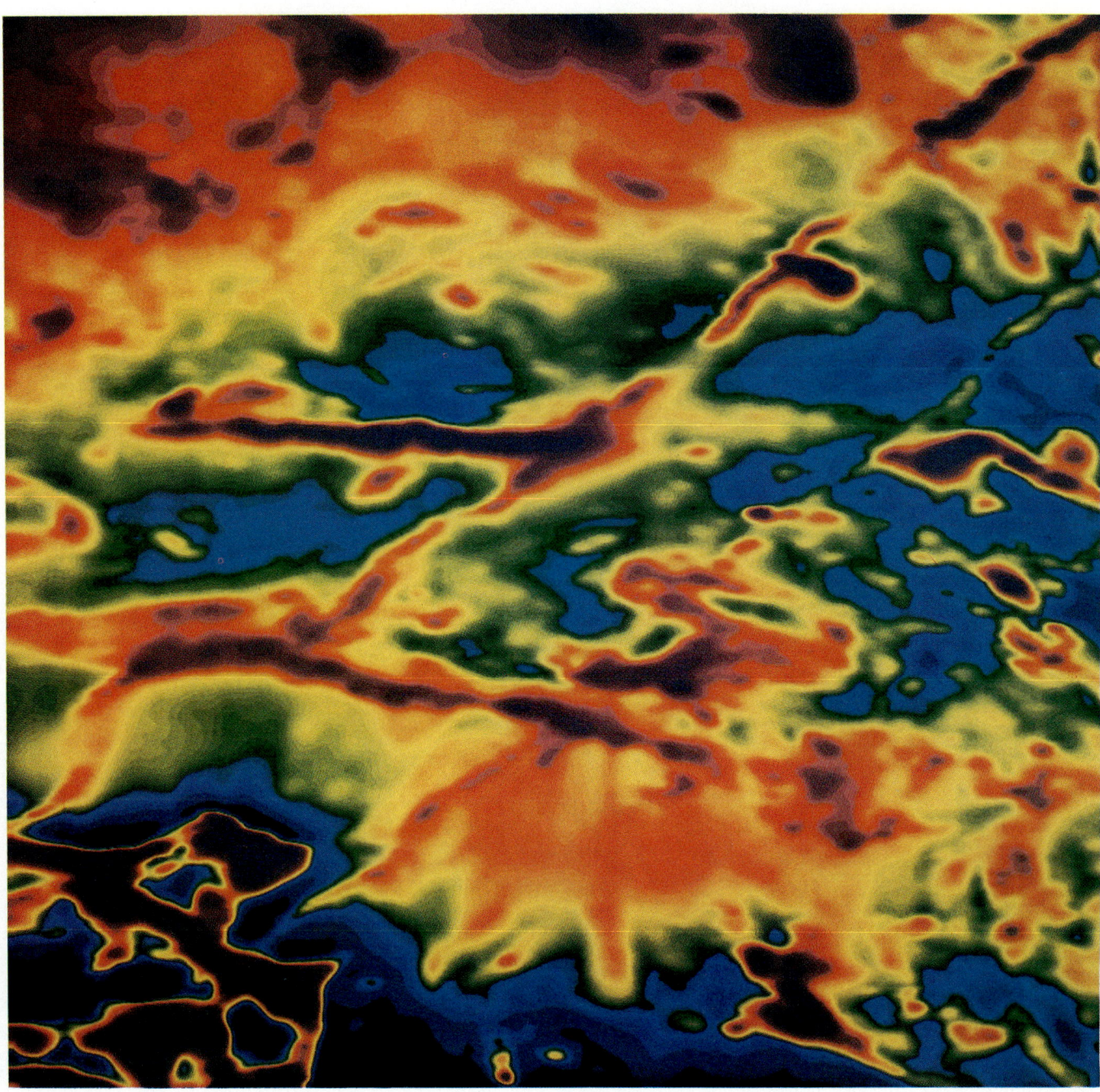

Fig. 8. Total-magnetic-field map of the study area (units are nT).

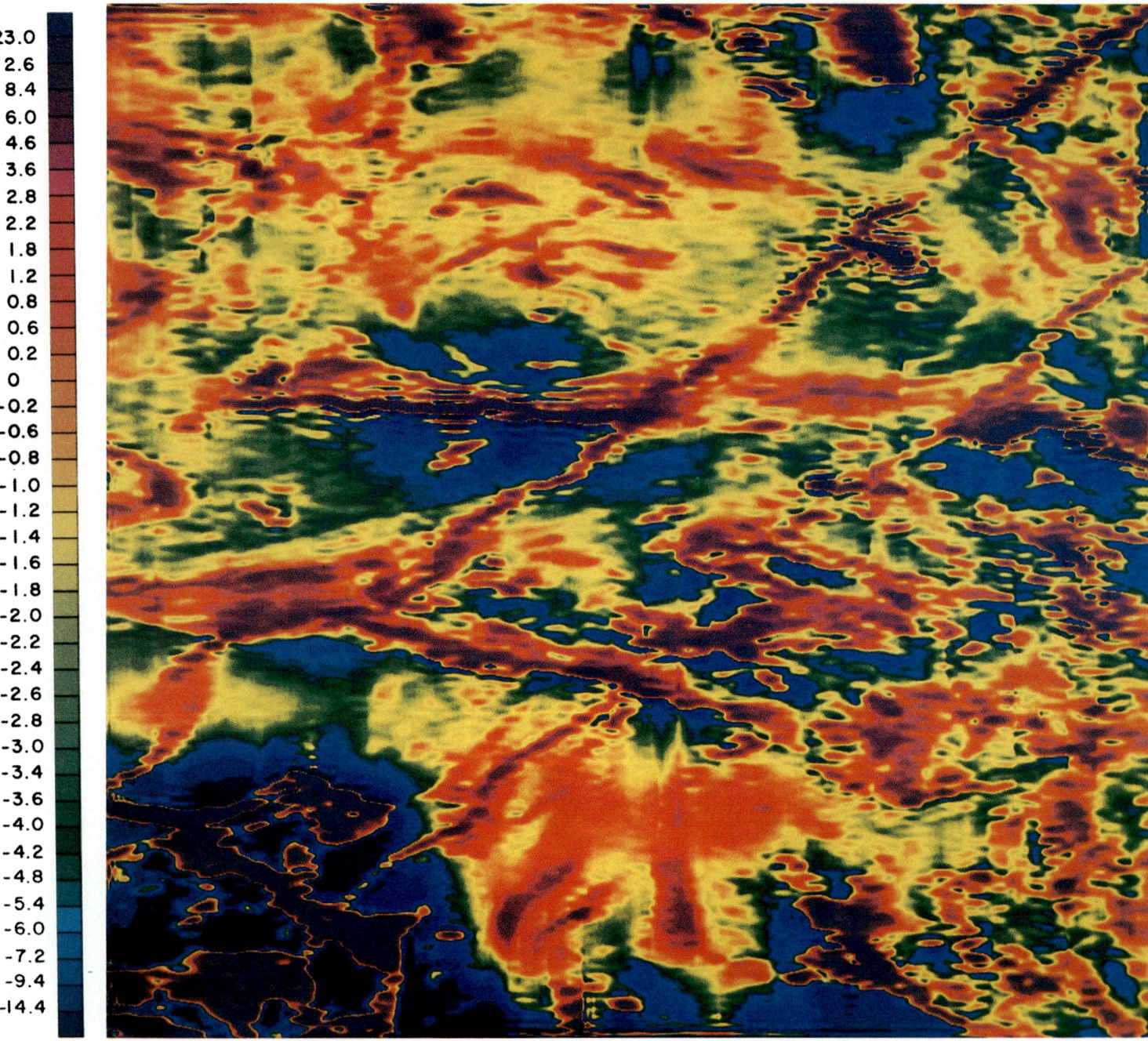

FIG. 9. Apparent-susceptibility map of the study area (units are SI).

Table 3. Magnetic properties of two batholiths.

Rock	No. of samples	K ($\times 10^{-4}$ SI)	No. of samples	J_r ($\times 10^{-3}$ SI)
Lacorne Batholith	6	38	5	20
LaMotte Batholith	7	8	3	2

detail on the distribution of mafic intrusive rocks and diabase dikes. In the northern part of the area, magnetic data indicate the existence of some small mafic intrusions. In the Lacorne Batholith, and north of the Duverny granite stock, two diabase dikes are interpreted and some of the dikes are extended. In the northeastern corner it is interpreted that banded volcanic–sedimentary sequences occur rather than a large area of sedimentary rocks.

CONCLUSION

A high-resolution aeromagnetic survey of the Amos–Barraute area, Quebec, was studied in detail. The total-field maps accurately portray the major magnetic features, but owing to anomaly superposition, much information is obscured. The apparent-susceptibility map, which is of much higher resolution than the total-field map, is a useful aid for surface geological interpretation. The contacts between rocks of different susceptibilities are more accurately outlined—i.e., the mafic and ultramafic intrusives and the diabase dikes are accurately located. In addition, the map values are in apparent-susceptibility units, which reflect a measurable rock property. A difference in apparent-susceptibility value was found, for example, between the Lacorne and LaMotte Batholiths in both the maps and the samples.

The second-derivative map also provides high-resolution information but not a comparison in level difference between different geologic rocks. Stratigraphic features in the volcanic–sedimentary rocks were detected, so that it was possible to distinguish between volcanic and granitic rocks, even though the magnetization level is similar.

ACKNOWLEDGMENTS

We express our appreciation to the Natural Sciences and Engineering Research Council (NSERC) of Canada for financial assistance, as well as to the Government of China. Dr. P. Hood of the Geological Survey of Canada kindly made the data tape available. We also express our appreciation to the anonymous reviewers who worked on the paper.

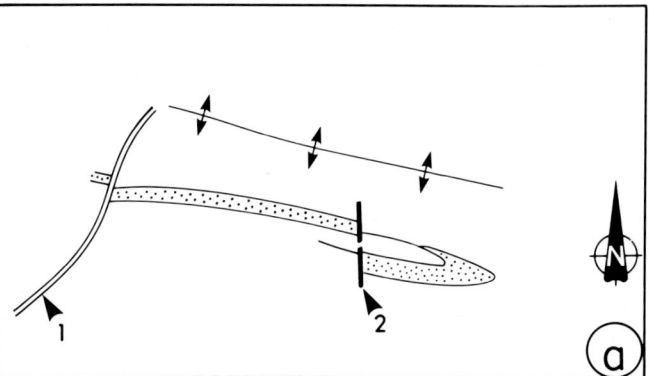
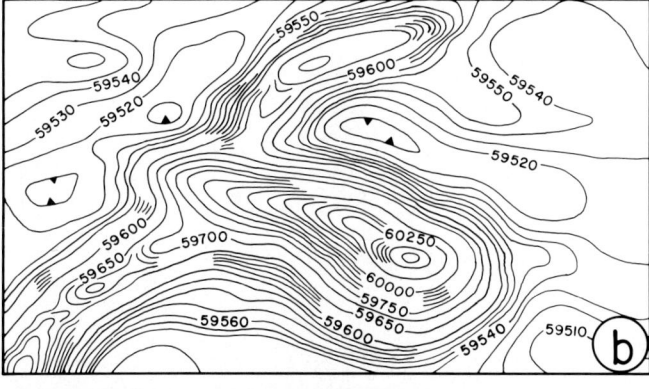
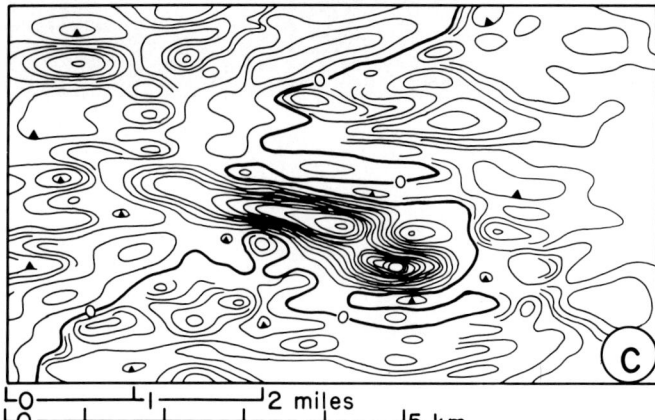

FIG. 10. Anomaly on a fault: a. Geology (1, dike; 2, fault). b. Total field. c. Apparent susceptibility.

REFERENCES

Bambrick, J., Letros, S., and Geissman, J., 1982, Apparent magnetic susceptibility (SUSC) mapping in theory and practice [abstr.]: 52nd Annu. Int. SEG Meet., Oct. 17–21, Dallas, Tex.

Bhattacharyya, B. K., 1966, Continuous spectrum of the total magnetic field anomaly due to a rectangular prismatic body: Geophysics, **31**, 97–121.

Brett, P. R., 1960, Southeast quarter of LaMotte township and the southwest quarter of Lacorne township, Electoral District of Abitibi-East: Prelim. Rep. 428, Que. Dep. Mines.

Brett, P. R., Jones, E. R., Leunier, W. R., and Latulippe, M., 1976, Lamotte township: Geol. Rep. 160, Que. Mineral Deposits Serv.

Brown, W. G., 1958, Northeast quarter of Fiedmont township, Electoral District of Abitibi-East: Prelim. Rep. 364, Que. Dep. Mines.

Grant, F. S., 1974, Magnetic susceptibility map (with accompanying notes): Open File Rep. 229(A), Geol. Surv. Can.

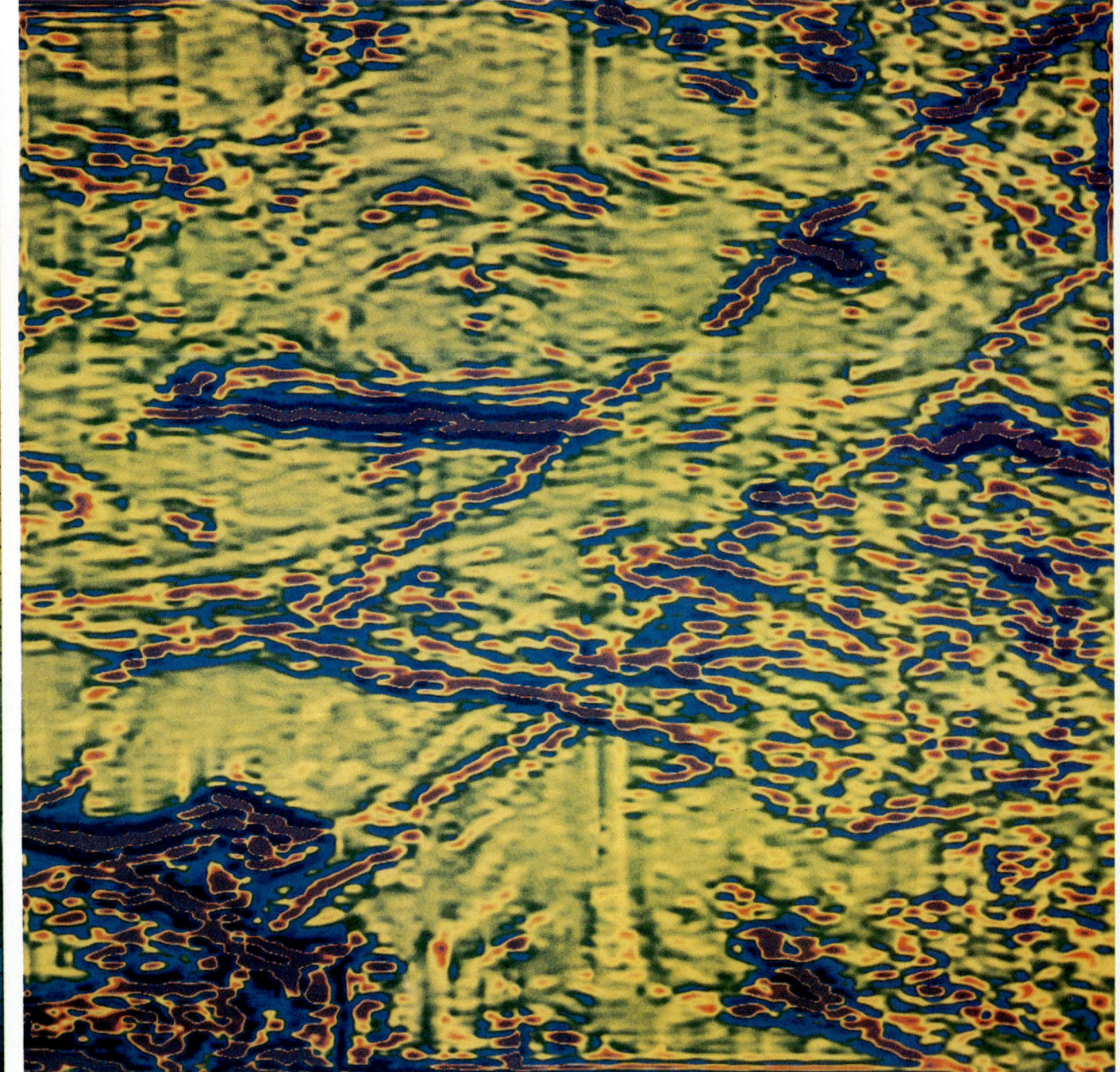

FIG. 11. Second-vertical-derivative map of the study area.

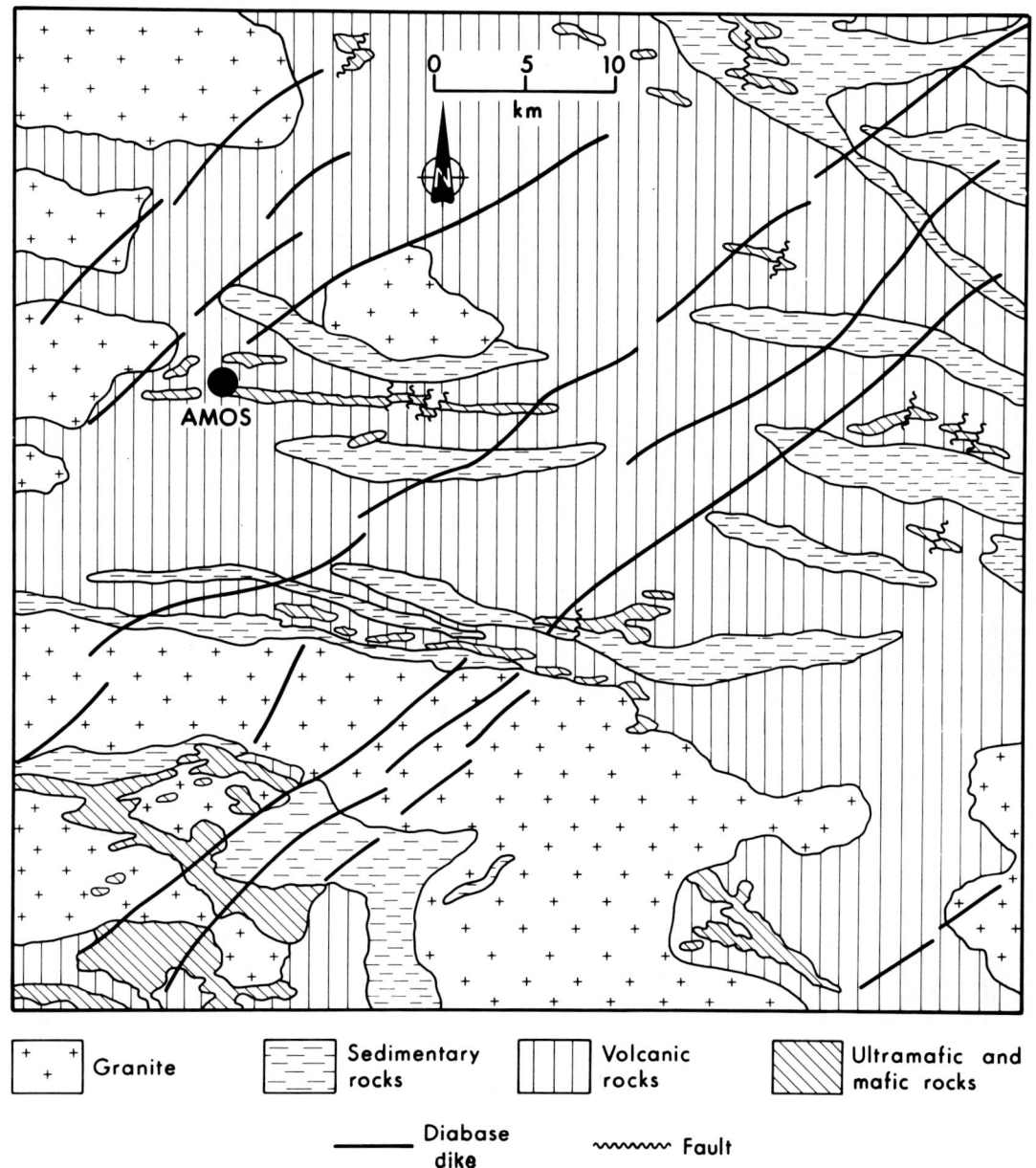

FIG. 12. Interpreted geologic map of the study area (compare with Figure 2).

Jolly, W. T., 1977, Metamorphic history of the Archean Abitibi belt: Pap. 77–1A, Geol. Surv. Can., 191–195.
———1978, Metamorphic history of the Archean Abitibi belt, in Metamorphism in the Canadian Shield: Pap. 78–10, Geol. Surv. Can., 63–78.
Jones, R. E., 1964, Northwest quarter of Fiedmont township, Abitibi-East County: Geol. Rep. 108, Que. Dep. Nat. Resources.
Kornik, L. J., McGrath, P. H., Holroyd, M. T., and Hood, P. J., 1975, Evaluation of high resolution aeromagnetic survey data over a test range in the Timmins area, Ontario: Pap. 75–1B, Geol. Surv. Can., 23–38.
Letros, S. W., 1980, Geological interpretation of high resolution aeromagnetic data in the Kirkland–Larder Lake area; M.S. thesis, Univ. Toronto, Dep. Phys.
Letros, S. W., Strangway, D. W., and Geissman, J., 1983, Apparent susceptibility mapping in the Kirkland Lake area, Ontario, Abitibi Greenstone Belt: Can. J. Earth Sci., 20, 548–560.
McDougall, D. J., 1965. Southeast quarter of Barraute township, Abitibi-East County: Geol. Rep. 114, Que. Dep. Nat. Resources.
Sawatzky, P., and Hood, P. J., 1972, High resolution aeromagnetic surveys: 1971: Pap. 72–1A, Geol. Surv. Can.

Sharpe, J. I., 1961, South half of Figuery and the southwest quarter of Landrienne township, Abitibi-East County: Prelim. Rep. 446, Que. Dep. Nat. Resources.
Spector, A., 1968, Spectral analysis of aeromagnetic data: Ph.D. thesis, Univ. Toronto.
Spector, A., and Grant, F. S., 1970, Statistical models for interpreting aeromagnetic data: Geophysics, 35, 293–302.
Spector, A., and Parker, W., 1979, Computer compilation and interpretation of geophysical data, in Hood, P. J., Ed., Geophysics and geochemistry in the search for metallic ores: Econ. Geol. Rep. 31, Geol. Surv. Can., 527–544.
Tremblay, L. P., 1950, Fiedmont map area, Abitibi County, Quebec: Mem. 253, Geol. Surv. Can.
Urquhart, W.E.S., 1976, Investigation of geological significance of the linear aeromagnetic anomalies in the English River gneiss belt: M.S. thesis, Univ. Toronto.
Urquhart, W.E.S., and West, G. F., 1979, Aeromagnetic anomaly pattern in the English River gneiss belt, Superior Province: Can. J. Earth Sci., 16, 1920–1932.
Weber, W. W., and Latulippe, M., 1964, Amos-Barraute area, Abitibi-East County: Geol. Rep. 109, Que. Dep. Nat. Resources.

Interpretation of part of an aeromagnetic survey in the Matagami area of Quebec

W.E.S. Urquhart* and D. W. Strangway*

ABSTRACT

Aeromagnetic data collected along 200-m-spaced lines at 120-m terrain clearance in the Matagami area of Quebec have been used to improve geological understanding in an area where bedrock is almost completely covered with overburden. The use of the apparent-susceptibility technique has reduced anomaly overlap and has led directly to the use of rock-type susceptibility as a mapping tool.

Volcanic rocks in the area are clearly divided into magnetic (6.0×10^{-2} SI) and weakly magnetic (4.0×10^{-3} SI) units. Chemical analysis of major elements shows that both units are iron-rich tholeiites with virtually the same bulk chemistry. The original oxygen fugacity and temperature of the source magma are thus proposed as the controlling factors influencing the magnetic character of the volcanic rocks.

INTRODUCTION

Of the many problems confronting the geologist, perhaps the most troublesome is the presence of overburden. In areas where there is good outcrop the masking effect of overburden may be only an aggravation. However, where the overburden coverage approaches 100 percent the geologist is faced with the hopeless task of piecing together small glimpses of the underlying geology. What can make the situation worse is that the collection of this small amount of data is not controlled by random processes, which would offer the luxury of statistical analysis. In general, outcrop is exposed to the geologist by geomorphological processes that will inevitably bias the sampling toward a few favored rock types. Thus, in areas of considerable overburden a geologic map will be the result of studying a few biased samples.

In overburden areas the geologist must turn to other indirect methods to assist the mapping exercise. These

*Department of Geology, University of Toronto, Toronto, Ont. M5S 1A1, Canada.

methods include geochemistry, biochemistry, air-photo interpretation, and geophysics.

The work documented here forms part of a larger project whose goal is to use airborne magnetic data to enhance the detailed geological knowledge of an area. Apparent-susceptibility maps were calculated from the magnetic data, leading directly to an inferred magnetic lithology map. Other studies have used the apparent-susceptibility technique, which has proved to be useful in Archean terranes (Grant, 1974; Urquhart and West, 1979; Letros et al., 1983; Yunsheng et al., this volume, a, b). Once a magnetic lithology map is developed, the known geology and field sampling data are reconciled with the magnetic lithologies to form a refined geologic map.

LOCATION

The larger project area is that covered by a Quebec Ministry of Natural Resources (1979a) Input®/magnetic survey (Figure 1). The survey was compiled in 12 map sheets (Figure 2) at a scale of 1:20 000 and covers approximately 3 000 km². It is accompanied by a detailed geologic compilation at the same scale (Quebec Ministry of Natural Resources, 1979b).

This report concerns the northwesternmost map sheet Figure 2, no. 1), although geologic/geophysical samples taken over the larger area were used in the sample studies.

Map sheet 1 was chosen because it contained the most geological information and control and offered a good opportunity to test the interpretation method before undertaking analysis of the remaining 11 map sheets. In addition, the magnetic data revealed a linear pattern in the volcanic units the cause of which was not apparent in the geologic mapping.

Access to and within the area is possible on roads ranging from a main highway to logging roads of which some are passable only with four-wheel-drive vehicles. Water travel is also possible along the Bell River and on Lake Matagami. The terrain in the area is gently rolling to flat muskeg except for 120 m relief in the Laurier Hills at the eastern edge of the map (Figure 3).

GENERAL GEOLOGIC DESCRIPTION

Exposure in the area is poor as a result of a mantle of glacial and glaciofluvial deposits. The best exposures are

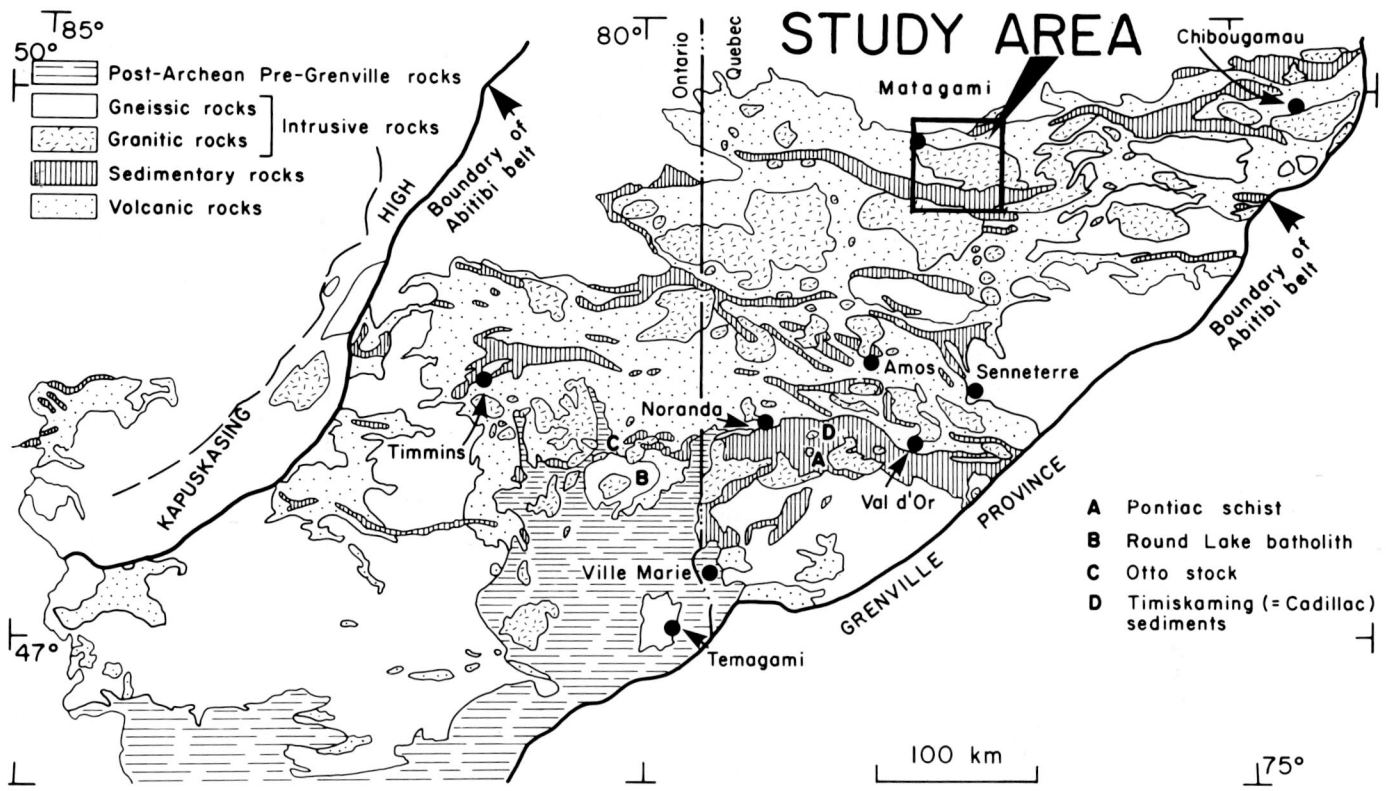

FIG. 1. Location map of the Matagami area, a part of the Abitibi greenstone belt (after Jolly, 1978).

seen along the shores of the lakes and rivers and in the Laurier Hills at the center of the eastern edge of the area. Muskeg virtually eliminates outcrop over large parts of the area. However, some holes have been drilled in the overburden areas, which are described on the geologic-compilation map (Quebec Ministry of Natural Resources, 1979b) and which indicate the underlying rock types.

The geology in the area has been examined by several workers over the years. The earliest mapping was done by Bell (1895, 1900). Bancroft (1912) first recognized the Bell River complex. More detailed work was published by Longley (1943), Freeman and Black (1944), and Beland (1953). More recently Sharpe (1968) mapped in and around the area at a scale of 1:12 000. All this work has been compiled at a scale of 1:20 000 by the Quebec Ministry of Natural Resources (1979b). The following discussion of the geology in the project area is based on the 1:20 000-scale compilation map, a report accompanying the maps prepared by Sharpe (1968), and observations made by Urquhart during field trips in 1981 and 1982.

The area (Figure 3, modified from Sharpe, 1968) is underlain by Precambrian rocks, which, on the basis of lithology and field relationships, can be divided into three major groups and a fourth minor group composed of diabase dikes.

Basal volcanic–sedimentary sequence

The oldest (basal) lithology consists of mafic to intermediate volcanic flows with a few rhyolites, related tuffs, intrusives, and sedimentary rocks, all of which are altered to greenschist facies. This sequence, which extends in an east–west direction, is bounded to the south by the Bell River complex and to the north by granitoid rocks. The northernmost units are primarily sedimentary in origin, whereas the southern part is composed of two volcanic sequences. The lower volcanic unit is known as the Watson Lake Group and is made up of rhyolite and altered volcanic rocks. The type locality is in the vicinity of Matagami Lake

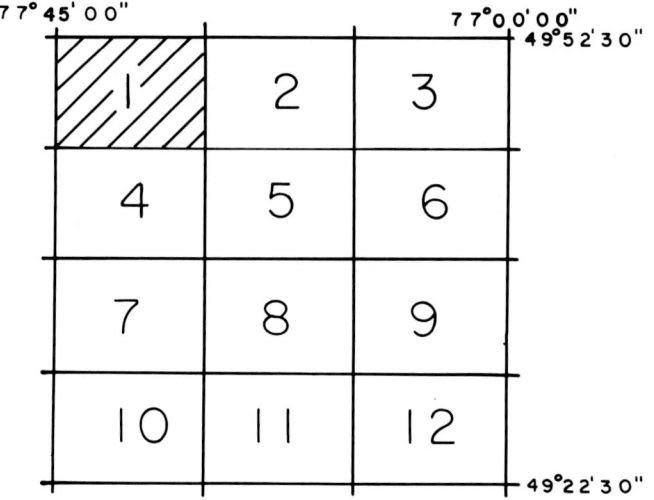

FIG. 2. Layout of map sheets covering study area.

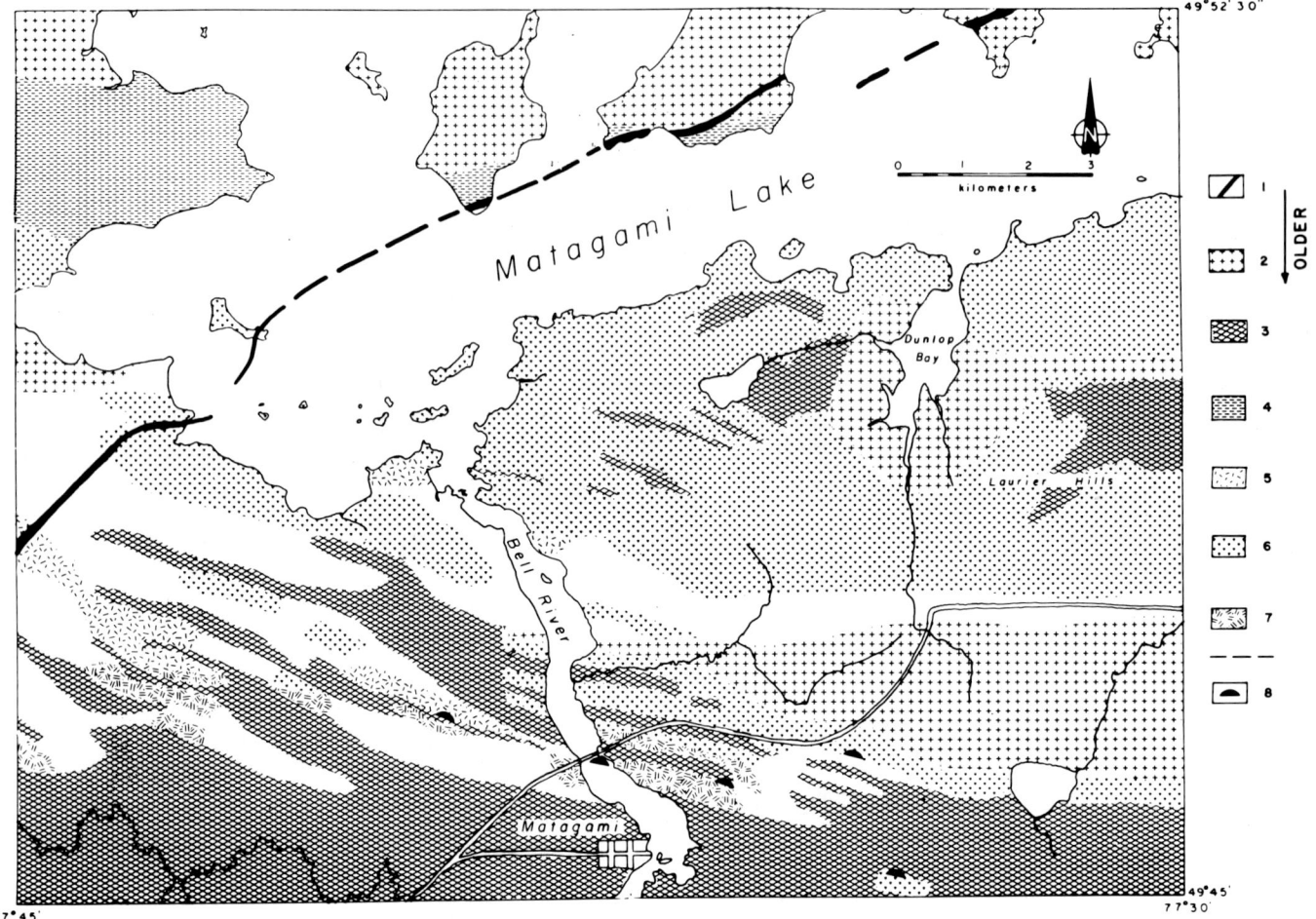

FIG. 3. General geologic map of map sheet 1 (Figure 2) (after Sharpe, 1968). 1 = diabase; 2 = granitoids; 3 = Bell River complex; 4 = metasedimentary rocks; 5 = feldspathic lava; 6 = mafic lava; 7 = rhyolite; 8 = sulfides (Wabassee: 5 and 6. Watson Lake: 7).

and Orchan mines (10 km southwest of the town of Matagami), where the top of the unit is well defined by a thin zone of laminated chert and tuffs ("key tuffite"). These beds separate porphyritic rhyolite from an overlying pillowed dacite lava of the Wabassee Group. In the study area the thin marker zone is not present, and thus the contact is less accurately drawn at the base of the spherulitic rhyolites (Figure 3; Sharpe, 1968). The Watson Lake Group is confined to a narrow band (Figure 3) that is intruded and disrupted by highly magnetic sills associated with the Bell River complex.

The Wabassee Group is typified by thick flows of contrasting dark mafic and light-colored feldspathic lavas. However, in the study area, andesitic-basalt lavas dominate. In addition, there appear to be sills that are similar petrographically to the massive gabbroic portions of the thicker mafic flows. As indicated in Figure 3, the Wabassee is the dominant group in the area.

Bell river complex

The Bell River complex covers a broad area that is oval in shape and about 17 km wide and 65 km long. The central part of the complex is composed of an anorthosite which exhibits no recognizable layering. The marginal zone of the anorthosite is intricately layered. The rocks are of the gabbro family, and the layers differ from one another in the relative proportions of pyroxene and feldspar, although some layers of almost pure titaniferous magnetite are present. The volcanics near the edge of the complex are intruded by sill-like layers of gabbroic composition that are highly magnetic in places and trend at 290 degrees. This unit is at the south edge of the map area.

Granitoid intrusives

The granitoid rocks range in composition from granitic (Dunlop Bay) to dioritic (northern parts of the area) and are considered intrusives because they tend to cut the main tectonic structures (Sharpe, 1968).

Mafic dikes

Only one diabase dike, which trends across the area in a northeasterly direction, has been mapped in the area. The dike is believed to be part of the Abitibi swarm.

GENERAL GEOPHYSICAL DESCRIPTION

Magnetic data

The area is covered by two government airborne geophysical surveys. The first was published in 1949 by the Canadian Federal Government. The data were collected with a Gulf fluxgate magnetometer along 800-m-spaced lines with a mean terrain clearance of 150 m. The data were compiled at a scale of 1:63 360 with a contour interval of 100 nT. The second data set is derived from a Quebec Government-sponsored Input® survey (DP-657; Quebec Ministry of Natural Resources, 1979a). The magnetic measurements were made using a proton magnetometer while flying along 200-m-spaced north–south lines at a mean terrain clearance of 120 m. The data were compiled digitally and contoured at 25 nT. The gridded data (grid-net size, 63.5 m) from the survey have been used for this project and are recontoured in Figure 4.

Magnetically, the area is dominated by a roughly east–west–trending anomaly pattern that, in general, agrees in style and complexity with the known geology. However, the direct relationship between magnetic anomalies and geology is not immediately clear except for that of a few diabase dikes and the highly magnetic gabbro sills associated with the Bell River complex.

Magnetic properties

Magnetic anomalies reflect changes in the concentration and habit of certain iron-rich minerals. The most commonly occurring magnetic minerals are the iron–titanium oxides. These include solid-solution series of magnetite (Fe_3O_4 and ulvöspinel (Fe_2TiO_4); hematite (α-Fe_2O_3) and ilmenite ($FeTiO_3$); exsolution phases of magnetite and ilmenite; and maghemite (γ-Fe_2O_3). Other magnetic phases, such as pyrrhotite ($Fe_{1-x}S$) and native iron, can also cause anomalies.

Changes in magnetic character are normally interpreted to correspond to changes in rock units. Unfortunately, existing geologic nomenclature is rather insensitive to changes in magnetic properties, making a simple conversion from a magnetic lithology map to a geologic map difficult. Not only can rocks of widely differing chemistries have similar magnetic properties, but rocks of similar chemistry can have widely differing magnetic properties.

The presence of primary magnetite in a particular rock is controlled by many factors, including the composition (iron and titanium) and the oxidation conditions. Haggerty (1979) summarized known information about the role of temperature and oxygen fugacity in determining the nature of the iron–titanium oxides in igneous rocks. These results are summarized with reference to the forsterite–magnetite–quartz buffer curve (Figure 5). The data show that mafic rocks tend to have formed under lower fugacity conditions than felsic rocks. This means that mafic rocks tend to have iron–titanium oxides in the range of the magnetite-ulvöspinel solid-solution series. These commonly occur as simple homogeneous grains without well-developed exsolution textures. If the titanium content is high, the susceptibility values are less than those of pure magnetite. Under somewhat more oxidizing conditions, an exsolution texture of magnetite and ilmenite tends to develop. This texture dramatically changes the magnetic properties by lowering the susceptibility and raising the remanence as the effective grain size becomes small. Under yet more oxidizing conditions, the composition of the iron–titanium oxides moves toward the ilmenite–hematite series. This series is generally weakly magnetic except for a limited composition range that is highly magnetic and is a common constituent of felsic rocks.

FIELD WORK

A rock-sampling program was carried out in the area to determine the susceptibility levels for each rock type. Rock outcrops were chosen from the existing geologic map to ensure that each rock unit was sampled. In addition, specific magnetic anomalies and their causative rocks were sought out to make certain that as few magnetic rock units as possible would go unsampled in the field work.

At each outcrop the rock unit was identified, and a representative hand sample taken and its susceptibility measured. In-situ susceptibility measurements were made and, depending on the uniformity of the susceptibilities encountered, included from 4 to 30 measurements. In all, 1 871 in-situ and 229 hand-sample susceptibility measurements were made. Ground magnetic profiles were measured and interpreted in places where particular magnetic anomalies were investigated to ensure that the actual causative body was sampled. Remanence measurements were not made, because the primary objective was to collect a large number of susceptibility data and most measurements were done in situ.

Fifty-one outcrops were visited within the map area (see Figure 17 for locations), and more than two hundred outcrops were sampled in the entire 12 map sheets.

ROCK-SAMPLE STUDIES

Geophysical parameters

In order to relate the magnetic lithologies observed on the magnetic and apparent-magnetic-susceptibility map with the mapped geology, the in-situ and hand-sample susceptibility values were analyzed by rock type. Because the rock units in the area extend beyond the sheet, the susceptibility data from the entire 12 sheets are used and presented in accumulated-percent form in Figures 6–13.

The use of the accumulated-percent plots is explained as follows. In Figure 8, for example, one can see that 0 percent of the samples had susceptibility values of more than 10×10^{-3} SI, and 95 percent of all the samples had susceptibility above 1.2×10^{-3} SI. The modal value of the sample set is marked by the steepest gradient and in Figure 8 is 3.5×10^{-3} SI.

Data below 1.2×10^{-3} SI are accumulated at the 1.2×10^{-3} SI level on the susceptibility diagrams. For example, 25 percent of the hand-specimen samples for diorites have susceptibility measurements at or below 1.2×10^{-3} SI. The truncation of the data at 1.2×10^{-3} SI was done for two reasons. First, the susceptibility-map minimum-contour level is 2.4×10^{-3} SI, and thus smaller values are not visible.

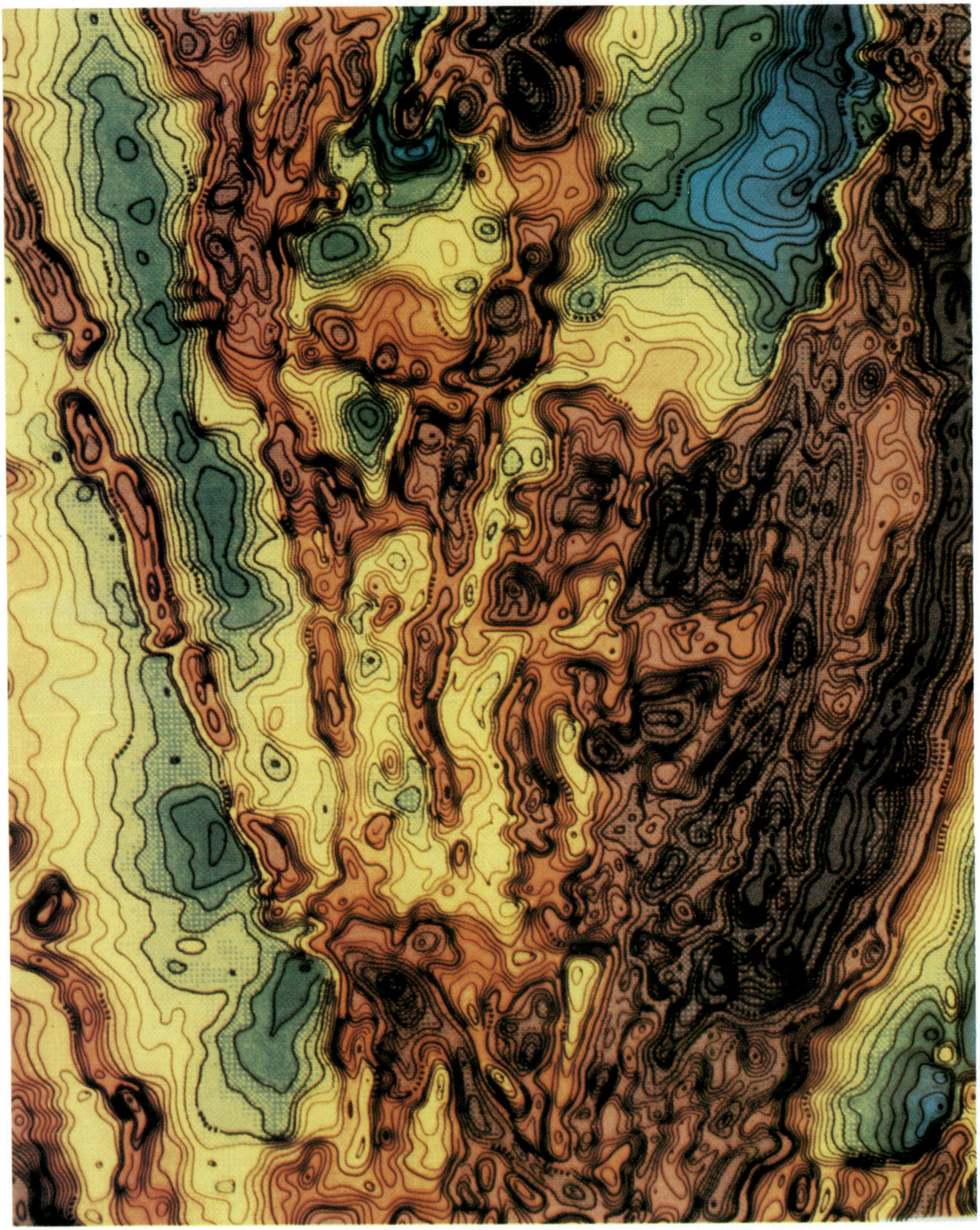

FIG. 4. Total-magnetic-anomaly map of map sheet 1 (Figure 2). Minimum contour interval, 25 nT.

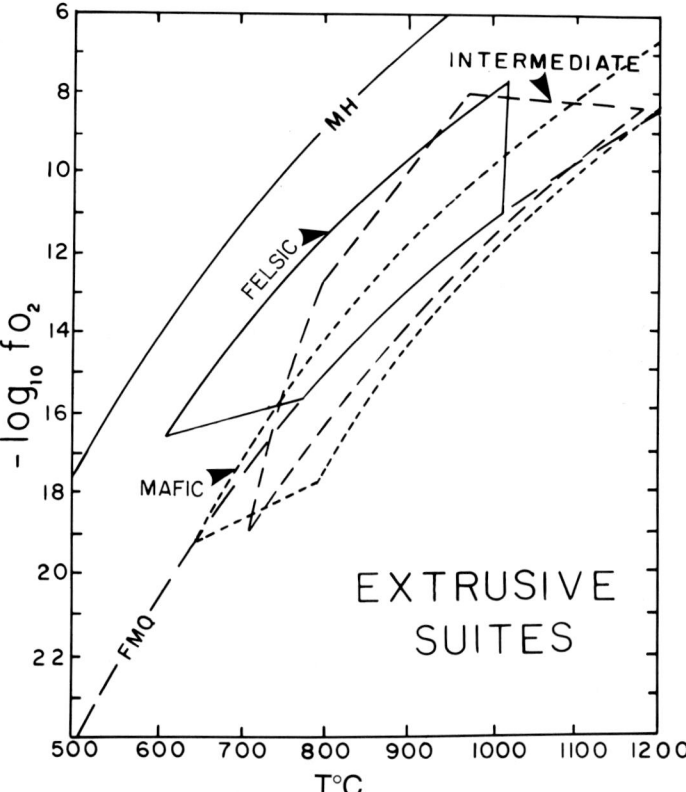

Fig. 5. Temperature and oxygen-fugacity (fO_2) conditions for different rock types. Values below the FMQ (at 10^5 Pa (1 bar)) buffer curve are less oxidized than values above the curve. Fayalite, magnetite, and quartz (FMQ) are in equilibrium along the curve (after Haggerty, 1976). MH defines the solid solution curve for magnetite$_{100}$, ulvöspinel.

Second, the field instruments were not accurate below 1.2×10^{-3} SI.

Too few samples were taken of the granitic rocks to establish clear modes (Figure 6). Apart from the samples of the Dunlop intrusion, at about 24.0×10^{-3} SI, the granites have values typically below 10.0×10^{-3} SI.

Only four outcrops of felsic volcanic rocks from the Wabassee and Watson Lake Groups were sampled; these samples yielded susceptibility values with an average of about 0.35×10^{-3} SI. This sampling is not representative but suggests that the level of susceptibilities for the group is low.

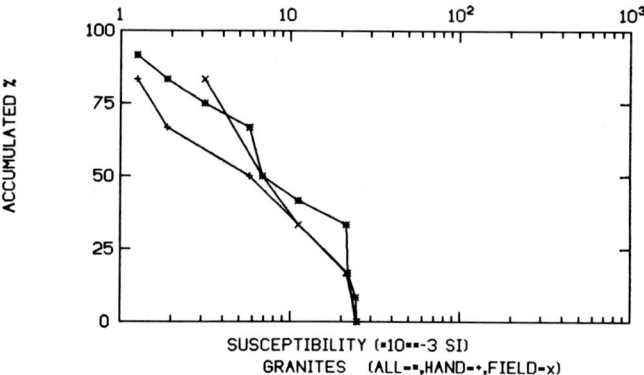

Fig. 6. Susceptibility data for granites in study area.

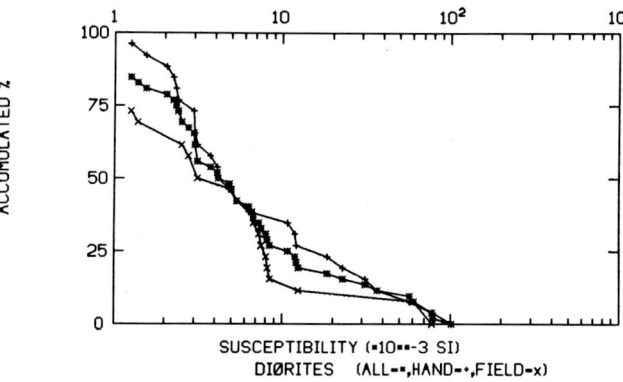

Fig. 7. Susceptibility data for diorites in study area.

Diorites are divided into two modes, one at approximately 5.0×10^{-3} SI and the other at about 70.0×10^{-3} SI (Figure 7). Andesite volcanic rocks form one well-defined unit at about 3.5×10^{-3} SI (Figure 8). Mafic volcanics have a bimodal distribution, with one mode being similar to the andesites with a susceptibility of about 4.0×10^{-3} SI (Figure 9), and the second mode having a susceptibility of about 60.0×10^{-3} SI.

Data for the gabbros reveal two different modes (Figure 10). The mode at about 60.0×10^{-3} SI appears to be caused by sills associated with the highly magnetic basaltic volcanics. The mode at about 4.0×10^{-3} SI represents gabbroic material that may be equivalent to weakly magnetic volcanic rocks with similar susceptibilities. Clearly, it will be difficult to distinguish between the sills and their corresponding volcanic rocks solely on the basis of susceptibility. The diabase-dike samples have one mode at about 60.0×10^{-3} SI (Figure 11).

The susceptibility data for the Bell River complex reveals one mode at about 180×10^{-3} SI (Figure 12). This mode is caused by the highly magnetic sills that intrude the volcanics. The rest of the data are not well clustered, reflecting the variable nature of the magnetic properties of the Bell River complex. The Bell River anorthosites are not included in the cumulative-percent data, as their susceptibilities all fall well below the 1.2×10^{-3} SI level. The anorthosite samples were taken from drill cores supplied by Noranda Mines, Ltd.

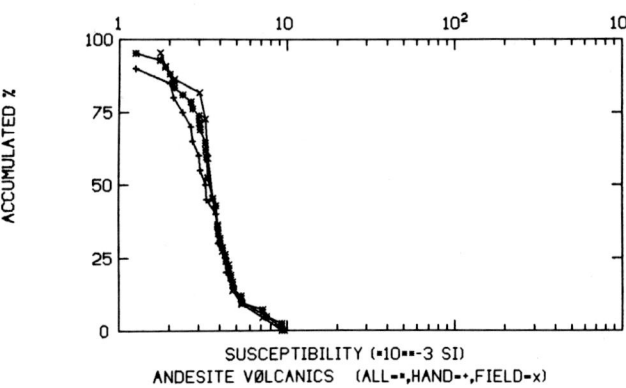

Fig. 8. Susceptibility data for andesite volcanic rocks in study area.

The data for metasedimentary rocks (Figure 13) represent samples taken entirely outside the map sheet, as exposures were not accessible within the area. However, the 4×10^{-3} SI mode indicated by the data is probably representative of the metasedimentary rocks in the area, as the sampled rocks are stratigraphically and mineralogically similar. The very high values are associated with iron formations south of this map area.

The susceptibility study has demonstrated graphically the difficulty in assigning rock type to given susceptibility levels. On the basis of susceptibility alone, it is impossible to distinguish between the gabbros, basaltic volcanic rocks, and diabases, which all have members at the 60.0×10^{-3} SI levels. Similarly, there are gabbros, basaltic volcanic rocks, andesitic volcanic rocks, or metasedimentary rocks in the 3.0–4.0×10^{-3} SI range. Clearly, structural information, such as the trend of a dike, or direct geologic correlation of anomalies with geologic units is necessary to assign any geologic names to the magnetic lithologic units. The magnetic pattern is likely to prove a better indication of rock type than susceptibility measurements in the absence of geologic control.

Geochemical data

Eleven mafic volcanic samples were taken for geochemical analysis. The primary objective was to determine if any definite chemical differences exist between the magnetic (60.0×10^{-3} SI) and weakly magnetic (3.5×10^{-3} SI) mafic volcanic units that could account for the banding seen on the magnetic map.

The samples were analyzed using standard x-ray-fluorescence techniques, with the NaO content being determined by neutron activation. The results that appear in Table 1 indicate that no major-element differences exist between the two magnetic groups. Figure 14 displays the data on an AFM diagram. (The AFM diagram is a plot of the percentages of alkali, total iron oxides, and magnesium oxides on a triangular plot.) The rocks fall within the tholeiitic region and are considered iron rich (Barker and Arth, 1976).

There is an indication that the magnetic samples have slightly more iron and magnesium, but the sampling is not adequate to establish this difference clearly. A plot of percentage of total iron (as Fe_2O_3) versus logarithmic susceptibility for the 11 samples (Figure 15) shows that no correlation exists between susceptibility and iron content.

The partitioning of the volcanic rocks into magnetic and weakly magnetic units is not associated with obvious compositional differences. Thus, oxygen fugacity and temperature are the factors which are most likely to explain why some units are magnetic and some weakly magnetic. The presence of parallel units of magnetic and weakly magnetic

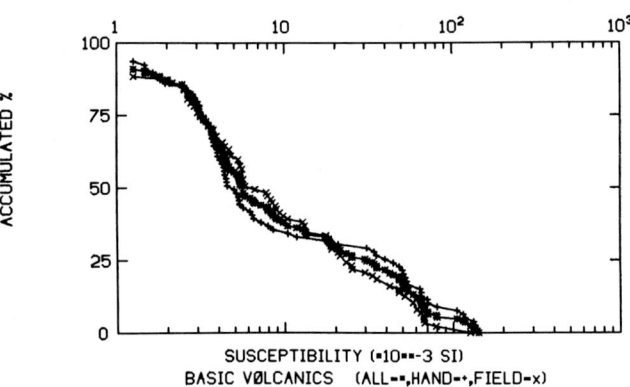

FIG. 9. Susceptibility data for mafic or basic volcanic rocks in study area.

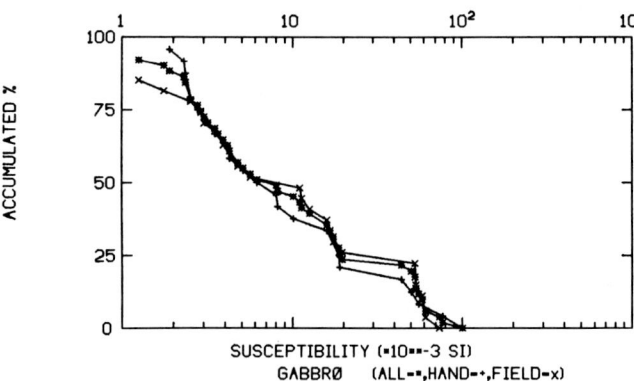

FIG. 10. Susceptibility data for gabbros in study area.

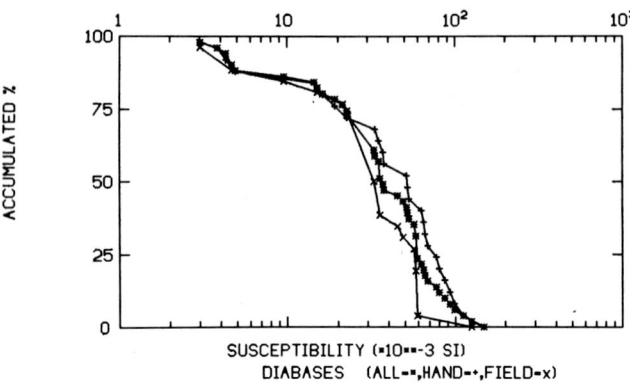

FIG. 11. Susceptibility data for diabases in study area.

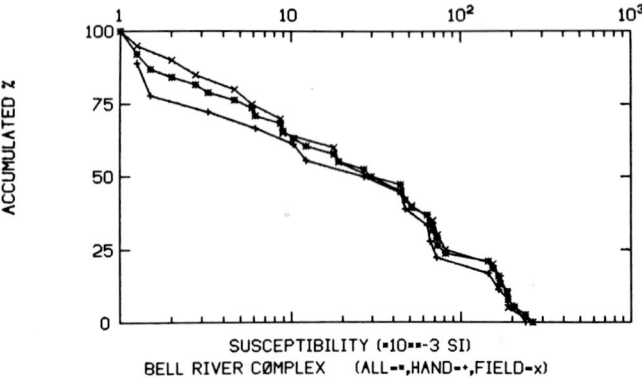

FIG. 12. Susceptibility data for Bell River complex in study area.

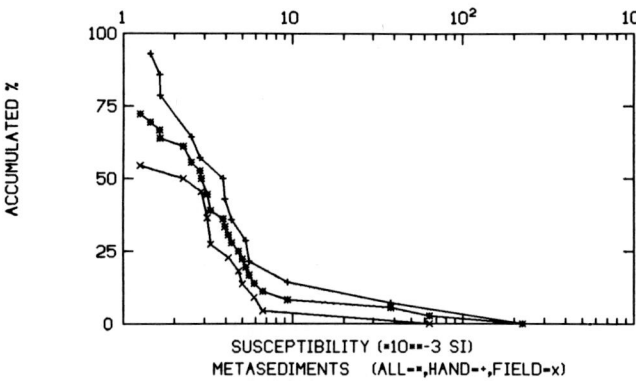

FIG. 13. Susceptibility data for metasedimentary rocks in study area.

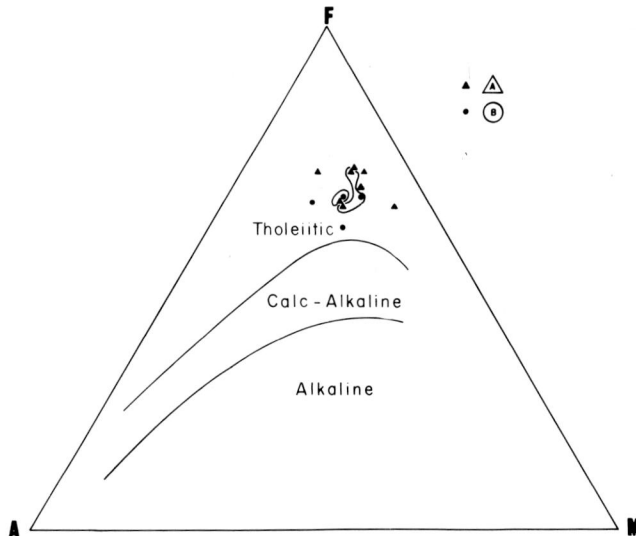

FIG. 14. AFM (alkali–total iron oxides–magnesium oxides) diagram, with data on volcanic-rock samples displayed. The A samples are the magnetic rocks, and the B samples are the weakly magnetic rocks. The samples grouped together came from the same outcrops.

volcanic rocks, probably corresponding to individual flows, and the presence of a magnetically similar suite of gabbroic rocks, suggest that the magnetic character of these rocks is primary. Thus, the magnetic properties of this volcanic suite are likely to be a function of changes in temperature and oxygen fugacity in the source magma. As summarized by Haggerty (1976), small changes in fugacity can lead to major changes in the magnetic phases and in their magnetic properties.

APPARENT-SUSCEPTIBILITY MAPPING

The magnetic data were converted to an apparent-susceptibility map (Figure 16) to aid in the interpretation (Grant, 1974; Urquhart and West, 1979; Letros et al., 1983). This type of map has four major benefits:

1. Anomaly overlap is substantially reduced.
2. Areas of uniform susceptibility appear relatively flat, and level differences appear between units with different magnetic properties.
3. Contacts between rocks of different susceptibility can be more accurately positioned.
4. The map values are in apparent-susceptibility units. This is a property that can be measured on the ground (by examining the susceptibility and remanence) and compared directly with those of the map.

Apparent-susceptibility mapping is a data-processing technique that converts a total-field magnetic map to an apparent-susceptibility map (Grant, 1974). This process can be

Table 1. Chemical analyses of 11 samples of mafic volcanic rocks in study area.

Sample no.	1	2	3	4	5	6	7	8	9	10	11
SiO_2	47.10	51.69	48.73	50.59	53.23	51.73	50.33	47.44	53.59	53.18	48.23
TiO_2	1.98	1.36	2.26	2.33	1.60	1.54	2.12	2.11	1.78	1.91	0.97
Al_2O_3	13.88	14.02	13.48	12.12	12.84	14.06	12.32	12.87	13.67	14.42	14.81
Fe_2O_3	18.11	14.34	17.80	18.33	14.56	15.44	18.28	16.84	15.56	14.20	10.48
FeO	0.00	0.00	0.00	0.00	0.00	0.00	0.00	0.00	0.00	0.00	0.00
MnO	0.27	0.19	0.20	0.20	0.19	0.17	0.22	0.23	0.21	0.25	0.17
MgO	5.29	4.40	5.85	5.01	4.70	7.24	4.85	5.91	4.79	3.22	4.15
CaO	7.40	8.43	6.46	5.82	7.56	4.51	7.21	8.25	6.34	7.01	9.36
Na_2O	1.73	2.86	3.73	2.23	1.88	1.30	2.17	1.98	2.61	4.22	2.48
K_2O	0.31	0.42	0.38	0.26	0.22	0.17	0.23	0.65	0.54	0.14	0.28
P_2O_5	0.32	0.17	0.28	0.34	0.21	0.22	0.24	0.26	0.16	0.95	0.11
Totals:	96.39	97.89	99.16	97.34	96.99	96.38	97.97	96.55	99.25	98.49	91.05
Susceptibility ($\times 10^{-3}$ SI):	142.0	133.0	103.0	74.0	34.0	21.0	5.3	5.1	3.7	3.1	3.0
Density (g/cm^3):	2.95	3.02	2.98	3.03	3.09	2.83	3.22	2.96	3.21	2.99	2.90

Standard x-ray-fluorescence techniques used in analyses, with NaO content having been determined by neutron activation.

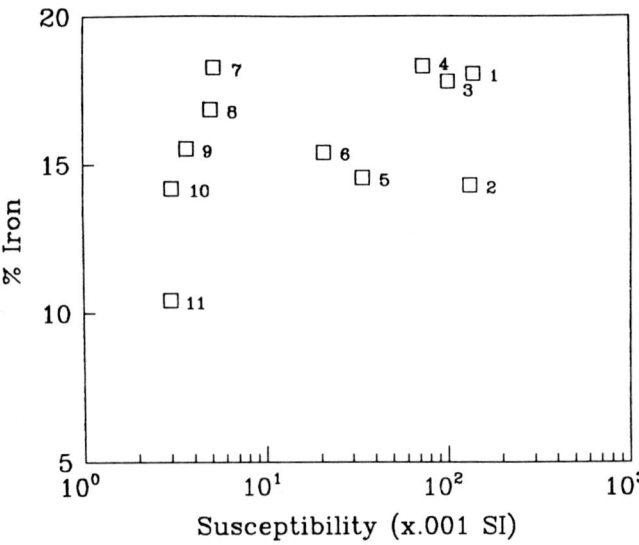

Fig. 15. Fe_2O_3 versus log susceptibility ($\times 10^{-3}$ SI). Numbers correspond to sample numbers in Table 1.

thought of as a combination of reduction to the pole (Baranov, 1975) and a stabilized downward continuation to the ground surface. If the dips of contacts in the area are vertical, the bottoms of magnetic bodies relatively deep, and the magnetization of the rocks results only from induction in the direction of the Earth's field, then this process yields an averaged susceptibility for the rock formations.

Deviation from vertical dip can cause errors in both the position of the body and the magnitude of the apparent susceptibility. The magnitude of these errors varies approximately as the cosine of the deviation from vertical.

In theory, an apparent-susceptibility map is calculated in the following way (Spector, 1968; Spector and Grant, 1970).

The anomalous total magnetic field (T), owing to an assemblage of bodies, can be written as a sum, i.e.,

$$\Delta T_{(x, y)} = \sum_{i=1}^{k} \Delta T_i(x, y). \quad (1)$$

The Fourier transform, being a linear operator, can likewise be written as

$$\Delta \mathbf{T}_{(u, v)} = \sum_{i=1}^{k} \Delta \mathbf{T}_i(u, v). \quad (2)$$

Each individual transform can be expressed as the product of four factors, which include the body parameters, i.e.,

$$\Delta \mathbf{T}_i(u, v) = h(h_{i, j, r}) \times S(a_i, b_i, \gamma_{ij} u, v)$$
$$\times m(\mathbf{T}_0, \mathbf{M}_{ij}, \theta) \times d(\xi_i, \eta_{ij}, u, v), \quad (3)$$

where h is the depth factor (e.g., in the case of a bottomless prism it is e^{-hr}, where

$r = (u^2 + v^2)^{1/2}$
$u = 2\pi f_x$
$v = 2\pi f_y$
$f =$ spectral frequency in cycles/data interval;

s is the size and orientation factor ($2a$, $2b$ are the body dimensions for a bottomless prism and γ its azimuth), and in the case of a rectangular prism it is

$$\frac{\sin [a(u \cos \gamma + v \sin \gamma)]}{a(u \cos \gamma + v \sin \gamma)} \times \frac{\sin [b(v \cos x - u \sin \gamma)]}{b(v \cos \gamma - u \sin \gamma)}; \quad (4)$$

m involves the orientation of $\mathbf{T}_0$, the local geomagnetic field vector, and also the intensity and orientation of magnetization, $\mathbf{M}$. In the case of only induced magnetization, $\mathbf{m}$ is

$$2\pi |\mathbf{M}| \times [n + i(\ell \sin \theta + m \cos \theta)]^2, \quad (5)$$

where (l, m, n) are the direction-cosines of $\mathbf{T}_0$, $\theta = \tan^{-1}(u, v)$, and d includes the horizontal position of the individual prisms (ξ, η), i.e.,

$$d = e^{-i(u\xi + v\eta)}.$$

Because all the bodies are assumed to be identical in shape, at the surface (constant h) and magnetized in the Earth's field direction only, the equation for the total-field-anomaly map in frequency space is

$$\Delta \mathbf{T}_{(u, v)} = \{h(h_j, v) \times S(a, b, \gamma_j, u, v)$$
$$\times 2\pi[n + i(\ell \sin \theta + m \cos \theta)]^2\}$$
$$\times \sum_{i=1}^{k} |\mathbf{M}_i| \times d(\xi_i, \eta_{i, j}, u, v). \quad (6)$$

The orientation factor is removed by reduction to the pole. This is done by dividing $\Delta \mathbf{T}(u, v)$ by $[n + i(l \sin \theta + m \cos \theta)^2]$;

$$\therefore \mathbf{T}_{\text{pole}}(u, v) = \{h(h_j, v) \times S(a, b, \gamma_j, u, v) \times 2\pi\}$$
$$\times \sum_{i=1}^{k} |\mathbf{M}_i| \times d(\xi_i, \eta_{i, j}, u, v). \quad (7)$$

The height factor (h) is removed by downward continuation. This is done by dividing $\Delta \mathbf{T}_{\text{pole}}(u, v)$ by e^{-hr} (for the case of the infinite vertical prism);

$$\therefore \Delta \mathbf{T}_{\text{surface plane}}(u, v) = \{S(a, b, \gamma_j, u, v) \times 2\pi\}$$
$$\times \sum_{i=1}^{k} |\mathbf{M}_i| \times d(\xi_i, \eta_{i, j}, u, v). \quad (8)$$

To reduce to a distribution of magnetization, one divides by the shape factor, equation (4), for vertical prisms:

$$\Delta \mathbf{T}_m(u, v) = 2\pi \sum_{i=1}^{k} |\mathbf{M}_i| \times d(\xi_i, \eta_{i, j}, u, v). \quad (9)$$

Now one transforms back to space domain to obtain:

$$\Delta \mathbf{T}_m(x, y) = \sum_{i=1}^{k} |\mathbf{M}_i| \times D(\xi_i, \eta_{i, j}, x, y), \quad (10)$$

where $D(\xi, \eta_j, x, y)$ is the spatial equivalent of d in the frequency domain, and is a δ-function. Therefore,

$$\Delta \mathbf{T}_m(x, y) = |\mathbf{M}_i|$$

for x, y at center of prism i (from δ-function).

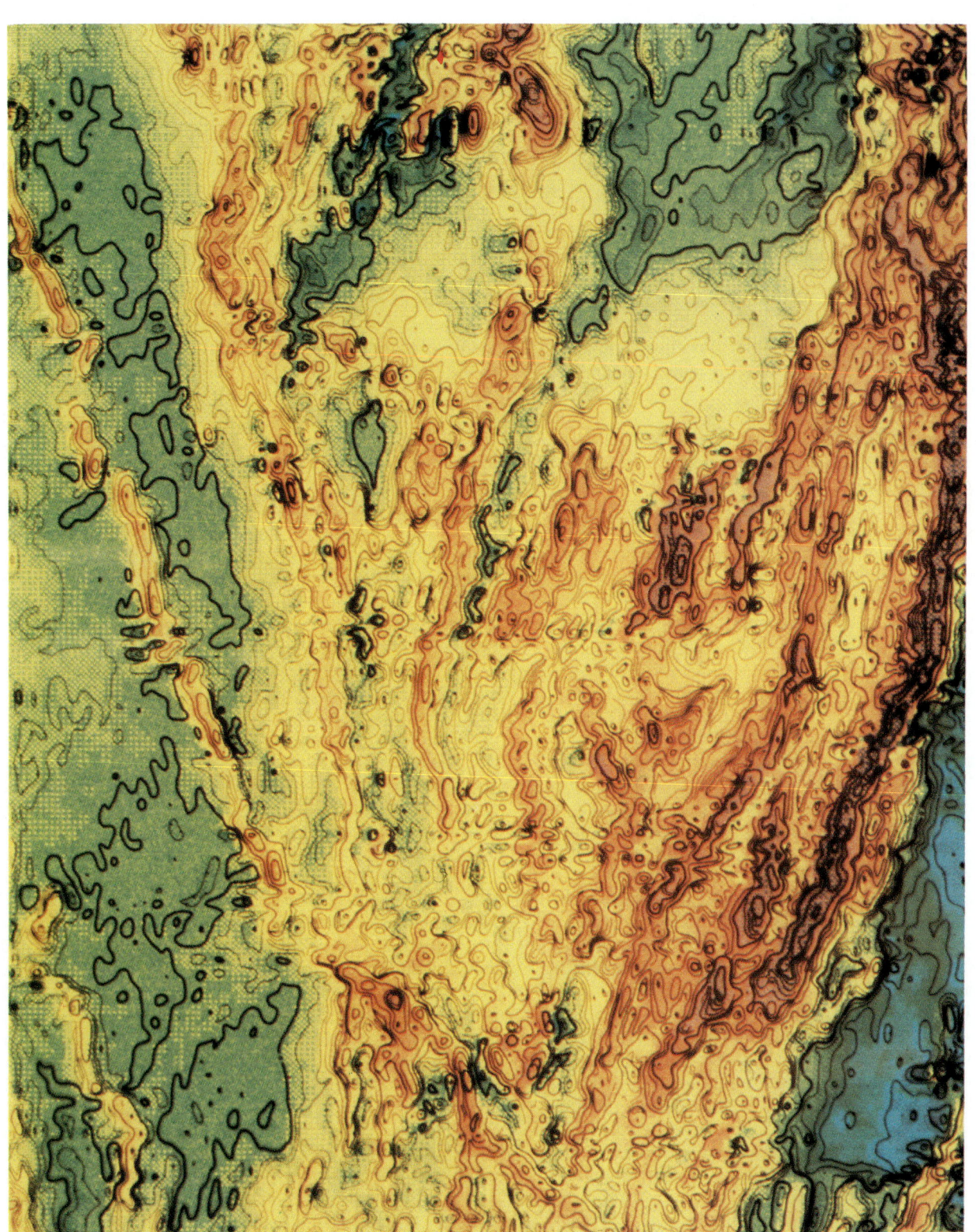

FIG. 16. Apparent-susceptibility map of map sheet 1 (Figure 2). Minimum contour interval, 2×10^{-3} emu (24×10^{-3} SI = 2×10^{-3} emu). Arbitrary zero contour is thick line in the green areas. Low is blue, high is red.

Thus each magnetization can be converted to an apparent-susceptibility value by dividing $T_m(x,y)$ by the geomagnetic-field strength.

In practice, two additional steps are required. High-frequency noise in the original data is enhanced, primarily by the downward-continuation operation, and must be suppressed. This can be done by using a Hanning-shaped cosine roll-off filter. For the map presented here, the roll-off started at 0.2 cycles per cell (CPC) and approached zero at 0.3 CPC. The second step is that of dealing with the arbitrary base level. In order to establish a meaningful base level, a tilted plane is added to the data after processing so that large areas which have very low fields will have susceptibility levels at or near zero. The entire 12 map sheets were "zeroed" at the same time with a single tilted plane "regional." Apparent-susceptibility values for sedimentary and dioritic units were used as zero-susceptibility points. The three zero-susceptibility points that define the plane are outside this map area.

The final result (Figure 16) of the process is an apparent-susceptibility map that has quantitative meaning only insofar as the following criteria are met:

1. The magnetic field can be described as due to an assemblage of bodies of rectangular cross-section, one grid cell in dimension.
2. The bodies are vertically sided.
3. The bodies are infinitely deep (bottomless).
4. The bodies are only magnetized in the direction of the Earth's magnetic field.

Qualitatively, the map is significant as a high-frequency-enhanced map even if the above criteria are not met.

MAGNETIC INTERPRETATION

The magnetic interpretation (Figure 17) was performed in a two-stage process. The first step was development of a magnetic lithology map. The magnetic-susceptibility map is ideally suited to this stage, as it separates the lithologies on the basis of magnetic properties. Lithologies are identified on the basis of linear continuity, susceptibility level, and general character. Disruptions in continuity, offsets, or abrupt changes of magnetic level are noted, and they commonly correspond to faults. The results of the first stage of interpretation is a magnetic lithology map that is presumably related to the geology but which has no geologic nomenclature attached.

The second stage of the interpretation is more difficult. It involves taking known geologic information and the magnetic lithology map and reconciling the two data sets to produce a meaningful geologic interpretation. At this stage, the resolution of the magnetic data becomes both a problem and an asset. The grid-cell dimension of the digital data is 63.5 m on the ground. Moreover, flight lines are 200 m apart, so that resolution normal to the flight lines is more limited. This means that at best every magnetic feature will appear to be no less than 63.5 m wide and that in practice, following noise-reduction filtering, 100 to 200 m is the best that can be seen. Under these conditions it is not possible to resolve narrow, closely spaced units. However, the field geologist does find recognizable units at a smaller scale, thus introducing an immediate difficulty in reconciling the detail between the magnetic lithology map and the mapped geology. On the other hand, because of poor exposure, it may be impossible for the geologist to join together units from outcrop to outcrop. Because of the larger scale of the magnetic data, it is particularly suited to identifying trends and assisting in joining together isolated outcrops. The magnetic lithology map portrays a broad picture of the geology in an area and results in a map that may better represent the larger lithologic features.

The process of transforming the magnetic lithology data into geologic terms must start with the analysis of the susceptibilities of the rock units in the area. Investigation of the rock susceptibilities has revealed that distinguishing between the magnetic dikes, sills, and flows or between the weakly magnetic basalt flows and sills is not possible. The granites and sedimentary rocks also have similar magnetic-susceptibility ranges. The only units that are identifiable on the basis of their susceptibilities are the sills that are related to the Bell River complex that have a susceptibility greater than 18.0×10^{-3} SI. Because of the difficulty in distinguishing rocks on the basis of their susceptibility alone, the transformation of a magnetic lithology map to a geologic interpretation rests heavily on the known outcrop geology and structure.

In the study area the diabase dikes strike at an angle that is not easily confused with the strike of the rest of the rock units. Thus the diabases are easily identified and can be interpreted with confidence. The sample susceptibility of 60.0×10^{-3} SI is consistent with the calculated-susceptibility data presented on the apparent-susceptibility map (varying between 20 and 60×10^{-3} SI).

The granitic rocks in the area occupy regions of low susceptibility (in accordance with the sample susceptibility data) which lack linear trends but have a mottled appearance, indicating a nonuniform susceptibility distribution. The one exception to this is an apparent-susceptibility anomaly between 30.0×10^{-3} and 60.0×10^{-3} SI south of the Dunlop intrusive. Field mapping indicates that granite underlies this particular anomaly.

Diorites and granodiorites are grouped together because they are not well distinguished on the geologic maps and appear to be related phases. The 5×10^{-3} SI susceptibility mode for these rock types is consistent with the calculated susceptibilities in the northern part of the map, where diorite and granodiorite outcrops are well documented. The contact between the diorite/granodiorite and metasedimentary rocks is based on the change from very flat, almost zero susceptibility for the metasedimentary rocks to slightly higher susceptibility values. Diorites and granodiorites elsewhere on the interpretation map have a wide range of magnetic-susceptibility values. East and west of the Dunlop intrusive (Dunlop Bay, Figure 3) the diorites appear as a wide zone with susceptibilities as high as 20.0×10^{-3} SI, particularly where they are in contact with volcanic rocks. A 70.0×10^{-3} SI mode is greater than that needed to explain these levels. A highly magnetic zone extending from the western part of the southern granitic body is interpreted to be dioritic or granodioritic, based on the mapped geology that indicates that dioritic rocks underlie the southernmost magnetic

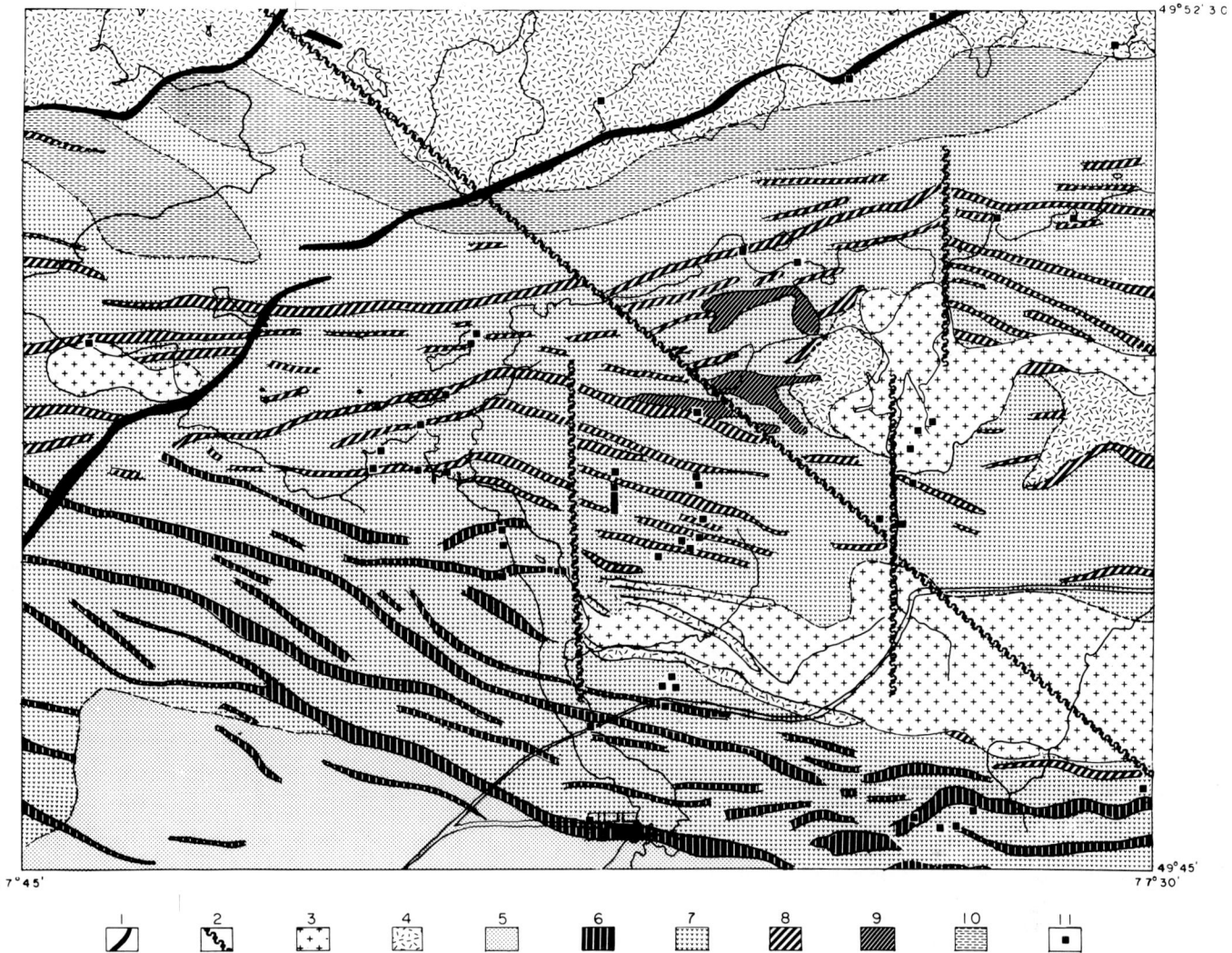

FIG. 17. Geologic map of map sheet 1 (Figure 2) constructed with the help of aeromagnetics. 1 = diabase dikes; 2 = faults; 3 = granite; 4 = diorite; 5 = Bell River complex anorthosites; 6 = Bell River complex gabbroic sills; 7 = undifferentiated volcanic rocks; 8 = magnetic volcanic rocks; 9 = gabbro sills; 10 = metasedimentary rocks; 11 = sample sites.

anomaly. The rest of the highly magnetic diorite is interpreted solely on the basis of the magnetic lithology. No outcrop is present in the area.

The metasedimentary rock units indicated in the northern part of the map occupy an area of uniformly low susceptibility and correlate with the map geology. The Bell River complex anorthosites are interpreted to occupy the area on the susceptibility map where particularly low and uniform susceptibilities are indicated. This is consistent with the measured susceptibility data ($>1 \times 10^{-3}$ SI).

The highly magnetic sills of the Bell River complex are easily recognized by the unusually high susceptibilities. In addition, they display a strike that is distinguishable from the trend of the volcanics. Susceptibility measurements made in situ and from hand samples are consistent with the calculated values. One measurement greater than 200.0×10^{-3} SI was made on the most highly magnetic sill, and measurements of 60.0×10^{-3} SI were made at points along the less magnetic sills. The nonmagnetic sills associated with the Bell River complex that intrude the volcanic rocks are, of course, not distinguishable from the nonmagnetic volcanic units. Thus, interpretation based on the susceptibilities alone is ambiguous in the region underlain by the nonmagnetic Bell River sills. It is for this reason that the area between the magnetic sills is arbitrarily displayed as volcanic on the interpretation map (Figure 17).

The volcanic sequences are the most difficult to interpret, because they contain a mix of magnetic and weakly magnetic flows and sills. It is not possible to use magnetic lithologic characteristics to distinguish the flows from the sills. Thus, only those units for which field-sample susceptibility data are available can be interpreted with confidence. The mapped geology was useful in a few places where the outcrop was sufficient to determine unambiguously the type of rock that coincides with a particular anomaly. No attempt was made to distinguish between the weakly magnetic basalts and the andesite units, because they are not well defined by the geologic mapping and appear to have similar susceptibilities.

The distinction between the Wabassee and the Watson Lake Groups, as seen near the bottom of Figure 3, is not possible on the basis of the magnetic data, because the contact lies within the region traversed by the highly magnetic sills. For this reason, all weakly magnetic volcanic rocks are indicated by unit 5 in Figure 17. An additional diabase dike has been interpreted at the northwest corner of Figure 17 on the basis of its strike length and direction (parallel to the known dike) and its susceptibility level.

Two distinct fault trends are indicated by the magnetic data, one striking at N 50° W and the other in a northerly direction (confirmed by geologic mapping). There appears to be some offset associated with the faulting, but it is not known if this indicates true strike-slip movement or simply apparent offsets caused by normal faulting cutting dipping beds.

SUMMARY AND CONCLUSIONS

The sample susceptibility study has shown that the calculated-susceptibility values on the apparent-susceptibility map agree well with measured susceptibilities. This suggests that the assumptions of magnetic induction and steep dip are approximately fulfilled in the area.

The uniform coverage offered by the magnetic data has allowed the recognition of significant linear continuity in the volcanic units. The linear character is revealed by a strong contrast in the magnetic susceptibility between flows. The susceptibility and geochemical data demonstrate that the flows are divided into two types, those that are magnetic (60.0×10^{-3} SI) and those that are weakly magnetic (4×10^{-3} SI), even though they have similar compositions. The presence of sills with the same susceptibility groupings and the close spatial relationship between magnetic and weakly magnetic units suggest that the magnetic characteristic is primary. Thus it would appear that varying oxygen fugacity and temperature within the source magma controlled the magnetic properties.

The geologic map prepared with the assistance of the interpreted magnetic data offers considerable improvement over the mapped geology. Contacts are extended along strike and more accurately placed. The overall flow pattern in the volcanic units is shown to be extensive and not localized. Two directions of faulting have been identified.

The use of high-resolution magnetic data, coupled with filter processing, has increased the geologic information about the area. It is clear that detailed magnetic interpretation can assist significantly in mapping the geology of the region.

ACKNOWLEDGMENTS

The authors are indebted to Noranda Mines, Ltd., for field assistance and for the award of a Bradfield scholarship to W.E.S. Urquhart. The National Science and Engineering Research Council of Canada supported the other aspects of the study. Able assistance was provided in the field and in sample work by S. Yunsheng and P. Halewood. Neutron-activation work was provided by the Slowpoke at the University of Toronto, and chemical analyses by M. Gorton and C. Cermignani.

The color maps were made by Data Plotting Services, Inc. The filtering and contouring software was provided by W.E.S. Urquhart Associates, Ltd.

REFERENCES

Bancroft, J. A., 1912, A report on the geology and natural resources of certain parts of the drainage basins of the Harricanaw and Nottaway Rivers, to the north of the National Transcontinental Railway in north-western Quebec: Rep. on Min. Oper., Que. Bur. Mines, 131–198.

Baranov, W., 1975, In potential fields and their transformations in applied geophysics: Geoexplor. Monogr. 6, Gebrüder Borntraeger.

Barker, F., and Arth, J. G., 1976, Generation of trondhjemitic–tonalitic liquids and Archean bimodal trondhjemite–basalt suites: Geology, **4**, 596–600.

Beland, R., 1953, Allard River area: Geol. Rep. 57, Que. Dep. Mines.

Bell, R., 1895, Geological Survey of Canada, annual report: v. 8, pt. A, 74–85.

———1900, Geology of the basins of the Nottaway River: Geol. Surv. Can., v. 13, pt. K.

Freeman B. C., and Black, J. M., 1944, The Opaoka River area: Geol. Rep. 16, Que. Dep. Mines.

Grant, F. S., 1974, Magnetic susceptibility map (with accompanying notes): Open File Rep. 229 (A), Geol. Surv. Can.

Haggerty, S. E., 1976, Opaque mineral oxides in terrestrial igneous rocks, in Oxide minerals: Short Course Notes, Mineral. Soc. Am., **3**.

———1979, The aeromagnetic mineralogy of igneous rocks: Can. J. Earth Sci., **16**, 1281–1293.

Jolly, W. T., 1978, Metamorphic history of the Archean Abitibi belt, in Metamorphism in the Canadian Shield: Pap. 78–10, Geol. Surv. Can., 68–78.

Letros, S. W., Strangway, D. W., Tasillo-Hirt, A., Geissman, J., and Jensen, L. S., 1983, Apparent susceptibility mapping in the Kirkland Lake area, Ontario, Abitibi greenstone belt: Can. J. Earth Sci., **20**, 548–560.

Longley, W. W., 1943, Kitchigama Lake area: Geol. Rep. 12, Que. Dep. Mines.

Quebec Ministry of Natural Resources, 1979a, Leve aeroporte input Matagami, contours du champ magnétique, DP-657.

———1979b, Leve aeroporte input Matagami, compilation géologique, DP-657.

Sharpe, J. I., 1968, Geology and sulfide deposits of the Matagami area, Abitibi-East County: Geol. Rep. 137, Que. Dep. Nat. Resources.

Spector, A., 1968, Spectral analysis of aeromagnetic data: Ph.D. thesis, Univ. Toronto.

Spector, A., and Grant, F. S., 1970, Statistical models for interpreting aeromagnetic data: Geophysics, **35**, 293–302.

Urquhart, W.E.S., and West, G. F., 1979, Aeromagnetic anomaly pattern in the English River gneiss belt, Superior Province: Can. J. Earth Sci., **16**, 1920–1932.

Yunsheng, S., Strangway, D. W., and Urquhart, W.E.S., Geological interpretation of a high-resolution aeromagnetic survey in the Amos–Barraute area of Quebec, this volume.

Yunsheng, S., Strangway, D. W., Urquhart, W.E.S., and Fengxing, S., Interpretation of an aeromagnetic survey of the Qian'an Archean metamorphic-rock series in China, this volume.

Interpretation of an aeromagnetic survey of the Qian'an Archean metamorphic-rock series in China

Sun Yunsheng,* D. W. Strangway,‡ W.E.S. Urquhart,‡ and Sun Fengxing§

ABSTRACT

An aeromagnetic survey was carried out over the Qian'an district of northeastern China. The area is underlain by a part of the Archean metamorphosed-rock series of Eastern Hebei province in China. From the total-field-intensity map, an apparent-susceptibility (SUSC) map and a regional-field map were produced. The regional-field map clearly indicates the structural trend of the Qian'an tectonic belt as well as the faults around it. The SUSC map was found to be useful in geological interpretation of the bedrock in the study area. Inferences from previous geological studies have been corrected. Subsequent field mapping in an area earlier mapped as sedimentary rocks has been shown to be an area of Archean crystalline rocks, and various units within the Archean have been reclassified. These changes have been confirmed by subsequent field work.

INTRODUCTION

The Qian'an district lies about 280 km east of Beijing (Figure 1) and is underlain by rocks of Precambrian age. These rocks are metamorphosed to granulite and amphibolite facies and are associated with complex structure. Tectonic activity during Proterozoic time folded the basement rocks. This was further complicated by later tectonic activity in the Mesozoic. Magnetite quartzite is the major resource of this area and is an important source of iron ore. Therefore, much geological and geophysical research has been conducted over several years (He, 1980; Shen et al., 1980).

Gravity and magnetic surveys have played an important role in both prospecting activity and in geologic mapping. In order to understand the geology better and to outline areas for prospecting, further mapping is under way. Therefore, it was considered useful to enhance interpretation of the aeromagnetic data using various data-processing techniques. The apparent-susceptibility method has been shown to be a useful tool for surface geological interpretation in regions where basement rocks are at the surface or are only buried to shallow depths (Letros, 1980; Letros et al., 1983; Bambrick et al., 1982; Urquhart and Strangway, this volume; Sun et al., this volume). The principal purpose of the present study is to obtain additional geological information by applying various processing techniques to the data from a standard-sensitivity aeromagnetic survey.

GEOLOGY

The Qian'an district of Eastern Hebei province belongs to the southern tectonic region of the Malanyu ancient continental nucleus at the northern margin of the Sino-Korean platform (Zhang, 1980). The area is primarily underlain by metamorphic rocks of Archean age. The Archean formations in Eastern Hebei have been divided into three types: granulite, amphibolite, and granite. It is believed that the metamorphic rocks were derived from a sequence of mafic to felsic volcanic rocks, with minor amounts of sedimentary rocks (Jin et al., 1980). The distribution of the formations is sketched in Figure 2.

Figure 3 is the geologic map of the area considered in this paper. Table 1 shows the stratigraphic sequence. The oldest rocks are metavolcanic rocks of Archean age, with Rb–Sr isotopic dates of about 3 to 3.6 b.y. (Shen et al., 1980). They are, in succession, the Shangchuan Formation (Arc) and the Santunying Formation (Ars) of the Qianxi Group and the Baimiaozi Formation (Arb) of the Dantazi Group. The inferred thicknesses of these formations are 400, 12 000, and 7 400 m, respectively. The rocks are classified as high-grade regional metamorphic, but generally they have locally undergone strong retrogressive metamorphism from granulite facies back to amphibolite facies. The rocks of these formations consist mainly of biotite-pyroxene-plagioclase gneisses, biotite-amphibolite-plagioclase gneisses, and amphibolite-plagioclase gneisses intercalated with lens-like or layered magnetite quartzites. Most of the rocks are slightly migmatized, although strong migmatization is ob-

---

*On leave, Department of Applied Geophysics, Changchun College of Geology, China.
‡Department of Geology, University of Toronto, Toronto, Ont. M5S 1A1, Canada.
§Department of Geology, Changchun College of Geology, China.

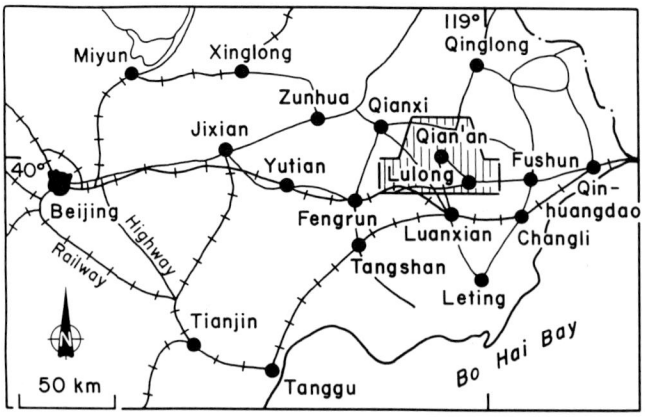

FIG. 1. Location map showing survey area approximately 150 km east of Beijing, northeastern China.

served locally. Since most of the rocks in the area are rich in ferromagnesian minerals, the original rocks are believed to have been mafic to intermediate volcanics.

The stratigraphic sequence strikes in a northeasterly direction. Surrounding the Qian'an block (except on the eastern side) are sedimentary rocks of Sinian age (Late Proterozoic), which consist of limestone, sandstone, and shale. At some places Cambrian sedimentary and Jurassic volcanic and sedimentary rocks unconformably overlie Precambrian rocks. In the southeastern corner and in the northern part, felsic to intermediate intrusive rocks, including granite, granodiorite, and diorite, are intruded into the volcanic and sedimentary rocks of older age. Some intrusive gabbro bodies occur in the southeastern corner. Pegmatite and alkaline dikes are distributed along the northeast-trending shear zone in the southeastern corner.

The study area is situated in the southeastern part of the Shanhaiguan uplift (Tianjin Institute of Geology, 1974). Rocks of Archean age form an uplifted block, the Qian'an block. Tectonic movements occurred at least twice during Archean time and later in Proterozoic time (Table 1), producing tight folds in the basement rocks. The axes of the folds strike north or northeast and are overturned to the west or to the northwest. Tectonic movements in the Mesozoic significantly affected the structure of the area, producing a broad anticlinorium that plunges to the west. Thus rocks of Precambrian age have an arc-like distribution.

Faults are highly developed in the area. Some of them were initiated in the Archean, but most date from the Mesozoic (Jin et al., 1980). The faults are divided into four sets: east–west, north–south, northeast, and northwest. Many faults—for example, the Lengkou fault and the Qinglong River fault—surround and define the Qian'an block. Both northeast-trending faults and northwest-trending faults are found within the block. In addition, in the southeastern corner, a northeast-trending shear zone separates the block from another tectonic unit that extends beyond the study area (Zhang, 1980).

The important resources of the area are iron deposits, which occur mainly in the upper series of the Santunying Formation. The ore bodies occur in lenses or layers that strike N 40° E to N 70° E and consist mainly of magnetite and martite (hematite replacing magnetite).

GEOPHYSICAL SURVEY AND DATA-PROCESSING TECHNIQUES

In connection with prospecting for iron and other ore deposits, various geophysical surveys have been undertaken in the past two decades. In 1964 a regional gravity map at a scale of 1:1 000 000 was prepared. During the 1973 and 1974 field season an aeromagnetic survey leading to a map at a scale of 1:50 000 was carried out in the area. In addition, ground magnetic surveys have been carried out in parts of the area. A great deal of information has been derived from these surveys, and many geologic and geophysical reports have been published (He, 1980).

The present study is based on an aeromagnetic survey performed in 1973 by the Baoding Company of Geophysics using a fluxgate magnetometer. The survey covered more than 5 000 km^2 in the Qian'an–Luanxian district at a line spacing of 500 m and an average terrain clearance of 150 m. The resulting total-magnetic-intensity maps were produced using manual compilation techniques and were published at a scale of 1:50 000. The present work is concerned with an area of about 3 000 km^2 in the northern part of the region.

In order to acquire a digital data set, the total-intensity contour maps were sampled along the contours using a Summagraphics digitizer. A program was written to convert the sampled data set into grid cells at intervals of 100 × 100 m using cubic spline interpolation. This is a useful interval for preparing contour maps and processing data, but it should be recognized that the actual data are limited by the original 500-m flight-line spacing. Figure 4 is part of the original map (Figure 4a) and of the map contoured from the interpolated grid set (Figure 4b). The comparison shows that the digitizing process led to a reasonable reproduction of the original hand-contoured maps. For interpretation, a regional-field map and a susceptibility (SUSC) map were derived from the digitized data.

Figure 5 is the radially averaged logarithmic energy-density spectra (LnEDS) of the aeromagnetic map. It clearly shows the existence of sources that contribute to the energy at low wavenumbers (lower than 0.01 cycle/digital interval, cpdi, or longer than 10 km). A low-pass filter was designed to extract the regional field. From the LnEDS a roll-off range of 0.01 to 0.031 cpdi was selected, and the Hanning function was applied to give the filter a smooth response from 1 to 0. In calculating the SUSC map, four filters (strike-sensitive filter, reduction to the pole, downward continuation, and size correction) were used. From the theory point of view, the following assumptions are made (Bambrick et al., 1982): (1) the magnetization is along the Earth's magnetic field, (2) the prism models are vertically sided, (3) the prism models have infinite depth extent, (4) the data are downward continued to the surface or to the top of the models, and (5) demagnetization effects are negligible. Deviation from these conditions will affect the results in different but characteristic ways, which were discussed by Bambrick et al. (1982). In the application of the method, corrections have to be made. Some of these have been analyzed by the authors and are discussed in a separate paper (Sun et al., this volume).

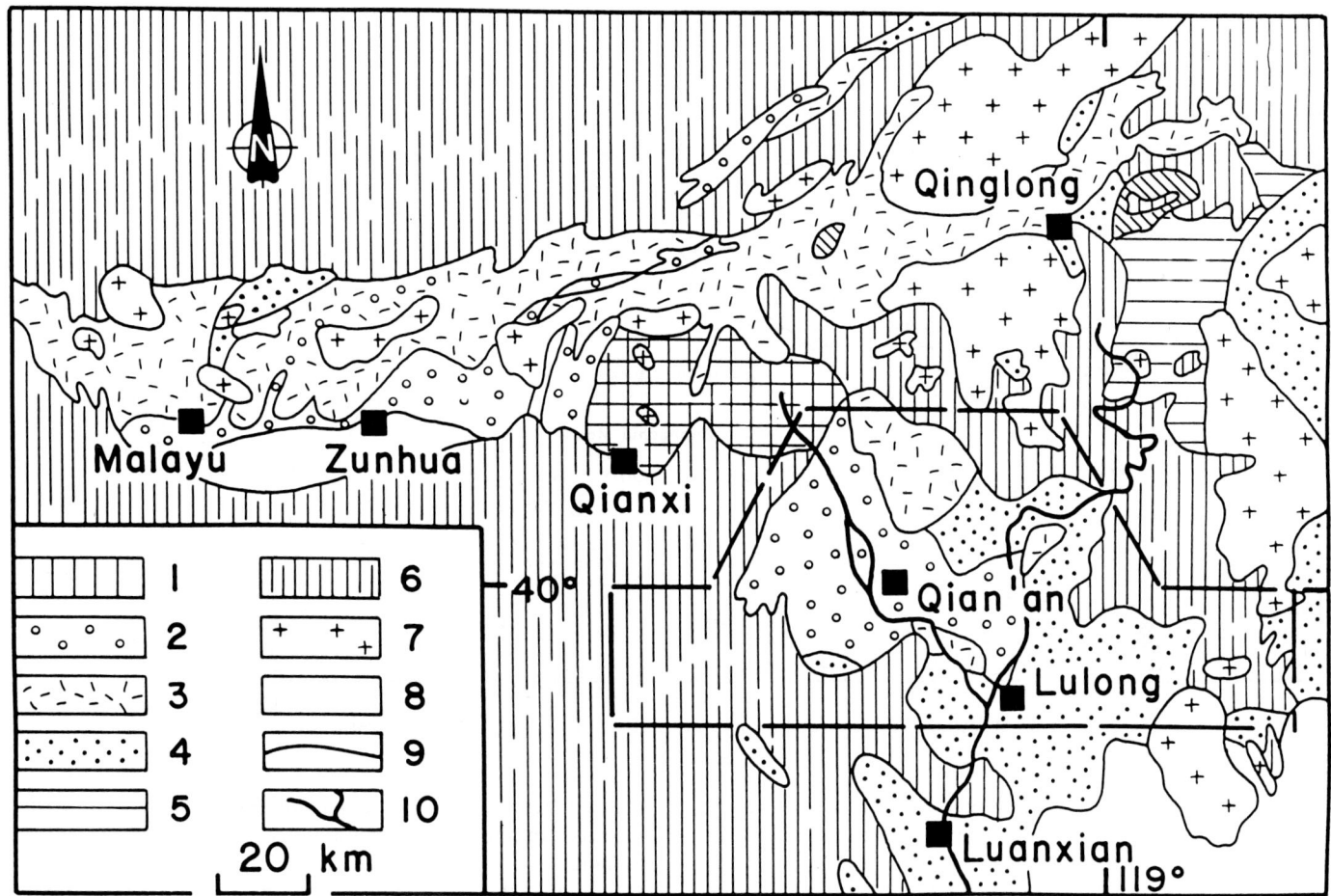

Fig. 2. Map of survey area and the Archean Qian'an block. Units 1–4 are Archean; 1 = pyroxenite-granulite; 2 = biotite-pyroxenite gneiss; 3 = plagioclase-amphibolite; 4 = biotite-plagioclase granulite and amphibolite; 5 = Lower Proterozoic; 6 = Sinian rocks, Upper Proterozoic; 7 = intrusives, mainly Mesozoic; 8 = Quaternary. Symbols 9 and 10 show contacts and rivers, respectively.

Among these effects, the assumption of induced magnetization is the most noticeable factor. Ten individual anomalies were selected to assess the effect of the remanent magnetization on anomalies using their maximum and minimum values (Sun, 1980). Figure 6 illustrates the test procedure on one of the selected anomalies. The results show that the average inclination of total polarization is about 60°, which is very close to that of the Earth's magnetic field (56.5°) for the area. That implies that the first assumption is essentially satisfied.

In addition to the existence of long wavelengths usually interpreted to mean deep sources (Spector and Grant, 1970), the LnEDS also show that significant energy is due to noise caused by the instrument and by compilation, digitization, and interpolation effects. To design the filters properly, three first-vertical-derivative maps with different high-wavenumber cutoff ranges (0.1–0.17, 0.12–0.20, 0.2–0.3 cpdi) were calculated. The results were used to select the range 0.1–0.2.

GEOLOGICAL INTERPRETATION

In the study, three filtered maps (regional field, apparent susceptibility, and first vertical derivative) were produced, and the total-field map and the apparent-susceptibility map were color-contoured to aid in better interpretation. The Capital Steel Works, the Wuhan College of Geology, and the Changchun College of Geology (1975) collected and measured the magnetic properties of more than 500 samples of various rock types in the area and vicinity, and the results were used as a control for the interpretation in this study (Table 2).

TOTAL-INTENSITY MAP AND MAJOR FEATURES

The anomalies in the total-intensity map (Figure 7) indicate the major geologic features. Pyroxene-plagioclase gneisses, amphibolite-plagioclase gneisses, and other rocks of the Qian'an block are characterized by highly irregular anomalies. In the Lulong area the scattered anomaly swarm is over the upper series of the Santunying (Ars) Formation. In the Qian'an area the slightly smooth and northeast-trending contours are related to the lower series of the Santunying Formation. In the east the flat appearance is associated with the Baimiaozi (Arb) Formation. The transition from one pattern to another is so gradual, however, that it is impos-

Table 2. Determination of magnetic properties of rock types in study area.

Rock Type	Number of samples	K ($\times 10^{-3}$ SI)	J_r ($\times 10^{-3}$ SI)	Q
Magnetite-quartzite	313	1307	4084	5.82
Ferruginous-amphibolite-plagioclase gneiss	23	82	54	1.23
Ferruginous amphibolite	2	15	45	5.41
Amphibolite-plagioclase gneiss	8	7	4	0.92
Granitic migmatite	15	<0.2	<0.2	No data
Sedimentary rocks	3	<0.2	<0.2	No data

Note: K = susceptibility; J_r = natural remanent magnetism; $Q = J_r/K\mathbf{H}$, where $\mathbf{H}$ is Earth's field.

amplitude anomalies of large areal extent coincide with the Mesozoic granite intrusives that were intruded into the earlier metavolcanic and sedimentary rocks. This analysis indicates that the major geologic features are reflected on the aeromagnetic map but that it is difficult to place the contacts accurately. Much of the information is obscured by anomaly superposition and noise. In order to improve the interpretation of the geology, maps of SUSC, regional field, and first derivatives were computed. These maps yield considerable additional information about the stratigraphic and structural features.

FILTERED MAPS AND INTERPRETATION

The radial LnEDS of the aeromagnetic data (Figure 5) indicate the existence of long-wavelength features. To extract this component, usually attributed to deep sources, a simple low-pass filter was employed, and the wavenumber interval of 0.01 to 0.031 cpdi was chosen to remove the residual field. The resulting regional map is shown as Figure 8. The map shows individual anomalies striking east. These anomalies are distributed along northeast and east-northeast trends. This regional orientation coincides with the strike of the Precambrian units. Field values, higher than 300 nT in the central part, correspond to the Qian'an block. In the north, the negative contours occur over the Sinian sedimentary rocks. The contact between the two regions is related to the west-northwest-trending Lengkou fault, which separates the Qian'an block from the sedimentary rocks.

In the western part of the area the positive anomaly (1 in Figure 8) is believed to be associated with another Precambrian block or with a Mesozoic intrusive which is separated from the Qian'an block by a north–south fault with low values. In the eastern part of the area (2), a broad zone of negative values is present at the eastern edge of the Qian'an block. South of the Qian'an block a series of east–west contours indicates east–west structural features (3). In addition, the map indicates many structural features within the Qian'an block. A northeast-trending anomaly (4) in the middle (about 100 nT) is related to the main anticlinorium.

Precambrian metamorphic formations containing metamorphosed ferruginous rocks are widely distributed in the Qian'an region and the Luanxian region, which is south of the study area. It had been formerly believed that the rock units strike in a north–south direction. Some geologists have proposed recently, however, that the rocks strike east–west and decrease in age from north to south (Zhang, 1980), while others propose that the rocks strike in a northeasterly direction. The distribution of the regional field in Figure 8 obviously strongly confirms the northeasterly strike.

For surface geological interpretation, the SUSC map can also be useful. In the study, four separate filters were applied to produce an apparent-SUSC map (Figure 9). Various levels of apparent susceptibilities and patterns appear to be related to the contrasting rock types and to structural features.

The magnetite quartzites are the most strongly magnetic rocks in the area, whereas metavolcanic rocks are relatively weakly magnetized. The magnetizations of granitic migmatites and sedimentary rocks are so low that the measuring instruments (sensitivity of about 2×10^{-4} SI) were not sensitive enough to measure them. Thus these rocks are nonmagnetic for all practical purposes.

At least seven anomaly patterns can be distinguished on the apparent-SUSC map. All the strongest anomalies, ranging in value from 600×10^{-3} (SI) to as high as $1\,400 \times 10^{-3}$ (SI), are related to the magnetite quartzites. Most of them are being exploited. The Shuichang Mine, the Gongdianzi Mine, the Qian'an Mine, the Baoguanying Mine, and other mines are all characterized by the highest apparent-SUSC values. According to geological reports (Tianjin Institute of Geology et al., 1974), ferruginous rocks exist mainly in the upper series of the Santunying Formation. On the SUSC map, the series themselves are characterized by irregular contour patterns except in the iron-formation areas. The apparent-SUSC values are about $50–100 \times 10^{-3}$ (SI). The typical patterns are seen in the south, the Mopan Mountain, and the Zhongchui Mountain area. This type of contour pattern is also observed in the Shuichang–Dashihe area.

In the central part of the map area the apparent-SUSC values with amplitudes of about 120×10^{-3} (SI) are a little higher than those discussed in the previous paragraph, but the contours are smoother. This same pattern occurs over the lower series of the Santunying Formation. It is apparent that the upper and lower series can be distinguished by this contrast. Using this criterion, it is suggested that the rocks west of Yulindian belong to the lower series rather than to the upper series as indicated on the geologic map. The appearance of these anomalies over the rocks west of

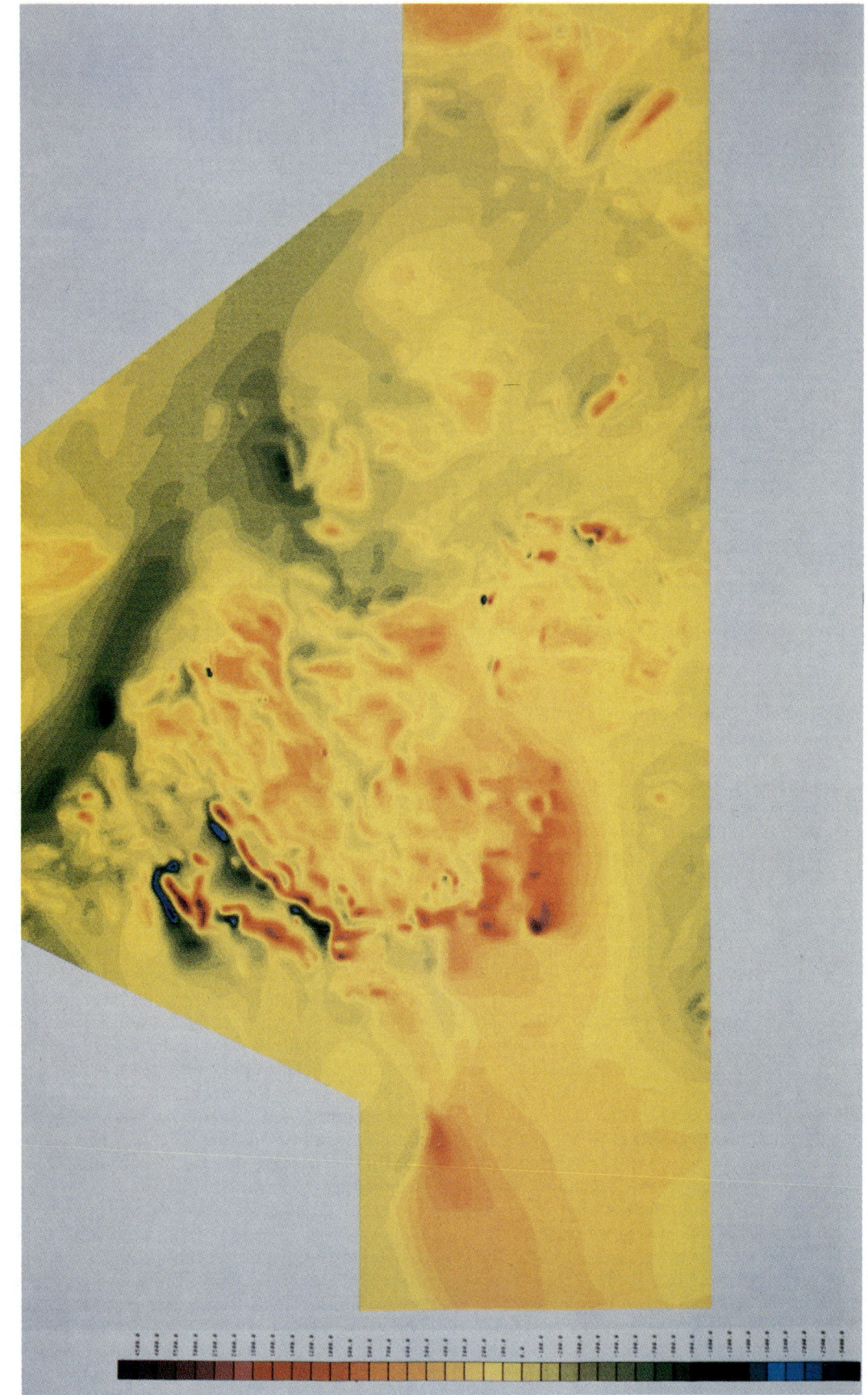

FIG. 7. Total-intensity anomaly map of study area. Contour interval, 100 nT.

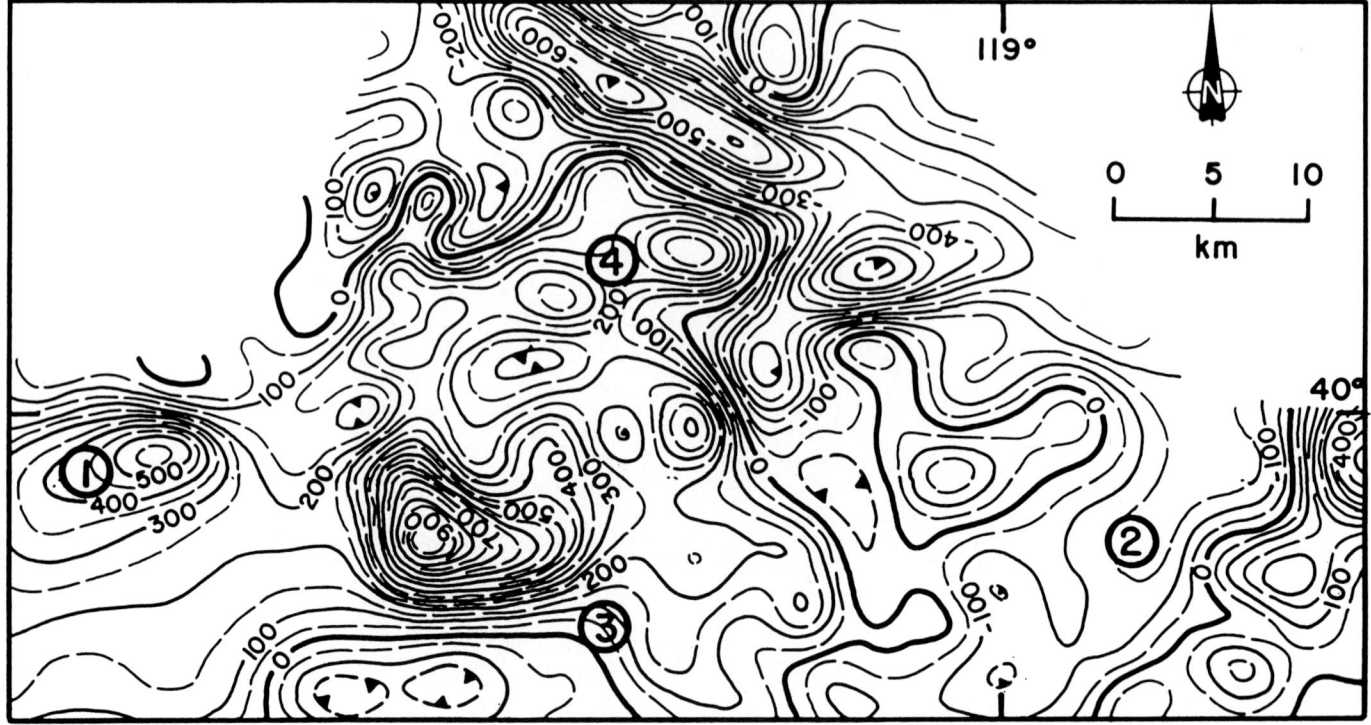

FIG. 8. Regional-field map of study area using the roll-off range from 0.01 to 0.031 cpdi. See text for explanation of numbered features.

Yulindian is similar to that of the lower series. In terms of lithology, the different SUSC values may suggest that the original volcanic rocks of the lower series are more mafic than those of the upper series. There is no field evidence to support this interpretation, however, as the rocks in the area have not been studied in detail. SUSC values of 120×10^{-3} (SI) or higher are present in the north. These values occur over the oldest rocks of the area, the Shangchuan Formation. The contours in the Caoguzhuang area and in the Jianchangying area are slightly different. The latter is featureless, while the expression of the former is higher and more irregular. They are associated with the upper and lower series, respectively, of the Shangchuan Formation. The different appearances may support a previous suggestion that the original rocks of the lower series are more mafic than those of the upper series. The correlations suggest that the Shangchuan Formation and the Santunying Formation represent two eruption cycles in which the rocks in each cycle change from mafic to intermediate to felsic.

Low to slightly negative SUSC values are associated with the metavolcanic rocks of the Baimiaozi Formation, which overlies the Santunying Formation in the eastern part of the area. A Precambrian fault has been mapped, which separates the Qianxi Group from the Dantazi Group along the Chaisho–Yulindian, Dahenghe–Lulong belt. The expressions of the two groups on the map are rather different. In the Lulong and Zhaojiawa areas the contour pattern is similar to that over rocks of the Baimiaozi Formation in the east, suggesting that the rocks of the Baimiaozi unit are present in this area. This interpretation, which was first suggested in this study, has been confirmed by subsequent detailed geologic mapping. The rocks were described as the Santunying (Ars) Formation on the earlier geologic map (1973).

North, west, and south of the Qian'an block, negative contours ranging from 120 to 250×10^{-3} (SI) correspond to Sinian sedimentary rocks that are essentially nonmagnetic. The negative apparent-SUSC values are partly caused by the choice of base level. Almost the same SUSC values are found over the same rock type in the northern and southern sedimentary zones, whereas the total-field values are different because of the influence of nearby anomalies and the regional field (as discussed previously). At the northern edge and the southeastern edge, SUSC values of about 120×10^{-3} (SI) are associated with diorite and granodiorite masses.

Near the southern border of the area the SUSC contour pattern is similar to that over rocks of Archean age. There is a sharp contact between this pattern and the negative contours to the north. It seems obvious that the contrasting patterns are associated with two different rock units. We conclude that the negative contours are related to the sedimentary rocks such as those discussed previously, and the positive contours are associated with Archean metavolcanics. This interpretation (reported here for the first time) has been confirmed by recent geological work at the Changchun College of Geology. This analysis indicates that the zero contours on the SUSC map roughly outline the contacts between metavolcanic and sedimentary rocks and that they can be used to delineate some intrusive bodies intruded into the sedimentary and metavolcanic rocks.

In addition to the lithology, numerous structural features are displayed on the apparent-SUSC map. A northeast-

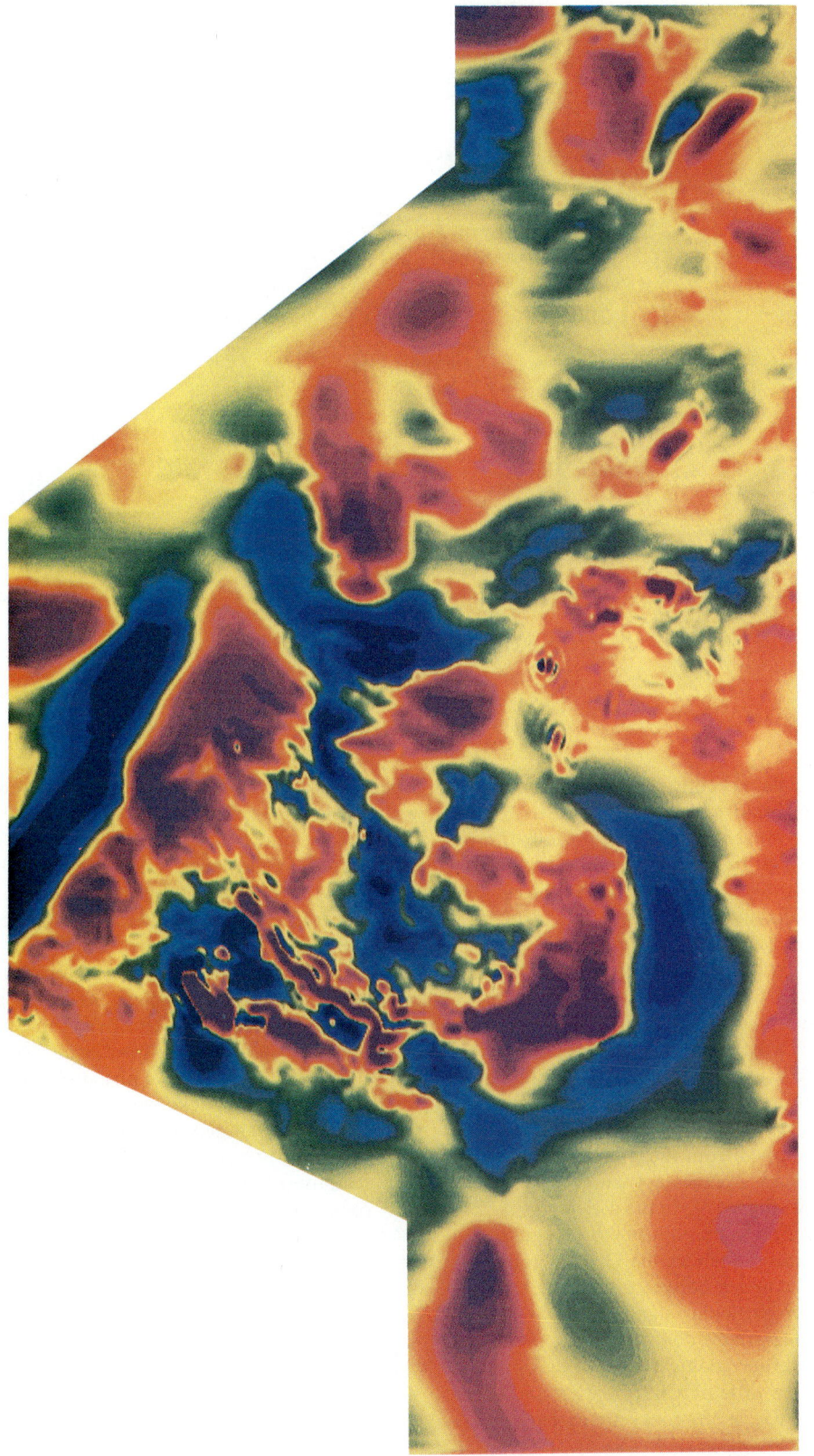

FIG. 9. Apparent-susceptibility map of study area. Contour interval in SI units.

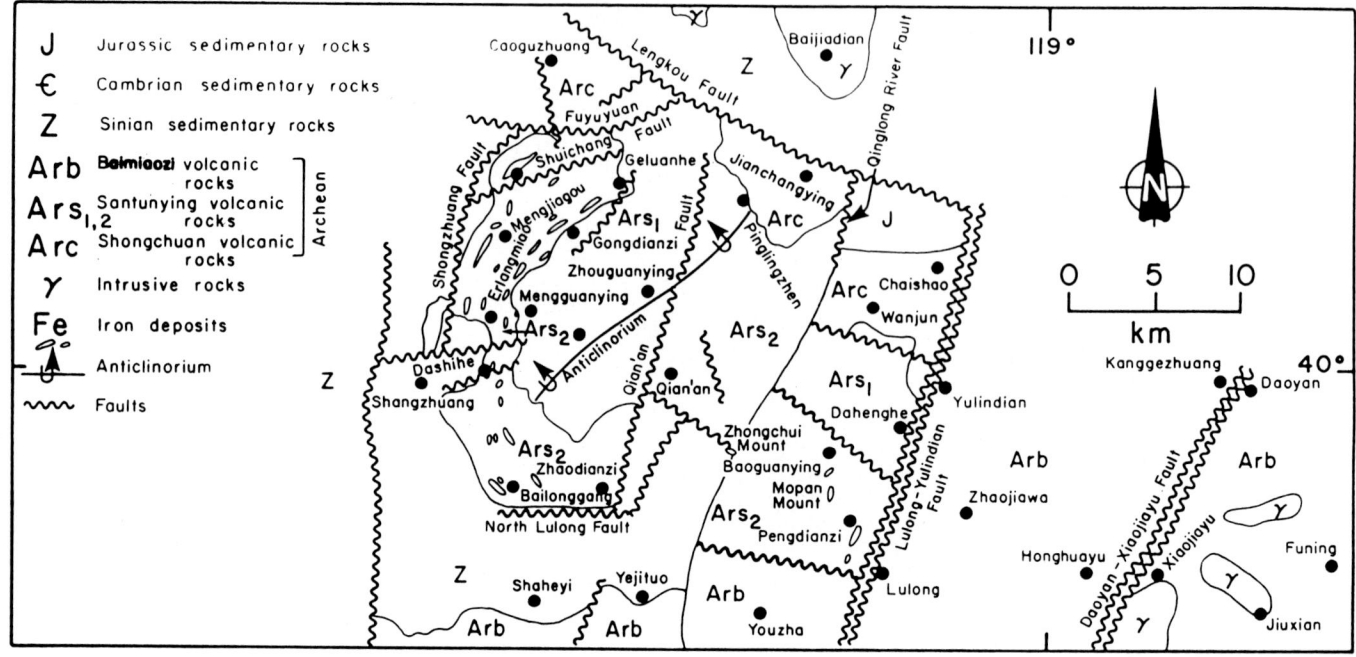

FIG. 10. Interpreted geologic map of study area.

trending trough in the SUSC map is consistent with a similar feature on the regional-field map and coincident with the anticlinorium discussed previously. In the eastern half, the SUSC values change from high to low and back to high in connection with two ancient, deep faults, the Lulong–Yulindian fault and the Xiaojiayu–Daoyan fault. The former separates the Qianxi Group (Ars, Arc) from the Dantazi (Arb) Group, and the latter divides the Qian'an block from a separate tectonic unit. The SUSC map appearance over the Qian'an block contrasts sharply with that over the region surrounding it. The borders of the Qian'an are defined by the Lengkou fault, the Qianlong River fault, the Fuyuyuan fault, and the Shongzhuang fault, respectively. The SUSC map also shows evidence of faults within the block (Figure 10).

A geologic map was developed from the geological interpretation (Figure 10). This map is consistent with the previous geological interpretation (Figure 3). In the areas of rocks of Archean age, this interpretation map gives more detail of both structure and stratigraphy, such as the Qian'an fault, the anticlinorium, and the contact between different subgroups. These results are of particular significance, as the area is largely covered by Quaternary deposits. Some incorrect interpretations on the earlier geologic map were found and corrected. This map gives little new information within the sedimentary rocks, however, as the rocks are essentially nonmagnetic.

SUMMARY AND CONCLUSIONS

An aeromagnetic survey of the Qian'an district, China, was studied in connection with a recent program of detailed geologic mapping. The total-field map reflects the dominant geologic features, but details are obscured. Various filters were employed to improve resolution of the data. The SUSC map was the most effective technique for surface geologic mapping. At least seven different patterns can be distinguished, which correspond to the different rock units in the area. Contrasts in SUSC patterns outline contacts between rocks of different types. The analysis of the SUSC map brought to light some incorrect interpretations on earlier geologic maps. For example, rocks at the southern edge of the area are Archean in age rather than Sinian sedimentary rocks. In the Lulong–Zhaojiawa area, the rocks are interpreted to be the Baimiaozi Formation rather than the Santunying Formation. These results have been confirmed by subsequent detailed geologic mapping. A regional field map produced by low-pass filtering proved useful for indicating deep structural features. Some folds and faults in the basement rocks are interpreted from the regional map.

An interpreted geologic map was developed by combining the aeromagnetic maps and the filtered maps, the sample susceptibility measurements, and existing geologic maps.

In general, this study demonstrates the utility of aeromagnetic surveying and various data-processing techniques for geologic mapping in Precambrian terranes.

ACKNOWLEDGMENTS

We express our appreciation to the Natural Sciences and Engineering Research Council (NSERC) of Canada for financial assistance. The Capital Steel Works and the Precambrian Research Office of Changchun College of Geology, China, kindly provided the aeromagnetic maps as well as other material. We also acknowledge Associate Professors

La Yugi and Yi Tinshong, Changchun College of Geology, for their helpful discussion. The Chinese/Canadian collaboration on this project was made possible by the June 7, 1979, arrangement between the Chinese Ministry of Education and the Council of Ministers of Education, Canada, to place Chinese scholars in Canadian universities. Detailed comments by anonymous reviewers were most helpful in rewriting this paper.

REFERENCES

Bambrick, J., Letros, S., and Geissman, J., 1982, Apparent magnetic susceptibility (SUSC) mapping in theory and practice [abstr.]: 52nd Annu. Int. SEG Meet., Oct. 17–21, Dallas, Tex.

Capital Steel Works, Wuhan College of Geology, Changchun College of Geology, Tianjin Institute of Geology, 1975, Analysis on the characteristics of iron deposit rock series in Qian'an mine area.

He, Shaoying, 1980, Preliminary analysis of the characteristics of regional gravitational and magnetic fields of pre-Sinian metamorphic rock series in Qian'an-Luanxian District, eastern Hebei: Bull. Chinese Acad. Geol. Sci. (ed. by Inst. Geol.), no. 2, 33–50.

Jin, Wenshan; Sun, Dazhong; Wang, Junlian; and Fang, Liancheng; 1980, Regional metamorphic formations of Archean rocks of eastern Hebei and their characteristics: Bull. Chinese Acad. Geol. Sci. (ed. by Tianjin Inst. Geol. Mineral Resources), no. 1, 38–43.

Letros, S. W., 1980, Geological interpretation of high-resolution aeromagnetic data in the Kirkland–Larder Lake area: M.S. thesis, Univ. Toronto, Dep. Phys.

Letros, S. W., Strangway, D. W., and Geissman, J., 1983, Apparent susceptibility mapping in the Kirkland Lake area, Ontario, Abitibi Greenstone Belt: Can. J. Earth Sci., 20, 548–560.

Shen, Qihan; Liu, Guohui; Zhang, Qinwen; He, Shaoying; and Gao, Jifeng; 1980, The correlation of the pre-Sinian metamorphosed ferruginous rock series of Qian'an and Luanxian, eastern Hebei: Acad. Sci. China, J. Geol., no. 2.

Spector, A., and Grant, F. S., 1970, Statistical models for interpreting aeromagnetic data: Geophysics, 35, 293–302.

Sun, Yunsheng, 1980, Analytical coefficients of characteristic points of magnetic anomalies due to several geometric bodies: J. Changchun Coll. Geol., no. 3, 65–76.

Sun, Yunsheng, Strangway, D. W., and Urquhart, W.E.S., Geological interpretation of a high-resolution aeromagnetic survey in the Amos–Barraute area of Quebec, this volume.

Tianjin Institute of Geology, Capital Steel Works, et al., 1974, Geological maps of eastern Hebei.

Urquhart, W.E.S., and Strangway, D. W., Interpretation of part of an aeromagnetic survey in the Matagami area of Quebec, this volume.

Zhang, Qinwen, 1980, Pre-Sinian east–west geotectonic belts and their inner structural pattern in Qian'an-Luanxian District, eastern Hebei: Bull. Chinese Acad. Geol. Sci. (ed. by Inst. Geol.), no. 2, 17–32.

INDEX

Abitibi greenstone belt, 392, 394, 411, 413
Adirondack Mountains, 13, 15, 17, 301
Alaska, 25, 27
Alaska compilation, 25
Aleutian active margin, 339, 343
alkaline stocks, 128
alluvial basins, Arizona, 164, 178
alluvial valleys, Arizona, 172, 176
Altamaha anomaly, 11
Amarillo uplift, 58
Amos-Barraute area, Quebec, 413, 417, 423
amplitude, 46, 60
Anadarko basin, 36
analytical continuation, 248, 265
Animikie Group, 380, 381, 382, 384
anorthosite, 428, 431
    gravity signatures, 117
    origin of, 109, 116
Antarctica, 104
Appalachian allochthon, 59
Appalachian fold belt, 11, 13, 17, 252
Appalachian Mountains, 1, 8, 11, 22, 36, 47, 52, 58, 288, 302, 308, 310, 343
Appalachian orogen, 11, 13, 17
apparent-density algorithm, 392, 412
apparent-density map, 392, 399
apparent-susceptibility maps, 413, 416, 417, 418, 419, 420, 423, 426, 429, 433, 434, 436, 438, 439, 441, 444, 447
aquifers, South Dakota, 233, 244, 246
arc-continent collision model, Michigan and Wisconsin, 267, 285
Archean rocks, China, 439, 440, 442, 446, 448
Arizona, 164, 165, 169, 171, 174, 175, 176, 178
Australia, 102, 103, 104, 105
Avalon terrane, 11

Baltimore Canyon trough, 325, 335, 336, 337
basement-lithology map, Ontario, 392
basement rift zones, 287, 294, 297, 298, 306
Basin and Range pattern, 33
Basin and Range province, 20, 164, 165, 169, 171, 172, 174, 176, 178, 198, 205, 206, 207, 208
basins, Early Proterozoic, 267, 268, 269, 271, 275, 277, 281, 283, 284
Belt basin, 59
Benioff zone, 343
Big Horn Mountains, 36
Black Hills, 13, 233, 236, 240, 241, 242, 243, 244, 245, 246
Black Mesa basin, 169
Blake Plateau, 36, 336, 337
Blake Plateau basin, 325, 335, 336, 337
Blake Spur Anomaly (BSA), 340, 343

block faulting, Kansas, 231
block structure, 102, 103, 104, 105
Blue Ridge, 308, 310
Botswana, 144, 147, 150, 153
Bouguer correction, 362
Bouguer reduction density, 362, 363, 365, 371, 372, 373
Brazil, 124, 126, 130
breccia pipes, Arizona, 176
Bright Angel anomaly, 169
Brunswick anomaly, 8, 11, 335, 336
Brunswick terrane, 11

calderas, 361, 363, 365, 366, 371, 372
Caledonian Range, 154
California, 347, 349, 351, 352, 356, 357, 359
Carlton County area, Minnesota, 375, 377, 380, 381, 384, 385, 390
Carolina-Mississippi fault, 11
Carolina Slate belt, 308, 310, 312, 320, 323
Carolina trough, 325, 335, 336
Cascade Range, 361, 363, 365, 370, 371, 372
Central North American rift system (CNARS) — see also Midcontinent rift system — 213, 225, 228, 231
central platforms, 59
Chaman fault zone, 141
Charleston magnetic terrane, 11
Charlotte belt, 308, 310, 312
chemical remanent magnetization (CRM), 321
Cheyenne belt, 20
China, 439
Churchill province, 17, 20, 72, 283, 291
Cincinnati arch, 15
Coastal Plain, 308, 310, 313, 314, 318
collision tectonics, 109, 112, 114, 115
collision zone, 59
Colorado lineament, 20
Colorado Plateau, 20, 33, 36, 164, 165, 169, 172, 176, 178, 182
Columbia Plateau, 4, 33, 36, 55
computer graphics, 377, 390
continental structure, 46
convergent-plate zones, 89, 96, 98
copper deposits, Arizona, 164, 169, 174, 175, 178
copper sulfide mineralization, Pakistan, 139
Cordillera, 1, 17, 22
Cordilleran system, 1, 17, 20, 22
core-mantle boundary, 88, 89, 95, 96, 98, 101
cratons, edges and interior, 59
Crosbytown anomaly, 6, 7
cross fractures, 102, 103, 105, 106, 108
Crozet geoid high, 95, 96, 98
crustal rifts, 58, 59, 60

crustal weakness, zones of, 59, 60
Curie isotherm, 1, 4, 20, 22, 262, 339, 343, 345
Curie point, 343
Curie-point geotherm, 67

Datil volcanic pile, 164, 171, 208
Decade of North American Geology (DNAG), 34, 69, 87
density units, 401
density zoning, 399
depth-to-bedrock map, Arizona, 164, 172
derivative maps of the magnetic field, 325
diabase dikes, Quebec, 427, 428, 431, 432, 436, 438
digital enhancement, 72
dike intrusion, Kansas, 231
dike swarms, 144, 147, 375, 380, 381, 384, 385, 389, 390, 428
downward continuation, 413, 417, 419, 434, 436, 440
downward continuation filters, 213
drape-to-level continuation, 181, 185, 186, 187, 188
draping, 248, 255
Duluth Complex, 375, 381, 387, 388, 389

Earth
  deep structure, 89
  gravity field, 88, 89, 95, 96, 98
  magnetic field, 88, 96
east–central Midcontinent, 287, 288, 294, 302, 305, 306
East Coast magnetic anomaly (ECMA), 8, 11, 325, 335, 336, 337, 340, 343
East continent gravity high, 294
economic applications of regional geophysics, Arizona, 175, 176
editing, gravity data, 39, 45
elevation versus gravity, 314
enhancement techniques, magnetic anomaly maps, 72
ensialic environment, 130
evaporite deposits, Arizona, 175, 176, 178
exploration strategies, Ontario, 392, 404, 411

failed-arm rift, 301
filtered maps of the magnetic field, 325
first-vertical-derivative filter, 255
France, 154
frequency-domain filtering, 375, 381, 390
frequency plots, 401

Gabbro Lake area, Minnesota, 375, 377, 380, 381, 387, 390
geoid anomalies, 88, 89, 94, 98, 101
geophysical terranes, 262, 265
Georges Bank basin, 325, 335, 336, 337
Georgia, 308, 310, 313, 314
geothermal gradients, 233, 238, 243, 244, 245, 246
geothermal system, South Dakota, 233, 243, 244
geothermometry, South Dakota, 233, 243, 244, 246
Giants Range batholith, 381
gneiss domes, Wisconsin, 267, 271, 272, 277, 284
gold deposits, Ontario, 393, 394
gold exploration, Ontario, 392, 404, 411
Gorda plate, 347, 356
granite–rhyolite terrane, 287, 291, 298, 306
granitic plutons, 228, 267, 268, 271, 272, 277, 280, 283, 284
Gravity Anomaly Map (GAM) Committee, 33
Gravity Anomaly Map of the United States, 33

gravity differentiation, 117
gravity "dipole" anomalies, 104, 105
gravity inversion, 154, 160, 162
gravity signature, 109, 113, 114, 115, 116, 122
gravity-to-geoid ratio (g/N), 88, 94, 95, 96
gravity-trend pattern, 103, 104, 105, 106
Great Basin, 36
Great Lakes tectonic zone (GLTZ), 268, 269, 275, 279, 281, 284, 285, 377, 380
Great Plains, 4, 8
Great Valley of California, 20, 37
greenstone belts, Africa, 144, 147
Grenville anomaly, 17
Grenville collision, 114, 115, 116
Grenville foreland zone, 291, 298
Grenville front, 17, 63, 109, 110, 111, 112, 113, 114, 115, 116, 121, 291, 297, 298, 301, 302, 306, 393, 399, 411
Grenville province, 13, 17, 109, 110, 111, 112, 114, 115, 116, 117, 121, 122, 287, 301, 306, 393, 395, 411
Grenville terrane, 17
grid of digital data, 38
ground water, 164, 172, 176, 178, 233, 244, 246
ground-water exploration, Pakistan, 132
Gulf Coast, 4

Hawaii, 25
Hawaii compilation, 25
heat flow, 164, 178, 244, 343, 345
Hercynian Range, 154
Hibbing area, Minnesota, 375, 377, 380, 381, 384, 390
high-frequency-pass filter, 213, 214, 225
high-pass filtering, 375, 382, 388, 390
high-resolution aeromagnetic data/survey, 325, 337, 375, 377, 389, 413, 420, 423, 438
hinge zones, 336, 337
Holbrook line, 169, 174, 178
horizontal compressive stress, 59, 60
horizontal gradients, 46, 47, 52, 55, 60, 181, 190, 192, 193, 196, 238, 255, 265
Humboldt fault, 213, 225, 228, 231
Humboldt zone, 20

Idaho batholith, 20, 33, 36, 55
Illinois basin, 60
induced magnetization, 164, 416, 420, 434, 441
International Geomagnetic Reference Field (IGRF), 2, 4, 25, 27, 62, 67, 165, 213, 225, 237, 249, 335, 377
inverse gravity problem, 239
iron formation, 375, 377, 380, 381, 382, 384, 387, 389, 390, 403, 432, 442, 444
isostasy, 176
isostatic compensation, 347, 349, 351, 356, 359, 373
isostatic residual gravity map, California, 347, 349, 351, 352, 356, 357, 359

Japan active margin, 339, 343
Japan trench, 343

Kalahari Desert, 144, 153
Kalahari line, 147, 150, 151, 153
Kansas, 213, 214, 225, 228, 231
Keweenawan, Minnesota, 375, 377, 380, 385, 388, 389, 390
Kheis belt, 151
kimberlite, 147, 153

kimberlite diatremes, 126, 128
Klamath Mountains, 347, 357
Kuril trench, 343

Lac Fournier massif, 118, 119, 120, 121
Laramide orogeny, 165, 175, 198, 208, 236, 237
Laramie Range, 36
level-to-drape continuation, 181, 185, 186, 187
level-to-level continuation, 181, 185, 187
Lexington–Bedford lineament, 302
lineaments, 128, 130, 174, 175
linear gravity highs, 59
Llano uplift, 6, 13, 17, 36
logarithmic energy-density spectra (LnEDS), 440, 441, 444
Long Island platform, 325, 335, 336
long-wavelength (LWL) anomalies, 46, 47, 52, 55, 59

magnetically calm areas, Brazil, 128
magnetically disturbed areas, Brazil, 128
Magnetic Anomaly Map (MAM) of Canada, 62, 63, 65, 67, 69, 72, 75
    boundary adjustments, 69
    compilation, 69
    continuation, 69
    digitization, 69
    filing, 69
    gridding, 69
    merging, 72
Magnetic Anomaly of the Paris Basin (MAPB), 154, 155, 160, 162, 163
magnetic basement, depth to, 141
magnetic data bank, 69, 75
magnetic inversion, 154, 160, 162
magnetic lineaments/lineations, 228, 231, 248, 250, 252, 254, 255, 265, 302, 380, 389
magnetic lithology map, Quebec, 426, 436
magnetic polarization, 4, 6
magnetic properties of rock types, China, 441
magnetic terranes, 248, 250, 254
magnetite quartzites, China, 439, 442, 444
magnetization boundaries, New Mexico, 181, 188, 189, 190, 192, 193, 196
magnetization-density ratio map, 248, 255, 262, 265
major discontinuities, Brazil, 124, 130
Matagami, Quebec, 426, 428
Mesa Butte magnetic anomaly, 164, 169, 178
metamorphic core complexes, Arizona, 164, 172, 174, 178
metamorphism, Carolina Slate belt, 320, 321, 322, 323
Michigan, 267, 269, 272, 274, 279, 281, 282, 284
Michigan basin, 13, 17, 60, 254, 255, 262, 265, 268, 269
Midcontinent geophysical (gravity and magnetic) anomaly (MGA), 13, 35, 47, 52, 55, 58, 60, 225, 250, 252, 254, 255
Midcontinent rift (MCR) system — see also Central North American rift system (CNARS) — 1, 7, 13, 17, 22, 33, 35, 250, 254, 255, 262, 267, 268, 269, 275, 282, 284, 285, 377, 385
Mid-Michigan geophysical anomaly, 35
mineral deposits, Brazil, 124, 125, 128, 130
mineral exploration, 144, 147, 153, 392, 411
Minnesota, 375, 377, 380, 381, 382, 385, 389, 390
Mississippi embayment, 1, 11, 15, 22, 36, 254, 255, 288, 297

Mississippi Valley graben, 36, 254, 262
Mississippi Valley–type mineralization, Pakistan, 138, 141
Missouri gravity low, 35
modeling, Paris basin, 154, 155, 162
Mohovoričić discontinuity (Moho), 4, 178, 314, 343, 345, 351, 359, 371

National Uranium Resource Evaluation (NURE), 2, 4, 27, 198, 200
Nemaha ridge, uplift, 213, 214, 225, 228, 231
New England sea mounts, 337
New Guinea geoid high, 88
New Madrid rift complex, 15, 287, 298, 301, 302, 306
New Mexico, 181, 182
New York–Alabama lineament, 11, 13, 17, 250, 252, 255, 262, 265, 301, 302, 306
nodes of regional metamorphism, Michigan and Wisconsin, 267, 269, 277, 279, 284
North Carolina, 320, 323
Northeast Atlantic passive margin, 339, 340

ocean–continent boundary, 339, 340, 343, 345
oil and gas, Arizona, 176
Ontario, 392, 393
Ophiolite belt, Pakistan, 138, 139, 141
Ouachita fold belt, 13
Ouachita front, 301
Ouachita orogenic belt, 36, 198, 207, 208, 209
oxygen fugacity, 426, 429, 432, 433, 438

Pacific Cordillera, 33, 36, 37
Pakistan, 132
paleorift, Arizona, 164, 178
Paleosubduction, 121
Paris basin, 154, 155, 162, 163
Peninsular Ranges batholith, 347, 356, 357
Penokean fold belt (PFB), 267, 268, 269, 272, 275, 277, 279, 280, 281, 282, 283, 284, 285
Phoenix arc, 164, 174, 175, 178
Piedmont, 308, 310, 312, 313
Piedmont gravity gradient, 310
plagioclase flotation, 109, 118, 122
plate tectonics, 109, 112, 132, 137
    driving force, 89
Poisson's equation, 255, 262
pole-correction map, Kansas, 213, 214, 225
polynomial surface fit, 39
porphyry copper deposits, Arizona, 164, 169, 174, 175, 178
potential fields, continuation of, 181, 185, 186, 190
Powder River basin, 244
Precambrian basement structure, Brazil, 124
Precambrian plate collision, Grenville province, 109, 116
Precambrian terranes, Kansas, 213, 214, 225, 231

Qian'an district, China, 439, 440, 441, 444, 448
Quebec, 413, 414, 424, 426

reduction to basement, New Mexico, 181, 185, 188, 196
reduction to the pole, 136, 139, 375, 381, 388, 390, 413, 417, 419, 434, 440, 442
Reelfoot rift, 1, 15, 22, 301
regional-field map, China, 439, 440, 441, 444, 448
regional-residual analysis, Ontario, 397

regional-residual separation, California, 347, 349, 356, 359
remanent component, Minnesota, 375
remanent magnetism, North Carolina, 321, 323
residual-gravity map, Ontario, 399
reverse faults, 59, 60
reverse polarization, 128
rift, arc–continent-collision model, Michigan and Wisconsin, 267, 285
Rio Grande rift, 20, 36, 200, 204, 206, 208, 246
rock-density measurements, Ontario, 393
rock magnetism, North Carolina, 320, 321, 322, 323
rock magnetization, Pakistan, 138, 141
rock-type susceptibility, Quebec, 426, 429, 431, 432, 433, 436, 437
Rocky Mountains, 47, 52, 59
Romaine River massif, 118, 119, 121
Rome trough, 15, 301
Rough Creek graben, 301
Round Lake Batholith, 392, 395, 397, 399, 401, 402, 404, 409, 411

Ste. Genevieve–Cookeville lineament, 302
St. Croix horst, 375, 385, 389
St. Francois Mountains, 13, 287, 288
St. Louis arm, 301
Salinian crustal block, 20
salt (halite) deposits, Arizona, 175
San Andreas fault, 20
San Juan basin, 181, 182, 187, 188, 192, 196
second vertical derivative, 375, 385, 390
second-vertical-derivative anomalies, 237
second-vertical-derivative map, 213, 228, 325, 335, 336, 337, 413, 418, 420, 423
sedimentary basins, 325
shaded relief, 69, 72, 75, 248, 255, 377, 382
Shining Tree greenstone belt, Ontario, 397, 399, 409, 410, 411
short-wavelength (SWL) anomalies, 46, 47, 52, 55, 59, 60
side lobes, 46, 52, 60
Sierra Nevada batholith, 37, 55, 347, 356, 357
silver–cobalt (Ag–Co) vein deposits, Ontario, 392, 393, 394, 411
Sinyala anomaly, 169
Snake River Plain, 8, 20
source-depths, 47
source geometry, 46
South Carolina, 314
South Dakota, 233
southeastern United States, 308
southern Africa, 144, 151
southern Indiana arm, 301
Southern Oklahoma aulacogen, 1, 13, 17, 22, 209, 255
Southern province, Canada, 393
Southern Rockies, 33
Spavinaw, Oklahoma, 288
Spearfish Formation, 236, 238, 239, 240, 242, 246
spectral enhancement, 213
spectrally filtered maps, Kansas, 213, 214, 231
spherical harmonic analysis, 89
spherical harmonic coefficients, 89, 95, 98
Sri Lanka geoid low, 88, 89, 95

strike-sensitive filtering, 375, 381, 382, 390, 440
strike-slip faults, Brazil, 125, 126, 128
structural-elements map, 337
structural relations, west Texas, 198, 209
subvolcanic intrusions, 361, 362, 363, 365, 366, 370, 371, 372, 373
Sudbury–Cobalt area, Ontario, 392, 393, 411
Superior province, 375, 393, 395, 399, 410, 411
susceptibility measurements, 429, 431, 432, 436, 437, 438
suture, Africa, 144, 151

Tallahassee/Suwannee terrane, 11
tectonic history, west Texas, 198, 209
tectonic terranes, Michigan and Wisconsin, 267, 268, 269, 271, 272, 274, 275, 277, 279, 280, 281, 285
Tennessee, 310
terrain corrections, 38, 46
Texas lineament, 36
Texas zone, Arizona, 174
38th-parallel lineament, 15, 302
Tibet plateau, 112
tilted-block pattern, Brazil, 130
total-field maps, 418, 423, 439, 448
total-intensity maps, China, 440, 441
transition zone, Arizona, 165
trend-pass filter, 213, 228, 231
triple junction, Africa, 147
two-dimensional gravity models, 267, 268, 269, 271, 281, 284
two-dimensional models, 320, 322, 323, 329

United States, conterminous, gravity data, 38
upward continuation filters, 213
U.S. Atlantic continental margin, 325
USSR, 103, 105

Ventura basin, 347, 357, 359
Venus, 88, 98, 101
Vermilion Granitic Complex, 375, 379, 389
vertical gradient mapping, Ontario, 392, 399
volcanic rocks, Pakistan, 134, 136, 138, 141
volcanic terrain, 361, 362, 363
volcanoes, 361, 362, 363, 365, 366, 370, 371, 372, 373

Walker lane, 36
wavelength-filtered gravity maps, 46
wavelength-filtered magnetic maps, 325, 336, 337
Western North Atlantic passive margin, 340
west Texas, 198, 209
White Mountains intrusive body, 164, 169, 171, 172, 174, 175, 178
Wichita-Arbuckle system, 36
Wichita system, 13
Wichita uplift, 58, 59, 60
Williston basin, 58
Wilson cycle tectonics, 130
Wisconsin, 267, 269, 272, 274, 277, 279, 280, 281, 282, 284
Wolf River batholith (WRB), 267, 268, 269, 272, 274, 280, 282, 284, 285

Zagros suture, 115
zone of failure, 59